ATHEROSCLEROSIS

ATHEROSCLEROSIS

Risks, Mechanisms, and Therapies

Edited by

HONG WANG
CAM PATTERSON

Co-Editors
DOMENICO PRATICÒ
MUKESH K. JAIN
XIAOFENG YANG
NICHOLAS E.S. SIBINGA
MARK W. FEINBERG

WILEY Blackwell

Published by John Wiley & Sons, Inc., Hoboken, New Jersey
Published simultaneously in Canada

Library of Congress Cataloging-in-Publication Data:

Atherosclerosis (Wang)
Atherosclerosis : risks, mechanisms, and therapies / edited by Hong Wang and Cam Patterson.
p. ; cm.
Includes bibliographical references and index.
ISBN 978-1-118-28591-6 (cloth)
I. Wang, Hong, 1956 September 19– , editor. II. Patterson, Cam, editor. III. Title.
[DNLM: 1. Coronary Artery Disease–physiopathology. 2. Coronary Artery Disease–prevention & control. 3. Coronary Artery Disease–therapy. WG 300]
RD598
617.412–dc23

2014034589

Cover image courtesy Hong Wang

Set in 10/12pt Times by SPi Publisher Services, Pondicherry, India

Printed in Singapore by C.O.S. Printers Pte Ltd

10 9 8 7 6 5 4 3 2 1

1 2015

CONTENTS

CONTRIBUTORS

CHIEF EDITORS

Cam Patterson, MD, MBA *New York-Presbyterian Hospital/Weill-Cornell Medical Center, New York, NY, USA*

Hong Wang, MD, PhD, EMBA *Cardiovascular Research Center, Department of Pharmacology, Temple University School of Medicine; Center for Metabolic Disease Research, Temple University School of Medicine, Philadelphia, PA, USA*

EDITORS

Mark W. Feinberg, MD *Cardiology Division, Brigham and Women's Hospital/Harvard Medical School, Boston, MA, USA*

Mukesh K. Jain, MD *Department of Medicine, Case Cardiovascular Research Institute, University Hospitals Case Medical Center, Case Western Reserve University; University Hospitals Harrington Heart and Vascular Institute, Cleveland, OH, USA*

Domenico Praticò, MD *Center for Translational Medicine and Department of Pharmacology, Temple University School of Medicine, Philadelphia, PA, USA*

Nicholas E. S. Sibinga, MD *Department of Medicine (Cardiology) & Department of Developmental and Molecular Biology, Albert Einstein College of Medicine, Bronx, NY, USA*

Xiaofeng Yang, MD, PhD *Cardiovascular Research Center, Department of Pharmacology; Center for Metabolic Disease Research, Temple University School of Medicine, Philadelphia, PA, USA*

CONTRIBUTORS

Ahsan Achtchi, DO *Emory University School of Medicine, Atlanta, GA, USA*

Chastain Anderson, MSc *Department of Cell and Molecular Biology, Tulane University, New Orleans, LA, USA*

G. Brandon Atkins, MD, PhD *Department of Medicine, Case Cardiovascular Research Institute, University Hospitals Case Medical Center, Case Western Reserve University; University Hospitals Harrington Heart and Vascular Institute, Cleveland, OH, USA*

Michael V. Autieri, PhD *Department of Physiology, Cardiovascular Research Center, Temple University School of Medicine, Philadelphia, PA, USA*

Jefferson Baer, MD, MPH *Department of Medicine, The Center for Heart Disease Prevention, Emory University School of Medicine, Atlanta, GA, USA*

Wayne E. Cascio, MD, FACC, FAHA *Department of Medicine/Division of Cardiology, University of North Carolina at Chapel Hill; Environmental Public Health Division, U.S. Environmental Protection Agency, Research Triangle Park, NC, USA*

Yuqing Eugene Chen, MD, PhD *Center for Advanced Models for Translational Sciences and Therapeutics, University of Michigan Medical Center, Ann Arbor, MI, USA*

David R. Clemmons, MD *Department of Medicine, UNC School of Medicine, University of North Carolina at Chapel Hill, Chapel Hill, NC, USA*

Caitlin Collins, *Department of Cell Biology and Physiology, University of North Carolina at Chapel Hill, Chapel Hill, NC, USA*

Adam Corken, *Department of Physiology and Biophysics, University of Arkansas for Medical Sciences, Little Rock, AR, USA*

Alan Daugherty, PhD, DSc, FAHA *Saha Cardiovascular Research Center, University of Kentucky, Lexington, KY, USA*

Sanjana Dayal, PhD *Department of Internal Medicine, University of Iowa Carver College of Medicine, Iowa City, IA, USA*

Arjun Deb, MD *Department of Medicine and Molecular, Cell and Developmental Biology, Broad Stem Cell Research Center, University of California, Los Angeles, CA, USA*

Joseph J. DeRose, MD *Department of Cardiovascular and Thoracic Surgery, Montefiore-Einstein Heart Center, Bronx, NY, USA*

Allen Dollar, MD *Department of Medicine, The Center for Heart Disease Prevention, Emory University School of Medicine, Atlanta, GA, USA*

Jie Du, PhD *Beijing An Zhen Hospital, Capital Medical University; The Key Laboratory of Remodeling-Related Cardiovascular Diseases, Ministry of Education; Beijing Institute of Heart Lung and Blood Vessel Diseases, Beijing, China*

William Durante, PhD *Department of Medical Pharmacology and Physiology, School of Medicine, University of Missouri-Columbia, Columbia, MO, USA*

Danny Eapen, MD *Department of Medicine, The Center for Heart Disease Prevention, Emory University School of Medicine, Atlanta, GA, USA*

Satoru Eguchi, MD, PhD *Department of Physiology, Cardiovascular Research Center, Temple University School of Medicine, Philadelphia, PA, USA*

Katherine J. Elliott, PhD *Department of Physiology, Cardiovascular Research Center, Temple University School of Medicine, Philadelphia, PA, USA*

Jianglin Fan, PhD *Department of Molecular Pathology, Interdisciplinary Graduate School of Medicine and Engineering, University of Yamanashi, Chuo-City, Yamanashi, Japan*

Pu Fang, PhD *Center for Metabolic Disease Research, Temple University School of Medicine, Philadelphia, PA, USA*

Justyna Fert-Bober, PhD, *Advanced Clinical Biosystems Research Institute, Heart Institute and Department of Medicine, Cedars-Sinai Medical Center, Los Angeles, CA, USA; Department of Medicine, Johns Hopkins University, Baltimore, MD, USA*

Edward A. Fisher, PhD, MD *Department of Medicine, Division of Cardiology, and the Marc and Ruti Bell Program in Vascular Biology, New York University School of Medicine; The Center for the Prevention of Cardiovascular Disease, NYU School of Medicine, New York, NY, USA*

Stephen J. Forest, MD *Department of Cardiovascular and Thoracic Surgery, Montefiore-Einstein Heart Center, Bronx, NY, USA*

Baiba K. Gillard, PhD *Atherosclerosis & Lipoprotein Research Laboratory, Cardiology Department, Houston Methodist Research Institute, Houston, TX, USA*

Antonio M. Gotto, Jr., MD, DPhil *Atherosclerosis & Lipoprotein Research Laboratory, Cardiology Department, Houston Methodist Research Institute, Houston, TX, USA*

David G. Harrison, MD *Division of Clinical Pharmacology, Department of Medicine, Vanderbilt University, Nashville, TN, USA*

Menno Hoekstra, PhD *Division of Biopharmaceutics, Leiden Academic Center for Drug Research, Gorlaeus Laboratories, Leiden University, Leiden, The Netherlands*

Ronald J. Holewinski, PhD *Advanced Clinical Biosystems Research Institute, Heart Institute and Department of Medicine, Cedars-Sinai Medical Center, Los Angeles, CA, USA; Department of Medicine, Johns Hopkins University, Baltimore, MD, USA*

Jonathan W. Homeister, MD, PhD *Department of Pathology and Laboratory Medicine, McAllister Heart Institute, University of North Carolina Medical School, Chapel Hill, NC, USA*

Paul L. Huang, MD, PhD *Cardiovascular Research Center and Cardiology Division, Department of Medicine, Massachusetts General Hospital and Harvard Medical School, Boston, MA, USA*

Ijeoma Isiadinso, MD, MPH *Emory University School of Medicine, Atlanta, GA, USA*

Farouc A. Jaffer, MD, PhD *Cardiovascular Research Center, Cardiology Division, Department of Medicine, Massachusetts General Hospital, Harvard Medical School; Department of Radiology, Massachusetts General Hospital, Boston, MA, USA*

Lixin Jia, PhD *Beijing An Zhen Hospital, Capital Medical University, The Key Laboratory of Remodeling-Related Cardiovascular Diseases, Ministry of Education, Beijing, China*

Paul Johnson, MD *Department of Medicine, University of North Carolina at Chapel Hill, Chapel Hill, NC, USA*

Golsa Joodi *Lipid and Prevention Clinic, University of North Carolina at Chapel Hill, Chapel Hill, NC, USA*

Christopher G. Kevil, PhD *Department of Pathology, School of Medicine in Shreveport, Louisiana State University; Departments of Cardiology, Pathology, and Molecular and Cellular Physiology, LSU Health-Shreveport, Shreveport, LA, USA*

Ahmed Khan, DO *Department of Cardiovascular and Thoracic Surgery, Montefiore-Einstein Heart Center, Bronx, NY, USA*

Annet Kirabo, DVM, MSc, PhD *Division of Clinical Pharmacology, Department of Medicine, Vanderbilt University, Nashville, TN, USA*

Joel A. Kooren, PhD *Advanced Clinical Biosystems Research Institute, Heart Institute and Department of Medicine, Cedars-Sinai Medical Center, Los Angeles, CA, USA; Department of Medicine, Johns Hopkins University, Baltimore, MD, USA*

Suzanne J.A. Korporaal, PhD *Department of Clinical Chemistry and Haematology, University Medical Center Utrecht, Utrecht, The Netherlands*

Jack Kuritzky, MD *Department of Medicine, University of North Carolina at Chapel Hill, Chapel Hill, NC, USA*

Steven R. Lentz, MD, PhD *Department of Internal Medicine, University of Iowa Carver College of Medicine, Iowa City, IA, USA*

Pin-Lan Li, MD, PhD *Department of Pharmacology and Toxicology, Medical College of Virginia Campus, Virginia Commonwealth University, Richmond, VA, USA*

Ming-Lin Liu, MD, PhD *Department of Dermatology, Perelman School of Medicine, University of Pennsylvania; Department of Medicine, Temple University School of Medicine, Philadelphia, PA, USA*

Hong Lu, PhD, MD *Saha Cardiovascular Research Center, University of Kentucky, Lexington, KY, USA*

Christopher Mack, PhD *Department of Pathology, University of North Carolina, Chapel Hill, NC, USA*

Lantz C. Mackey, PhD *Department of Pathology and Laboratory Medicine, McAllister Heart Institute, University of North Carolina Medical School, Chapel Hill, NC, USA*

Jietang Mai, PhD *Center for Metabolic Disease Research; Cardiovascular Research Center, Department of Pharmacology, Temple University School of Medicine, Philadelphia, PA, USA*

Kreton Mavromatis, MD *Emory University School of Medicine, Atlanta, GA, USA*

Phil Mendys, Pharm D *Cardiology Division, University of North Carolina, Chapel Hill, NC, USA*

Shu Meng, MD, PhD *Methodist Research Institute, Department of Cardiovascular Sciences, Houston, TX, USA*

Diana Neamtu *Department of Physiology and Biophysics, School of Medicine, University of Louisville, Louisville, KY, USA*

Reynaria Nieva, MD *University of Colorado, Denver, CO, USA*

Eric N. Olson, PhD *Department of Molecular Biology, University of Texas Southwestern Medical Center, Dallas, TX, USA*

Gabriela Orasanu, MD *Department of Medicine, Case Cardiovascular Research Institute, University Hospitals Case Medical Center, Case Western Reserve University; University Hospitals Harrington Heart and Vascular Institute, Cleveland, OH, USA*

Susmita Parashar, MD, MPH, MS *Emory University School of Medicine, Atlanta, GA, USA*

Aalok Patel, MD *Mayo Clinic, Rochester, MN, USA*

Timir Paul, MD, PhD *Division of Cardiology, University of North Carolina, Chapel Hill, NC, USA*

Elvis A. Peter, MD *Department of Pathology, School of Medicine in Shreveport, Louisiana State University; Departments of Cardiology, Pathology, and Molecular and Cellular Physiology, LSU Health-Shreveport, Shreveport, LA, USA*

Terrence Pong, PhD *School of Medicine, University of California, San Francisco, CA, USA*

Henry J. Pownall, PhD *Atherosclerosis & Lipoprotein Research Laboratory, Cardiology Department, Houston Methodist Research Institute, Houston, TX, USA*

Daniel J. Rader, MD *Preventive Cardiovascular Medicine and Lipid Clinic, University of Pennsylvania Health System; Institute for Translational Medicine and Therapeutics, University of Pennsylvania School of Medicine; Cardiovascular Metabolism Unit, Institute for Diabetes, Obesity and Metabolism, University of Pennsylvania School of Medicine; General Clinical Research Center, University of Pennsylvania Medical Center, Philadelphia, PA, USA*

Ravichandran Ramasamy, PhD *Diabetes Research Program, Division of Endocrinology, Department of Medicine, New York University Langone Medical Center, New York, NY, USA*

Srinivasa T. Reddy, PhD *Department of Medicine/Cardiology, University of California Los Angeles, Los Angeles, CA, USA*

Dario F. Riascos-Bernal, MD, PhD *Department of Medicine (Cardiology), and Department of Developmental and Molecular Biology, Albert Einstein College of Medicine, Bronx, NY, USA*

Victor Rizzo, PhD *Department of Anatomy and Cell Biology, Cardiovascular Research Center, Temple University School of Medicine, Philadelphia, PA, USA*

Peggy Robinet, PhD *Department of Cellular and Molecular Medicine, Cleveland Clinic Lerner Research Institute/NC10, Cleveland, OH, USA*

Corina Rosales, PhD *Atherosclerosis & Lipoprotein Research Laboratory, Cardiology Department, Houston Methodist Research Institute, Houston, TX, USA*

Marschall Runge, MD, PhD *NC TraCS Institute, University of North Carolina School of Medicine, Chapel Hill, NC, USA*

Andrew J. Sampson, MD *Division of Cardiology, University of North Carolina, Chapel Hill, NC, USA*

Ann Marie Schmidt, MD *Diabetes Research Program, Division of Endocrinology, Department of Medicine, New York University Langone Medical Center, New York, NY, USA*

Nicholas E.S. Sibinga, MD *Department of Medicine (Cardiology) and Department of Developmental and Molecular Biology, Albert Einstein College of Medicine, Bronx, NY, USA*

Ross J. Simpson, Jr., MD, PhD *Cardiology Section, Lipid and Prevention Clinic, University of North Carolina at Chapel Hill, Chapel Hill, NC, USA*

Jonathan D. Smith, PhD *Department of Cellular and Molecular Medicine, Cleveland Clinic Lerner Research Institute/NC10, Cleveland, OH, USA*

Ping Song, PhD *Section of Molecular Medicine, Department of Internal Medicine; Department of Biochemistry and Molecular Biology, University of Oklahoma Health Sciences Center, Oklahoma City, OK, USA*

Laurence Sperling, MD *Department of Medicine, The Center for Heart Disease Prevention, Emory University School of Medicine, Atlanta, GA, USA*

George A. Stouffer, MD *Division of Cardiology, University of North Carolina, Chapel Hill, NC, USA*

Arun K. Thukkani, MD, PhD *Cardiovascular Division, Brigham and Women's Hospital, Harvard Medical School, Boston, MA, USA*

Uwe J.F. Tietge, MD *Department of Pediatrics, University of Groningen, University Medical Center Groningen, Groningen, The Netherlands*

Suresh C. Tyagi, PhD *Department of Physiology and Biophysics, School of Medicine, University of Louisville, Louisville, KY, USA*

Ellie Tzima, PhD *Department of Cell Biology and Physiology, University of North Carolina at Chapel Hill, Chapel Hill, NC, USA*

Thomas P. Vacek *Department of Physiology and Biophysics, School of Medicine, University of Louisville, Louisville, KY, USA*

Theo J.C. Van Berkel, PhD *Division of Biopharmaceutics, Leiden Academic Centre for Drug Research, Gorlaeus Laboratories, Leiden University, Leiden, The Netherlands*

Miranda Van Eck, PhD *Division of Biopharmaceutics, Leiden Academic Centre for Drug Research, Gorlaeus Laboratories, Leiden University, Leiden, The Netherlands*

Jennifer E. Van Eyk, PhD *Advanced Clinical Biosystems Research Institute, Heart Institute and Department of Medicine, Cedars-Sinai Medical Center, Los Angeles, CA, USA; Department of Medicine, Johns Hopkins University; Departments of Biomedical Engineering and Biological Chemistry, JHU-Bayview Proteomics Center, Johns Hopkins University, Baltimore, MD, USA*

Kiran Valiani, MD *Emory University School of Medicine, Atlanta, GA, USA*

Jai Varma, MD *Department of Pathology, School of Medicine in Shreveport, Louisiana State University; Departments of Cardiology, Pathology, and Molecular and Cellular Physiology, LSU Health-Shreveport, Shreveport, LA, USA*

Anthony Virtue, PhD *Center for Metabolic Disease Research; Cardiovascular Research Center, Department of Pharmacology, Temple University School of Medicine, Philadelphia, PA, USA*

Shusheng Wang, PhD *Department of Cell and Molecular Biology, Tulane University; Department of Ophthalmology, Tulane University, New Orleans, LA, USA*

Ying Wang, MD, PhD *Beijing An Zhen Hospital, Capital Medical University; The Key Laboratory of Remodeling-Related Cardiovascular Diseases, Ministry of Education, Beijing, China*

Jerry Ware, PhD *Department of Physiology and Biophysics, University of Arkansas for Medical Sciences, Little Rock, AR, USA*

Kevin Jon Williams, MD *Section of Endocrinology, Diabetes, & Metabolism, Departments of Medicine and Physiology, Cardiovascular Research Center, Temple University School of Medicine; Department of Molecular and Clinical Medicine, Sahlgrenska Center for Cardiovascular and Metabolic Research, University of Gothenburg, Göteborg, Sweden*

Mei Mei Wong, PhD *Cardiovascular Division, King's College London, London, UK*

Xin-Yu Xiong *Department of Pharmacology, Cardiovascular Research Center; Thrombosis Research Center, Temple University School of Medicine, Philadelphia, PA, USA*

Qingbo Xu, MD, PhD *Cardiovascular Division, King's College London, London, UK*

Shi Fang Yan, MD *Diabetes Research Program, Division of Endocrinology, Department of Medicine, New York University Langone Medical Center, New York, NY, USA*

Faresa Zarreen, MD *Emory University School of Medicine, Atlanta, GA, USA*

Jifeng Zhang, PhD *Center for Advanced Models for Translational Sciences and Therapeutics, University of Michigan Medical Center, Ann Arbor, MI, USA*

Yang Zhang, PhD *Department of Pharmacology and Toxicology, Medical College of Virginia Campus, Virginia Commonwealth University, Richmond, VA, USA*

Ying Zhao, PhD *Division of Biopharmaceutics, Leiden Academic Center for Drug Research, Gorlaeus Laboratories, Leiden University, Leiden, The Netherlands*

Ming-Hui Zou, MD, PhD *Department of Medicine, University of Oklahoma Health Science Center, Oklahoma City, OK, USA*

FOREWORD

Atherothrombosis is a pathogenic process that affects all vascular beds, most notably the cerebral, coronary, and peripheral vasculatures. This disease is widespread in the United States, where it accounted for 917,000 deaths in 2010 and is increasing rapidly worldwide as well: in 2010, there were 9,914,500 deaths from atherothrombotic disease, a 32.3% increase over the number of deaths reported in 1990.

Owing to its prevalence and consequences for global mortality and morbidity, atherothrombosis remains an important disease process that warrants a careful assessment of its molecular and cellular underpinnings. Over the past three decades, there have been many advances in our knowledge of the mechanisms of atherogenesis. From the molecular and cellular determinants of the conventional risk factors, hyperlipidemia, and diabetes mellitus to increased understanding of the intermediate pathophenotypes of inflammation and thrombosis, this disease process has been progressively dissected and the elegance of its complexity increasingly appreciated. With knowledge of mechanisms comes better strategies for disease prevention and treatment. Since 1968, the death rate for atherothrombotic disorders has decreased progressively in the United States, a clear testimony of the benefits of rigorous and relentless biomedical research and research investment.

In this book, Hong Wang and Cam Patterson and their section editors, Mark W. Feinberg, Mukesh K. Jain, Domenico Pratico, Xiaofeng Yang, and Nicholas E.S. Sibinga, have assembled a marvelous compilation of chapters that focus on a contemporary view of the cellular, molecular, and biochemical mechanisms of atherothrombogenesis. The topics contained within these chapters range widely and include discussions of the roles of hyperlipidemia, endothelial dysfunction, and the renin–angiotensin system in the disease process, as well as cutting-edge presentations on the roles of microRNAs, hydrogen sulfide, epigenesis, and posttranslational modification of the proteome. These detailed mechanistic sections are followed by a section on clinical management, which presents up-to-date information on topics ranging from molecular imaging to contemporary pharmacotherapy to stem cell–based therapies.

In all, *Atherosclerosis* is a comprehensive, authoritative treatise on a highly prevalent, complex, chronic disease process. The text is a most useful compendium of information for basic, translational, and clinical investigators alike and is presented in a rigorous and thoughtful way. The editors and authors have provided the scientific community with what will, no doubt, quickly become a standard reference for this dynamic, important field of disease research.

Joseph Loscalzo

Department of Medicine
Brigham and Women's Hospital
Boston, MA, USA

PREFACE

Atherosclerosis, the primary antecedent to heart disease, is among the most deeply understood human disorders at the mechanistic level. This foundation of knowledge has served as a platform for the development of some of the most effective and widely prescribed drugs of our time. Nevertheless, heart disease continues to increase in prevalence worldwide, and current projections show no sign of abating. Thus, many opportunities and obstacles remain in the quest for deeper insights and additional approaches for prevention and treatment of this disorder.

Atherosclerotic vascular disease results from the pathological deposition of lipids, such as cholesterol and triglycerides, within the walls of arterial blood vessels, leading to calcification, inflammation, recruitment of macrophages and foam cells, and eventual thrombosis. Elegant mechanistic dissection of the cellular processes and molecular mechanisms responsible for the initiation and progression of atherosclerosis has revealed the central involvement of low-density lipoprotein (LDL) metabolism in the disease. Molecular dissection of the disease process in cultured cells and animal models, combined with population studies and human genetics, has unveiled key therapeutic targets within the biochemical pathways of lipoprotein metabolism, inflammatory signaling, and thrombosis. These discoveries have led to the development of statins, as lipid-lowering agents, as well as drugs that target other regulatory steps in the atherosclerotic disease process. Surgical intervention through angioplasty procedures also provides benefit in late stages of the disease.

Atherosclerosis occurs over a lifetime and is influenced by numerous factors, including excessive dietary fat, smoking, diabetes, sedentary lifestyle, and genetic factors. A variety of cell types influence the disease process, including endothelial cells, vascular smooth muscle cells, fibroblasts, inflammatory cells, and platelets. Each of these cell types participates in a complex dialog of inter- and intracellular signaling to modulate the disease phenotype and clinical outcome.

The discovery of gene regulatory and signaling pathways involved in atherosclerotic heart disease continues to reveal new mechanisms of disease regulation and potential therapeutic targets. Particular excitement has surrounded the discovery of microRNAs as regulators of lipoprotein metabolism, vascular remodeling, and inflammation, pointing to unexplored inroads into the disease. Epigenetic regulatory mechanisms impose additional regulation on disease manifestation. Population studies have also had a transformative impact on understanding of atherosclerosis through the identification of new disease-modifying genes, such as *PCSK9*, which controls processing of the LDL receptor.

Few, if any, areas of contemporary medicine have been so profoundly influenced by detailed scientific inquiry into the mechanistic basis of disease as the field of atherosclerosis. In *Atherosclerosis*, Hong Wang and Cam Patterson, together with their section editors, Mark Feinberg, Mukesh K. Jain, Domenico Pratico, Xiaofeng Yang, and Nicholas E.S. Sibinga, bring together articles by the leaders in the field to provide a timely, insightful, and comprehensive view of this complex disorder. This collection of articles should serve as the go-to source of information and a guide for the field for years to come.

Eric N. Olson

Dallas, TX, USA

SECTION I

RISK FACTORS: OLD AND NEW

1

LIPOPROTEINS

MING-LIN LIU[1,2] AND DANIEL J. RADER[3,4,5,6]

[1] *Department of Dermatology, Perelman School of Medicine, University of Pennsylvania, Philadelphia, PA, USA*

[2] *Department of Medicine, Temple University School of Medicine, Philadelphia, PA, USA*

[3] *Preventive Cardiovascular Medicine and Lipid Clinic, University of Pennsylvania Health System, Philadelphia, PA, USA*

[4] *Institute for Translational Medicine and Therapeutics, University of Pennsylvania School of Medicine, Philadelphia, PA, USA*

[5] *Cardiovascular Metabolism Unit, Institute for Diabetes, Obesity and Metabolism, University of Pennsylvania School of Medicine, Philadelphia, PA, USA*

[6] *General Clinical Research Center, University of Pennsylvania Medical Center, Philadelphia, PA, USA*

BRIEF HISTORY OF CHOLESTEROL AND ATHEROSCLEROSIS

There is a long history of a relationship between cholesterol and atherosclerosis. Vogel first identified the deposits of cholesterol in the artery wall in 1840s [1, 2]. The German pathologist Felix Jacob Marchand first introduced the descriptive term "atherosclerosis" in 1904 to describe the "gruel-like" fatty substance build-up inside a hardened artery and suggested that atherosclerosis is responsible for obstructive processes in the arteries [3]. However, the causal relationship between cholesterol and atherosclerosis was not clarified until 1910s by a series of laboratory experiments that were initiated by an accidental discovery when they studied the role of diet in aging [4, 5]. Feeding rabbits with meat or egg yolk induced atherosclerosis, while feeding the animals with egg white did not produce atherosclerosis [5]. Based on these observations, the investigators concluded that meat and egg yolk play a critical role in atherosclerosis formation. Sobolev et al. performed a systematic survey to evaluate factors in meat and egg yolk that could lead to atherosclerosis [5], and Anitschkow and Chalatow in 1913 continued the work by feeding rabbits with purified cholesterol that resulted in the development of typical atheromatous lesions in rabbits [3, 5, 6]. These and subsequent studies led Anitschkow to the conclusion that "there is no atherosclerosis without cholesterol" [7]. These series of experimental studies established the causal relationship of dietary cholesterol to the development of atherosclerosis in rabbits and marked the beginning of the modern era of atherosclerosis research.

Cholesterol is a hydrophobic lipid that is not solubilized by the aqueous environment of the

Atherosclerosis: Risks, Mechanisms, and Therapies, First Edition. Edited by Hong Wang and Cam Patterson.

plasma and is therefore transported by macromolecular complexes known as lipoproteins. In 1929, Macheboeuf first isolated a stable, water-soluble lipoprotein by precipitation from horse serum [8, 9]. This lipoprotein contains 59% protein and 41% lipid, which consisted of 18% cholesterol and 23% phospholipid and could be re-dissolved in water to form a clear solution. Macheboeuf called it lipid-containing α-globulin at the time, and it is now recognized as high-density lipoprotein (HDL). The development of the ultracentrifuge by Svedberg in the 1920s provided an important laboratory instrument for the flotation and isolation of lipoproteins. In the 1930s, by using ultracentrifugation, McFarlane first noticed a time- and salt-dependent layer that existed between the albumin and globulin layers. He named it "X-fraction" [10], which was later named low-density lipoprotein (LDL) [9, 11]. In 1941, Blix separated α- and β-globulins by paper electrophoresis [12], which is now replaced by more advanced technology with agarose gel electrophoresis.

Among those who tried to characterize the full spectrum of lipoproteins in blood, the American medical physicist John W. Gofman was the first to successfully achieve this goal [13]. Later Gofman and colleagues also provided important epidemiologic evidence that cholesterol-carrying molecules (lipoproteins) in the blood predicted the risk of heart disease [14]. Gofman's work opened the window on the complexity of the lipoproteins and sparked the explosive increase in the attention of the research on the plasma lipoproteins and their relationship to atherosclerosis. Ancel Keys postulated a correlation between cholesterol levels and cardiovascular disease (CVD) and initiated a study of Minnesota businessmen (the first prospective study of CVD), demonstrating an association of blood cholesterol with CVD [15]. He subsequently extended this observation to several other countries in a study known as the Seven Countries Study [16]. In 1965, Frederickson and Lees in the National Institutes of Health (NIH) used more advanced technology for lipoprotein electrophoresis and developed a system of lipoprotein phenotypes [17], later refined by Frederickson and Levy [18] to a classification of hyperlipoproteinemias that remains in use to this day. The Framingham Heart Study, a long-term prospective study of cardiovascular risk factors definitively established that incident coronary artery disease (CAD) risk is highest in groups with highest blood cholesterol levels at baseline [19]. Participants with total cholesterol greater than 245 mg/dl at baseline had a threefold increased risk of incident coronary heart disease (CHD) compared with participants with a total cholesterol less than 210 mg/dl.

LIPOPROTEINS

Lipoproteins are large macromolecular complexes that transport hydrophobic lipids in the blood. Lipoproteins consist of a core of hydrophobic lipids (triglycerides [TGs] and cholesteryl esters) surrounded by a shell of hydrophilic lipids (phospholipids and unesterified cholesterol) and proteins. The plasma lipoproteins are divided into five major classes based on their relative density: chylomicrons (Figure 1.1), very-low-density lipoproteins (VLDLs), intermediate-density lipoproteins (IDLs), LDL, and HDL. The density of a lipoprotein is determined by the amount of lipid per particle. The proteins associated with lipoproteins, called apolipoproteins, are required for the assembly, structure, and function of lipoproteins. Apolipoproteins activate enzymes, mediate lipid transfer, and act as ligands for cell surface receptors. Apolipoprotein-B (apoB) is the major structural protein of chylomicrons, VLDL, IDL, and LDL; whereas apoA-I is the major protein in HDL particles [20].

LDL AND ATHEROSCLEROSIS

Metabolism of LDL

LDL is derived from the lipolysis of TGs in VLDLs, which are secreted by the liver (Figure 1.2). The rates of VLDL production

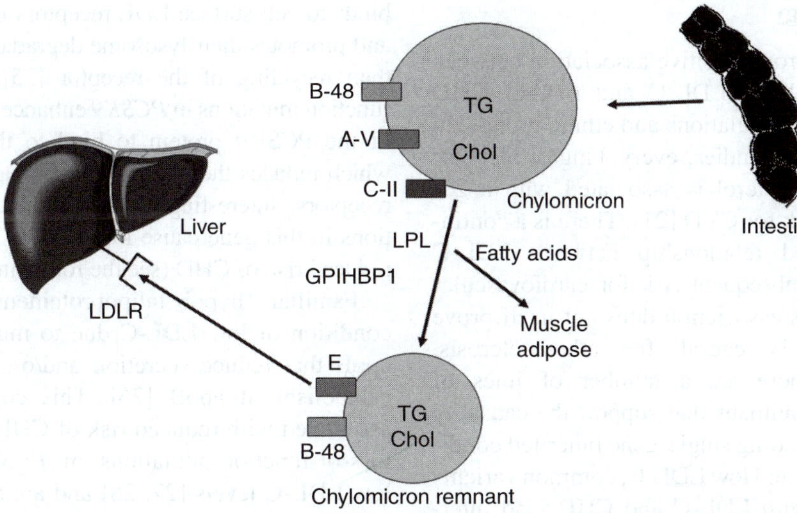

Figure 1.1 Schematic diagram of chylomicron metabolism. A-V, apolipoprotein A-V; B-48, apoliporotein B48; C-II, apolipoprotein CII; Chol, cholesterol; E, apolipoprotein E; LDLR, LDL receptor; LPL, lipoprotein lipase; TG, triglyceride; GPIHBP1, glycosylphosphatidylinositol-anchored high-density lipoprotein-binding protein 1.

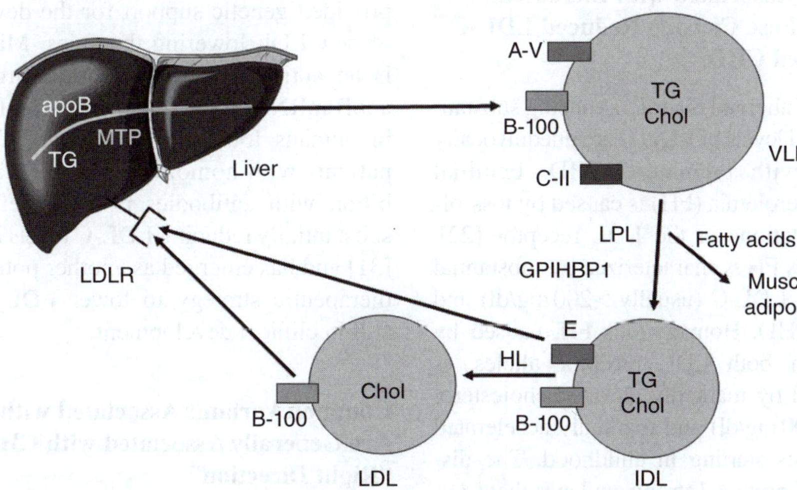

Figure 1.2 Schematic diagram of VLDL, IDL, and LDL metabolism. apoB, apolipoprotein B; B-100, apolipoprotein B100; HL, hepatic lipase; MTP, microsomal triglyceride transfer protein.

and lipolysis of VLDL TG influence plasma levels of LDL cholesterol (LDL-C). However, the most important determinant of plasma LDL-C is the rate of LDL catabolism by the liver. This is primarily influenced by the abundance and activity of the LDL receptors on the surface of the hepatocytes. The LDL receptor is regulated by cellular cholesterol content at both transcriptional and post-transcriptional levels. In addition, the protein proprotein convertase subtilisin/kexin type 9 (PCSK9) is secreted by the liver and targets the LDL receptor for degradation, thereby providing another layer of regulation. High levels of LDL-C are generally a result of reduced LDL receptor expression due to genetic or environmental influences or both.

Epidemiology

There is a strong positive association between plasma levels of LDL-C and risk of CHD across many populations and ethnic groups. In epidemiologic studies, every 1 mg/dl increase in LDL cholesterol is associated with a 2% increased risk for CVD [21]. There is a continuous, graded relationship between LDL-C levels and subsequent risk for cardiovascular disease. This association does not itself prove that LDL is causal for atherosclerosis. However, there are a number of lines of evidence in humans that support the causality of LDL, including single-gene inherited conditions of high and low LDL-C, common variants associated with LDL-C and CHD, and intervention studies of LDL-C reduction leading to reduction in CV events.

Inherited Conditions Causing Elevated LDL-C Are Associated with Increased CHD and Those Causing Reduced LDL-C with Reduced CHD

Single-gene inherited disorders causing substantially elevated levels of LDL-C are unequivocally associated with premature CHD. Familial hypercholesterolemia (FH) is caused by loss-of-function mutations in the LDL receptor [22]. Heterozygous FH is characterized by substantial elevations in LDL-C (usually >200 mg/dl) and premature CHD. Homozygous FH, caused by mutations in both LDL receptor alleles, is characterized by markedly elevated cholesterol (usually >400 mg/dl) and markedly accelerated atherosclerosis starting in childhood. The discovery of FH provided strong evidence that LDL is causally linked to atherosclerotic disease.

Other single-gene causes of elevated LDL-C also increase the risk of CHD. Familial defective apoB-100 (FDB) [23] is caused by mutations in the receptor-binding region of apoB-100 that impair its binding to the LDL receptor and delay clearance of LDL. Like heterozygous FH, FDB is associated with elevated LDL-C and premature coronary disease [23]. Gain-of-function mutations in the *PCSK9* gene also cause elevated LDL-C and premature CHD [24]. PCSK9 binds to cell surface LDL receptors in the liver and promotes their lysosome degradation rather than recycling of the receptor [25]. Gain-of-function mutations in *PCSK9* enhance the ability of the PCSK9 protein to bind to the LDLR, which reduces the number of cell surface LDL receptors. Interestingly, loss-of-function mutations in this gene cause low LDL-C levels and reduced risk of CHD (see the following text).

Familial hypobetalipoproteinemia is a condition of low LDL-C due to mutations in apoB that reduce secretion and/or accelerate catabolism of apoB [26]. This condition is associated with reduced risk of CHD. Further, loss-of-function mutations in *PCSK9* cause low LDL-C levels [27, 28] and are associated with a lifelong reduction in CHD [29]. These two conditions strongly support a causal role for LDL in atherosclerosis and that lower levels of LDL-C over a lifetime substantially reduces CHD risk. These conditions have also provided genetic support for the development of new LDL-lowering therapies. Mipomersen is an antisense oligonucleotide targeting the apoB mRNA in liver. It reduces LDL-C levels in humans [30] and is approved for use in patients with homozygous FH. PCSK9 inhibition with antibodies is highly effective in substantially reducing LDL-C levels in humans [31] and has emerged as another potential new therapeutic strategy to lower LDL-C that is still in clinical development.

Common Variants Associated with LDL-C Are Generally Associated with CHD in the "Right Direction"

Genome-wide association studies (GWASs) of common DNA variants have identified dozens of loci associated with LDL-C levels, including both "known" genes and "novel" genes. These studies have substantially reinforced the evidence that LDL-C plays a strong causal role in CHD. For example, the majority of LDL-C loci were significantly associated with CHD with concordant directionality, that is, variants associated with higher LDL-C were associated with increased risk of CHD [32]. Further, a

GWAS meta-analysis for CHD, detailed a total of 46 independent loci that were genome-wide significant and 20% of these loci were also genome-wide significantly associated with LDL-C in the direction concordant with CHD risk [33, 34]. This discovery of a major enrichment of "LDL-C genes" in an experiment focused on genes associated with CAD further confirms the indisputable causal relationship between LDL-C levels and CAD.

Interventions That Reduce LDL-C Levels Reduce CHD Events

There is now a very large body of evidence based on randomized controlled trials indicating that intervention to reduce LDL-C levels reduces cardiovascular events [35]. Most of these studies involve HMG-CoA reductase inhibitors (statins), although other approaches to reducing LDL-C have also been shown to reduce events. A detailed review of these studies is beyond the scope of this chapter. However, these data strongly support the observational epidemiology, single-gene conditions, and common variant genetics in supporting the causal role of plasma LDL-C levels in atherosclerotic CVD.

TRIGLYCERIDE-RICH LIPOPROTEINS AND ATHEROSCLEROSIS

Metabolism of Triglyceride-rich Lipoproteins

Triglyceride-rich lipoproteins (TRLs) are synthesized by both the intestine and the liver. Dietary lipids are absorbed in the small intestine and packaged with apoB-48 to form chylomicrons that are secreted into the intestinal lymph and delivered directly to the systemic circulation (Figure 1.1). In peripheral tissues such as adipose and muscle, lipoprotein lipase (LPL) hydrolyzes the TGs in chylomicrons—an effect that creates chylomicron remnants that are rapidly removed from circulation by the liver through a process that requires apoE as a ligand for receptors in the liver. The liver esterifies free fatty acids to TGs and incorporates them into VLDL particles that contain apoB-100 and are secreted into the plasma (Figure 1.2). As with chylomicrons, the TGs of VLDL are hydrolyzed by LPL in muscle and adipose tissue creating VLDL remnants or IDLs. The liver removes approximately half of IDL particles by LDL receptor-mediated endocytosis via binding to apoE. The remainder of the IDL is remodeled by hepatic lipase (HL) to form LDL.

Epidemiology

The independent relationship of TG levels and cardiovascular disease has been a topic of debate over the years; but recently, a general consensus has developed that TGs are an independent predictor of cardiovascular risk [36]. In addition, single-gene conditions and common variant studies support a causal role for at least some types of TRLs in atherosclerosis. However, interpretation of the observational epidemiology and genetic studies is complicated by the heterogeneity of TRL and the likelihood that certain TRLs (such as chylomicrons) are not atherogenic, whereas other TRLs (such as VLDL remnants) are highly atherogenic.

Inherited Conditions Influencing TGs Are Variably Associated with CHD

Single-gene disorders associated with elevation in chylomicrons alone are probably not associated with increased CHD. The familial hyperchylomicronemia syndrome (FCS) is caused by homozygosity for loss-of-function mutations in one of two genes encoding the protein LPL and apoC-II [37]. FCS is characterized by extreme hypertriglyceridemia (>1000 mg/dl) usually presenting in childhood with acute pancreatitis, eruptive xanthomas, lipemia retinalis, and/or hepatosplenomegaly. Interestingly, despite markedly elevated TG (and cholesterol) levels, premature atherosclerotic cardiovascular disease is not generally a feature of this disease, suggesting that chylomicrons themselves are not atherogenic.

On the contrary, genetic conditions that increase chylomicron and VLDL-remnant lipoproteins appear to be associated with increased CHD risk. Familial dysbetalipoproteinemia (FD), or type III hyperlipoproteinemia, is caused by mutations in the gene for apoE [38]. ApoE on chylomicron and VLDL remnants normally mediate their catabolism by binding to receptors in the liver. FD is usually caused by homozygosity for a common variant called apoE2, which differs from the wild-type apoE3 form by a substitution of a cysteine for an arginine at position 158. ApoE2 has impaired binding to lipoprotein receptors such as the LDL receptor, resulting in defective removal of chylomicron and VLDL remnants. Importantly, premature atherosclerotic CVD is common in this disorder, an observation that helped to clarify that remnant lipoproteins are highly atherogenic.

Common Variants Associated with TG Levels Are Variably Associated with CHD

There are a number of loci in the genome that are strongly associated with plasma TG levels. Many are also associated with LDL-C, and the majority of these are also associated with CHD. One TG locus not associated with LDL-C but strongly associated with CHD is the LPL locus. The gain-of-function variant (rs264 leading to S447X) is associated with reduced TG, elevated HDL-C, and significantly reduced risk of CHD [32, 33]. Alternatively, another variant (rs12678919) is associated with increased TG, reduced HDL-C, and increased CHD risk [39]. Thus, the actions of LPL modulate TG (and HDL-C) levels, thereby directly influencing CHD risk. While this does not prove that TRLs are causally pro-atherogenic, it supports the concept.

Interventions That Reduce TG Levels Do Not Consistently Reduce CHD Events

There are a much smaller number of randomized controlled trials with interventions that reduce TG levels (primarily fibrates and omega-3 fatty acids or fish oils). A detailed review of these studies is beyond the scope of this chapter. The fibrate studies have yielded mixed results but on balance have been disappointing [40]. The fish oil trials have not been designed specifically as TG-lowering trials. Importantly, none of the trials with fibrates or fish oils were specifically performed in subjects selected for elevated TG levels at baseline. Ultimately, trials of TG-lowering drugs in individuals with elevated TG levels will be needed to address the causal role of plasma TG levels in atherosclerotic CVD.

HDL AND ATHEROSCLEROSIS

Metabolism

HDL metabolism is complex. Nascent HDL particles are synthesized by the intestine and the liver (Figure 1.3). Newly secreted apoA-I rapidly acquires phospholipids and unesterified cholesterol from its site of synthesis (intestine or liver) via efflux promoted by the membrane protein ATP-binding cassette protein A1 (ABCA1). Within the HDL particle, the cholesterol is esterified by lecithin-cholesterol acyltransferase (LCAT) to cholesteryl ester and the HDL becomes spherical. HDL cholesterol is transported to hepatoctyes through two pathways. HDL cholesteryl esters can be transferred to apoB-containing lipoproteins in exchange for TG by the cholesteryl ester transfer protein (CETP). The cholesteryl esters are then removed from the circulation by LDL receptor-mediated endocytosis. HDL cholesterol can also be taken up directly by hepatocytes via the scavenger receptor class BI (SR-BI), a cell surface receptor that mediates the selective transfer of lipids to cells. HDL particles also undergo remodeling by a variety of lipid transfer proteins and lipases. The phospholipid transfer protein (PLTP) has the net effect of transferring phospholipids from other lipoproteins to HDL. After CETP-mediated lipid exchange, the TG-enriched HDL becomes a much better substrate for hepatic lipase (HL), which hydrolyzes the TGs and phospholipids to generate smaller HDL particles. Endothelial

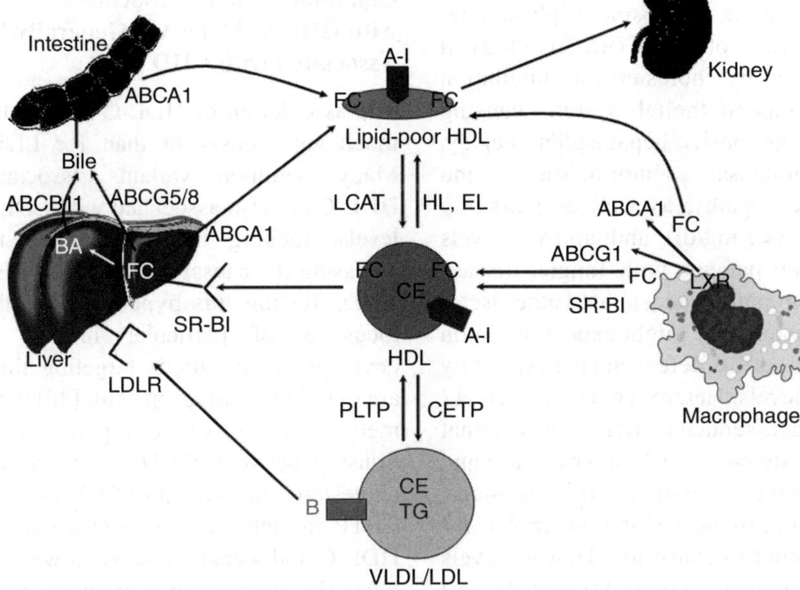

Figure 1.3 Schematic diagram of HDL metabolism. ABCA1, ATP-binding cassette transporter A1; ABCB11, ATP-binding cassette transporter B11; ABCG1, ATP-binding cassette transporter G1; ABCG5/8, ATP-binding cassette transporter G5 and G8; A-I, apolipoprotein A-I; CE, cholesterol ester; CETP, cholesteryl ester transfer protein; EL, endothelial lipase; FC, free cholesterol; LCAT, lecithin-cholesterol acyltransferase; PLTP, phospholipid transfer protein; SR-BI, Scavenger receptor class B member 1.

lipase (EL) hydrolyzes HDL phospholipids, generating smaller HDL particles that are catabolized faster.

HDLs are very heterogeneous with regard to size and density, lipid composition, and apolipoprotein composition. While there has been considerable interest over the years in the relationship of subfractions of HDL based on size and density, there are no compelling data to support any HDL subfraction having significantly greater predictive power than HDL-C itself. The nature of the apolipoproteins on HDL could be a determinant of differential relationship to CHD. For example, apoC-III-containing HDLs have been reported to lack the inverse association with CHD and have properties *in vitro* consistent with a pro-atherogenic profile.

Epidemiology

The inverse association between HDL-C and CHD has been repeatedly and consistently noted in scores of observational studies across the world and is one of the most consistent epidemiologic observations with regard to cardiovascular risk. In epidemiologic studies, every 1 mg/dl increase in HDL cholesterol is associated with a 2–3% decreased risk of CHD [21]. While the association is unquestioned, there remain substantial questions about the causal nature of this association. Neither the inherited conditions nor the common genetic variant studies support a causal role for HDL in influencing CHD risk.

Inherited Disorders Causing Low HDL-C Are Generally Not Associated with Increased CHD

Missense or nonsense mutations that result in structurally abnormal or truncated apoA-I proteins cause low HDL-C but are not generally associated with an increased risk of atherosclerosis [41]. Only complete genetic deficiency of apoA-I has been associated with increased CHD. Tangier disease is caused by loss-of-function mutations in both alleles encoding the gene adenosine triphosphate-binding cassette protein A1 (*ABCA1*) [42]. It is characterized by cholesterol accumulation in the reticuloendothelial system causing enlarged orange tonsils, hepatosplenomegaly, intestinal mucosal abnormalities, and peripheral neuropathy, as well as markedly low HDL-C (<5 mg/dl) and apoA-I levels [42]. However, patients with Tangier disease do not develop rapidly accelerated atherosclerosis to the extent one might expect based on the cholesterol efflux defect and the extremely low HDL-C levels. Heterozygotes for *ABCA1* mutations have reduced HDL-C levels that are intermediate between Tangier disease and normal but have no evidence of cholesterol accumulation in tissues. Mutations in *ABCA1* have been found to cause low HDL-C levels in some families in which Tangier disease homozygotes are not found [43], and these individuals are also not at increased risk of CHD. Finally, LCAT deficiency is caused by loss-of-function mutations in both alleles of the *LCAT* gene [44]. LCAT is the enzyme that esterifies the free cholesterol present in HDL to cholesteryl ester, creating a cholesteryl ester core and resulting in maturation of HDL. In the absence of functional LCAT and cholesterol esterification, mature HDL particles are not formed and nascent HDL particles containing apoA-I are rapidly catabolized [45]. Interestingly, LCAT deficiency is not clearly associated with premature coronary disease, despite the markedly reduced HDL-C levels [44].

CETP deficiency is caused by loss-of-function mutations in both alleles of the *CETP* gene [46]. CETP transfers cholesteryl esters from HDL to apoB-containing lipoproteins in exchange for TGs. Lack of functional CETP results in markedly elevated HDL-C levels due to lack of HDL remodeling, accumulation of cholesteryl esters in HDL, and slower turnover of apoA-I [47]. Whether homozygous or heterozygous, CETP deficiency is associated with increased, decreased, or unchanged cardiovascular risk that remains to be resolved.

Common Variants Associated with HDL-C Alone Are Generally Not Associated with CHD

The association of HDL-C loci with CHD is much less consistent than for LDL-C loci. Many common variants associated with HDL-C are also associated with LDL-C or TG levels, making them unsuitable tools for assessing the causal relationship of HDL-C to CHD. Testing this hypothesis for the CETP locus is of particular interest as drug development programs targeting this protein are ongoing. Thompson et al. [48] conducted a meta-analysis of studies reporting on the relationship between *CETP* gene variants, HDL cholesterol, and risk for CHD. Three common CETP variants were associated with increased HDL-C and weakly associated with decreased CAD risk to the degree predicted by observational epidemiology data. However, multiple variants at this locus were also found to be significantly associated with LDL-C and TGs in addition to HDL-C; therefore, the possibility that the CAD-associated CETP variants exert their effect on cardiovascular disease risk through changes in LDL-C or TG levels cannot be excluded. As noted earlier, ABCA1 is an important regulator of plasma HDL-C levels. Interestingly, variants in ABCA1 associated with lower HDL-C levels were not found to be associated with increased risk of CHD [49]. However, many of these variants are also associated with lower LDL-C levels.

Thus, the efforts to address this question have focused on genetic variants that are specifically associated with HDL-C and not LDL-C or TG. As noted earlier, LCAT is a key enzyme that regulates HDL metabolism and HDL-C levels. Low-frequency variants at the LCAT locus associated with reduced HDL-C were found to have no association with increased CHD risk [50]. Endothelial lipase (EL; gene name *LIPG*) hydrolyzes HDL phospholipids and accelerates HDL catabolism and has been shown in mice to be an important regulator of HDL-C levels [51]. Deep candidate gene re-sequencing of the *LIPG* gene in subjects with extremes of HDL-C identified rare, nonsynonymous loss-of-function variants in subjects with extremely high HDL-C levels not found in persons with low HDL-C [52]. In addition, a low-frequency (~2% minor allele frequency) loss-of-function variant, N396S, was shown to be associated with significantly elevated HDL-C levels [52]. In a very large analysis, this variant was confirmed to be associated with a significant increase in HDL-C but no association with other lipid or nonlipid risk factors [32]. While observational epidemiology would have predicted a significant reduction in myocardial infarction (MI) risk based on the HDL-C increase, there was no association of this variant with CHD risk. In the same analysis, 14 additional HDL-C variants that were not associated with changes in other lipids were also found to have no association with CHD [32]. Overall, these aggregated results provide poor support for the concept that plasma HDL-C levels are causally associated with atherosclerotic disease.

Interventions That Increase HDL-C Levels Have Not Been Shown to Reduce CHD Events

There are only a few randomized controlled trials with interventions that raise HDL-C levels (primarily niacin and CETP inhibitors), but they have not supported the "HDL cholesterol hypothesis." A detailed review of these studies is beyond the scope of this chapter. Two studies of niacin added to a statin both showed increases in HDL-C but failed to demonstrate a reduction of CV events [53]. Two different CETP inhibitors, torcetrapib and dalcetrapib, raised HDL-C substantially but failed to reduce CV events [54, 55]. There remains substantial interest in the concept that HDL composition, for example, content of apoC-III [56], or HDL function, for example, capacity to promote cholesterol efflux, may be causally more relevant than HDL-C levels *per se* [57] and that therapeutically targeting "HDL flux" might reduce CV risk [58]. In addition, the post-translational modification of apoA-I can directly lead to HDL dysfunction as observed, for example, in HDL from diabetic subjects [59]. Ultimately, trials of interventions that promote cholesterol efflux and reverse cholesterol transport will be needed to address this hypothesis.

SUMMARY

Lipoproteins are important risk factors for atherosclerotic CVD. LDL-C is clearly an independent risk factor and all available evidence indicates that LDL-C is a causal factor in promoting atherosclerosis and cardiovascular events. Plasma TG levels are probably an independent risk factor as well, and the evidence suggests that certain TRLs, particularly remnant lipoproteins, are causal whereas others, particularly chylomicrons, are not. Finally, HDL-C is a major independent inverse risk factor for CHD. However, available evidence suggests that this strong association may not be causal in nature. LDL-C, TG, and HDL-C can all be used to predict risk of CHD, but data currently exist only for intervention targeted toward reducing LDL-C as a strategy to reduce risk of cardiovascular events.

REFERENCES

1. Besterman EM. (1957) Lipoproteins in coronary artery disease. *British Heart Journal* 19, 503–515.
2. Hoefner DM. (2008) The cholesterol controversy continues. *MLO: Medical Laboratory Observer* 40, 24–28.
3. Konstantinov IE, Mejevoi N, Anichkov NM. (2006) Nikolai n. Anichkov and his theory of atherosclerosis. *Texas Heart Institute Journal* 33, 417–423.
4. Ignatowsky A. (1909) Über die wirkung des ferischen eiweisses auf die aorta and parenchymatösen organe der kaninchen. *Virchows Archives Pathological Anatomy and Histology* 198, 248–270.
5. Hoeg JM, Klimov AN. (1993) Cholesterol and atherosclerosis: The new is the old rediscovered. *American Journal of Cardiology* 72, 1071–1072.

6. Anitschkow N, Chalatow S. (1913) Über experimentelle cholesterinsteatose und ihre bedeutung für die entstehung einiger pathologischer prozesse. *Zentralblatt ffr Allgemeine Pathologie und Pathologische Anatomie* 24, 1–9.
7. Anichkov N. (1915) New data on the pathology and etiology of arteriosclerosis (atherosclerosis). *Russkii Vrach [Russian Physician]* 9, 207–211.
8. Macheboeuf MA. (1929) Recherches sur les phosphoaminolipides et les sterides du serum et du plasma sanguins. *Bulletin de la Société de Chimie Biologique* 11, 268–293.
9. Steinberg D. (2004) Thematic review series: The pathogenesis of atherosclerosis. An interpretive history of the cholesterol controversy: Part I. *Journal of Lipid Research* 45, 1583–1593.
10. McFarlane AS. (1935) Lxxix: The ultracentrifugal protein sedimentation diagram of normal human, cow, and horse serum. *The Biochemical Journal* 29, 660–693.
11. Pedersen KO. (1947) On a low-density lipoprotein appearing in normal human plasma. *Journal of Physical and Colloid Chemistry* 51, 156–163.
12. Blix G, Tiselius A, Svensson H. (1941) Lipids and polysaccharides in electrophoretically separated blood serum proteins. *The Journal of Biological Chemistry* 137, 485–494.
13. Gofman JW, Lindgren FT, Elliott H. (1949) Ultracentrifugal studies of lipoproteins of human serum. *The Journal of Biological Chemistry* 179, 973–979.
14. Gofman JW. (1956) Serum lipoproteins and the evaluation of atherosclerosis. *Annals of the New York Academy of Sciences* 64, 590–595.
15. Keys A, Taylor HL, Blackburn H, Brozek J, Anderson JT, Simonson E. (1963) Coronary heart disease among Minnesota business and professional men followed fifteen years. *Circulation* 28, 381–395.
16. Keys A. (1970) Coronary heart disease in seven countries. *Circulation* 41, 1–211.
17. Fredrickson DS, Lees RS. (1965) A system for phenotyping hyperlipoproteinemia. *Circulation* 31, 321–327.
18. Levy RI, Fredrickson DS. (1968) Diagnosis and management of hyperlipoproteinemia. *The American Journal of Cardiology* 22, 576–583.
19. Kannel WB, Dawber TR, Kagan A, Revotskie N, Stokes J, 3rd. (1961) Factors of risk in the development of coronary heart disease—Six year follow-up experience. The Framingham study. *Annals of Internal Medicine* 55, 33–50.
20. Rader DJ. (2006) Molecular regulation of HDL metabolism and function: Implications for novel therapies. *The Journal of Clinical Investigation* 116, 3090–3100.
21. Gordon DJ, Rifkind BM. (1989) High-density lipoprotein—The clinical implications of recent studies. *The New England Journal of Medicine* 321, 1311–1316.
22. Hobbs HH, Brown MS, Goldstein JL. (1992) Molecular genetics of the LDL receptor gene in familial hypercholesterolemia. *Human Mutation* 1, 445–466.
23. Tybjaerg-Hansen A, Humphries SE. (1992) Familial defective apolipoprotein b-100: A single mutation that causes hypercholesterolemia and premature coronary artery disease. *Atherosclerosis* 96, 91–107.
24. Abifadel M, Varret M, Rabes JP, et al. (2003) Mutations in pcsk9 cause autosomal dominant hypercholesterolemia. *Nature Genetics* 34, 154–156.
25. Horton JD, Cohen JC, Hobbs HH. (2007) Molecular biology of pcsk9: Its role in LDL metabolism. *Trends in Biochemical Sciences* 32, 71–77.
26. Linton MF, Farese RV, Jr., Young SG. (1993) Familial hypobetalipoproteinemia. *Journal of Lipid Research* 34, 521–541.
27. Cohen J, Pertsemlidis A, Kotowski IK, Graham R, Garcia CK, Hobbs HH. (2005) Low LDL cholesterol in individuals of African descent resulting from frequent nonsense mutations in pcsk9. *Nature Genetics* 37, 161–165.
28. Kotowski IK, Pertsemlidis A, Luke A, et al. (2006) A spectrum of pcsk9 alleles contributes to plasma levels of low-density lipoprotein cholesterol. *American Journal of Human Genetics* 78, 410–422.
29. Cohen JC, Boerwinkle E, Mosley TH, Jr., Hobbs HH. (2006) Sequence variations in pcsk9, low LDL, and protection against coronary heart disease. *The New England Journal of Medicine* 354, 1264–1272.
30. Raal FJ, Santos RD, Blom DJ, et al. (2010) Mipomersen, an apolipoprotein b synthesis inhibitor, for lowering of LDL cholesterol concentrations in patients with homozygous familial hypercholesterolaemia: A randomised,

double-blind, placebo-controlled trial. *Lancet* 375, 998–1006.

31. Stein EA, Mellis S, Yancopoulos GD, et al. (2012) Effect of a monoclonal antibody to pcsk9 on LDL cholesterol. *The New England Journal of Medicine* 366, 1108–1118.
32. Voight BF, Peloso GM, Orho-Melander M, et al. (2012) Plasma HDL cholesterol and risk of myocardial infarction: A mendelian randomisation study. *Lancet* 380(9841), 572–580.
33. Schunkert H, Konig IR, Kathiresan S, et al. (2011) Large-scale association analysis identifies 13 new susceptibility loci for coronary artery disease. *Nature Genetics* 43, 333–338.
34. Coronary Artery Disease (C4D) Genetics Consortium. (2011) A genome-wide association study in Europeans and south Asians identifies five new loci for coronary artery disease. *Nature Genetics* 43, 339–344.
35. Gotto AM, Jr. (2005) Review of primary and secondary prevention trials with lovastatin, pravastatin, and simvastatin. *The American Journal of Cardiology* 96, 34F–38F.
36. Sarwar N, Sandhu MS, Ricketts SL, et al. (2010) Triglyceride-mediated pathways and coronary disease: Collaborative analysis of 101 studies. *Lancet* 375, 1634–1639.
37. Santamarina-Fojo S. (1992) Genetic dyslipoproteinemias: Role of lipoprotein lipase and apolipoprotein c-ii. *Current Opinion in Lipidology* 3, 186–195.
38. Mahley RW, Huang Y, Rall SC, Jr. (1999) Pathogenesis of type iii hyperlipoproteinemia (dysbetalipoproteinemia). Questions, quandaries, and paradoxes. *Journal of Lipid Research* 40, 1933–1949.
39. Teslovich TM, Musunuru K, Smith AV, et al. (2010) Biological, clinical and population relevance of 95 loci for blood lipids. *Nature* 466, 707–713.
40. Jun M, Foote C, Lv J, et al. (2010) Effects of fibrates on cardiovascular outcomes: A systematic review and meta-analysis. *Lancet* 375, 1875–1884.
41. Chiesa G, Sirtori CR. (2003) Apolipoprotein a-i(milano): Current perspectives. *Current Opinion in Lipidology* 14, 159–163.
42. Hobbs HH, Rader DJ. (1999) Abc1: Connecting yellow tonsils, neuropathy, and very low HDL. *The Journal of Clinical Investigation* 104, 1015–1017.
43. Marcil M, Brooks-Wilson A, Clee SM, et al. (1999) Mutations in the abc1 gene in familial HDL deficiency with defective cholesterol efflux. *Lancet* 354, 1341–1346.
44. Kuivenhoven JA, Pritchard H, Hill J, et al. (1997) The molecular pathology of lecithin: Cholesterol acyltransferase (LCAT) deficiency syndromes. *Journal of Lipid Research* 38, 191–205.
45. Rader DJ, Ikewaki K, Duverger N, et al. (1994) Markedly accelerated catabolism of apolipoprotein A-II (apoA-II) and high density lipoproteins containing apoA-II in classic lecithin: Cholesterol acyltransferase deficiency and fish-eye disease. *The Journal of Clinical Investigation* 93, 321–330.
46. Brown ML, Inazu A, Hesler CB, et al. (1989) Molecular basis of lipid transfer protein deficiency in a family with increased high-density lipoproteins. *Nature* 342, 448–451.
47. Ikewaki K, Nishiwaki M, Sakamoto T, et al. (1995) Increased catabolic rate of low density lipoproteins in humans with cholesteryl ester transfer protein deficiency. *The Journal of Clinical Investigation* 96, 1573–1581.
48. Thompson A, Di Angelantonio E, Sarwar N, et al. (2008). Association of cholesteryl ester transfer protein genotypes with CEPT mass and activity, lipid levels, and coronary risk. *JAMA* 299, 2777–2788.
49. Frikke-Schmidt R, Nordestgaard BG, Stene MC, et al. (2008) Association of loss-of-function mutations in the abca1 gene with high-density lipoprotein cholesterol levels and risk of ischemic heart disease. *JAMA* 299, 2524–2532.
50. Haase CL, Tybjaerg-Hansen A, Qayyum AA, Schou J, Nordestgaard BG, Frikke-Schmidt R. (2012) LCAT, HDL cholesterol and ischemic cardiovascular disease: A mendelian randomization study of HDL cholesterol in 54,500 individuals. *The Journal of Clinical Endocrinology and Metabolism* 97, E248–E256.
51. Degoma EM, Rader DJ. (2011) Novel HDL-directed pharmacotherapeutic strategies. *Nature Reviews Cardiology* 8(5), 266–277.
52. Edmondson AC, Brown RJ, Kathiresan S, et al. (2009) Loss-of-function variants in endothelial lipase are a cause of elevated HDL cholesterol in humans. *The Journal of Clinical Investigation* 119, 1042–1050.
53. Boden WE, Probstfield JL, Anderson T, et al. (2011) Niacin in patients with low HDL

cholesterol levels receiving intensive statin therapy. *The New England Journal of Medicine* 365, 2255–2267.

54. Barter PJ, Caulfield M, Eriksson M, et al. (2007) Effects of torcetrapib in patients at high risk for coronary events. *The New England Journal of Medicine* 357(21), 2109–2122.
55. Schwartz GG, Olsson AG, Abt M, et al. (2012) Effects of dalcetrapib in patients with a recent acute coronary syndrome. *The New England Journal of Medicine* 367(22), 2089–2099.
56. Jensen MK, Rimm EB, Furtado JD, Sacks FM. (2012) Apolipoprotein c-iii as a potential modulator of the association between HDL-cholesterol and incident coronary heart disease. *Journal of American Heart Association*. 1, pii: jah3-e000232.
57. Khera AV, Cuchel M, de la Llera-Moya M, et al. (2011) Cholesterol efflux capacity, high-density lipoprotein function, and atherosclerosis. *The New England Journal of Medicine* 364, 127–135.
58. Rader DJ, Tall AR. (2012) The not-so-simple HDL story: Is it time to revise the HDL cholesterol hypothesis? *Nature Medicine* 18, 1344–1346.
59. Fisher EA, Feig JE, Hewing B, Hazen SL, Smith JD. (2012) High-density lipoprotein function, dysfunction, and reverse cholesterol transport. *Arteriosclerosis, Thrombosis, and Vascular Biology* 32, 2813–2820.

2

ROLE OF DYSGLYCEMIA IN ATHEROSCLEROSIS

DAVID R. CLEMMONS

Department of Medicine, UNC School of Medicine, University of North Carolina at Chapel Hill, Chapel Hill, NC, USA

INTRODUCTION

Hyperglycemia is a major risk factor for accelerating the rate of development of multiple cardiovascular complications of diabetes. Dysglycemia is a term that describes abnormal glucose regulation that occurs both prior to and following the development of overt diabetes mellitus. The available data support the conclusion that there is a direct effect of blood glucose on blood vessel function as well as indirect effects that modify the responses to other cardiovascular risk factors such as hypertension, hyperlipidemia, and cytokines [1]. Furthermore, the effect of hyperglycemia is strong enough that even minor deviations from normal glucose levels can lead to increased cardiovascular risk; however, the proportion of patients who are affected increases substantially with worsening severity of dysglycemia [2]. Therefore, attention has been focused on the assessment of cardiovascular risk in several different subpopulations of patients across the entire spectrum of severity of dysglycemia from those who exhibit only abnormal regulation of blood glucose dynamics to patients who have severe insulin-dependent diabetes. Patients with abnormal blood glucose regulation prior to the development of overt diabetes have been classified as those with insulin resistance, impaired fasting glucose, impaired glucose tolerance, and "wide variability" in blood glucose. Since each of these groups overlaps, clinical studies that attempt to stratify subjects often are unable to include pure populations and simply attempt to define the degree overlap that exists. Patients with overt diabetes have the highest risk of subsequent development of cardiovascular disease with an overall risk for myocardial infarction that is equal to the Framingham risk equivalent of individuals with established coronary artery disease (CAD) [3]. Studies using multifactorial linear regression to take into account the effects of other risk factors have shown that diabetes alone is an independent risk factor for cardiovascular

Atherosclerosis: Risks, Mechanisms, and Therapies, First Edition. Edited by Hong Wang and Cam Patterson.

disease. Plaque burden is more extensive in coronary vessels in young people with diabetes. Vascular events represent the leading cause of death in patients with diabetes and 75% of all hospitalizations for people with diabetes are for CAD. The variables that are major determinants of the percentage of patients with diabetes who have detectible vascular disease are age and the presence of nephropathy [4]. CAD was only present in 9% of men in type I diabetics aged between 15 and 29 years, whereas it was detected in 25% of patients between 45 and 59 years who had a similar duration of disease. In the presence of nephropathy, 43% of patients with type I diabetes of 20 years duration had CAD as compared to 7% of similar patients without nephropathy [4].

Patients who have not yet developed overt diabetes are also at increased risk for vascular diseases. However, patients with prediabetes and overt diabetes are part of a continuum since at the time, they develop either impaired fasting glucose or glucose intolerance significant beta cell destruction has already occurred [5]. Therefore, although they may not carry the classification of overt diabetes, prediabetics clearly have underlying pathophysiologic changes that are likely to lead to the development of overt diabetes in the future. This is supported by epidemiologic studies that show that greater than 40% of patients with impaired fasting glucose will progress to type II diabetes within 5 years [6]. Studies that have attempted to stratify subjects based on impaired fasting glucose versus glucose intolerance have shown different levels of cardiovascular disease risk. An extensive Japanese study concluded that impaired glucose tolerance was a risk factor for cardiovascular disease but not impaired fasting glucose, and a Finnish study indicated that impaired glucose tolerance at baseline was an independent predictor of incidence of cardiovascular disease [7]. The Chicago heart study also emphasized the importance of postprandial hyperglycemia. The DECODE study analyzed data from 10 prospective European cohort studies that included 22,000 subjects. Death rates were higher for CAD (relative risk ratio 1.56:1) in subjects who had 2-hour postprandial plasma glucose values that were elevated [8]. There was also increased cardiovascular morbidity in this group, but not in the group with impaired fasting glucose. These differences no doubt reflect, in part, that subjects with abnormal postprandial glucose are more likely to have elevated blood glucose throughout the day as compared to those with normal postprandial glucose but elevated fasting glucose. This conclusion is supported by the observation that subjects with abnormal glucose tolerance have higher hemoglobin A1c (HbA1c) concentrations than subjects who have impaired fasting glucose [9]. In the DECODE study, both abnormal fasting blood glucose and postprandial glucose were correlated with increasing risk (relative risk 1.09:1 and 1.56:1, respectively) although the degree of risk for CAD was greater in subjects with elevated 2-hour blood glucose values. Neither group showed a degree of risk of vascular events that was as severe as the subjects with overt diabetes. In the patients who had vascular events, the prevalence of impaired glucose tolerance was 31%; whereas for overt diabetes, it was 60%. There is also a linear relationship between excess cardiovascular mortality and the degree of abnormality in postprandial glucose concentrations [10]. More recently, HbA1c has been used to make these comparisons. The UKPDS study showed an 11% increase in cardiovascular risk for every percentage point increase in HbA1c [11]. Data from the Swedish National Registry also verified the importance of the degree of abnormality based on HbA1c since they showed the lowest risk n patients with a HbA1c of 6.5% (20% decrease), and a substantially higher risk (relative risk ratio 1.15:1) in patients with HbA1c values between 8 and 9% when compared to patients with a mean value of 7.5% [12]. These findings emphasize the importance of screening patients with dysglycemia for the presence of other cardiovascular risk factors including hypertension, hyperlipidemia, nephropathy, or retinopathy.

Given that hyperglycemia predisposes to increased cardiovascular risk, several studies have attempted to determine whether lowering glucose in patients with overt diabetes has an effect on cardiovascular disease outcomes. The strongest findings come from patients with type I diabetes who were treated with intensive insulin therapy in the DCCT study. Subjects were divided into a group that received intensive therapy and a group that received conventional treatment. The study duration was 7 years. The two groups achieved very different HbA1c values of 7.4 and 9.1%, respectively. Importantly, a follow-up study was conducted after the 7-year trial ended. In this study, termed the EDIC trial, 75% of the subjects elected to be followed for an additional 11 years [13]. Because there was no attempt made to enforce tight control, the HbA1c values for both groups converged; and within a year, they were 8.0 and 8.2%, respectively. Both groups received the same treatment for other risk factors such as hypertension and hyperlipidemia. By year 11, the group that had been treated conventionally had a greater prevalence of microalbuminuria and serum creatinine greater than 2.0 mg/dl. The relative risk of cardiovascular events was 42% less in the intensively treated group, and the occurrence of nonfatal myocardial infarction or stroke was reduced by 57% [13]. These findings strongly suggest that intensive lowering of glucose in patients with type I diabetes results in a sustained benefit that persists for several years after this intervention. The UKPDS trial also compared conventional with intensive therapy for 10 years in patients with type II diabetes [11]. Subjects were then followed for an additional 10 years. Both groups had a convergence of their HbA1c values during follow-up. The intensively treated group showed a significantly lower cumulative incidence of myocardial infarction (e.g., 15% decrease) and death (13% decrease).

More recent studies in patients with longstanding type II diabetes and the presence of established CAD or additional cardiovascular risk factors have not supported that intensive glucose lowering is as effective. The ACCORD trial demonstrated a higher rate of mortality in subjects in the intensive therapy group that targeted a HbA1c level below 6.0% compared to the standard therapy group that targeted a HbA1c level from 7 to 7.9% [14]. The increased deaths in the intensive therapy group remain unclear but do not appear to be associated with nonfatal myocardial infarction that was significantly reduced (hazard ratio 0.76). The ADVANCE trial, which also compared the effects of tighter control in older subjects, showed a major reduction in microvascular events (hazard ratio 0.86) but not macrovascular events in subjects who had intensive control. In that study, there was no difference in overall mortality [15]. The Veterans Affairs Diabetes Trial also addressed the same question of intensive or standard glucose control therapy. This study showed that there was no significant difference for major cardiovascular events, deaths, or microvascular complications (with the exception for microalbuminuria) [16].

In summary, these findings suggest that dysglycemia is a major predisposing factor for the development and progression of CAD, and that the risk is amplified with increasing age, the presence of cardiovascular risk factors, or nephropathy. Significant intervention at younger ages in subjects who have minimal or no vascular disease burden may result in a long-term benefit. Aggressive glycemic control in type II diabetic subjects who have well-established cardiovascular disease or risk factors is not associated with further benefit and may result in increased overall mortality. These findings strongly suggest that major interventions that attempt to inhibit the effect of hyperglycemia on cardiovascular disease progression should be undertaken in younger patients who are relatively asymptomatic and who are likely to have minimal atherosclerotic disease compared to more advanced lesion burden. These studies also emphasize the need to individualize therapy and to pursue more modest treatment objectives in older patients with established CAD.

MOLECULAR MECHANISMS BY WHICH DYSGLYCEMIA ACCELERATES ATHEROSCLEROSIS

The molecular mechanisms by which hyperglycemia leads to accelerated progression of atherosclerosis are complex. Clearly, hyperglycemia is an independent risk factor, but it also interacts with other risk factors. When the animal models that have been utilized to determine the effects of dysglycemia have been evaluated, the results are often confounded by attempting to determine the effects that are directly dependent on hyperglycemia versus those that are indirectly mediated through induction of other pathophysiologic pathways such as increased cytokine secretion, nonenzymatic glycosylation of proteins, modification of lipoprotein signaling through various pathways, modification of peptide growth factor signaling, and alterations in extracellular matrix protein composition leading to changes in integrin receptor function. Furthermore, some animal models are susceptible to the development of complications such as hypertension and abnormal lipoprotein metabolism simply as a function of raising glucose. Therefore, interpretation of experiments in animal models should consider if each of those factors in isolation are due to changes in glucose as compared to those factors that are changed due to exacerbation of other pathophysiological variables. Nevertheless, despite these challenges several observations regarding the pathogenesis of alterations in vascular function that are due to hyperglycemia alone have been forthcoming.

The effects of glucose on vascular function have been examined in multiple cell types that are involved in atherosclerotic lesion formation within the vessel wall. These include macrophages, endothelial cells, smooth muscle cells, and immune cells such as dendritic cells and T lymphocytes. Both the direct effects of exposure of each of the cell type to high glucose and indirect effects wherein high glucose mediates changes in signaling proteins that are responsive to other stimuli, such as growth factors and cytokines, have been analyzed. Furthermore, in many cases the changes in cellular function have been analyzed *in vitro* following *in vivo* exposure to hyperglycemia. These experimental approaches have allowed the identification of several aberrant signaling pathways by which hyperglycemia mediates changes in cellular function (Figure 2.1).

One of the major pathways that has been analyzed is oxidative stress. Oxidative stress has been shown to occur in endothelium, smooth muscle cells and macrophages in response to high glucose exposure. The common mechanism that has been proposed is that diacylglycerol is activated, which leads to the activation of the enzyme PKC. Several PKC isoforms are induced and linked to further increases in oxidative stress [17]. The increase in reactive oxygen species (ROS) that occurs in mitochondria is due to accumulation of multiple metabolites; whereas in other subcellular compartments, it is due to activation of NAPDH oxidases. Both endothelial and vascular smooth muscle cells (VSMCs) generate ROS following exposure to hyperglycemia through activation of NAPDH oxidases and mitochondrial ROS generation. The exact biochemical steps that lead to activation of NAPDH oxidases have not been extensively delineated. Diaclyglycerol generation with subsequent PKC activation is involved; however, the initiating steps in this pathway and the signaling elements through which these are propagated are still under active investigation. ROS release by NAPDH oxidases can have several consequences that often depend on the source of the ROS and the subcellular location wherein the ROS release occurs.

The generation of ROS in mitochondria leads to a decrease in the activity of GAPDH, which increases glycolytic intermediates that are upstream [18]. These intermediates increase flux through the polyl pathway and the methylglyoxal level, which induces glycation of proteins. They also increase PKC activity as well as hexosamine pathway activation. Inhibition of ROS production in mitochondria inhibits the reduction in GAPDH activity and the activation of these pathways.

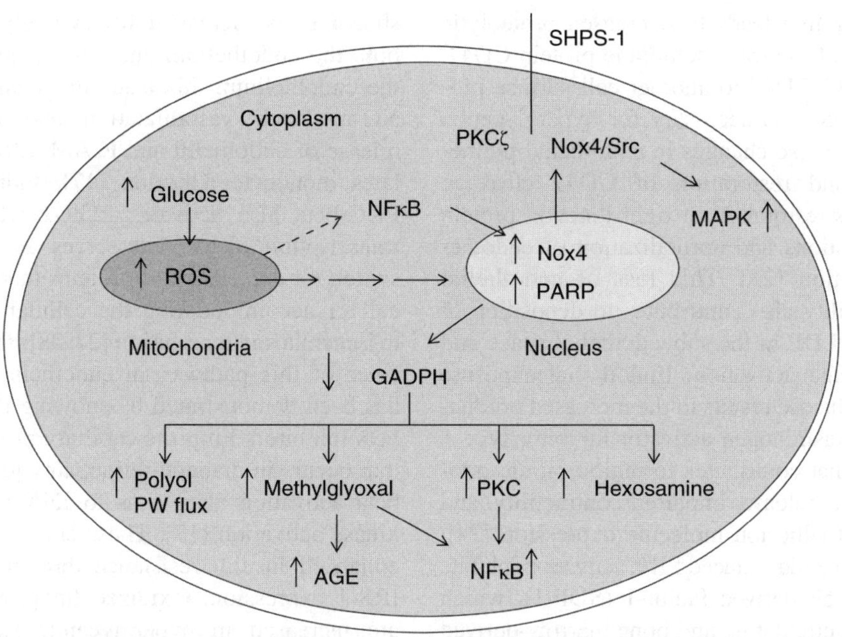

Figure 2.1 Hyperglycemia-induced activation of mitochondrial ROS suppresses GAPDH activity. Hyperglycemia acts on mitochondria to increase ROS production and PARP expression that suppresses GAPDH activity. Suppression of GAPDH activity increases accumulation of intermediary metabolites that stimulate multiple signaling pathways aberrantly.

DYSGLYCEMIA AND ENDOTHELIAL CELLS

In the vascular endothelium (VE), increased ROS leads to enhancement of redox-sensitive transcription factors such as NF-kappaB (NF-κB that impairs nitric oxide synthase activity [19]. This results in a decrease in nitric oxide bioavailability, decreased vasodilation, and increased endothelial stiffness. Disruption of eNOS activation also generates superoxide that reacts with NO to form peroxynitrite that generates further ROS. The two primary forms of NAPDH oxidases in endothelium are Nox2 and Nox4. These oxidases can act directly to enhance endothelial permeability and Nox-dependent changes in gene expression. Under nondiabetic conditions, Nox2 may be the dominant endothelial isoform mediating endothelial dysfunction since endogenous Nox4 protected mice during ischemic or inflammatory stress [20]. Insulin also plays a major role in enhancing NOS activity in endothelial cells. Mice in which the insulin receptor was selectively eliminated from endothelial cells exhibited reduced eNOS activity [21]. Induction of insulin resistance leads to decreased eNOS activation. Inhibition of NAPDH oxidases has been shown to improve vascular relaxation in arteries from diabetic animals. Mitochondrial ROS generation in the endothelium is also an important variable. Under hyperglycemic conditions, it contributes to the formation of advanced glycation end (AGE) products, as well as PKC and NF-κB activation. All of these events lead to further endothelial dysfunction. Exposure of endothelium to hyperglycemia also has direct effects on vascular permeability. Specifically, the junctional proteins such as VE cadherin, claudin and occludin are translocated away from tight junctions leading to enhanced cellular permeability [22]. The exact mechanisms that account for dysregulation of these proteins have not been definitively determined; however, increased oxidative stress is one variable that has been shown to contribute, hyperglycemia leads to decreased proteolytic cleavage of the transmembrane protein CD47. Binding of CD47 to another cell surface protein, SHPS-1, is necessary for hyperglycemia to induce these changes in endothelial permeability, and restoration of CD47 cleavage facilitates restoration of tight junction protein concentrations and normalization of endothelial function [23]. This loss of endothelial permeability also contributes to deposition of oxidized LDL in the subendothelial space and macrophage activation. Endothelial exposure to high glucose results in the increased production of plasminogen activator inhibitor type 1 (PAI-1) that contributes to inhibition of endothelial NO release, impaired contractility, and increased adhesion molecule expression [24]. Hyperglycemia-induced PKC activation inhibits stromal cell derived factor-1 (SDF-1), which regulates circulating and bone marrow-derived endothelial progenitor cell mobilization, and the interaction with its receptor, CXCR4 [25].

DYSGLYCEMIA AND VASCULAR SMOOTH MUSCLE CELLS

Vascular smooth muscle cells (VSMCs) also undergo several biochemical changes in response to hyperglycemia. ROS generation is also important for smooth muscle cell function. Increased glucose flux leads to increased formation of AGE products bind to the RAGE receptor and activate pathophysiologic events that are described in Chapter 3 and include ROS generation. Mitochondrial overproduction of ROS leads to changes that are similar to those described for endothelial cells. In VSMCs, hyperglycemia increases phospholipase C activation, which leads to diacylglycerol formation that directly activates PKC isoenzymes. Activation of PKC-zeta has been shown to lead directly to increased synthesis of Nox4 that enhances growth factor-induced signaling [26]. Mitochondrial ROS also activates PKC-beta—an effect leading to induction of the early growth response-1 (EGR-1) pathway [27]. PKC-beta activation can also occur in response to hypoxia. PKC-beta II has been shown to be activated by hyperglycemia in both the endothelium and smooth muscle. In the endothelium, this leads to impairment of NO-mediated vasodilatation and increased release of endothelin and ICAM-1 that stimulates monocyte adhesion [17]. Induction of PKC-beta also activates EGR-1 which is a transcription factor that serves as a master switch for expression of proteins that are critical for accommodating the cellular response to ischemia and reperfusion [27, 28]. The importance of this pathway in endothelial biology has been demonstrated by showing that PKC-beta inhibitors limit the capillary proliferation that occurs in diabetic retinopathy [29]. PKC-beta activation also leads to JNK p38 MAP kinase activation [17]. These enzymes lead to enhanced insulin resistance through reduced IRS-1 expression. Oxidized lipoproteins that are increased in hyperglycemic states also contribute to overexpression of EGR-1 in VSMCs and macrophages. Human atherosclerotic lesions exhibit increased EGR-1 expression associated with increased expression of TNFα, ICAM-1, PAI-1, and M-CSF [27]. In addition, compound EGR-1- and apoE-deficient mice displayed reduced expression for these markers and atherosclerotic lesion formation [27]. Similar changes were noted in compound PKC-beta- and apoE-deficient mice suggesting that many of the changes were mediated through induction of EGR-1 [27].

Nox1 and Nox4 are the functionally active NAPDH oxidases in VSMCs. The activity of these two enzymes is regulated by several molecules that also regulate vascular function. Angiotensin II exposure results in oxidation of Src kinase through a Nox4-dependent activity [30]—an effect that increases Src enzymatic activity and VSMC proliferation and migration. Other studies that are relevant to diabetes have shown that hyperglycemia enhances insulin-like growth factor I (IGF-1)-stimulated Src activation by increasing Nox4 expression [26]. Under these conditions, IGF-1 stimulation leads to the recruitment of the p22 phox subunit to NOX4, resulting in enhanced Src oxidation and IGF-1-stimulated VSMC migration and proliferation. Blocking Nox4 activation in diabetic mice reversed these changes [26]. NAPDH oxidases have also been shown to work cooperatively with other growth factors such as VE growth factor and transforming growth factor beta [31].

A major mechanism by which hyperglycemia induces insulin resistance is by a reduction in IRS-1 signaling. Upon activation, IRS-1 binds to the insulin receptor and acts as a scaffolding protein to activate both the PI-3 and MAP kinase pathways. IRS-1 binds directly to the p85 subunit of PI3 kinase leading to enhanced activation. Because glucose suppresses IRS signaling, this pathway is a critical mediator of insulin resistance [32]. In addition, cytokines increased in diabetes, such as IL-6 and TNFα, enhance JNK activity and reduces IRS-1 expression. The importance of hyperglycemia-mediated reduction of IRS-1 signaling was examined in compound IRS-1- and apoE-deficient mice that exhibited accelerated atherosclerotic lesion formation [33]. Nonenzymatic glycosylation of LDL oxidizes LDL, which may directly stimulates VSMC proliferation is mediated through the lipoprotein receptor CD36 independent of the RAGE receptor [34]. Chronic exposure to high glucose leads to an imbalance in insulin action in VSMCs, in part, due to protein tyrosine phosphatase PTP1B and decreased AKT phosphorylation [35]. Blocking PKC-beta 2 activation also inhibits high glucose-induced phosphorylation of Rac 1B, p47 phox, p67 phox, and ROS production in VSMCs [36]. High glucose also enhances CD36 expression and suppresses the expression of the ABC transporter cassette ABCG1, thereby inhibiting cholesterol efflux—an effect mediated, in part, through NF-κB activation [37]. High glucose also enhances VSMC chemotactic responses to platelet-derived growth factor by increasing PDGF receptor tyrosine kinase activation [38]. Finally, hyperglycemic rats demonstrate a Rho kinasedependent reduction in eNOS gene expression and vasoconstriction [39]. Therefore, there are multiple glucose dependent mechanisms by which insulin signaling can be disrupted.

VSMCs also respond to hyperglycemic stress by activating the alpha V beta 3 integrin (αVβ3) (Figure 2.2). This integrin mediates

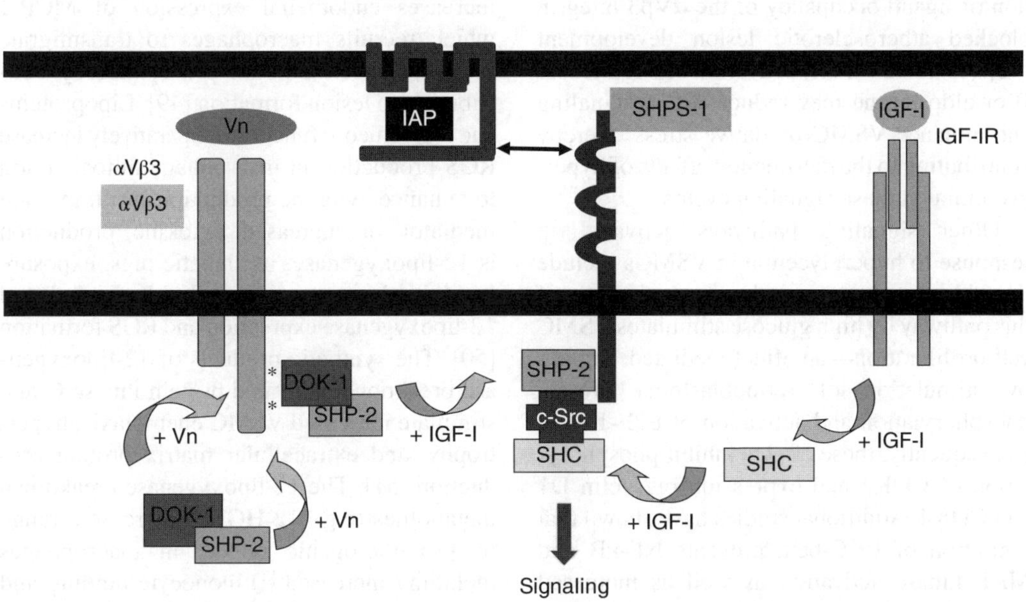

Figure 2.2 Hyperglycemia induces integrin αVβ3 activation that forms an alternative signaling scaffold. Following exposure to hyperglycemia, endothelial, and smooth muscle cells secrete increased concentrations of αVβ3 ligands. This results in stimulation of β3 tyrosine phosphorylation of the β3 subunit which recruits SHP-2 to the cell surface. Subsequently, the IGF-1 receptor phosphorylates tyrosines on the scaffolding protein SHPS-1. SHP-2 is transferred to SHPS-1 that allows assembly of a signaling complex that stimulates activation of both the MAP kinase and PI-3 kinase signaling pathways.

VSMC attachment to a transitional matrix that activates VSMC proliferation and migration. Following exposure to hyperglycemia, VSMCs synthesize substantially greater quantities of αVβ3 ligands such as osteopontin and vitronectin [40]. These ligands interact with αVβ3 to stimulate phosphorylation of the β3 subunit, which results in recruitment of the signaling protein SHP-2 from the cytoskeleton to the plasma membrane. If cells are stimulated with growth factors such as IGF-1, then the scaffolding protein SHPS-1 is tyrosine phosphorylated and a signaling complex is formed which can activate both the PI3 and MAP kinase pathways [41]. Since IRS-1 mediates the ability of IGF-1 to inhibit VSMC apoptosis and is markedly reduced in response to hyperglycemia, SHPS-1 activation serves as an alternative signaling mechanism for activation of these pathways [41]. However, in addition to preventing apoptosis, SHPS-1 activation enhances VSMC migration and proliferation, thereby potentially contributing to atherosclerotic lesion formation. In diabetic pigs, inhibition of ligand occupancy of the αVβ3 integrin blocked atherosclerotic lesion development [42]. As noted earlier, activation of angiotensin II or aldosterone may reduce IRS-1 signaling and enhance VSMC oxidative stress, thereby contributing to the detrimental effects of hyperglycemia on these signaling events.

Other signaling pathways activated in response to hyperglycemia in VSMCs include the aldose reductase pathway. Activation of this pathway by high glucose stimulates VSMC cell proliferation—an effect mediated, in part, by stimulation of retinoblastoma protein phosphorylation and activation of E2F-1 [43]. Consequently, these effects inhibit phosphorylation of CDK2 and expression of cyclin D1 and CDK4. Additional studies have shown that inhibition of PKC-beta prevents NF-κB and MAP kinase activation as well as increased AGE formation in SMC. High glucose-mediated induction of AGE in VSMCs increases IL-6 and monocyte chemoattractant protein (MCP)-1 expression, effects that may lead to lesional macrophage accumulation and VSMC proliferation [44]. Finally, high glucose induces SDF-1 and CXCR4 expression that can stimulate PI3 kinase-mediated VSMC proliferation in rats.

DYSGLYCEMIA AND MACROPHAGES

Hyperglycemia also strongly influences macrophage function. Exposure to high glucose in macrophages increases diacylglycerol formation with activation of PKC-alpha and PKC-beta. Inhibition of PKC-beta has been shown to reduce ROS generation in macrophages as well as cytokine expression [45] (please enter reference for this). Glucose also stimulates LDL and oxLDL uptake by macrophages in diabetic mice, in part, by increased expression [46, 47]. Additional studies in diabetic mice have shown that there is decreased synthesis of ABCA1. Consequently, macrophages may accumulate more cholesterol, a common feature of macrophages that have accumulated in the subendothelial space [48]. Further, high glucose increases endothelial expression of MCP-1 which recruits macrophages to transmigrate into, the subendothelial space, thereby contributing to lesion formation [49]. Lipoproteins and high glucose function cooperatively increase ROS production in macrophage mitochondria to enhance cytokine production. An important mediator of increased cytokine production is 12-lipoxygenase. In diabetic pigs, exposure to high glucose and hyperlipidemia induced 12-lipoxygenase expression and ROS formation [50]. The synthetic products of 12-lipoxygenase breakdown activated protein kinase C and stimulate increased VSMC chemotaxis, hypertrophy, and extracellular matrix protein production [51]. The 12-lipoxygenase breakdown metabolites (e.g., 12 HGTE) increase a range of pro-atherogenic effects in macrophages including increased (i) monocyte binding and uptake by the vessel wall; (ii) CD36 expression and cholesterol efflux; and (iii) activation the NF-κB pathway and production of multiple cytokines including IL-1 beta, IL-6, IL-8, and TNF-alpha in diabetic mice [52]. A central role for hyperglycemia was observed in atherosclerotic-prone LDL-receptor –/– "Reversa" mice that harbor a conditional knockout gene encoding the microsomal triglyceride transfer protein to lower elevated plasma LDL levels. In the Reversa mice rendered diabetic by streptozotocin, plaque cholesterol, and macrophage numbers were retained longer than controls and exhibited higher oxidant stress and proinflammatory markers [53]. These findings underscore the notion that hyperglycemia cooperatively functions with hyperlipidemia to enhance macrophage activation and lesion formation.

CONCLUSION

In summary, multiple molecular mechanisms converge to enhance lesion progression in the presence of hyperglycemia. While most animal models require simultaneous exposure to hyperlipidemia and hyperglycemia, the severity of lesion progression is clearly accelerated during exposure to chronic hyperglycemia. Furthermore, many of the biochemical events induced by activated lipids are also induced by glucose. Therefore, in the presence of both stimuli, the subsequent pathophysiologic events and the anatomic changes in the vessel wall that ensue are likely to be a result of increases in both types of metabolites on downstream signaling pathways. The extent to which hyperglycemia alone can activate atherosclerosis in animal models has not been extensively studied. This may be due, in part, to the time necessary to generate lesion progression that can be for periods as long as 5 years in models such as obese, diabetic monkeys. Therefore, due to time and resource limitations, most investigators have accelerated atherosclerotic lesion development by utilizing simultaneous stimulation with hyperglycemia and hyperlipidemia. However, the available data would indicate that hyperglycemia can increase the sensitivity to deleterious effects of normal lipoprotein concentrations suggesting that even hyperglycemic patients who are maintaining LDL-cholesterol in a reasonable range may be at risk for accelerated lesion development. Further, the role of non-HDL cholesterol that is bound to lipoprotein particles other than LDL in accelerating atherosclerosis may be important. Since this cholesterol fraction is frequently elevated in patients with diabetes, understanding the mechanistic interactions between hyperlipidemia and hyperglycemia will continue to be an important priority for research and offers major potential for the development of new therapeutic approaches.

ACKNOWLEDGMENTS

The author thanks Ms Laura Lindsey for her help in preparing the manuscript. This work was supported by grant AG02331 from the National Institutes of Health.

REFERENCES

1. Hanna-Moussa, A., Gardner, M.J., Kurukulasuriva, L.R., Sowers, J.R. (2009) Dysglycemia/prediabetes and cardiovascular risk. Rev Cardiovasc Med 10(4):202–208.
2. Ryden, L., Standle, E., Bartnik, M., et al. (2007) Guidelines on diabetes: pre-diabetes, and cardiovascular diseases: executive summary. The task force on Diabetes and Cardiovascular Diseases of the European Society of Cardiology (ESC) and the European Association for the Study of Diabetes (EASD). Eur Heart J 28:88–136.
3. Haffner, S.M., Lehto, S., Ronnemaa, T., Pyorala, K., Laakso, M. (1998) Mortality from coronary heart disease in subjects with type 2 diabetes and in nondiabetic subjects with and without prior myocardial infarction. N Engl J Med 339: 229–234.
4. Tuomilehto, J., Borch-Johnsen, K., Molarius, A., et al. (1998) Incidence of cardiovascular disease in type 1 (insulin-dependent) diabetic subjects with and without diabetic nephropathy in Finland. Diabetologia 41(7):784–790.
5. Butler, A.E., Janson, J., Bonner-Weir, S., Ritzel, R., Rizza, R.A., Butler, P.C. (2003) β-Cell deficit and increased β-cell apoptosis in humans with type 2 diabetes. Diabetes 52:102–100.

6. Knowler, W.C., Barrett-Conner, E., Fowler, S.E., et al. (2002) Reduction in the incidence of type 2 diabetes with lifestyle intervention or metformin. N Engl J Med 346:393–403.
7. Anto, T., Okada, S., Niijima, Y., et al. (2010) Impaired glucose tolerance, but not impaired fasting glucose, is a risk factor for early-stage atherosclerosis. Diabetic Med 27(12):1430–1435.
8. DECODE Study Group (2001) Glucose tolerance and cardiovascular mortality, comparison of fasting and 2-hour diagnostic criteria. Arch Intern Med 161:397–405.
9. Woerle, H.J., Pimenta, W.P., Meyer, C., et al. (2004) Diagnostic and therapeutic implications of relationships between fasting, 2 hour postchallenge plasma glucose and hemoglobin A1C values. Arch Intern Med 164:1627–1632.
10. Qiao, Q., Pyorala, K., Pyorala, M., et al. (2002) Two-hour glucose is a better risk predictor for incident coronary heart disease and cardiovascular mortality than fasting glucose. Eur Heart J 23:1267–1275.
11. Homan, R.R., Paul, S.K., Bethel, M.A., et al. (2008) 10 yr follow-up of intensive glucose control in type 2 diabetes. N Engl J Med 359:15771589.
12. Eeg-Olofsson K., Cederholm J., Nilsson P.M., et al. (2010) New aspects if HbA1c as a risk factor for cardiovascular diseases in type 2 diabetes: an observational study from the Swedish National Diabetes Register (NDR). J Intern Med 268: 471–482.
13. Nathan, D.M., Cleary, P.A., Backlund, J.Y., et al. (2005) Intensive diabetes treatment and cardiovascular disease in patients with type 2 diabetes. N Engl J Med 353(25):2643–2653.
14. Gerstein, H.C., Miller, M.E., Byington, R.P., et al. (2008) Effects of intensive glucose lowering in type 2 diabetes. N Engl J Med 358: 2545–2559.
15. Patel, A., MacMahon, S., Chalmers, J., et al. (2008) Intensive blood glucose control and vascular outcomes in patients with type 2 diabetes. N Engl J Med 358:2560–2572.
16. Duckworth, W., Abraira, C., Mortiz, T., et al. (2009) Glucose control and vascular complications in veterans with type 2 diabetes. N Engl J Med 360:129–139.
17. Geradles, P., King, G.L. (2010) Activation of protein kinase C isoforms and its impact on diabetic complications. Circ Res 106(8):1319–1331.
18. Giacco, F., Brownlee, M. (2010) Oxidative stress and diabetic complications. Circ Res 107: 1058–1070.
19. Eugenio, C., Ralph, A.D. (2006) Insulin resistance and endothelial dysfunction: the road map to cardiovascular diseases. Diabetes/Metab Res Rev 22:423–436.
20. Schroder, K., Zhang, M., Benkhoff, S., et al. (2012) Nox4 is a protective reactive oxygen species generating vascular NADPH oxidase. Circ Res 110:1217–1225.
21. Rask-Madsen, C., Li, Q., Freund, B., et al. (2010) Loss of insulin signaling in vascular endothelial cells accelerates atherosclerosis in apolipoprotein E null mice. Cell Metab 11: 379–389.
22. Leal, E.C., Martins, J., Voabil, P., et al. (2010) Calcium dobesilate inhibits the alterations in tight junction proteins and leukocyte adhesion to retinal endothelial cells induced by diabetes. Diabetes 59:2637–2645.
23. Maile, L.A., Gollahon, K., Wai, C., Byfield, G., Hartnett, M.E., Clemmons, D.R. (2012) Disruption of the association of integrin-associated protein (IAP) with tyrosine phosphatase non-receptor type substrate-1 (SHPS-1 inhibits pathophysiological changes in retinal endothelial function in a rat model of diabetes. Diabetologia 55(3):835–844.
24. Fatehl-Hassanabad, Z., Chan, C.B., Furman, B.L. (2010) Reactive oxygen species and endothelial function in diabetes. Eur J Pharmacol 636:8–17.
25. Egan, C.C., Lavery, R., Caporali, F., et al. (2008) Generalized reduction of putative endothelial progenitors and CXCR4-positive peripheral blood cells in type 2 diabetes. Diabetologia 51:1296–1305.
26. Xi, G., Shen, X., Maile, L.A., Wai, C., Gollahon, K., Clemmons D.R. (2012) Hyperglycemia enhances IGF-I-stimulated Src activation via increasing Nox4-derived reactive oxygen species in a PKCζ-dependent manner in vascular smooth muscle cells. Diabetes 61(1):104–113.
27. Yan, S., Harja, E., Andrassy, M., Fugita, T., Schmidt, A.M. (2006) Protein kinase C β/early growth response-1 pathway. A key player in ischemia, atherosclerosis and restenosis. JACC 48(9):A47–A55.
28. Fujita, T., Asaim, T, Andrassy, M., et al. (2004) PKC beta regulates schemia/reperfusion injury in the lung. J Clin Invest 113(11):1615–1623.
29. Harja, E., Bucciarelli, G., Lu, Y., et al. (2004) Early growth response-1 promotes atherogenesis: mice deficient in early growth response-1 and apolipoprotein E display decreased atherosclerosis and vascular inflammation. Circ Res 94(3):333–339.

30. Block, K., Eid, A., Griendling, K.K., et al. (2008) Nox4 NAD(P)H oxidase mediates Src-dependent tyrosine phosphorylation of PDK-1 in response to angiotensin II: role in mesangial cell hypertrophy and fibronectin expression. J Biol Chem 283:24061–24076.
31. Ghouleh, I.A., Kohh, N., Knaus, U.G., et al. (2011) Oxidases and peroxidases in cardiovascular and lung disease: new concepts in reactive oxygen species signaling. Free Rad Biomed 51:1271–1288.
32. Carvalho, E., Rondinone, C., Smith, U. (2000) Insulin resistance in fat cells from obese Zucker rats- evidence for an impaired activation and translocation of protein kinase B and glucose transporter 4. Mol Cell Biochem 206:7–16.
33. Galkina, E.V., Butcher, M., Keller, S.R., et al. (2012) Accelerated atherosclerosis in Apoe$^{-/-}$ mice heterozygous for the insulin receptor and the insulin receptor substrate-1. Arterioscler Thromb Vasc Biol 32:247–256.
34. Allen, L.B., Capps, B.E., Miller, E.C., Clemmons, D.R., Maile, L.A. (2009) Glucose-oxidized low-density lipoproteins enhance insulin-like growth factor I-stimulated smooth muscle cell proliferation by inhibiting integrin-associated protein cleavage. Endocrinology 150(3):1321–1329.
35. Pandey, N.R., Benkirane, K., Amiri, F., Schiffrin, E.L. (2007) Effects of PPAR-gamma knockdown and hyperglycemia on insulin signaling in vascular smooth muscle cells from hypertensive rats. J Cardiovasc Pharmacol 49(6): 346–354.
36. Shaw, S., Wang, X., Redd, H., Alexander, G.D., Isales, C.M., Marrero, M.B. (2003) High glucose augments the angiotensin II-induced activation of JAK2 in vascular smooth muscle cells via the polyol pathway. J Biol Chem 278(33):30634–30641.
37. Xue, J.H., Yuan, Z., Wu, Y., et al. (2010) High glucose promotes intracellular lipid accumulation in vascular smooth muscle cells by impairing cholesterol influx and efflux balance. Cardiovasc Res 86(1):141–150.
38. Campbell, M., Allen, W.E., Silversides, J.A., Trimble, E.R. (2003) Glucose-induced phosphatidylinositol 3-kinase and mitogen-activated protein kinase-dependent upregulation to the platelet-derived growth factor-beta receptor potentiates vascular smooth muscle cell chemotaxis. Diabetes 52(2):519–526.
39. Ishiko, K., Sakoda, T., Akagami, T., et al. (2010) Hyperglycemia induced cell growth and gene expression via the serum response element through RhoA and Rho-kinase in vascular smooth muscle cells. Prep Biochem Biotechnol 40(2):139–151.
40. Nilsson-Berglund, L.M., Zetterqvist, A.V., Nilsson-Ohman, J., et al. (2010) Nuclear factor of activated T cells regulates osteopontin expression in arterial smooth muscle in response to diabetes-induced hyperglycemia. Arterioscler Thromb Vasc Biol 30(2):218–224.
41. Radhakrishnan, Y., Shen, X., Maile, L.A., Xi, G., Clemmons, D.R. (2011) IGF-I stimulates cooperative interaction between the IGF-I receptor and CSK homologous kinase that regulates SHPS-1 phosphorylation in vascular smooth muscle cells. Mol Endocrinol 25(9):1636–1649.
42. Maile, L.A., Busby, W.H., Nichols, T.C., et al. (2010) A monoclonal antibody against alphaVbeta3 integrin inhibits development of atherosclerotic lesions in diabetic pigs. Sci Transl Med 2(18):18ra11.
43. Tammall, R., Saxena, A., Srivastava, S.K., Ramana, K.V. (2010) Aldose reductase regulates vascular smooth muscle cell proliferation by modulating G1/S phase transition of cell cycle. Endocrinology 151(5):2140–2150.
44. Meng, L., Park, J., Cai, Q., Lanting, L., Reddy, M.A., Natarajan, R. (2010) Diabetic conditions promote binding of monocytes to vascular smooth muscle cells and their subsequent differentiation. Am J Physiol Heart Circ Physiol 298(3): H736–H745.
45. Carnevale, K.A., Cathcart, M.K. (2003) Protein kinase C beta is required for human monocyte chemotaxis to MCP-1. J Biol Chem 278: 25317–25322.
46. Lamharzi, N., Renard, C.B., Kramer, F., et al. (2004) Hyperlipidemia in concert with hyperglycemia stimulates the proliferation of macrophages in atherosclerotic lesions: potential role of glucose-oxidized LDL. Diabetes 53(12): 3217–3225.
47. Graessler, J., Pietzsch, J., Westendorf, T., Julius, U., Bornstein, S.R., Kopprasch, S. (2007) Glycoxidized LDL isolated from subjects with impaired glucose tolerance increases CD36 and peroxisome proliferator-activator receptor gamma gene expression in macrophages. Diabetologia 50(5):1080–1088.
48. Tang, C., Kanter, J.E., Bornfeldt, K.E., Leboeuf, R.C., Oram, J.F. (2010) Diabetes reduces the cholesterol exporter ABCA1 in mouse macrophages and kidneys. J Lipid Res 51(7):1719–1728.

49. Piga, R., Naito, Y., Kokura, S., Handa, O., Yoshikawa, T. (2007) Short-term high glucose exposure induces monocyte-endothelial cells adhesion and transmigration by increasing VCAM-1 and MCP-1 expression in human aortic endothelial cells. Atherosclerosis 193(2): 328–334.
50. Natarajan, R., Gerrity, R.G., Gu, J.L., Lanting, L., Thomas, L., Nadler, J.L. (2002) Role of 12-lipoxygenase and oxidant stress in hyperglycaemia-induced acceleration of atherosclerosis in a diabetic pig model. Diabetologia 45: 125–133.
51. Dobrian, A.D., Lieb, D.C., Cole, B.K., et al. (2011) Functional and pathological roles of the 12-and 15-lipoxygenases. Prog Lipid Res 50(1): 115–131.
52. Li, S., Reddy M.A., Cai, Q., et al. (2006) Enhanced proatherogenic responses in macrophages and vascular smooth muscle cells derived from diabetic db/db mice. Diabetes 55:2611–2619.
53. Parathath, S., Grauer, L., Huang, L.S., et al. (2011) Diabetes adversely affects macrophages during atherosclerotic plaque regression in mice. Diabetes 60:1759–1769.

3

GLYCATION, INFLAMMATION AND RAGE: MECHANISMS CONTRIBUTING TO THE ACCELERATED ATHEROSCLEROSIS OF DIABETES

RAVICHANDRAN RAMASAMY, SHI FANG YAN, AND ANN MARIE SCHMIDT

Diabetes Research Program, Division of Endocrinology, Department of Medicine, New York University Langone Medical Center, New York, NY, USA

INTRODUCTION

In both type 1 and type 2 diabetes, a major cause of morbidity and mortality is accelerated cardiovascular disease, manifested as increased incidence and severity of heart attacks and strokes [1]. A chief challenge in the field has been to discern mechanisms that specifically contribute to accelerated atherosclerosis in the diabetic state, so that in addition to therapies that target the well-known risk factors, such as hypertension, hyperlipidemia, and smoking, diabetes-specific accelerating factors might be identified.

In this chapter, we consider the role of glycation in the development of vascular inflammation and accelerated atherosclerosis in diabetes.

ANIMAL MODELS OF DIABETIC ATHEROSCLEROSIS

A chief experimental strategy that has shed light on mechanisms of diabetic atherosclerosis has been the use of animal models. One of the most expedient animal models for the study of in-depth mechanistic insights into disease is the mouse. In some of the very earliest studies, Kunjathoor et al. [2] examined the impact of streptozotocin-induced diabetes on atherosclerosis in C57BL/6 and BALB/c mice fed an atherogenic diet. In both strains of mice, diabetes resulted in increased atherosclerotic lesions; but, overall, the lesions were relatively small and limited largely to the fatty streak. Both macrophages and smooth muscle cells were present in the lesions. A caveat raised about this study was

Atherosclerosis: Risks, Mechanisms, and Therapies, First Edition. Edited by Hong Wang and Cam Patterson.

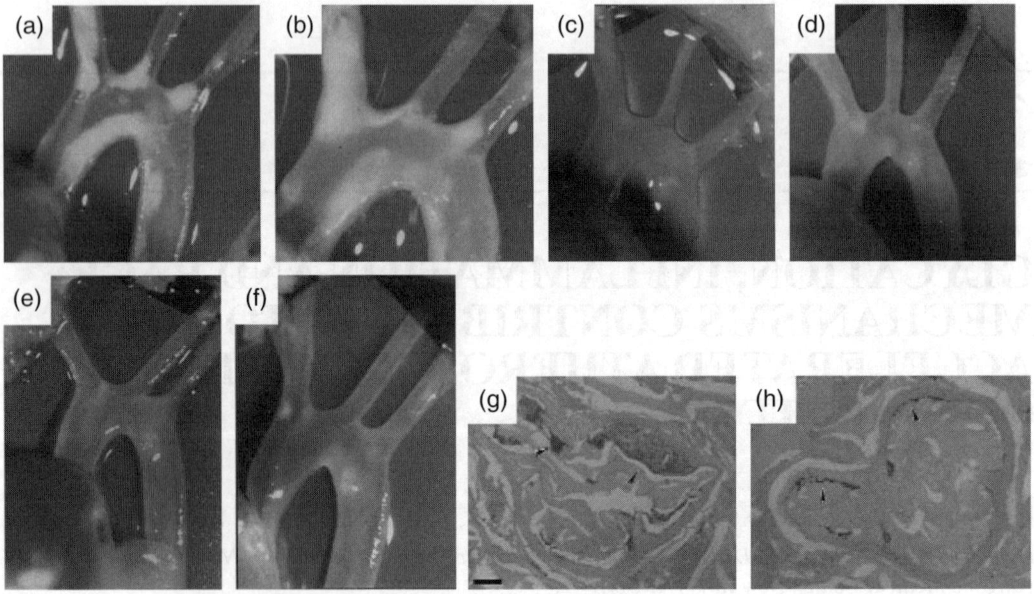

Figure 3.1 Accelerated atherosclerosis in diabetic apoE-null mice. (a–f) Dissection microscopy. ApoE-null mice were either made diabetic with streptotozocin (STZ) or treated with citrate buffer as control. After 6 weeks of diabetes or control treatment, the heart and aorta were dissected under microscopy and photographed. Diabetic mice treated with murine serum albumin (MSA) at 80 μg/day (a) or sRAGE at 3 μg/day (b), 20 μg/day (c) or 40 μg/day (d); control without diabetes treated with MSA at 40 μg/day (e) or sRAGE at 20 μg/day (f). (g, h) Histologic analysis. Typical histologic sections from the aortic sinus are shown from diabetic (g) and nondiabetic (h) apoE-null mice. *Arrows* indicate areas of atherosclerotic lesion as determined by staining with Oil Red O. Scale bar represents 33 μm (Adapted from Park et al. [3]. © Nature). (*See insert for color representation of the figure.*)

that the lesions did not show significant evidence of complexity, such as necrotic cores or calcification. As human diabetes is characterized by the presence of advanced lesions, this model recapitulated only some of the features of diabetes relevant to human pathology.

Hence, a major advance in the study of diabetic atherosclerosis was the use of genetically modified mice. The two most commonly used mouse models to study diabetic atherosclerosis have been mice devoid of apolipoprotein E (apoE) and mice devoid of low-density lipoprotein receptor (LDLR). Park et al. reported on the impact of induction of diabetes in apoE-null mice. Male mice were rendered diabetic with streptozotocin or treated with vehicle; all mice were fed normal chow. By approximately 6 weeks after the development of diabetes, the hyperglycemic mice displayed significantly increased atherosclerotic lesion area at the aortic sinus, in parallel with greater degrees of lesion complexity (Figure 3.1). The diabetic mice did display increased levels of total cholesterol, suggesting that at least, in part, higher levels of cholesterol contributed to the acceleration of atherosclerosis [3]. In that study, the authors reported that levels of advanced glycation endproducts (AGEs) were higher in the plasma and tissues of the diabetic mice compared to the nondiabetic control animals [3] (Figure 3.2). In this chapter, we will discuss glycation and its roles in diabetic atherosclerosis—from basic biochemical mechanisms, to receptors—and the impact in vascular inflammation and acceleration of diabetic atherosclerosis.

In addition to murine models, other animal models of accelerated diabetic atherosclerosis

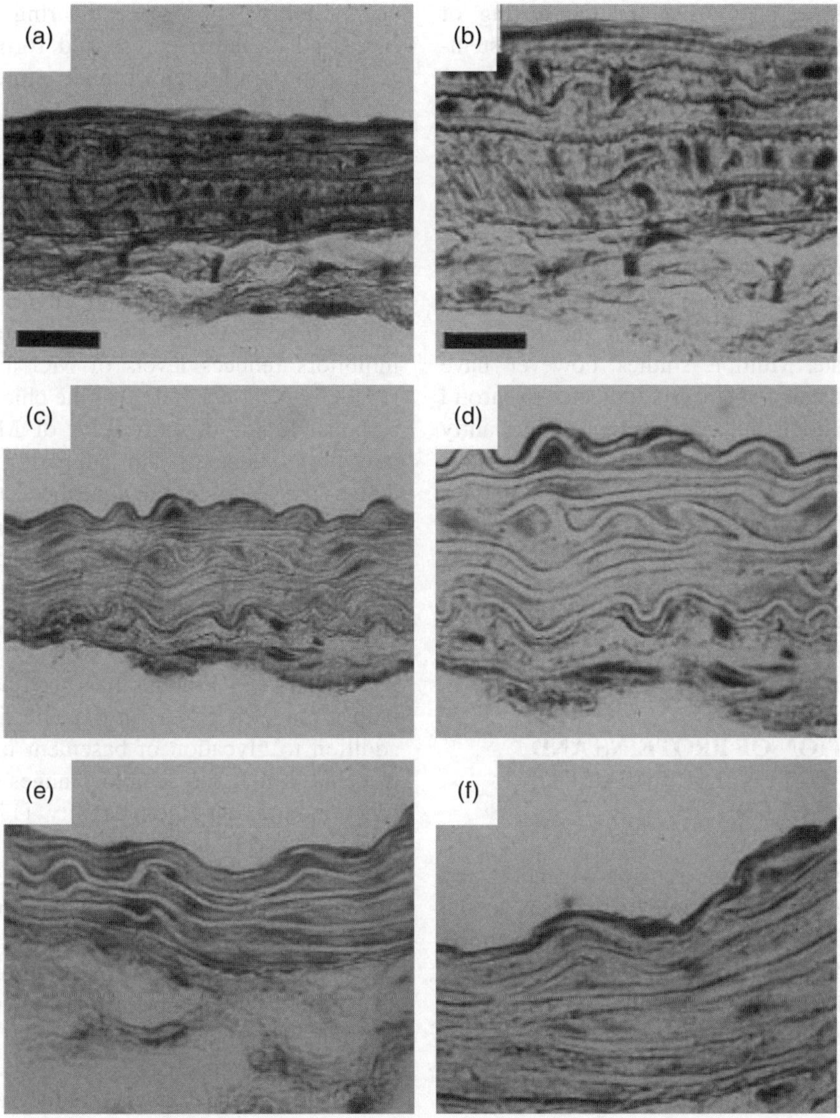

Figure 3.2 Increased AGE-reactive epitopes in diabetic atherosclerosis: studies in ApoE-null mice. (a–f) Vascular sections from the indicated ApoE-null mice were subjected to immunohistochemistry with anti-AGE IgG. a and b, diabetic mice; c and d, control mice; e and f, diabetes treated with sRAGE (20 µg/day). Scale bars represent 33 µm (a, c, and e) and 20 µm (b, d, and f). In all cases, controls with appropriate nonimmune IgG were negative (not shown) (Adapted from Park et al. [3]. © Nature). (*See insert for color representation of the figure.*)

and vascular disease have been developed. Nonhuman primates [4] and swine [5] models are particularly relevant since atherosclerotic lesions in these animals are often advanced comparable to human atherosclerotic lesions, and they display many of the mechanistic features linked to accelerated vascular disease in rodents, such as accumulation of AGEs and their precursors, and evidence of increased inflammation and oxidative stress. Swine models have been used for the testing of vascular intervention strategies such as stenting [6]. The study of alloxan-treated (type 1 diabetes) rabbits has proved useful for the study of accelerated diabetic atherosclerosis and basic mechanisms [7]. Interestingly, a recent study reported on the use of hypercholesterolemic zebrafish for the testing of antiglycation agents such as cinnamon and clove agents [8]. Like mice, rats do not develop accelerated atherosclerosis in the hyperglycemic state. Multiple studies, however, have illustrated that diabetic rats subjected to carotid artery balloon injury develop significantly increased restenosis compared to their nondiabetic counterparts [9]. Taken together, studies from zebrafish to nonhuman primates have made important contributions to the delineation of mechanisms and potential treatment strategies for diabetes and accelerated atherosclerosis.

GLYCATION OF PROTEINS AND LIPIDS: POTENTIAL IMPACT ON ATHEROSCLEROSIS

The process of nonenzymatic glycation and oxidation of proteins and lipids leads to the formation of AGEs. Lysine and arginine amino acid residues are particularly vulnerable to glycation [10]. Major AGE precursors formed particularly in the presence of high glucose include 3-deoxyglucosone (3-DG) and the alpha-oxoaldehyde methylglyoxal (MG) [11]. MG is a very potent and highly reactive glycation precursor and is increased in the blood of diabetic subjects [12]. In addition to forming adducts on proteins, MG may also form adducts on other species such as lipids, DNA, and RNA [12]. It has been hypothesized that such MG-adducts may disrupt antigen presentation mechanisms with implications for the biology of atherosclerotic plaques, particularly in diabetes. MG feeding to diabetic rats resulted in a plethora of complications, including impaired endothelial function, as demonstrated by reduced vasorelaxation responses to acetylcholine in vascular ring studies, increased oxidative stress and formation of AGEs, in parallel with higher expression of the chemokine monocyte chemoattractant peptide 1 (MCP-1) and receptor for AGE (RAGE) [13].

Ramasamy et al. demonstrated that flux via the polyol pathway in hyperglycemia contributes to driving formation of MG and oxidative stress. In an AGE-enriched environment of aging, treatment with aldose reductase (AR) inhibitors reduces levels of MG and AGEs [14]. Glyoxalase-1 (Glo1) is the chief enzyme responsible for detoxification of MG; Glo1 may be modulated by inflammation and oxidative stress [15]. RAGE contributes to regulation of Glo1, as diabetic mice devoid of RAGE display lower levels of MG and AGEs despite equivalent degrees of hyperglycemia, in parallel with increased levels of Glo1 mRNA, protein, and activity [16].

Glycation of proteins has been associated with a range of effects in diabetic tissues. In addition to glycation of basement membrane proteins, which may lead to changes in barrier function and hyperpermeability [17], AGEs may bind to cell surface receptors and mediate changes in gene expression patterns that increase inflammation, oxidative stress, and cellular demise.

Importantly, glycation of lipoproteins has also been linked to diabetic atherosclerosis. For example, Isoda et al. found that AGE-LDL-exposed human macrophages exhibited higher expression of the CC chemokine receptor 2 (CCR2). This AGE effect was mediated by RAGE, as pretreatment of the monocytes with anti-RAGE IgG antibodies (Abs) blocked the increase of CCR2 mRNA expression [18]. Lam et al. [19] showed that glycoxidized LDL increased expression of scavenger receptor-A (SR-A) and CD36, and decreased expression of scavenger receptor class B type 1 (SR-B1) in THP-1 macrophages. Harja et al. showed that *in vitro*-prepared oxidized LDL contained AGE epitopes. When native and oxidized LDLs were subjected to SDS-PAGE followed by Western blotting with an antibody to heterogeneous

AGEs, the AGE-immunoreactive epitopes were substantially higher in the oxLDL versus the native LDL material [20]. Finally, glycation of LDL by MG results in the formation of small, dense LDL with enhanced atherogenicity [21].

In addition to endogenously formed AGEs, dietary AGEs may also contribute to acceleration of atherosclerosis in apoE-null mice. These mice were fed high- or low-AGE diets. After 2 months of diet, atherosclerotic lesions were significantly lower in the low-AGE-fed mice compared with the high-AGE-fed mice by more than 50%. In parallel, inflammation and levels of AGE-receptor 1 and 2, and RAGE were significantly lower in the low- versus high-AGE-fed apoE-null mice [22].

Given the heterogeneous nature of glycation products and AGEs, it is not surprising that AGEs may interact with multiple receptors.

NON-RAGE GLYCATION RECEPTORS AND IMPLICATIONS FOR ATHEROSCLEROSIS

There are multiple reported receptors for AGEs. AGEs bind to lectin-like oxidized LDL receptors and AGEs produced from three different types of reducing sugars—including glucose, fructose, and ribose—bind in a dose-dependent and saturable manner to LOX-1-expressing cells [23]. Chen et al. showed that diabetes upregulates expression of LOX-1 in the diabetic rat vessel wall. In aortic endothelial cells, upregulation of LOX-1 was induced by rat serum and by AGE-BSA (bovine serum albumin) [24]. Others suggested that FEEL-1 and FEEL-2 were endocytic receptors for aGEs [25].

AGEs may bind to multiple classes of scavenger receptors, including scavenger receptor-A, CD36 and scavenger receptor-B1. SR-B1 plays key roles in reverse cholesterol transport by promoting selective uptake of cholesteryl esters in HDL by hepatocyes. AGE-BSA blocks selective uptake of HDL-cholesteryl esters and inhibits cholesterol efflux from CHO-SR-B1 cells to HDL [26]. Such findings implicate AGE interaction with these scavenger receptors in atherosclerosis.

Vlassara et al. identified AGER-1, -2, and -3 as receptors for AGEs. Of these, AGER1 displays protective effects against vascular perturbation. As circulating AGEs may produce reactive oxygen species (ROS) via the NADPH oxidase system, expression of AGER1 reduces ROS production [27]. Hence, such studies indicate that specific AGE receptors mediate protection against AGEs, likely via internalization of the AGE ligands. One of the best-characterized AGE receptors is receptor for AGE or RAGE.

RAGE AND IMPLICATIONS FOR DIABETIC ATHEROSCLEROSIS

The finding that RAGE was a member of the immunoglobulin superfamily lent first insights into the roles RAGE might play in atherosclerosis. Extracellular RAGE comprises one V-type immunoglobulin-like domain followed by two C-type immunogloblulin-like domains. AGEs and AGE peptides bind to the V-type domain; recent published structures of RAGE reveal a large hydrophobic "patch" in the V-type domain that serves as highly charged surface for AGE ligand binding [28]. Importantly, RAGE is a multiligand receptor [29]. In addition to AGEs, RAGE is also a receptor for pro-inflammatory S100/calgranulins, high mobility group box 1, amyloid-β peptide, and β-sheet fibrils. Most of the ligands bind to RAGE at the V-domain, but some of the S100/calgranulin ligands are reported to bind via the C2-type Ig domain [29]. Beyond the extracellular domain is a single transmembrane spanning domain followed by a short, highly charged cytoplasmic domain. The cytoplasmic domain of RAGE lacks endogenous tyrosine kinase activity; however, ligand binding to RAGE still activates signal transduction in a manner requiring the cytoplasmic domain. The cytoplasmic domain of RAGE binds to the mammalian diaphanous-related formin-1 (mDia1) [30]. In transformed cells, macrophages, and vascular smooth muscle cells, mDia1 is required for the effects of RAGE ligands on signal transduction.

RAGE is expressed on a broad range of cell types that are implicated in diabetic atherosclerosis, such as endothelial cells, vascular smooth muscle cells, monocytes/macrophages, T and B lymphocytes, and dendritic cells. Ligand–RAGE interaction induces inflammatory cell mediators and activates signal transduction cascades that eventuate in activation of the transcription factors, Nuclear factor-kappaB (NF-κB) and early growth response protein-1 (Egr-1) [31]. Of note, the promoter for the gene encoding *RAGE* (or *AGER*) contains functional NF-κB-binding sites. Hence, AGEs lead to the activation of NF-κB, at least in part via RAGE, as part of a positive-feedback mechanism. These experiments provide mechanistic support for the finding that from human subjects to rodents, diabetes promotes the expression of RAGE in the vasculature.

RAGE and Accelerated Atherosclerosis in Diabetes

Atherosclerotic plaques of diabetic human subjects undergoing carotid endarterectomy [32] or from autopsy specimens [33] displayed larger necrotic cores and greater plaque load compared to nondiabetic plaques. Diabetic atherosclerotic plaques also exhibited greater staining for macrophages, T lymphocytes, and HLA-DR [33]. Diabetic lesions also exhibited increased expression of RAGE and S100A12 (localized to apoptotic vascular smooth muscle cells and macrophages), activated NF-κB, cyclooxygenase-2 (cox-2), matrix metalloproteinases (MMPs), gelatinolytic activity, and increased lipid and oxidized LDL content. By contrast, collagen content was reduced in the diabetic plaques, suggestive of less plaque stability [32].

To test the concept that increased MG-AGE generation in diabetic atherosclerosis would accelerate atherosclerosis, Goldberg and Ramasamy bred transgenic mice expressing human-relevant levels of AR into the LDL receptor null background. These mice and nontransgenic LDL receptor-null mice were rendered diabetic or maintained in the nondiabetic state. Although there were no differences in atherosclerosis between the transgenic and nontransgenic LDL receptor-null mice in the nondiabetic state, diabetic transgenic human AR mice displayed significantly greater degrees of atherosclerosis compared to the nontransgenic diabetic LDL receptor-null mice. In parallel, diabetic AR transgenic mouse macrophages displayed higher expression of scavenger receptors compared to nontransgenic macrophages, and levels of antioxidant stress defenses (such as genes regulating glutathione) were reduced in the diabetic transgenic AR LDL receptor-null mice aortas [34]. In a later study, these same authors bred apoE-null mice into two distinct lines of transgenic human AR mice: one in which AR was expressed globally and the other using the Tie-2 promoter, in which AR was expressed largely in endothelial cells. Compared to nontransgenic diabetic apoE-null mice, both of these lines of mice displayed significantly increased atherosclerosis. In the transgenic human AR apoE-null mice with diabetes, treatment with an AR inhibitor, zopolrestat, attenuated accelerated atherosclerosis. There was evidence of AR-induced endothelial dysfunction, as aortic rings prepared from the transgenic human AR diabetic apoE-null mice demonstrated the greatest degree of impaired relaxation in response to acetylcholine versus other groups. Inflammatory gene expression in the aortas was also significantly greater in the AR transgenic versus nontransgenic diabetic mice devoid of apoE [35].

To test the hypothesis that a chief AGE receptor RAGE participated in accelerated atherosclerosis in diabetic mice, we employed soluble RAGE (sRAGE), the extracellular ligand-binding domain of RAGE as a test of the concept that binding RAGE ligands and preventing their interaction with and activation of the cell surface receptor would mitigate diabetic atherosclerosis. Administration of sRAGE to diabetic apoE-null mice beginning immediately after the induction of diabetes resulted in a dose-dependent suppression of accelerated diabetic atherosclerosis (Figure 3.1) and reduced AGE levels (Figure 3.2). sRAGE

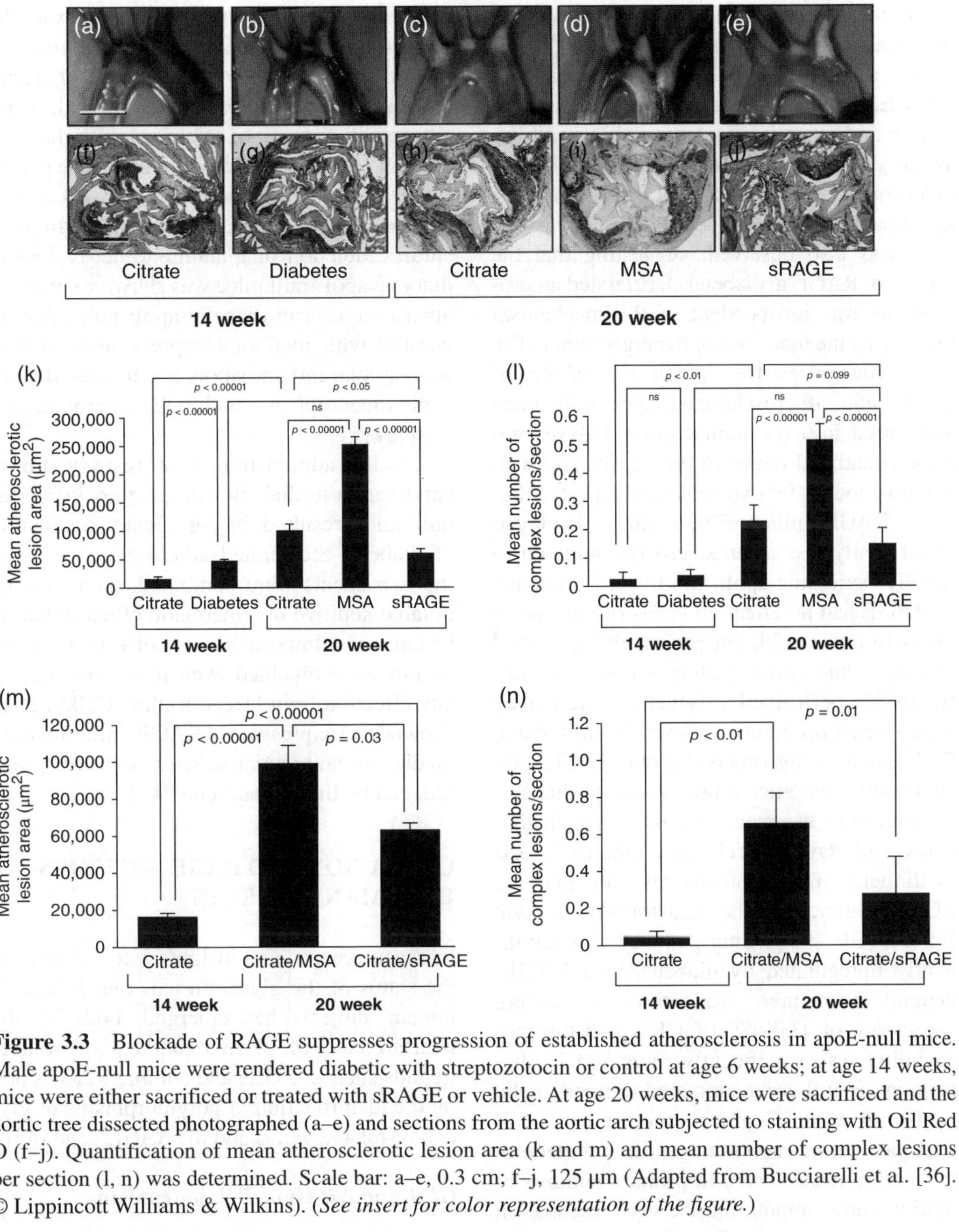

Figure 3.3 Blockade of RAGE suppresses progression of established atherosclerosis in apoE-null mice. Male apoE-null mice were rendered diabetic with streptozotocin or control at age 6 weeks; at age 14 weeks, mice were either sacrificed or treated with sRAGE or vehicle. At age 20 weeks, mice were sacrificed and the aortic tree dissected photographed (a–e) and sections from the aortic arch subjected to staining with Oil Red O (f–j). Quantification of mean atherosclerotic lesion area (k and m) and mean number of complex lesions per section (l, n) was determined. Scale bar: a–e, 0.3 cm; f–j, 125 μm (Adapted from Bucciarelli et al. [36]. © Lippincott Williams & Wilkins). (*See insert for color representation of the figure.*)

had no impact on levels of glucose or lipids in the diabetic animals [3]. In a later study, sRAGE was administered to diabetic apoE-null mice beginning 6 weeks after the documentation of diabetes and continued for an additional 6 weeks. Compared to baseline (i.e., accelerated diabetic atherosclerosis compared to nondiabetic apoE-null mice), sRAGE-treated diabetic mice displayed a significant reduction in progression of accelerated diabetic atherosclerosis compared to the vehicle-treated (murine serum albumin) diabetic mice (Figure 3.3). Diabetes accelerated vascular inflammation and infiltration of lipids and macrophages, increased oxidative stress, and increased activation of proinflammatory NF-κB in the aorta tissue. In animals treated with sRAGE, significant reduction in vascular inflammation and oxidative stress was observed [36]. When sRAGE was administered to apoE-null/db/db mice, a mouse model of type 2 diabetes, significant reduction in accelerated atherosclerosis was also observed, suggesting that the impact of RAGE in diabetes-accelerated atherosclerosis was independent of the mechanism underlying the induction of hyperglycemia [37].

In order to test the specific role of RAGE in diabetic atherosclerosis, apoE-null mice were bred into the homozygous RAGE null background and rendered type 1 diabetic with streptozotocin. In two separate reports, diabetic RAGE-null/apoE-null mice displayed significantly less atherosclerosis compared to RAGE-expressing apoE-null mice. Deficiency of RAGE had no effect on levels of glucose or lipids in this model, but profoundly impacted oxidative and inflammatory stress [38, 39]. Bu et al. performed affymetrix microarray experiments on aortas retrieved from diabetic RAGE-expressing versus diabetic RAGE-null/apoE-null mice after a few weeks of diabetes to determine the mechanisms by which diabetes and RAGE-accelerated atherosclerosis in diabetes. Data analysis revealed that the ROCK1 branch of the transforming growth factor (TGF)-β signaling pathway was prominently upregulated by diabetes in a RAGE-dependent manner. Transcript and protein expression of TGF-β2, ROCK1, and thromospondin1 was significantly increased in diabetic apoE-null aorta compared to nondiabetic apoE-null aorta. Confocal microscopy localized the sites of expression of these molecules to smooth muscle cells. In primary cultures of murine aortic smooth muscle cells, incubation with the prototypic RAGE ligand S100B resulted in increased proliferation and migration of smooth muscle cells in a manner significantly reduced by pretreatment with the ROCK inhibitor fasudil or Y-27632 [39] (Figure 3.4).

Investigators have used these models of diabetes-accelerated atherosclerosis to test the effects of distinct interventions. For example, in the context of the hypothesized role for oxidative stress in the pathogenesis of diabetes-accelerated atherosclerosis, partial, aorta site-specific benefit of treatment with the antioxidant ebselen (a mimic of glutathione peroxidase) in reducing atherosclerosis was shown in type 1 diabetic apoE-null mice [40]. In other studies, deletion of glutathione peroxidase in diabetic apoE-null mice was shown to increase atherosclerosis in diabetic apoE-null mice, in parallel with increased expression of RAGE and vascular inflammation, and increased numbers of macrophages within the atherosclerotic plaques [41].

Finally, administration of rosuvastatin or candesartan to diabetic (streptozotocin) apoE-null mice resulted in significant attenuation of diabetes-accelerated atherosclerosis; when the two agents were combined, there was no notable additive or synergistic effect noted on lesion areas. Importantly, the effects of rosuvastatin were obtained even in the absence of any effect on lipid levels. Rather, Calkin et al. showed that expression of RAGE, inflammatory mediators, and oxidative stress molecules was reduced by these treatments [42].

GLYCATION AND RAGE—STUDIES IN HUMAN SUBJECTS

In recent years, new insights into monitoring the status of the glycation axis and RAGE in human subjects has emerged, both by the measurement of plasma markers of soluble ligands such as AGEs and soluble RAGEs and by the identification of polymorphisms of glycation-related genes and the *RAGE* gene [43].

Glo1 and *AR* Gene Polymorphisms

Wu et al. examined a study population of 362 patients with type 2 diabetes and 301 nondiabetic patients for association of polymorphisms of *Glo1* gene (A419C and C-7T) and the *AR* gene (C-106T) with type 2 diabetes and carotid atherosclerosis in a Chinese Han population. Carotid intima-media thickness (CIMT) and the

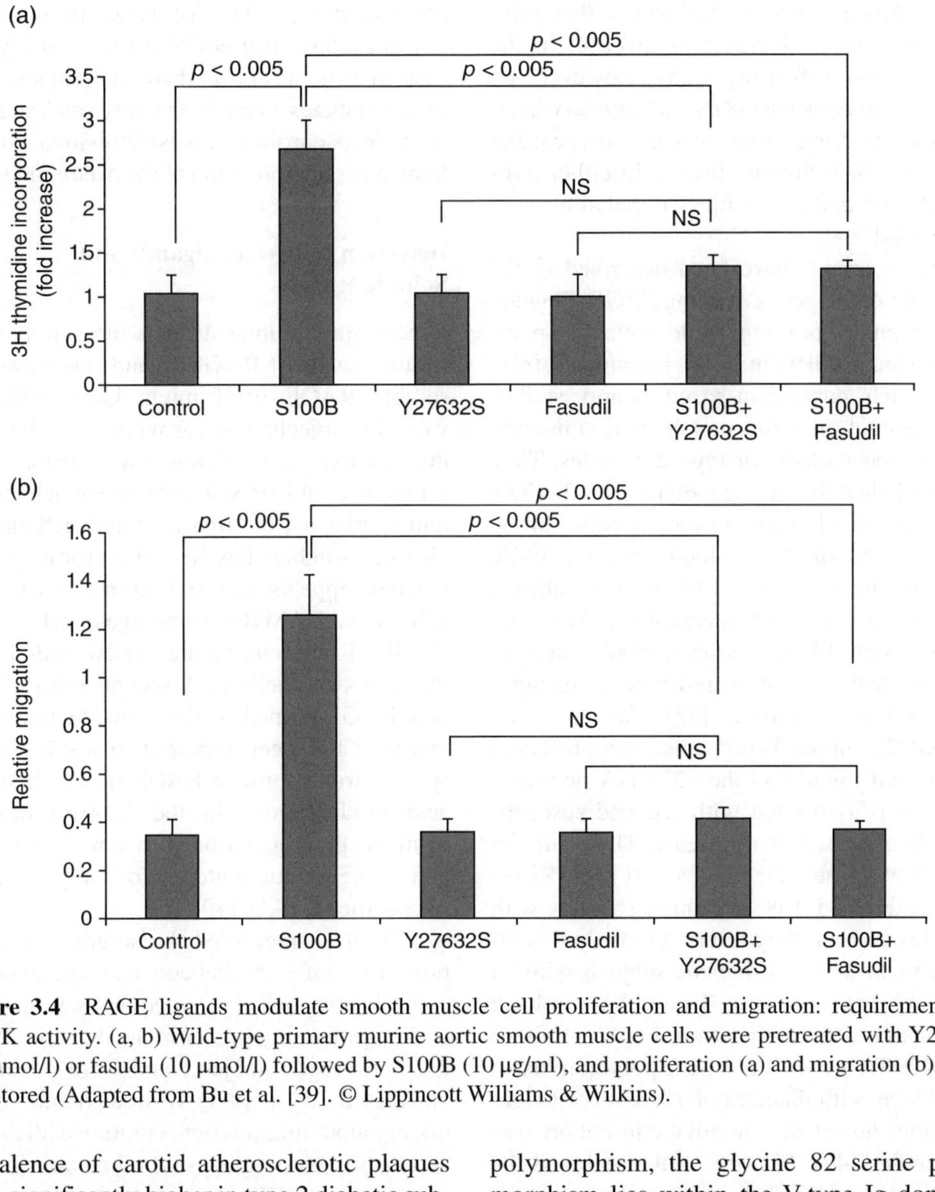

Figure 3.4 RAGE ligands modulate smooth muscle cell proliferation and migration: requirement for ROCK activity. (a, b) Wild-type primary murine aortic smooth muscle cells were pretreated with Y27632 (10 μmol/l) or fasudil (10 μmol/l) followed by S100B (10 μg/ml), and proliferation (a) and migration (b) were monitored (Adapted from Bu et al. [39]. © Lippincott Williams & Wilkins).

prevalence of carotid atherosclerotic plaques were significantly higher in type 2 diabetic subjects with the Glo1-7CC genotype versus Glo1-7CT or Glo1-7TT genotypes [43]. No other associations with the distinct polymorphisms were found in these specific population.

RAGE Gene Polymorphisms

Polymorphisms of the gene encoding *RAGE* have been described, both in the coding and promoter regions. A particularly intriguing polymorphism, the glycine 82 serine polymorphism lies within the V-type Ig domain; most of the ligands bind within the V-type domain; and hence, it was hypothesized that such a mutation might impact ligand binding. Compared to the wild-type allele, the G82S when transfected into low-RAGE-expressing cells imparted a significant increase in affinity for the S100/calgranulin ligands. In parallel, ligand-stimulated cytokine generation and activation of MMPs was increased in the presence of the G82S versus the wild-type allele. Hence, it was logical to test this polymorphism in cardiovascular disease. In the Framingham offspring study, however, no relationship between G82S and coronary heart disease, cardiovascular disease, myocardial infarction or ischemic disease in either nondiabetic or diabetic subject populations was found [44].

Other variants have been described in the promoter of the gene encoding *RAGE*. Engelen et al. examined nine tag single nucleotide polymorphisms (SNPs) in 1291 individuals from two Dutch population-based cohort studies with normal glucose metabolism, impaired glucose metabolism, or type 2 diabetes. They reported that the AA genotype of –374T/A was associated with lower systolic blood pressure and diastolic blood pressure, pulse pressure, and less arterial stiffness in subjects with normal glucose metabolism, but with higher systolic blood pressure and pulse pressure in individuals with impaired glucose metabolism or type 2 diabetes [45]. Falcone et al. studied 259 nondiabetic subjects in an Italian cohort and found that the –374T/A polymorphism was associated with reduced susceptibility to coronary artery disease. These authors did not examine a diabetic cohort [46]. Kirbis et al. examined 168 Slovenian subjects with diabetes and coronary artery disease and compared them with 241 diabetic subjects without coronary artery disease. The authors did not demonstrate an association between either the –429T/C or –374T/A gene polymorphisms in subjects with diabetes of 10 years or longer duration; however, a nondiabetic cohort was not studied [47]. Falcone et al. examined the links between –374T/A polymorphism and the severity of coronary artery disease; 14.3% of the AA subjects had type 2 diabetes and 15.5% of the subjects bore AT or TT genotype. The authors reported that the mean number of diseased vessels was significantly lower in the AA genotype subjects versus those bearing AT or TT, even after consideration of confounding variables [48]. Taken together, although no impact of the RAGE G82S was identified in one Caucasian (Framingham offspring study) population, the –374 AA genotype might be linked to lower degrees of coronary artery disease in general and perhaps in subjects with type 2 diabetes. Overall, the very small number of subjects enrolled in these individual studies limited the interpretation of the overall findings.

Detection of RAGE Ligands and Soluble RAGEs

Recent studies indicate that measurement of soluble levels of RAGE ligands and levels of soluble RAGE itself might have predictive value in subjects with (or without) diabetes in the context of cardiovascular disease. Two forms of soluble RAGEs have been detected in human plasma; the first is "total" cell-surface cleaved soluble RAGE. This form of the receptor appears to result from the action of MMPs or ADAM10 on cleavage of cell surface RAGE from circulating cells and/or the vascular endothelium. A second form of soluble RAGE termed "endogenous secretory" or esRAGE has been reported to result from a splice variant form of RAGE in which amino acid modifications in the C2-type domain result in the generation of a novel sequence, and hence unique epitopes for the production of specific ELISAs [49].

Aso et al. measured the serum concentrations of AGEs in diabetic and age-matched control subjects and reported that serum AGEs were significantly higher in the diabetic versus the nondiabetic subjects and were positively correlated with glycosylated hemoglobin, urinary albumin excretion, duration of diabetes, and fasting blood glucose. In the diabetic subjects, serum AGE levels were higher in the group with versus without coronary artery disease [50]. Kilhovd et al. studied AGE levels in 874 Finish diabetic subjects followed for 18 years and reported that serum levels of AGEs were significantly positively associated with total and cardiovascular disease mortality in women, but not in men [51]. Steine et al. studied 26 subjects with type 1 diabetes and 16 age-matched controls. In multiple regression analyses, they found that serum AGE levels and duration of type 1 diabetes significantly predicted systolic strain by Doppler imaging of the heart [52].

Others studied non-AGE RAGE ligands and the relationship to coronary artery disease. Yan et al. [53] reported that serum levels of HMGB1 in 512 subjects were independently associated with coronary artery disease in the diabetic and nondiabetic subjects. Mahajan et al. examined peripheral blood mononuclear cell (PBMC) levels of S100A12 in nondiabetic subjects with angiographically proven coronary artery disease and a control group with risk factors but no evidence of coronary artery disease. Levels of PBMC S100A12 and RAGE were higher in the group with premature coronary artery disease compared with without premature disease. Further, the severity of disease was positively correlated with the expression of RAGE and S100A12. S100A12 expression was positively correlated with RAGE expression and negatively associated with levels of total soluble RAGE [54].

Levels of soluble RAGE have been examined in human subject plasma to test the premise that such levels might correlate with the presence, absence, or degree of atherosclerosis in populations. Falcone et al. reported on levels of total sRAGE in 328 nondiabetic men with angiographically proven coronary artery disease and 328 matched subjects without coronary artery disease. Overall, the levels of plasma sRAGE were significantly lower in coronary artery disease versus control subjects. The authors divided the sRAGE levels into quartiles and found that the odds ratio for the presence of coronary artery disease was 6.719 when the lowest quartile of sRAGE was compared with the highest level of sRAGE [55]. Nakamura et al. examined total levels of AGEs and sRAGE in type 2 diabetic subjects and found that levels of AGEs and sRAGE were positively associated with serum levels of MCP-1 and TNF-alpha in type 2 diabetic subjects [56]. Tan et al. showed that in diabetic subjects, levels of serum AGEs (and creatinine levels) were main determinations of serum total sRAGE concentration [57]. These data suggested that in parallel with measurement of sRAGEs, it is important to assess levels of renal function.

Gohda et al. studied 107 type 2 diabetic subjects divided into three groups (A–B–C) based on their renal function [no nephropathy (A), nephropathy without hemodialysis (B), and hemodialysis (C), respectively]. Serum esRAGE and serum AGEs were higher in Group C compared with Group A or B. After multiple regression analysis, the levels of serum creatinine and duration of diabetes were independently associated with serum esRAGE levels [58].

Katakami et al. assessed levels of total and esRAGE in type 1 diabetic subjects with respect to coronary artery disease and found that esRAGE, but not total sRAGE, was inversely correlated with CIMT, a surrogate for the presence coronary artery disease [59]. Colhoun et al. studied subjects from the CARDS trial and analyzed levels of total sRAGE and esRAGE in type 2 diabetic subjects treated with atorvastatin and followed for 3.9 years. Levels of sRAGE and esRAGE were positively correlated with incident coronary heart disease in that trial but not with stroke. These authors found that treatment with atorvastatin (10 mg/day) had no effect on sRAGE levels [60].

The data from the CARDS trial stand in contrast to a study by Tam et al. in which they treated diabetic subjects with atorvastatin (10 mg/day for 3 months then 20 mg/day for an additional 3 months). In the type 2 diabetic subjects, there was a significant increase in both sRAGE and esRAGE levels after 6 months' treatment, but there was no change in placebo group [61]. A correlation was found between the increase of serum esRAGE and the reduction of LDL. Santilli et al. studied a nondiabetic patient population with hypercholesterolemia who received atorvastatin or pravastatin; these were subjects free of vascular disease. After 8 weeks treatment, subjects who received atorvastatin but not pravastatin displayed a significant increase in sRAGE levels [62]. Finally, Forbes et al. treated type 1 diabetic subjects with 24 months perindopril, an angiotensin-converting enzyme inhibitor. The authors found that there was a significant increase in plasma sRAGE in subjects treated with perindopril, and that complexes between sRAGE and the AGE carboxymethyl lysine (CML) were observed in the plasma of the diabetic subjects [63]. Taken together, these data suggest that pharmacological interventions might, by reducing overall oxidative and inflammatory stress, reduce RAGE ligands and, therefore, potentially increase the available levels of sRAGE.

Overall, however, much more work is needed to settle the issue of whether low or high levels of total sRAGE and esRAGE are adaptive, harmful or merely predictive of the state of cardiovascular disease in the diabetic or nondiabetic state. In this context, it is essential to monitor renal function, as studies suggest that sRAGE levels may rise with renal insufficiency. Prospective, multiple collection studies are needed in newly diagnosed type 1 and type 2 diabetic subjects, versus age- and gender-matched controls, to resolve the potential of sRAGE and ligand levels as biomarkers in diabetes and atherosclerosis.

CONCLUSIONS

In conclusion, with the burgeoning increase in both types 1 and 2 diabetes worldwide, the complications of diabetes, particularly in the macrovasculature, pose a great threat to the health and well-being of the world's population. Evidence is accruing linking glycation products and their interaction with cellular receptors to the pathogenesis of diabetic accelerated atherosclerosis. Efforts to target glycation and the impact of glycation products interaction with their key receptors may provide a novel means to suppress the increased incidence and severity of heart attacks and strokes that occur in this population. Finally, data suggest that measurement of AGEs and the proinflammatory ligands of RAGE (S100/calgranulins and HMGB1), as well as soluble levels of RAGE, might provide novel means to identify and track the course of progression of diabetic atherosclerosis and the response to therapeutic intervention in human subjects.

REFERENCES

1. Anand, D.V., Lim, E., Darko, D., et al. (2007) Determinants of progression of coronary artery calcification in type 2 diabetes role of glycemic control and inflammatory/vascular calcification markers. *J Am Coll Cardiol* 50, 2218–2225.
2. Kunjathoor, V.V., Wilson, D.L., LeBoeuf, R.C. (1996) Increased atherosclerosis in streptozotocin-induced diabetic mice. *J Clin Invest* 97, 1767–1773.
3. Park, L., Raman K.G., Lee, K.J., et al. (1998) Suppression of accelerated diabetic atherosclerosis by the soluble receptor for advanced glycation endproducts. *Nat Med* 4, 1025–1031.
4. Rowe, P.A., Kavanagh, K., Zhang, L., Harwood, H.J. Jr., Wagner JD. (2011) Short term hyperglycemia increases arterial superoxide production and iron dysregulation in atherosclerotic monkeys. *Metabolism* 60, 1070–1080.
5. Gerrity, R.G., Natarajan, R., Nadler, J.L., Kimsey, T. (2001) Diabetes-accelerated atherosclerosis in swine. *Diabetes* 50, 1654–1665.
6. Granada, J.F, Kaluza, G.L., Wilensky, R.L., Biedermann, B.C., Schwartz, R.S., Falk, E. (2009) Porcine models of coronary atherosclerosis and vulnerable plaque for imaging and interventional research. *EuroIntervention* 5, 140–148.
7. Roy, H., Bhardwaj, S., Babu, M., et al. (2006) VEGF-A, VEGF-D, VEGF receptor-1, VEGF receptor-2, NF-kappaB, and RAGE in atherosclerotic lesions of diabetic Watanabe heritable hyperlipidemic rabbits. *FASEB J* 20, 2159–2161.
8. Jin S., Cho, K.H. (2011) Water extracts of cinnamon and clove exhibits potent inhibition of protein glycation and anti-atherosclerotic activity in vitro and in vivo hypolipidemic activity in zebrafish. *Food Chem Toxicol* 49, 1521–1529.
9. Park, S.H., Marso, S.P., Zhou, Z., Foroudi, F., Topol, E.J., Lincoff, A.M. (2001) Neointimal hyperplasia after arterial injury is increased in a rat model of non-insulin dependent diabetes mellitus. *Circulation* 104, 815–819.
10. Baynes, J.W., Thorpe, S.R. (2000). Glycoxidation and lipoxidation in atherogenesis. *Free Radic Biol Med* 28, 1708–1716.

11. Westwood, M.E., Thornalley, P.J. (1995) Molecular characteristics of methylglyoxal-modified bovine and human serum albumins. Comparisons with glucose-derived advanced glycation endproduct-modified serum albumins. *J Protein Chem* 14, 359–372.
12. Price, C.L., Knight, S.C. (2007) Advanced glycation: a novel outlook on atherosclerosis. *Curr Pharm Des* 13, 3681–3687.
13. Sena, C.M., Matafome, P., Crosostomo, J., et al. (2012) Methylglyoxal promotes oxidative stress and endothelial dysfunction. *Pharmacol Res* 65, 497–506.
14. Hallam, K.M., Li, Q., Ananthakrishnan, R., et al. (2010) Aldose reductase and AGE-RAGE pathways: central roles in the pathogenesis of vascular dysfunction in aging rats. *Aging Cell* 9, 776–784.
15. Thornalley, P.J. (2003) Glyoxalase I: structure, function and a critical role in the enzymatic defence against glycation. *Biochem Soc Trans* 31(Pt6), 1343–1348.
16. Reiniger, N., Lau, K., McCalla, D., et al. (2010) Deletion of the receptor for advanced glycation endproducts reduces glomerulosclerosis and preserves renal function in the diabetic OVE26 mouse. *Diabetes* 59, 2043–2054.
17. Monnier, V.M., Sell, D.R., Nagaraj, R.H., et al. (1992) Maillard reaction-mediated molecular damage to extracellular matrix and other tissue proteins in diabetes, aging and uremia. *Diabetes* 41(Suppl 2), 36–41.
18. Isoda, K., Folco, E., Marwali, M.R., Ohsuzu, F., Libby, P. (2008) Glycated LDL increases monocyte CC chemokine receptor 2 expression and monocyte chemoattractant protein-1 mediated chemotaxis. *Atherosclerosis* 198, 307–312.
19. Lam, M.C., Tan, K.C., Lam, K.S. (2004) Glycoxidized low density lipoprotein regulates the expression of scavenger receptors in THP-1 macrophages. *Atherosclerosis* 177, 313–320.
20. Harja, E., Bu, D., Hudson, B.I., et al. (2008) Vascular and inflammatory stresses mediate atherosclerosis via RAGE and its ligands in apoE null mice. *J Clin Invest* 118, 183–194.
21. Rabbani, N., Godfrey, L., Xue, M., et al. (2011) Glycation of LDL by methylglyoxal increases arterial atherogenicity: a possible contributor to increased risk of cardiovascular disease in diabetes. *Diabetes* 60, 1973–1980.
22. Lin, R.Y., Choudhury, R.P., Cai, W., et al. (2003) Dietary glycotoxins promote diabetic atherosclerosis in apolipoprotein E deficient mice. *Atherosclerosis* 168, 213–220.
23. Chen, X., Zhang, T., Du, G. (2008) Advanced glycation end products serve as ligands for lectin-like oxidized low-density lipoprotein receptor-1 (lox-1): biochemical and binding characterizations assay. *Cell Biochem Funct* 26, 760–770.
24. Chen, M., Nagase, M, Fujita, T., Narumiya, S., Masaki, T., Sawamura, T. (2001) Diabetes enhances lectin-like oxidized LDL receptor-1 expression in the vascular endothelium: possible role of LOX-1 ligand and AGE. *Biochem Biophys Res Commun* 287, 962–968.
25. Tamura, Y., Adachi, H., Osuga, J., et al. (2003) FEEL-1 and FEEL-2 are endocytic receptors for advanced glycation endproducts. *J Biol Chem* 278, 12613–12617.
26. Miyazaki, A., Nakayama, H., Horiuchi, S. (2002) Scavenger receptors that recognize advanced glycation end products. *Trends Cardiovasc Med* 12, 258–262.
27. Cai, W., Torreggiani, M., Zhu, L., et al. (2010) AGER1 regulates endothelial cell NADPH oxidase-dependent oxidant stress via PCK delta: implications for vascular disease. *Am J Physiol Cell Physiol* 298, C624–C634.
28. Xue, J., Rai, V., Singer D., et al. (2011) Advanced glycation endproduct recognition by the receptor for AGEs. *Structure* 19, 722–732.
29. Leclerc, E., Fritz, G., Vetter, S.W., Heizmann, C.W. (2009) Binding of S100 proteins to RAGE: an update. *Biochim Biophys Acta* 1793, 993–1007.
30. Hudson, B.I., Kalea, A.Z., Del Mar Arriero, M., et al. (2008) Interaction of the RAGE cytoplasmic domain with diaphanous-1 is required for ligand-stimulated cellular migration through activation of rac1 and cdc42. *J Biol Chem* 283, 34457–34468.
31. Ramasamy, R., Yan, S.F., Schmidt, A.M. (2011) Receptor for AGE: signaling mechanisms in the pathogenesis of diabetes and its complications. *Ann NY Acad Sci* 1243, 88–102.
32. Cipollone, F., Iezzi, A., Fazia M.L., et al. (2003) The receptor RAGE as a progression factor amplifying arachidonate-dependent inflammatory and proteolytic response in human atherosclerotic plaques: role of glycemic control. *Circulation* 108, 1070–1077.
33. Burke, A.P., Kolodgie, F.D., Zieske, A., et al. (2004) Morphologic findings of coronary

atherosclerotic plaques in diabetics: a postmortem study. *Arterioscler Thromb Vasc Biol* 24, 1266–1271.

34. Vikramadithyan, R.K., Hu, Y., Noh, H.L., et al. (2011) Human aldose reductase expression accelerates diabetic atherosclerosis in transgenic mice. *J Clin Invest* 115, 2434–2443.
35. Vedantham S., Noh, H., Ananthakrishnan, R., et al. (2011) Human aldose reductase expression accelerates atherosclerosis in diabetic apolipoprotein E null mice. *Arterioscler Thromb Vasc Biol* 31, 1805–1813.
36. Bucciarelli, L.G., Wendt, T., Qu, W., et al. (2002) RAGE blockade stabilizes established atherosclerosis in diabetic apolipoprotein E null mice. *Circulation* 106, 2827–2835.
37. Wendt, T., Harja, E., Bucciarelli, L., et al. (2006) RAGE modulates vascular inflammation and atherosclerosis in a murine model of type 2 diabetes. *Atherosclerosis* 185, 70–77.
38. Soro-Paavonen, A., Watson, A.M., Li, J., et al. (2008) Receptor for advanced glycation endproducts (RAGE) deficiency attenuates the development of atherosclerosis in diabetes. *Diabetes* 57, 2461–2469.
39. Bu, D.X., Rai, V., Shen, X., et al. (2010) Activation of the ROCK1 branch of the transforming growth factor beta signaling pathway contributes to RAGE-dependent acceleration of atherosclerosis in diabetic apoE null mice. *Circ Res* 106, 1040–1051.
40. Chew, P., Yuen, D.Y., Koh, P., et al. (2009) Site specific antiatherogenic effect of the antioxidant ebselen in the diabetic apolipoprotein E deficient mouse. *Arterioscler Thromb Vasc Biol* 29, 823–830.
41. Lewis, P., Stefanovic, N., Pete, J., et al. (2007) Lack of the antioxidant enzyme glutathione peroxidase 1 accelerates atherosclerosis in diabetic apolipoprotein E deficient mice. *Circulation* 115, 2178–2187.
42. Calkin, A.C., Giunti, S., Sheehy, K.J., et al. (2008) The HMG CoA reductase inhibitor rosuvastatin and the angiotensin receptor antagonist candesartan attenuate atherosclerosis in an apolipoprotein E deficient mouse model of diabetes via effects on advanced glycation, oxidative stress and inflammation. *Diabetologia* 51, 1731–1740.
43. Wu, J.C., Li, X.H., Wang, J.B., Tang, J.F., Wang, Y.F., Peng, Y.D. (2011) Glyoxalase I and aldose reductase gene polymorphisms and susceptibility to carotid atherosclerosis in type 2 diabetes. *Genet Test Mol Biomarkers* 15, 273–279.
44. Hofmann, M.A., Yang, Q., Harja, E., et al. (2005) The RAGE Gly82Ser polymorphism is not associated with cardiovascular disease in the Framingham offspring study. *Atherosclerosis* 182, 301–305.
45. Engelen, L., Ferreira, I., Gaens, K.H., et al. (2010) The association between the -374T/A polymorphism of the receptor for advanced glycation endproducts gene and blood pressure and arterial stiffness is modified by glucose metabolism status: the Hoorn and CoDAM studies. *J Hypertens* 28, 285–293.
46. Falcone, C., Campo, I., Emanuele, E., et al. (2004) Relationship between the -374T/A RAGE gene polymorphism and angiographic coronary artery disease. *Int J Mol Med* 14, 1061–1064.
47. Kirbis, J., Milutinovic, A., Steblovnik, K., Teran, N., Terzic, R., Zorc, M. (2004) The -429T/C and -374T/A gene polymorphisms of the receptor of advanced glycation endproducts gene (RAGE) are not risk factors for coronary artery disease in Slovene population with type 2 diabetes. *Coll Antropol* 28, 611–616.
48. Falcone, C., Campo, I., Emanuele, E., et al. (2005) -374T/A polymorphism of the RAGE gene promoter in relation to severity of coronary atherosclerosis. *Clin Chim Acta* 354, 111–116.
49. Santilli, F., Vazzana, N., Bucciarelli, L.G., Davi, G. (2009) Soluble forms of RAGE in human diseases: clinical and therapeutical implications. *Curr Med Chem* 16, 940–952.
50. Aso, Y., Inukai, T., Tayama, K., Takemura, Y. (2000) Serum concentration of advanced glycation endproducts are associated with the development of atherosclerosis as well as diabetic microangiopathy in patients with type 2 diabetes. *Acta Diabetol* 37, 87–92.
51. Kilhovd, B.K., Juutilainen, A., Lehto, S., et al. (2007) Increased serum levels of advanced glycation endproducts predict total, cardiovascular and coronary mortality in women with type 2 diabetes: a population-based 18 year follow up study. *Diabetologia* 50, 1409–1417.
52. Steine, K., Larsen, J.R., Stugaard, M., Berg, T.J., Brekke, M., Dah-Jorgensen, K. (2007) LV systolic impairment in patients with asymptomatic coronary heart disease and type 1 diabetes is related to coronary atherosclerosis, glycaemic control and advanced glycation endproducts. *Eur J Heart Fail* 9, 1044–1050.

53. Yan, X.X., Lu, L., Peng, W.H., et al. (2009) Increased serum HMGB1 level is associated with coronary artery disease in nondiabetic and type 2 diabetic patients. *Atherosclerosis* 205, 544–548.
54. Mahajan, N., Malik, N., Bahl, A., Dhawan, V. (2009) Receptor for advanced glycation end-products (RAGE) and its inflammatory ligand EN-RAGE in non-diabetic subjects with premature coronary artery disease. *Atherosclerosis* 207, 597–602.
55. Falcone, C., Emanuele, E., D'Angelo, A., et al. (2005) Plasma levels of soluble receptor for advanced glycation endproducts and coronary artery disease in non-diabetic men. *Arterioscler Thromb Vasc Biol* 25, 1032–1037.
56. Nakamura, K., Yamagishi, S., Adachi, H., et al. (2007) Serum levels of sRAGE, the soluble form of receptor for advanced glycation end-products, are associated with inflammatory markers in patients with type 2 diabetes. *Mol Med* 13, 185–189.
57. Tan, K.C., Shiu, S.W., Chow, W.S., Leng, L., Bucala, R., Betteridge, D.J. (2006) Association between serum levels of soluble receptor for advanced glycation endproducts and circulating advanced glycation endproducts in type 2 diabetes. *Diabetologia* 49, 2756–2762.
58. Godha, T., Tanimoto, M., Moon, J.Y., et al. (2008) Increased serum endogenous secretory receptor for advanced glycation end product (esRAGE) levels in type 2 diabetic patients with decreased renal function. *Diabetes Res Clin Pract* 81, 196–201.
59. Katakami, N., Matsuhisa, M., Kaneto, H., et al. (2008) Endogenous secretory RAGE but not soluble RAGE is associated with carotid atherosclerosis in type 1 diabetes patients. *Diabetes Vasc Dis Res* 5, 190–197.
60. Colhoun, H.M., Betteridge, D.J., Durrington, P., et al. (2011) Total soluble and endogenous secretory receptor for advanced glycation end-products as predictive biomarkers of coronary heart disease risk in patients with type 2 diabetes: an analysis from the CARDS trial. *Diabetes* 60, 2379–2385.
61. Tam, H.L., Shiu, S.W., Wong, Y., Chow, W.S., Betteridge, D.J., Tan, K.C. (2010) Effects of atorvastatin on serum soluble receptors for advanced glycation endproducts in type 2 diabetes. *Atherosclerosis* 209, 173–177.
62. Santilli, F., Bucciarelli, L., Noto, D., et al. (2007) Decreased plasma soluble RAGE in patients with hypercholesterolemia: effect of statins. *Free Radic Biol Med* 43, 1255–1262.
63. Forbes, J.M., Thorpe, S.R., Thallas-Bonke, V., et al. (2005) Modulation of soluble receptor for advanced glycation endproducts by angiotensin-converting enzyme-1 inhibition in diabetic nephropathy. *J Am Soc Nephrol* 16, 2363–2372.

4

INFLAMMATION

CAITLIN COLLINS[1], ELLIE TZIMA[1], AND CAM PATTERSON[2]

[1] *Department of Cell Biology and Physiology, University of North Carolina at Chapel Hill, Chapel Hill, NC, USA*

[2] *New York-Presbyterian Hospital/Weill-Cornell Medical Center, New York, NY, USA*

INTRODUCTION

Atherosclerosis is characterized by the accumulation of lipids and fibrous elements in the inner layer of the vessel wall. The disease has long been associated with hypercholesterolemia. However, in recent decades, we have begun to appreciate the critical role of inflammation in the development of the atherosclerosis. Indeed, one hallmark of atherosclerosis is endothelial dysfunction, or chronic inflammation of the endothelium, and accompanying infiltration of immune cells such as leukocytes and monocytes, into the inner layers of the vessel wall. After exiting the circulation, the newly resident monocytes transition into a macrophage-like identity begin to express scavenger receptors that internalize modified lipoprotein particles present within the developing atheroma. As a result of the internalization of resident lipid particles, the macrophage once again takes on a new identity, transitioning into a foam cell that further contributes to lesion formation by secreting pro-inflammatory cytokines that recruit additional immune cells to the vessel wall. Inflammatory signaling pathways that promote activation of the endothelium and adhesion of leukocytes to endothelial cells are some of the earliest molecular cascades that contribute to the development and progression of atherosclerosis. Thus, mechanisms of endothelial activation and the factors that contribute to infiltration of immune cells into the vessel wall will be the focus of this chapter.

MONOCYTE RECRUITMENT TO THE VESSEL WALL

Despite their constant presence in the circulating blood, immune cells rarely adhere to the endothelium. In fact, in normal blood flow conditions, endothelial cells are largely resistant to the adhesion of monocytes. However, during inflammation, recruitment of immune

Atherosclerosis: Risks, Mechanisms, and Therapies, First Edition. Edited by Hong Wang and Cam Patterson.

cells to sites of injury is a critical component of the immune response. This recruitment is facilitated by the release of two types of inflammatory signaling molecules by vascular and immune cells: chemokines and cytokines.

CHEMOKINES

Chemokines are part of a large family of low-molecular-weight proteins that have been shown to play numerous roles in the vascular system. As their name suggests, chemokines function as chemotactic cytokines, recruiting monocytes, and other immune cells to sites of inflammation, such as atherosclerotic plaques. Chemokines can also activate platelets [1, 2], a process that is linked with progression of atherosclerosis. Given their role in recruiting inflammatory cells to the vessel wall, it is no surprise that chemokines have a well-established role in both the onset and progression of atherosclerosis. In fact, we now know that expression of cytokines and their receptors increase with plaque progression in mouse models of atherosclerosis [3]. As leukocyte recruitment to sites of inflammation and migration through the endothelium to the developing plaque is a critical step in the development and progression of atherosclerosis, identification of the molecular players that specifically recruit leukocytes to the sites of inflammation remains an intense area of research.

MCP-1/CCL2

The most well-documented chemokine to date is monocyte chemoattractant protein-1 (MCP-1) or CCL2. MCP-1 is a chemokine that recruits monocytes, memory T cells, and dendritic cells to sites of inflammation, such as atherosclerotic lesions. MCP-1 is upregulated in mouse models of atherosclerosis and has been directly implicated in the progression of the disease [4]. MCP-1 knockout mice display reduced plaque development, associated with decreased leukocyte infiltration in mouse models of atherosclerosis [5, 6].

CCL5/RANTES

CCL5, or regulated and normal T cell expressed and secreted (RANTES), is a chemokine deposited by activated platelets onto the endothelium [7]. Expression of CCL5 positively correlates with the presence of atherosclerotic plaques. Similar to MCP-1, CCL5 functions to recruit leukocytes to the activated endothelium [8]. Therefore, it is not surprising that expression of CCL5 promotes lesion development in mouse models of atherosclerosis. Furthermore, manipulation of CCL5-mediated signaling has shown some success in minimizing development of plaques, as antagonism of the CCL5 receptor, CCR5, reduces plaque formation in the apoE$^{-/-}$ atherosusceptible mouse model [9].

CYTOKINES

Once immune cells are recruited to vessel wall, they initiate the release of other pro-inflammatory factors, such as cytokines, that perpetuate the local inflammatory response. Cytokines are low-molecular-weight proteins that function to mediate local inflammatory signaling. Importantly, cytokines can be produced by and act on all cell types involved in atherosclerosis (endothelial cells, vascular smooth muscle cells, macrophages, and T cells). This chapter will focus on three cytokines that have been shown to promote inflammation in the endothelium: interleukins (IL), tumor necrosis factors (TNF), and interferon-γ (IFN-γ).

Interleukins

ILs are a large class of proteins that contribute to inflammatory signaling in numerous cell types and diseases. In atherosclerosis, multiple ILs (IL-1, IL-2, IL-3, IL-4, IL-5, IL-6, IL-10, IL-12, IL-18, and IL-20) are expressed in atherosclerotic plaques and contribute to inflammatory signaling during the progression of the disease. ILs are expressed by all cell types within the plaque and have multiple functions in the progression of the disease [10]. However, within the endothelium, ILs have been shown to increase expression of adhesion molecules that facilitate stable leukocyte–endothelial interactions. Interestingly, some ILs have documented anti-inflammatory effects in atherosclerosis [11, 12]. Therefore, the role of individual ILs in the development or prevention of atherosclerosis must be considered when defining their role in atherosclerosis. For this reason, elucidation of specific molecular events mediated by each interleukin family member remains under intense investigation.

Tumor Necrosis Factor

TNFα, released by macrophages and vascular smooth muscle cells, has a profound pro-inflammatory effect on endothelial cells. Stimulation of endothelial cells with TNFα affects the organization of junctional proteins and leads to weakening of endothelial cell–cell junctions [13, 14]. These effects influence permeability of the monolayer and facilitate leukocyte transendothelial migration *in vivo*. Furthermore, TNFα$^{-/-}$/apoE$^{-/-}$ mice develop smaller atherosclerotic plaques when compared with their apoE$^{-/-}$ counterparts [15], providing direct evidence of the contribution of TNFα to lesion development.

Interferon

IFN-γ, produced by macrophages within the plaque, also significantly influences the endothelial cell phenotype. Treatment of endothelial cells with IFN-γ modulates endothelial permeability by altering the organization of junctional proteins, such as cadherins and catenins [13]. IFN-γ also contributes to inflammatory pathways activated in other cell types within the plaque. The role of IFN-γ in foam cell formation has been well documented, and IFN-γ regulates the immune system by influencing the recruitment and activity of immune cells, such as monocytes and T-lymphocytes [16]. Not only does IFN-γ influence the development of plaques, but it has also been shown to influence destabilization of the plaque [16]. In agreement with its multiple roles in the progression of atherosclerosis, therapies that target IFN-γ have shown some promise. Blocking IFN-γ function in the apoE$^{-/-}$ model of atherosclerosis reduced plaque size, while also promoting stabilization of advanced plaques [17].

SHEAR STRESS AND INFLAMMATION

Risk factors associated with atherosclerosis, such as high cholesterol, elevated lipoprotein levels, and coagulation factors are systemic in nature and uniformly circulate throughout the entire vasculature. However, it has been recognized for decades that atherosclerosis is a highly focal disease. Atherosclerotic plaques develop in distinct regions of the vasculature, such as vessel branch points and regions of curvature, which coincide with regions of low and disturbed blood flow. Conversely, few plaques develop in regions of the vasculature that experience unidirectional laminar blood flow. These observations suggest that the shear force of blood flow directly influences the health of the vessel. Endothelial cells lining the lumen of the blood vessel are constantly exposed to shear stress, or the parallel frictional force of flowing blood. Numerous *in vitro* and *in vivo* studies have identified that the magnitude of the force produced by blood flow, as well as the direction of flow, greatly influence physiological processes and contribute to the overall health of the vessel.

High (10–70 dyn/cm^2) laminar flow with a net forward movement in straight regions of the vessel elicits an anti-inflammatory and atheroprotective gene expression profile that promotes the integrity and maintenance of the vessel wall [18]. Specifically, activation of anti-inflammatory signaling molecules, such as endothelial nitric oxide synthase (eNOS), promotes vessel health, vasorelaxation, and endothelial cell quiescence. Nitric oxide (NO), a product of eNOS activation, is a potent vasodilator. Importantly, NO also inhibits numerous processes that are involved in the formation of atherosclerotic plaques, such as inflammation, aberrant proliferation, and thrombotic events.

Other anti-inflammatory pathways that promote vessel health include activation of the Krüppel-like family (KLF) family of transcription factors and subsequent expression of anti-apoptotic and anti-inflammatory signaling molecules that protect against the development of lesions [19]. Activation of these pathways, and others, are critical for maintenance of vessel integrity and actively protect against the formation of atherosclerotic plaques in these regions of the vasculature.

In contrast to regions of high laminar shear, regions of low ($<5\,dyn/cm^2$) and disturbed shear stress are sites that are susceptible to chronic inflammation and plaque development. Endothelial cells in these regions of the vessel display chronic activation of pro-inflammatory signaling pathways that elicit numerous deleterious effects in endothelial cells and other layers of the vessel wall. The molecular players involved in the pro-inflammatory signaling cascade will be discussed in detail later in the chapter.

In light of the observations indicating that shear stress is a major determinant of atherosclerotic plaque development, there has been a significant push to identify endothelial mechanosensors or proteins and subcellular structures that have the ability to convert the physical force of blood flow into various intracellular signaling events. Over the years, numerous mechanosensors have been proposed, and it is thought that multiple mechanosensors decorating all surfaces of the endothelial cell cooperatively work together to regulate endothelial cell phenotype. We will highlight only a subset of these proposed mechanosensors, and it should be noted that the list of endothelial mechanosensors will likely grow, as mechanotransduction of shear stress remains an intense area of research.

Ion channels

The apical surface of the endothelium provides the interface between the flowing blood and underlying tissue. Therefore, it is not surprising that numerous apical surface proteins have been proposed as mechanosensors over the past several decades. Numerous studies have identified rapid activation of multiple ion channels on the surface of endothelial cells. For instance, shear stress elicits rapid hyperpolarization of the endothelial membrane due to activation of potassium channels [20]. Shear stress also results in activation of plasma membrane bound P2X4 purinoceptors, resulting in a rapid influx of calcium [21]. Calcium signaling has also been shown to mediate a subset of shear stress-induced signaling pathways, such as activation of eNOS. Taken as a whole, these results suggest that flow-induced changes in activity of multiple ion channels cooperatively regulate and influence the cellular response to shear stress.

GPCRs

Membrane-bound G-protein-coupled receptors (GPCRs) have also been proposed as potential mechanotransducers. One such example is the bradykinin B_2 GPCR. Recent work has revealed that fluid shear stress elicits a conformational change in the B_2 GPCR independent of ligand binding [22]. These observations led to the hypothesis that these receptors may be important for mediating mechanotransduction signaling events in endothelial cells. However, the exact role of the B_2 bradykinin receptor in endothelial cell mechanotransduction remains elusive.

Caveolae

Caveolae are plasma membrane microdomains that are rich in cholesterol and the structural protein caveolin-1. These microdomains provide important docking sites for numerous signaling molecules, including eNOS, Ras, and Src family kinases [23]. Molecular studies have revealed a requirement of caveolin-1 in a subset of fluid shear stress signaling pathways, such as activation of extracellular signal-related kinase (ERK) [24]. These results suggest that caveolin-1 is required for proper organization of caveolin-like microdomains that contain shear-responsive signaling molecules.

Integrins

It has been proposed that shear stress is transmitted from the apical surface of endothelial cells to distal regions of the cell through tensile cytoskeletal elements, such as actin fibers. Therefore, proteins at the basal surface of the cell have also been proposed as potential mechanostransducers. Integrins are heterodimeric proteins that connect the internal actin cytoskeleton to the underlying extracellular matrix (ECM). Therefore, they are in a prime position to experience transmission of shear forces. Integrins on the basal surface of endothelial cells are activated in response to shear stress, and integrin ligation with the underlying ECM is required for a number of flow-responsive events, such as activation of the small GTPase RhoA and alignment of the actin cytoskeleton [25, 26]. Furthermore, direct application of force on integrins results in intracellular signaling events, including calcium release and activation of RhoA [26, 27].

Junctional Proteins

In recent years, a mechanosensory complex comprising platelet endothelial cell adhesion molecule-1 (PECAM-1), vascular endothelial (VE)-cadherin, and vascular endothelial growth factor receptor 2 (VEGFR2) was shown to be required for a subset of endothelial responses to shear stress [28]. Within this complex, PECAM-1 functions as the mechanosensor that first initiates intracellular signaling events. Indeed, direct force application on PECAM-1 results in activation of intracellular signaling pathways and changes in the actin cytoskeleton [29, 30]. Furthermore, PECAM-$1^{-/-}$ endothelial cells do not activate a number of shear-stress responsive pathways, and PECAM-1 knockout animals are impaired in a number of flow-mediated pathways [28, 31, 32]. In order to investigate the role of PECAM-1-mediated shear stress signaling in the development of atherosclerosis, the PECAM-$1^{-/-}$ mouse has been crossed to the $LDLR^{-/-}$ mouse and the $apoE^{-/-}$ mouse models of atherosclerosis. Both of these mouse models revealed decreased atherosclerotic lesion formation in regions of disturbed flow, such as the aortic arch [33–35]. Interestingly, the PECAM-$1^{-/-}$/$LDLR^{-/-}$ mouse also displayed increased atherosclerotic plaque development in the descending aorta that experiences laminar flow and is usually protected from plaque development. These results well displayed the dual effects of shear stress on the development of atherosclerosis *in vivo*: Disturbed shear stress promotes lesion formation, while laminar shear stress (such as in the descending aorta) actively protects against the development of atherosclerotic plaque formation.

While numerous *in vitro* and *in vivo* studies have demonstrated that disturbed flow correlates with sites of atherosclerotic plaque development, only recently was direct causality between low and disturbed shear stress and plaque formation established [36]. Acute onset of disturbed flow in an atherosusceptible mouse model is sufficient to induce plaque formation in a region of the vessel that is normally protected from lesion formation. Thus, this study provided the first direct evidence that shear stress-mediated inflammation is sufficient to induce plaque formation. Similarly, the development of the perivascular shear stress modifier cuff provided evidence directly linking disturbed flow to atherogenesis [37, 38].

MECHANISMS OF FLOW-INDUCED INFLAMMATION

Inflammation of the endothelium promotes adhesion of leukocytes to the endothelial monolayer, and disturbed shear is a potent activator of numerous inflammatory signaling cascades. Thus, flow-induced inflammatory signaling pathways that promote stable leukocyte–endothelial interactions have been under intense investigation in recent years.

MAP Kinases

Identification of complex flow patterns by endothelial mechanosensors triggers activation of numerous signaling cascades, including initiation of the mitogen-activated protein (MAP) kinase pathway. Several MAP kinases have been identified as pro-inflammatory signaling molecules in the endothelium, including extracellular signal-regulated kinase (ERK), p38, and c-Jun-NH2-terminal kinase (JNK). These MAP kinases demonstrate sustained activation in response to disturbed flow and contribute to the inflammatory phenotype. Evidence of the involvement of MAP kinases in the inflammation and progression of atherosclerosis has also been further supported by *in vivo* studies. $JNK^{-/-}$ mice on an atherosusceptible background show reduced lesion formation compared with their wild-type counterparts [39]. Importantly, the MAP kinases, such as JNK, mediate their effects, in part through activation of pro-inflammatory transcription factors, such as factor nuclear kappa-light-chain-enhancer of activated B cells (NF-κB) that initiate elaborate pro-inflammatory gene expression profiles.

NF-κB

The pro-inflammatory transcription factor NF-κB is a central mediator of endothelial cell dysfunction, and therefore also plays a crucial role in the development of atherosclerosis. NF-κB initiates a number of inflammatory signaling pathways in numerous cell types and contexts, and is activated in endothelial cells in response to disturbed shear stress. The NF-κB heterodimer, consisting of the p50/p65 subunits, is held inactive in the cytoplasm via association with the inhibitor of kappa B (IκB), which masks the nuclear localization signal (NLS) contained with the protein. Upon inflammatory insult, phosphorylation of IκB results in degradation of the inhibitor. In the absence of IκB binding, the p50/p65 heterodimer translocates to the nucleus, where it is able to bind to the promoters of several pro-inflammatory genes. Thus, activation of NF-κB results in increased expression of multiple inflammatory proteins.

In the context of atherosclerosis, shear stress is a major contributor to activation of NF-κB in the endothelium. Upon activation, NF-κB binds to shear stress response elements (SSREs) of multiple pro-inflammatory genes, including the chemokine MCP-1 and adhesion molecules such as endothelial–leukocyte adhesion molecule-1 (ELAM-1 or E-selectin), intracellular adhesion molecule-1 (ICAM-1), and vascular cell adhesion molecule-1 (VCAM-1) [40–43]. Expression of these NF-κB responsive genes collectively facilitates recruitment and adhesion of monocytes to the endothelium.

E-selectin

Leukocyte rolling along the activated endothelium is the first step of adhesion to endothelial cells, and is therefore one of the earliest steps in the development of atherosclerosis. The selectin family of adhesion receptors is critical for leukocyte–endothelial interactions. ELAM-1, or E-selectin, is rapidly upregulated and presented at the surface of activated endothelial cells. Increased levels of E-selectin are mediated by NF-κB, as the E-selectin promoter contains three NF-kB consensus sites that allow maximal expression during inflammation. In agreement with its role in inflammation, E-selectin expression is upregulated in endothelial cells in response to disturbed shear stress and is locally upregulated in regions of disturbed flow *in vivo*. Animal studies have also supported its role in the initiation of atherosclerosis. Loss of E-selectin in an atherosusceptible mouse decreases plaque formation [44]. These results provide specific evidence highlighting the pro-inflammatory role of the E-selectin in the development of atherosclerosis.

ICAM-1

Once circulating leukocytes and monocytes are recruited to sites of inflammation in the vessel wall by chemokines and cytokines, they must be able to form stable interactions with the activated endothelium in order to exit the bloodstream and infiltrate into the developing plaque. In order to accomplish this task, both leukocytes and endothelial cells rapidly upregulate expression of adhesion molecules that facilitate binding between the opposing cell types. Stable interactions between the leukocyte and endothelial cell eventually enable transmigration through the endothelium to the developing plaque in the vessel wall. One such adhesion molecule is ICAM-1. ICAM-1 is constitutively expressed at low levels in endothelial cells and has long been recognized for its role in stabilizing leukocyte–endothelial interactions during leukocyte transendothelial migration. The adhesion molecule is upregulated in response to oscillatory shear stress, and site-specific upregulation of ICAM-1 is observed in regions of disturbed flow *in vivo* [45, 46]. Given its role in stabilizing leukocyte–endothelial interactions and mediating important inflammatory processes, the protein was hypothesized to facilitate plaque progression in atherosclerosis. This hypothesis was supported by a study performed in recent years, in which researchers demonstrated that ICAM-1-null mice show a reduction in plaque burden in an apoE$^{-/-}$ model of atherosclerosis [47].

VCAM-1

VCAM-1 expression is mediated, in part, through activation of the NF-κB pathway. Expression of the protein is nearly undetectable in unstimulated endothelial cells. However, there is rapid upregulation of the protein following inflammatory insult, such as cytokine stimulation or disturbed blood flow. In agreement with numerous *in vitro* studies, VCAM-1 is specifically upregulated in regions of disturbed flow *in vivo* [46]. Furthermore, the adhesion molecule is highly upregulated in endothelial cells in diseases of inflammation, such as atherosclerosis. VCAM-1 mediates the adhesion of multiple circulating blood cells, including leukocytes, monocytes, eosinophils, and basophils to the activated endothelium. Similar to ICAM-1, VCAM-1 also has a documented role in facilitating leukocyte transendothelial migration. The involvement of the adhesion molecule is also well documented in atherosclerosis, VCAM-1$^{-/-}$/apoE$^{-/-}$ mice display reduced monocyte infiltration in the vessel wall and reduced atherosclerotic plaque formation compared with VCAM-1$^{+/+}$/apoE$^{-/-}$ mice [48, 49].

CONCLUSIONS

Atherosclerosis is a multifaceted and chronic inflammatory disease. Multiple inflammatory pathways cooperatively contribute to progression of the disease. Disturbed shear stress promotes inflammatory signaling in the endothelium, including activation the pro-inflammatory transcription factor, NF-κB, and increased expression of adhesion molecules, such as E-selectin, VCAM-1, and ICAM-1. Chemokine signaling recruits circulating monocytes to activation regions of the endothelium, while adhesion molecules facilitate leukocyte rolling and adhesion. Finally, once monocytes have infiltrated into the plaque, the release of cytokines further perpetuate inflammation within the lesion and progression of the disease. While this chapter focused on endothelial dysfunction and the role of shear stress in the initiation of atherosclerosis, there are many other inflammatory signaling processes occurring concomitantly in other cell types such as the vascular smooth muscle cells and the infiltrated immune cells. Knowledge of the molecular pathways involved in the onset and progression of the disease is growing rapidly. The identification of new cellular players and pathways, both soluble and mechanical, in the progression of the disease will facilitate the development of new therapeutic approaches to minimize inflammation and progression of atherosclerosis.

REFERENCES

1. Kowalska, M.A., Ratajczak, M.Z., Majka, M. et al. (2000) Stromal cell-derived factor-1 and macrophage-derived chemokine: 2 chemokines that activate platelets. *Blood.* 96, 50–57.
2. Abi-Younes, S., Sauty, A., Mach, F., Sukhova, G.K., Libby, P., Luster, A.D. (2000) The stromal

cell-derived factor-1 chemokine is a potent platelet agonist highly expressed in atherosclerotic plaques. *Circ Res.* 86, 131–138.

3. Veillard, N.R., Steffens, S., Burger, F., Pelli G., Mach, F. (2004) Differential expression patterns of proinflammatory and antiinflammatory mediators during atherogenesis in mice. *Arterioscler Thromb Vasc Biol.* 24, 2339–2344.
4. Veillard, N.R., Kwak, B., Pelli, G. et al. (2004). Antagonism of rantes receptors reduces atherosclerotic plaque formation in mice. *Circ Res.* 94, 253–261.
5. Gosling, J., Slaymaker, S., Gu, L. et al. (1999) Mcp-1 deficiency reduces susceptibility to atherosclerosis in mice that overexpress human apolipoprotein b. *J Clin Invest.* 103, 773–778.
6. Gu, L., Okada, Y., Clinton, S.K. et al. (1998) Absence of monocyte chemoattractant protein-1 reduces atherosclerosis in low density lipoprotein receptor-deficient mice. *Mol Cell.* 2, 275–281.
7. Schober, A., Manka, D., von Hundelshausen, P. et al. (2002) Deposition of platelet rantes triggering monocyte recruitment requires p-selectin and is involved in neointima formation after arterial injury. *Circulation.* 106, 1523–1529.
8. Huo, Y., Schober, A., Forlow, S.B. et al. (2003) Circulating activated platelets exacerbate atherosclerosis in mice deficient in apolipoprotein e. *Nat Med.* 9, 61–67.
9. Braunersreuther, V., Steffens, S., Arnaud, C. et al. (2008) A novel rantes antagonist prevents progression of established atherosclerotic lesions in mice. *Arterioscler Thromb Vasc Biol.* 28, 1090–1096.
10. Kleemann, R., Zadelaar, S., Kooistra, T. (2008) Cytokines and atherosclerosis: A comprehensive review of studies in mice. *Cardiovasc Res.* 79, 360–376.
11. Pinderski, L.J., Fischbein, M.P., Subbanagounder, G. et al. (2002) Overexpression of interleukin-10 by activated t lymphocytes inhibits atherosclerosis in ldl receptor-deficient mice by altering lymphocyte and macrophage phenotypes. *Circ Res.* 90, 1064–1071.
12. Mallat, Z., Besnard, S., Duriez, M. et al. (1999) Protective role of interleukin-10 in atherosclerosis. *Circ Res.* 85, e17–e24.
13. Kleinbongard, P., Heusch, G., Schulz, R. (2010) Tnfalpha in atherosclerosis, myocardial ischemia/reperfusion and heart failure. *Pharmacol Ther.* 127, 295–314.
14. Komarova, Y., Malik, A.B. (2010) Regulation of endothelial permeability via paracellular and transcellular transport pathways. *Annu Rev Physiol.* 72, 463–493.
15. Ohta, H., Wada, H., Niwa, T. et al. (2005) Disruption of tumor necrosis factor-alpha gene diminishes the development of atherosclerosis in apoe-deficient mice. *Atherosclerosis.* 180, 11–17.
16. Harvey, E.J., Ramji, D.P. (2005) Interferon-gamma and atherosclerosis: Pro- or anti-atherogenic? *Cardiovasc Res.* 67, 11–20.
17. Koga, M., Kai, H., Yasukawa, H. et al. (2007) Postnatal blocking of interferon-gamma function prevented atherosclerotic plaque formation in apolipoprotein e-knockout mice. *Hypertens Res.* 30, 259–267.
18. Davies, P.F. (1995) Flow-mediated endothelial mechanotransduction. *Physiol Rev.* 75, 519–560.
19. Dekker, R.J., van Soest, S., Fontijn, R.D. et al. (2002) Prolonged fluid shear stress induces a distinct set of endothelial cell genes, most specifically lung kruppel-like factor (klf2). *Blood.* 100, 1689–1698.
20. Hoger, J.H., Ilyin, V.I., Forsyth, S., Hoger, A. (2002) Shear stress regulates the endothelial kir2.1 ion channel. *Proc Natl Acad Sci U S A.* 99, 7780–7785.
21. Ando, J., Ohtsuka, A., Korenaga, R., Sakuma, I., Kamiya, A. (1993) Flow-induced calcium transients and release of endothelium-derived relaxing factor in cultured vascular endothelial cells. *Front Med Biol Eng.* 5, 17–21.
22. Chachisvilis, M., Zhang, Y.L., Frangos, J.A. (2006) G protein-coupled receptors sense fluid shear stress in endothelial cells. *Proc Natl Acad Sci U S A.* 103, 15463–15468.
23. Park, H., Go, Y.M., Darji, R. et al. (2000) Caveolin-1 regulates shear stress-dependent activation of extracellular signal-regulated kinase. *Am J Physiol Heart Circ Physiol.* 278, H1285–H1293.
24. Goetz, J.G., Minguet, S., Navarro-Lerida, I. et al. (2011) Biomechanical remodeling of the microenvironment by stromal caveolin-1 favors tumor invasion and metastasis. *Cell.* 146, 148–163.
25. Tzima, E., del Pozo, M.A., Shattil, S.J., Chien, S., Schwartz, M.A. (2001) Activation of integrins in endothelial cells by fluid shear stress mediates rho-dependent cytoskeletal alignment. *EMBO J.* 20, 4639–4647.

26. Matthews, B.D., Overby, D.R., Mannix, R., Ingber, D.E. (2006) Cellular adaptation to mechanical stress: Role of integrins, rho, cytoskeletal tension and mechanosensitive ion channels. *J Cell Sci.* 119, 508–518.
27. Guilluy, C., Swaminathan, V., Garcia-Mata, R., O'Brien, E.T., Superfine, R., Burridge, K. (2011) The rho gefs larg and gef-h1 regulate the mechanical response to force on integrins. *Nat Cell Biol.* 13, 722–727.
28. Tzima, E., Irani-Tehrani, M., Kiosses, W.B. et al. (2005) A mechanosensory complex that mediates the endothelial cell response to fluid shear stress. *Nature.* 437, 426–431.
29. Osawa, M., Masuda, M., Harada, N., Lopes, R.B., Fujiwara, K. (1997) Tyrosine phosphorylation of platelet endothelial cell adhesion molecule-1 (pecam-1, cd31) in mechanically stimulated vascular endothelial cells. *Eur J Cell Biol.* 72, 229–237.
30. Collins, C., Guilluy, C., Welch, C. et al. (2012) Localized tensional forces on pecam-1 elicit a global mechanotransduction response via the integrin-rhoa pathway. *Curr Biol.* 22, 2087–2094.
31. Chen, Z., Tzima, E. (2009) Pecam-1 is necessary for flow-induced vascular remodeling. *Arterioscler Thromb Vasc Biol.* 29, 1067–1073.
32. Liu, Y., Bubolz, A.H., Shi, Y., Newman, P.J., Newman, D.K., Gutterman, D.D. (2006) Peroxynitrite reduces the endothelium-derived hyperpolarizing factor component of coronary flow-mediated dilation in pecam-1-knockout mice. *Am J Physiol Regul Integr Comp Physiol.* 290, R57–R65.
33. Stevens, H.Y., Melchior, B., Bell, K.S., Yun, S., Yeh, J.C., Frangos, J.A. (2008) Pecam-1 is a critical mediator of atherosclerosis. *Dis Model Mech.* 1, 175–181; discussion 179.
34. Goel, R., Schrank, B.R., Arora, S. et al. (2008) Site-specific effects of pecam-1 on atherosclerosis in ldl receptor-deficient mice. *Arterioscler Thromb Vasc Biol.* 28, 1996–2002.
35. Harry, B.L., Sanders, J.M., Feaver, R.E. et al. (2008) Endothelial cell pecam-1 promotes atherosclerotic lesions in areas of disturbed flow in apoe-deficient mice. *Arterioscler Thromb Vasc Biol.* 28, 2003–2008.
36. Nam, D., Ni, C.W., Rezvan, A. et al. (2010) A model of disturbed flow-induced atherosclerosis in mouse carotid artery by partial ligation and a simple method of rna isolation from carotid endothelium. *J Vis Exp.* 40, pii 1861.
37. Cheng, C., Tempel, D., van Haperen, R. et al. (2006) Atherosclerotic lesion size and vulnerability are determined by patterns of fluid shear stress. *Circulation.* 113, 2744–2753.
38. Cheng, C., van Haperen, R., de Waard, M. et al. (2005) Shear stress affects the intracellular distribution of enos: Direct demonstration by a novel in vivo technique. *Blood.* 106, 3691–3698.
39. Wang, J., An, F.S., Zhang, W. et al. (2011) Inhibition of c-jun n-terminal kinase attenuates low shear stress-induced atherogenesis in apolipoprotein e-deficient mice. *Mol Med.* 17, 990–999.
40. Collins, T., Read, M.A., Neish, A.S., Whitley, M.Z., Thanos, D., Maniatis, T. (1995) Transcriptional regulation of endothelial cell adhesion molecules: Nf-kappa b and cytokine-inducible enhancers. *FASEB J.* 9, 899–909.
41. Schindler, U., Baichwal, V.R. (1994) Three nf-kappa b binding sites in the human e-selectin gene required for maximal tumor necrosis factor alpha-induced expression. *Mol Cell Biol.* 14, 5820–5831.
42. Marui, N., Offermann, M.K., Swerlick, R. et al. (1993) Vascular cell adhesion molecule-1 (vcam-1) gene transcription and expression are regulated through an antioxidant-sensitive mechanism in human vascular endothelial cells. *J Clin Invest.* 92, 1866–1874.
43. Sakurada, S., Kato, T., Okamoto, T. (1996) Induction of cytokines and icam-1 by proinflammatory cytokines in primary rheumatoid synovial fibroblasts and inhibition by n-acetyl-l-cysteine and aspirin. *Int Immunol.* 8, 1483–1493.
44. Dong, Z.M., Chapman, S.M., Brown, A.A., Frenette, P.S., Hynes, R.O., Wagner, D.D. (1998) The combined role of p- and e-selectins in atherosclerosis. *J Clin Invest.* 102, 145–152.
45. Iiyama, K., Hajra, L., Iiyama, M. et al. (1999) Patterns of vascular cell adhesion molecule-1 and intercellular adhesion molecule-1 expression in rabbit and mouse atherosclerotic lesions and at sites predisposed to lesion formation. *Circ Res.* 85, 199–207.
46. Nakashima, Y., Raines, E.W., Plump, A.S., Breslow, J.L., Ross, R. (1998) Upregulation of vcam-1 and icam-1 at atherosclerosis-prone sites on the endothelium in the apoe-deficient mouse. *Arterioscler Thromb Vasc Biol.* 18, 842–851.

47. Bourdillon, M.C., Poston, R.N., Covacho, C., Chignier, E., Bricca, G., McGregor, J.L. (2000) Icam-1 deficiency reduces atherosclerotic lesions in double-knockout mice (apoe(-/-)/icam-1(-/-)) fed a fat or a chow diet. *Arterioscler Thromb Vasc Biol.* 20, 2630–2635.
48. Ley, K., Huo, Y. (2001) Vcam-1 is critical in atherosclerosis. *J Clin Invest.* 107, 1209–1210.
49. Cybulsky, M.I., Iiyama, K., Li, H. et al. (2001) A major role for vcam-1, but not icam-1, in early atherosclerosis. *J Clin Invest.* 107, 1255–1262.

5

HOMOCYSTEINE: A CONTROVERSIAL CARDIOVASCULAR RISK FACTOR

SANJANA DAYAL AND STEVEN R. LENTZ

Department of Internal Medicine, University of Iowa Carver College of Medicine, Iowa City, IA, USA

INTRODUCTION

Homocysteine is a naturally occurring nondietary amino acid that is found in low micromolar concentrations in blood plasma. The synthesis and metabolism of homocysteine is dependent on two major cyclic pathways: the methionine and folate cycles. As depicted in Figure 5.1, homocysteine is formed from methionine, and then undergoes one of two metabolic fates: it is either metabolized to cystathionine by the vitamin B6-dependent enzyme cystathionine β-synthase (CBS) or remethylated back to methionine by one of two enzymes: betaine homocysteine methyl transferase (BHMT) or methionine synthase (MS). BHMT utilizes betaine, which is derived from dietary choline, as the methyl donor for remethylation of homocysteine to methionine. The methyl donor for MS, 5-methyltetrahydrofolate, is generated by 5,10-methylene tetrahydrofolate reductase (MTHFR) in the folate cycle. Thus, tight regulation of several key enzymes, including CBS, BHMT, MS, MTHFR, and methionine synthase reductase (which maintains MS activity) is necessary to prevent excessive accumulation of plasma homocysteine. Many of these enzymes are allosterically regulated by methionine or *S*-adenosylmethionine, an essential substrate for intracellular transmethylation reactions [1].

The recognition of homocysteine as a cardiovascular risk factor dates back to the 1960s, when adverse vascular events were first reported in patients with excessive levels of urinary homocysteine, a condition termed "homocystinuria." Homocystinuria was first described in 1962 [2], and an enzymatic defect in CBS in a patient with homocystinuria was reported by Mudd and colleagues in 1964 [3]. Clinical abnormalities reported in these patients included mental retardation, musculoskeletal abnormalities, dislocation of the ocular lens, and major cardiovascular and thromboembolic events. While it was recognized from these early cases that premature cardiovascular disease was a component of the disease phenotype, it was not until 1969 that homocysteine was proposed to be a general cardiovascular risk factor [4]. In this

Atherosclerosis: Risks, Mechanisms, and Therapies, First Edition. Edited by Hong Wang and Cam Patterson.

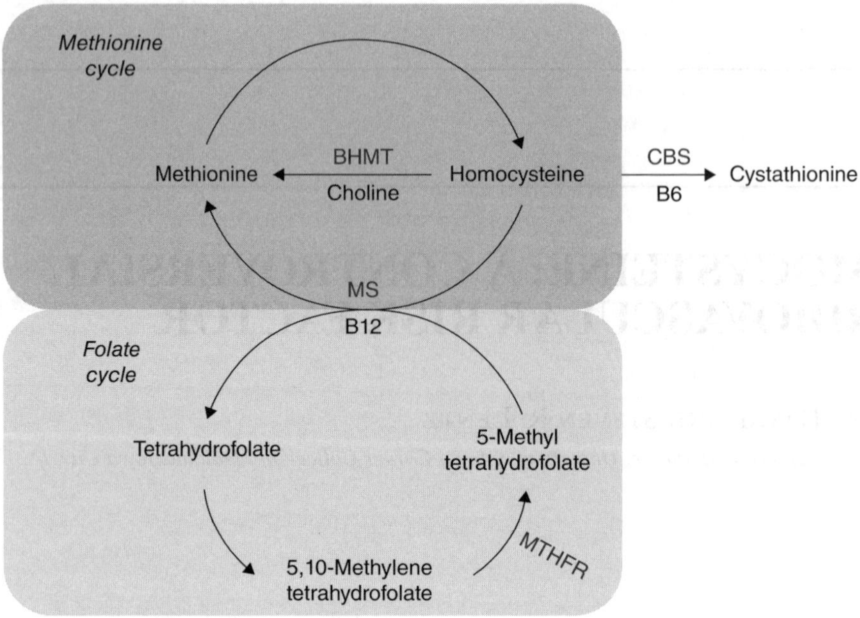

Figure 5.1 Homocysteine metabolic pathways. Homocysteine is formed in the methionine cycle by demethylation of methionine. Homocysteine then either undergoes transsulfuration to cystathionine or is remethylated to methionine. In the transsulfuration reaction, homocysteine is condensed with serine to form cystathionine by the enzyme cystathionine β-synthase (CBS), which uses vitamin B6 as a cofactor. In the remethylation pathway, which couples homocysteine metabolism to the folate cycle, the enzyme methionine synthase (MS) transfers a methyl group from 5-methyltetrahydrofolate in a reaction that requires vitamin B12 as a cofactor. 5-methyltetrahydrofolate is generated by reduction of 5,10-methylene tetrahydrofolate through the action of 5,10-methylene tetrahydrofolate reductase (MTHFR). Homocysteine remethylation is alternatively carried out by betaine:homocysteine methyl transferase (BHMT), which uses betaine (which is ultimately derived from choline) as the methyl donor.

landmark paper, McCully reported autopsy findings from two homocystinuric children who had suffered from separate enzymatic defects in the methionine and folate cycles. The similarity of the arterial lesions in these children led McCully to propose that homocysteine itself may be a causative mediator of arteriosclerosis and thrombosis, regardless of the specific metabolic defect responsible for elevation of homocysteine. McCully also suggested that elevation in homocysteine might be a risk factor for arteriosclerosis in patients with nongenetic causes of hyperhomocysteinemia.

Homocysteine is normally present in plasma as a mixture of molecular species (Figure 5.2). The total concentration of the thiol and disulfide forms of homocysteine is termed "total homocysteine" (tHcy) [5]. In addition, a variable amount of homocysteine is linked to plasma proteins through isopeptide bonds (rather than disulfide bonds) [6]. This isopeptide (N-linked) protein-bound homocysteine is not detected by clinical homocysteine assays, which measures only tHcy. The development in the 1980s and 1990s of reliable clinical laboratory techniques to accurately determine tHcy concentrations in human plasma facilitated the diagnosis of hyperhomocysteinemia and helped to establish its association with cardiovascular disease.

Normal reference levels of plasma homocysteine vary in different populations depending on dietary intake of folate and other nutritional factors. In North America, where

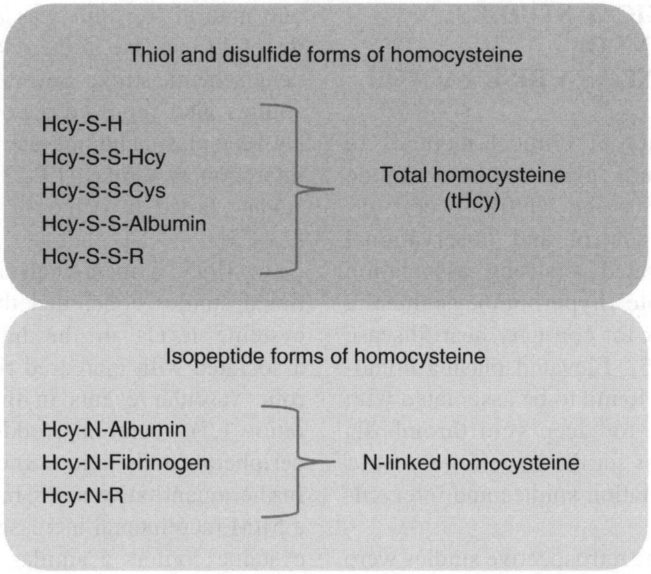

Figure 5.2 Molecular forms of homocysteine found in plasma. The total concentration of the thiol and disulfide forms of homocysteine is termed "total homocysteine" (tHcy) [5]. The largest amount is present as a mixed disulfide with albumin (Hcy-S-S-albumin) with lesser amounts in the free thiol form (Hcy-S-H), the homocystine form (Hcy-S-S-Hcy), or in the form of a mixed disulfides with cysteine (Hcy-S-S-Cys) or other sulfur-containing molecules (Hcy-S-R). In addition, some homocysteine is linked to albumin (Hcy-N-albumin), fibrinogen (Hcy-N-fibrinogen), or other proteins or peptides (Hcy-N-R) through isopeptide bonds (rather than disulfide bonds) [6]. These N-linked forms of homocysteine are not detected by clinical homocysteine assays, which measure only tHcy.

the food supply is fortified with folic acid, most healthy individuals now have plasma homocysteine levels less than 10 μmol/l [7]. Hyperhomocysteinemia can be broadly classified into two categories: moderate and severe. Severe hyperhomocysteinemia (plasma homocysteine >50 μmol/l) is relatively rare, is most often genetic, and usually occurs due to inborn errors in methionine, folate, or vitamin B12 metabolism [1] or in cases of severe vitamin B12 deficiency due to pernicious anemia [8]. Moderate hyperhomocysteinemia (plasma homocysteine >10 but <50 μmol/l), which is more prevalent than severe hyperhomocysteinemia, is often caused by nutritional deficiencies in folate or vitamin B12 or by renal disease [9, 10], though it can also be caused by polymorphisms in the CBS or MTHFR genes.

A number of interesting parallels exist between the emergence of homocysteine and cholesterol as cardiovascular risk factors. For example, both hyperhomocysteinemia and hypercholesterolemia were initially observed to be associated with cardiovascular disease in patients with inborn errors of metabolism that led to marked increases in homocysteine or cholesterol (hereditary homocystinuria and familial hypercholesterolemia, respectively). Both conditions span a wide range of severity; and in both cases, the lessons learned from studying patients with the severe condition have been applied to cases of moderate disease. However, key differences exist between moderate and severe hyperhomocysteinemia that have raised important questions regarding whether homocysteine is a causative mediator or rather a biomarker of increased susceptibility to vascular disease. In subsequent sections, we describe the pertinent epidemiological studies, discoveries made using animal models of hyperhomocysteinemia, and findings from homocysteine-lowering intervention trials.

EPIDEMIOLOGICAL STUDIES: ESTABLISHMENT OF HOMOCYSTEINE AS A RISK FACTOR

As the availability of clinical methods to accurately measure plasma homocysteine spread in the 1980s and 1990s, several retrospective case control and observational studies demonstrated a strong association between moderate hyperhomocysteinemia and increased risk for coronary heart disease and stroke [11–13]. Elevated plasma homocysteine was also found to be associated with an increased risk for deep vein thrombosis [14]. These studies set the stage for a host of prospective association studies and intervention trials.

Results from the retrospective studies were confirmed by several large prospective cohort studies that evaluated risk for vascular events in subjects with elevated plasma homocysteine. For example, a nested case control study with 14,916 US physician participants prospectively assessed the risk of cardiovascular heart disease associated with plasma levels of homocysteine in subjects with no prior myocardial infarction or stroke at baseline [15]. A striking increase in the relative risk (with an odds ratio of 3.4) of developing cardiovascular disease was observed within 5 years in subjects whose homocysteine levels were in the highest 5% as compared with those who had homocysteine levels in the lowest 90%. Similarly, the British United Provident Association study found a relative risk of 2.9 for fatal coronary heart disease in men whose homocysteine levels were in the highest quartile compared with those in lowest quartile [16]. Furthermore, a prospective study [17] that investigated the association of plasma homocysteine with incident stroke identified plasma homocysteine as a significant independent predictor of new atherothrombotic brain infarcts in older subjects; with each 1 μM increase in plasma homocysteine, the estimated risk increased by 7.9%. Likewise, in a nested case control study by Tanne and colleagues [18], an increase in homocysteine by one natural log unit was associated with a threefold increase in the relative odds of incident ischemic stroke. Several other prospective studies also reported a positive association between plasma homocysteine and myocardial infarction or stroke [19–22], although a few studies failed to demonstrate an association [23–25].

In 1995, a meta-analysis of 27 observational studies concluded that plasma homocysteine levels in the highest 5–10% are associated with increased risk for atherosclerotic vascular events in the coronary (odds ratio 1.7), cerebral (odds ratio 2.5), or peripheral (odds ratio 6.8) circulation [26]. A further quantitative assessment suggested that a 5 μM incremental increase in plasma homocysteine confers a similar risk for coronary artery disease as that of a 0.5 mM incremental increase in plasma cholesterol [26]. A subsequent meta-analysis of individual participant data from 12 prospective studies found that, after adjusting for other known risk factors, a 25% elevation in plasma homocysteine is associated with a ~10% higher risk of cardiovascular disease and a ~20% higher risk of stroke [27]. Finally, a meta-analysis of 10 case control studies [28] also supported hyperhomocysteinemia as a risk factor for venous thrombosis. Taken together, these results provided support for the hypothesis that homocysteine is a clinically relevant, independent risk factor for cardiovascular disease, stroke, and venous thrombosis.

ANIMAL MODELS OF HYPERHOMOCYSTEINEMIA

Much has been learned about the vascular consequences of elevated homocysteine by studying animal models [29]. As shown in Figure 5.3, hyperhomocysteinemia can be produced in animals through a variety of different dietary or genetic approaches. Some of the dietary interventions that have been used to induce hyperhomocysteinemia include the following: (i) addition of excess methionine to

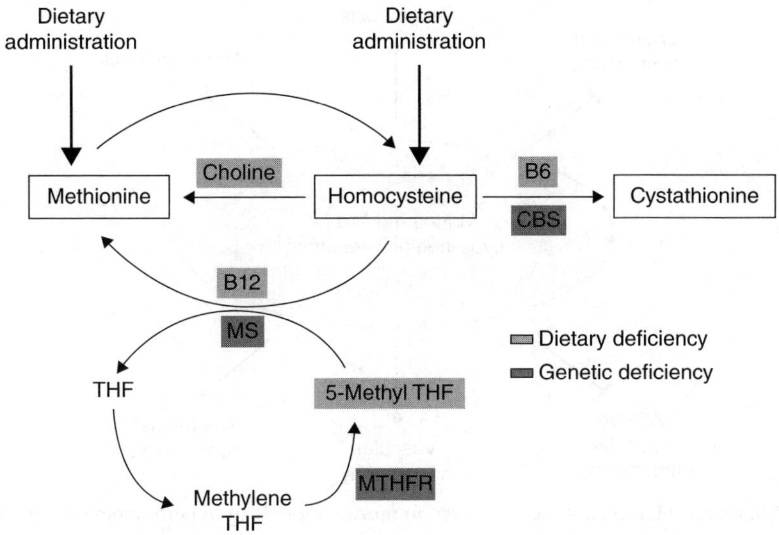

Figure 5.3 Experimental strategies for the development of genetic and dietary models of hyperhomocysteinemia. Dietary interventions that have been used to induce hyperhomocysteinemia include addition of excess methionine to increase metabolic flux through the methionine cycle, addition of homocysteine itself to the solid diet or drinking water, or utilization of diets deficient in vitamin B6, B12, choline, and/or folate. The most widely used dietary strategy is a combination of a high methionine and low folate diet. Genetic approaches include the use of mice that have targeted mutations or deletions in one of the key homocysteine metabolic enzymes such as CBS, MTHFR, or MS. Some studies have adopted a combined dietary and genetic strategy.

the diet to increase metabolic flux through the methionine cycle, (ii) diets deficient in vitamin B6 to limit the CBS-dependent conversion of homocysteine to cystathionine, (iii) diets deficient in folate and/or vitamin B12 to limit the MS-mediated remethylation of homocysteine to methionine, (iv) diets deficient in choline to limit BHMT-mediated homocysteine remethylation, and (v) addition of homocysteine itself to the diet or drinking water. These dietary approaches have been used to produce experimental hyperhomocysteinemia in nonhuman primates, mini pigs, rabbits, rats, and mice [30]. A majority of the animal studies have been conducted in mice because murine models allow for genetic as well as dietary strategies to manipulate homocysteine levels. The genetic approaches include use of mice that have targeted mutations or deletions in one of the key homocysteine metabolic enzymes such as CBS [31], MTHFR [32], and MS [33]. A particularly useful murine model is one in which mice lacking an intact endogenous murine CBS gene express an inducible wild-type or mutant human *CBS* transgene [34]. The following paragraphs highlight some recent scientific advances using these murine models.

Murine models of hyperhomocysteinemia mimic many of the vascular abnormalities observed in patients with elevated plasma homocysteine, including endothelial dysfunction, accelerated atherosclerosis and thrombosis, neointimal hyperplasia, vascular hypertrophy, and altered vascular mechanics (Figure 5.4) [29, 30, 35]. Although these effects have been observed in several arterial vascular beds, cerebral blood vessels appear to be particularly sensitive to the adverse effects of hyperhomocysteinemia [36, 37]. The hypersensitivity of the cerebral circulation to hyperhomocysteinemia in mice is of interest in light of the clinical

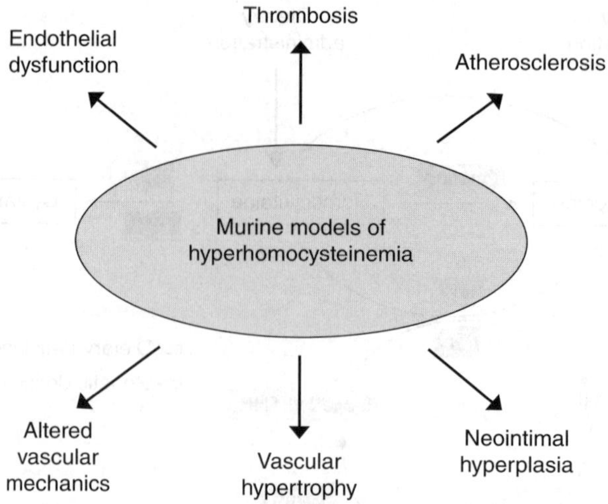

Figure 5.4 Major vascular phenotypes observed in murine models of hyperhomocysteinemia.

observation that elevated plasma homocysteine is associated with a higher relative risk for stroke than for myocardial infarction [27].

In general, the findings from murine models support the hypothesis that hyperhomocysteinemia is a causative risk factor for cerebrovascular and cardiovascular diseases. However, there are some limitations and inconsistent findings from the murine models which suggest that elevation of homocysteine itself is not sufficient to reproduce the entire spectrum of vascular abnormalities associated with hyperhomocysteinemia in humans. Although some vascular phenotypes, such as endothelial dysfunction, accelerated atherosclerosis, and vascular hypertrophy, have been seen in both dietary and genetic models of hyperhomocysteinemia, other phenotypes have not consistently been found to occur in all models. For example, increased susceptibility to thrombosis has been observed in murine models of diet-induced hyperhomocysteinemia [38], yet a genetic model of severe hyperhomocysteinemia demonstrated a complete lack of a prothrombotic phenotype [39]. The paradoxical absence of effects on arterial or venous thrombosis in this genetic model suggests that increased susceptibility to thrombosis requires factors other than, or in addition to elevation of homocysteine. Furthermore, because levels of homocysteine are regulated by two cyclic metabolic pathways (Figure 5.1), all of the dietary and genetic approaches to produce hyperhomocysteinemia inevitably result in alterations of other metabolites, some of which may be responsible for adverse vascular effects. Despite these limitations, the availability of a growing number of murine models with targeted mutations in specific genes within the methionine and folate cycles should help clarify the relative contributions of homocysteine and related metabolites in the vascular phenotype of hyperhomocysteinemia.

CLINICAL INTERVENTION TRIALS

The rationale for the investigation of homocysteine-lowering therapy to prevent cardiovascular and cerebrovascular events was established by numerous epidemiological studies implicating hyperhomocysteinemia as a cardiovascular risk factor (see Section "Epidemiological studies: establishment of homocysteine as a risk factor"). Further support for large-scale homocysteine-lowering intervention trials was provided by a landmark publication in 2001 demonstrating the clinical benefit of homocysteine-lowering therapy in patients with severe hyperhomocysteinemia. In this retrospective study, homocysteine-lowering therapy (supplementation with B vitamins and betaine in conjunction with restriction of dietary methionine) decreased the apparent risk of vascular events by an estimated 85% [40].

Beginning in the late 1990s, several large, prospective, randomized clinical trials were conducted to examine the potential benefits of homocysteine-lowering therapy for the secondary prevention of adverse vascular outcomes. These trials assessed the effects of supplementation with folic acid and other B vitamins on plasma homocysteine concentration and subsequent vascular outcomes in patients with a history of a prior vascular event such as myocardial infarction, ischemic stroke, or venous thromboembolism. All of the intervention trials showed a significant reduction in plasma homocysteine concentration. Despite the reduction in plasma tHcy, however, none of the trials demonstrated a significant overall clinical benefit of homocysteine-lowering therapy for the secondary prevention of vascular events. The negative results of these trials may be related to several factors: (i) inadequate sample size and lack of sufficient statistical power to demonstrate the expected 10–20% improvement in clinical vascular outcomes, (ii) confounding effects due to folate fortification and inadequate vitamin B12 supplementation in some of the trials, (iii) short trial duration, and (iv) possible adverse effects of B vitamins [41–43]

Subsequent meta-analyses of individual participant data from the randomized intervention trials was performed to address some of the above concerns, including sample size, event rates, and duration of follow-up [44–46]. The results of these meta-analyses confirmed a lack of clinical benefit of homocysteine-lowering therapy for the secondary prevention of coronary heart disease but suggested a possible benefit in stroke prevention. The largest meta-analyses, which analyzed data from over 45,000 subjects, estimated a small benefit in the prevention of stroke, with a relative risk estimate of 0.88–0.93 [46, 47]. This degree of protection is consistent with the benefit that might be expected based on the results of prospective association studies [27]. Several other reports have suggested a possible benefit of homocysteine-lowering therapy for stroke prevention in certain subgroups of subjects [45–49], but all of these analyses suffer from the inherent limitations of subgroup analysis.

SUMMARY AND PERSPECTIVES

Since the discovery of patients with hyperhomocysteinemia, a great deal of clinical, epidemiologic, and basic science research has attempted to define the link between elevated homocysteine levels and the risk for developing cardiovascular events. Despite over 6000 publications in the field, it is still controversial whether homocysteine plays a causal role in vascular disease.

Several important questions still need to be answered before a definitive conclusion can be reached about homocysteine as a treatable cardiovascular risk factor. First, since cardiovascular disease is associated with multiple risk factors, each of which may make a small relative contribution to clinical vascular events, it is not surprising that intervention trials may require very large sample sizes to demonstrate the protective effects of modifying a single risk factor such as homocysteine. This is particularly true for secondary intervention trials, since outcomes are likely to be affected by additional therapeutic interventions directed at other risk factors. For example, many subjects in the homocysteine-lowering trials were also treated with antiplatelet agents, lipid-lowering drugs, and/or anti-hypertensives in addition to B vitamin supplementation. In this regard, it is interesting to note that one subgroup analysis suggested a greater benefit of homocysteine-lowering therapy in preventing stroke in subjects who were not taking aspirin or other antiplatelet drugs [45].

Second, there is growing evidence that homocysteine-lowering therapy may be more beneficial for the primary prevention of vascular events than for secondary prevention. Indirect evidence for that lowering of homocysteine is beneficial in the primary prevention setting was obtained from a large population-based cohort study, which suggested that homocysteine-lowering via folic acid fortification may prevent stroke [50]. A subsequent meta-analysis of eight randomized trials confirmed these observations [48]. It is important to note that all of the reported primary prevention trials only examined stroke as the vascular endpoint; therefore, the effect of homocysteine-lowering on primary prevention of other vascular endpoints remains undefined.

Clearly, further work is needed to clarify whether there is a subset of vascular disease patients who may benefit from homocysteine-lowering therapy. The new murine genetic models of hyperhomocysteinemia provide an excellent platform to ask mechanistic questions regarding the role of homocysteine in vascular pathophysiology. Reconciling the discrepant results between genetic and dietary models of hyperhomocysteinemia may help resolve the seemingly contradictory results of homocysteine-lowering interventions in patients with severe versus moderate hyperhomocysteinemia. Likewise, a more complete understanding of the precise role of homocysteine itself in vascular pathophysiology may suggest specific therapeutic strategies, as opposed to nonspecific interventions such as vitamin supplementation, to target homocysteine as a potentially modifiable cardiovascular risk factor.

REFERENCES

1. Finkelstein JD. Inborn errors of sulfur-containing amino acid metabolism. J. Nutr. 2006;136: 1750S–1754S.
2. Carson NA, Neill DW. Metabolic abnormalities detected in a survey of mentally backward individuals in Northern Ireland. Arch. Dis. Child. 1962;37:505–513.
3. Mudd SH, Finkelstein JD, Irreverre F, Laster L. Homocystinuria: an enzymatic defect. Science. 1964;143:1443–1445.
4. McCully KS. Vascular pathology of homocysteinemia: implications for the pathogenesis of arteriosclerosis. Am. J. Pathol. 1969;56: 111–128.
5. Mudd SH, Finkelstein JD, Refsum H, et al. Homocysteine and its disulfide derivatives: a suggested consensus terminology. Arterioscler. Thromb. Vasc. Biol. 2000;20:1704–1706.
6. Jakubowski H, Glowacki R. Chemical biology of homocysteine thiolactone and related metabolites. Adv. Clin. Chem. 2011;55:81–103.
7. Yang Q, Cogswell ME, Hamner HC, et al. Folic acid source, usual intake, and folate and vitamin B-12 status in US adults: National health and nutrition examination survey (NHANES) 2003-2006. Am. J. Clin. Nutr. 2010;91:64–72.
8. Limal N, Scheuermaier K, Tazi Z, Sene D, Piette JC, Cacoub P. Hyperhomocysteinaemia, thrombosis and pernicious anaemia. Thromb. Haemost. 2006;96:233–235.
9. Selhub J, Jacques PF, Wilson PW, Rush D, Rosenberg IH. Vitamin status and intake as primary determinants of homocysteinemia in an elderly population. JAMA. 1993;270: 2693–2698.
10. vanGuldener C, Stehouwer CD. Homocysteine and methionine metabolism in renal failure. Semin. Vasc. Med. 2005;5:201–208.
11. Clarke R, Daly L, Robinson K, et al. Hyperhomocysteinemia: an independent risk factor for vascular disease. N. Engl. J. Med. 1991;324:1149–1155.
12. Evers S, Koch HG, Grotemeyer KH, Lange B, Deufel T, Ringelstein EB. Features, symptoms, and neurophysiological findings in stroke associated with hyperhomocysteinemia. Arch. Neurol. 1997;54:1276–1282.
13. Montalescot G, Ankri A, Chadefauxvekemans B, et al. Plasma homocysteine and the extent of atherosclerosis in patients with coronary artery disease. Int. J. Cardiol. 1997;60:295–300.
14. denHeijer M, Koster T, Blom HJ, et al. Hyperhomocysteinemia as a risk factor for deep-vein thrombosis. N. Engl. J. Med. 1996;334: 759–762.
15. Stampfer MJ, Malinow MR, Willett WC, et al. A prospective study of plasma homocyst(e)ine and risk of myocardial infarction in US physicians. JAMA. 1992;268:877–881.

16. Wald NJ, Watt HC, Law MR, Weir DG, McPartlin J, Scott JM. Homocysteine and ischemic heart disease: results of a prospective study with implications regarding prevention. Arch. Intern. Med. 1998;158:862–867.
17. Aronow WS, Ahn C, Gutstein H. Increased plasma homocysteine is an independent predictor of new atherothrombotic brain infarction in older persons. Am. J. Cardiol. 2000;86: 585–586, A510.
18. Tanne D, Haim M, Goldbourt U, et al. Prospective study of serum homocysteine and risk of ischemic stroke among patients with preexisting coronary heart disease. Stroke. 2003;34:632–636.
19. Stehouwer CD, Weijenberg MP, van den Berg M, Jakobs C, Feskens EJ, Kromhout D. Serum homocysteine and risk of coronary heart disease and cerebrovascular disease in elderly men: a 10-year follow-up. Arterioscler. Thromb. Vasc. Biol. 1998;18:1895–1901.
20. Arnesen E, Refsum H, Bonaa KH, Ueland PM, Forde OH, Nordrehaug JE. Serum total homocysteine and coronary heart disease. Int. J. Epidemiol. 1995;24:704–709.
21. Perry IJ, Refsum H, Morris RW, Ebrahim SB, Ueland PM, Shaper AG. Prospective study of serum total homocysteine concentration and risk of stroke in middle-aged British men. Lancet. 1995;346:1395–1398.
22. Bots ML, Launer LJ, Lindemans J, et al. Homocysteine and short-term risk of myocardial infarction and stroke in the elderly: the Rotterdam study. Arch. Intern. Med. 1999;159: 38–44.
23. Chasan-Taber L, Selhub J, Rosenberg IH, et al. A prospective study of folate and vitamin B6 and risk of myocardial infarction in US physicians. J. Am. Coll. Nutr. 1996;15:136–143.
24. Verhoef P, Hennekens CH, Malinow MR, Kok FJ, Willett WC, Stampfer MJ. A prospective study of plasma homocyst(e)ine and risk of ischemic stroke. Stroke. 1994;25:1924–1930.
25. Folsom AR, Nieto FJ, McGovern PG, et al. Prospective study of coronary heart disease incidence in relation to fasting total homocysteine, related genetic polymorphisms, and B vitamins: the atherosclerosis risk in communities (ARIC) study. Circulation. 1998;98:204–210.
26. Boushey CJ, Beresford SA, Omenn GS, Motulsky AG. A quantitative assessment of plasma homocysteine as a risk factor for vascular disease. Probable benefits of increasing folic acid intakes. JAMA. 1995;274:1049–1057.
27. Homocysteine Studies Collaboration. Homocysteine and risk of ischemic heart disease and stroke: a meta-analysis. JAMA. 2002; 288:2015–2022.
28. denHeijer M, Rosendaal FR, Blom HJ, Gerrits WB, Bos GM. Hyperhomocysteinemia and venous thrombosis: a meta-analysis. Thromb. Haemost. 1998;80:874–877.
29. Dayal S, Lentz SR. Murine models of hyperhomocysteinemia and their vascular phenotypes. Arterioscler. Thromb. Vasc. Biol. 2008;28: 1596–1605.
30. Dayal S, Lentz SR. Role of redox reactions in the vascular phenotype of hyperhomocysteinemic animals. Antioxid. Redox. Signal. 2007; 9:1899–1909.
31. Watanabe M, Osada J, Aratani Y, et al. Mice deficient in cystathionine beta-synthase: animal models for mild and severe homocyst(e)inemia. Proc. Natl. Acad. Sci. U. S. A. 1995;92:1585–1589.
32. Chen Z, Karaplis AC, Ackerman SL, et al. Mice deficient in methylenetetrahydrofolate reductase exhibit hyperhomocysteinemia and decreased methylation capacity, with neuropathology and aortic lipid deposition. Hum. Mol. Genet. 2001;10:433–443.
33. Swanson DA, Liu ML, Baker PJ, et al. Targeted disruption of the methionine synthase gene in mice. Mol. Cell. Biol. 2001;21:1058–1065.
34. Wang L, Chen X, Tang B, Hua X, Klein-Szanto A, Kruger WD. Expression of mutant human cystathionine beta-synthase rescues neonatal lethality but not homocystinuria in a mouse model. Hum. Mol. Genet. 2005;14:2201–2208.
35. Cheng Z, Jiang X, Kruger WD, et al. Hyperhomocysteinemia impairs endothelium-derived hyperpolarizing factor-mediated vasorelaxation in transgenic cystathionine beta synthase-deficient mice. Blood. 2011;118: 1998–2006.
36. Dayal S, Arning E, Bottiglieri T, et al. Cerebral vascular dysfunction mediated by superoxide in hyperhomocysteinemic mice. Stroke. 2004; 35:1957–1962.
37. Baumbach GL, Sigmund CD, Bottiglieri T, Lentz SR. Structure of cerebral arterioles in cystathionine beta-synthase-deficient mice. Circ Res. 2002;91:931–937.
38. Dayal S, Wilson KM, Leo L, Arning E, Bottiglieri T, Lentz SR. Enhanced susceptibility

to arterial thrombosis in a murine model of hyperhomocysteinemia. Blood. 2006;108:2237–2243.

39. Dayal S, Chauhan AK, Jensen M, et al. Paradoxical absence of a prothrombotic phenotype in a mouse model of severe hyperhomocysteinemia. Blood. 2012;119:3176–3183.
40. Yap S, Boers GH, Wilcken B, et al. Vascular outcome in patients with homocystinuria due to cystathionine beta-synthase deficiency treated chronically: a multicenter observational study. Arterioscler. Thromb. Vasc. Biol. 2001;21: 2080–2085.
41. Bostom AG, Selhub J, Jacques PF, Rosenberg IH. Power shortage: clinical trials testing the "homocysteine hypothesis" against a background of folic acid-fortified cereal grain flour. Ann. Int. Med. 2001;135:133–137.
42. Spence JD. Homocysteine: call off the funeral. Stroke. 2006;37:282–283.
43. Spence JD. B vitamin therapy for homocysteine: renal function and vitamin B12 determine cardiovascular outcomes. Clin. Chem. Lab. Med. 2013;51:633–637.
44. Clarke R, Halsey J, Lewington S, et al. Effects of lowering homocysteine levels with B vitamins on cardiovascular disease, cancer, and cause-specific mortality: meta-analysis of 8 randomized trials involving 37,485 individuals. Arch. Int. Med. 2010;170:1622–1631.
45. Lee M, Hong KS, Chang SC, Saver JL. Efficacy of homocysteine-lowering therapy with folic acid in stroke prevention: a meta-analysis. Stroke. 2010;41:1205–1212.
46. Huang T, Chen Y, Yang B, Yang J, Wahlqvist ML, Li D. Meta-analysis of B vitamin supplementation on plasma homocysteine, cardiovascular and all-cause mortality. Clin. Nutr. 2012;31: 448–454.
47. Ji Y, Tan S, Xu Y, et al. Vitamin B supplementation, homocysteine levels, and the risk of cerebrovascular disease: a meta-analysis. Neurology. 2013;81: 1298–1307.
48. Wang X, Qin X, Demirtas H, et al. Efficacy of folic acid supplementation in stroke prevention: a meta-analysis. Lancet. 2007;369:1876–1882.
49. Hankey GJ. Is homocysteine a causal and treatable risk factor for stroke? Lancet Neurol. 2007;6:751–752.
50. Yang Q, Botto LD, Erickson JD, et al. Improvement in stroke mortality in Canada and the United States, 1990 to 2002. Circulation. 2006;113:1335–1343.

6

HYPERTENSION AS A RISK FACTOR FOR ATHEROSCLEROSIS

ANNET KIRABO AND DAVID G. HARRISON

Division of Clinical Pharmacology, Department of Medicine, Vanderbilt University, Nashville, TN, USA

INTRODUCTION

Hypertension represents an enormous health care burden in Western societies. Approximately 30% of the Western population and 70% of adults over 70 years of age are afflicted with this disease [1]. Another 30% of the population has prehypertension, defined as systolic blood pressure of 120–139 mm Hg, and these individuals commonly progress to have overt hypertension within 4 years [2]. While the overall prevalence of hypertension in the United States has remained constant over the past decade, its incidence in young adults (aged 18–39 years) has increased from 27.6% in 1999 to 49.1% in 2008. The overall annual healthcare cost of hypertension in the United States was recently estimated to be over $93.5 billion [3].

The clinical sequelae of hypertension are largely related to end-organ insults including renal failure, stroke, myocardial hypertrophy, heart failure, and vascular disease. The vascular consequences of hypertension include alterations of vascular reactivity, vascular hypertrophy, fibrosis, dropout or rarefaction of smaller vessels, and a striking predilection to atherosclerosis. The mechanisms by which hypertension promotes atherosclerosis remain a matter of continued investigation. In this chapter, we will review the molecular, cellular, and hemodynamic factors that link hypertension with atherosclerosis and discuss future directions for research and potential therapeutic implications.

EVIDENCE LINKING ATHEROSCLEROSIS AND HYPERTENSION

Clinical and epidemiology studies strongly support the concept that hypertension is a major risk factor for atherosclerosis. In large populations, elevated values of blood pressure predict future coronary heart disease mortality beyond other risk factors such as smoking, age, and lipoprotein levels [4]. In the Multiple Risk

Atherosclerosis: Risks, Mechanisms, and Therapies, First Edition. Edited by Hong Wang and Cam Patterson.

Factor Intervention Trial (MRFIT), a systolic pressure exceeding 142 mm Hg increased cardiovascular disease-related mortality by almost 3-fold, while diastolic pressures greater than 92 mm Hg increased risk by 2.25-fold [5]. Autopsies performed in the Bogalusa Heart study showed that the incidence of fatty streaks and fibrous plaques correlates significantly with premortem systolic blood pressure [6]. Likewise, in the Pathological Determinants of Atherosclerosis in the Youth (PDAY) study, the extent of coronary artery surface area covered with atherosclerotic lesions was almost twice as high in men greater than 30 years of age with renal changes of hypertension compared to those without such changes [7]. Aortic plaque volume, as estimated by magnetic resonance imaging, is increased by the presence of hypertension by threefold in men and by 25% in women [8]. The incidence of major adverse events in atherosclerosis, including ischemic heart disease, ischemic stroke, and other vascular diseases are linearly related to both systolic and diastolic blood pressure over all age groups [9]. Among individuals aged 50–59 years of age, stroke mortality is increased by 16-fold and ischemic heart disease mortality by 8-fold in patients with blood pressures of 180 mm Hg [9].

There have been efforts to identify genetic variants responsible for hypertension. Several genome-wide association studies have identified correlations between single nucleotide polymorphisms and blood pressure [10]. In the International Consortium for Blood Pressure Genome-Wide Association Studies, a genetic risk score based on 29 significant variants was associated with not only hypertension but also with stroke and coronary artery disease [11]. A recent study showed that additional genetic variants are more closely associated with pulse pressure and mean arterial pressure than with systolic blood pressure [12]. One variant identified in this analysis lies near the gene encoding ADAMTS8, which is highly expressed in macrophage-rich regions of atherosclerotic plaques. The HYPERGENES project identified a hypertension susceptibility locus in the promoter region of the endothelial nitric oxide synthase (NOS) gene in a European population and performed meta-analysis of this locus in several other existing databases of almost 22,000 individuals [13]. Overall, the odds ratio of hypertension in subjects with this variation was 1.34 (95% CI: 1.25–1.44; $P = 1.032 \times 10^{-14}$). This locus is adjacent to an ETS-transcription factor-binding site, known to modulate eNOS transcription. Another recent genome-wide meta-analysis of over 86,000 individuals revealed single nucleotide polymorphisms in the beta-1 adrenergic receptor, angiotensinogen, and the angiotensin I-converting enzyme (ACE) genes that reached genome wide significance [14]. Given the role of nitric oxide, catecholamines and angiotensin II in modulating atherosclerosis, it is possible that these gene variations associated with hypertension influence vascular lesion formation. These genome-wide association studies not only point to new potential gene targets for treatment but also have the potential to describe gene networks that interact to cause hypertension.

In addition to these clinical studies, there is strong experimental evidence supporting a link between hypertension and atherosclerosis. Induction of hypertension by infusion of angiotensin II, deoxycorticosterone (DOCA)-salt challenge or suprarenal constriction dramatically enhances atherosclerosis in apoE$^{-/-}$ mice [15–17]. Angiotensin II also increases large artery stiffening and collagen deposition in hyperlipidemic mice [18]. Induction of hypertension by renal artery stenosis not only augments the severity and distribution of atherosclerosis but also causes transition of lesions to an unstable phenotype, with thin fibrous caps and reduced quantities of vascular smooth muscle cells [19]. Thus, there is ample evidence from clinical, epidemiological, and experimental studies to support the concept that hypertension markedly enhances atherosclerotic lesion formation, promotes adverse events, and increases cardiovascular mortality.

MOLECULAR AND CELLULAR MECHANISMS LINKING HYPERTENSION AND ATHEROSCLEROSIS

Various factors encountered in hypertension including increased vascular stretch, angiotensin II, catecholamines, reactive oxygen species (ROS), and excess salt have effects on the vessel wall that create a pathological setting favorable to the development of atherosclerosis. While there can be direct effects of these stimuli, many are mediated by an increase in oxidative stress and a loss of biologically active nitric oxide (Figure 6.1). We will review each of these components and the cellular mechanisms by which they induce atherosclerotic lesion formation in the following paragraphs.

Oxidative Stress

Oxidative stress is covered in detail elsewhere in this book. In brief, during normal cellular metabolism, several enzymes generate electrons that reduce molecular oxygen to form ROS. In vascular cells, the major sources of ROS include the NADPH oxidases, the mitochondria, xanthine oxidase, and the endothelial NOS (eNOS). Biologically relevant ROS include superoxide (O_2^-), hydrogen peroxide (H_2O_2), peroxynitrite ($OONO^-$), and lipid radicals including lipid hydroperoxides. These are formed at low levels normally and have important signaling roles in cell growth and function; however when produced in excess, lead to cellular injury, often referred to as oxidative injury or oxidative stress. An important consequence of excessive

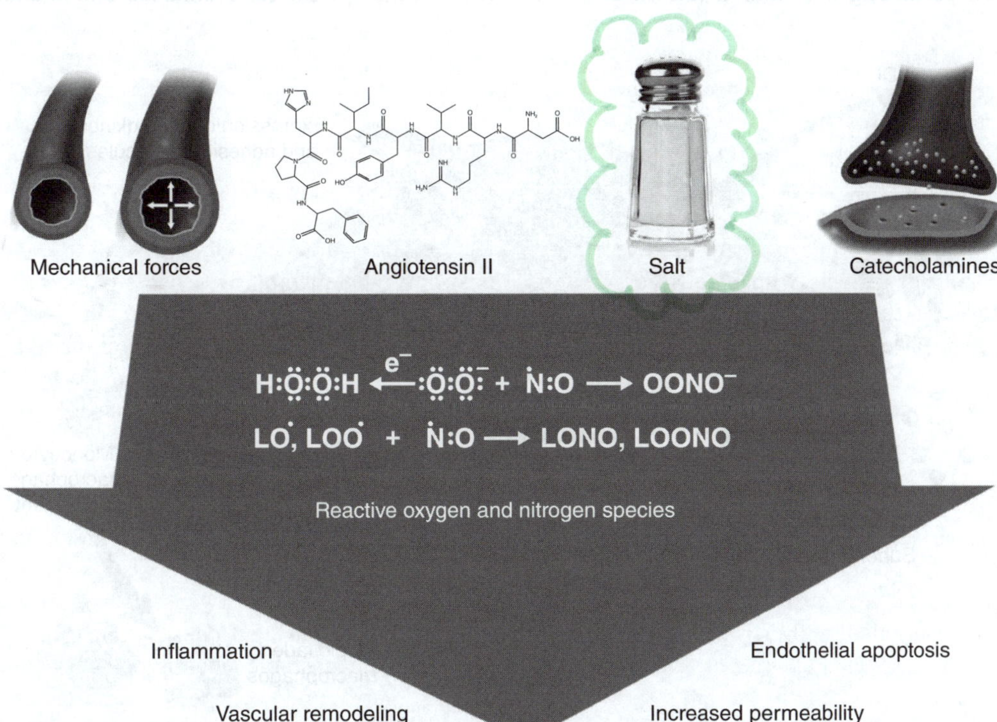

Figure 6.1 Various hypertensive stimuli increase vascular production of reactive oxygen species (ROS) to mediate vascular disease. In hypertension, altered mechanical forces such as stretch, angiotensin II, increased sodium, and catecholamines stimulate the production of ROS. Increased production of superoxide leads to a loss of nitric oxide and formation of the strong oxidant peroxynitrite. Lipid radicals including lipid alkoxy- (LO•) and peroxy-radicals (LOO•) also react with NO. These ROS stimulate various events that promote atherosclerosis.

O_2^- production is oxidative inactivation of NO. This reaction occurs at a diffusion-limited rate leading to formation of the strong oxidant $OONO^-$ and a reduction in bioavailable NO.

During the past two decades, it has become clear that ROS play an important role in the genesis of hypertension [20]. Hypertensive stimuli such as angiotensin II, salt, and increased catecholamines activate the enzymes mentioned earlier in various cells. As an example, angiotensin II, by acting on angiotensin II type 1 (AT_1) receptors, activates the NADPH oxidases. ROS produced by the NADPH oxidases can affect mitochondrial function augmenting ROS production in this organelle in a feed-forward fashion. Likewise, tetrahydrobiopterin, a critical cofactor of the NOS enzymes, is oxidized by ROS generated from the NADPH oxidases, promoting a condition known as NOS uncoupling, in which these enzymes begin to produce O_2^- rather than NO. These feed-forward mechanisms lead to a situation in which low levels of ROS (i.e., kindling radicals) lead to high levels of ROS (bonfire radicals) in various cells. Both in cell culture and *in vivo*, angiotensin II can thereby increase endothelial and vascular smooth muscle superoxide production by more than threefold. ROS produced in the central nervous system (CNS) signal an increase in sympathetic outflow. In the kidney, ROS promote glomerular injury, afferent arteriolar vasoconstriction and sodium, and volume retention. In the vasculature, ROS alter vasomotion and promote vascular smooth muscle hypertrophy and proliferation.

Vascular formation of ROS has been implicated in virtually every step of atherosclerosis (Figure 6.2). ROS enhance endothelial

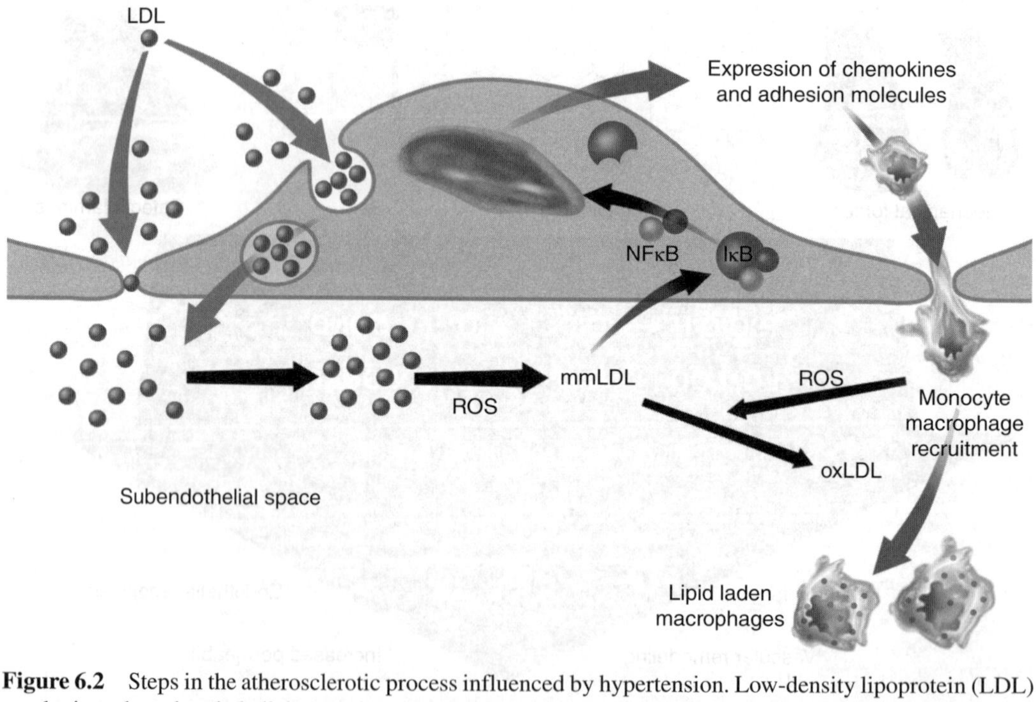

Figure 6.2 Steps in the atherosclerotic process influenced by hypertension. Low-density lipoprotein (LDL) uptake into the subendothelial space is enhanced by mechanical stretch and ROS generated in response to hypertension. The oxidative environment of the vessel wall promotes the formation of minimally modified LDL (mmLDL), which enhances activation of proinflammatory transcription factors such as NFκB. This leads to endothelial cell production of adhesion molecules and chemokines that recruit monocytes to the subendothelium, where they become macrophages. Subendothelial macrophages generate additional ROS that lead to further oxidation of LDL. Extensively oxidized LDL is taken up by the scavenger receptor on macrophages, leading to formation of lipid-laden macrophages or foam cells.

permeability to low-density lipoprotein (LDL) [21], and promote lipid peroxidation, leading to the formation of oxidatively modified LDL in the subendothelium. Oxidized LDL is more readily taken up by macrophages to form foam cells via the scavenger LDL receptor when compared with native LDL. Oxidatively modified lipoproteins stimulate production of adhesion molecules and chemokines, enhancing recruitment of inflammatory cells to the lesion during atherogenesis. In many cases, expression of these pro-inflammatory molecules is governed by the redox sensitive transcription factor NFkB. ROS also activate matrix metalloproteinases, which are involved in plaque remodeling, plaque rupture, and endothelial cell apoptosis.

As illustrated in Figure 6.1, ROS including O_2^- and lipid radicals rapidly react with endothelium-derived NO, leading to a loss of its biological activity. NO has several antiatherosclerotic properties, including inhibition of NFkB activation, lipid peroxidation, vascular smooth muscle growth, and platelet adhesion. The loss of the beneficial effects of NO, therefore, promotes a proatherogenic vascular environment.

Inflammation

Inflammation is a critical mediator of both hypertension and atherosclerosis. Lipoproteins entering the subendothelial space become oxidatively modified, particularly when the oxidative environment of the vessel wall is increased by conditions such as hypertension, cigarette smoking, and diabetes. The modified lipoproteins are proinflammatory, in that they signal endothelial expression of leukocyte adhesion molecules and chemokines, including the vascular cell adhesion molecule-1 (VCAM1), the intracellular adhesion molecule-1 (ICAM1), the monocyte chemoattractant peptide-1 (MCP1), RANTES, and others [22–25]. These promote rolling, adhesion, and ultimately diapedesis of inflammatory cells, including monocyte/macrophages, T cells, and mast cells to the subendothelial space, which in turn release a variety of mediators, including interleukin (IL)-1β, interferon gamma, IL-17, and TNFα that promote plaque development, remodeling, and ultimately rupture [26]. Gene targeting or pharmacological blockade of these various steps in atherosclerosis-prone mice have shown that many of these are critical in lesion development [26]. As an example, the cross of mice lacking functional lymphocytes with $apoE^{-/-}$ mice leads to lesions that are markedly reduced compared with $apoE^{-/-}$ mice with lymphocytes [27]. Adoptive transfer of $CD4^+$ T cells to SCID/SCID X $apoE^{-/-}$ mice restores their lesion size to that observed in immunocompetent $apoE^{-/-}$ mice, showing that $CD4^+$ cells play a critical role in lesion development.

In addition to the subendothelium, substantial inflammation also occurs in the adventitia in both experimental and human atherosclerosis [28, 29]. T lymphocytes are present in the adventitia of human atherosclerotic coronary arteries in both $apoE^{-/-}$ mice and human atherosclerosis. There is evidence that these can worsen experimental atherosclerosis. In a fascinating study, Ohman et al. performed allogeneic transplants of visceral fat adjacent to carotid arteries of $apoE^{-/-}$ mice and found that these markedly worsened atherosclerosis and endothelium-dependent vasodilatation, while subcutaneous fat transplants had minimal effects [30]. It is interesting to speculate that cytokines released from inflammatory cells or activated adipocytes present in the perivascular fat diffuse to the adjacent vessel and enhance lesion formation (Figure 6.3).

Interestingly, many of these inflammatory processes are recapitulated in hypertension. Mechanical stretch enhances the uptake of LDL in large arteries [31]. Various aspects of the hypertensive milieu, including angiotensin II, catecholamines, salt, oxidative stress, and altered mechanical forces are associated with increased vascular expression of the same proinflammatory genes observed in atherosclerosis, including ICAM1, VCAM1, P-selectin, RANTES, and MCP1 [32–35]. In many cases, a cascade of signaling molecules including NFkB, MAP kinases, and Rho kinase coordinately regulate these inflammatory genes. Inhibition of NFkB via either pharmacological or genetic means prevents leukocyte accumulation in target tissues and end-organ damage

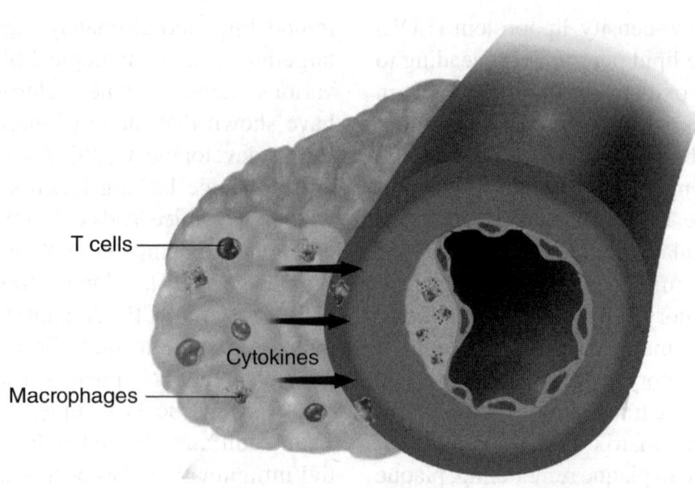

Figure 6.3 Inflammatory cells in the perivascular adipose tissue and adventitia contribute to vascular disease in hypertension. Hypertension promotes accumulation of macrophages and activated T cells in the perivascular tissues. These cells produce inflammatory cytokines including IL-17, TNFα and IFN-γ, which in turn promote vascular disease and atheroma formation.

in experimental hypertension [36, 37]. Unlike experimental atherosclerosis, there is limited accumulation of inflammatory cells in the subendothelial space of large vessels in hypertension. There is, however, a striking accumulation of macrophages and T cells in the perivascular space, including the adventitia and perivascular fat of large vessels. In several studies, we have found that the number of total leukocytes and T cells are increased three- to fourfold in response to angiotensin II, norepinephrine, or DOCA-salt-induced hypertension [34, 38, 39]. In the kidney, T cells and macrophages accumulate adjacent to arterioles and the glomerulus, where they likely promote renal injury commonly observed in long-term hypertension [40]. The perivascular regions are richly invested with sympathetic nerve endings, and it is possible that locally released catecholamines and perhaps other neurotransmitters mediate some of these inflammatory responses.

This accumulation of inflammatory cells in the perivascular fat and in the kidney contributes to vascular dysfunction and hypertension. We have found that mice lacking lymphocytes (RAG-$1^{-/-}$ mice) are protected against several forms of experimental hypertension, including that caused by angiotensin II, norepinephrine, DOCA-salt, and stress [34, 38, 41]. The alterations in endothelium-dependent vasodilatation, increases in vascular superoxide production and vascular hypertrophy, commonly observed in hypertension, are blunted in RAG-$1^{-/-}$ mice. Adoptive transfer of T cells, but not B cells, restores hypertension and vascular dysfunction in these animals. Several lines of evidence suggest that signals from the CNS are essential for this inflammatory process. Lesions in the anteroventral third ventricular (AV3V) region, which are known to prevent most forms of experimental hypertension, prevent T-cell activation and vascular inflammation in response to angiotensin II infusion [38]. By contrast, induction of oxidative stress in the circumventricular organs by deletion of the extracellular superoxide dismutase promotes hypertension, T-cell activation, and vascular inflammation [39]. We have proposed that signals arising from the CNS modestly elevate blood pressure to prehypertensive levels, which in turn leads to barotrauma and formation of neoantigens that promote T-cell activation (Figure 6.4). Activated T cells accumulate in the kidney and perivascular regions where they release cytokines that lead to altered renal sodium handling and vascular dysfunction.

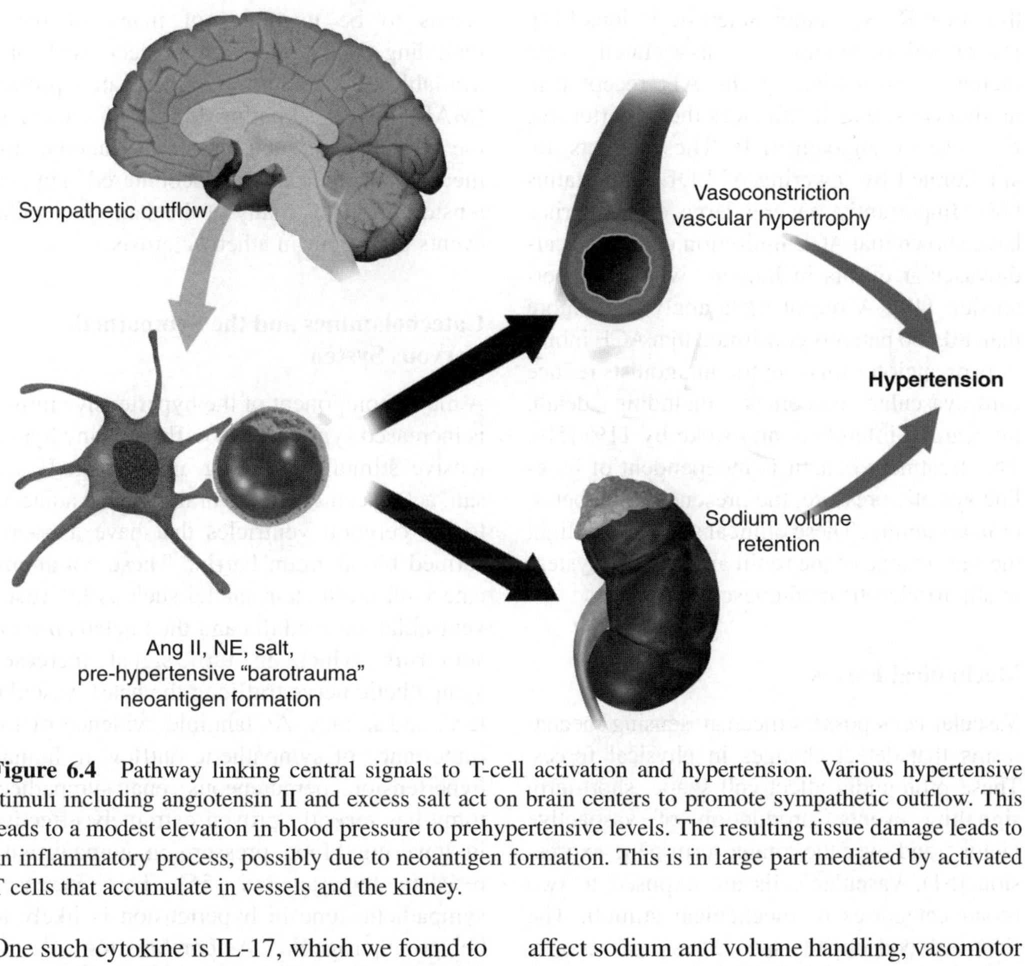

Figure 6.4 Pathway linking central signals to T-cell activation and hypertension. Various hypertensive stimuli including angiotensin II and excess salt act on brain centers to promote sympathetic outflow. This leads to a modest elevation in blood pressure to prehypertensive levels. The resulting tissue damage leads to an inflammatory process, possibly due to neoantigen formation. This is in large part mediated by activated T cells that accumulate in vessels and the kidney.

One such cytokine is IL-17, which we found to be produced by T cells of angiotensin II infused mice [42]. Mice lacking IL-17 are protected against the hypertension and vascular dysfunction caused by angiotensin II. Moreover, we found that circulating plasma IL-17 levels are threefold higher in humans with hypertension and metabolic syndrome compared to those with metabolic syndrome without hypertension. Other cytokines that have been implicated in the end-organ dysfunction hypertension include TNFα, IL-6, and IL-10 [43–45]. The last of these has protective effects.

The Renin Angiotensin System

A major contributor to hypertension is the octapeptide angiotensin II and its related metabolites. Angiotensin II has pleiotropic actions that affect sodium and volume handling, vasomotor tone, vascular and cardiac hypertrophy, and sympathetic outflow. Some of these actions are protective in the setting of hemorrhage or dehydration; however, they become injurious in pathological conditions such as hypertension. In the majority of patients with hypertension, treatment with angiotensin I-converting enzyme (ACE) inhibitors or angiotensin II-receptor antagonists lower blood pressure, perhaps not to normal, but significantly. These responses occur in individuals with "essential" hypertension in whom the plasma renin activity is not elevated [46]. This is in part because angiotensin II is produced locally within specific tissues and cells, and not reflected by circulating angiotensin II or renin activity. Of note, macrophages in atherosclerotic lesions produce angiotensin II, emphasizing the importance of this local RAS in atherosclerotic lesions [47]. Hypercholesterolemia is associated with increased expression of the AT_1 receptor in human cells, which enhances the hypertensive response to angiotensin II. These effects are ameliorated by lowering of LDL with statins [48]. Importantly, several large clinical trials have shown that ACE inhibition decreases cardiovascular events in humans without hypertension [49]. A recent meta-analysis of more than 80,000 patients confirmed that ACE inhibitors or angiotensin receptor antagonists reduce cardiovascular outcomes, including death, myocardial infarction, and stoke by 11% [50]. This treatment benefit is independent of baseline systolic pressure, the presence of diabetes, or heart failure. These clinical studies highlight the importance of the renin angiotensin system in atherosclerotic cardiovascular disease.

Mechanical Forces

Vascular cells possess mechanosensing mechanisms that detect changes in physical forces. These profoundly affect cell shape, short-term signaling events, production of vasoactive factors, and over the longer term gene expression [51]. Vascular cells are exposed to two broad categories of mechanical stimuli. The flow of blood imparts shear stress, particularly to the endothelial cell surface, while the pressure exerted within the vessel leads to stretch of the endothelium and vascular smooth muscle. In general, regions of the circulation exposed to unidirectional laminar shear are protected, while regions exposed to disturbed or oscillatory shear stress are predisposed to atherosclerosis.

As the pressure rises during a normal cardiac cycle from 80 to 120 mm Hg, the endothelium is exposed to a stretch of about 5%. In the setting of hypertension, this often doubles to greater than 10% [52]. Studies of cultured cells have shown that excessive stretch induces an inflammatory response, characterized by increased production of ROS, chemokines, and adhesion molecules [53, 54]. While multiple signaling pathways likely contribute to these responses, the increase in ROS production seems to be upstream of many of these, including NFkB activation, decreased bioavailable NO, and mitogen-activated protein (MAP) kinase signaling [55]. Thus, even in the absence of neurohumoral factors, the mechanical environment encountered in hypertension provokes many of the same molecular events that occur in atherosclerosis.

Catecholamines and the Sympathetic Nervous System

A major component of the hypertensive milieu is increased sympathetic outflow. Many hypertensive stimuli, including angiotensin II and salt, act on centers of the brain located adjacent to the cerebral ventricles that have a poorly formed blood–brain barrier. These communicate with brain stem nuclei such as the rostal ventral lateral medulla and the nucleus *tractus solitarius*, which in turn signal increased sympathetic nerve traffic to the heart, vasculature, and kidney. As tangible evidence of the importance of sympathetic outflow in human hypertension, percutaneous renal sympathectomy has recently proven extremely effective in lowering blood pressure in humans with resistant hypertension [56]. This increased sympathetic tone in hypertension is likely to be proatherogenic. A combination of the alpha antagonist prazosin and the AT_1 receptor antagonist Losartan markedly reduces atherosclerosis in $apoE^{-/-}$ mice, while either of these alone does not [57]. Beta-blockade also reduces atherosclerosis in cholesterol-fed rabbits, and in primates exposed to a stress paradigm [58]. Treatment with the beta-adrenergic antagonist metoprolol reduces the rate of progression of carotid intimal medial thickness in humans [59]. These effects of adrenergic blockade might simply be due to alterations of hemodynamics; however, immune cells involved in atherosclerosis could also be affected. Antigen presenting cells, macrophages, and T cells possess both alpha- and beta-adrenergic receptors that can modulate their functions. Stimulation of $alpha_2$-adrenergic receptors on dendritic cells enhances their antigen uptake [60]. By contrast, beta adrenergic stimulation inhibits dendritic cell function [61]. Both alpha- and beta-adrenergic receptors are present on T cells, and these seem to modify polarization, proliferation, and surface marker expression. Adrenergic receptors on macrophages modulate proinflammatory cytokine production in response to toll-like receptor (TLR) activation [62, 63]. Norepinephrine also enhances entry of T cells and macrophages in the perivascular regions [38], promoting the proatherogenic inflammatory events described earlier. Thus, increased sympathetic outflow in hypertension might not only raise blood pressure but also promote atherosclerosis.

Micro-RNAs, Hypertension, and Atherosclerosis

Micro-RNAs are small noncoding RNAs formed as hairpin precursors that are further processed by the endonuclease Dicer to form 21–24 nucleotide sequences. These are subsequently incorporated into an RNA-induced silencing complex (RISC) that targets mRNAs. Micro-RNAs can either suppress RNA translation or lead to mRNA degradation. The ultimate degree of regulation is often subtle, with only modest changes in target mRNA and protein levels. Frequently, micro-RNAs target several mRNAs involved in a common cellular program and therefore confer executive functions [64]. Over one-half of human genes are potential targets of micro-RNAs and several have been identified that are important in hypertension [65]. One example is mir143/145, the genes of which reside within 2 kb on murine chromosome 18. Mir145 is the most prevalent micro-RNA in vascular smooth muscle cells [66], and its deletion in mice results in these cells displaying a synthetic phenotype [67]. These animals have reduced blood pressure, and their vessels exhibit reduced medial thickness and are hypocontractile to angiotensin II, phenylephrine, and calcium. Interestingly, these mice exhibit a fivefold increase in ACE expression in vascular smooth muscle and enhanced hypotensive responses to ACE inhibitors. Conversely, overexpression of mir145 using an adenovirus in rat carotid arteries inhibits neointimal formation following balloon injury. Thus, mir143/145 seems to affect both blood pressure and vascular lesions. The precise role of this micro-RNA complex on atherosclerosis has not been studied. Another micro-RNA potentially related to hypertension is mir155, which targets AT_1R. Overexpression of this blunts angiotensin II signaling in primary lung fibroblasts [68]. This mir is prevalent in atherosclerotic plaques and proinflammatory macrophages and its absence reduces atherosclerotic lesion formation in apoE$^{-/-}$ mice [69]. Further studies of micro-RNAs are likely to provide insight into the link between hypertension and atherosclerosis.

IMPLICATIONS FOR TREATMENT AND FUTURE DIRECTIONS

As can be understood from the discussion above, the disease of hypertension is substantially more complex than simply an elevation of blood pressure. It comprises a "hypertensive milieu" that includes oxidative stress, inflammation, excess catecholamines, and activation of the renin angiotensin system. Beyond these biochemical and humoral factors, hypertension seldom occurs as an isolated finding in the clinic. The majority of hypertensive patients in Western societies exhibit components of the metabolic syndrome, including obesity, insulin resistance, and dyslipidemia. Many hypertensive patients do not obtain adequate exercise and consume unhealthy diets. These factors coalesce and conspire to cause both hypertension and atherosclerosis. Hypertension physicians and clinics cannot simply address blood pressure, but need to emphasize weight loss, exercise, correction of hyperlipidemia, and attention to insulin resistance. The precise mechanisms by which these factors interact remain an important focus of research. Moreover, while inflammation and oxidative stress clearly contribute to hypertension and atherosclerosis in experimental models, it is unclear how important they are in human disease or how they could be treated clinically. It is conceivable that short-term treatment with immune modulating agents might be helpful in selected patients with resistant hypertension or atherosclerosis. Methotrexate is currently being studied in a trial to determine if it reduces cardiovascular events in patients with stable coronary artery disease and high C-reactive proteins [70]. In addition, an ongoing clinical trial is examining the use of an IL-1β antagonist on cardiovascular events in patients with elevated hs-CRP, a marker of inflammation [71]. Several large trials, including the HOPE study, have not shown benefit of vitamin E and other antioxidants in prevention of cardiovascular disease; and in some cases, vitamin E has proven deleterious [72–74]. It is possible that these trials have been negative because they have not targeted individuals who truly have oxidative stress. In this regard, the effect of vitamin E on cardiovascular outcomes was studied in almost 1500 diabetic individuals aged over 55 years with the haptaglobin 2/2 genotype [75]. Haptaglobin has antioxidant properties, and the 2/2 genotype is associated with a reduction of this function. This study was stopped prematurely because of a striking benefit of vitamin E. Compared to placebo, vitamin E reduced all cardiovascular events, myocardial infarction, and stroke by 50%. It is also possible that more potent, catalytic agents such as Tempol or targeted agents such as mitoTempol, which scavenges radicals specifically in the mitochondria, would have greater efficacy. In experimental studies, these agents have proven extremely effective [76, 77]. These directions of research are paramount for understanding the link between hypertension and atherosclerosis and for reducing the untoward consequences of these commonly coexisting conditions.

REFERENCES

1. Chobanian AV, Bakris GL, Black HR, et al. The Seventh Report of the Joint National Committee on prevention, detection, evaluation, and treatment of high blood pressure: the JNC 7 report. JAMA. 2003;289:2560–72.
2. Julius S, Nesbitt SD, Egan BM, et al. Feasibility of treating prehypertension with an angiotensin-receptor blocker. N Engl J Med. 2006;354: 1685–97.
3. Heidenreich PA, Trogdon JG, Khavjou OA, et al. Forecasting the future of cardiovascular disease in the United States: a policy statement from the American Heart Association. Circulation. [Consensus Development Conference]. 2011;123: 933–44.
4. Multiple Risk Factor Intervention Trial Research Group. Relationship between baseline risk factors and coronary heart disease and total mortality in the Multiple Risk Factor Intervention Trial. Prev Med. 1986;15:254–73.
5. Domanski M, Mitchell G, Pfeffer M, et al. Pulse pressure and cardiovascular disease-related mortality: follow-up study of the Multiple Risk Factor Intervention Trial (MRFIT). JAMA. 2002; 287:2677–83.
6. Berenson GS, Srinivasan SR, Bao W, Newman WP, 3rd, Tracy RE, Wattigney WA. Association between multiple cardiovascular risk factors and atherosclerosis in children and young adults. The Bogalusa Heart Study. N Engl J Med. 1998;338:1650–6.
7. McGill HC, Jr., McMahan CA, Tracy RE, et al. Relation of a postmortem renal index of hypertension to atherosclerosis and coronary artery size in young men and women. Pathobiological Determinants of Atherosclerosis in Youth (PDAY) Research Group. Arterioscler Thromb Vasc Biol. 1998;18:1108–18.
8. Oyama N, Gona P, Salton CJ, et al. Differential impact of age, sex, and hypertension on aortic atherosclerosis: the Framingham Heart Study. Arterioscler Thromb Vasc Biol. 2008;28: 155–9.
9. Lewington S, Clarke R, Qizilbash N, Peto R, Collins R. Age-specific relevance of usual blood pressure to vascular mortality: a meta-analysis of individual data for one million adults in 61 prospective studies. Lancet. 2002;360:1903–13.
10. Rafiq S, Anand S, Roberts R. Genome-wide association studies of hypertension: have they been fruitful? J Cardiovasc Transl Res. 2010;3: 189–96.
11. Ehret GB, Munroe PB, Rice KM, et al. Genetic variants in novel pathways influence blood pressure and cardiovascular disease risk. Nature. 2011;478:103–9.
12. Wain LV, Verwoert GC, O'Reilly PF, et al. Genome-wide association study identifies six

new loci influencing pulse pressure and mean arterial pressure. Nat Genet. 2011;43:1005–11.
13. Salvi E, Kutalik Z, Glorioso N, et al. Genomewide association study using a high-density single nucleotide polymorphism array and case-control design identifies a novel essential hypertension susceptibility locus in the promoter region of endothelial NO synthase. Hypertension. 2012;59: 248–55.
14. Johnson AD, Newton-Cheh C, Chasman DI, et al. Association of hypertension drug target genes with blood pressure and hypertension in 86,588 individuals. Hypertension. 2011;57:903–10.
15. Daugherty A, Manning MW, Cassis LA. Angiotensin II promotes atherosclerotic lesions and aneurysms in apolipoprotein E-deficient mice. J Clin Invest. 2000;105:1605–12.
16. Weiss D, Taylor WR. Deoxycorticosterone acetate salt hypertension in apolipoprotein E-/- mice results in accelerated atherosclerosis: the role of angiotensin II. Hypertension. 2008;51:218–24.
17. Wu JH, Hagaman J, Kim S, Reddick RL, Maeda N. Aortic constriction exacerbates atherosclerosis and induces cardiac dysfunction in mice lacking apolipoprotein E. Arterioscler Thromb Vasc Biol. 2002;22:469–75.
18. Tham DM, Martin-McNulty B, Wang YX, et al. Angiotensin II injures the arterial wall causing increased aortic stiffening in apolipoprotein E-deficient mice. Am J Physiol Regul Integr Comp Physiol. 2002;283:R1442–9.
19. Mazzolai L, Duchosal MA, Korber M, et al. Endogenous angiotensin II induces atherosclerotic plaque vulnerability and elicits a Th1 response in ApoE-/- mice. Hypertension. 2004;44: 277–82.
20. Harrison DG, Gongora MC. Oxidative stress and hypertension. Med Clin North Am. 2009;93:621–35.
21. Alexander JJ, Graham DJ, Miguel R. Oxygen radicals alter LDL permeability and uptake by an endothelial-smooth muscle cell bilayer. J Surg Res. 1991;51:361–7.
22. Cybulsky MI, Gimbrone MA, Jr. Endothelial expression of a mononuclear leukocyte adhesion molecule during atherogenesis. Science. 1991;251:788–91.
23. Nakashima Y, Raines EW, Plump AS, Breslow JL, Ross R. Upregulation of VCAM-1 and ICAM-1 at atherosclerosis-prone sites on the endothelium in the ApoE-deficient mouse. Arterioscler Thromb Vasc Biol. 1998;18: 842–51.
24. Liao F, Berliner JA, Mehrabian M, et al. Minimally modified low density lipoprotein is biologically active in vivo in mice. J Clin Invest. 1991;87:2253–7.
25. Gleissner CA, Leitinger N, Ley K. Effects of native and modified low-density lipoproteins on monocyte recruitment in atherosclerosis. Hypertension. 2007;50:276–83.
26. Hansson GK. Inflammatory mechanisms in atherosclerosis. J Thromb Haemost. 2009;7 Suppl 1:328–31.
27. Zhou X, Nicoletti A, Elhage R, Hansson GK. Transfer of CD4(+) T cells aggravates atherosclerosis in immunodeficient apolipoprotein E knockout mice. Circulation. 2000;102:2919–22.
28. Galkina E, Kadl A, Sanders J, Varughese D, Sarembock IJ, Ley K. Lymphocyte recruitment into the aortic wall before and during development of atherosclerosis is partially L-selectin dependent. J Exp Med. 2006;203:1273–82.
29. Watanabe M, Sangawa A, Sasaki Y, et al. Distribution of inflammatory cells in adventitia changed with advancing atherosclerosis of human coronary artery. J Atheroscler Thromb. 2007; 14:325–31.
30. Ohman MK, Luo W, Wang H, et al. Perivascular visceral adipose tissue induces atherosclerosis in apolipoprotein E deficient mice. Atherosclerosis. 2011;219:33–9.
31. Meyer G, Merval R, Tedgui A. Effects of pressure-induced stretch and convection on low-density lipoprotein and albumin uptake in the rabbit aortic wall. Circ Res. 1996;79:532–40.
32. Haller H, Park JK, Dragun D, Lippoldt A, Luft FC. Leukocyte infiltration and ICAM 1 expression in two-kidney one-clip hypertension. Nephrol Dial Transplant. 1997;12:899–903.
33. Luft FC. Angiotensin, inflammation, hypertension, and cardiovascular disease. Curr Hypertens Rep. 2001;3:61–7.
34. Guzik TJ, Hoch NE, Brown KA, et al. Role of the T cell in the genesis of angiotensin II induced hypertension and vascular dysfunction. J Exp Med. 2007;204:2449–60.
35. Funakoshi Y, Ichiki T, Shimokawa H, et al. Rho-kinase mediates angiotensin II-induced monocyte chemoattractant protein-1 expression in rat vascular smooth muscle cells. Hypertension. 2001;38:100–4.
36. Muller DN, Dechend R, Mervaala EM, et al. NF-kappaB inhibition ameliorates angiotensin II-induced inflammatory damage in rats. Hypertension. 2000;35:193–201.

37. Henke N, Schmidt-Ullrich R, Dechend R, et al. Vascular endothelial cell-specific NF-kappaB suppression attenuates hypertension-induced renal damage. Circ Res. 2007;101:268–76.
38. Marvar PJ, Thabet SR, Guzik TJ, et al. Central and peripheral mechanisms of T-lymphocyte activation and vascular inflammation produced by angiotensin II-induced hypertension. Circ Res. 2010;107:263–70.
39. Lob HE, Marvar PJ, Guzik TJ, et al. Induction of hypertension and peripheral inflammation by reduction of extracellular superoxide dismutase in the central nervous system. Hypertension. 2010;55:277–83.
40. Rodriguez-Iturbe B, Quiroz Y, Nava M, et al. Reduction of renal immune cell infiltration results in blood pressure control in genetically hypertensive rats. Am J Physiol Renal Physiol. 2002;282:F191–201.
41. Marvar PJ, Vinh A, Thabet S, et al. T lymphocytes and vascular inflammation contribute to stress-dependent hypertension. Biol Psychiatry. 2012;71:774–82.
42. Madhur MS, Lob HE, McCann LA, et al. Interleukin 17 promotes angiotensin II-induced hypertension and vascular dysfunction. Hypertension. 2010;55:500–7.
43. Venegas-Pont M, Manigrasso MB, Grifoni SC, et al. Tumor necrosis factor-alpha antagonist etanercept decreases blood pressure and protects the kidney in a mouse model of systemic lupus erythematosus. Hypertension. 2010;56:643–9.
44. Lee DL, Sturgis LC, Labazi H, et al. Angiotensin II hypertension is attenuated in interleukin-6 knockout mice. Am J Physiol Heart Circ Physiol. 2006;290:H935–40.
45. Didion SP, Kinzenbaw DA, Schrader LI, Chu Y, Faraci FM. Endogenous interleukin-10 inhibits angiotensin II-induced vascular dysfunction. Hypertension. 2009;54:619–24.
46. Muiesan G, Alicandri CL, Agabiti-Rosei E, et al. Angiotensin-converting enzyme inhibition, catecholamines and hemodynamics in essential hypertension. Am J Cardiol. 1982;49:1420–4.
47. Potter DD, Sobey CG, Tompkins PK, Rossen JD, Heistad DD. Evidence that macrophages in atherosclerotic lesions contain angiotensin II. Circulation. 1998;98:800–7.
48. Nickenig G, Baumer AT, Temur Y, Kebben D, Jockenhovel F, Bohm M. Statin-sensitive dysregulated AT1 receptor function and density in hypercholesterolemic men. Circulation. 1999; 100:2131–4.
49. Yusuf S, Sleight P, Pogue J, Bosch J, Davies R, Dagenais G. Effects of an angiotensin-converting-enzyme inhibitor, ramipril, on cardiovascular events in high-risk patients. The Heart Outcomes Prevention Evaluation Study Investigators. N Engl J Med. 2000;342:145–53.
50. McAlister FA. Angiotensin-converting enzyme inhibitors or angiotensin receptor blockers are beneficial in normotensive atherosclerotic patients: a collaborative meta-analysis of randomized trials. Eur Heart J. 2012;33:505–14.
51. Li YS, Haga JH, Chien S. Molecular basis of the effects of shear stress on vascular endothelial cells. J Biomech. 2005;38:1949–71.
52. Estrada R, Giridharan GA, Nguyen MD, et al. Endothelial cell culture model for replication of physiological profiles of pressure, flow, stretch, and shear stress in vitro. Anal Chem. 2011;83: 3170–7.
53. Hishikawa K, Luscher TF. Pulsatile stretch stimulates superoxide production in human aortic endothelial cells. Circulation. 1997;96:3610–6.
54. Ando J, Yamamoto K. Effects of shear stress and stretch on endothelial function. Antioxid Redox Signal. 2011;15:1389–403.
55. Lehoux S. Redox signalling in vascular responses to shear and stretch. Cardiovasc Res. 2006;71:269–79.
56. Esler MD, Krum H, Sobotka PA, Schlaich MP, Schmieder RE, Bohm M. Renal sympathetic denervation in patients with treatment-resistant hypertension (The Symplicity HTN-2 Trial): a randomised controlled trial. Lancet. 2010;376: 1903–9.
57. Makaritsis KP, Gavras H, Du Y, Chobanian AV, Brecher P. Alpha1-adrenergic plus angiotensin receptor blockade reduces atherosclerosis in apolipoprotein E-deficient mice. Hypertension. 1998;32:1044–8.
58. Kaplan JR, Pettersson K, Manuck SB, Olsson G. Role of sympathoadrenal medullary activation in the initiation and progression of atherosclerosis. Circulation. 1991;84:VI23–32.
59. Ostling G, Goncalves I, Wikstrand J, Berglund G, Nilsson J, Hedblad B. Long-term treatment with low-dose metoprolol CR/XL is associated with increased plaque echogenicity: the Beta-blocker Cholesterol-lowering Asymptomatic Plaque Study (BCAPS). Atherosclerosis. 2011; 215:440–5.

60. Yanagawa Y, Matsumoto M, Togashi H. Enhanced dendritic cell antigen uptake via alpha2 adrenoceptor-mediated PI3K activation following brief exposure to noradrenaline. J Immunol. 2010;185:5762–8.
61. Maestroni GJ. Adrenergic modulation of dendritic cells function: relevance for the immune homeostasis. Curr Neurovasc Res. 2005;2:169–73.
62. Grisanti LA, Woster AP, Dahlman J, Sauter ER, Combs CK, Porter JE. alpha1-adrenergic receptors positively regulate Toll-like receptor cytokine production from human monocytes and macrophages. J Pharmacol Exp Ther. 2011;338: 648–57.
63. Grisanti LA, Evanson J, Marchus E, et al. Proinflammatory responses in human monocytes are beta1-adrenergic receptor subtype dependent. Mol Immunol. 2010;47:1244–54.
64. Pasquinelli AE. MicroRNAs and their targets: recognition, regulation and an emerging reciprocal relationship. Nat Rev Genet. 2012;13:271–82.
65. Batkai S, Thum T. MicroRNAs in hypertension: mechanisms and therapeutic targets. Curr Hypertens Rep. 2012;14:79–87.
66. Ji R, Cheng Y, Yue J, et al. MicroRNA expression signature and antisense-mediated depletion reveal an essential role of MicroRNA in vascular neointimal lesion formation. Circ Res. 2007;100:1579–88.
67. Boettger T, Beetz N, Kostin S, et al. Acquisition of the contractile phenotype by murine arterial smooth muscle cells depends on the Mir143/145 gene cluster. J Clin Invest. 2009;119:2634–47.
68. Martin MM, Lee EJ, Buckenberger JA, Schmittgen TD, Elton TS. MicroRNA-155 regulates human angiotensin II type 1 receptor expression in fibroblasts. J Biol Chem. 2006; 281:18277–84.
69. Nazari-Jahantigh M, Wei Y, Noels H, et al. MicroRNA-155 promotes atherosclerosis by repressing Bcl6 in macrophages. J Clin Invest. 2012;122:4190–202.
70. Ridker PM. Testing the inflammatory hypothesis of atherothrombosis: scientific rationale for the cardiovascular inflammation reduction trial (CIRT). J Thromb Haemost. 2009;7 Suppl 1:332–9.
71. Ridker PM, Thuren T, Zalewski A, Libby P. Interleukin-1beta inhibition and the prevention of recurrent cardiovascular events: rationale and design of the Canakinumab Anti-inflammatory Thrombosis Outcomes Study (CANTOS). Am Heart J. 2011;162:597–605.
72. Gruppo Italiano per lo Studio della Sopravvivenza nell'Infarto miocardico. Dietary supplementation with n-3 polyunsaturated fatty acids and vitamin E after myocardial infarction: results of the GISSI-Prevenzione trial. Lancet. 1999;354:447–55.
73. Yusuf S, Dagenais G, Pogue J, Bosch J, Sleight P. Vitamin E supplementation and cardiovascular events in high-risk patients. The Heart Outcomes Prevention Evaluation Study Investigators. N Engl J Med. 2000;342:154–60.
74. MRC/BHF Heart Protection Study of antioxidant vitamin supplementation in 20,536 high-risk individuals: a randomised placebo-controlled trial. Lancet. 2002;360:23–33.
75. Milman U, Blum S, Shapira C, et al. Vitamin E supplementation reduces cardiovascular events in a subgroup of middle-aged individuals with both type 2 diabetes mellitus and the haptoglobin 2-2 genotype: a prospective double-blinded clinical trial. Arterioscler Thromb Vasc Biol. 2008;28:341–7.
76. Schnackenberg CG, Welch WJ, Wilcox CS. Normalization of blood pressure and renal vascular resistance in SHR with a membrane-permeable superoxide dismutase mimetic: role of nitric oxide. Hypertension. 1998;32:59–64.
77. Dikalova AE, Bikineyeva AT, Budzyn K, et al. Therapeutic targeting of mitochondrial superoxide in hypertension. Circ Res. 2010;107:106–16.

7

THE GENETICS OF ATHEROSCLEROSIS: FROM POLYMORPHISMS TO POPULATIONS

PAUL JOHNSON[1], JACK KURITZKY[1], AND MARSCHALL RUNGE[2]

[1] *Department of Medicine, University of North Carolina at Chapel Hill, Chapel Hill, NC, USA*

[2] *NC TraCS Institute, University of North Carolina School of Medicine, Chapel Hill, NC, USA*

INTRODUCTION

In 1953, nearly a century after Gregor Mendel's landmark genetic plant studies, James Watson and Francis Crick identified the structure of DNA, ushering in the modern era of genetics. Since then, and particularly with the sequencing of the human genome, our knowledge of the genetic basis of disease has grown exponentially. The role of heredity in heart attack and stroke—the most significant risks in individuals with atherosclerosis—has long been appreciated. Understanding the contribution of specific genetic abnormalities at the molecular level, however, has been both elusive and complicated.

Brown and Goldstein's Nobel Prize-winning discovery of the importance of low-density lipoprotein (LDL)-cholesterol receptor expression and function and the potential to regulate LDL-cholesterol levels (discussed in more detail later) is an excellent case in point. Their work was built upon the contemporaneous observation of a monogenic autosomal-dominant disease, familial hypercholesterolemia (hereafter FH), that was and remains quite rare. Brown and Goldstein's work identified a handful of mutations that characterize this disorder in expression of functional LDL receptors. Based on these initial observations, they postulated an LDL receptor-based therapy that led to the development of a new class of drugs, the statins. Without question, their studies resulted in the development of one of the farthest-reaching therapies in all of clinical medicine.

Now, 25 years later, it is clear that like most common diseases, atherosclerosis is a complicated process that can involve any of a large number of biologic mechanisms. Although a handful of examples exist of Mendelian diseases (like LDL receptor abnormalities) with monogenic mechanisms, as illustrated in Figure 7.1, the vast majority of cardiovascular disease represents a complex interplay between polygenic risk alleles and environmental triggers [2, 3].

Atherosclerosis: Risks, Mechanisms, and Therapies, First Edition. Edited by Hong Wang and Cam Patterson.

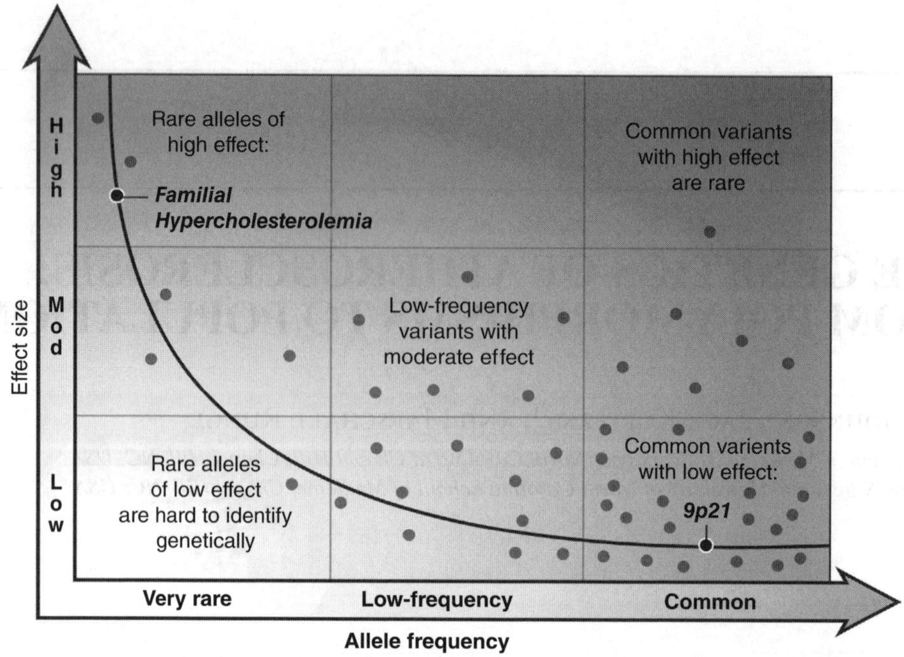

Figure 7.1 Relationship between effect size and allele frequency. While there are some genes with great impact on disease, such as rare mutations leading to familial hypercholesterolemia, a majority of cardiovascular disease can be attributed to multifactorial contributions of genetic and environmental factors. It is hypothesized that the common variants that contribute to cardiovascular disease risk individually exert little effect on disease. However, in combination, they may significantly increase risk (Adapted with permission from Manolio et al. [1]. © Nature).

Indeed, genetic variants that have only minimal impact on disease risk are common (Figure 7.1) [1], and several low-level genetic risk factors may exist in a single individual or family. Twin studies have highlighted genetic predisposition of atherosclerosis; for example, the Danish Twin Registry revealed a higher incidence of atherosclerosis and death in monozygotic twins (44%) compared with dizygotic twins (14%) [4]. These genetic abnormalities may exert their effect through known atherosclerosis risk factors, various steps in the atherosclerotic process itself, thrombosis, and likely through molecular pathways that have yet to be identified.

A practical side of this reality is that despite the hope for personalized medicine, genetic screening for atherosclerosis risk is in its infancy. Given the diversity of potential risk genes and their individual impact, scientists and care providers must be cautious in their approach. The list of polymorphisms and mutations linked to disease is incomplete, and the translation of genetic knowledge to clinical care is at a very early stage. Except in unusual circumstances or when conducted for research purposes, genetic testing for atherosclerosis typically introduces more questions than it answers.

This chapter focuses on two key issues regarding the genetics of atherosclerosis. First, rather than attempting to provide an exhaustive compendium of polymorphisms and mutations that may be important in atherosclerosis, we provide illustrative examples of specific diseases and genetic populations that are of current or future clinical importance. Second, because rapid advances in technology are changing the landscape in genetic analysis, we describe the most important technologies today and for the future.

LESSONS FROM MENDELIAN DISEASES—FAMILIAL HYPERCHOLESTEROLEMIA

Genetic disorders that are inherited in Mendelian patterns are typically monogenic, that is, caused by mutations in single genes. Mendelian diseases that dramatically affect a particular disease but are not lethal prior to reproduction are important because the disease can be transmitted to subsequent generations. By their very nature, related to dramatically increased mortality and morbidity, severe Mendelian diseases appear in the general population at low frequency, as is the case for FH. FH is a hereditary disorder characterized by severely elevated levels of LDL cholesterol. Phenotypic manifestations of FH include xanthalasmas, tendenous xanthomata, arcus senilis, premature symptomatic atherosclerosis, and cardiovascular disease. These physical findings led to the relatively early identification of FH as a hereditary disease.

FH was described by Kachadurian in the 1960s, when he identified families in Lebanon who had both significantly elevated LDL cholesterol (LDL-C) levels and the phenotypes noted earlier [5]. He determined that their hypercholesterolemia was inherited in an autosomal codominant pattern with LDL-C levels four times higher than normal in homozygotes and twice as high as normal in heterozygotes. In the general population, the severe phenotype (total cholesterol generally 650–1000 mg/dl) is present with a prevalence of one per million and the more common mild phenotype (total cholesterol 350–550 mg/dl) occurs in only one in 500 persons [6]. Nonetheless, FH is a very important causative factor for atherosclerosis in young individuals. In one report, FH was reported to be a causative factor in 20% of individuals who present with myocardial infarctions under the age of 45 years [6].

FH is seen with higher frequency in select populations—most prominently Lebanese, French Canadian, and South Afrikaner [7]. This phenomenon exemplifies the *founder effect,* meaning that a mutation in an initial founder has been transmitted through many generations. Even in these known populations and despite the presence of characteristic physical findings (which are not always present in Mendelian diseases), only a portion of those with the disease are identified, and even fewer receive appropriate treatment.

Initially, hypercholesterolemia in those with FH was thought to be caused by overproduction of LDL-C. The discovery by Brown and Goldstein that altered expression of LDL-receptor was responsible for the dramatically increased LDL-C levels in FH led to their interest in characterizing the LDL receptor pathway [8, 9] and ultimately to their Nobel Prize in Physiology or Medicine. Once the LDL receptor gene was cloned (*LDLR*), mutations were identified that cause defects in receptor function by several different mechanisms. It was then clear that the genetic etiology of FH on a molecular level was not overproduction of LDL-C but rather mutations in the LDL receptor pathway that resulted in reduced clearance of LDL-C from the bloodstream [10].

The LDL Receptor Pathway

The LDL receptor (LDLR) pathway is complex, and genetic defects in any of a number of steps in this pathway can lead to phenotypic FH (Figure 7.2). The LDLR is a glycoprotein found on the cell surface of hepatocytes that binds specifically to apolipoprotein B-100 on the surface of LDL particles. Once bound, the LDLR and its ligand, LDL-C, enter the cell through endocytosis. LDLR adaptor protein 1 (LDLRAP1) and other proteins interact with the receptor–ligand complex forming the endosome. After endocytosis, the acidic environment of the endosomal compartment causes release of LDL-C from the LDLR. The LDLR is then recycled to the cell wall, and free LDL-C can then be used within the cell. The hydrolysis of cholesterol esters within LDL-C leads to accumulation of free cholesterol. Excess intracellular cholesterol inactivates sterol regulatory element-binding protein

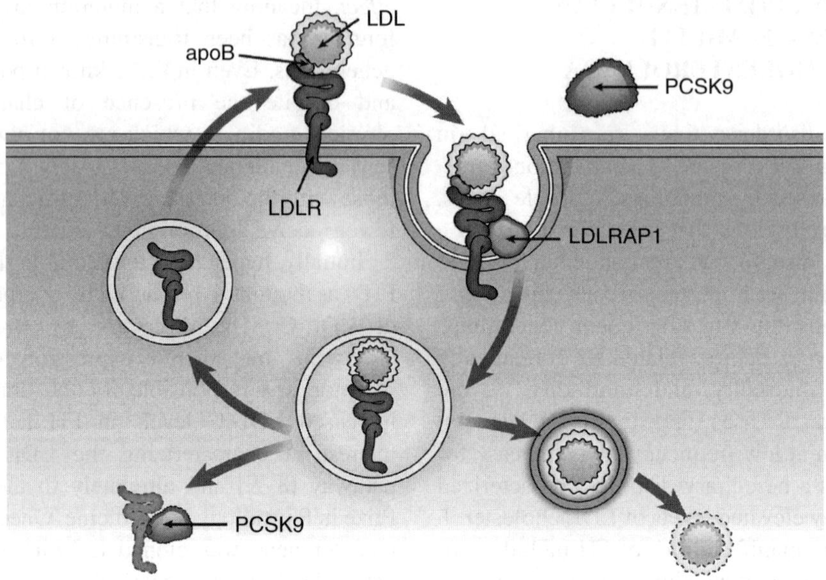

Figure 7.2 LDL receptor pathway. The LDL receptor (LDLR), found on the cell surface of hepatocytes, binds to apoB 100 within the LDL particle. This complex then binds to LDLR adaptor protein 1, which facilitates endocytosis, and the complex enters the cell. Once inside the cell, LDL is released for use and LDLR is recycled to the cell surface. PCSK9 likely acts within the cell to target LDLR for degradation. It may also bind LDLR on the cell surface, where the LDLR–PCSK9 complex is then internalized and LDLR is either degraded or prevented from recycling to the cell surface.

(SREBP), a transcription factor that increases expression of genes involved in cholesterol and LDLR synthesis, which in turn reduces uptake of LDL from the blood. Finally, proprotein convertase subtilisin kexin type 9 (PCSK9), whose complete function is not yet fully understood [11], likely blocks recycling of LDL receptor to the cell membrane and promotes its degradation. It is this multistep feedback mechanism that regulates LDL-C levels in the blood stream under normal conditions.

Mutations within the genes encoding each of the proteins involved in the pathway, but most commonly the LDLR, have been shown to cause FH (see Table 7.1). To date, over a thousand mutations that cause FH by perturbations in the LDLR pathway have been identified [12]. The most common mutations in the key proteins of the pathway are discussed below.

LDLR Mutations

Mutations in *LDLR*, the gene that codes for the LDL receptor, cause 85–90% of classic FH [6]. The *LDLR* gene is located on chromosome 19p13.2. It is a large, complex gene, 45 kb long and comprises 18 exons that code for different domains within the 839 amino acid LDLR protein. Since the discovery of the first mutations, which were large deletions [10], over 1100 unique allelic variants have been identified [13], and additional mutations and polymorphisms continue to be reported. The mutations are spread throughout the gene, affecting multiple domains and causing multiple defects within the LDL receptor. Rearrangements, missense mutations, single aminoacid substitutions, and variants that affect splicing—all of which cause defects in important functions of the LDLR—have been described. With the exception of a few variants found in large

TABLE 7.1 LDL receptor cycle defects in familial hypercholesterolemia

Defect	Gene	Percentage of FH [6, 12, 14, 54]	Approximate reported gene mutations* [55]	Phenotype	Prevalence
LDLR is not synthesized at all (null alleles) or is defective	*LDLR*	85–90%	>1200	Heterozygote: LDL 2–3 × normal Homozygote: LDL >3–4 × normal Early atherosclerosis	Heterozygotes: 1 in 500 Homozygotes: 1 in 10^6
LDL degradation (either before or after reaching cell surface)	*PCSK9*	<5%	47	Mild LDL elevation to 3–4 × normal	Unknown
Ineffective binding of ApoB100 and LDLR	*APOB*	5–10%	110 (Arg3500Gln is most common)	Variable penetrance Mild LDL elevation to 2–3 × normal Early atherosclerosis seen in severe phenotypes	Heterozygotes: 1 in 1000
Ineffective endocytosis of LDLR–LDL through clathrin-coated pits	*LDLRAP1*	Only 100s of cases	15	Mild LDL elevation to 3–4 × normal	Extremely rare

The number of reported mutations causing a familial hypercholesterolemia phenotype continues to grow, with most of these occurring in the LDL receptor.

*The reported mutations have variable effects on the proteins listed, and thus have variable effects of serum cholesterol levels.

numbers of patients in founder populations, the majority of mutations are uncommon, often found only in a single family [14].

APOB Mutations

In contrast to the many mutations in the *LDLR* gene, only a few significant mutations of the apoB gene (*APOB*) have been reported to cause phenotypic FH. Families with *APOB* mutations are said to have the familial-defective apoB 100 syndrome [15]. ApoB 100 is encoded by a gene on chromosome 2p24–p23 that comprises 29 exons and spans approximately 43 kb. The most common mutation, seen in approximately 5–10% of Europeans with FH, is a single amino acid substitution of arginine with glutamine at amino acid 3500 in the protein. This Arg3500Gln mutation results in a defect in the binding domain of apoB 100 with LDL-C [12]. Insufficient LDL binding to apoB 100 means that LDL will not bind efficiently to LDLR, leading to elevated levels of LDL-C in the serum. Patients with familial-defective apoB 100 have similar clinical findings and increased risk of premature myocardial infarction, as do FH patients; but in general, LDL-C levels are lower [16].

PCSK9 Mutations

The gene for *PCSK9* is found on chromosome 1p32 and is 3.6 kb in size. Its 12 exons encode a 695 amino acid protein. The mechanism by which *PCSK9* gene defects cause autosomal-dominant FH is likely due to increased PCSK9 function [17]. The proposed mechanism is that the liver secretes PCSK9 into the bloodstream where it can bind to the extracellular domain of the LDLR [18]. Upon internalization of the PCSK9–LDLR complex, LDLR recycling and LDLR numbers on the cell surface are decreased [19]. Consequently, LDL-C is less rapidly cleared from the circulation, and LDL-C concentrations in blood are markedly increased.

Although *PCSK9* mutations are rare, comprising only a small portion (up to 3%) of cases of FH [12], this proposed mechanism of action has generated significant interest in the gene as a drug target because, in animal models, loss of function mutations of the *PCSK9* gene leads to significant reductions in LDL-C levels [11]. It is likely that this is relevant in humans, as loss of function mutations of PCSK9 in the African American population have been associated with an 88% reduction in risk of coronary heart disease compared to those without the mutation [20]. In a manner analogous to knowledge of the LDLR pathway leading to development of statin therapies, these observations have spawned therapeutic interest to target PCSK9 to reduce LDL-C and cardiovascular events.

ROLE FOR GENETIC DIAGNOSIS AND SCREENING

Identification of mutations in the LDLR pathway is one area in which genetic screening is both clinically useful and cost effective. Although clinical diagnostic criteria for identification of FH patients exist and are inexpensive, the physical examination fails to identify a significant portion of those who have not yet had phenotypic expression of the disease or who have a mild form [6]. In FH, early diagnosis is of great importance because it allows pharmacologic intervention before cardiovascular disease develops. Genetic testing is less available and more expensive than clinical screening; but given the cost of treating late-stage atherosclerosis, it may prove to be cost-effective in the future on a population-wide basis.

However, identifying the specific molecular etiology in a family can have profound cascade effects, since presymptomatic testing for a known familial mutation can then be offered to unaffected family members. In the cascade screening method, first-degree relatives of a patient identified with a specific mutation are screened for the mutation. If the mutation is found, all of their first-degree relatives are screened, and so on. Those who test positive for the known familial mutation can then pursue more aggressive surveillance and pharmacotherapy, while those who test negative can be presumed to have the normal population risk for hypercholesterolemia and atherosclerosis. Because cascade screening only involves first-degree relatives of affected patients, the prevalence of mutations is much higher than in the general population [21]. Challenges to this strategy include (i) failure to identify a genetic cause—despite advances in genetic testing, approximately 15% of those with a clinical diagnosis of FH have no identifiable genetic cause [12]; (ii) the implications for medical insurance; and (iii) uncertainty about who will actually have phenotypic expression. As these molecular pathways are more fully elucidated and the cost of genetic analysis decreases due to improved technology, genetic testing for FH will improve in specificity and sensitivity at reduced cost.

GENETICALLY ISOLATED POPULATION STUDIES: IMPLICATIONS FOR ATHEROSCLEROSIS IN THE GENERAL POPULATION

An important tool in the study of genetics and human disease involves evaluating genetically isolated populations, which have both decreased genetic diversity and limited environmental heterogeneity. While much more time-consuming than studies of Mendelian diseases, studies of isolated populations have the potential to identify cellular/molecular pathways and genetic abnormalities that are broadly relevant to the general population. In addition, studies of genetically isolated populations can offer unique insights into the role of environmental factors in disease progression (see the example of the Pima Indians).

One interesting example is the Hutterite communities in the prairies of North America. This group, currently estimated to number approximately 40,000, is relatively genetically homogeneous, the majority having descended from fewer than 90 ancestors with little out-marriage [22]. Common variant susceptibility alleles for cardiovascular diseases have been identified in the Hutterite population. Interestingly, as outbreeding (marriage of Hutterites to non-Hutterites) has become more common for this population, it has been reported that the susceptibility alleles for cardiovascular disease identified in inbred Hutterites also carry increased risk of disease in outbred populations [23], suggesting that these candidate genes may well be applicable to the general population. Over time, two different methods—linkage analysis and association studies—have been applied to the Hutterite population in an attempt to identify cardiovascular risk genes.

Linkage analysis was developed first and has been used in families and founder populations to identify loci containing candidate genes. During gametogenesis, the formation of a sperm or egg in germ cells, recombination occurs between the paired chromosomes from each parent, such that the chromosome passed to the offspring contains a mixture of alleles from both grandparents. Segments of the genome located in close proximity are more likely to remain together during recombination, while segments that are further apart on the chromosome are more likely to segregate.

In linkage analysis, the recombination patterns of chromosomal markers mapped to a specific region are evaluated in families in which several members have the disease in question. Markers that segregate at higher rates in family members with the disease state are likely located near a gene involved in the disease. Further mapping can then be done to identify specific genes located near the marker. The *ABCA1* gene, the causative gene for hypercholesterolemia in Tangier disease, was identified using linkage analysis in one family [24]. This method works well for Mendelian diseases with genes that have a large effect; but in the population-wide search for the genetic contribution to complex diseases such as atherosclerosis, linkage analysis has largely been replaced by more efficient methods [25].

Association studies, a second method to search for candidate genes, utilize a case-control design to compare allele frequencies in those with disease and those without. Initially, association studies used a method called linkage disequilibrium to identify candidate genes. Linkage disequilibrium depends on nonrandom segregation of alleles over many generations. Important findings have been made in populations such as the Hutterites using linkage disequilibrium analysis, including identification of a major locus on chromosome 2q14 associated with lower triglyceride levels [26].

The technological advances that led to the ability to perform genome-wide association studies (GWAS) resulted in much more rapid and detailed association studies than were possible using earlier forms of linkage disequilibrium analysis. Modern GWAS (described in more detail later) leverage the technical capacity to assay millions of single nucleotide polymorphisms simultaneously, thus greatly increasing the coverage of the genome and the resolution of the linkage disequilibrium. The true benefit of GWAS is believed to be identification of novel molecular pathways and targets for pharmacologic intervention.

A second genetically isolated population of interest to atherosclerosis research is the Pima Indians, in this case because of the potential to study gene/environment interactions. Within the past century, a time during which there can be little genetic drift, this genetically homogenous group of Native Americans based in the Gila River Indian Community in Arizona has experienced profound phenotypic changes. The Pima Indian population has developed the highest reported prevalence of diabetes in the world: almost half of all Pima adults in the United States have diabetes [27, 28]. This occurred in conjunction with—and almost certainly because of—a dramatic change in lifestyle. Over about a century, the Pima Indians' lifestyle changed from a nomadic existence, high on physical activity and low on readily available high-calorie food, to a much more sedentary lifestyle and a diet rich in processed foods and animal fats.

In a manner far too rapid to be attributable to genetic causes, the Pima population transitioned from a society with low body weight and no diabetes to one plagued by obesity, insulin resistance, and an epidemic of diabetes completely out of proportion to other populations with the same lifestyle and diet. The search for causative genetic factors centers on identifying genetic variations that confer no or limited risk prior to their modern lifestyle and significant risk when exposed. Whole genome linkage analyses and more recently GWAS have been conducted in this population, and studies are ongoing to identify candidate genes that would explain this observation [29]. The Pima serve as a prime example of how environmental exposures can have various effects on a patient's disease state given genetic makeup.

Although GWAS and other association studies are now being performed in the general population, the importance of studying genetically isolated populations cannot be overstated. Novel mutations conferring increased risk for cardiovascular diseases continue to emerge from studies comparing genetically isolated populations and outbred families. With the continued mixing of populations throughout the world, special populations with retained genetic and environmental homogeneity will continue to play an important role in gene discovery and will likely not be supplanted in the near future.

GWAS TO IDENTIFY ATHEROSCLEROSIS RISK GENES IN THE GENERAL POPULATION

Tight associations between causative genes and disease are much more challenging in the general population, where the preponderance of cardiovascular disease does not result from Mendelian disorders and does not occur within a genetically homogeneous population. With GWAS becoming widely available in the past 10 years, there has been a tremendous interest in using this technology to identify single nucleotide polymorphisms (SNPs) that are associated with risk for multifactorial disease in the general population. It has been speculated for more than a century that, in contrast to rare single-gene disorders such as FH, most common diseases cannot be attributed to an isolated genetic defect. Generally, common diseases are polygenic, meaning that an array of susceptibility alleles, combined with certain environmental triggers and random chance events, impart risk for developing disease. These types of abnormalities are referred to as common variants [30]. The odds ratios conferred by these SNPs generally range from 1.1 to 1.5, consistent with a low effect size [31]. Successful sequencing of the human genome in 2001, the publication of the International Haplotype Map project, and advances in array-based platforms that enable massive genetic sequencing have enabled researchers to obtain a much better understanding of common variants through GWAS [31].

GWAS are predicated upon identifying strong case-control associations. In these studies, the frequency of a genetic marker is compared in cases versus controls. A marker found at higher frequency in cases may indicate the presence of an allele conferring increased risk in close proximity to the marker. Markers that are present at higher frequency in the control population may indicate the presence of a protective allele located in close proximity to the marker. This type of study allows researchers to identify disease associations in real-world populations. Because these studies do not target gene identification according to a biologic hypothesis, GWAS may facilitate identification of novel gene–disease associations.

Due to the immense number of comparisons being made between cases and controls in GWAS, it is important to recognize that there are stringent criteria for declaring associations significant. To illustrate this point, if one compares genotypes for 100,000 SNPs at a *P*-value of 0.05, one would expect to see 5000 associations that are due to chance; when 1,000,000 genotypes are assessed, there will be 50,000 chance associations. Thus, researchers apply a correction for multiple testing depending on the number of SNPs assayed on their microarray in order to evaluate for genome-wide significance. The minimal *P*-value is thus more stringent (5×10^{-7} to 5×10^{-8} depending on the assay) before an association is declared statistically significant [31, 32].

Thus, GWAS can identify common variants that are associated with disease. To date, many risk variants have been identified by GWAS in diseases across different organ systems, from coronary artery disease to Crohn's disease to bipolar disorder [33].

An important limitation of case-control studies in general, and GWAS in particular, is that these studies are observational and only imply a relationship. Another important drawback is that GWAS can only identify genetic markers and may not identify the allele that is directly responsible for the disease risk—they may simply be in strong linkage disequilibrium with one another [31]. Some variants identified through GWAS directly affect the translated protein and thus could affect the function of the gene. Other variants may exert their effect by modifying or regulating the expression of nearby genes or even genes a great distance away, thus greatly complicating further investigation. Indeed, in the spectrum of variants associated with coronary artery disease identified by GWAS, more than 70% are found in sequences that do not code for proteins, suggesting that they may exert their effect by regulating gene expression [32]. Therefore, identification of disease-associated genetic variants by GWAS must be followed by extensive investigation to determine the biologic and/or prognostic relevance of the newly identified variant.

For these reasons, GWAS are most appropriately considered hypothesis-generating studies. As an illustration of this principle in atherosclerosis, in studies reported between 2007 and 2011, GWAS identified 30 variants strongly associated with CAD and MI (see Table 7.2). Further investigation revealed evidence for disease-causing mechanisms in only 8 of the 30 variants, and the relevance of these 8 remains to be proven. GWAS has inspired a flurry of research to learn more about how these variants affect atherosclerosis.

TABLE 7.2 Common variants for CAD identified by GWAS

Band	Year of publication	Neighboring genes	Odds ratio for CAD
9p21.3	2007	*CDKN2A, CDKN2B*	1.25 (1.18–1.31)
1p13.3	2007	*SORT1*	1.29 (1.18–1.40)
1q41	2007	*MIA3*	1.20 (1.12–1.30)
10q11.21	2007	*CXCL12*	1.33 (1.20–1.48)
1p32.3	2009	*PCSK9*	1.15 (1.10–1.21)
2q33.1	2009	*WDR12*	1.16 (1.10–1.22)
3q22.3	2009	*MRAS*	1.15 (1.11–1.19)
6p24.1	2009	*PHACTR1*	1.13 (1.09–1.17)
6q25.3	2009	*LPA*	1.92 (1.48–2.49)
12q24.12	2009	*SH2B3*	1.13 (1.08–1.18)
19p13.2	2009	*LDLR*	1.14 (1.09–1.19)
21q22.11	2009	*MRPS6*	1.19 (1.13–1.27)
1p32.2	2011	*PPAP2B*	1.17 (1.13–1.22)
6p21.31	2011	*ANKS1A*	1.07 (1.05–1.10)
6q23.2	2011	*TCF21*	1.08 (1.06–1.10)
7q32.2	2011	*ZC3HC1*	1.09 (1.07–1.12)
9q34.2	2011	*ABO*	1.10 (1.07–1.13)
10q24.32	2011	*CYP17A1, CNNM2, NT5C2*	1.12 (1.08–1.16)
11q23.3	2011	*ZNF259, APOA5, APOA4, APOC3, APOA1*	1.13 (1.10–1.16)
13q34	2011	*COL4A1, COL4A2*	1.07 (1.05–1.09)
14q32.2	2011	*HHIPL1*	1.07 (1.05–1.10)
17p13.3	2011	*SMG6, SRR*	1.07 (1.05–1.09)
17p11.2	2011	*RASD1, SMCR3, PEMT*	1.07 (1.05–1.09)
17q21.32	2011	*UBE2Z, GIP, ATP5G1, SNF8*	1.06 (1.04–1.08)
10q23.31	2011	*LIPA*	1.09 (1.07–1.12)
11q22.3	2011	*PDGF*	1.07 (1.04–1.09)
7q22.3	2011	*BCAP29*	1.08 (1.05–1.11)
10p11.23	2011	*KIAA1462*	1.07 (1.04–1.09)
6p24.1	2011	*C6orf105*	1.65 (1.44–1.90)
15q25.1	2011	*ADAMTS7*	1.19 (1.13–1.24)

From 2007 to 2011, 30 risk loci for CAD were identified through GWAS. Of these, 30 risk variants, nearly 75% (22/30) exert their effect via unknown mechanisms. (Table and information adapted with permission from Roberts and Stewart [32]).

IDENTIFICATION OF A RISK ALLELE LOCATED AT 9P21 BY GWAS

The location of a risk allele on chromosome 9 (9p21) was the first and most robust variant identified through GWAS that portends risk for atherosclerosis. In many ways, it is prototypical of other variants associated with multifactorial disease risk. The 9p21 risk allele is common, estimated to be present in more than four billion individuals worldwide [31]; it exerts its effect via a novel pathway, potentially leading to new therapeutic targets; and, although it is consistently associated with disease, its specific disease-causing mechanism remains elusive. For these reasons, 9p21 is an interesting and illustrative example of the power and pitfalls of gene variant identification by GWAS.

9p21 was first described in 2007 when multiple investigators simultaneously uncovered and confirmed its significance, independent of known risk factors [32, 34–36]. Cases were variably defined in these studies, ranging from individuals who had known CAD by coronary angiography to individuals who carried the diagnosis of myocardial infarction based on any of a number of criteria. The study sizes ranged from 322 to 4587 affected cases and

312–12,767 controls. After case sample analysis, it became readily apparent that the 9p21 variant occurred with greater frequency in case patients than in controls. Across all of these studies, it was found that 50% of cases had one copy of the 9p21 risk allele and 20–25% had two copies. The presence of a single copy of the 9p21 risk allele conferred a 20–30% increased risk for disease. Initial studies were performed primarily in Caucasian populations, an important issue because risk variants important in one genetic background may not be important in other genetic (or environmental) backgrounds. In the case of the 9p21 risk allele, the association with cardiovascular disease has been validated in most but not all studies across ethnicities [32]. To date, one important study in an African American population failed to find an association of the 9p21 risk allele with cardiovascular disease [37]. Interestingly, while initial studies focused on its association with atherosclerosis, 9p21 has since been associated with other forms of vascular disease: cerebrovascular disease, dementia, peripheral vascular disease, and even intracranial and abdominal aortic aneurysm [38–41]. Its association with aneurysms, combined with the observation that it can increase risk for atherosclerosis without MI, has been interpreted as evidence that the 9p21 haplotype exerts its effect in the vessel wall rather than modulating risk for thrombosis.

The risk-conferring 9p21 domain contains many SNPs that are in strong linkage disequilibrium with one another. Together, they form a string of alleles in the p21 region of chromosome 9 that is approximately 53 kb long, creating a risk haplotype. The majority of the 9p21 variant consists of noncoding regions. It is likely that the 9p21 variant/haplotype does not cause disease as a result of an abnormality in one or more genes or even by transcription factor-mediated regulation of an adjacent gene [42]. The 9p21 haplotype does occur in a region of high linkage disequilibrium with the tumor-suppressor genes *CDKN2A* and *CDKN2B*, located 100 kb away. These genes are abundantly expressed in atherosclerotic tissue [43] and regulate the cell cycle, apoptosis, and senescence. *CDKN2A* and *CDKN2B* have been reported to promote smooth muscle cell and leukocyte proliferation, both of which have been reported to be important steps in atherogenesis. Nevertheless, cases (patients) with 9p21 variants do not consistently demonstrate altered expression of *CDKN2A* or *CDKN2B*, bringing into question this general hypothesis for the biologic mechanism of the 9p21 variant in atherosclerosis [43].

An alternative proposed mechanism for the 9p21 variant rests in *ANRIL*, also known as *CDKN2BAS*, a large noncoding RNA sequence that overlaps *CDKN2B*. While *ANRIL*'s precise biologic function remains a topic of current investigation, its sequence contains repetitive elements and binding sites likely to be important in transcriptional regulation. *ANRIL* has been shown to be expressed in numerous tissues involved in atherosclerosis—coronary arteries, carotid arteries, aortic aneurysms, vascular endothelial cells, and even macrophages, as well as in the myocardium [44]. Increased *ANRIL* expression has also been reported to directly correlate with atherosclerotic plaque burden [45]. Finally, unlike *CDKN2A* and *CDKN2B*, *ANRIL* expression is highly associated with the 9p21 haplotype. For all these reasons, there is great interest in further characterization of the potential biologic role of *ANRIL* variants in atherosclerosis.

Even in the example of the 9p21 haplotype, a risk allele highly correlated with cardiovascular risk across populations, it is clear that much more mechanistic data are required. The development of targeted therapies awaits further investigation. Even application of these findings to assess cardiovascular risk is premature because it is unclear which of these very common SNPs are of greatest importance and, even should a high-risk profile of SNPs be identified, how to counsel individual patients is unclear as well. Indeed, the 9p21 variant carries a similar level of risk for atherosclerosis for some traditional cardiovascular risk factors such as smoking, high cholesterol, or hypertension [46, 47].

There is no question that GWAS have changed the landscape with regard to the study of cardiovascular risk. Whether these studies will have an impact on treatment, prevention, or even genetic screening for cardiovascular diseases remains to be determined.

TECHNOLOGIC ADVANCES AND THE FUTURE

The past two decades have witnessed amazing advances in our ability to sequence DNA, opening the door to more efficient and accurate analysis of the human genome. For decades, DNA sequencing relied on Sanger biochemistry [48], using DNA fragments incorporated into plasmid vectors and then transformed into *Escherichia coli* for cloning and amplification. The rate-limiting step in Sanger sequencing is the use of gel electrophoresis to identify individual sequences, making this approach time-consuming and cost prohibitive for large sequencing projects. Newer techniques that have been developed over the past decade, identified as next-generation sequencing, allow massively parallel sequencing of multiple parts of the genome, drastically reducing the cost and time required to obtain high-quality sequence information [49]. While several unique techniques exist, each of the next-generation methods relies on development of an array of thousands of genomic fragments ligated to a solid substrate. Fluorescently labeled nucleotides are then incorporated into the genomic fragments with PCR. With each cycle, an image of the array is taken to determine the sequence (see Figure 7.3). In this way, tens of millions of sequences can be identified in parallel.

Already this new technology has permitted a more thorough exploration of the specific mutations leading to Mendelian diseases such as FH through targeted sequencing. Since 2010, for example, genome sequencing has identified *ABCG5* mutations as a cause of severe hypercholesterolemia [50] and novel *ANGPTL3* mutations in familial combined hypolipidemia [51]. Furthermore, next-generation sequencing can serve a complementary role to GWAS by allowing rapid sequencing of regions of the genome identified by GWAS. Sequencing can also identify rare polymorphisms that could be utilized in future GWAS to identify new loci associated with cardiovascular disease, in addition to those discussed earlier at the 9p21 locus.

While these new genetic technologies have generated great excitement in the literature, there are significant challenges to transforming this new data into clinically useful information. With a few exceptions (principally, the FH example and genetic screening for long-QT syndrome as a cause of sudden cardiac death), the vast majority of genetic research in atherosclerosis has yet to translate into clinical practice. GWAS are, as noted earlier, primarily hypothesis generating. These limitations notwithstanding, there is reason for optimism. Brown and Goldstein's work on familial hypercholesterolemia in the 1970s and 1980s required an additional 15 years to be translated into the development, testing, and implementation of statin therapy for prevention of major adverse cardiac events [52]. Even if one new pathway of this significance is identified by genetic approaches, the impact for those at risk for cardiovascular disease will be high.

For years, genetic counselors have sought better tools to advise individuals on their risk of cardiovascular disease. Many researchers in this field have proposed utilizing a genetic panel that combines hundreds of variants that portend the highest risk [30]. However, in order for a genetic risk profile to have a predictive performance similar to current risk profiles, such as the Framingham risk score, a large number of variants, 50–100, with high odds ratios would be required [53].

Today's tools are of great relevance in familial disorders and for research on human populations. However, while it was hoped that the results of GWAS and other approaches would be useful for screening of the general population, this is almost certainly not the case at present. The scientific and medical

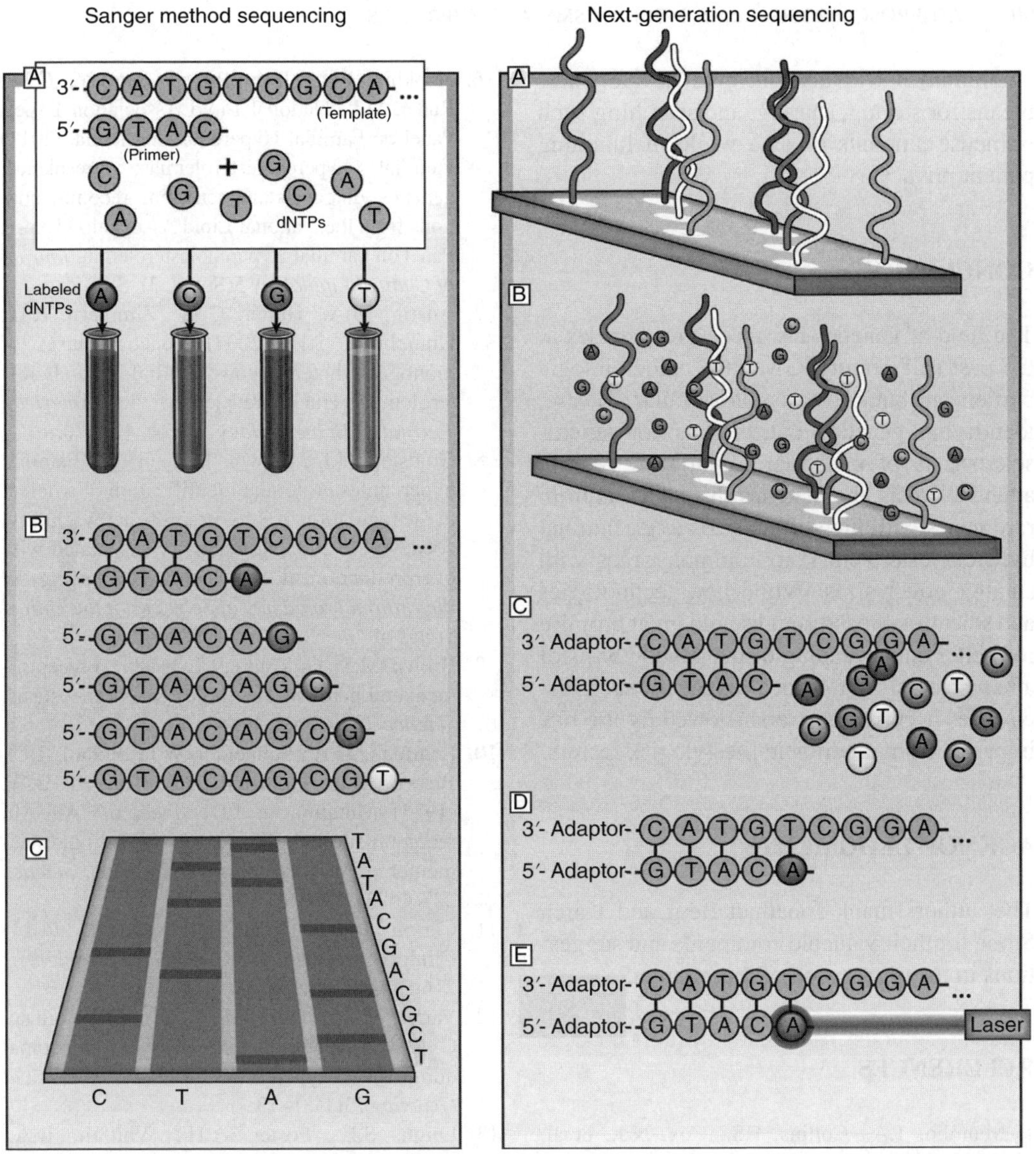

Figure 7.3 Next-generation DNA sequencing. In shotgun Sanger sequencing, depicted on the left-hand side of the figure, random fragments of genomic DNA are cloned into bacterial plasmids and amplified. These clonal DNA fragment templates are then placed into four separate mixtures [A], which include the DNA fragment template, all four deoxynucleotides nucleotides (dNTP), and one type of fluorescently labeled dideoxynucleotide (ddNTP), one of G, A, T, or C. Each tube completes cycle sequencing with extension of the primer until incorporation of a ddNTP, stopping the reaction [B]. The fragments, which are now of variable lengths, are separated in a high-resolution capillary-based polymer gel electrophoresis. Because each fragment terminates with a known nucleotide, the DNA sequence can be determined by examining the results of the electrophoresis [C].While several specific techniques are used in next-generation sequencing, depicted on the right-hand side of the figure, each is based on massively parallel sequencing of DNA fragments. Random DNA fragments are ligated onto a solid substrate with common adaptor sequences [A]. Each fragment is copied so that millions of fragments are clustered on the array. Sequencing reagents and fluorescently labeled dNTPs are then added to the array [B, C]. The sequence reaction is stopped after the addition of one fluorescently labeled nucleotide to each fragment [D]. With each addition, an image of the array is taken that identifies the terminal nucleotide and thus the sequence of the fragment [E]. With each cycle, millions of copies of fragments are sequenced in parallel (Adapted with permission from Aparicio and Huntsman [56]. © Wiley).

communities will also have to develop new means for storing, sharing, and searching such immense amounts of data while maintaining patient privacy.

CONCLUSION

The field of genetics has made great strides in the past half century. Given the epidemiologic burden attributable to cardiovascular disease, identifying genetic contributions for atherosclerosis is of particular importance. Within atherosclerosis, the field has shifted focus from rare genes with high effect sizes (e.g., familial hypercholesterolemia) to common genes with smaller effect sizes. While new technologies and scientific breakthroughs hold great promise for better understanding of the genetic basis of atherosclerosis, the clinical impact of such discoveries is currently overshadowed by the risk imparted from modifiable lifestyle risk factors.

ACKNOWLEDGMENTS

The authors thank Jonathan Berg and Karen Stone for their valuable comments and suggestions in the preparation of this chapter.

REFERENCES

1. Manolio, T.A., Collins, F.S., Cox, N.J., et al. (2009) Finding the missing heritability of complex diseases. *Nature* 461(7265), 747–53.
2. Reich, D.E., Lander, E.S. (2001) On the allelic spectrum of human disease. *Trends in Genetics: TIG* 17(9), 502–10.
3. Pritchard, J.K. (2001) Are rare variants responsible for susceptibility to complex diseases? *American Journal of Human Genetics* 69(1), 124–37.
4. Allen, G., Harvald, B., Shields, J. (1967) Measures of twin concordance. *Acta Genetica et Statistica Medica* 17(6), 475–81.
5. Khachadurian, A.K. (1964) The inheritance of essential familial hypercholesterolemia. *The American Journal of Medicine* 37, 402–7.
6. Hopkins, P.N., Toth, P.P., Ballantyne, C.M., Rader, D.J.; National Lipid Association Expert Panel on Familial Hypercholesterolemia. (2011) Familial hypercholesterolemias: prevalence, genetics, diagnosis and screening recommendations from the National Lipid Association Expert Panel on Familial Hypercholesterolemia. *Journal of Clinical Lipidology* 5(Suppl. 3), S9–17.
7. Austin, M.A., Hutter, C.M., Zimmern, R.L., Humphries, S.E. (2004) Genetic causes of monogenic heterozygous familial hypercholesterolemia: a HuGE prevalence review. *American Journal of Epidemiology* 160(5), 407–20.
8. Goldstein, J.L., Brown, M.S. (1973) Familial hypercholesterolemia: identification of a defect in the regulation of 3-hydroxy-3-methylglutaryl coenzyme A reductase activity associated with overproduction of cholesterol. *Proceedings of the National Academy of Sciences of the United States of America* 70(10), 2804–8.
9. Brown, M.S., Goldstein, J.L. (1986) A receptor-mediated pathway for cholesterol homeostasis. *Science* 232(4746), 34–47.
10. Lehrman, M.A., Schneider, W.J., Sudhof, T.C., Brown, M.S., Goldstein, J.L., Russell, D.W. (1985) Mutation in LDL receptor: Alu-Alu recombination deletes exons encoding transmembrane and cytoplasmic domains. *Science* 227(4683), 140–6.
11. Cariou, B., Le May, C., Costet, P. (2011) Clinical aspects of PCSK9. *Atherosclerosis* 216(2), 258–65.
12. Varret, M., Abifadel, M., Rabes, J.P., Boileau, C. (2008) Genetic heterogeneity of autosomal dominant hypercholesterolemia. *Clinical Genetics* 73(1), 1–13.
13. Leigh, S.E., Foster, A.H., Whittall, R.A., Hubbart, C.S., Humphries, C.S. (2008) Update and analysis of the University College London low density lipoprotein receptor familial hypercholesterolemia database. *Annals of Human Genetics* 72(4), 485–98.
14. Soutar, A.K. (2010) Rare genetic causes of autosomal dominant or recessive hypercholesterolaemia. *IUBMB Life* 62(2), 125–31.
15. Soutar, A.K., Naoumova, R.P. (2007) Mechanisms of disease: genetic causes of familial hypercholesterolemia. *Nature Clinical Practice Cardiovascular Medicine* 4(4), 214–25.
16. Ejarque, I., Real, J.T., Martinez-Hervas, S., et al. (2008) Evaluation of clinical diagnosis criteria of familial ligand defective apoB 100 and

lipoprotein phenotype comparison between LDL receptor gene mutations affecting ligand-binding domain and the R3500Q mutation of the apoB gene in patients from a South European population. *Translational Research: The Journal of Laboratory and Clinical Medicine* 151(3), 162–7.

17. Abifadel, M., Rabes, J.P., Devillers, M., et al. (2009) Mutations and polymorphisms in the proprotein convertase subtilisin kexin 9 (PCSK9) gene in cholesterol metabolism and disease. *Human Mutation* 30(4), 520–9.
18. Maxwell, K.N., Fisher, E.A., Breslow, J.L. (2005) Overexpression of PCSK9 accelerates the degradation of the LDLR in a post-endoplasmic reticulum compartment. *Proceedings of the National Academy of Sciences of the United States of America* 102(6), 2069–74.
19. Horton, J.D., Cohen, J.C., Hobbs, H.H. (2009) PCSK9: a convertase that coordinates LDL catabolism. *Journal of Lipid Research* 50 Suppl:S172–7.
20. Cohen, J.C., Boerwinkle, E., Mosley, T.H., Jr., Hobbs, H.H. (2006) Sequence variations in PCSK9, low LDL, and protection against coronary heart disease. *The New England Journal of Medicine* 354(12), 1264–72.
21. Marks, D., Wonderling, D., Thorogood, M., Lambert, H., Humphries, S.E., Neil, H.A. (2002) Cost effectiveness analysis of different approaches of screening for familial hypercholesterolaemia. *BMJ* 324(7349), 1303.
22. Martin, A.O. (1970) The founder effect in a human isolate: evolutionary implications. *American Journal of Physical Anthropology* 32(3), 351–367.
23. Newman, D.L., Hoffjan, S., Bourgain, C., et al. (2004) Are common disease susceptibility alleles the same in outbred and founder populations? *European Journal of Human Genetics: EJHG* 12(7), 584–90.
24. Brooks-Wilson, A., Marcil, M., Clee, S.M., et al. (1999) Mutations in ABC1 in Tangier disease and familial high-density lipoprotein deficiency. *Nature Genetics* 22(4), 336-45.
25. Pollex, R.L., Hegele, R.A. (2005) Complex trait locus linkage mapping in atherosclerosis: time to take a step back before moving forward? *Arteriosclerosis, Thrombosis, and Vascular Biology* 25(8), 1541–4.
26. Newman, D.L., Abney, M., Dytch, H., Parry, R., McPeek, M.S., Ober, C. (2003) Major loci influencing serum triglyceride levels on 2q14 and 9p21 localized by homozygosity-by-descent mapping in a large Hutterite pedigree. *Human Molecular Genetics* 12(2), 137–44.
27. Schulz, L.O., Bennett, P.H., Ravussin, E., et al. (2006) Effects of traditional and western environments on prevalence of type 2 diabetes in Pima Indians in Mexico and the U.S. *Diabetes Care* 29(8), 1866–71.
28. Knowler, W.C., Pettitt, D.J., Saad, M.F., Bennett, P.H. (1990) Diabetes mellitus in the Pima Indians: incidence, risk factors and pathogenesis. *Diabetes/Metabolism Reviews* 6(1), 1–27.
29. Rong, R., Hanson, R.L., Ortiz, D., et al. (2009) Association analysis of variation in/near FTO, CDKAL1, SLC30A8, HHEX, EXT2, IGF2BP2, LOC387761, and CDKN2B with type 2 diabetes and related quantitative traits in Pima Indians. *Diabetes* 58(2), 478–88.
30. O'Donnell, C.J., Nabel, E.G. (2011) Genomics of cardiovascular disease. *The New England Journal of Medicine* 365(22), 2098–109.
31. Lotta, L.A. (2010) Genome-wide association studies in atherothrombosis. *European Journal of Internal Medicine* 21(2), 74–8.
32. Roberts, R., Stewart, A.F. (2012) 9p21 and the genetic revolution for coronary artery disease. *Clinical Chemistry* 58(1), 104–12.
33. Wellcome Trust Case Control Consortium. (2007) Genome-wide association study of 14,000 cases of seven common diseases and 3,000 shared controls. *Nature* 447(7145), 661–78.
34. Helgadottir, A., Thorleifsson, G., Manolescu, A., et al. (2007) A common variant on chromosome 9p21 affects the risk of myocardial infarction. *Science* 316(5830), 1491–3.
35. McPherson, R., Pertsemlidis, A., Kavaslar, N., et al. (2007) A common allele on chromosome 9 associated with coronary heart disease. *Science* 316(5830), 1488–91.
36. Samani, N.J., Erdmann, J., Hall, A.S., et al. (2007) Genomewide association analysis of coronary artery disease. *The New England Journal of Medicine* 357(5), 443–53.
37. Assimes, T.L., Knowles, J.W., Basu, A., et al. (2008) Susceptibility locus for clinical and subclinical coronary artery disease at chromosome 9p21 in the multi-ethnic ADVANCE study. *Human Molecular Genetics* 17(15), 2320–8.
38. Helgadottir, A., Thorleifsson, G., Magnusson, K.P., et al. (2008) The same sequence variant on 9p21 associates with myocardial infarction,

abdominal aortic aneurysm and intracranial aneurysm. *Nature Genetics* 40(2), 217–24.

39. Cluett, C., McDermott, M.M., Guralnik, J., et al. (2009) The 9p21 myocardial infarction risk allele increases risk of peripheral artery disease in older people. *Circulation Cardiovascular Genetics* 2(4), 347–53.
40. Emanuele, E., Lista, S., Ghidoni, R., et al. (2011) Chromosome 9p21.3 genotype is associated with vascular dementia and Alzheimer's disease. *Neurobiology of Aging* 32(7), 1231–5.
41. Gschwendtner, A., Bevan, S., Cole, J.W., et al. (2009) Sequence variants on chromosome 9p21.3 confer risk for atherosclerotic stroke. *Annals of Neurology* 65(5), 531–9.
42. Cunnington, M.S., Keavney, B. (2011) Genetic mechanisms mediating atherosclerosis susceptibility at the chromosome 9p21 locus. *Current Atherosclerosis Reports* 13(3), 193–201.
43. Holdt, L.M., Sass, K., Gabel, G., Bergert, H., Thiery, J., Teupser, D. (2011) Expression of Chr9p21 genes CDKN2B (p15(INK4b)), CDKN2A (p16(INK4a), p14(ARF)) and MTAP in human atherosclerotic plaque. *Atherosclerosis* 214(2), 264–70.
44. Jarinova, O., Stewart, A.F., Roberts, R., et al. (2009) Functional analysis of the chromosome 9p21.3 coronary artery disease risk locus. *Arteriosclerosis, Thrombosis, and Vascular Biology* 29(10), 1671–7.
45. Holdt, L,M., Beutner, F., Scholz, M., et al. (2010) ANRIL expression is associated with atherosclerosis risk at chromosome 9p21. *Arteriosclerosis, Thrombosis, and Vascular Biology* 30(3), 620–7.
46. Emerging Risk Factors Collaboration. (2009) Major lipids, apolipoproteins, and risk of vascular disease. *JAMA* 302(11), 1993–2000.
47. Sesso, H.D., Stampfer, M.J., Di Angelantonio, E., et al. (2000) Systolic and diastolic blood pressure, pulse pressure, and mean arterial pressure as predictors of cardiovascular disease risk in men. *Hypertension* 36(5), 801–7.
48. Sanger, F., Air, G.M., Barrell, B.G., et al. (1977) Nucleotide sequence of bacteriophage phi X174 DNA. *Nature* 265(5596), 687–95.
49. Shendure, J., Ji, H. (2008) Next-generation DNA sequencing. *Nature Biotechnology* 26(10), 1135–45.
50. Rios, J., Stein, E., Shendure, J., Hobbs, H.H., Cohen, J.C. (2010) Identification by whole-genome resequencing of gene defect responsible for severe hypercholesterolemia. *Human Molecular Genetics* 19(22), 4313–8.
51. Musunuru, K., Pirruccello, J.P., Do, R., et al. (2010) Exome sequencing, ANGPTL3 mutations, and familial combined hypolipidemia. *The New England Journal of Medicine* 363(23), 2220–7.
52. Libby, P., Ridker, P.M., Hansson, G.K. (2011) Progress and challenges in translating the biology of atherosclerosis. *Nature* 473(7347), 317–25.
53. Van der Net, J.B., Janssens, A.C., Sijbrands, E.J., Steyerberg, E.W. (2009) Value of genetic profiling for the prediction of coronary heart disease. *American Heart Journal* 158(1), 105–10.
54. Hobbs, H.H., Russell, D.W., Brown, M.S., Goldstein, J.L. (1990) The LDL receptor locus in familial hypercholesterolemia: mutational analysis of a membrane protein. *Annual Review of Genetics* 24, 133–70.
55. Stenson, P.D., Mort, M., Ball, E.V., et al. (2009) The Human Gene Mutation Database: 2008 update. *Genome Medicine* 1(1), 13.
56. Aparicio, S.A., Huntsman, D.G. (2010) Does massively parallel DNA resequencing signify the end of histopathology as we know it? *Journal of Pathology* 220, 307–15.

8

LIFE STYLE

WAYNE E. CASCIO[1,2]

[1]Environmental Public Health Division, U.S. Environmental Protection Agency, Research Triangle Park, NC, USA

[2]Department of Medicine/Division of Cardiology, University of North Carolina at Chapel Hill, Chapel Hill, NC, USA

INTRODUCTION

Lifestyle behaviors such as physical activity, smoking and diet, and environmental factors contribute to cardiovascular health and well-being in the United States and abroad. The Global Burden of Disease Study 2010 showed that 8 of the leading 12 risk factors for global disease burden are lifestyle behaviors or environmental factors, namely tobacco smoking, including second-hand smoke, alcohol use, household air pollution from solid fuels, diets low in fruits, ambient particulate air pollution, physical inactivity, diets high in sodium, and diets low in nuts and seeds [1]. Individual behavioral choices modify traditional risk factors such as lipid profiles, body mass index (BMI), and blood pressure and are believed to play a role in the development and progression of atherosclerosis. The living environment with its attendant social stresses and toxin exposures also modifies atherosclerosis.

This chapter provides an overview of lifestyle behaviors and environmental factors that impact the development and progression of atherosclerosis either directly or indirectly. Because lifestyle behaviors such as regular physical activity, a healthy diet, and the absence of smoking prolong life and enhance well-being, it is important for health care providers and public health advocates to promote healthy lifestyle behaviors and to encourage research seeking to understand the relationship between lifestyle behaviors and mechanisms of atherogenesis and atherothrombosis. By contrast, relatively little is known about the influence of social stressors and environmental exposures on mechanistic links to

Disclaimer

This work is not a product of the United States Government or the United States Environmental Protection Agency, and the author is not doing this work in any governmental capacity. The views expressed are those of the author only and do not necessarily represent those of the United States or the US EPA.

Atherosclerosis: Risks, Mechanisms, and Therapies, First Edition. Edited by Hong Wang and Cam Patterson.

atherosclerosis, or their interaction with traditional risk factors.

LIFESTYLE FACTORS AND CARDIOVASCULAR DISEASE

Healthy lifestyle behaviors benefit cardiovascular health and extend life. The European SENECA Study showed that regular physical activity, the absence of smoking, and a healthy diet improved survival [2], even among those with advanced age. By contrast, physical inactivity, smoking, and obesity hastened the onset of disability by more than 8 years and death relative to a middle-aged cohort lacking these risk factors [3]. Many other studies also show that the presence of fewer lifestyle risk factors during middle age improves survival. Much of the health benefit attributable to healthy lifestyle behaviors relates to the delay in the onset of vascular disease [4]. As shown in the Physicians Health Study, a higher prevalence of healthy behaviors delayed the onset of coronary artery disease [4].

Short and excessively long sleep durations [5, 6], and disordered sleep [7] are associated with inflammation, insulin resistance, obesity, diabetes, and coronary heart disease. Studies utilizing prolonged sleep restriction and circadian disruption decrease resting metabolic rate and postprandial insulin secretion [8].

LIFESTYLE FACTORS AND CARDIOVASCULAR DISEASE IN THE YOUNG

Subclinical atherosclerosis exists in children with cardiovascular risk factors. The longitudinal Cardiovascular Risk in Young Finns Study followed nearly 3600 children aged 3 through 18 years for the subsequent 30 years and offers insight into the impact of early childhood experiences on cardiovascular risk in later life [9]. Observations from this study show that traditional risk and lifestyle factors when present in children predict subclinical atherosclerosis. Consistent with other studies hypertension, obesity, and lipid abnormalities in adulthood were associated with childhood increased BMI, socioeconomic status, parental risk factor status, and certain genetic polymorphisms. These observations highlight the importance of early identification and correction of risk factors driving the initiation of atherosclerosis and its progression.

PHYSICAL ACTIVITY

Sedentary lifestyle increases the risk for hypertension, metabolic syndrome, diabetes, cardiovascular disease, and mortality. By contrast, cohort studies demonstrate that moderate to high levels of physical activity are associated with lower risk of coronary heart disease (CHD) [10]. In a large cohort, healthy middle-aged men having a high level of fitness were 40% less likely to have advanced coronary artery calcium (CAC) when compared to the lowest level of fitness [11]. Likewise, in the Prospective Army Coronary Calcium (PACC) project, habitual physical activity was associated with less CAC [12]. Therefore, habitual exercise and greater aerobic fitness was protective and associated with less prevalent advanced coronary atherosclerosis as assessed by CAC in two asymptomatic middle-aged male populations. Health benefits have also been documented after increasing fitness among those less fit [13]. In a cohort of 14,345 men with a mean age of 44 years during 11.4 years of follow-up, changes in fitness and BMI were correlated to all-cause and cardiovascular mortality. For each 1 metabolic equivalent (MET) improvement in fitness, all-cause and cardiovascular disease mortality decreased by 15 and 19%, respectively [13].

The mechanisms accounting for physical activity's cardiovascular health benefit are understood incompletely [14]. One of the challenges is that physical activity is associated with increased caloric expenditures and weight loss. Consequently, in small clinical studies, the effects of physical activity are difficult to differentiate from risk reduction attendant to weight loss. Yet, numerous mechanisms are proposed including positive effects on sympathetic–parasympathetic balance and O_2 demand, improved vascular function, vascular collateralization, remodeling and risk of plaque rupture, decreased risk of thrombosis, infarct sparing due to myocardial preconditioning, inflammatory mediator release from skeletal muscle and adipose tissue [14], and improved mood and chronic mental stress [15]. Sedentary behavior associates with adiposity and biomarkers of systemic inflammation. For example, lifestyle behaviors, physical characteristics, and biomarkers of systemic inflammation were assessed in an ethnically diverse middle-aged population in the Multiethnic Study of Atherosclerosis (MESA) Study cohort and sedentary behavior was associated with increased leptin, TNF-α, and a lower adiponectin-to-leptin ratio [16]. Interestingly, regular exercise does not appear to confer its health benefit through effects on atherogenenic lipids. Regular exercise apart from its effects on weight loss does not change total cholesterol [17], although daily regular exercise can increase low-density lipoprotein (LDL) particle size [18]. Other studies point to an increase in high-density lipoprotein (HDL) and lowering of triglycerides [19].

In the Studies of a Targeted Risk Reduction Intervention through Defined Exercise (STRRIDE) Study moderate intensity exercise lowered triglycerides and improved insulin sensitivity more effectively than vigorous exercise [20]. The observations in STRRIDE suggest that habitual moderate intensity exercise might modify cardiovascular risk factors favorably. Moderate intensity exercise training for 8 months results in significant changes in very-low-density lipoprotein (VLDL) triglyceride values that are maintained for 15 days after exercise ceases.

Benefits of exercise are also documented among individuals with diabetes and hypertension. A 6-month, supervised, exercise program involving exercise three times each week yielded improvement in BMI, percentage body fat, fitness, and glycemic control as determined by HbA1c in a sedentary middle-aged cohort with type 2 diabetes and hypertension. Neither endothelial biomarkers (E-selectin, P-selectin, ICAM, VCAM, or tPA) nor endothelial function improved in the exercise group compared to the sedentary control. Yet, improvement in risk factors did correlate with some endothelial biomarkers. The associations of changes in cardiovascular risk factors and endothelial biomarkers suggest that the exercise-induced improvements in endothelial function might be mediated by improvement in cardiovascular risk factors [21].

DIET AND NUTRITION

Dietary Patterns

Dietary patterns in the United States are generally characterized by excessive amounts of sugar-containing beverages, sweets, baked goods, and salt associated with increases in cardiovascular risk factors [22] and the development of atherosclerosis. For example, in the Atherosclerosis Risk in Communities (ARIC) Study, a "Western" dietary pattern (red meats, fried potatoes, refined grains, high-fat dairy, desserts, sugar-sweetened beverages, candy, white potatoes, eggs, pizza, and butter) was associated positively with markers of systemic inflammation. By contrast, a dietary pattern featuring more healthy choices (vegetables, fruit, legumes, fish, tomatoes, whole grains, nuts, and poultry) was associated inversely with CD14, monocyte toll-like receptor-1, and platelet glycoprotein IIb (CD41) [23], supporting a direct effect of dietary pattern on markers of inflammation and endothelial activation.

The Mediterranean, Japanese, and the Dietary Approaches to Stop Hypertension (DASH) diets are associated with less cardiovascular disease or in the case of the DASH diet designed to prevent heart disease. Clinical evidence shows that Mediterranean and Japanese diets lessen the risk of cardiovascular disease. Mediterranean diets are rich in fruits, vegetables, olive oil, nuts, and breads made from wheat, and have been demonstrated in clinical trials to reduce the risk of metabolic syndrome [24]. Japanese diets are also rich in fruits, vegetables, but also include soybean, green tea, and seaweed. The Mediterranean and Japanese diets share some common characteristics. They include fish and are low in saturated fats, meats, and cholesterol. The DASH diet emphasizes fruits, vegetables, and low-fat dairy products; includes whole grains, poultry, fish, and nuts; and is lower in red meat, sweets, and sugar-sweetened beverages [22]. Despite the documented benefits of diets rich in whole grains, fruits, vegetables, nuts, legumes and seeds, fish and shellfish, and dietary fiber, the majority of US adults fail to meet the daily recommendations for such foods [22].

Diet and Nutrients

Atherosclerosis involves a process mediated in part by oxidative stress and inflammation. Diets rich in calories lead to obesity, insulin resistance, systemic inflammation, increased risk of diabetes, hypertension, and atherosclerotic cardiovascular disease. Whereas diets rich in antioxidants including vitamins C, E, and A that presumably counter the free radicals and prevent oxidative injury to cells might offer protection from the development of atherosclerosis.

Diets rich in fruits, vegetables, whole grains, nuts, and fish reduced the incidence of CHD [22]. Associations with milk and dairy findings were mixed. Eggs neither benefited nor worsened risk of CHD. Whereas, diets rich in meats were associated with high incidence of CHD, and processed meats appeared worse [22].

Fruits and vegetables are a natural source of dietary antioxidants. Because oxidative stress is postulated to underlie atherogenesis, dietary intake of fruits and vegetables is expected to reduce the risk of cardiovascular disease. Epidemiological data demonstrating the protective effects of fruits and vegetables against CHD are limited [25]. Yet, the health benefits of antioxidants observed in epidemiological studies have not been substantiated in interventional trials [26], and antioxidants might even be harmful. For example, in the large prospective randomized placebo-controlled Women's Health Study and the Physicians' Health Study II, vitamins E did not improve risk for cardiovascular events or mortality [27, 28]. Likewise long-term supplementation with vitamin C in the Physicians' Health Study II [28] and betacarotene [29] in the Physicians' Health Study did not improve cardiovascular disease risk or mortality.

Hypovitaminosis D has been associated with endothelial dysfunction, hypertension, atherosclerosis, peripheral vascular disease, diabetes, myocardial infarction, stroke, and congestive heart failure [30]. Yet, studies designed to determine the benefit of vitamin D replacement on cardiovascular disease have produced outcomes that are mixed emphasizing the need for randomized controlled trials to establish whether vitamin D supplementation will slow the progression of atherosclerosis and reduce cardiovascular events [31, 32].

Genomic processes affect atherogenesis. Recently, epigenetic influences have been studied. Epigenetic effects are modulated, in part, by methylation of DNA and histones and micro-RNA-mediated effects on mRNA and protein expression. Lifestyle-related risk factors have been associated with methylation of DNA. Numerous nutrients such as folate, polyphenols, and flavonoids may serve as a source for DNA methylation. Such changes modulate genomic stability and transcription of key pathways [33]. Epigenetic research appears to be a promising field that is likely to contribute significantly to understanding the role of lifestyle behaviors and their interaction with more traditional risk factors and genetic determinants of cardiovascular risk.

Smoking

Mainstream tobacco smoke Smoking increases the development of atherosclerosis, triggers myocardial infarction, and increases cardiovascular morbidity and mortality. Tobacco smoke contains many pro-oxidants and reactive free radicals, promoting chronic inflammation and vascular injury. CAC images show the acceleration of vascular aging among smokers when compared with nonsmokers. Measurements of carotid intima-media thickness (CIMT), as a measure of carotid artery atherosclerosis showed that smokers develop atherosclerosis at a younger age [34]. In the Heinz Nixdorf Recall Study, a 10-year difference in vascular age based on the extent of CAC was found between middle-aged never-smokers and smokers. Discontinuation of smoking slowed the progression of CAC, but it did not affect the burden of calcium accumulated previously. Of note is that traditional risk factors for the development of atherosclerosis such as blood pressure, LDL cholesterol, BMI, and diabetes contributed to only a small share of the risk among smokers to the CAC in this population [35]. Positive associations observed between the degree of CAC and fibrinogen further support a mechanistic link between chronic inflammation and progression of atherosclerosis [35].

ENVIRONMENTAL EFFECTS

Life Stress and Alcohol Use

Stress is a maladaptive response to real or perceived threats and can activate the sympathetic nervous system and modify known risk factors for atherosclerosis such as hypertension via the rennin–angiotensin system, as well as insulin resistance, platelet aggregation, adiposity, inflammation, shear stress, and vascular remodeling [36]. Life stress arising from psychosocial stress is implicated as a risk factor for the development of cardiovascular disease and poor long-term outcomes [37]. Recently, the relationship between measures of chronic life stress and unhealthy lifestyle characteristics was assessed in the MESA Study cohort without cardiovascular disease [38]. This study suggests that unhealthy coping behaviors to chronic life stress such as smoking, excessive alcohol use, high caloric intake, sedentary lifestyle, and obesity rather than life stress itself contribute to subclinical atherosclerosis measured by CAC. Individual personality and coping strategies might determine responses to external stressors and possibly predict cardiovascular risk. Such influences are likely to be important determinants of risk in populations more susceptible to social stressors such as minority populations, the impoverished, and those of low educational attainment.

Alcohol use contributes significantly to the global burden of disease and can indicate a maladaptive coping strategy for stress with attendant increases in blood pressure, triglycerides, and cardiovascular risk. While epidemiological studies show adverse health effects of primarily heavy alcohol use, low-to-moderate levels of alcohol ingestion are associated with cardiovascular health benefits. Low-to-moderate alcohol consumption is associated with decreased risk of cardiovascular disease mortality, incident ischemic heart disease, and incident stroke [39], type 2 diabetes [40], and progression of coronary atherosclerosis [41]. The beneficial effects of moderate alcohol consumption might be related to higher HDL and adiponectin [42] improved insulin resistance [22], and lower levels of pro-inflammatory cytokines, IL-6 and TNF-alpha [43], and other inflammatory markers such as white blood cell count and fibrinogen [42, 44].

Secondhand Tobacco Smoke

Exposure to secondhand smoke remains an important public health problem. Like mainstream tobacco smoke, secondhand smoke contains particulate matter (PM), air toxins, and carcinogens known to have adverse long-term health effects. In a cross-sectional study, the association of secondhand smoke on CAC was assessed in never smokers between 45 and 75 years of age and without clinical evidence of CHD. The history of secondhand tobacco exposure was acknowledged in 21.1% of the cohort. A person exposed to secondhand smoke was more likely to have a calcium score greater than zero indicating subclinical atherosclerosis. After adjustment for other CV risk factors, the strength of the association decreased, yet it was still 15.4% greater in those exposed compared to those not exposed [45].

Secondhand tobacco smoke exposure is a risk factor for children and perhaps the fetus, as early life exposures to secondhand smoke contribute to risk for asthma. Yet, early life exposures can also adversely impact the cardiovascular system. Early life exposures to passive smoking has been associated with decreased aortic elastic properties [46], altered markers of lipid metabolism and inflammation [47], increased common carotid artery intima-media thickness [48], and decreased flow-mediated dilation [49].

Ambient Air Pollution

Short- and long-term exposures to air particle pollution cause adverse health effects and increase cardiopulmonary mortality. Epidemiological studies show *short-term* daily exposures to airborne PM, a heterogeneous collection of solid and liquid particles largely formed from products of combustion, associate with acute coronary syndrome, ischemic stroke, exacerbation of heart failure, and possibly arrhythmia. A number of epidemiological studies provide evidence that *long-term* exposure to PM, and traffic-related or combustion-related air pollution at ambient levels is associated with increased ischemic heart disease, vascular disease, and cardiovascular mortality. Progressive reductions in ambient PM levels throughout the United States since the implementation of the Clean Air Act have resulted in less PM-related cardiopulmonary mortality and all-cause mortality [50]. Continued, yet smaller reductions in annual average PM levels achieved between 2000 and 2007 have also been associated with increased longevity [51].

Short-term clinical effects, those occurring within minutes to hours of exposure are likely caused by inhaled PM activating pulmonary chemoreceptors and altering autonomic balance, producing systemic inflammation, activating endothelial cells, modulating endothelial function, and altering hemostasis, thrombosis, and fibrinolysis [50] as shown in Figure 8.1.

Long-term clinical outcomes caused by PM inhalation are likely related to activation of the sympathetic nervous system, systemic inflammation, endothelial dysfunction, and the development and progression of atherosclerosis [50]. Noninvasive vascular function [52] and imaging studies [53, 54] demonstrate that endothelial dysfunction and carotid artery intima-media thickness, a surrogate marker of atherosclerosis, are proportional to the annual levels of ambient PM exposure.

The Veterans Administration Normative Aging Study has served as a rich source of human data to investigate relationships between air pollutant exposure and health effects. Of particular interest are those effects that bear directly to either risk factors or signaling pathways driving atherogenesis. For example, estimates of long-term exposure to black carbon, a surrogate marker of traffic-related air pollution, was found in the Boston area to be associated with higher blood pressure. A recent prospective population-based cohort study of middle-aged adults conducted in Europe, the German Heinz Nixdorf Recall study showed that long-term residential exposure to traffic and PM increases atherosclerosis as assessed by CAC [55]. Among this cohort individuals living within 100 m of a major road were more likely to have high CAC when compared to individuals living more than 200 m from a major road. In the ARIC Study, long-term exposure to traffic-related air pollution was associated with an increase in the development of CHD in a cohort of middle-aged men and women without CHD. Smoking status, sex, obesity, LDL, hypertension, and education did not modify the effect [56].

The previous studies show that *long-term* exposure to ambient PM is probably associated with the progression of atherosclerosis and vascular calcium deposition. The mechanisms accounting for these observations are now under study. The MESA Air Pollution Study (MESA Air) is an ancillary study to MESA and

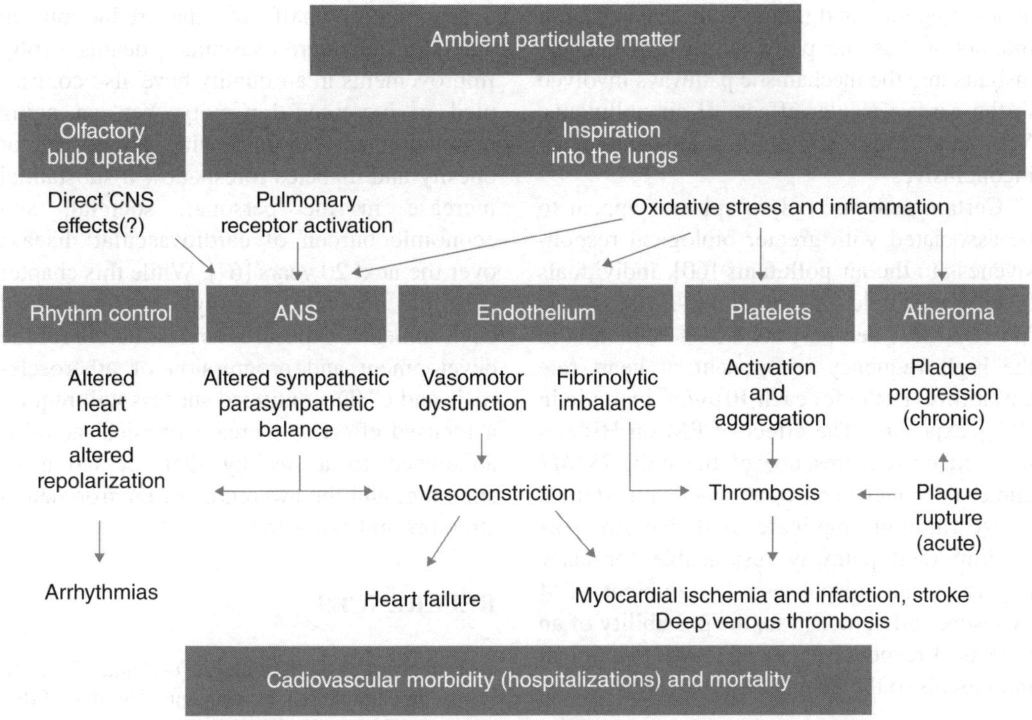

Figure 8.1 Possible mechanisms by which ambient particulate matter cause acute and chronic cardiovascular disease and clinical events. ANS, autonomic nervous system.

was designed to investigate the relationship between long-term exposure to ambient $PM_{2.5}$ on the development and progression of subclinical atherosclerosis and cardiovascular events in a healthy ethnically diverse middle-aged population [57]. The postulated mechanisms include oxidative stress, activation of the innate and adaptive immune system, or the actions of macrophages and dendritic cells that can sequester particles and subsequently leave the lung through the systemic circulation where they might affect signaling in lymph nodes and blood vessels. The smallest of inhaled ambient air particles may translocate from the lungs to the circulation or traverse the blood–brain barrier and enter the brain where they could have direct effects [58]. In MESA Air, *long-term* exposure to PM was associated with endothelial dysfunction as evidenced by a 0.3% decrease in flow-mediated dilation for each 3 μg/m^3 increase in $PM_{2.5}$ after adjustments for traditional risk factors, and other covariants [52]. These data provide new insights into the long-term risk for the development of atherosclerosis.

The relationship between air pollutant exposure and DNA methylation of tissue factor, Toll-like receptor 2 (TLR-2), and ICAM-1 was measured in the Normative Aging Study [59]. Interestingly, different types of pollutants affected different pathways. Pollutants generally associated with traffic, that is, particle number, black carbon, NO_2, and CO were associated with an increase in fibrinogen. The secondary pollutant ozone was associated with increased CRP and ICAM-1. Particle number, black carbon, NO_2, CO, $PM_{2.5}$, and sulfates were associated with increased ICAM-1 and VCAM-1 [59] biomarkers associated with cardiovascular risk. Reduced methylation of DNA encoding tissue factor, TLR-2 yielded stronger effects of air pollution implicating epigenetic states as a determinant of susceptibility to air pollutants. Study of gene-, epigene-, and metabolome-environment interaction has the potential to provide new insights into the mechanistic pathways involved in the cardiovascular effects of air pollutants. Yet, data from such studies are limited or inconclusive.

Certain genomic polymorphisms appear to be associated with greater biological responsiveness to the air pollutants [60]. Individuals lacking the allele for glutathione-*S*-transferase M1 (*GSTM1*) demonstrate a 34% reduction in the high-frequency component of heart rate variability (HRV) for each 10 μg/m³ increase in $PM_{2.5}$ exposure. The effect of PM on HRV is blocked by the presence of the ~~null~~ *GSTM1* allele or concurrent treatment with statins. These findings implicate oxidative stress as an important pathway responsible for early physiological responses to ambient PM exposure and demonstrate the possibility of an increased responsiveness of *GSTM1*-deficient individuals to the effects of inhaled PM.

Other Environmental Factors

Other environmental exposures are known to drive vascular disease and atherosclerosis. The oxidative properties of arsenic contribute to vascular disease, hypertension, and atherosclerosis [61]. Environmental lead exposure is associated with hypertension, and positively but less convincingly associated with cardiovascular outcomes such as coronary artery disease, peripheral vascular disease, and stroke mortality [62]. Exposure to mercury might be associated with cardiovascular disease [63]. Endocrine disruptors and pesticides might contribute indirectly to atherogenesis through modulation of cardiovascular risk factors such as obesity, metabolic syndrome, and diabetes [64, 65].

SUMMARY

Over the past 30 years, improvement of lifestyle behaviors, namely reductions in smoking and increased physical activity account for approximately half of the reduction in age-adjusted cardiovascular deaths [66]. Improvements in air quality have also contributed to increased longevity. Yet, an aging population and an increasing prevalence of obesity and diabetes foreshadow a substantial increase in the personal, societal, and economic burden of cardiovascular disease over the next 20 years [67]. While this chapter describes individual lifestyle behaviors and environmental exposures as risk factors for the development and progression of atherosclerosis and CHD, continued success will require a focused effort to increase physical activity, adherence to a healthy diet, cessation of smoking, and the awareness of environmental stressors and exposures.

REFERENCES

1. Lim SS, Vos T, Flaxman AD, Danaei G, et al. A comparative risk assessment of burden of disease and injury attributable to 67 risk factors and risk factor clusters in 21 regions, 1990–2010: a systematic analysis for the Global Burden of Disease Study 2010. *Lancet* 2012;380:2224–60.
2. Haveman-Nies A, de Groot L, Burema J, et al. Dietary quality and lifestyle factors in relation to 10-year mortality in older Europeans: the SENECA study. *Am J Epidemiol* 2002;156:962–8.
3. Chakravarty EF, Hubert HB, Krishnan E, Bruce BB, Lingala VB, Fries JF. Lifestyle risk factors predict disability and death in healthy aging adults. *Am J Med* 2012;125:190–7.
4. Yates LB, Djoussé L, Kurth T, Buring JE, Gaziano JM. Exceptional longevity in men: modifiable factors associated with survival and function to age 90 years. *Arch Intern Med* 2008;168:284–90.
5. Buxton OM, Marcelli E. Short and long sleep are positively associated with obesity, diabetes, hypertension, and cardiovascular disease among adults in the United States. *Soc Sci Med* 2010;71:1027–36.
6. Cappuccio FP, D'Elia L, Strazzullo P, Miller MA. Quantity and quality of sleep and incidence of type 2 diabetes: a systematic review and meta-analysis. *Diabetes Care* 2010;33:414–20.
7. Punjabi NM, Shahar E, Redline S, et al. Sleep-disordered breathing, glucose intolerance, and insultin resistance: the Sleep Heart Health Study. *Am J Epidemiol* 2004;160:521–30.

8. Buxton OM, Cain SW, O'Connor SP, et al. Adverse metabolic consequences in humans of prolonged sleep restriction combined with circadian disruption. *Sci Transl Med* 2012;4: 129ra43.
9. Juonala M, Viikari JS, Raitakari OT. Main findings from the prospective Cardiovascular Risk in Young Finns Study. *Curr Opin Lipidol* 2012;24(1):57–64.
10. Sofi F, Capalbo A, Cesari F, Abbate R, Gensini GF. Physical activity during leisure time and primary prevention of coronary heart disease: an updated meta-analysis of cohort studies. *Eur J Cardiovasc Prev Rehabil* 2008;15:247–57.
11. Sung J, Cho SJ, Choe YH, Choi YH, Hong KP. Prevalence of coronary atherosclerosis in asymptomatic middle-age men with high aerobic fitness. *Am J Cardiol* 2012;109:839–43.
12. Simprini LA, Villines TC, Rich M, Taylor AJ. The relationship between subclinical atherosclerosis, non-high-density lipoprotein cholesterol, exercise, and diet among male participants of the PACC Project. *J Clin Lipidol* 2012;6:174–9.
13. Lee DC, Sui X, Artero EG, et al. Long-term effects of changes in cardiorespiratory fitness and body mass index on all-cause and cardiovascular disease mortality in men: the Aerobics Center Longitudinal Study. *Circulation* 2011;124: 2483–90.
14. Bowles DK, Laughlin MH. Mechanism of beneficial effects of physical activity on atherosclerosis and coronary heart disease. *J Appl Physiol* 2011;111:308–310.
15. Hamer M, Endrighi R, Poole L. Physical activity, stress reduction, and mood: insight into immunological mechanisms. *Methods Mol Biol* 2012;934:89–102.
16. Allison MA, Jensky NE, Marshall SJ, Bertoni AG, Cushman M. Sedentary behavior and adiposity-associated inflammation: the Multi-Ethnic Study of Atherosclerosis. *Am J Prev Med* 2012;42:8–13.
17. Ahmed HM, Blaha MJ, Nasir K, Rivera JJ, Blumenthal RS. Effects of physical activity on cardiovascular disease. *Am J Cardiol* 2012;109: 288–95.
18. Kraus WE, Houmard JA, Duscha BD, et al. Effects of the amount and intensity of exercise on plasma lipoproteins. *N Engl J Med* 2002;347: 1483–92.
19. Carroll S, Dudfield M. What is the relationship between exercise and metabolic abnormalities? A review of the metabolic syndrome. *Sports Med* 2004;34:371–418.
20. Kraus WE, Slentz CA. Exercise training, lipid regulation, and insulin action: a tangled web of cause and effect. *Obesity (Silver Spring)* 2009;17 Suppl 3:S21–6.
21. Barone Gibbs B, Dobrosielski DA, Bonekamp S, Stewart KJ, Clark JM. A randomized trial of exercise for blood pressure reduction in type 2 diabetes: effect on flow-mediated dilation and circulating biomarkers of endothelial function. *Atherosclerosis* 2012;224:446–53.
22. Mozaffarian D, Appel LJ, Van Horn L. Components of a cardioprotective diet: new insights. *Circulation* 2011;123:2870–91.
23. Nettleton JA, Matijevic N, Follis JL, Folsom AR, Boerwinkle E. Associations between dietary patterns and flow cytometry-measured biomarkers of inflammation and cellular activation in the Atherosclerosis Risk in Communities (ARIC) Carotid Artery MRI Study. *Atherosclerosis* 2010;212:260–7.
24. Kastorini CM, Milionis HJ, Esposito K, Giugliano D, Goudevenos JA, Panagiotakos DB. The effect of Mediterranean diet on metabolic syndrome and its components: a meta-analysis of 50 studies and 534,906 individuals. *J Am Coll Cardiol* 2011;57:1299–313.
25. Dauchet L, Amouyel P, Dallongeville J. Fruits, vegetables and coronary heart disease. *Nat Rev Cardiol* 2009;6:599–608.
26. Núñez-Córdoba JM, Martínez-González MA. Antioxidant vitamins and cardiovascular disease. *Curr Top Med Chem* 2011;11:1861–9.
27. Lee IM, Cook NR, Gaziano JM, et al. Vitamin E in the primary prevention of cardiovascular disease and cancer: the Women's Health Study: a randomized controlled trial. *JAMA* 2005; 294:56–65.
28. Sesso HD, Buring JE, Christen WG, et al. Vitamins E and C in the prevention of cardiovascular disease in men: the Physicians' Health Study II randomized controlled trial. *JAMA* 2008;300:2123–33.
29. Hennekens CH, Buring JE, Manson JE, et al. Lack of effect of long-term supplementation with beta-carotene on the incidence of malignant neoplasms and cardiovascular disease. *N Eng J Med* 1996;334:1145–9.
30. McGreevy C, Williams D. New insights about vitamin D and cardiovascular disease: a narrative review. *Ann Intern Med* 2011;155:820–6.

31. Brewer LC, Michos ED, Reis JP. Vitamin D in atherosclerosis, vascular disease, and endothelial function. *Curr Drug Targets* 2011;12:54–60.
32. Lavie CJ, Lee JH, Milani RV. Vitamin D and cardiovascular disease will it live up to its hype? *J Am Coll Cardiol* 2011;58:1547–56.
33. Lim U, Song MA. Dietary and lifestyle factors of DNA methylation. *Methods Mol Biol* 2012;863:359–76.
34. Giudice R, Izzo R, Manzi MV, et al. Lifestyle-related risk factors, smoking status and cardiovascular disease. *High Blood Press Cardiovasc Prev* 2012;19:85–92.
35. Jöckel KH, Lehmann N, Jäger BR, et al. Smoking cessation and subclinical atherosclerosis – results from the Heinz Nixdorf Recall Study. *Atherosclerosis* 2009;203:221–227.
36. Hamer M, Malan L. Psychophysiological risk markers of cardiovascular disease. *Neurosci Biobehav Rev* 2010;35:76–83.
37. Brotman DJ, Golden SH, Wittstein IS. The cardiovascular toll of stress. *Lancet* 2007;370: 1089–100.
38. Mainous AG 3rd, Everett CJ, Diaz VA, Player MS, Gebregziabher M, Smith DW. Life stress and atherosclerosis: a pathway through unhealthy lifestyle. *Int J Psychiatry Med* 2010;40:147–61.
39. Ronksley PE, Brien SE, Turner BJ, Mukamal KJ, Ghali WA. Association of alcohol consumption with selected cardiovascular disease outcomes: a systematic review and meta-analysis. *BMJ* 2011;342:d671.
40. Roerecke M, Rehm J. Alcohol intake revisited: risks and benefits. *Curr Atheroscler Rep* 2012; 14:556–62.
41. Janszky I, Mukamal KJ, Orth-Gomér K, et al. Alcohol consumption and coronary atherosclerosis progression–the Stockholm Female Coronary Risk Angiographic Study. *Atherosclerosis* 2004;176:311–9.
42. Brien SE, Ronksley PE, Turner BJ, Mukamal KJ, Ghali WA. Effect of alcohol consumption on biological markers associated with risk of coronary heart disease: systematic review and meta-analysis of interventional studies. *BMJ* 2011; 342:d636.
43. Marques-Vidal P, Bochud M, Bastardot F, et al. Associations between alcohol consumption and selected cytokines in a Swiss population-based sample (CoLaus study). *Atherosclerosis* 2012; 222:245–50.
44. Mukamal KJ, Cushman M, Mittleman MA, Tracy RP, Siscovick DS. Alcohol consumption and inflammatory markers in older adults: the Cardiovascular Health Study. *Atherosclerosis* 2004;173:79–87.
45. Peinemann F, Moebus S, Dragano N, et al. Secondhand smoke exposure and coronary artery calcification among nonsmoking participants of a population-based cohort. *Environ Health Perspect* 2011;119:1556–61.
46. Kallio K, Jokinen E, Hämäläinen M, et al. Decreased aortic elasticity in healthy 11-year-old children exposed to tobacco smoke. *Pediatrics* 2009;123:e267–73.
47. Nagel G, Arnold FJ, Wilhelm M, Link B, Zoellner I, Koenig W. Environmental tobacco smoke and cardiometabolic risk in young children: results from a survey in south-west Germany. *Eur Heart J* 2009;30:1885–93.
48. Geerts CC, Bots ML, Grobbee DE, Uiterwaal CS. Parental smoking and vascular damage in young adult offspring: is early life exposure critical? The atherosclerosis risk in young adults study. *Arterioscler Thromb Vasc Biol* 2008;28:2296–302.
49. Kallio K, Jokinen E, Saarinen M, et al. Arterial intima-media thickness, endothelial function, and apolipoproteins in adolescents frequently exposed to tobacco smoke. *Circ Cardiovasc Qual Outcomes* 2010;3:196–203.
50. Brook RD, Rajagopalan S, Pope CA 3rd, et al. Particulate matter air pollution and cardiovascular disease: an update to the scientific statement from the American Heart Association. *Circulation* 2010;121:2331–78.
51. Correia AW, Pope CA III, Dockery DW, Wang Y, Ezzati M, Dominici F. Effect of air pollution control on life expectancy in the United States. An analysis of 545 U.S. Counties for the period from 2000 to 2007. *Epidemiology* 2013;24:1–9.
52. Krishnan RM, Adar SD, Szpiro AA, et al. Vascular responses to long- and short-term exposure to fine particulate matter: MESA Air (Multi-Ethnic Study of Atherosclerosis and Air Pollution). *J Am Coll Cardiol* 2012;60:2158–66.
53. Bauer M, Moebus S, Möhlenkamp S, et al. Urban particulate matter air pollution is associated with subclinical atherosclerosis: results from the HNR (Heinz Nixdorf Recall) study. *J Am Coll Cardiol* 2010;56:1803–8.

54. Künzli N, Jerrett M, Mack WJ, et al. Ambient air pollution and atherosclerosis in Los Angeles. *Environ Health Perspect* 2005;113:201–6.
55. Hoffmann B, Moebus S, Möhlenkamp S, et al. Residential exposure to traffic is associated with coronary atherosclerosis. *Circulation* 2007; 116:489–96.
56. Kan H, Heiss G, Rose KM, Whitsel EA, Lurmann F, London SJ. Prospective analysis of traffic exposure as a risk factor for incident coronary heart disease: the Atherosclerosis Risk in Communities (ARIC) study. *Environ Health Perspect* 2008;116:1463–8.
57. Kaufman JD, Adar SD, Allen RW, et al. Prospective study of particulate air pollution exposures, subclinical atherosclerosis, and clinical cardiovascular disease: the Multi-Ethnic Study of Atherosclerosis and Air Pollution (MESA Air). *Am J Epidemiol* 2012;176: 825–37.
58. Campen MJ, Lund A, Rosenfeld M. Mechanisms linking traffic-related air pollution and atherosclerosis. *Curr Opin Pulm Med* 2012; 18:155–60.
59. Bind MA, Baccarelli A, Zanobetti A, et al. Air pollution and markers of coagulation, inflammation, and endothelial function: associations and epigene-environment interactions in an elderly cohort. *Epidemiology* 2012;23:332–40.
60. Zanobetti A, Baccarelli A, Schwartz J. Gene-air pollution interaction and cardiovascular disease: a review. *Prog Cardiovasc Dis* 2011; 53:344–52.
61. Jomova K, Jenisova Z, Feszterova M, et al. Arsenic: toxicity, oxidative stress and human disease. *J Appl Toxicol* 2011;31:95–107.
62. Navas-Acien A, Guallar E, Silbergeld EK, Rothenberg SJ. Lead exposure and cardiovascular disease – a systematic review. *Environ Health Perspect* 2007;115:472–82.
63. Virtanen JK, Rissanen TH, Voutilainen S, Tuomainen TP. Mercury as a risk factor for cardiovascular diseases. *J Nutr Biochem* 2007;18:75–85.
64. De Coster S, van Larebeke N. Endocrine-disrupting chemicals: associated disorders and mechanisms of action. *J Environ Public Health* 2012:713696.
65. Thayer KA, Heindel JJ, Bucher JR, Gallo MA. Role of environmental chemicals in diabetes and obesity: a National Toxicology Program workshop review. *Environ Health Perspect* 2012;120:779–89.
66. Ford ES, Ajani UA, Croft JB, et al. Explaining the decrease in U.S. deaths from coronary disease, 1980-2000. *N Engl J Med* 2007;356: 2388–98.
67. Heidenreich PA, Trogdon JG, Khavjou OA, et al. Forecasting the future of cardiovascular disease in the United States: a policy statement from the American Heart Association. *Circulation* 2011;123:933–44.

SECTION II

CELLULAR MECHANISMS

9

ENDOTHELIAL CELLS

G. Brandon Atkins[1,2], Gabriela Orasanu[1,2], and Mukesh K. Jain[1,2]
[1] *Department of Medicine, Case Cardiovascular Research Institute, University Hospitals Case Medical Center, Case Western Reserve University, Cleveland, OH, USA*
[2] *University Hospitals Harrington Heart and Vascular Institute, Cleveland, OH, USA*

INTRODUCTION

The vascular endothelium is a vital organ that has been demonstrated to play a critical role in regulating vascular health and disease. It consists of a single layer of endothelial cells that line the blood vessel lumen and therefore is uniquely positioned to transduce and integrate both biochemical (e.g., inflammatory cytokines) and biomechanical (e.g., shear stress) stimuli from the blood to the underlying blood vessel wall. The primary homeostatic functions of the normal endothelium include regulation of selective permeability, blood coagulation, and homing of immune cells to specific sites of the body. Significant alteration of these endothelial homeostatic functions is a major factor in the process of atherosclerosis, contributing to the progression of this disease and its attendant complications. This chapter will focus on highlighting some of the more recent major advancements in the study of endothelial cells and their role in the development of atherosclerosis.

ENDOTHELIAL NITRIC OXIDE

Furchgott and colleagues made the initial discovery that endothelial cells release a factor that relaxes vascular smooth muscle cells resulting in vasodilation [1]. This factor initially called endothelium-derived relaxing factor (EDRF) was subsequently identified to be the gas, nitric oxide (NO) [2]. Since those initial discoveries, NO has come to be recognized as the "third gas" of the respiratory cycle and is increasingly understood to be as important as O_2 and CO_2 in normal physiology and pathophysiology of disease [3].

Endothelial Dysfunction

NO in the vasculature is predominantly produced in endothelial cells by endothelial nitric oxide synthase (eNOS), which uses nicotinamide adenine dinucleotide phosphate (NADPH), tetrahydrobiopterin, and O_2 to convert L-arginine to L-citrulline resulting in the release of NO. Until recently, endothelial NO's primary mechanism

Atherosclerosis: Risks, Mechanisms, and Therapies, First Edition. Edited by Hong Wang and Cam Patterson.

of action in the vasculature was believed to be through diffusion to the underlying smooth muscle cells and binding to its intracellular receptor, soluble guanylyl cyclase (sGC). NO's binding to sGC results in an increase in cyclic guanosine monophosphate (cGMP) and activation of protein kinase (PK)G with subsequent smooth muscle cell relaxation, ultimately resulting in vasodilation and improved blood flow. Impaired endothelium-dependent (i.e., NO-dependent) vasodilation, through a defect in NO production or activity, is the hallmark of endothelial dysfunction and has been proposed as a major a contributor to atherosclerosis [4]. Ludmer et al. made the original discovery, using invasive methods, that endothelial dysfunction in the setting of atherosclerosis can be clinically assessed [5]. Since this finding, several more recent studies demonstrate that noninvasive clinical assessment of endothelial dysfunction in conduit arteries using brachial artery flow-mediated dilation (FMD) correlates with atherosclerosis and predicts cardiovascular disease events [6].

S-Nitrosylation

More recently, the effects of NO beyond cGMP-dependent vasodilation have begun to be recognized. eNOS has been demonstrated to be an anti-inflammatory, antiadhesive, and antithrombotic factor in the endothelium. In an effort to identify other molecular mechanisms by which NO may exert its effects, a novel post-translational protein modification process, called *S*-nitrosylation, was discovered to be the likely factor responsible for these cGMP-independent effects. *S*-nitrosylation is the covalent modification of a protein cysteine thiol by an NO group to generate *S*-nitrosothiol (SNO) [7]. As with phosphorylation, numerous proteins can be *S*-nitrosylated, thereby explaining the multiple effects of NO beyond the cGMP axis. Notably, *S*-nitrosylation can both activate and inhibit many NO regulatory factors, including eNOS itself, thereby making major contributions to the mechanisms of endothelial dysfunction [8].

In vivo evidence indicates that protein *S*-nitrosylation is not only dependent on the rates of NO production via NO synthases but rather is highly dependent on the delicate balance between *S*-nitrosylation and denitrosylation pathways [9]. Specifically, this balance between *S*-nitrosylation and denitrosylation has been demonstrated to play key roles in atherosclerosis via regulation of superoxide production [10, 11], endothelial protein trafficking, and proinflammatory cytokine, and adhesion molecule expression [12]. Interestingly, statins have been proposed to exhibit their vasculoprotective effects in the endothelium by an *S*-nitrosylation mechanism [13].

FLOW AND ARTERIAL SHEAR STRESS

It has long been recognized that the earliest lesions in atherosclerosis occur in a nonuniform fashion with lesions occurring at branch points and areas of major curvature in the vasculature. These initial observations led to the hypothesis that local events and/or intrinsic differences in endothelial cells at specific regions of the vascular tree may contribute to the focal nature of vascular disease. Accumulating evidence suggests various forms of shear stress can differentially modulate various aspects of endothelial cell function. Sustained laminar flow with high shear stress (seen in unbranched portions of vessels) upregulates expression of atheroprotective endothelial cell factors, whereas disturbed or turbulent flow with reciprocal low shear stress (seen near branch points and major curves) upregulates endothelial cell factors that promote atherogenesis [14].

Mechanosensors and Intracellular Signaling Pathways

Over the past decade, research has led to a better understanding of the factors that are positioned on the endothelial cell surface and sense shear stress as well as the intracellular signaling pathways involved in shear stress-mediated regulation of endothelial cell function. One notable discovery was the identification of a mechanosensory complex consisting of platelet endothelial cell adhesion molecule (PECAM)1 (which directly transmits mechanical force), vascular endothelial cell (VE) cadherin (which functions as an adaptor), and vascular endothelial cell growth factor receptor (VEGFR)2 (which activates phosphatidylinositol-3-OH kinase). This complex has been demonstrated to be sufficient to confer responsiveness to flow *in vitro* and required *in vivo* for activation of nuclear factor kappa-B (NFκB) and downstream genes in regions of disturbed flow, one of the earliest events in atherogenesis [15]. Interestingly, endothelial cell primary cilium, an important component of the endothelial cell apical membrane, has recently been shown to contain mechanosensing proteins that may play roles in regulating endothelial cell response to disturbed flow [16]. In addition to the previously characterized flow-mediated induction of the mitogen/extracellular signal-regulated kinasekinase (MEK)5/extracellular signal-regulated kinase (ERK)5/myocyte enhancer factor (MEF)2 intracellular signaling pathway, recent evidence demonstrates the requirement of adenosine monophosphate (AMP)-activated protein kinase (AMPK) for the activation of ERK5/MEF2, implicating AMPK as an upstream signaling molecule of this pathway, thereby playing a significant role in the regulation of vascular homeostasis [17].

Kruppel-Like Factor 2

Kruppel-like factor 2 (KLF2) is a member of the zinc finger family of transcription factors that is highly expressed in endothelial cells and has been implicated as a "molecular switch" regulating endothelial function in health and disease [18]. Studies have shown KLF2 to have the ability to differentially regulate the expression of factors that confer anti-inflammatory, antithrombotic, and antiproliferative effects in endothelial cells. Notably, KLF2 strongly upregulates eNOS and thrombomodulin (TM) and inhibits cytokine-mediated induction of cell adhesion molecules. KLF2 is upregulated by laminar shear stress and inhibited by turbulent shear stress and inflammatory cytokines. KLF2 has been shown to be regulated by MEF2 and the well-characterized MEK5/ERK5/MEF2 flow pathway and more recently identified AMPK/ERK5/MEF2 pathway [17], thereby establishing a molecular link between flow and its downstream effects in endothelial cells. Atkins and colleagues were the first to establish KLF2 as an antiatherogenic factor *in vivo*, demonstrating that KLF2 hemizygous mice have increased plaque burden in the apolipoprotein (apo)E-deficient model of atherosclerosis [19]. Interestingly, the statin effects in the vasculature have been shown to be KLF2 dependent [20]. Similar to KLF2, KLF4 is another endothelial KLF that has been demonstrated to play an important role in regulating vascular function [18].

Nuclear Factor (Erythroid-Derived 2)-Like-2

Nuclear factor (erythroid-derived 2)-like-2 (Nrf-2) is a shear stress-induced transcription factor in endothelial cells that has been shown to play a role in antioxidant gene expression [21]. The redox state of endothelial cells is regulated through phosphoinositol 3-kinase/Akt-dependent activation of Nrf2 and its downstream transcriptional targets in response to different flow patterns. The nuclear localization of Nrf2 is facilitated by KLF2, and these two factors are involved in the regulation of approximately 70% of the shear stress-induced genes in ECs [22].

Thioredoxin-Interacting Protein

Thioredoxin-interacting protein (TXNIP) plays a pivotal role in cardiovascular disorders, functioning as a sensor for biomechanical and oxidative stress and as regulator of tumor necrosis factor (TNF) signaling and inflammation in the endothelium. TXNIP expression in endothelial cells is low but highly induced in endothelial cells exposed to disturbed flow, inflammatory cytokines, and hyperglycemia. TXNIP binding to thioredoxin (Trx)1 relieves inhibition of apoptosis signal-regulating-kinase (ASK)-1 enabling c-Jun N-terminal kinase (JNK)-dependent induction of vascular cell adhesion molecule (VCAM)-1 [23]. TXNIP–Trx1 complex acts as a redox-sensitive mediator to control VEGFR2 signaling, thereby promoting cell survival under oxidative stress. TXNIP mediates Trx1 translocation to the plasma membrane in response to TNFα: a key mechanism for VEGFR2 transactivation by reactive oxygen species [24]. Physiologic fluid shear stress decreases TXNIP expression and limits proinflammatory events mediated by the TNF–ASK1–JNK/p38 pathway [23].

Histone Deacetylases

Histone deacetylases (HDACs) serve as important mechanosensitive molecules that regulate the endothelial cell response to flow and shear stress. *In vitro* and *in vivo* studies have demonstrated that oscillatory shear stress (or turbulent flow) induces the expression and nuclear accumulation of both class I (HDAC-1/2/3) and class II (HDAC-5/7) HDACs in endothelial cells. Turbulent flow induced the association of HDAC-1/2/3 with Nrf2 and the association of HDAC-3/5/7 with MEF2, leading to downregulation of NADPH dehydrogenasequinone 1 (NQO1) and KLF2 in endothelial cells. By contrast, pulsatile shear stress (or laminar flow) induces phosphorylation-dependent nuclear export of class II HDACs [25]. These results confirm earlier studies that demonstrate that laminar flow induces phosphorylation-dependent nuclear export of HDAC5 and its dissociation from MEF2, leading to inductions of KLF2 and eNOS [26]. These findings have elucidated the roles of different classes of HDACs in modulating vascular endothelial oxidative, inflammatory, and proliferative responses to different flow patterns and shear stress.

Sirtuin1

Sirtuin1 (SIRT1) catalyses the NAD$^+$-dependent deacetylation of ϵ-amino-acetylated lysine residues from protein substrates and represents a key regulator of oxidative stress in the cardiovascular system by targeting multiple redox genes. SIRT1 improves endothelium-dependent vascular function. Expression of SIRT1 is higher in vessel areas under pulsatile flow (thoracic aorta) than in areas under disturbed flow (aortic arch) in C57BL/6 mice. Human umbilical vein endothelial cells (HUVECs) exposed to laminar flow *in vitro* enhance the expression of AMPK and SIRT1 and display increased NAD$^+$ levels and SIRT1 activity, thereby activating eNOS and inducing NO generation [27]. Endothelial overexpression of SIRT1 prevents atherosclerosis by improving vascular function in atherosclerosis-prone apoE$^{-/-}$ mice kept on a high-fat diet [28]. SIRT1 deacetylates RelA/p65 at K310 and suppresses its binding to naked DNA in human aortic endothelial cells (HAECs), thereby interfering with a crucial step in NFκB signaling activation and reducing the expression of endothelial adhesion molecules. Suppression of NFκB signaling by SIRT1 also plays an antiatherogenic role *in vivo* as intraperitoneal injection of lipopolysaccharide (LPS) to boost NFκB signaling leads to increased aortic endothelial expression of intracellular adhesion molecule (ICAM)-1 and VCAM-1 in apoE$^{-/-}$ SIRT1$^{+/-}$ compared to apoE$^{-/-}$ SIRT1$^{+/+}$ mice [29].

Micro-RNA

Micro-RNAs (miRNAs) control the inflammatory responses and are important mediators in vascular remodeling induced by the shear stress, angiotensin II, oxidized low-density lipoprotein, or apoptosis in endothelial cells. miRNAs are noncoding single-stranded RNA molecules of approximately 22 nucleotides that bind to target sites in 3′-untranslated regions (3′-UTRs) of mRNAs to cause their degradation or translation repression [30]. miRNA profiles of endothelial cells revealed that specific miRNAs are enriched in endothelial cells including let-7b, miR-16, miR-21, miR-23a, miR-29, miR-100, miR-126, miR-221, and miR-222 [31].

Turbulent flow induces activator protein (AP)-1-dependent miR-21 expression, which directly targets PPARα to inhibit its expression, thereby allowing activation of AP-1 and the promotion of monocyte contributing to the proinflammatory responses of vascular endothelium under oscillatory shear stress [32]. miR-21 is a regulator of neointima lesion formation and downregulation of aberrantly expressed miR-21 decreased neointima formation in rat carotid artery after angioplasty [33]. MiR-10a is downregulated in endothelial cells from atherosusceptible regions of the aortic arch and aortorenal branches, which leads to activation of the proinflammatory NFκB pathway. Other miRs, such as miR-155, miR-221, and miR-222, play important roles in restricting the proinflammatory response of endothelial cells upon angiotensin II stimulation [34].

MiR-126 is not only specifically expressed in endothelial cells, but it is also one of the most abundant endothelial miRs. miR-126 plasma levels were found to be decreased in patients with coronary atherosclerosis, which can be explained by impaired packaging of miR-126 in microvesicles [34]. Furthermore, miR-126 reduces leukocyte adhesion to endothelial cells by suppressing the expression of VCAM-1, suggesting a crucial role in vascular inflammation [35]. MiR-126 was also shown to inhibit atherosclerosis in mice through the regulator of G-protein signaling 16 (RGS16) that results in an increased expression of C–X–C chemokine receptor type (CXCR)4 in endothelial cells that leads to plaque stabilization [36].

Recent studies have demonstrated the importance of the KLF2-miR axis in endothelial cells. miR-92a is downregulated by atheroprotective laminar flow resulting in a subsequent increase in KLF2 expression and augmentation of eNOS and TM, two potent endothelial homeostatic factors [37]. *In vivo* studies confirm miR-92a's ability to regulate KLF2 as well as KLF4 in a site-specific manner in the atherosusceptible endothelium [38]. Interestingly, endothelial KLF2 is necessary for shear stress-mediated upregulation and packaging of miR-143/145 into extracellular vesicles for secretion to vascular smooth muscle cells and reduction of atherosclerotic lesion formation. These findings link shear-responsive endothelial KLF2 expression to the regulation of vascular smooth muscle cell phenotype in atherosclerosis [39].

PEROXISOME PROLIFERATOR-ACTIVATED RECEPTOR NUCLEAR RECEPTOR FAMILY

Peroxisome proliferator-activated receptors (PPARs) are key regulators of inflammation, fatty acid and lipoprotein metabolism, glucose homeostasis, cellular proliferation/differentiation, and the immune response, and have been demonstrated to be important targets in atherosclerosis. There are three distinct PPARs—PPARα, PPARδ/β, and PPARγ—encoded by a separate gene and with specific tissue distribution. All three PPARs are expressed in endothelial cells. Activation of PPAR is initiated by binding of an agonist, either an endogenous ligand (e.g., fatty acids, eicosanoids, or oxidized phospholipids) or synthetic ligand to the ligand-binding domain of the nuclear receptor, which then forms heterodimers with retinoid X receptor (RXR), another nuclear receptor activated by its own ligand (purported to be 9-*cis*-retinoic acid). The PPAR–RXR complex subsequently recognizes and binds to DNA at sequence-specific regions on target gene promoters, known as PPAR response elements (PPREs), thereby activating their expression [40].

PPARα

PPARα agonists inhibit transcriptional expression of VCAM-1 induced by inflammatory cytokines such as TNFα, deriving at least in part from NFκB inhibition. The basal levels of VCAM-1 in microvascular endothelial cells from PPARα-deficient mice are increased as compared to endothelial cells isolated from wild-type mice, implicating PPARα as a tonic inhibitor of inflammation [41]. Endogenous pathways of lipid metabolism, such as hydrolysis of very low-density lipoprotein (VLDL) by lipoprotein lipase or high-density lipoprotein (HDL) hydrolysis by endothelial lipase, that activate PPARα can also limit cytokine-induced VCAM-1 expression [42, 43]. PPARα activation may confer a vasoprotective effect and improve vascular reactivity reported in hypertriglyceridemic human subjects in response to fibrate treatment [44].

PPARγ

PPARγ agonists inhibit inflammation in endothelial cells and suppress the expression of vasoconstrictors endothelin-1 and angiotensin-II, but enhance the expression and activity of vasodilator nitric oxide [45]. Endothelial PPARγ is involved in blood pressure regulation, including the possibility that certain angiotensin receptor blockers may activate PPARγ [46]. PPARγ agonists inhibit the expression of chemokine genes interferon gamma-induced protein (IP)-10, Mig, and interferon-inducible T-cell alpha chemoattractant (I-TAC) in human endothelial cells. PPARγ activation inhibits the expression of proinflammatory adhesion molecules (ICAM-1, VCAM-1, and E-selectin), leading to decreased adherence of monocytes to the activated endothelial cell *in vitro* and *in vivo* [45]. PPARγ-specific agonist pioglitazone represses inflammatory responses involving endothelial VCAM-1 in a PPARα-dependent manner in mice and expands the potential mechanisms of action for this agent [47].

LDL receptor $(LDLR)^{-/-}$ mice with endothelial cell-specific disruption of PPARγ develop atherosclerosis as early as 4 weeks on a high-cholesterol diet, suggesting an important vascular-protective role of endothelial PPARγ against atherogenesis. In these mice disruption of endothelial cell, PPARγ exacerbated dyslipidemia, raised systolic BP, increased endothelial permeability, enhanced expression of proinflammatory cytokines, and increased monocyte/macrophage recruitment and infiltration into vessel walls [48].

PPARδ

In endothelial cells, PPARδ activation reportedly limits adhesion molecule expression through various mechanisms, including protection against oxidative stress, inducing expression of target genes such as thioredoxin and catalase. PPARδ-specific agonist GW0742 was found to inhibit LPS-induced expression of proinflammatory targets, including cyclooxygenase (COX)-2 and inducible nitric oxide synthase (iNOS). Interestingly, unliganded PPARδ sequester BCL6, a repressor of inflammation; whereas in the presence of bound ligand, B-cell lymphoma (BCL)6 is released from PPARδ and able to limit inflammation suggesting BCL6 involvement in PPARδ-mediated inhibition of inflammation. PPARδ may protect the endothelium from shear stress, an effect that may involve PPARγ coactivator (PGC)1 as a PPARδ accessory molecule [41]. PPARδ agonists limited atherosclerosis in apoE-deficient mice as well as in an angiotensin II-accelerated model [49].

ENDOTHELIAL PROGENITOR CELLS

Endothelial progenitor cells (EPCs) play an integral role in the cellular repair mechanisms for endothelial regeneration and maintenance. EPC reduction is considered one mechanism whereby risk factors negatively affect cardiovascular function and promote cardiovascular disease. Classic cardiovascular disease risk factors, smoking, hypertension, hypercholesterolemia, obesity, diabetes, hyperhomocysteinemia, microalbuminuria, inflammation, and insulin resistance have been consistently associated with reduced circulating EPC, while protective factors have been correlated with increased numbers of EPCs, such as HDL, statins, and angiotensin II inhibitors, thiazolidines, granulocyte colony-stimulating factor (G-CSF) and erythropoietin [50, 51].

In patients with angiographically confirmed coronary artery disease, the $CD34^{+}KDR^{+}$ EPCs, as well as $CD133^{+}$ cells and CFU-endothelial cells were predictive of a first major cardiovascular event, independent of potential confounders and that patients with high numbers of EPCs had a reduced risk of revascularization [52]. The subjects with high cell counts have preserved flow-mediated brachial reactivity irrespective of whether they had a high- or low-risk score and, similarly, those with low cell counts had depressed flow-mediated brachial reactivity; thereby, the activity of EPCs is a better predictor of endothelial function than the presence or absence of conventional risk factors [53]. Intracoronary injection of EPCs may improve left ventricular function after acute myocardial infarction [52]. In patients with severe coronary stenosis, those with increased circulating EPCs had better collateral formation compared with those with lower EPC counts [54].

Aging is a major risk factor for the onset and progression of cardiovascular disease, and EPCs are subject to age-associated changes that diminish their number in circulation and function, thereby enhancing vascular disease risk [55]. Chronic TNFα treatment (a key player in the pathogenesis of atherosclerosis) has been found to induce premature senescence of highly proliferative EPCs, which was related to an increase in expression of the cell cycle inhibitor $p16^{INK4a}$ [56].

EPCs can also contribute to atherogenesis by secreting proinflammatory cytokines, inducing smooth muscle cell recruitment to the plaque, and increasing adventitia neovascularization that could be attenuated by pretreatment with statins [57]. Autologous EPCs overexpressing eNOS can rapidly restore endothelial integrity when transplanted into mice after balloon carotid artery injury [58]. Infusion of EPCs into $apoe^{-oe}$ mice, which are prone to the development of atherosclerosis, improves endothelial function and reduce atherosclerosis [59].

FUTURE DIRECTIONS

Interesting perspectives for the treatment of vascular diseases in the future consist of the development of techniques to deliver antiatherogenic genes specifically to the vascular wall. Despite disappointing results of various oral antioxidant treatment trials, promising findings have been reported using gene delivery of enzymes to improve NO bioavailability and decrease oxidative stress in animal models for vascular diseases. PPAR-directed therapies often have unexpected effects because of their pleiotropism and wide distribution. Promising future developments undoubtedly lie in the field of selective PPARα modulators (SPPARMs). Similarly, augmenting KLF2 expression and/or activity may also be a viable strategy that can be exploited for therapeutic gain.

Knowledge about miRNAs provides an opportunity to use antagomirs (oligonucleotide inhibitors to silence endogenous miRNAs) and miRNA mimics to modulate biological function in the pathology of diseases. However, stringent research is needed to efficiently deliver these miRNA inhibitors to the specific target organs or cells. Circulating miRNAs could be measured in body fluids as early biomarkers for disease diagnosis, prognosis, and response to treatment, but large-scale studies are required to confirm potential usefulness as predictive markers.

EPC-based therapies for tissue ischemia and blood vessel repair appear promising. Genetic engineering of EPC may provide an important strategy to enhance EPC mobilization, survival, engraftment, and function, thereby rendering these cells efficient therapeutic modalities for cardiovascular diseases [60].

Modulation of SIRT1, a molecular switch that is able to inhibit O_2^- production from uncoupled eNOS, while also augmenting free radical metabolism by intracellular enzymes, may be a promising strategy for treating chronic inflammatory diseases. Polyphenolic compounds, such as resveratrol, a known SIRT1 activator and flavonols, effective vasodilators and antioxidants, are cardioprotective and have potential as an adjunctive therapeutic agent in atherosclerosis, diabetes, and myocardial ischemia reperfusion. These biological activities have generated great interest in searching for and designing flavonols for potential use as therapeutic agents in the prevention and treatment of CVD.

REFERENCES

1. Furchgott R.F., Zawadzki J.V. (1980) The obligatory role of endothelial cells in the relaxation of arterial smooth muscle by acetylcholine. *Nature* 288, 373–376.
2. Ignarro L.J., Buga G.M., Wood K.S., Byrns R.E., Chaudhuri G. (1987) Endothelium-derived relaxing factor produced and released from artery and vein is nitric oxide. *Proc Natl Acad Sci U S A* 84, 9265–9269.
3. McMahon T.J., Moon R.E., Luschinger B.P. et al. (2020) Nitric oxide in the human respiratory cycle. *Nat Med* 8, 711–717.
4. Davignon J., Ganz P. (2004) Role of endothelial dysfunction in atherosclerosis. *Circulation* 109, 27–32.
5. Ludmer P.L., Selwyn A.P., Shook T.L. et al. (1986) Paradoxical vasoconstriction induced by acetylcholine in atherosclerotic coronary arteries. *N Engl J Med* 315, 1046–1051.
6. Vita J.A. (2011) Endothelial function. *Circulation* 124, e906–e912.
7. Hess D.T., Matsumoto A., Kim S.O., Marshall H.E., Stamler J.S. (2005) Protein S-nitrosylation: purview and parameters. *Nat Rev Mol Cell Biol* 6, 150–166.
8. Lima B., Forrester M.T., Hess D.T., Stamler J.S. (2010) S-nitrosylation in cardiovascular signaling. *Circ Res* 106, 633–646.
9. Benhar M., Forrester M.T., Stamler J.S. (2009) Protein denitrosylation: enzymatic mechanisms and cellular functions. *Nat Rev Mol Cell Biol* 10, 721–732.
10. Selemidis S., Dusting G.J., Peshavariya H., Kemp-Harper B.K., Drummond G.R. (2007) Nitric oxide suppresses NADPH oxidase-dependent superoxide production by S-nitrosylation in human endothelial cells. *Cardiovasc Res* 75, 349–358.
11. Kim J.H., Bugaj L.J., Oh Y.J., et al. (2009) Arginase inhibition restores NOS coupling and reverses endothelial dysfunction and vascular stiffness in old rats. *J Appl Physiol* 107, 1249–1257.
12. Lowenstein C.J. (2007) Nitric oxide regulation of protein trafficking in the cardiovascular system. *Cardiovasc Res* 75, 240–246.
13. Haendeler J., Hoffmann J., Zeiher A.M., Dimmeler S. (2004) Antioxidant effects of statins via S-nitrosylation and activation of thioredoxin in endothelial cells: a novel vasculoprotective function of statins. *Circulation* 110, 856–861.
14. Chiu J.J., Chien S. (2011) Effects of disturbed flow on vascular endothelium: pathophysiological basis and clinical perspectives. *Physiol Rev* 91, 327–387.
15. Tzima E., Irani-Tehrani M., Kiosses W.B. et al. (2005) A mechanosensory complex that mediates the endothelial cell response to fluid shear stress. *Nature* 437, 426–431.
16. Van der Heiden K., Hierck B.P., Krams R. et al. (2008) Endothelial primary cilia in areas of disturbed flow are at the base of atherosclerosis. *Atherosclerosis* 196, 542–550.
17. Young A., Wu W., Sun W. et al. (2009) Flow activation of AMP-activated protein kinase in vascular endothelium leads to Kruppel-like factor 2 expression. *Arterioscler Thromb Vasc Biol* 29, 1902–1908.
18. Atkins G.B., Jain M.K. (2007) Role of Kruppel-like transcription factors in endothelial biology. *Circ Res* 100, 1686–1695.
19. Atkins G.B., Wang Y., Mahabeleshwar G.H. et al. (2008) Hemizygous deficiency of Kruppel-like factor 2 augments experimental atherosclerosis. *Circ Res* 103, 690-693.
20. Sen-Banerjee S., Mir S., Lin Z. et al. (2005) Kruppel-like factor 2 as a novel mediator of statin effects in endothelial cells. *Circulation* 112, 720–726.
21. Dai G., Vaughn S., Zhang Y., Wang E.T., Garcia-Cardena G., Gimbrone M.A., Jr. (2007) Biomechanical forces in atherosclerosis-resistant vascular regions regulate endothelial redox balance via phosphoinositol 3-kinase/Akt-dependent activation of Nrf2. *Circ Res* 101, 723–733.
22. Fledderus J.O., Boon R.A., Volger O.L. et al. (2008) KLF2 primes the antioxidant transcription factor Nrf2 for activation in endothelial cells. *Arterioscler Thromb Vasc Biol* 28, 1339–1346.
23. Yamawaki H., Pan S., Lee R.T., Berk B.C. (2005) Fluid shear stress inhibits vascular inflammation by decreasing thioredoxin-interacting protein in endothelial cells. *J Clin Invest* 115, 733–738.
24. World C., Spindel O.N., Berk B.C. (2011) Thioredoxin-interacting protein mediates TRX1 translocation to the plasma membrane in response to tumor necrosis factor-alpha: a key

mechanism for vascular endothelial growth factor receptor-2 transactivation by reactive oxygen species. *Arterioscler Thromb Vasc Biol* 31, 1890–1897.

25. Lee D.Y., Lee C.I., Lin T.E. et al. (2012) Role of histone deacetylases in transcription factor regulation and cell cycle modulation in endothelial cells in response to disturbed flow. *Proc Natl Acad Sci U S A* 109, 1967–1972.
26. Wang W., Ha C.H., Jhun B.S., Wong C., Jain M.K., Jin Z.G. (2010) Fluid shear stress stimulates phosphorylation-dependent nuclear export of HDAC5 and mediates expression of KLF2 and eNOS. *Blood* 115, 2971–2979.
27. Chen Z., Peng I.C., Cui X., Li Y.S., Chien S., Shyy J.Y. (2010) Shear stress, SIRT1, and vascular homeostasis. *Proc Natl Acad Sci U S A* 107, 10268–10273.
28. Zhang Q.J., Wang Z., Chen H.Z. et al. (2008) Endothelium-specific overexpression of class III deacetylase SIRT1 decreases atherosclerosis in apolipoprotein E-deficient mice. *Cardiovasc Res* 80, 191–199.
29. Stein S., Schafer N., Breitenstein A. et al. (2010) SIRT1 reduces endothelial activation without affecting vascular function in ApoE-/- mice. *Aging (Albany NY)* 2, 353–360.
30. Bartel D.P. (2004) MicroRNAs: genomics, biogenesis, mechanism, and function. *Cell* 116, 281–297.
31. Urbich C., Kuehbacher A., Dimmeler S. (2008) Role of microRNAs in vascular diseases, inflammation, and angiogenesis. *Cardiovasc Res* 79, 581–588.
32. Zhou J., Wang K.C., Wu W. et al. (2011) MicroRNA-21 targets peroxisome proliferators-activated receptor-alpha in an autoregulatory loop to modulate flow-induced endothelial inflammation. *Proc Natl Acad Sci U S A* 108, 10355–10360.
33. Ji R., Cheng Y., Yue J. et al.(2007) MicroRNA expression signature and antisense-mediated depletion reveal an essential role of MicroRNA in vascular neointimal lesion formation. *Circ Res* 100, 1579–1588.
34. Nazari-Jahantigh M., Wei Y., Schober A. (2012) The role of microRNAs in arterial remodelling. *Thromb Haemost* 107, 611–618.
35. Harris T.A., Yamakuchi M., Ferlito M., Mendell J.T., Lowenstein C.J. (2008) MicroRNA-126 regulates endothelial expression of vascular cell adhesion molecule 1. *Proc Natl Acad Sci U S A* 105, 1516–1521.
36. Zernecke A., Bidzhekov K., Noels H. et al. (2009) Delivery of microRNA-126 by apoptotic bodies induces CXCL12-dependent vascular protection. *Sci Signal* 2, ra81.
37. Wu W., Xiao H., Laguna-Fernandez A. et al. (2011) Flow-dependent regulation of kruppel-like factor 2 is mediated by microRNA-92a. *Circulation* 124, 633–641.
38. Fang Y., Davies P.F. (2012) Site-specific microRNA-92a regulation of Kruppel-like factors 4 and 2 in atherosusceptible endothelium. *Arterioscler Thromb Vasc Biol* 32, 979–987.
39. Hergenreider E., Heydt S., Treguer K. et al. (2012) Atheroprotective communication between endothelial cells and smooth muscle cells through miRNAs. *Nat Cell Biol* 14, 249–256.
40. Fruchart J.C. (2009) Peroxisome proliferator-activated receptor-alpha (PPARalpha): at the crossroads of obesity, diabetes and cardiovascular disease. *Atherosclerosis* 205, 1–8.
41. Plutzky J. (2011) The PPAR-RXR transcriptional complex in the vasculature: energy in the balance. *Circ Res* 108, 1002–1016.
42. Ziouzenkova O., Perrey S., Asatryan L. et al. (2003) Lipolysis of triglyceride-rich lipoproteins generates PPAR ligands: evidence for an antiinflammatory role for lipoprotein lipase. *Proc Natl Acad Sci U S A* 100, 2730–2735.
43. Ahmed W., Orasanu G., Nehra V. et al. (2006) High-density lipoprotein hydrolysis by endothelial lipase activates PPARalpha: a candidate mechanism for high-density lipoprotein-mediated repression of leukocyte adhesion. *Circ Res* 98, 490–498.
44. Capell W.H., DeSouza C.A., Poirier P. et al. (2003) Short-term triglyceride lowering with fenofibrate improves vasodilator function in subjects with hypertriglyceridemia. *Arterioscler Thromb Vasc Biol* 23, 307–313.
45. Duan S.Z., Usher M.G., Mortensen R.M. (2008) Peroxisome proliferator-activated receptor-gamma-mediated effects in the vasculature. *Circ Res* 102, 283–294.
46. Kurtz T.W. Beyond the classic angiotensin-receptor-blocker profile. *Nat Clin Pract Cardiovasc Med* 5(Suppl 1), S19–S26.
47. Orasanu G., Ziouzenkova O., Devchand P.R. et al. (2008) The peroxisome proliferator-activated receptor-gamma agonist pioglitazone represses inflammation in a peroxisome proliferator-activated receptor-alpha-dependent manner in vitro and in vivo in mice. *J Am Coll Cardiol* 52, 869–881.

48. Qu A., Shah Y.M., Manna S.K., Gonzalez F.J. (2012) Disruption of endothelial peroxisome proliferator-activated receptor gamma accelerates diet-induced atherogenesis in LDL receptor-null mice. *Arterioscler Thromb Vasc Biol* 32, 65–73.
49. Bishop-Bailey D., Bystrom J. (2009) Emerging roles of peroxisome proliferator-activated receptor-beta/delta in inflammation. *Pharmacol Ther* 124, 141–150.
50. Fadini G.P., Losordo D., Dimmeler S. (2012) Critical reevaluation of endothelial progenitor cell phenotypes for therapeutic and diagnostic use. *Circ Res* 110, 624–637.
51. Sirker A.A., Astroulakis Z.M., Hill J.M. (2009) Vascular progenitor cells and translational research: the role of endothelial and smooth muscle progenitor cells in endogenous arterial remodelling in the adult. *Clin Sci (Lond)* 116, 283–299.
52. Werner N., Kosiol S., Schiegl T. et al. (2005) Circulating endothelial progenitor cells and cardiovascular outcomes. *N Engl J Med* 353, 999–1007.
53. Hill J.M., Zalos G., Halcox J.P. et al.(2003) Circulating endothelial progenitor cells, vascular function, and cardiovascular risk. *N Engl J Med* 348, 593–600.
54. Tokgozoglu L., Yorgun H., Gurses K.M. et al. (2011) The association between circulating endothelial progenitor cells and coronary collateral formation. *Atherosclerosis* 219, 851–854.
55. Williamson K., Stringer S.E., Alexander M.Y. (2012) Endothelial progenitor cells enter the aging arena. *Front Physiol* 3, 30.
56. Zhang Y., Herbert B.S., Rajashekhar G. et al. (2009) Premature senescence of highly proliferative endothelial progenitor cells is induced by tumor necrosis factor-alpha via the p38 mitogen-activated protein kinase pathway. *FASEB J* 23, 1358–1365.
57. Zhang Y., Ingram D.A., Murphy M.P. et al. (2009) Release of proinflammatory mediators and expression of proinflammatory adhesion molecules by endothelial progenitor cells. *Am J Physiol Heart Circ Physiol* 296, H1675–H1682.
58. Kong D., Melo L.G., Mangi A.A. et al. (2004) Enhanced inhibition of neointimal hyperplasia by genetically engineered endothelial progenitor cells. *Circulation* 109, 1769–1775.
59. Cubbon R.M., Kahn M.B., Wheatcroft S.B. (2009) Effects of insulin resistance on endothelial progenitor cells and vascular repair. *Clin Sci (Lond)* 117, 173–190.
60. Dzau V.J., Gnecchi M., Pachori A.S., Morello F., Melo L.G. (2005) Therapeutic potential of endothelial progenitor cells in cardiovascular diseases. *Hypertension* 46, 7–18.

10

VASCULAR SMOOTH MUSCLE CELLS

Dario F. Riascos-Bernal and Nicholas E. S. Sibinga
Department of Medicine (Cardiology) and Department of Developmental and Molecular Biology, Albert Einstein College of Medicine, Bronx, NY, USA

INTRODUCTION

The vascular smooth muscle cell (SMC) is the main component of the vessel wall throughout the circulatory system, with the exception of the microvasculature, where mural cells are called pericytes. In normal adult medium and large arteries, quiescent SMCs provide structural support and regulate vessel tone and diameter. In arterioles, SMC contractile function determines local and systemic vascular resistance, as well as systemic blood pressure. Remarkably, differentiated adult SMCs can proliferate, migrate, and synthetize abundant protein during vascular injury or disease, in essence re-assuming a dedifferentiated phenotype like that found during embryonic development [1]. This profound SMC plasticity is centrally important in the pathogenesis of vascular diseases such as atherosclerosis, restenosis after angioplasty, in-stent stenosis, and transplant arteriosclerosis. This chapter addresses extrinsic signals that affect SMC phenotype, and intrinsic cellular and molecular processes that affect SMC activities during atherogenesis.

SMCS IN ATHEROMA PROGRESSION

SMCs participate in all stages of atheroma formation. In early lesions, endothelial dysfunction and intimal lipid accumulation trigger inflammation and promote recruitment of leukocytes, including neutrophils and monocytes [2]. In the intima, monocytes differentiate into macrophages that phagocytose modified lipids and lipoprotein particles and become foam cells. Cytokines and growth factors such as platelet-derived growth factor (PDGF)-BB released from macrophages and foam cells induce dedifferentiation (phenotypic modulation or switching) (Figure10.1) and migration of SMCs from media to intima. In the intima, SMCs contribute to atheroma expansion and complexity by proliferating, synthesizing extracellular matrix (ECM)

Atherosclerosis: Risks, Mechanisms, and Therapies, First Edition. Edited by Hong Wang and Cam Patterson.

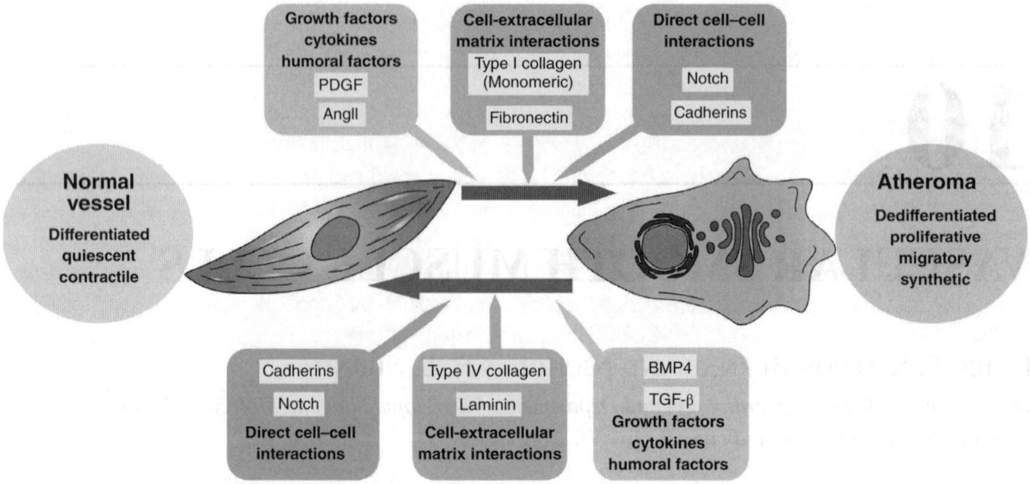

Figure 10.1 Phenotypic switching of smooth muscle cells. Adult SMCs possess remarkable plasticity that allows them to go from a quiescent, contractile, and differentiated state in the normal vessel wall to a dedifferentiated, proliferative, migratory, and synthetic state during atherosclerosis, and *viceversa*. Soluble factors, cell-extracellular matrix interactions, and direct cell-cell interactions provide signals that direct SMCs to one or the other state.

(mainly collagen), engulfing lipids and lipoprotein particles, and by releasing inflammatory cytokines. SMCs are centrally involved in the progression from early fatty streak to advanced fibrofatty atheroma [3]; this lesion expansion can lead to stenosis. In addition to SMCs that migrate from the media, resident intimal SMCs have been described in humans; these cells respond to environmental cues, undergo phenotypic modulation, and contribute to atheroma formation. Sca1$^+$ adventitial progenitors can contribute to neointimal formation [4], and such cells are increased in the adventitia of human atherosclerotic vessels [5]. Alternative sources of SMCs that may contribute to atheroma formation have been described, including bone marrow (via circulating progenitor cells) and endothelium (via cellular transdifferentiation); the importance of these pathways for clinical disease is controversial [6].

Plaque stability becomes critical as atherosclerosis progresses. Rupture of an unstable plaque can initiate thrombosis, leading to vascular occlusion and downstream tissue infarction. SMCs support plaque stability by synthesizing ECM proteins, lending mechanical strength to the fibrous cap overlying the plaque. Cap thickness and composition depend on the balance of ECM synthesis and degradation; expression and activity of matrix metalloproteinases (MMPs), which degrade ECM, and tissue inhibitors of metalloproteinases (TIMPs), which inhibit MMPs, become significant. SMC death is also important, as loss of SMCs limits ECM synthesis and leads to a thin and acellular fibrous cap [7].

EXTRINSIC CONTROL OF SMC BIOLOGY

Soluble Factors

Platelet-derived growth factors The PDGF family consists of five different disulfide-linked dimers (PDGF-AA, PDGF-AB, PDGF-BB, PDGF-CC, and PDGF-DD) that act via two receptor tyrosine kinases (PDGFRs α and β). In vascular biology, the classical PDGF polypeptide chains (PDGF-A and PDGF-B) have been studied extensively; the newer PDGFs, PDGF-C, and PDGF-D, are expressed in SMCs in human atherosclerotic lesions, but their functions have not been determined [8].

PDGF-BB, produced mainly by platelets and macrophages, induces SMC migration and proliferation and inhibits expression of SMC marker genes including smooth muscle (SM) α-actin, SM-myosin heavy chain (MHC), and SM22α. Multiple mechanisms defined in cultured SMCs mediate this inhibition [9], including (i) induction of Kruppel-like factor 4 (KLF4) expression, which recruits histone deacetylase 2 (HDAC2), leading to histone 4 (H4) deacetylation, chromatin compaction, and inhibition of myocardin/serum response factor (SRF) binding to SMC genes; (ii) mitogen-activated protein kinase kinase (MEK1/2)-mediated phosphorylation of E26 (ETS)-like transcription factor-1 (Elk-1), which competes for binding of SRF to CArG elements in SMC marker gene promoters; (iii) inhibition of myocardin-dependent SMC marker gene transactivation through the HES-related repressor protein-1 (HERP1) transcription factor, which prevents SRF binding to CArG box through

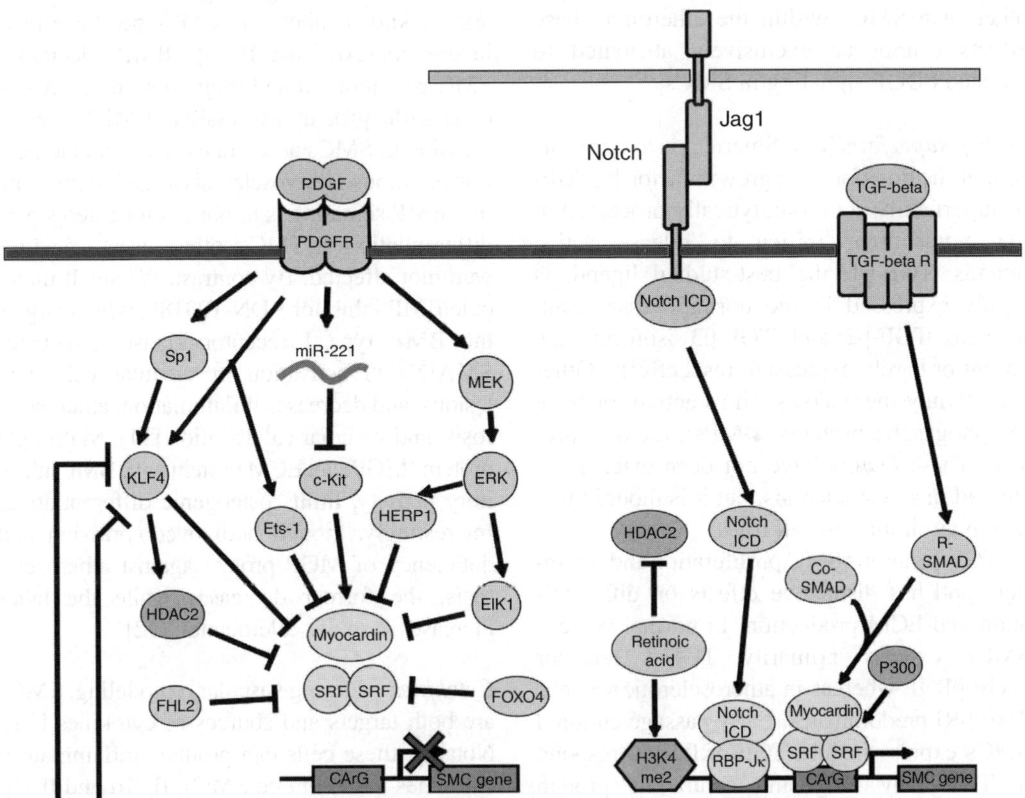

Figure 10.2 Regulation of SMC gene expression. Regulation occurs at different levels: (a) interaction of *cis*-acting elements and *trans*-acting factors, (b) chromatin remodeling and histone posttranslational modifications, (c) microRNAs (mRNAs) enhance or decrease SMC gene expression by modulating levels of important regulators. PDGF reduces SMC gene expression, whereas TGF and jagged1 enhance it (see text for details). The *arrows* indicate stimulatory interactions, whereas the perpendicular lines indicate inhibition. Interactions are not necessarily direct.

physical association with SRF; and (iv) induction of specificity protein 1 (Sp1) and ets-1 [V-ets erythroblastosis virus E26 oncogene homolog 1 (avian)] transcription factors that suppress SMC marker gene expression (Figure10.2).

Interestingly, antibody-based PDGFR blockade reduces lesion size by 67% and SMC population in the neointima by 80% in a mouse model of atherosclerosis [1]. This effect was not seen with selective PDGFR-α blockade, suggesting that PDGFR-β function is more important in atherosclerosis. However, because PDGFRs are expressed in cell types other than SMCs within the atheroma, these effects cannot be exclusively attributed to reduced PDGF signaling in SMCs.

TGF-β superfamily Dimeric inactive precursors of the transforming growth factor-β (TGF-β) superfamily are proteolytically processed in the extracellular milieu to release active ligands. TGF-β1, the best-studied ligand, is highly expressed in the normal vessel wall, whereas TGF-β2 and TGF-β3 isoforms are absent or barely expressed, respectively. Other superfamily members, such as activin or bone morphogenetic proteins (BMPs), are also present. These factors have not been extensively studied in atherosclerosis, but it is thought that they play similar overall roles.

TGF-β inhibits SMC proliferation and migration, and has distinctive effects on differentiation and ECM production. In normal vessels, SMCs express primarily TGF-β receptor (TGF-βR)II; whereas in atherosclerotic vessels, TGF-βRI predominates. Early passage cultured SMCs express mainly TGF-βRII, and respond to TGF-β by increasing contractile protein expression; later-passage SMCs express more TGF-βRI and respond to TGF-β by markedly increasing ECM production [10].

Adult *tgfb1*$^{+/-}$ mice show reduced SMC differentiation, consistent with the idea that TGF-β contributes to vascular homeostasis. Inactivation of TGF-β signaling in a mouse model of atherosclerosis exacerbates lipid-rich lesion development and recruitment of inflammatory cells, including macrophages, and neutralizing TGF-β activity decreases ECM deposition in atherosclerotic lesions. These observations support the idea that TGF-βRII activation opposes fatty streak formation at early stages, while TGF-βRI signaling promotes plaque stability in advanced atheromas by increasing SMC-dependent ECM production in the fibrous cap [10].

Both endothelial and SMCs are significant sources of BMPs, and BMP signaling is active in endothelium, SMCs, and subintimal macrophages of early atherosclerotic lesions. Several lines of evidence support a significant role for endothelial BMP signaling in atherosclerosis; less is known about its SMC-specific effects in this context. Like TGF-β, BMP4 decreases SMC proliferation and migration and increases contractile protein expression. BMP2 overexpression in SMCs accelerates vascular calcification in a mouse atherosclerosis model, suggesting that BMP signaling is important for osteogeneic differentiation of SMCs; other disease features were not affected. By contrast, the small molecule BMP inhibitor LDN-193189, which targets the BMP type I receptor kinase, attenuates SMAD1/5/8 activation in vascular cells and lesions, and decreases inflammation, atherosclerosis, and vascular calcification [11]. Matrix gla protein (MGP), an ECM protein with BMP inhibitory activity, limits osteogenic differentiation. Interestingly, though both overexpression and deficiency of MGP protect against atherosclerosis, the former decreases while the latter increases vascular calcification [12].

Cytokines During vascular remodeling, SMCs are both targets and sources of cytokines [13]. Notably, these cells can produce inflammatory cytokines—in cultured SMCs, IL-1α and IL-1β induce IL-1β production, TNF-α induces TNF-α production, and IL-6 is constitutively released. They also produce chemokines, including monocyte chemotactic protein-1 (MCP-1), eotaxin, or monokine induced by γ-interferon (MIG), and constitutively express mRNA encoding chemokine receptors CCR1 and CCR2.

Some cytokines affect SMC phenotype [13]. IL-1 is a potent SMC mitogen, while interferon-γ upregulates PDGFR-β expression and potentiates PDGF-BB proliferative effects. IL-1 and TNF-α promote SMC production of collagen types I and III, whereas interferon-γ represses collagen synthesis. MMP activity is relevant for ECM homeostasis; IL-1 and TNF-α induce MMPs types 1, 3, 8, and 9, while TIMPs types 1, 2, and 3, endogenous MMP inhibitors are constitutively expressed.

Angiotensin II Angiotensin II (AngII) activity is increased in atherosclerosis and aggravates the disease at all stages. AngII administration accelerates atherosclerosis in animal models, whereas pharmacologic inhibition slows atherosclerosis both in animal models and in humans [14]. The multiple effects of AngII on SMCs are particularly relevant for atherosclerosis.

AngII promotes SMC migration, proliferation, and hypertrophy via multiple mechanisms [14]. The angiotensin type 1 receptor (AT1R) mediates most of these effects, which may be partially opposed by signaling through the type 2 receptor (AT2R). AngII activates several serine-threonine and tyrosine kinases, generates oxygen radicals, and releases arachidonic acid by phospholipase A2 to produce SMC-activating metabolites, for example, 12-hydroxyeicosatetraenoic acid (12-HETE) and 20-HETE. AngII also increases cytochrome P450 1B1-dependent generation of reactive oxygen species (ROS), and activates extracellular signal-regulated kinase (ERK)1/2 and p38 mitogen-activated protein kinase (MAPK). The full AngII proliferative response requires ERK1/2 and phosphatidylinositol 3-kinase (PI3K) activation and functional epidermal growth factor receptor (EGFR) signaling. AngII proliferative effects are ultimately mediated through induction of proto-oncogenes (e.g., *c-fos, c-myc,* and *c-jun*) and increased expression of growth factors, including PDGF.

AngII also promotes SMC apoptosis. In rats, AngII administration increases Fas ligand (FasL) expression and apoptosis in carotid artery medial SMCs. AngII binds AT2R to initiate signaling through MAPK kinase (MEK), ERK1/2, and c-Jun N-terminal kinase (JNK), which activates the GATA-6 transcription factor. GATA-6 promotes FasL expression, stimulating apoptosis in a FasL- and caspase 8-dependent manner [15]. AngII-induced oxidative stress also increases FasL expression. FasL is expressed in human atherosclerotic plaques, and its serum levels increase in patients with acute myocardial infarction or unstable angina [15]; it is plausible that AngII promotes both SMC apoptosis and plaque instability.

AngII promotes production of oxidized lipoprotein particles and indirectly increases SMC foam cell formation [14]. AngII increases NADPH oxidase activation and superoxide radical (O_2) production. These enzymes, along with lipooxygenases and myeloperoxidases, oxidize subendothelial low-density lipoprotein (LDL) particles. Resulting oxidized LDL particles increase foam cell formation and cytokine release. In addition, SMC exposure to LDL cholesterol significantly induces AT1R mRNA expression and AngII binding, augmenting AngII signaling.

ECM-Cell Interactions

During atherogenesis, vessel wall ECM changes significantly. Since pericellular matrix exerts important effects on SMC phenotype, it has been proposed that ECM changes antecede the phenotypic modulation observed in SMCs during disease. ECM affects SMC differentiation state and responsiveness to growth factors such as PDGF [16]. A differentiated phenotype is preserved better when SMCs are plated on laminin or type IV collagen rather than fibronectin or monomeric type I collagen. Monomeric and fibrillar type I collagens have opposite effects on PDGF mitogenic signaling and proliferation of SMCs: while monomeric type I collagen promotes PDGF-induced proliferation, the fibrillar form prevents it.

The effect of ECM on SMC phenotype is mediated by interactions between ECM proteins and integrins and other adhesion molecules (Table10.1) [16]. Thus, ECM proteins and their receptors that favor a differentiated state of SMCs should be abundant in normal

TABLE 10.1 Extracellular matrix protein-cell interactions and SMC phenotype

ECM protein	Integrin/ receptor	SMC phenotype
Collagen I	$\alpha1\beta1/\alpha2\beta1/$ DDR1	Differentiated
Collagen IV	$\alpha1\beta1/\alpha2\beta1$	Differentiated
Laminin	$\alpha1\beta1/\alpha3\beta1/\alpha7\beta1$	Differentiated
COMP	$\alpha7\beta1$	Differentiated
Collagen VIII	$\alpha1\beta1/\alpha2\beta1$	Dedifferentiated
Fibronectin	$\alpha5\beta1$	Dedifferentiated
Vitronectin	$\alpha v\beta1$	Dedifferentiated

DDR1, discoidin domain receptor 1.

arteries, but decreased upon injury or disease. For example, expression of cartilage oligomeric matrix protein (COMP) [17], also called thrombospondin 5, is high in normal vessels but falls significantly after balloon-mediated injury. COMP expression normally promotes SMC differentiation via Integrin α7β1; when COMP levels decrease in cultured, PDGF-treated SMCs, so do SMC differentiation markers. The specific role of COMP in atherosclerosis is not known, although its overexpression limits neointimal formation after injury.

Integrin signaling Transmembrane integrin proteins mediate ECM-cell interactions (focal adhesions) and regulate activation of multiple cell types, including SMCs. Migration requires dynamic interaction between SMCs and surrounding ECM, and this process is important in atheroma formation. Collagen and fibronectin promote SMC migration through β1 integrins, whereas vitronectin, osteopontin, and periostin promote SMC migration mainly through β3 integrin [18].

Focal adhesion kinase (FAK) associates constitutively with β integrins, is activated by integrin-ECM, and transduces migration, survival, and growth signals. Integrin and FAK inactivation *in vivo* lead to vascular defects and embryonic lethality, preventing their study in atherosclerosis models. However, some evidence suggests that SMC integrin/FAK signaling is important during atherosclerosis. Integrin β3 is induced in the vessel wall in response to vascular injury. Osteopontin, enriched in human atheroscleorotic plaques, promotes SMC migration through interaction with β3 integrin. PDGF and TGF-β, both implicated in atherosclerotic plaque formation, induce β3 integrin expression in SMCs. In addition, peptide antagonists that selectively bind to β3 integrin reduce neointima formation after balloon injury [19].

Signals From Direct Cell-Cell Interactions

Adhesion molecules are important in leukocyte trafficking in the vessel wall, but comparatively little is known about direct cell-cell interactions involving SMCs in the development and progression of atherosclerosis. Notch and cadherin family proteins mediate such interactions and have important vascular functions.

Notch signaling Notch signaling, a major developmental pathway affecting specification and differentiation, acts in the endothelium to regulate vasculogenesis and angiogenesis; in SMCs, Notch supports SMC investment and multilayer vessel wall formation. How Notch affects SMCs in the postnatal period is not well understood.

Notch1, Notch2, and Notch3 are expressed in adult arteries. Mutations in Notch3 cause an autosomal-dominant inherited disorder known as cerebral autosomal-dominant arteriopathy with subcortical infarcts and leukoencephalopathy (CADASIL) syndrome, with multiple characteristic pathological findings indicating abnormal SMC function. *In vitro*, Notch3 regulates SMC proliferation, migration, and survival; *in vivo*, Notch3 expression decreases after balloon injury in carotid arteries [20]. In vascular remodeling after artery ligation, Notch1, not Notch3, promotes SMC proliferation and neointima formation through a HERP-mediated process [21]. These observations suggest that SMC Notch signaling may be important in atherosclerosis.

Interestingly, pharmacologic inhibition of gamma-secretase—the enzyme that releases the Notch intracellular domain via regulated intramembrane proteolysis (RIP) to mediate subsequent transcriptional regulation—shows beneficial effects in a mouse model of atherosclerosis [22]. Although this has been attributed to impaired macrophage activation resulting from decreased Notch1/HERP signaling, such studies cannot exclude effects on Notch in other cell types, including SMCs, or on other RIP targets.

Cadherins Cadherins, along with selectins, integrins, and immunoglobulin-domain containing cell adhesion molecules, constitute the four major groups of cell adhesion proteins. Cadherins belong to a transmembrane protein superfamily that in vertebrates comprises more than 100 members. Classical and desmosomal cadherins function in direct cell-cell adhesion, but other cadherins (e.g., fat and protocadherins) appear not to have strong adhesive properties and more likely work as signaling molecules.

Cadherins also contribute to cell recognition and sorting, polarity, proliferation, migration, survival, cytoskeletal organization, and other functions [23].

SMCs express N-, R-, T-, and Fat1 cadherins. N-cadherin loss promotes SMC proliferation via decreased β-catenin sequestration and consequent increase in nuclear β-catenin/T-cell factor (TCF)-mediated Wnt target gene expression. N-cadherin also enhances SMC survival *in vitro* via activation of FGF and PI3K/Akt signaling. Interestingly, expression of a soluble N-cadherin extracellular fragment in atherosclerotic mice promotes stable plaque features, including decreased SMC apoptosis and macrophage infiltration, and increased SMC-rich fibrous cap thickness [24].

R-cadherin can also limit proliferation. R-cadherin levels decrease after carotid arterial injury; its expression in atherosclerosis is not known. By contrast, T-cadherin expression increases in intimal SMCs during atheroma formation. T-cadherin attenuates SMC growth and motility, adding to its protective role as a receptor for adiponectin, an antiatherosclerotic adipokine with multiple salutary metabolic effects.

The atypical cadherin Fat1 is expressed robustly in cultured SMCs; it inhibits proliferation and promotes migration. Fat1 is induced after arterial injury [25], and may regulate SMC distribution during vessel healing. Finally, although E-cadherin is not expressed in normal arteries, it is induced in human atherosclerotic lesions; its significance in this setting is not known.

INTRINSIC CONTROL OF SMC BIOLOGY

Integration of Signals—The mTOR Pathway

Extracellular signals including mitogens and hormones promote protein translation and growth by activating the highly conserved mTOR serine/threonine kinase. mTOR promotes cell growth and proliferation, primarily as an effector of the PI3K pathway, by activating the eIF4E-binding protein (4E-BP1) and p70S6 kinase that are involved in protein translation and growth [26].

Pharmacologic PI3K or mTOR inhibition regulates several aspects of SMC phenotype [27]. mTOR inhibition by rapamycin inhibits proliferation, and everolimus—a rapamycin derivative—causes G0/G1 phase cell cycle arrest by inhibiting cyclin-dependent kinases. Although rapamycin reduces total protein synthesis, it concurrently promotes SMC marker gene expression—translation of a subset of 5′ structured mRNAs is inhibited, but residual activity allows contractile protein translation. Both rapamycin and everolimus attenuate atherosclerotic plaque progression in murine models, delaying transition from early, macrophage-rich lesions to more complex advanced atheromas [28, 29], likely due in part to effects on SMCs [3].

Reactive Oxygen Species

ROS are small, oxygen-derived molecules including radicals (superoxide, hydroxyl, peroxyl, and alkoxyl) and nonradicals (hypochlorous acid, ozone, singlet oxygen, and hydrogen peroxide). ROS may be primary reaction products, as mediated by NADPH oxidases, or may arise as reaction by-products in mitochondria, peroxisomes, or microsomes (via cytochrome P450) [30]. Balanced production and scavenging permits physiologic ROS function in signaling. With increased cell stress in atherosclerosis, for example, ROS levels can rise significantly, allowing these highly reactive molecules to modify other oxygen species, proteins, or lipids. Macrophages and SMCs are the main cellular sources of oxidative molecules and ROS in the atherosclerotic plaque [31].

ROS come from several sources in SMCs, including NADPH oxidases, xanthine oxidase, the mitochondrial respiratory chain, lipooxygenases, and nitric oxide synthases. NADPH oxidase is a multiprotein complex of five subunits: $p40^{PHOX}$ (PHOX=phagocyte oxidase), $p47^{PHOX}$, $p67^{PHOX}$, $p22^{PHOX}$, and Nox. The transmembrane Nox proteins transport electrons across membranes to reduce oxygen to superoxide. SMCs express Nox1, 2, and 4; the latter accounts for basal NADPH oxidase activity. Factors relevant to atherosclerosis, including AngII, thrombin, PDGF, TNF-α, and mechanical wall stress, activate these oxidases mainly by modulating expression of Nox subunits. For example, prolonged treatment with AngII, PDGF, or serum induces Nox1 expression, but downregulates Nox4 in aortic SMCs [31].

Nox-derived ROS are thought to participate in nitric oxide inactivation, neointimal formation, SMC growth, proliferation, migration, ECM regulation, kinase activation, secretion of inflammatory cytokines, and control of differentiation. Nox4 knockdown, for example, reduces SMC marker gene expression, suggesting that Nox4 acts to maintain a differentiated phenotype [31].

Regulation of SMC Marker Gene Expression

Regulation of SMC marker gene transcription involves complex interactions of *trans*-acting factors with *cis*-acting elements in the gene promoters (Figure10.2). SRF interaction with CArG ($CC(AT)_6GG$) elements activates expression of many SMC marker genes [1, 9, 32], and is strongly enhanced by cofactors myocardin and myocardin-related transcription factor A and B (MRTF-A and MRTF-B).

Decreased marker gene expression is a hallmark of SMC phenotypic switching. A conserved G/C repressor element in many marker gene promoters mediates their suppression during atherosclerotic lesion development. Moreover, binding of KLF4, a factor that maintains stem cell pluripotency, regulates G/C repressor inhibitory function. KLF4 expression increases in atherosclerotic lesions; its overexpression reduces myocardin and SMC marker gene expression, and induces epigenetic silencing via histone deacetylase recruitment and histone hypoacetylation [9]. Other factors can limit SMC marker gene expression by perturbing the functional SRF-myocardin or MRTF complex. Examples include the forkhead box protein O4 (FOXO4), homeodomain-only protein (HOP), FHL2/DRAL/Slim3 [9], and Yap1 [33].

SMC differentiation is also regulated at the epigenetic level through chromatin remodeling, histone post-translational modifications, DNA methylation, exchange of histone and histone variants, and small RNA molecules. For example, retinoic acid promotes SRF binding and transcriptional activation of SMC marker genes by hyperacetylation and H3K4 dimethylation (H3K4diMe) of histones within the CArG-containing regions of SMC marker genes [34]. H3K4diMe, specific to SMCs, persists when SMCs undergo phenotypic switching, suggesting that H3K4diMe may serve as an SMC lineage epigenetic marker.

Micro-RNAs (miRNAs), about 22 nucleotide-long noncoding RNAs, regulate diverse biological functions through post-transcriptional repression of target genes. miRNAs are broadly implicated in atherosclerosis through effects on endothelial cells, SMCs, macrophages, and lipoprotein metabolism [35]. Specific miRNAs can promote SMC differentiation (miR-143/145, miR-1, and miR-21) or dedifferentiation (miR-221, miR-146a, miR-24, and miR-26a) [36]. miR-143/145 induce SMC differentiation mainly by repressing KLF4, but are also required for prodifferentiation effects of jagged-1 (Figure10.2). miR-145 overexpression reduces neointimal formation after injury; interestingly, miR-143/145 levels are reduced in patients with aortic aneurysm or coronary artery disease. Similarly, myocardin induces miR-1 expression, which inhibits proliferation by repressing Pim-1 and KLF4, known promoters of SMC proliferation. By contrast, PDGF induces miR-221, which enhances proliferation and migration and reduces contractile protein expression. miR-221 increases proliferation by repressing the cyclin-dependent kinase inhibitor p27Kip, and limits expression of contractile proteins by indirectly inhibiting myocardin expression via repression of the tyrosine kinase c-kit, a known stimulator of myocardin expression. Differential miRNA expression in SMCs during atherosclerosis may mediate diverse cellular outcomes by regulating multiple targets.

AUTOPHAGY AND CELL DEATH

The balance of SMC survival and death bears critically on atherosclerotic plaque progression and stability. At basal levels, autophagy recycles damaged cellular components and promotes survival, and therefore may serve to stabilize atheromas. Both apoptotic and necrotic cell death have been identified in atherosclerotic plaques, but potential beneficial or detrimental roles of these processes are not fully understood. Programmed cell death in early lesions supports efficient clearance of cellular corpses. On the other hand, apoptosis is higher in unstable plaques than in stable plaques or normal vessels; and in preclinical models, induction of SMC apoptosis promotes the formation of vulnerable plaques [7].

Autophagy

Autophagy, a conserved adaptive cellular response to environmental stresses such as oxidative stress or nutrient deprivation, promotes cell survival via recycling of substrates such as amino acids, free fatty acids, and nucleotides derived from degradation of endogenous cellular components. The role of autophagy in atherosclerosis is not settled, but ultrastructural analysis of advanced plaques reveals features of autophagy in SMCs within the fibrous cap, including formation of myelin figures and vacuolization. Proposed autophagy inducers are inflammation, ROS, oxidized lipoproteins, hypoxia, and endoplasmic reticulum (ER) stress. Suggested consequences of autophagy may be beneficial or detrimental. Basal levels of autophagy protect against oxidative stress by degrading damaged cellular material (e.g., impaired mitochondria), and thus promote cell survival. However, excessive autophagy will induce autophagic SMC death, decrease collagen synthesis, and promote thinning of the fibrous cap and plaque instability [37]. *In vivo* studies are still necessary to test the significance of SMC autophagy in plaque evolution.

Apoptosis

Numerous factors associated with atherosclerosis may induce SMC apoptosis, including oxidative stress, hypoxia, interferon-γ, cholesterol overload, inflammatory cells, cytokines, altered blood flow, and pressure. These factors are counterbalanced by so-called survival factors, including growth factors, integrin-mediated cell-matrix interactions, and cadherin-mediated cell-cell contacts. SMC loss due to apoptosis leads to decreased collagen synthesis, fibrous cap thinning, and plaque instability. In addition, apoptotic cells that are not cleared (due to a failure in efferocytosis) undergo secondary necrosis, with consequent leakage of cytokines such as IL1-α; this in turn acts on neighboring cells and induces IL-6 and MCP-1 expression, contributing to a chronic inflammatory state. Moreover, apoptotic SMCs increase plaque thrombogenicity through translocation of phosphatidylserine to the plasma membrane outlet leaflet, which promotes thrombin formation [38, 39].

Ultrastructural analysis of dedifferentiated SMCs shows an abundance of ER and Golgi apparatus. The ER normally assures proper folding of secreted proteins; this key function can be compromised by increased secreted protein demand, greater misfolded protein load, and/or decreased folding capacity, all of which contribute to ER stress. ER stress is sensed by resident ER proteins including inositol requiring protein 1 (IRE1), protein kinase RNA-like ER kinase (PERK), and activating transcription factor-6 (ATF6); these factors act to relieve stress through a corrective action termed the unfolded protein response (UPR). With persistent ER stress, prolonged UPR activation leads to apoptosis through activity of IRE1 or downstream effectors such as C/EBP-homologous protein (CHOP) [40]. ER stress occurs in atherosclerotic plaques, especially at advanced stages. Moreover, markers of ER stress such as increased CHOP or activation of the IRE1 branch of the UPR have been identified in SMCs in human lesions; however, a causal relationship between activation of the UPR and SMC apoptosis has not been established.

Necrosis

Dying SMCs in human atherosclerotic plaques show ultrastructural changes typical of necrotic cell death, and experimental evidence suggests that necrosis promotes atherogenesis or plaque instability. Dying cells release their cytoplasmic contents to a large lipid mass known as the necrotic core. Proposed necrosis inducers are oxidative stress (excess ROS, highly oxidized LDL, toxic concentrations of mildly oxidized LDL, or high oxysterol levels), cellular ATP depletion (caused by low glucose and oxygen availability in the lesion center due to limited diffusion), lack of apoptotic cell clearance (defective efferocytosis), and increased intracellular calcium (influx of calcium across the plasma membrane).

Necrotic cell death in the atheroma provokes inflammation by releasing molecules such as HMGB1, cyclophilin A, heat shock proteins, DNA-chromatin complexes, uric acid, thioredoxin, galectins, adenosine, and ATP. Passively released ECM-degrading proteases further undermine plaque stability. Accumulating cellular debris softens and expands the necrotic core, adding to mechanical stress on the overlying cap and increasing likelihood of its rupture. Finally, high levels of tissue factor render the necrotic core highly thrombogenic [41].

SUMMARY

The remarkable plasticity of the SMC shapes its contribution to atherosclerosis. This plasticity is mediated by a multitude of extrinsic factors and intrinsic mechanisms that control differentiation, proliferation, migration, ECM production and degradation, and cell death. These SMC phenotypic changes are critical to plaque evolution from early fatty streak to advanced fibrofatty atheroma, and in maintenance of plaque stability. Greater understanding of the mechanisms underlying SMC activities in atherosclerosis may point to novel strategies for intervention.

REFERENCES

1. Owens GK, Kumar MS, Wamhoff BR. Molecular regulation of vascular smooth muscle cell differentiation in development and disease. *Physiol Rev*. 2004;84:767–801.
2. Weber C, Noels H. Atherosclerosis: Current pathogenesis and therapeutic options. *Nat Med*. 2011;17:1410–1422.
3. Falk E. Pathogenesis of atherosclerosis. *J Am Coll Cardiol*. 2006;47:C7–C12.
4. Hu Y, Zhang Z, Torsney E, et al. Abundant progenitor cells in the adventitia contribute to atherosclerosis of vein grafts in apoe-deficient mice. *J Clin Invest*. 2004;113:1258–1265.
5. Torsney E, Mandal K, Halliday A, Jahangiri M, Xu Q. Characterisation of progenitor cells in human atherosclerotic vessels. *Atherosclerosis*. 2007;191:259–264.
6. De Geest B. The origin of intimal smooth muscle cells: Are we on a steady road back to the past? *Cardiovasc Res*. 2009;81:7–8.
7. Clarke MC, Figg N, Maguire JJ, et al. Apoptosis of vascular smooth muscle cells induces features of plaque vulnerability in atherosclerosis. *Nat Med*. 2006;12:1075–1080.

8. Fredriksson L, Li H, Eriksson U. The PDGF family: Four gene products form five dimeric isoforms. *Cytokin Growth Factor Rev*. 2004; 15:197–204.
9. Kawai-Kowase K, Owens GK. Multiple repressor pathways contribute to phenotypic switching of vascular smooth muscle cells. *Am J Physiol Cell*. 2007;292:C59–C69.
10. Grainger DJ. Transforming growth factor beta and atherosclerosis: So far, so good for the protective cytokine hypothesis. *Arterioscler Thromb Vasc Biol*. 2004;24:399–404.
11. Derwall M, Malhotra R, Lai CS, et al. Inhibition of bone morphogenetic protein signaling reduces vascular calcification and atherosclerosis. *Arterioscler Thromb Vasc Biol*. 2012;32: 613–622.
12. Yao Y, Bennett BJ, Wang X, et al. Inhibition of bone morphogenetic proteins protects against atherosclerosis and vascular calcification. *Circ Res*. 2010;107:485–494.
13. Tedgui A, Mallat Z. Cytokines in atherosclerosis: Pathogenic and regulatory pathways. *Physiol Rev*. 2006;86:515–581.
14. Ferrario CM, Richmond RS, Smith R, Levy P, Strawn WB, Kivlighn S. Renin-angiotensin system as a therapeutic target in managing atherosclerosis. *Am J Therapeut*. 2004;11:44–53.
15. Tan NY, Li JM, Stocker R, Khachigian LM. Angiotensin II-inducible smooth muscle cell apoptosis involves the angiotensin II type 2 receptor, GATA-6 activation, and FasL-Fas engagement. *Circ Res*. 2009;105:422–430.
16. Raines EW, Bornfeldt KE. Integrin alpha(7) beta(1) compels smooth muscle cells to maintain their quiescence. *Circ Res*. 2010;106:427–429.
17. Wang L, Zheng J, Du Y, et al. Cartilage oligomeric matrix protein maintains the contractile phenotype of vascular smooth muscle cells by interacting with alpha7beta1 integrin. *Circ Res*. 2010;106:514–525.
18. Li G, Jin R, Norris RA, et al. Periostin mediates vascular smooth muscle cell migration through the integrinsalphavbeta3 and alphavbeta5 and focal adhesion kinase (FAK) pathway. *Atherosclerosis*. 2010;208:358–365.
19. Srivatsa SS, Fitzpatrick LA, Tsao PW, et al. Selective alphavbeta3 integrin blockade potently limits neointimal hyperplasia and lumen stenosis following deep coronary arterial stent injury: Evidence for the functional importance of integrin alphavbeta3 and osteopontin expression during neointima formation. *Cardiovasc Res*. 1997;36:408–428.
20. Wang T, Baron M, Trump D. An overview of Notch3 function in vascular smooth muscle cells. *Prog Biophys Mol Biol*. 2008;96:499–509.
21. Li Y, Takeshita K, Liu PY, et al. Smooth muscle Notch1 mediates neointimal formation after vascular injury. *Circulation*. 2009;119:2686–2692.
22. Aoyama T, Takeshita K, Kikuchi R, et al. Gamma-secretase inhibitor reduces diet-induced atherosclerosis in apolipoprotein E-deficient mice. *Biochem Biophys Res Commun*. 2009; 383:216–221.
23. Wheelock MJ, Johnson KR. Cadherins as modulators of cellular phenotype. *Ann Rev Cell Dev Biol*. 2003;19:207–235.
24. Lyon CA, Johnson JL, Williams H, Sala-Newby GB, George SJ. Soluble N-cadherin overexpression reduces features of atherosclerotic plaque instability. *Arterioscler Thromb Vasc Biol*. 2009;29:195–201.
25. Hou R, Liu L, Anees S, Hiroyasu S, Sibinga NES. The Fat1 cadherin integrates vascular smooth muscle cell growth and migration signals. *J Cell Biol*. 2006;173:417–429.
26. Wullschleger S, Loewith R, Hall MN. TOR signaling in growth and metabolism. *Cell*. 2006;124:471–484.
27. Martin KA, Rzucidlo EM, Merenick BL, et al. The mTOR/p70 S6K1 pathway regulates vascular smooth muscle cell differentiation. *Am J Physiol Cell Physiol*. 2004;286:C507–C517.
28. Mueller MA, Beutner F, Teupser D, Ceglarek U, Thiery J. Prevention of atherosclerosis by the mTOR inhibitor everolimus in LDLR-/- mice despite severe hypercholesterolemia. *Atherosclerosis*. 2008;198:39–48.
29. Pakala R, Stabile E, Jang GJ, Clavijo L, Waksman R. Rapamycin attenuates atherosclerotic plaque progression in apolipoprotein-E knockout mice: Inhibitory effect on monocyte chemotaxis. *J Cardiovasc Pharmacol*. 2005; 46:481–486.
30. Bedard K, Krause KH. The Nox family of ROS-generating NADPH oxidases: Physiology and pathophysiology. *Physiol Rev*. 2007; 87:245–313.
31. Clempus RE, Griendling KK. Reactive oxygen species signaling in vascular smooth muscle cells. *Cardiovasc Res*. 2006;71:216–225.
32. Xie C, Ritchie RP, Huang H, Zhang J, Chen YE. Smooth muscle cell differentiation in vitro: Models and underlying molecular

mechanisms. *Arterioscler Thromb Vasc Biol.* 2011;31:1485–1494.

33. Xie C, Guo Y, Zhu T, Zhang J, Ma PX, Chen YE. Yap1 protein regulates vascular smooth muscle cell phenotypic switch by interaction with myocardin. *J Biol Chem* 2012;287:14598–14605.
34. Gomez D, Owens GK. Smooth muscle cell phenotypic switching in atherosclerosis. *Cardiovasc Res.* 2012;95(2):156–164.
35. Vickers KC, Remaley AT. MicroRNAs in atherosclerosis and lipoprotein metabolism. *Curr Opin Endocrinol Diab Obes.* 2010;17:150–155.
36. Davis-Dusenbery BN, Wu C, Hata A. Micromanaging vascular smooth muscle cell differentiation and phenotypic modulation. *Arterioscler Thromb Vasc Biol.* 2011;31: 2370–2377.
37. Martinet W, De Meyer GR. Autophagy in atherosclerosis: A cell survival and death phenomenon with therapeutic potential. *Circ Res.* 2009;104:304–317.
38. Van Vre EA, Ait-Oufella H, Tedgui A, Mallat Z. Apoptotic cell death and efferocytosis in atherosclerosis. *Arterioscler Thromb Vasc Biol.* 2012;32:887–893.
39. Martinet W, Schrijvers DM, De Meyer GR. Pharmacological modulation of cell death in atherosclerosis: A promising approach towards plaque stabilization? *Br J Pharmacol.* 2011; 164:1–13.
40. Scull CM, Tabas I. Mechanisms of ER stress-induced apoptosis in atherosclerosis. *Arterioscler Thromb Vasc Biol.* 2011;31:2792–2797.
41. Martinet W, Schrijvers DM, De Meyer GR. Necrotic cell death in atherosclerosis. *Bas Res Cardiol.* 2011;106:749–760.

11

FIBROBLASTS

Christopher Mack

Department of Pathology, University of North Carolina, Chapel Hill, NC, USA

INTRODUCTION

Blood vessels are exposed to a variety of pathologic stimuli that lead to tissue damage including hemodynamic forces, atherogenic lipids, and inflammatory cytokines. Although fibroblasts play a ubiquitous role in the process of wound healing, most studies examining the vessel injury response have focused on the role of smooth muscle cells (SMCs). However, many studies have now demonstrated that fibroblasts within the adventitia contribute to the development and progression of atherosclerosis and postangioplasty restenosis. This chapter will highlight advancements in this area and will discuss the implications of recent findings on the developmental origins of vessel fibroblasts and the existence of a population of adventitial cells that may serve as vascular progenitors in adult animals.

FIBROBLAST-DEPENDENT WOUND HEALING

The process of fibroblast-dependent wound healing and connective tissue remodeling has been well-characterized (see Refs. [1, 2] for reviews). Under normal conditions, fibroblasts are mostly quiescent, have a relatively less-ordered cortical actin cytoskeleton, and basally express a variety of extracellular matrix (ECM) components that contribute to organ structure. They are identified by a combination of features including the expression of selective fibroblast marker genes (i.e., vimentin, fibroblast-specific protein-1, PDGF receptor β), localization to interstitial regions with high ECM content, and by the absence of markers for other cell lineages. Upon tissue damage, fibroblasts are activated by inflammatory mediators and other growth factors released from resident lymphocytes or parenchymal cells leading to a significant rise in

Atherosclerosis: Risks, Mechanisms, and Therapies, First Edition. Edited by Hong Wang and Cam Patterson.

fibroblast proliferation indices, migratory activity, and synthesis of ECM components. Another major event in wound healing is the differentiation of fibroblasts into myofibroblasts. This process is characterized by increased actin polymerization and stress fiber formation, and by the activation of SM α-actin, an early marker of differentiated SMC. These cytoskeletal changes provide the increase in cellular tension that is important for proper wound closure. SM α-actin expression is considered a major hallmark of myofibroblast differentiation and is frequently used to identify myofibroblasts in injured tissue. It is important to note that all fibroblast/myofibroblast markers recognize other mesenchymally derived cell types, and that fibroblasts even within the same tissue are fairly heterogenous in regard to embryonic origin, marker gene expression profiles, and functional characteristics [2]. Thus, it is critical to use caution when identifying fibroblasts/myofibroblasts within a tissue and when interpreting results from different fibroblast populations. Although myofibroblasts typically apoptose upon completion of wound repair, they can persist under chronic inflammatory conditions or repeated physical injury. The extensive tissue fibrosis that occurs under these circumstances can mechanically interfere with normal tissue function effectively aggravating the underlying pathology.

ROLE OF THE ADVENTITIA IN THE VESSEL INJURY RESPONSE

The major source of fibroblasts in the vessel wall is the tunica adventitia, a loosely structured layer of collagen- and proteoglycan-rich connective tissue that surrounds larger vessels, and is separated from the medial layer of differentiated SMCs by the outer elastic lamina. In addition to fibroblasts, the adventitia contains resident lymphocytes, perivascular nerves, and the vasa vasorum—a capillary network that supplies the outer layers of the vessel wall. Recent studies suggest that the adventitia is also a source of vascular progenitor cells (see the following text). The involvement of adventitial fibroblasts in the development of atherosclerosis and postangioplasty restenosis is relatively underappreciated. Extensive studies have shown that phenotypically modulated medial SMC migrate into the intimal space where they proliferate and secrete matrix components to stabilize the developing plaque (see Chapter 10 for more details). However, many pathologic insults including atherosclerosis, hypertension, and balloon injury have been shown to activate adventitial fibroblasts in a time course that parallels, and often precedes, changes in SMC phenotype and intimal growth [3, 4]. It is now becoming clear that these activated adventitial cells contribute significantly to vessel remodeling by several mechanisms.

Soluble factors released from activated fibroblasts and inflammatory cells within the adventitia affect the function of SMC within the media and intima. Several components of the nicotinamide adenine dinucleotide phosphate (NADPH) oxidase system are upregulated in adventitial fibroblasts in response to vascular injury or atherosclerosis [3], and an increase in reactive oxygen species (ROS) primarily in the adventitia has been detected under these conditions [5, 6]. ROS stimulate SMC growth and migration [3, 5], and importantly, adenoviral delivery of a dominant-negative NADPH oxidase to the adventitia significantly decreased neointimal hyperplasia in a rat carotid injury model [7]. Adventitial fibroblasts also express a number of cytokines in an injury-dependent manner including monocyte chemotactic protein-1 [8]. The accumulation of inflammatory cells that can occur as early as 2 hours after balloon injury then potentiates ROS generation [9]. Adhesion receptor expression in the vasa vasorum facilitates inflammatory cell recruitment to injured vessels [9], and it has been hypothesized that increased angiogenesis within the vessel wall plays a role in the development and progression of atherosclerosis [10]. Supporting this idea, an expansion of the vaso vasorum in apolipoprotein E (apoE)/low-density lipoprotein (LDL) double knockout mice strongly correlated with lesion development [11], while treatment of LDL knockout mice with the angiogenesis inhibitor, angiostatin, decreased plaque angiogenesis and lesion size [12]. Increased adventitial angiogenesis and lymphocyte recruitment also correlated with vessel thickening during the development of hypertension [13, 14].

Interestingly, adventitial fibroblasts have been shown to migrate through the medial SMC layer and to directly contribute to intima formation. In these studies, rat carotid artery adventitial cells were infected with a LacZ expressing adenovirus before balloon injury. A substantial number of LacZ-positive cells were observed in the intima and media at 7 and 14 days post injury, while LacZ-expressing cells were confined to the adventitia in uninjured vessels [15, 16]. In addition, perivascular transfer of inhibitory SMAD7 [17] or a dominant-negative variant of platelet-derived growth factor (PDGF) receptor β [18], attenuated the appearance of LacZ-labeled cells in the intima strongly suggesting that transforming growth factor (TGF) and PDGF signaling pathways were critical for adventitial cell activation and/or migration in this model. It is important to note that not all studies have demonstrated a significant contribution of adventitial fibroblasts to intima formation [19–21] and that the precise origins of SM α-actin expressing intimal cells are still somewhat controversial. Species and model-specific differences in the type and extent of injury may partially explain these discrepancies. However, the similarity between phenotypically modulated SMC and activated fibroblasts has also made it difficult to trace the origin of cells within complex atherosclerotic lesions. Separate groups have now used inducible SMC-specific Cre-dependent mouse lines crossed to the Rosa26 LacZ line to fate map SMC in $apoE^{-/-}$ and wire injury models [22, 23]. Importantly, both studies demonstrated that the majority of intimal SMCs derive from the differentiated cells within the medial layer. Although these results do not rule out the presence of adventitial fibroblasts within the intima, they suggest that fibroblasts may play a vital role as a source of paracrine factors that secondarily affect SMC phenotype and lymphocyte recruitment.

In addition to effects on intimal growth, the adventitia-dependent injury response also leads to vessel wall remodeling. In early atherosclerotic lesions, increased matrix metalloprotease production by fibroblasts and SMC results in the breakdown of matrix components [24]. The eventual thinning of the medial and adventitial layers allows for outward expansion of the vessel wall and the maintenance of lumen diameter, a process known as positive remodeling. Prolonged insults especially those associated with hypertension result in negative remodeling. This process is mediated in large part by adventitial fibroblast proliferation, myofibroblast-dependent cell contracture, and increased matrix synthesis [2]. Indeed, results from restenosis models indicate that constrictive remodeling, usually defined by a decrease in the circumference of the outer elastic lamina, contributes as much or more to reduction in lumen diameter as intimal growth [25]. Importantly, since new covalent cross-links form between matrix components, negative remodeling can lead to long-lasting changes in vessel function. Negative remodeling also decreases vessel wall compliance interfering with vasodilatation and causing changes in blood flow dynamics that can further enhance the atherosusceptibility of endothelial cells [26].

DEVELOPMENTAL ORIGINS OF ADVENTITIAL FIBROBLASTS

Although mesenchymal in origin, fibroblasts are distinct from the primary mesenchymal cells that arise during gastrulation. By the time fibroblasts appear in the mouse around E9, they have gone through several important transitions (see Ref. [2] for review). During early organ development, primary mesenchyme undergoes a mesenchymal–epithelial transition (MET) to create the epithelial layers that line the surfaces of most organs. In later stages of organ maturation, fibroblasts arise from the epithelial layer following an epithelial–mesenchymal transition (EMT). The down-regulation of cell–cell junction proteins (e.g., E-cadherin, ZO-1, occludins) coupled with the activation of migratory pathways (Rho GTPases, cytoskeletal regulators, etc.) allows these cells to escape the epithelial layer and take up residence in the interstitial space.

Although we are only beginning to understand the specific origins of adventitial cells, the development of the adventitial layers of the coronary and mesenteric vessels seems to follow this paradigm fairly closely. Early studies in quail-chick transplantation chimeras demonstrated that the SMC and adventitial layers that surround the coronary arteries originate from the epicardium [27], a mesothelial layer of cells that forms around the heart at around E9.5. These findings have since been confirmed by fate mapping studies in a mouse line that expresses Cre under the control of the mesothelial-specific Wilms tumor 1 (Wt1) promoter [28]. The process begins around E12.5 when epicardial cells undergo an EMT, delaminate from the epicardium, and move into the subepicardial space. A major portion of these epicardial-derived mesenchymal cells (EPDCs) are recruited to the myocardial vascular plexus, perhaps in response to endothelial-derived PDGF-BB [29], and then differentiate into SMC. Some EPDCs remain within the myocardium as cardiac fibroblasts, while others end up as fibroblasts in the adventitial layer [27]. Whether these cell types represent subpopulations of EPDCs or whether their fate is controlled by local cues is currently unknown. Wilm et al. demonstrated that a similar EMT occurs in serosal mesothelium that lines the intestines, and most of the SMCs that invest the mesenteric arteries originate from this structure [30]. Like the epicardial layer, the serosal mesothelium also contributes fibroblasts to the intestines and to the mesenteric adventitia (personal communication, David Bader, Vanderbilt University). Interestingly, the fibroblasts within the aortic arch adventitia do not originate in the cardiac neural crest like the SMCs in the medial layer beneath them [31]. Since the adventitial injury response could depend on adventitial cell origin, it will be important to fully delineate the origins of this and other adventitial fibroblast populations.

SMC are clearly distinguished from adventitial fibroblasts by the expression of the late and relatively more specific SMC differentiation markers such as SM myosin heavy chain and smoothelin. However, given their similar origins in at least some vascular beds and the fact that they both express SMC-specific proteins, significant questions remain as to whether these cell types represent different stages of SMC development. The characterization of heterozygous elastin knockout mice provides some interesting insight into the relationship between these two cell types [32]. It is well known that the number of SMC layers within muscular arteries correlates with the circumferential wall stress to which these vessels are subjected, and in the mouse thoracic aorta, SMC layering is established by around E16. In heterozygous elastin mice that have much thinner laminae, additional layers of SMCs are added to the inner adventitial layer during late gestation and the early postnatal period [32]. These results suggest that a mechanosensitive mechanism exists that determines the location of the SMC–adventitia border. A similar addition of SMC layers at the media–adventitia border occurs during the vessel remodeling associated with pulmonary hypertension [33].

VASCULAR PROGENITORS IN THE VESSEL WALL

Recent advancements in the identification and analysis of vascular stem cells have added to our understanding of injury response within the vessel wall. Hu et al. identified a population of vascular progenitor cells within the adventitia just outside of the outer elastic lamina [34]. These cells express the stem cell antigen-1 (Sca1) surface marker, and when isolated from the aortic root could activate the entire SMC differentiation marker gene profile in response to PDGF-BB. Interestingly, when LacZ labeled $Sca1^+$ cells were seeded onto the outer surface of a vein graft that was then transplanted into the arterial circulation, LacZ-expressing cells that were positive for SMC markers but negative for Sca1 were found within the intima at 4 weeks. Although $Sca1^+$ cells express PDGF-β receptor (a fibroblast marker), it is not completely clear whether $Sca1^+$ cells share a lineage with adventitial fibroblasts. Sonic hedgehog expression at the SMC–adventitia border was shown to be required for the maintenance of this progenitor niche [31], but the molecular signaling and transcriptional pathways involved remain to be described. Of interest, Sainz et al. identified a "side population" of $Sca1^+$ vascular progenitor cells in the medial layer of mouse aortas using Hoechst-dye exclusion methods [35]. These cells did not express SMC markers but could differentiate into SMC cells when treated with TGF-β or PDGF-BB. Taken together, these results suggest that the adventitia (and perhaps the media) contain precursors that can be mobilized into SMC (or SMC-like myofibroblasts) at least under pathologic or *in vitro* conditions. $Sca1^+$ cells could be induced to differentiate into endothelial cells by treatment with VEGF [34, 35], and Zengin et al. identified $CD34^+$ endothelial cell precursors at the SMC–adventitia border in human vessels [36]. Thus, it is possible that these progenitor cells contribute to the angiogenic response in the adventitia during the development of atherosclerosis or pulmonary hypertension [11, 13].

The bone marrow has also been shown to contain fibroblast and/or SMC precursors [37], and early bone marrow transplant studies suggested that a significant number of SM α-actin positive cells within the injured vessel wall were derived from this source [38]. However, the contributions of bone marrow-derived cells to SM α-actin expressing cells in the intima are minimal and can be overestimated by using imaging methods that lack single-cell resolution of marker gene expression. Thus, while bone marrow-derived cells certainly contribute to the vessel injury response, most of these are monocytes/macrophages that are present only during the transient inflammatory period. Several studies have shown that endothelial cells can transdifferentiate *in vitro* into fibroblast-like cells that express SM α-actin [39, 40]. This process is referred to as endothelial-mesenchymal transition (EndoMT) and plays a critical role in heart valve and ventricular septum development [41]. Zeisberg et al. used endothelial cell-specific Tie1-Cre mice to demonstrate that a large percentage of the SM α-actin-positive cardiac fibroblasts that emerge following aortic banding are of endothelial origin [42]. Whether EndoMT plays a significant role during the vessel injury response will be an important area for future studies.

SIGNALING MECHANISMS THAT INFLUENCE MYOFIBROBLAST DIFFERENTIATION

The most commonly known inducer of myofibroblast differentiation is TGF-β. This cytokine is released from a variety of cell types present within the injured vessel wall including platelets, macrophages, EC, and SMC, and autocrine production of TGF-β seems to be important for maintaining the fibrotic response after the initial inflammatory stimulus has subsided [43]. TGF-β signals through a heteromeric complex of type I and type II receptors, and many of its effects are mediated by Smad-dependent transcriptional activation (see Ref. [44] for review). Major downstream TGF-β targets in fibroblasts include SM α-actin, matrix components (collagens and elastin), matrix metallaproteases (MMPs 2 and 9), and matrix-synthesizing enzymes (lysyloxidase). TGF-β also alters fibronectin mRNA splicing to produce an ED-A variant that includes an "extra domain" [45]. Interestingly, fibronectin ED-A expression precedes SM α-actin expression in TGF-β-treated fibroblasts, and a blocking antibody specific to the ED-A domain inhibits TGF-β-dependent myofibroblast differentiation [45]. Although the precise mechanisms by which fibronectin ED-A regulates myofibroblast differentiation are not completely clear, the ED-A domain interacts with the alpha9/beta1 and alpha4/beta1 integrins [46] and likely signals through focal adhesions.

Extensive evidence indicates that mechanical tension is also critical for myofibroblast differentiation. By plating cells on collagen matrices with different compliances, Arora et al. demonstrated that myofibroblast differentiation correlated strongly with matrix rigidity [47]. Hinz et al. obtained similar results *in vivo*, showing that myofibroblasts appeared earlier and persisted longer in wounds that were splinted with a plastic frame that increased mechanical tension across granulation tissue [48]. Recent studies have shown that SM α-actin is only incorporated into stress fibers in cells that exhibit high levels of cell tension and "super mature" focal adhesions that help support excessive mechanical loads [49]. Since the presence of SM α-actin in stress fibers significantly increases mechanical force generation, SM α-actin expression seems to be part of a feed forward loop that helps maintain myofibroblast differentiation.

Taken together, these results support a model in which adventitial fibroblasts sense and respond to changes in mechanical tension by altering their differentiation state. Under this paradigm, cell death or matrix degradation in the vessel wall transfers mechanical load to adventitial fibroblasts stimulating myofibroblast differentiation. The increased tension generated by newly differentiated myofibroblasts provides support for the vessel wall, while new matrix components are been laid down and a new lattice established. Interestingly, the release of mechanical tension is an important apoptotic signal for myofibroblasts [50]. Thus, near the end of the injury response, the waning of inflammatory signals and the completion of vessel remodeling combine to promote myofibroblast cell death.

Many cell signaling pathways that are critical for myofibroblast function are regulated by RhoA, the founding member of the Rho subfamily of small GTPases. RhoA stimulates actin polymerization and stress fiber formation by activating downstream effectors including Rho-kinase and the diaphanous formins, a highly conserved family of proteins that catalyze actin nucleation (see Ref. [51] for review). Rho-kinase also regulates actin–myosin-dependent force generation by inhibiting myosin phosphatase. RhoA signaling enhances the expression of many contractile-associated genes including SM α-actin (see the following text), and recent studies have shown that RhoA activity was required for myofibroblast differentiation in response to TGF-β, mechanical force, and thrombin [52–54]. RhoA activation has also been implicated in the regulation of the epicardial EMT that occurs during coronary vessel development [55] and with EMTs that occur during pathologic fibrosis [56].

TRANSCRIPTION MECHANISMS THAT REGULATE MYOFIBROBLAST DIFFERENTIATION

Much of our knowledge on the transcriptional control of myofibroblast differentiation comes from studies on the regulation of SM α-actin expression, and it is clear that fibroblasts and SMC share similar mechanisms in this regard (see Chapter 10). However, the fact that vessel injury activates SM α-actin expression in adventitial fibroblasts but dramatically downregulates SM α-actin expression in media and intimal SMC strongly suggests that mechanisms exist that differentially regulate this gene in these cell types. A number of positive and negative cis regulatory elements have been identified within the SM α-actin promoter (see Figure 11.1). Two CArG elements within the first 125 base pairs of the proximal promoter and one within the first intron are particularly important and have been shown to bind the ubiquitously expressed transcription factor

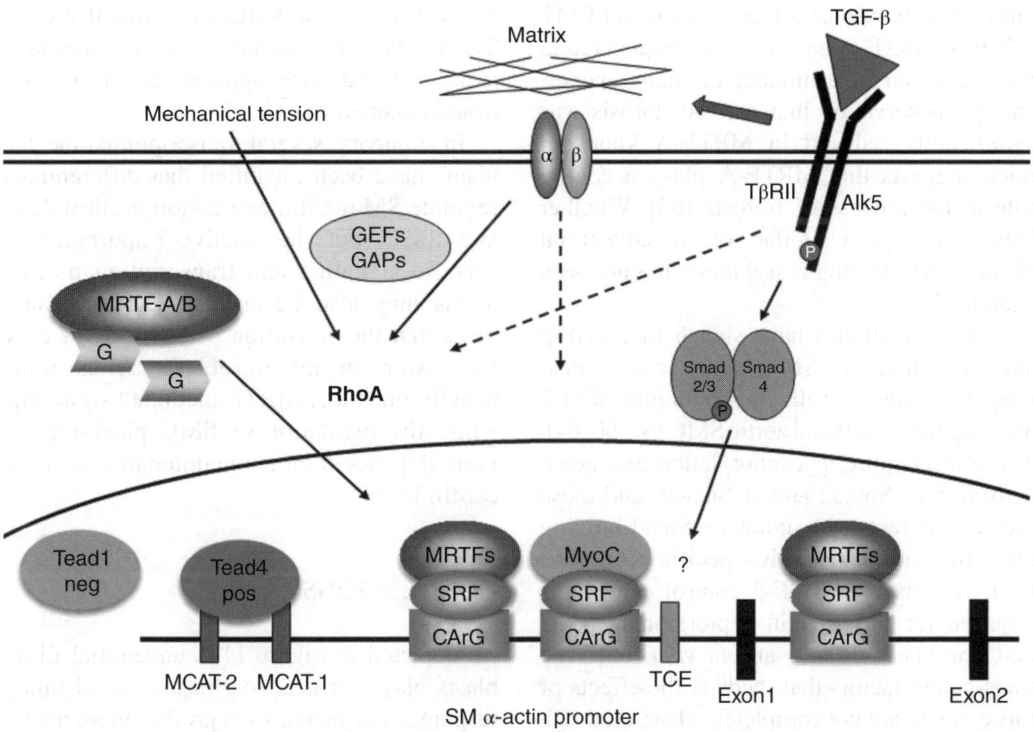

Figure 11.1 Transcription mechanisms that regulate SM α-actin expression. Serum response factor (SRF) and the myocardin family of SRF cofactors interact with three conserved CArG elements within the proximal promoter and first intron. While myocardin is a critical activator of SM α-actin expression in SMC, the myocardin-related factors A and B are likely more important in fibroblasts. The nuclear localization of the MRTFs is regulated by RhoA-dependent actin polymerization, and many extrinsic cues that stimulate SM α-actin expression in fibroblasts activate this pathway. The effects of TGF-β are mediated, at least in part, by phosphorylation of SMAD2/3 although the precise cis elements to which these factors bind have not been clearly defined. Finally, two MCAT elements upstream of the CArG region have positive transcriptional activity in fibroblasts, but act as inhibitory elements in SMC. These effects are thought to be mediated by differential expression of TEF transcription factor subtypes, TEAD4 in fibroblasts and TEAD1 in SMC.

serum response factor (SRF) [57]. Transgenic promoter analyses demonstrated that all three CArGs were required for SM α-actin expression in SMC and myofibroblasts, while only the proximal CArGs were required for SM α-actin expression in cardiac and skeletal muscle during development [57].

The SMC- and cardiac-selective SRF cofactor, myocardin, powerfully transactivates SM α-actin gene expression and has been shown to be required for SMC differentiation *in vitro* and *in vivo* [58]. Myocardin expression is significantly downregulated in intimal SMC and in primary SMC cultures strongly suggesting that its presence is a critical determinant of the differentiated SMC phenotype. Two myocardin-related transcription factors—MRTF-A and MRTF-B—have also been identified that are expressed more widely including in some epithelial layers and in many fibroblasts [59]. These factors have similar transcription properties to myocardin and have been shown to be important for SM α-actin expression in fibroblast cell lines [52, 60]. Interestingly, the nuclear localization of the MRTFs (but not myocardin) is controlled by RhoA-dependent actin polymerization [61], and this mechanism was shown to promote SM α-actin expression during myofibroblast differentiation and EMT [60, 62]. MRTF-A has also been implicated in the regulation of a number of matrix genes, and the observation that cardiac fibrosis was significantly reduced in MRTF-A knockout mice suggests that MRTF-A plays a central role in the process of fibrosis [63]. Whether MRTF-A plays a similar role in adventitial fibroblasts following vessel injury has not been examined.

Extensive studies have shown that TGF-β strongly stimulates SM α-actin gene expression in a number of cell types including 10T1/2 and lung fibroblasts and aortic SMC [52, 60, 64]. This effect requires phosphorylation-dependent activation of Smad2 and/or Smad3, and these factors interact with putative Smad-binding elements within the SMC-specific promoters [64]. A conserved TGF-β control element is required for SM α-actin expression in SMC and fibroblasts *in vitro* and *in vivo* [65]. The trans-acting factors that mediate the effects of this element are not completely clear, although some evidence suggests that it may involve competitive interactions between Kruppel-like and Sp1/Sp3 transcription factors [66]. TGF-β also upregulates the expression of many proteins that have been shown to have secondary effects on SM α-actin expression including matrix and matrix remodeling proteins [44], RhoA signaling molecules [67], and the NADPH oxidase, Nox4 [68].

Members of the TEAD/TEF transcription factor family interact with two MCAT elements within the proximal SM α-actin promoter and have been shown to differentially regulate SM α-actin in fibroblasts and SMC. For example, mutation of the MCAT elements significantly inhibited SM α-actin promoter activity in AKR-2B fibroblasts, but increased SM α-actin promoter activity in primary SMC [69]. Additional studies in transgenic mice by Gan et al. demonstrated that the MCAT mutations had no effect on SM α-actin promoter activity in adult SMC, but inhibited the SM α-actin promoter activity in myofibroblasts [70]. These authors went on to shown that the MCAT elements interacted with TEAD4 in fibroblasts, but with TEAD1 in SMC suggesting that these TEAD family members are differentially expressed and have opposite effects on SM α-actin expression.

In summary, several transcriptional mechanisms have been identified that differentially regulate SM α-actin expression in fibroblasts and SMC, but the relative importance of specific signaling and transcription mechanisms may also be critical. Evidence suggests that the activation of SM α-actin gene expression in myofibroblasts relies more heavily on RhoA/MRTF-mediated signaling, while the regulation of SMC phenotype is more dependent on the maintenance of myocardin levels.

CONCLUSIONS

As depicted in Figure 11.2, adventitial fibroblasts play a critical role in the vessel injury response. The increases in proliferation, matrix production, actin polymerization, and contractility that occur upon myofibroblast activation add mechanical support to the injured vessel. Although the primary therapeutic goal for the treatment of cardiovascular disease is to prevent the injury stimulus that initiates the process of atherosclerosis, an alternative strategy is to promote vessel repair in a manner that stabilizes a vulnerable plaque. An important consideration for such an approach would be to limit the inflammatory and fibrotic responses that can exacerbate the underlying pathology. Clearly, the vessel injury response is modified by inflammatory cells, humoral signals, and mechanical stresses, and a better understanding of the signals that regulate myofobroblast differentiation and apoptosis will be critical. Given that RhoA signaling regulates cytoskeletal reorganization, actin–myosin-based cell contractility, and myoblast differentiation, interventions that affect this pathway may be particularly useful.

The discovery of vascular stem cells within the adventitia is an exciting advance. While the importance of these cells during vascular

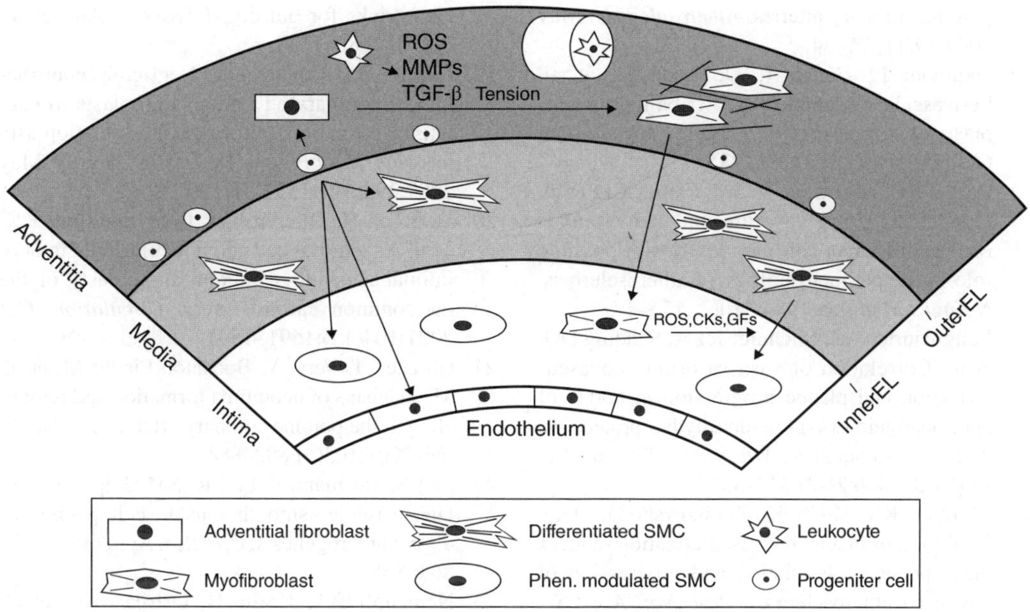

Figure 11.2 Contribution of the adventitia to vessel remodeling. Upon vessel injury, the release of reactive oxygen species (ROS), cytokines, growth factors, and matrix metalloproteases (MMPs) from adventitial fibroblasts and leukocytes facilitates the phenotypic modulation of SMC and the expansion of the intimal layer. These factors also stimulate the differentiation of adventitial fibroblasts into myofibroblasts. This process is important for vessel stabilization under these conditions; but when prolonged, it can lead to significant negative remodeling. Cell-tracking studies have demonstrated that some cells within the intima originated from the adventitia. When coupled with the discovery of progenitor cells at the adventia–media border, these data suggest that vessel homeostasis is the result of complex interactions between the many different cell types within the entire vessel wall.

development or the injury response is currently unknown, the ability to influence their function and fate could be a novel approach for treating cardiovascular disease.

REFERENCES

1. Desmouliere A, Chaponnier C, Gabbiani G. Tissue repair, contraction, and the myofibroblast. *Wound Repair Regen*. Jan-Feb 2005;13(1):7–12.
2. Neilson EG. Mechanisms of disease: Fibroblasts–a new look at an old problem. *Nat Clin Pract Nephrol*. Feb 2006;2(2):101–108.
3. Haurani MJ, Pagano PJ. Adventitial fibroblast reactive oxygen species as autacrine and paracrine mediators of remodeling: Bellwether for vascular disease? *Cardiovasc Res*. Sep 2007; 75(4):679–689.
4. Shi Y, Pieniek M, Fard A, et al. Adventitial remodeling after coronary arterial injury. *Circulation*. Jan 1996;93(2):340–348.
5. Szocs K, Lassegue B, Sorescu D, et al. Upregulation of Nox-based NAD(P)H oxidases in restenosis after carotid injury. *Arterioscler Thromb Vasc Biol*. Jan 2002;22(1):21–27.
6. Shi Y, Niculescu R, Wang D, et al. Increased NAD(P)H oxidase and reactive oxygen species in coronary arteries after balloon injury. *Arterioscler Thromb Vasc Biol*. May 2001; 21(5):739–745.
7. Weaver M, Liu J, Pimentel D, et al. Adventitial delivery of dominant-negative p67phox attenuates neointimal hyperplasia of the rat carotid artery. *Am J Physiol Heart Circ Physiol*. May 2006;290(5):H1933–H1941.
8. Jabs A, Okamoto E, Vinten-Johansen J, et al. Sequential patterns of chemokine- and chemokine receptor-synthesis following vessel wall injury in

porcine coronary arteries. *Atherosclerosis*. May 2007;192(1):75–84.

9. Okamoto E, Couse T, De Leon H, et al. Perivascular inflammation after balloon angioplasty of porcine coronary arteries. *Circulation*. Oct 2001;104(18):2228–2235.
10. Barger AC, Beeuwkes R, 3rd, Lainey LL, et al. Hypothesis: Vasa vasorum and neovascularization of human coronary arteries. A possible role in the pathophysiology of atherosclerosis. *N Engl J Med*. Jan 1984;310(3):175–177.
11. Langheinrich AC, Michniewicz A, Sedding DG, et al. Correlation of vasa vasorum neovascularization and plaque progression in aortas of apolipoprotein E(-/-)/low-density lipoprotein(-/-) double knockout mice. *Arterioscler Thromb Vasc Biol*. Feb 2006;26(2):347–352.
12. Moulton KS, Vakili K, Zurakowski D, et al. Inhibition of plaque neovascularization reduces macrophage accumulation and progression of advanced atherosclerosis. *Proc Natl Acad Sci U S A*. Apr 2003;100(8):4736–4741.
13. Davie NJ, Gerasimovskaya EV, Hofmeister SE, et al. Pulmonary artery adventitial fibroblasts cooperate with vasa vasorum endothelial cells to regulate vasa vasorum neovascularization: A process mediated by hypoxia and endothelin-1. *Am J Pathol*. Jun 2006;168(6):1793–1807.
14. Bush E, Maeda N, Kuziel WA, et al. CC chemokine receptor 2 is required for macrophage infiltration and vascular hypertrophy in angiotensin II-induced hypertension. *Hypertension*. Sep 2000; 36(3):360–363.
15. Li G, Chen SJ, Oparil S, et al. Direct in vivo evidence demonstrating neointimal migration of adventitial fibroblasts after balloon injury of rat carotid arteries. *Circulation*. Mar 2000; 101(12):1362–1365.
16. Siow RC, Mallawaarachchi CM, Weissberg PL. Migration of adventitial myofibroblasts following vascular balloon injury: Insights from in vivo gene transfer to rat carotid arteries. *Cardiovasc Res*. Jul 2003;59(1):212–221.
17. Mallawaarachchi CM, Weissberg PL, Siow RC. Smad7 gene transfer attenuates adventitial cell migration and vascular remodeling after balloon injury. *Arterioscler Thromb Vasc Biol*. Jul 2005; 25(7):1383–1387.
18. Mallawaarachchi CM, Weissberg PL, Siow RC. Antagonism of platelet-derived growth factor by perivascular gene transfer attenuates adventitial cell migration after vascular injury: New tricks for old dogs? *Faseb J*. Aug 2006; 20(10):1686–1688.
19. Fleenor BS, Bowles DK. Negligible contribution of coronary adventitial fibroblasts to neointimal formation following balloon angioplasty in swine. *Am J Physiol Heart Circ Physiol*. May 2009;296(5):H1532–H1539.
20. De Leon H, Ollerenshaw JD, Griendling KK, et al. Adventitial cells do not contribute to neointimal mass after balloon angioplasty of the rat common carotid artery. *Circulation*. Oct 2001;104(14):1591–1593.
21. Christen T, Verin V, Bochaton-Piallat M, et al. Mechanisms of neointima formation and remodeling in the porcine coronary artery. *Circulation*. Feb 2001;103(6):882–888.
22. Feil S, Hofmann F, Feil R. SM22alpha modulates vascular smooth muscle cell phenotype during atherogenesis. *Circ Res*. Apr 2004;94(7): 863–865.
23. Nemenoff RA, Horita H, Ostriker AC, et al. SDF-1alpha induction in mature smooth muscle cells by inactivation of PTEN is a critical mediator of exacerbated injury-induced neointima formation. *Arterioscler Thromb Vasc Biol*. Jun 2011;31(6):1300–1308.
24. Shi Y, Patel S, Niculescu R, et al. Role of matrix metalloproteinases and their tissue inhibitors in the regulation of coronary cell migration. *Arterioscler Thromb Vasc Biol*. May 1999;19(5): 1150–1155.
25. Michel JB, Thaunat O, Houard X, et al. Topological determinants and consequences of adventitial responses to arterial wall injury. *Arterioscler Thromb Vasc Biol*. Jun 2007;27(6): 1259–1268.
26. Aird WC. Phenotypic heterogeneity of the endothelium: I. Structure, function, and mechanisms. *Circ Res*. Feb 2007;100(2):158–173.
27. Dettman RW, Denetclaw W, Jr., Ordahl CP, et al. Common epicardial origin of coronary vascular smooth muscle, perivascular fibroblasts, and intermyocardial fibroblasts in the avian heart. *Dev Biol*. Jan 1998;193(2):169–181.
28. Zhou B, von Gise A, Ma Q, et al. Genetic fate mapping demonstrates contribution of epicardium-derived cells to the annulus fibrosis of the mammalian heart. *Dev Biol*. Feb 2010; 338(2):251–261.
29. Mellgren AM, Smith CL, Olsen GS, et al. Platelet-derived growth factor receptor beta signaling is required for efficient epicardial cell

migration and development of two distinct coronary vascular smooth muscle cell populations. *Circ Res*. Dec 2008;103(12):1393–1401.

30. Wilm B, Ipenberg A, Hastie ND, et al. The serosal mesothelium is a major source of smooth muscle cells of the gut vasculature. *Development*. Dec 2005;132(23):5317–5328.
31. Passman JN, Dong XR, Wu SP, et al. A sonic hedgehog signaling domain in the arterial adventitia supports resident Sca1+ smooth muscle progenitor cells. *Proc Natl Acad Sci U S A*. Jul 2008;105(27):9349–9354.
32. Wagenseil JE, Ciliberto CH, Knutsen RH, et al. The importance of elastin to aortic development in mice. *Am J Physiol Heart Circ Physiol*. Aug 2010;299(2):H257–H264.
33. Stenmark KR, Davie N, Frid M, et al. Role of the adventitia in pulmonary vascular remodeling. *Physiology (Bethesda)*. Apr 2006;21:134–145.
34. Hu Y, Zhang Z, Torsney E, et al. Abundant progenitor cells in the adventitia contribute to atherosclerosis of vein grafts in ApoE-deficient mice. *J Clin Invest*. May 2004;113(9): 1258–1265.
35. Sainz J, Al Haj Zen A, Caligiuri G, et al. Isolation of "side population" progenitor cells from healthy arteries of adult mice. *Arterioscler Thromb Vasc Biol*. Feb 2006;26(2):281–286.
36. Zengin E, Chalajour F, Gehling UM, et al. Vascular wall resident progenitor cells: A source for postnatal vasculogenesis. *Development*. Apr 2006;133(8):1543–1551.
37. Abe R, Donnelly SC, Peng T, et al. Peripheral blood fibrocytes: Differentiation pathway and migration to wound sites. *J Immunol*. Jun 2001; 166(12):7556–7562.
38. Sata M, Saiura A, Kunisato A, et al. Hematopoietic stem cells differentiate into vascular cells that participate in the pathogenesis of atherosclerosis. *Nat Med*. Apr 2002;8(4):403–409.
39. Kumarswamy R, Volkmann I, Jazbutyte V, et al. Transforming growth factor-beta-induced endothelial-to-mesenchymal transition is partly mediated by microRNA-21. *Arterioscler Thromb Vasc Biol*. Feb 2012;32(2):361–369.
40. Arciniegas E, Sutton AB, Allen TD, et al. Transforming growth factor beta 1 promotes the differentiation of endothelial cells into smooth muscle-like cells in vitro. *J Cell Sci*. Oct 1992;103(Pt 2):521–529.
41. Liebner S, Cattelino A, Gallini R, et al. Beta-catenin is required for endothelial-mesenchymal transformation during heart cushion development in the mouse. *J Cell Biol*. Aug 2004;166(3): 359–367.
42. Zeisberg EM, Tarnavski O, Zeisberg M, et al. Endothelial-to-mesenchymal transition contributes to cardiac fibrosis. *Nat Med*. Aug 2007; 13(8):952–961.
43. Tomasek JJ, Gabbiani G, Hinz B, et al. Myofibroblasts and mechano-regulation of connective tissue remodelling. *Nat Rev Mol Cell Biol*. May 2002;3(5):349–363.
44. Massague J, Seoane J, Wotton D. Smad transcription factors. *Genes Dev*. Dec 2005;19(23): 2783–2810.
45. Serini G, Bochaton-Piallat ML, Ropraz P, et al. The fibronectin domain ED-A is crucial for myofibroblastic phenotype induction by transforming growth factor-beta1. *J Cell Biol*. Aug 1998;142(3):873–881.
46. Shinde AV, Bystroff C, Wang C, et al. Identification of the peptide sequences within the EIIIA (EDA) segment of fibronectin that mediate integrin alpha9beta1-dependent cellular activities. *J Biol Chem*. Feb 2008;283(5):2858–2870.
47. Arora PD, Narani N, McCulloch CA. The compliance of collagen gels regulates transforming growth factor-beta induction of alpha-smooth muscle actin in fibroblasts. *Am J Pathol*. Mar 1999;154(3):871–882.
48. Hinz B, Mastrangelo D, Iselin CE, et al. Mechanical tension controls granulation tissue contractile activity and myofibroblast differentiation. *Am J Pathol*. Sep 2001;159(3): 1009–1020.
49. Goffin JM, Pittet P, Csucs G, et al. Focal adhesion size controls tension-dependent recruitment of alpha-smooth muscle actin to stress fibers. *J Cell Biol*. Jan 2006;172(2):259–268.
50. Grinnell F, Zhu M, Carlson MA, et al. Release of mechanical tension triggers apoptosis of human fibroblasts in a model of regressing granulation tissue. *Exp Cell Res*. May 1999;248(2):608–619.
51. Heasman SJ, Ridley AJ. Mammalian Rho GTPases: New insights into their functions from in vivo studies. *Nat Rev Mol Cell Biol*. Sep 2008;9(9):690–701.
52. Sandbo N, Lau A, Kach J, et al. Delayed stress fiber formation mediates pulmonary myofibroblast differentiation in response to TGF-beta. *Am J Physiol Lung Cell Mol Physiol*. Nov 2011; 301(5):L656–L666.

53. Chan MW, Chaudary F, Lee W, et al. Force-induced myofibroblast differentiation through collagen receptors is dependent on mammalian diaphanous (mDia). *J Biol Chem*. Mar 2010; 285(12):9273–9281.
54. Bogatkevich GS, Tourkina E, Silver RM, et al. Thrombin differentiates normal lung fibroblasts to a myofibroblast phenotype via the proteolytically activated receptor-1 and a protein kinase C-dependent pathway. *J Biol Chem*. Nov 2001; 276(48):45184–45192.
55. Lu J, Landerholm TE, Wei JS, et al. Coronary smooth muscle differentiation from proepicardial cells requires rhoA-mediated actin reorganization and p160 rho-kinase activity. *Dev Biol*. Dec 2001;240(2):404–418.
56. Fan L, Sebe A, Peterfi Z, et al. Cell contact-dependent regulation of epithelial-myofibroblast transition via the rho-rho kinase-phospho-myosin pathway. *Mol Biol Cell*. Mar 2007;18(3): 1083–1097.
57. Mack CP, Owens GK. Regulation of smooth muscle alpha-actin expression in vivo is dependent on CArG elements within the 5' and first intron promoter regions. *Circ Res*. Apr 1999; 84(7):852–861.
58. Wang Z, Wang DZ, Pipes GC, et al. Myocardin is a master regulator of smooth muscle gene expression. *Proc Natl Acad Sci U S A*. Jun 2003; 100(12):7129–7134.
59. Wang DZ, Li S, Hockemeyer D, et al. Potentiation of serum response factor activity by a family of myocardin- related transcription factors. *Proc Natl Acad Sci U S A*. Nov 2002; 99(23):14855–14860.
60. Hinson JS, Medlin MD, Lockman K, et al. Smooth muscle cell-specific transcription is regulated by nuclear localization of the myocardin-related transcription factors. *Am J Physiol Heart Circ Physiol*. Feb 2007;292(2):H1170–H1180.
61. Miralles F, Posern G, Zaromytidou AI, et al. Actin dynamics control SRF activity by regulation of its coactivator MAL. *Cell*. May 2003; 113(3):329–342.
62. Sebe A, Masszi A, Zulys M, et al. Rac, PAK and p38 regulate cell contact-dependent nuclear translocation of myocardin-related transcription factor. *FEBS Lett*. Jan 2008;582(2): 291–298.
63. Small EM, Thatcher JE, Sutherland LB, et al. Myocardin-related transcription factor-a controls myofibroblast activation and fibrosis in response to myocardial infarction. *Circ Res*. Jul 2010;107(2):294–304.
64. Hu B, Wu Z, Phan SH. Smad3 mediates transforming growth factor-beta-induced alpha-smooth muscle actin expression. *Am J Respir Cell Mol Biol*. Sep 2003;29(3 Pt 1):397–404.
65. Tomasek JJ, McRae J, Owens GK, et al. Regulation of alpha-smooth muscle actin expression in granulation tissue myofibroblasts is dependent on the intronic CArG element and the transforming growth factor-beta1 control element. *Am J Pathol*. May 2005;166(5):1343–1351.
66. Adam PJ, Regan CP, Hautmann MB, et al. Positive- and negative-acting Kruppel-like transcription factors bind a transforming growth factor beta control element required for expression of the smooth muscle cell differentiation marker SM22alpha in vivo. *J Biol Chem*. Dec 2000;275(48):37798–37806.
67. Vardouli L, Vasilaki E, Papadimitriou E, et al. A novel mechanism of TGFbeta-induced actin reorganization mediated by Smad proteins and Rho GTPases. *Febs J*. Aug 2008;275(16): 4074–4087.
68. Cucoranu I, Clempus R, Dikalova A, et al. NAD(P)H oxidase 4 mediates transforming growth factor-beta1-induced differentiation of cardiac fibroblasts into myofibroblasts. *Circ Res*. Oct 2005;97(9):900–907.
69. Swartz EA, Johnson AD, Owens GK. Two MCAT elements of the SM alpha-actin promoter function differentially in SM vs. non-SM cells. *Am J Physiol*. Aug 1998;275(2 Pt 1): C608–C618.
70. Gan Q, Yoshida T, Li J, et al. Smooth muscle cells and myofibroblasts use distinct transcriptional mechanisms for smooth muscle alpha-actin expression. *Circ Res*. Oct 2007; 101(9):883–892.

12

MONOCYTES AND MACROPHAGES IN ATHEROSCLEROSIS

Shu Meng[1], Pu Fang[3], Xiaofeng Yang[2,3], and Hong Wang[2,3]

[1] *Methodist Research Institute, Department of Cardiovascular Sciences, Houston, TX, USA*

[2] *Cardiovascular Research Center, Department of Pharmacology, Temple University School of Medicine, Philadelphia, PA, USA*

[3] *Center for Metabolic Disease Research, Temple University School of Medicine, Philadelphia, PA, USA*

INTRODUCTION

Atherosclerosis is a chronic inflammatory disease with atherosclerotic plaques predominantly consisting of infiltrated monocytes and macrophages. Monocyte infiltration and differentiation into macrophage and foam cells in the vessel wall are the critical processes of atherosclerosis initiation and progression. Recent advancements in immunology research have indicated that monocytes and macrophages are heterogeneous in both humans and mice with three distinct monocyte subsets and two major macrophage subsets identified. These subsets have great heterogeneity in terms of expression of cell surface marker, adhesion, and chemotactic molecules. They have distinct cytokine production profile and infiltrating capacity. Emerging evidence has implicated inflammatory subsets of monocyte and macrophage promotes atherosclerosis. This chapter reviews the role of monocytes and macrophages in atherogenesis including cell origin, differentiation, and transcriptional regulation. This chapter also emphasizes monocyte and macrophage heterogeneity, subset characterization, and functional implication in pathological conditions, such as hyperlipidemia and hyperhomocysteinemia.

MONOCYTE AND MACROPHAGE IN ATHEROSCLEROSIS

Atherosclerosis is featured by monocyte and T-cell infiltration, macrophage differentiation, lipid deposition, foam cell formation, and atherosclerotic plaque development in the vessel wall. Monocytes and macrophages have been found to be the major cellular components in the atherosclerotic plaque both morphologically and functionally. Using bone marrow transplantation approach by transfusing bone marrow cells from enhanced green fluorescent protein (EGFP) transgenic mice after sublethal irradiation, we (Figure 12.1) and others

Atherosclerosis: Risks, Mechanisms, and Therapies, First Edition. Edited by Hong Wang and Cam Patterson.

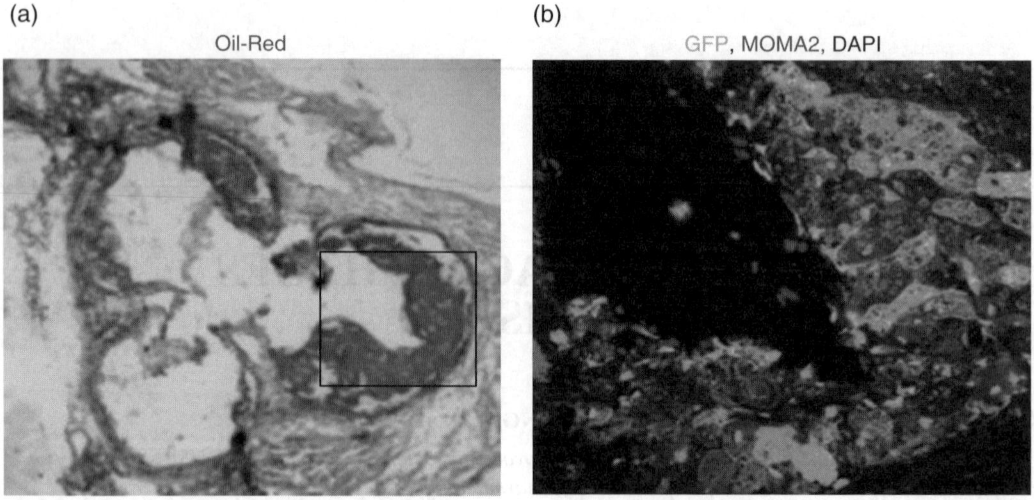

Figure 12.1 Monocyte/macrophage infiltration in atherosclerotic lesion. *Ldlr*$^{-/-}$ mice were lethally irradiated, received bone marrow transplantation from GFP mice at the age of 8 weeks and fed with high-fat diet for 6 weeks. (a) Oil Red O staining of mouse aortic sinus cross-section. Oil Red O staining is a classic staining method to detect triglyceride and lipids. The red staining indicates lipid deposition in the atherosclerotic lesion. (b) Immunostaining of aortic sinus cross-section. Only the section within the black box in (a) was shown. Antibodies against GFP (*green*), MOMA-2 (*red*, monocyte/macrophage marker), and DAPI (*blue*, nuclear marker). *(See insert for color representation of the figure.)*

demonstrated that bone marrow derived monocytes and macrophages (green, GFP$^+$ and red, MOMA-2$^+$) are dominant populations in atherosclerotic lesion [1]. Monocytes are present in the fatty steak, the earliest visible lesion in human arteries [2]. Circulating monocytes could infiltrate into the peripheral tissues and differentiate into macrophages, inflammatory dendritic cells (DCs), tumor necrosis factor-α (TNF-α) and inducible nitric oxide synthase (iNOS)-producing DC (TIP DCs) [3] or other monocyte/macrophage lineage cells such as osteoclasts [4]. In human atherosclerosis, CD68 expressing macrophages are enriched within the arterial wall [5]. In addition, T cells may be involved in the atherosclerosis formation, especially in the early phase of plaque formation [6].

Experimental and clinical studies strongly suggest that monocytes and macrophages are not only present in atherosclerosis but also significantly contribute to its initiation and progression. Atherosclerosis is often accompanied with monocytosis, an abnormal increase in the numbers of circulating monocytes. Several lines of evidence support a contributory role for monocyte and macrophage in the development of atherosclerosis. First, M280 polymorphism of CX3C chemokine receptor 1 (CX3CR1), a proinflammatory leukocyte receptor expressed in most monocytes, NK cells, and both CD4$^+$ and CD8$^+$ cytotoxic T lymphocytes are associated with reduced prevalence of acute coronary events, atherosclerosis, and coronary endothelial dysfunction in humans [7]. Second, depleting macrophages by clodronate-containing liposomes reduced neointima formation in balloon-injured coronary in rats and rabbits [8]. Third, deficiency of macrophage colony-stimulating factor (M-CSF) gene, the key molecule in monocyte differentiation and trafficking, largely abolishes atherosclerosis in male apolipoprotein E-deficient (apoE$^{-/-}$)/osteopetrotic mutant (op/op) mice [9]. Fourth, functional inhibition of monocyte recruitment into the vessel wall by the interference of chemokine receptors including chemokine (C–C motif) receptor 2 (CCR2), CX3CR1, and CCR5, drastically reduced atherosclerosis in mouse models and was associated with reduced monocyte infiltration. Finally, macrophages may exacerbate myocardial infarction, as both the results from animal [10] and clinical [11] studies suggest the detrimental roles of monocyte and macrophage recruitment in myocardial infarction and heart failure.

MONOCYTE ORIGIN AND DIFFERENTIATION

Developmental commitment to the monocyte lineage in bone marrow is determined at the stage of the macrophage and DC progenitor (MDP), at which point, lymphoid, erythroid, megakaryocyte, and granulocyte fates have been excluded [12] (Figure 12.2). MDPs are originally from hematopoietic stem cells (HSCs) through several intermediary progenitor stages. HSCs give rise to common myeloid progenitors (CMPs) with the help of the Ets family transcription factor PU.1. CMPs further differentiate into granulocyte and macrophage progenitors (GMPs). GMPs give rise to MDPs. MDPs give rise to monocytes and common DC progenitors (CDP) [13]. Genetic studies with knockout mice have revealed the crucial role of transcription factors such as PU.1 and CCAAT enhancer-binding protein α (C/EBPα) in monocyte/macrophage lineage commitment [14, 15]. The transcription factor Kruppel-like factor 4 (KLF4) plays a significant role in monocyte differentiation from earlier myeloid progenitors and can rescue PU.1$^{-/-}$ fetal liver cells along the monocytic lineage [16]. Gene-deficient mice whose monocyte progenitor cells have defective cholesterol efflux due to the deficiency of ABCA1 and ABCG1 transporters have increased monocyte numbers and increased atherosclerosis [17]. Monocytes can directly differentiate into macrophages, DCs [12], or TIP DCs. It is now well accepted that monocyte and macrophage differentiation occurs in a stepwise manner through a strictly time-dependent change of transcriptional programs [18]. However, it remains unclear how cell fate is decided and what transcription factors drive MDP differentiation. A recent report showed monocyte-predisposed hematopoietic progenitor cells in the adventitia of mouse postnatal aorta [19].

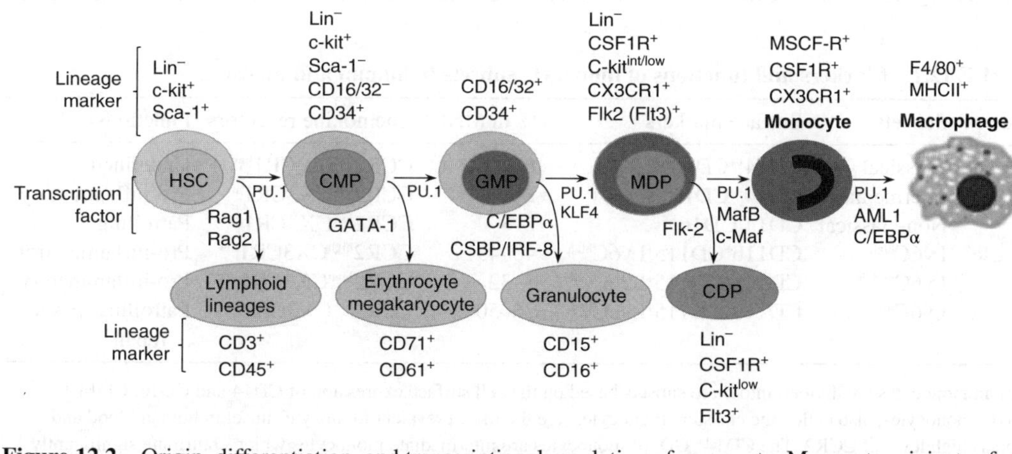

Figure 12.2 Origin, differentiation, and transcriptional regulation of monocyte. Monocyte originates from MDP through the activation of transcription factor PU.1, MafB and c-Maf. MDP is originally from HSC. HSC gives rise to CMP and then further differentiates into GMP. GMP gives rise to MDP with the help of transcription factor PU.1 and KLF4. Monocytes will differentiate into tissue macrophages, and this process involves transcription factor PU.1, AML1, and C/EBPα. HSC, hematopoietic stem cells; CMP, common myeloid progenitor; GMP, granulocyte and macrophage progenitor; MDP, macrophage and dendritic cells progenitors; and CDP, common dendritic cell progenitor. Lineage markers and transcription factors suggested for lineage differentiation are indicated.

MONOCYTE SUBSETS AND TRANSCRIPTIONAL REGULATION

Circulating monocytes constitute 5–10% of the peripheral blood leukocytes in humans. It becomes increasingly clear that monocytes are not homogenous populations but discrete functional subsets. CD14 has been used as a cell surface marker to distinguish human monocytes. Based on the cell surface expression of CD16, human monocytes are further characterized as three distinct populations, including $CD14^{++}CD16^{-}$, $CD14^{++}CD16^{+}$, and $CD14^{+}CD16^{++}$ monocyte subsets [3, 20] (Table 12.1). The majority of circulating human monocytes (80–95%) are $CD14^{++}$ $CD16^{-}$ cells, termed "classical monocytes" that express high levels of CCR2 and lymphocyte selectin (L-selectin or CD62L), low levels of CX3CR1, and major histocompatibility complex class II (MHC II). The function of human $CD14^{++}CD16^{-}$ classical monocytes remains undefined. The $CD14^{++}CD16^{+}$ subset is termed "intermediate monocytes" and consists of 2–11% human circulating monocytes. These populations express CCR5 [21], secret proinflammatory cytokines such as interleukin-1β (IL-1β) and TNF-α [22], and is associated with atherosclerosis development in human. Another subset is the $CD14^{+}CD16^{++}$ nonclassical monocytes (2–8%), which express low levels of CCR2 and CD62L, and high levels of CX3CR1 and MHC II [23, 24]. Human $CD14^{+}CD16^{++}$ nonclassical monocytes fulfill patrolling and antiviral function [3]. $CD14^{+}$ $CD16^{++}$ monocytes secrete proinflammatory cytokines IL-1β and TNF-α, mediated by a toll-like receptor (TLR) 7—adaptor MyD88—MEK kinase signaling in response to bacterial lipopolysaccharide (LPS), the cell wall component of gram-negative bacteria and a TLR4 ligand [25]. However, the detailed function of these monocyte subsets is not fully understood.

Mouse monocyte counterparts of human monocyte subsets are termed $Ly6C^{low}$, $Ly6C^{middle}$, and $Ly6C^{high}$ monocyte subsets based on the levels of their cell surface expression of Ly6C, a chemokine receptor (Table 12.1). The surface markers for $Ly6C^{high}$ and $Ly6C^{middle}$ monocytes are $CD11b^{+}CD115^{+}Ly6C^{high}$ and $CD11b^{+}$ $CD115^{+}Ly6C^{middle}$, respectively. These two subsets express the same chemokine receptors—$CCR2^{high}$ $CX3CR1^{low}$. By contrast, the surface

TABLE 12.1 Markers and functions of monocyte subsets in human and mouse

Species	Subsets	Surface markers	% in MNC	Chemokine receptors	Functions
Human	Classical	$CD14^{++}CD16^{-}$	80–95	$CCR2^{hi}CX3CR1^{low}$	Undefined
	Intermediate	$CD14^{++}CD16^{+}$	2–11	$CCR2^{low}CX3CR1^{hi}$	Pro-inflammatory
	Non-classical	$CD14^{+}CD16^{++}$	2–8	$CCR2^{low}CX3CR1^{hi}$	Patrolling
Mouse	$Ly6C^{high}$	$CD11b^{+}CD115^{+}Ly6C^{high}$	40–45	$CCR2^{high}CX3CR1^{low}$	Pro-inflammatory
	$Ly6C^{middle}$	$CD11b^{+}CD115^{+}Ly6C^{middle}$	5–32	$CCR2^{high}CX3CR1^{low}$	Pro-inflammatory
	$Ly6C^{low}$	$CD11b^{+}CD115^{+}Ly6C^{low}$	26–50	$CCR2^{low}CX3CR1^{high}$	Patrolling; tissue repair

Human monocytes are divided into three subsets based on the cell surface expression of CD14 and CD16. $CD14^{++}$ $CD16^{-}$ monocytes, also called the classical monocytes, are the most prevalent monocyte subset in human blood and express high level of CCR2. The $CD14^{++}CD16^{+}$ monocytes are intermediate monocytes which contribute significantly to atherosclerosis. The $CD14^{+}CD16^{++}$ monocytes are referred to as non-classical monocytes which perform *in vivo* patrolling function. Mouse monocytes are divided into two subsets based on their cell surface expression of Ly6C. The $Ly6C^{high}$ and $Ly6C^{middle}$ subsets perform pro-inflammatory function and express high level of CCR2, which is considered as the counterpart of human classical monocytes. The $Ly6C^{low}$ subsets express low level of CCR2, majorly patrol along the vascular endothelium and are involved in tissue repair, functionally similar to non-classical monocytes. CD, cluster of differentiation; CCR2, chemokine (C-C motif) receptor 2; CX3CR1, CX3C chemokine receptor 1; Ly6C, lymphocyte antigen 6 complex.

markers and chemokine receptors for Ly6C^{low} monocytes are CD11b^{+}CD115^{+} Ly6C^{low} and CX3CR1^{high}CCR2^{low}, respectively [26].

All three monocyte subsets are generated in the bone marrow and released into circulation. Spleen is the major repertoire of the blood monocytes. Unlike human monocyte subsets, Ly6C^{high} and Ly6C^{low}, monocytes have a 1:1 ratio in the blood. The mouse Ly6C^{high} monocytes are considered as the counterparts of human CD14^{++}CD16^{-} classical monocytes for their high level of CCR2 and low level of CX3CR1 [23]. Ly6C^{high} monocytes also express higher levels of CD11a, CD31, CD49b, and CD62L, and have lower expressions of CX3CR1, CD11c, and CD43 compared with Ly6C^{low} monocytes [4].

Ly6C^{high} and Ly6C^{middle} monocytes are classified as "inflammatory" monocytes due to their high expression of chemokine receptors, high infiltration capacities, and secretion of proinflammatory cytokines in response to stimulation in the cultured system (Figure 12.3). Based on gene expression profile analysis from

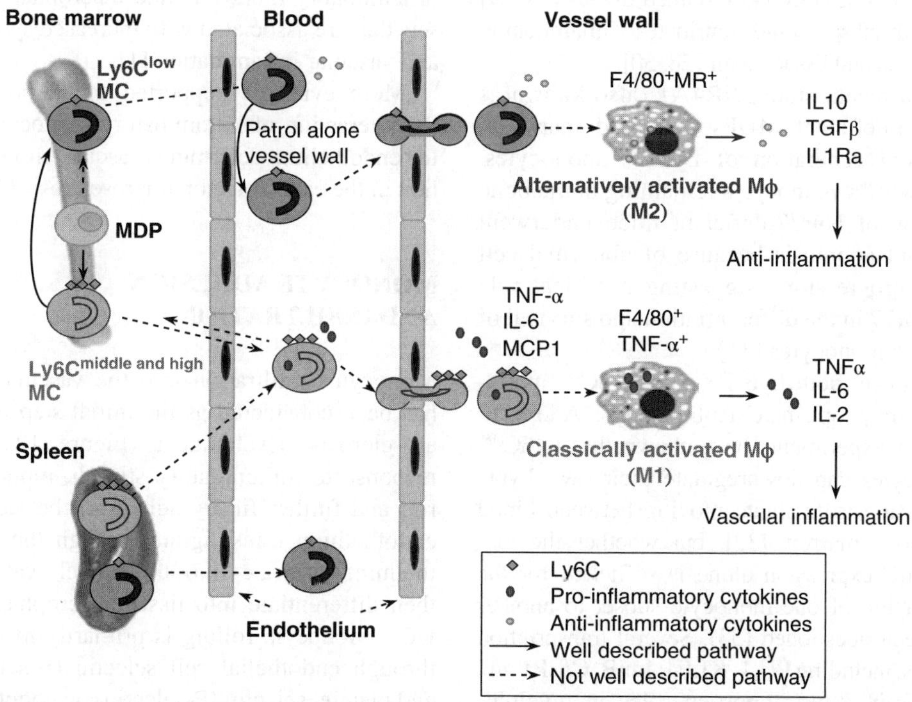

Figure 12.3 Trafficking and differentiation of circulating MC and Mϕ subsets. MDPs in the bone marrow give rise to Ly6C^{high} MC, which can serve as an intermediate for Ly6C^{low} MC. Ly6C^{high} MCs exit the bone marrow in a CC-chemokine receptor 2 (CCR2)-dependent manner. In the absence of inflammation, some circulating Ly6C^{high} MCs could return to the bone marrow. In inflammatory conditions, Ly6C^{hi} MCs migrate through the inflamed endothelium and potentially differentiate into classically activated Mϕ (M1). M1 Mϕ can be distinguished as F4/80^{+}TNF-α^{+}, that secret proinflammatory cytokines such as TNF-α, IL-6, and IL-2 and contribute to vascular inflammation and atherosclerosis formation. Ly6C^{low} MCs patrol along the blood vessel lumen by associating with the vascular endothelium. A small fraction of Ly6C^{low} MCs are also recruited to vessel wall and can differentiate into alternatively activated Mϕ (M2). M2 Mϕ are F4/80^{+}MR^{+} and secret anti-inflammatory cytokines such as IL-10, TGF-β, and IL-1 receptor antagonist (IL-1Rα) and is suggested to play a role in tissue repair. The spleen functions as a reservoir for blood MC. MDP, macrophage and dendritic cell precursor; MC, monocyte; and Mϕ, macrophages. *(See insert for color representation of the figure.)*

microarray study, mouse Ly6C^{high} monocytes are similar to human CD14^{++}CD16^{-} classical monocytes, but the inflammatory property of human CD14^{++}CD16^{-} monocytes has not been defined.

Mouse Ly6C^{low} monocytes are analogous to CD14^{+}CD16^{++} human monocytes based on their high expression of CX3CR1 and low level of CCR2. Ly6C^{low} monocytes are also called residential monocytes as their major behavior is patrolling along the vessel wall [27]. *In vitro* study has demonstrated a lower migration capacity of Ly6C^{low} monocytes compared with Ly6C^{high}. However, both subsets could be recruited to atherosclerotic lesions and inflamed tissues [28, 29] and both subsets could contribute to inflammation resolution and tissue repair [28, 30].

In a recent study, NR4A1 (also known as Nur77)-deficient MDPs exhibited markedly impaired generation of Ly6C^{low} monocytes. The Ly6C^{low} monocytes remaining in the bone marrow of Nur77-deficient mice underwent aberrant apoptosis because of abnormal cell cycle progression, suggesting a critical role for Nur77 in the differentiation and survival of Ly6C^{low} monocytes [31].

In mice, both Ly6C^{high} and Ly6C^{low} monocytes may originate from MDPs. Adoptive-transfer experiments have shown that Ly6C^{high} monocytes can downregulate their own Ly6C expression to facilitate moving between blood and bone marrow [32], but whether the loss of Ly6C expression alone is sufficient for the conversion of one monocyte subset to another has been questioned [33]. Several transcription factors, including PU.1, KLF4, JunB, C/EBPα/β, and IRF8, have important roles in myeloid-lineage differentiation. However, specific transcription factors that drive monocyte subset differentiation remain to be identified [34].

INFLAMMATORY MONOCYTE SUBSETS AND VASCULAR INFLAMMATION

Ly6C^{high} and Ly6C^{middle} inflammatory monocyte subsets have been linked with inflammatory disease, including hypercholesterolemia and atherosclerosis [35]. These subsets can produce inflammatory cytokines and infiltrate through the endothelium and contribute to chronic vascular disease [35, 36]. Ly6C^{high} monocytes show strong inflammatory response to LPS *in vitro*, and they are recruited to atherosclerotic lesions or inflamed tissues [23] and can differentiate into macrophages or DCs. It was found that hyperhomocysteinemia (HHcy), an independent risk factor of cardiovascular disease, elevates circulating Ly6C^{high} and Ly6C^{middle} monocyte independent of hyperlipidemia [36]. We reported that HHcy increases vessel wall inflammatory monocyte and macrophage subsets that are associated with increased systemic and vascular inflammation [1].

More evidence supported the notion that the interaction of inflammatory monocyte and the endothelium determines vascular inflammation in the early stage of atherosclerosis [4, 35].

MONOCYTE ADHESION AND INFILTRATION

Monocytes infiltration into the vascular wall has been considered as the initial step in atherosclerosis development (Figure 12.3). In response to inflammatory stimuli, monocytes roll and further firmly adhere to the vascular endothelium, transmigrate through the endothelium, infiltrate into the vessel wall, and then differentiate into tissue macrophages or DCs. Monocyte rolling is primarily mediated through endothelial cell selectin (E-selectin) and platelet selectin (P-selectin) on endothelial cells and their ligand P-selectin glycoprotein ligand-1 (PSGL-1) on monocytes. The Ly6C^{high} population, which is the major infiltrating monocyte subset, expresses high levels of PSGL-1. PSGL-1-deficiency has shown to decrease atherosclerosis in mice [37]. *ApoE*$^{-/-}$ mice deficient in both E-selectin and P-selectin have less severe atherosclerosis [38]. L-selectin is also considered critical for monocyte rolling. Ly6C^{high} monocyte subset expresses high level of L-selectin. Surprisingly, the L-selectin-deficient mice had promoted atherosclerotic lesion formation, suggesting a protective role of L-selectin [39]. Adhesion of monocytes are mediated by their cell surface expression of lymphocyte function-associated antigen 1 (LFA-1) and very late antigen 4 (VLA4), which bind to their endothelial cell ligands intercellular adhesion molecule 1 (ICAM-1) and vascular cell-adhesion molecule 1 (VCAM-1) on endothelial cells. Low-density lipoprotein receptor deficient (*Ldlr*$^{-/-}$) mice that express a truncated nonfunctional VCAM-1 develop less atherosclerotic plaques [40]. However, conflicting results have been reported regarding pro-atherosclerotic effects of ICAM-1 [40, 41].

Chemokine-mediated monocyte trafficking is a critical process involved in monocyte accumulation in the vessel wall [3]. Monocyte chemoattractant protein 1 (MCP-1) and its receptor, CCR2 are noted to play a significant role in the initiation of atherosclerosis, probably due to their role in monocyte infiltration and atherosclerosis development. Crossing *apoE*$^{-/-}$ or *Ldlr*$^{-/-}$ mice with mice lacking MCP-1 or CCR2 leads to significant lesion decrease [42, 43]. Other chemokines are also detected in atherosclerotic plaques, such as the cell-surface anchored CX3-chemokine ligand 1 (CX3CL1), which is expressed on vascular smooth muscle cells (VSMCs). CX3CR1 is expressed on monocytes and macrophages. Observations from the compound-deficient *apoE*$^{-/-}$ and *CX3CR1*$^{-/-}$ mice suggest that CX3CR1 may also be involved in monocyte recruitment to the vessel wall, thereby promoting atherosclerotic plaque formation [44]. It is unknown that how CX3CL1 expressed on VSMCs mediates the transendothelial migration of CX3CR1^{+} monocytes. In human populations, the M280 polymorphism of CX3CR1 has been associated with protection from coronary artery disease [7]. Taken together, these studies suggest that chemokines and chemokine receptors are strongly involved in atherogenesis, at least partially through their facilitation of monocyte trafficking and infiltration into atherosclerotic lesions.

In response to conditions such as inflammation, oxidative stress, and immune stimuli, monocytes transmigrate from the blood circulation to tissues and differentiate into macrophages to perform host defense or tissue repair functions. This process also holds true in the blood vessel during atherosclerosis where monocytes and macrophages participate in a maladaptive nonresolving inflammation [17]. Recently it was found that HHcy accelerates atherosclerosis by promoting inflammatory monocyte differentiation and macrophage accumulation in atherosclerotic lesions of an HHcy mouse model consisting of an inducible human cystathionine β-synthase (*CBS*) transgene in the background of *apoE*$^{-/-}$ and *Cbs*$^{-/-}$ compound knockout (Tg-h *CBSCbs*$^{-/-}$/*apoE*$^{-/-}$ mice) fed a high-fat diet [36]. Although both Ly6C^{high} and Ly6C^{low} monocytes could be found in the atherosclerotic plaque of these mice, the inflammatory Ly6C^{high} monocytes possess a higher infiltration capacity and represent a major population of infiltrated monocytes. Similarly, in mice with hypercholesterolemia, the development of atherosclerosis is in accordance with increased numbers of monocytes. It is suggested that the increase of circulating Ly6C^{high} monocytes may be due to its impaired conversion into less inflammatory Ly6C^{low} monocytes [35]. Although not highly infiltrating as Ly6C^{high} monocytes, Ly6C^{low} monocytes serve as important source to generate DCs [45].

VESSEL WALL MONOCYTE ORIGIN/DIFFERENTIATION AND INFLAMMATION

After infiltration, promoted by M-CSF and other differentiating factors, most monocytes undergo differentiation into macrophages at the early atherosclerotic lesions and then foam cells that contain a lipid-loaded cytosol after ingestion and processing apoB-LPs. In addition, it is shown that monocytes, especially Ly6C^{high} monocytes, directly and quickly differentiate into macrophages in the atheroma in aged *apoE*$^{-/-}$ mice [35]. In human atherosclerosis, CD68-expressing macrophages are found within the arterial wall of atherosclerosis [5]. It is very interesting that monocytes/macrophages in the atherosclerotic plaque are mainly confined to the subendothelial region above the elastin lamina. Beside apoptosis, it is speculated that monocyte/macrophage in the atherosclerotic lesion could escape the location through two possible ways. One is directly going back into the blood, which is supported by only a few experimental evidences, and the other one is to egress though the nearby lymph systems. In graft studies, after the atheroma aorta from *apoE*$^{-/-}$ mice was transplanted to wild-type mice, monocytes regressed from the donor plaque can be detected from the local lymph nodes near the transplantation site [46], indicating that infiltrated monocytes in the vessel wall could migrate further to the adventitia and then return to the blood circulation via the lymph system.

MACROPHAGE AND ATHEROSCLEROSIS

Macrophages are well known for their functions in phagocytosis, cytokine secretion, and antigen presentation to trigger adaptive immune response. Tissue macrophage could directly derive from circulating monocytes. Macrophages can be further activated by extracellular signals and manifest distinct functional patterns depending on the different cytokines and bacterial products present in the environment.

The roles of macrophages in atherogenesis are featured in several important aspects as shown by the following findings:

1. *Cholesterol loading*: Macrophages could phagocytose and degrade lipid, LPS, and apoptotic cells through scavenger receptors [47] including CD36, CD68, CXCL16, scavenger receptor A (SR-A), SR-B1, and lectin-type oxidized low-density lipoprotein receptor 1 (LOX1). Hence, scavenger receptors are considered crucial for lipid accumulation in the macrophages to finally form the foam cells, which significantly contribute to the formation of fatty streak and established lesions. Recent studies suggest that SR-A and CD36 [48] have important functions in mediating uptake of oxLDL and promotion of atherosclerotic development. Oxidation of LDL has been proposed to be atherogenic particle [49]. It has been shown that macrophages may promoter LDL oxidation *in vitro* through lipoxygenases (LO), iNOS, myeloperoxidase, and NADPH oxidases. These enzymes are expressed in human atherosclerotic lesions. In *apoE*$^{-/-}$ mice, disruption of the 12/15 LO gene diminishes atherosclerosis [50]. There are conflicting results regarding *iNOS*$^{-/-}$ mice [51, 52]. Many studies have shown the deficiency of NADPH oxidase components lead to significant decrease of atherosclerosis [53, 54].
2. *Inflammatory activation*: Macrophages play a critical role in sensing endogenous metabolic danger signals including cholesterol crystals via a NALP3 inflammasome manner, activate caspase-1, and initiate secretion of proinflammatory cytokines IL-1β and IL-18 [55]. The leukotrienes are lipid mediators of inflammation, and the leukotriene pathways have been proposed to play a significant role in atherosclerosis. 5-LO is the rate-limiting enzyme in leukotriene synthesis and *Ldlr*$^{-/-}$ mice that missed only one of the two allelic copies of the 5-LO gene had a dramatic decrease in atherosclerosis [56].
3. *Cell death and thinning of fibrous cap*: The accumulation of lipids in macrophages may finally contribute to the apoptosis and death of macrophage itself, which is shown as the necrotic core and thinning of fibrous cap of the atherosclerotic plaque. When apoptotic macrophages cannot be efficiently cleared by phagocytes, necrosis happens in the plaque. Efficient efferocytosis of

apoptotic macrophages results in protective effects.

4. *Antigen presentation and T-cell activation*: The antigen presentation function of macrophages may trigger adaptive immunity and potentially contributes to the autoimmune character of atherosclerosis. For example, IFN-γ, a typical type 1 $CD4^+$ T helper cell (Th1) cell-secreted cytokine, is expressed in the atherosclerotic plaque. CD40 and CD40 ligand (CD40L), critically involved in the costimulation of T-cell immunity, is also widely expressed in the lesion cells [57]. Inhibition of CD40 signaling, including genetic disruption of CD40L in *apoE*$^{-/-}$ mice or treating *Ldlr*$^{-/-}$ mice with CD40L antibodies, reduces atherosclerotic lesion formation [58]. Treatment with CD40L antibodies also inhibits the progression of established plaques and maintains their stable phenotype [59].

MACROPHAGE SUBSETS AND FUNCTION

Macrophages have been classified into two major types based on their activation pathways: classically activated (M1) macrophages and alternatively activated (M2) macrophages [60] (Figure 12.3). M1 and M2 macrophages have been proposed to derive from different subsets of circulating monocytes. Ly6C^{high} monocytes, the inflammatory monocytes, are thought to be precursors for M1 macrophages, whereas, Ly6C^{low} monocytes, the resident monocytes, are considered as precursors for M2 macrophages [27]. Studies from several independent groups indicate that Ly6C^{low} monocytes may differentiate into M2 macrophages [61] (Figure 12.3). In response to Th1 cytokine such as IFN-γ and LPS, M1 macrophages are polarized from "unpolarized" M0 macrophages and display a classical activation phenotype and produce mainly proinflammatory cytokines including TNF-α and IL-12 [60]. These macrophages promote inflammation. The Jak/Stat and activator protein 1 (AP-1)/NF-κB signaling pathways have been shown to play critical roles in classical activation of macrophages [60]. IFN-γ accumulated in the atherosclerotic plaque may promote M1 macrophage formation. M1 macrophages are conversely associated with robust intracellular killing of bacteria and support of Th1 cells [62]. Alternatively, M2 macrophages are either differentiated in the presence of M-CSF and IL-4 or IL-13 (M2a), immune complexes and IL-1β or LPS (M2b), or in the presence of IL-10, transforming growth factor-β (TGF-β) or glucocorticoids (M2c) [63]. M2 macrophages exhibit anti-inflammatory and tissue repair properties [64, 65]. M2 macrophages are typically associated with parasitic infections, humoral immune responses favored by type 2 $CD4^+$ T-helper cells (Th2), angiogenesis, and wound healing [62]. M2 macrophages play an essential role in inflammation resolution due to their function of suppressing inflammation and phagocytosing cellular debris [66]. It has been reported that M2 macrophage may stabilize the regressing atherosclerotic plaques and contribute to recovery of myocardial infarction [28, 67]. It has been reported that M1 macrophages could express IL-12, iNOS, CCL15, CCL20, CXCL9, CXCL10, and CXCL11, while M2 macrophages express CCL18, YM1, RELMα, CCL17, IL-27Rα, IGF1, CCL22, DCIR, stabilin-1, and factor XIII-A [64]. The imbalance of M1 and M2 macrophages in the atherosclerotic lesion may contribute to atherosclerosis progression.

In addition, two more subsets of macrophages have been characterized: M4 macrophages are defined as monocyte-derived macrophages that can proliferate in the presence of platelet factor 4 without requirement of M-CSF [68]. Mox macrophages are defined as mouse bone marrow-derived macrophages that respond to oxidized phospholipid ox PAPC stimulation and express a unique Mox transcriptome including heme oxygenase-1 gene [69].

Evidently, the same population of monocytes can be reversibly differentiated into either type of the two macrophages depending on the cytokines present [62]. Moreover, macrophage can exist in either M1 or M2 state depending on the polarization conditions [62].

MACROPHAGE TRANSCRIPTIONAL REGULATION

Previous studies on human monocyte-to-macrophage differentiation have largely relied on myeloid leukemic-like cell lines such as U937 and THP-1 and may not necessarily represent physiologic conditions. Recent global transcriptome analysis revealed profound changes in gene expression during monocyte-to-macrophage terminal differentiation [70]. Macrophages can be further activated by extracellular signals and manifest distinct functional patterns depending on the different cytokines and bacterial products present in the environment. It has been shown that IRF5 [71] and STAT1 [72] are crucially involved in M1 macrophage polarization, while IRF4 [73] and STAT6 [74] regulate M2 macrophage polarization. However, the key transcriptional mechanism controlling primary human monocyte-to-macrophage differentiation under atherosclerosis condition remains poorly defined. Recently, VentX, a human homolog of the Xenopus homeobox transcriptional factor Xom, was shown to be predominantly expressed in hematopoietic cells and was reported to play an essential role in human primary monocyte-to-macrophage terminal differentiation. VentX is also required for optimal proinflammatory response during classical macrophage activation [60].

FUTURE DIRECTIONS AND THERAPEUTIC IMPLICATIONS

Although the control of risks factors are still the major prevention of atherosclerosis, clinical treatments are mainly focused on lowering blood lipid levels, maintaining blood pressure, and managing thrombotic complications. A better understanding of atherosclerosis as an inflammatory disease, and the biological role of monocyte/macrophage in this process will provide a variety of potential therapeutic targets [75]. For example, investigation into monocyte foam cell formation has generated interest in novel therapies such as vaccines that specifically target oxLDL. Other promising work has been involved in disrupting various cytokine pathways to decrease monocyte adhesion and infiltration [76]. Since Ly6C^{high} monocytes are the major type of infiltrating monocytes and differentiate further into the inflammatory M1 macrophages in the atherosclerotic plaque, additional in-depth understanding of the molecular and cellular mechanisms on how Ly6C^{high} monocytes are generated in the bone marrow and what signals are responsible or their differentiation into M1 macrophages may unravel the enigma of this cell surface antigen. Finally, inhibition of recruitment of Ly6C^{high} monocytes or depletion of this unique population in the circulation may also potentially serve as antiatherosclerosis therapies.

REFERENCES

1. Zhang, D., Fang, P., Jiang, X. et al. (2012) Severe hyperhomocysteinemia promotes bone marrow-derived and resident inflammatory monocyte differentiation and atherosclerosis in LDLr/CBS-deficient mice. *Circ Res* 111, 37–49.
2. Napoli, C., D'Armiento, F. P., Mancini, F. P. et al. (1997) Fatty streak formation occurs in human fetal aortas and is greatly enhanced by maternal hypercholesterolemia. Intimal accumulation of low density lipoprotein and its oxidation precede monocyte recruitment into early atherosclerotic lesions. *J Clin Invest* 100, 2680–90.
3. Shi, C. & Pamer, E. G. (2011) Monocyte recruitment during infection and inflammation. *Nat Rev Immunol* 11, 762–74.
4. Gordon, S. & Taylor, P. R. (2005) Monocyte and macrophage heterogeneity. *Nat Rev Immunol* 5, 953–64.
5. Spanbroek, R., Grabner, R., Lotzer, K. et al. (2003) Expanding expression of the 5-lipoxygenase pathway within the arterial wall during human atherogenesis. *Proc Natl Acad Sci U S A* 100, 1238–43.

6. Song, L., Leung, C. & Schindler, C. (2001) Lymphocytes are important in early atherosclerosis. *J Clin Invest* 108, 251–9.
7. McDermott, D. H., Fong, A. M., Yang, Q. et al. (2003) Chemokine receptor mutant CX3CR1-M280 has impaired adhesive function and correlates with protection from cardiovascular disease in humans. *J Clin Invest* 111, 1241–50.
8. Danenberg, H. D., Fishbein, I., Gao, J. et al. (2002) Macrophage depletion by clodronate-containing liposomes reduces neointimal formation after balloon injury in rats and rabbits. *Circulation* 106, 599–605.
9. Qiao, J. H., Tripathi, J., Mishra, N. K. et al. (1997) Role of macrophage colony-stimulating factor in atherosclerosis: studies of osteopetrotic mice. *Am J Pathol* 150, 1687–99.
10. Panizzi, P., Swirski, F. K., Figueiredo, J. L. et al. (2010) Impaired infarct healing in atherosclerotic mice with Ly-6C(hi) monocytosis. *J Am Coll Cardiol* 55, 1629–38.
11. Tsujioka, H., Imanishi, T., Ikejima, H. et al. (2009) Impact of heterogeneity of human peripheral blood monocyte subsets on myocardial salvage in patients with primary acute myocardial infarction. *J Am Coll Cardiol* 54, 130–8.
12. Chow, A., Brown, B. D. & Merad, M. (2011) Studying the mononuclear phagocyte system in the molecular age. *Nat Rev Immunol* 11, 788–98.
13. Fogg, D. K., Sibon, C., Miled, C. et al. (2006) A clonogenic bone marrow progenitor specific for macrophages and dendritic cells. *Science* 311, 83–7.
14. Yeamans, C., Wang, D., Paz-Priel, I. et al. (2007) C/EBPalpha binds and activates the PU.1 distal enhancer to induce monocyte lineage commitment. *Blood* 110, 3136–42.
15. Scott, E. W., Simon, M. C., Anastasi, J. & Singh, H. (1994) Requirement of transcription factor PU.1 in the development of multiple hematopoietic lineages. *Science* 265, 1573–7.
16. Feinberg, M. W., Wara, A. K., Cao, Z. et al. (2007) The Kruppel-like factor KLF4 is a critical regulator of monocyte differentiation. *Embo J* 26, 4138–48.
17. Moore, K. J. & Tabas, I. (2011) Macrophages in the pathogenesis of atherosclerosis. *Cell* 145, 341–55.
18. Friedman, A. D. (2007) Transcriptional control of granulocyte and monocyte development. *Oncogene* 26, 6816–28.
19. Psaltis, P. J., Harbuzariu, A., Delacroix, S. et al. (2012) Identification of a monocyte-predisposed hierarchy of hematopoietic progenitor cells in the adventitia of postnatal murine aorta. *Circulation* 125, 592–603.
20. Zawada, A. M., Rogacev, K. S., Schirmer, S. H. et al. (2012) Monocyte heterogeneity in human cardiovascular disease. *Immunobiology* 217, 1273–84.
21. Ancuta, P., Rao, R., Moses, A. et al. (2003) Fractalkine preferentially mediates arrest and migration of CD16+ monocytes. *J Exp Med* 197, 1701–7.
22. Cros, J., Cagnard, N., Woollard, K. et al. (2010) Human CD14dim monocytes patrol and sense nucleic acids and viruses via TLR7 and TLR8 receptors. *Immunity* 33, 375–86.
23. Geissmann, F., Jung, S. & Littman, D. R. (2003) Blood monocytes consist of two principal subsets with distinct migratory properties. *Immunity* 19, 71–82.
24. Woollard, K. J. & Geissmann, F. (2010) Monocytes in atherosclerosis: subsets and functions. *Nat Rev Cardiol* 7, 77–86.
25. Belge, K. U., Dayyani, F., Horelt, A. et al. (2002) The proinflammatory CD14+CD16+DR++ monocytes are a major source of TNF. *J Immunol* 168, 3536–42.
26. Si, Y., Tsou, C. L., Croft, K. & Charo, I. F. (2010) CCR2 mediates hematopoietic stem and progenitor cell trafficking to sites of inflammation in mice. *J Clin Invest* 120, 1192–203.
27. Auffray, C., Fogg, D., Garfa, M. et al. (2007) Monitoring of blood vessels and tissues by a population of monocytes with patrolling behavior. *Science* 317, 666–70.
28. Nahrendorf, M., Swirski, F. K., Aikawa, E. et al. (2007) The healing myocardium sequentially mobilizes two monocyte subsets with divergent and complementary functions. *J Exp Med* 204, 3037–47.
29. Tacke, F., Alvarez, D., Kaplan, T. J. et al. (2007) Monocyte subsets differentially employ CCR2, CCR5, and CX3CR1 to accumulate within atherosclerotic plaques. *J Clin Invest* 117, 185–94.
30. Shechter, R., London, A., Varol, C. et al. (2009) Infiltrating blood-derived macrophages are vital cells playing an anti-inflammatory role

in recovery from spinal cord injury in mice. *PLoS Med* 6, e1000113.

31. Hanna, R. N., Carlin, L. M., Hubbeling, H. G. et al. (2011)The transcription factor NR4A1 (Nur77) controls bone marrow differentiation and the survival of Ly6C- monocytes. *Nat Immunol* 12, 778–85.
32. Yrlid, U., Jenkins, C. D. & MacPherson, G. G. (2006) Relationships between distinct blood monocyte subsets and migrating intestinal lymph dendritic cells in vivo under steady-state conditions. *J Immunol* 176, 4155–62.
33. Geissmann, F., Manz, M. G., Jung, S. et al. (2010) Development of monocytes, macrophages, and dendritic cells. *Science* 327, 656–61.
34. Auffray, C., Sieweke, M. H. & Geissmann, F. (2009) Blood monocytes: development, heterogeneity, and relationship with dendritic cells. *Annu Rev Immunol* 27, 669–92.
35. Swirski, F. K., Libby, P., Aikawa, E. et al. (2007) Ly-6Chi monocytes dominate hypercholesterolemia-associated monocytosis and give rise to macrophages in atheromata. *J Clin Invest* 117, 195–205.
36. Zhang, D., Jiang, X., Fang, P. et al. (2009) Hyperhomocysteinemia promotes inflammatory monocyte generation and accelerates atherosclerosis in transgenic cystathionine beta-synthase-deficient mice. *Circulation* 120, 1893–902.
37. An, G., Wang, H., Tang, R. et al. (2008) P-selectin glycoprotein ligand-1 is highly expressed on Ly-6Chi monocytes and a major determinant for Ly-6Chi monocyte recruitment to sites of atherosclerosis in mice. *Circulation* 117, 3227–37.
38. Dong, Z. M., Chapman, S. M., Brown, A. A. et al. (1998) The combined role of P- and E-selectins in atherosclerosis. *J Clin Invest* 102, 145–52.
39. Rozenberg, I., Sluka, S. H., Mocharla, P. et al. (2011) Deletion of L-selectin increases atherosclerosis development in ApoE-/- mice. *PLoS One* 6, e21675.
40. Cybulsky, M. I., Iiyama, K., Li, H. et al. (2001) A major role for VCAM-1, but not ICAM-1, in early atherosclerosis. *J Clin Invest* 107, 1255–62.
41. Collins, R. G., Velji, R., Guevara, N. V. et al. (2000) P-Selectin or intercellular adhesion molecule (ICAM)-1 deficiency substantially protects against atherosclerosis in apolipoprotein E-deficient mice. *J Exp Med* 191, 189–94.
42. Boring, L., Gosling, J., Cleary, M. & Charo, I. F. (1998) Decreased lesion formation in CCR2-/- mice reveals a role for chemokines in the initiation of atherosclerosis. *Nature* 394, 894–7.
43. Gu, L., Okada, Y., Clinton, S. K. et al. (1998) Absence of monocyte chemoattractant protein-1 reduces atherosclerosis in low density lipoprotein receptor-deficient mice. *Mol Cell* 2, 275–81.
44. Combadiere, C., Potteaux, S., Gao, J. L. et al. (2003) Decreased atherosclerotic lesion formation in CX3CR1/apolipoprotein E double knockout mice. *Circulation* 107, 1009–16.
45. Peng, Y., Latchman, Y. & Elkon, K. B. (2009) Ly6C(low) monocytes differentiate into dendritic cells and cross-tolerize T cells through PDL-1. *J Immunol* 182, 2777–85.
46. Llodra, J., Angeli, V., Liu, J. et al. (2004) Emigration of monocyte-derived cells from atherosclerotic lesions characterizes regressive, but not progressive, plaques. *Proc Natl Acad Sci U S A* 101, 11779–84.
47. Peiser, L., Mukhopadhyay, S. & Gordon, S. (2002) Scavenger receptors in innate immunity. *Curr Opin Immunol* 14, 123–8.
48. Febbraio, M., Hajjar, D. P. & Silverstein, R. L. (2001) CD36: a class B scavenger receptor involved in angiogenesis, atherosclerosis, inflammation, and lipid metabolism. *J Clin Invest* 108, 785–91.
49. Gaut, J. P. & Heinecke, J. W. (2001) Mechanisms for oxidizing low-density lipoprotein. Insights from patterns of oxidation products in the artery wall and from mouse models of atherosclerosis. *Trends Cardiovasc Med* 11, 103–12.
50. Cyrus, T., Witztum, J. L., Rader, D. J. et al. (1999) Disruption of the 12/15-lipoxygenase gene diminishes atherosclerosis in apo E-deficient mice. *J Clin Invest* 103, 1597–604.
51. Detmers, P. A., Hernandez, M., Mudgett, J. et al. (2000) Deficiency in inducible nitric oxide synthase results in reduced atherosclerosis in apolipoprotein E-deficient mice. *J Immunol* 165, 3430–5.
52. Niu, X. L., Yang, X., Hoshiai, K. et al. (2001) Inducible nitric oxide synthase deficiency does not affect the susceptibility of mice to atherosclerosis but increases collagen content in lesions. *Circulation* 103, 1115–20.
53. Judkins, C. P., Diep, H., Broughton, B. R. et al. (2010) Direct evidence of a role for Nox2 in superoxide production, reduced nitric oxide bioavailability, and early atherosclerotic plaque

formation in ApoE-/- mice. *Am J Physiol Heart Circ Physiol* 298, H24–32.

54. Barry-Lane, P. A., Patterson, C., van der Merwe, M. et al. (2001) p47phox is required for atherosclerotic lesion progression in ApoE(-/-) mice. *J Clin Invest* 108, 1513–22.
55. Swirski, F. K. & Nahrendorf, M. (2013) Leukocyte behavior in atherosclerosis, myocardial infarction, and heart failure. *Science* 339, 161–6.
56. Mehrabian, M., Allayee, H., Wong, J. et al. (2002) Identification of 5-lipoxygenase as a major gene contributing to atherosclerosis susceptibility in mice. *Circ Res* 91, 120–6.
57. Mach, F., Schonbeck, U., Sukhova, G. K. et al. (1997) Functional CD40 ligand is expressed on human vascular endothelial cells, smooth muscle cells, and macrophages: implications for CD40-CD40 ligand signaling in atherosclerosis. *Proc Natl Acad Sci U S A* 94, 1931–6.
58. Mach, F., Schonbeck, U., Sukhova, G. K., Atkinson, E. & Libby, P. (1998) Reduction of atherosclerosis in mice by inhibition of CD40 signalling. *Nature* 394, 200–3.
59. Lutgens, E., Cleutjens, K. B., Heeneman, S. et al. (2000) Both early and delayed anti-CD40L antibody treatment induces a stable plaque phenotype. *Proc Natl Acad Sci U S A* 97, 7464–9.
60. Wu, X., Gao, H., Ke, W., Giese, R. W. & Zhu, Z. (2011) The homeobox transcription factor VentX controls human macrophage terminal differentiation and proinflammatory activation. *J Clin Invest* 121, 2599–613.
61. Arnold, L., Henry, A., Poron, F. et al. (2007) Inflammatory monocytes recruited after skeletal muscle injury switch into antiinflammatory macrophages to support myogenesis. *J Exp Med* 204, 1057–69.
62. Randolph, G. J. (2011) Immunology. No need to coax monocytes. *Science* 332, 1268–9.
63. Martinez, F. O., Sica, A., Mantovani, A. & Locati, M. (2008) Macrophage activation and polarization. *Front Biosci* 13, 453–61.
64. Mosser, D. M. & Edwards, J. P. (2008) Exploring the full spectrum of macrophage activation. *Nat Rev Immunol* 8, 958–69.
65. Schroder, K., Sweet, M. J. & Hume, D. A. (2006) Signal integration between IFNgamma and TLR signalling pathways in macrophages. *Immunobiology* 211, 511–24.
66. Tabas, I. (2009) Macrophage death and defective inflammation resolution in atherosclerosis. *Nat Rev Immunol* 10, 36–46.
67. Feig, J. E., Rong, J. X., Shamir, R. et al. (2011) HDL promotes rapid atherosclerosis regression in mice and alters inflammatory properties of plaque monocyte-derived cells. *Proc Natl Acad Sci U S A* 108, 7166–71.
68. Gleissner, C. A., Shaked, I., Little, K. M. & Ley, K. (2010) CXC chemokine ligand 4 induces a unique transcriptome in monocyte-derived macrophages. *J Immunol* 184, 4810–8.
69. Kadl, A., Meher, A. K., Sharma, P. R. et al. (2010) Identification of a novel macrophage phenotype that develops in response to atherogenic phospholipids via Nrf2. *Circ Res* 107, 737–46.
70. Martinez, F. O., Gordon, S., Locati, M. & Mantovani, A. (2006) Transcriptional profiling of the human monocyte-to-macrophage differentiation and polarization: new molecules and patterns of gene expression. *J Immunol* 177, 7303–11.
71. Krausgruber, T., Blazek, K., Smallie, T. et al. (2011) IRF5 promotes inflammatory macrophage polarization and TH1-TH17 responses. *Nat Immunol* 12, 231–8.
72. Darnell, J. E., Jr., Kerr, I. M. & Stark, G. R. (1994) Jak-STAT pathways and transcriptional activation in response to IFNs and other extracellular signaling proteins. *Science* 264, 1415–21.
73. Satoh, T., Takeuchi, O., Vandenbon, A. et al. (2010) The Jmjd3-Irf4 axis regulates M2 macrophage polarization and host responses against helminth infection. *Nat Immunol* 11, 936–44.
74. Ohmori, Y. & Hamilton, T. A. (1997) IL-4-induced STAT6 suppresses IFN-gamma-stimulated STAT1-dependent transcription in mouse macrophages. *J Immunol* 159, 5474–82.
75. Tabas, I. & Glass, C. K. (2013) Anti-inflammatory therapy in chronic disease: challenges and opportunities. *Science* 339, 166–72.
76. Charo, I. F. & Taub, R. (2011) Anti-inflammatory therapeutics for the treatment of atherosclerosis. *Nat Rev Drug Discov* 10, 365–76.

13

LYMPHOCYTES AND ATHEROSCLEROSIS

Anthony Virtue, Jietang Mai, Hong Wang, and Xiaofeng Yang
Center for Metabolic Disease Research and Cardiovascular Research Center, Department of Pharmacology, Temple University School of Medicine, Philadelphia, PA, USA

As highlighted by the text of this book, atherosclerosis is a chronic inflammatory disease involving a wide array of residential vascular, inflammatory, and immune cells. Clinical research has found that patients affalicted with other autoimmune diseases have up to a 50-fold higher risk of developing atherosclerosis/cardiovascular disease [1]. This suggests a strong association between immune hyperactivity and atherosclerosis progression [2]. In this chapter, we will provide a general overview on lymphocytes and how they impact atherosclerosis initiation, progression, and aggravation. We will examine how each major lymphocyte cell type contributes to or protects against atherosclerosis development, how this is accomplished, and any therapeutic potential these subsets possess. Specifically, we will discuss the role of traditional CD4$^+$T cell subsets including type 1T-helper cells (Th1), type 2T-helper cells (Th2), and natural killer T (NKT) cells. Then, we will examine the contributions of emerging CD4$^+$ T-cell subsets such as T regulatory cells (Tregs), T-helper 17 cells (Th17), and γδ T cells. Finally, we will explore how other lymphocyte cell types, such as B cells and NK cells, also participate in atherosclerosis. The continued improvement of our understanding of how these lymphocyte subsets participate in atherosclerotic pathogenesis could eventually lead to the development of novel therapeutics for atherosclerosis and atherosclerosis-related myocardial infarction and stroke.

THE CONTRIBUTIONS OF B AND T CELLS IN ATHEROSCLEROSIS

Lymphocytes are an integral part of the adaptive and innate immune system that are responsible for differentiating self from nonself, antigen

Atherosclerosis: Risks, Mechanisms, and Therapies, First Edition. Edited by Hong Wang and Cam Patterson.

presentation, immunological memory, as well as the destruction of abnormal cells and foreign bodies. Constituting 20–40% of the body's white blood cell population, lymphocytes can be subdivided into three broad categories: T cells, B cells, or NK cells. Within these three groups, more specialized subsets exist that are unique in their attributes and capabilities. Throughout this chapter, we will discuss in greater detail these subsets and their individual contributions to atherosclerosis.

As a chronic inflammatory disease with the feature of lymphocyte infiltration, it is logical to extrapolate that lymphocytes would play a role in atherosclerosis. In fact, the use of mouse models has confirmed this postulation. Severe combined immunodeficiency (SCID) mice contain a rare recessive mutation that alleviates the function of protein kinase, DNA-activated, catalytic polypeptide. This enzyme is critical for DNA repair and is required in variable, diversity, and joining (VDJ) recombination of immunoglobulin and T-cell receptor genes. As a result, both B cells and T cells fail to fully mature. It should be noted that NK cells, the third lymphocyte group, maintain their function in SCID mice since VDJ recombination is not required for their maturation. The crossing of SCID mice into an atherosclerosis-susceptible genetic background allows for the assessment of B- and T-cell contributions in atherosclerosis. Evaluation of SCID mice with an apolipoprotein E (apoE) gene deficient (–/–) background shows that fatty streak and lesion formation are significantly inhibited with a 73% reduction in lesion size [3]. These findings are supported by similar discoveries gathered in an alternative immunological-deficient model. Recombination-activating genes (RAGs) encode for two VDJ recombination enzymes: RAG-1 and RAG-2. Similar to SCID mice, a deficiency of either RAG enzyme results in impaired VDJ recombination and therefore, immature B- and T-cell populations. Expectantly, crossing RAG-1-deficient mice with apoE$^{-/-}$ or low-density lipoprotein receptor (Ldlr)$^{-/-}$ mice leads to a reduction in total atherosclerosis, whereas in RAG-2$^{-/-}$ mice, this reduction appears to be location dependent [4]. Taken together, these immunological-deficient mouse models clearly demonstrate that lymphocytes play an important role in atherosclerosis development.

T cells are present in atherosclerotic lesions at all stages of development, while prominently exhibiting activation markers at sites of plaque rupture. Aside from migration to lesions, the strongly skewed pattern of the complementarity-determining region 3 in the T-cell antigen receptor (TCR) is indicative of oligoclonal expansions of T cells and suggests the occurrence of antigen-driven T-cell proliferation in atherosclerosis (see Figure 13.1) [5]. This is significant since the activation and clonal expansion of lymphocytes is critical to the establishment of both a humoral and cell-mediated immune response. T cells in the peripheral blood are recruited to developing atherosclerotic plaques by the production of several chemokines by vascular cells and other inflammatory cells within atherosclerotic plaques (see Figure 13.2). Mouse knockout models for either monocyte chemotactic protein 1 (MCP-1) or its receptor demonstrate diminished monocyte and T-cells numbers within plaques and an overall reduction in atherosclerosis [6]. Meanwhile, both chemokine receptor 7 (CCR7)-dependent T-cell priming in secondary lymphoid organs and CCR7-dependent recirculation of T cells between secondary lymphoid organs and inflamed tissue are crucially involved in atherosclerotic plaque development [7]. In addition, macrophages and vascular cells also produce CC-chemokine ligand 5 (CCL5), CXC-chemokine ligand 10 (CXCL10), and CXCL11, which are all known attractants for T cells [6].

Analysis of individual T cells in atherosclerotic plaques reveals a high level of activation, suggesting the presence of antigen(s) that trigger the infiltration and activation of these cells [8]. Traditionally, the activation of mature T cells in the periphery is dependent on the interaction of a TCR with a specific antigen presented by a professional antigen-presenting cell (APC). Therefore, the identification of

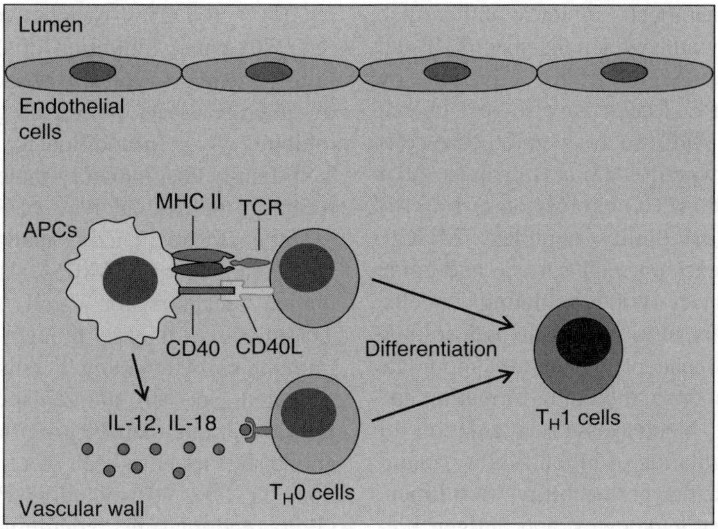

Figure 13.1 Inside the vessel wall, naïve T lymphocytes physically interact with professional antigen-presenting cells (APCs) such as macrophages and dendritic cells through major histocompatibility complex II (MHC II)/T cell receptor (TCR) and CD40/CD40L interactions. In addition to direct cell–cell communication, naïve T lymphocytes are also activated by locally released cytokines such as IL-12 and IL-18. These direct and indirect associations lead to the differentiation of naïve T lymphocytes into the proinflammatory T helper 1 (Th1) subset.

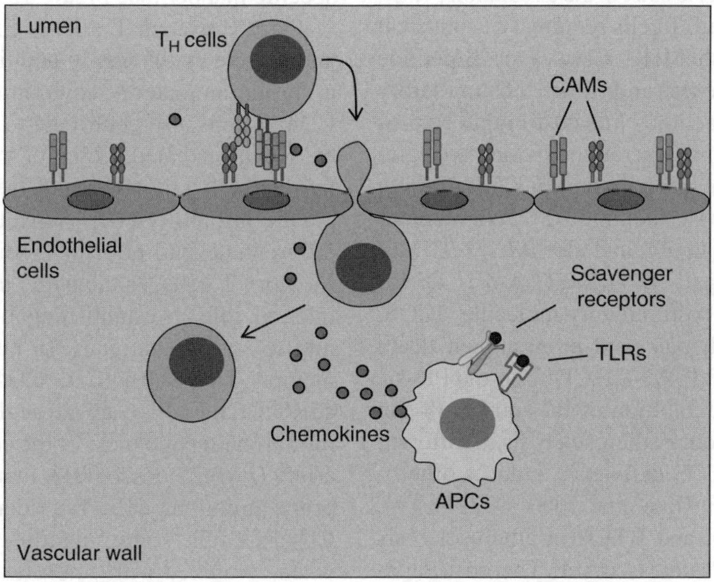

Figure 13.2 The augmented expression of cell adhesion molecules (CAMs) by activated endothelial cells leads to increased rolling and adhesion of circulating T lymphocytes. Meanwhile, the production of proinflammatory chemokines by antigen-presenting cells (APCs) causes the transmigration of the adhered T cells into the vessel wall.

bona fide dendritic cells in aortic and cardiac valves suggests atherosclerosis-specific T-cell activation [9]. Aside from activation by professional APC, T cells can also be activated by conditional APCs such as endothelial cells stimulated by hyperlipidemia. Human vascular endothelial cells (ECs) express class I and II major histocompatibility complex (MHC)–peptide complexes on their surface and come in regular contact with circulating T cells. Activated T cells, in turn, provide both soluble and contact-dependent signals to modulate normal EC functions, including formation and remodeling of blood vessels, regulation of blood flow, regulation of blood fluidity, maintenance of vascular permeability, recruitment of inflammatory leukocytes, and antigen presentation leading to activation of T cells [10].

$CD4^+$ T cells from atherosclerotic plaques have specificity for several atherosclerosis-specific autoantigens including oxidized LDL (oxLDL), heat shock protein 60 (HSP60), β2-glycoprotein I (β2-GPI), and other potential exogenous and endogenous antigens. As much as 10% of lesion T cells are ox-LDL-specific, which also exist in the circulation [4]. To be fully activated, T cells require TCR physical interaction with MHC Class I or II/peptide antigen complexes, and T cell costimulatory molecule interaction. Nine costimulatory proteins have been characterized, which are categorized to two protein families, B7 family and tumor necrosis factor (TNF)/TNF receptor family. The B7 family includes B7-1/2 (CD80/CD86)-CD28 pair, B7-1/2-CTLA-4 (CD152) pair, inducible costimulatory molecule (ICOS, CD278)-ICOSL pair, and programmed death (PD)-PD ligand 1 (PD-L1, CD274), and PD-L2 (CD273) pair. Inhibition of B7 and CD28 can aggravate or reduce atherosclerosis in different settings and ICOS deficiency induces atherosclerosis. Meanwhile, the roles of CTLA-4, B7-H3, B7-H4, and BTLA in atherosclerosis have not been addressed yet. The costimulatory molecules in the TNF superfamily include CD40–CD154 (CD40 ligand) pair, OX40–OX40 ligand pair, CD137–CD137 ligand pair, CD70–CD27 pair, GITR–GITRL pair, and LIGHT (CD258)–lymphotoxin β-HVEM (CD270) pairs. Inhibition of the TNF superfamily members typically results in a decrease of atherosclerosis unless Tregs' function is inhibited [11]. In addition to B7 family and TNF family costimulatory molecules, Toll-like receptors (TLRs), a type of innate immune receptors for recognizing pathogen-associated molecular patterns (PAMPs), also act as costimulation receptors for T-cell activation [12]. Traditionally, it was thought that adaptive immune cells including T cells could only be activated in an antigen-specific manner. However, more than 14 proinflammatory metabolic stresses have been identified as endogenous TLR ligands including HSP60, HSP70, GP96 (endoplasmic reticulum-resident chaperone), high-mobility group B1 (HMGB1), apo CIII (apoCIII), mRNA, fibrogen, fibronectin extra type III domain A (EDA), heparin sulfate, hyaluronan fragment, β-defensin 2, oxidized phospholipid, mild oxidized LDL, and nucleic acids [13]. This suggests that T cells can also be activated by proatherogenic metabolic stresses via TLRs in a nonantigen-specific manner (see Figure 13.3).

Once activated, T cells begin to proliferate and secrete cytokines. In addition to fostering a proinflammatory environment, activated $CD4^+$ T cells also contribute to atherosclerosis by T cell-mediated killing of vascular smooth muscle cells (VSMCs) and the promotion of plaque instability [14]. Activated T cells can differentiate into effector or memory T cells. Memory T cells, as the name implies, play an integral role in establishing life-long immunity to various antigens. In humans, effector memory T cells (T_{EM}:$CD3^+CD4^+CD45RA^-CD45RO^+CCR7^-$cells) are strongly related with intima-media thickness of the common carotid artery ($r = 0.27$; $P < 0.001$). In atherosclerosis-prone animal models, T_{EM} (identified as $CD4^+ CD44^+CD62L^-$) are significantly increased when compared with controls and can be correlated with aortic root lesion size ($r = 0.56$; $P < 0.01$) [15]. Effector lymphocytes activate other cells through secretion of cytokines and direct physical interactions.

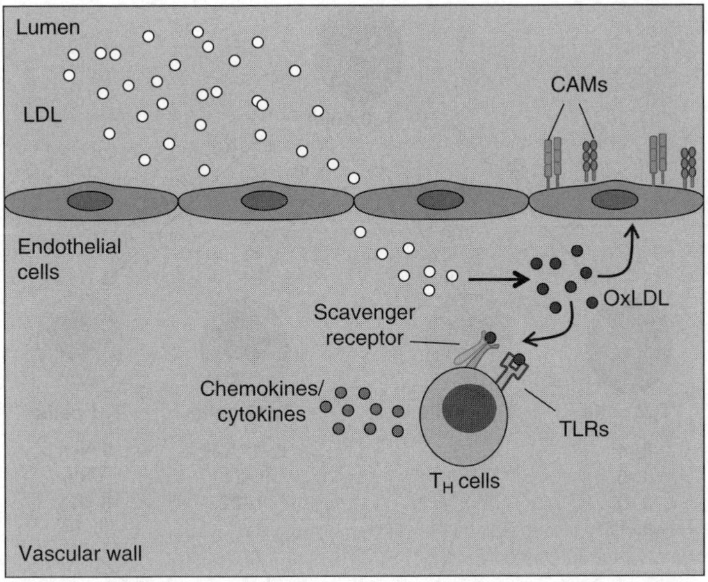

Figure 13.3 Low-density lipoprotein (LDL) diffuses from the blood into the vessel wall where interactions with enzymes and oxygen radicals lead to chemical modifications of LDL into biologically active lipids such as oxidized LDL (oxLDL). These active lipids interact with endothelial cells resulting in their activation and the elevated expression of cell adhesion molecules (CAMs) like vascular cell adhesion molecule 1 (VCAM1) and intercellular adhesion molecule 1 (ICAM1). OxLDL also interacts with T helper (Th) cells through scavenger receptors leading to oxLDL uptake and through toll-like receptors (TLRs), causing the production of proinflammatory cytokines/chemokines.

Meanwhile, the activation of mature B cells is antigen dependent. Naïve B cells circulate the blood and lymph until they encounter an antigen specific to its membrane-bound antibody. Following activation, B cells undergo proliferation, clonal expansion, secretion of antigen-specific antibodies, as well as processing and presenting antigens for T-cell recognition. The activity of the third lymphocyte group, NK cells, is quite different from T- and B cells in that NK cells do not require antigen specificity or harbor immunological "memory." In fact, NK cells do not require previous activation to carry out their cytotoxic function since NK cells constitutively maintain intracellular granules that contain perforin and granzymes. As a member of the innate immune system, NK cells are responsible for fostering an early response to immunological triggers through the recognition of abnormal antigen patterns on the surface of cells. The expansion and activity of NK cells can be induced by various cytokines such as interferons (IFNs) and interleukin-12 (IL-12). The activation and population expansion of lymphocyte cell types work in concert to drive, maintain, and regulate inflammatory and immune responses.

THE CONTRIBUTIONS OF PATHOGENIC Th1 CELLS TO ATHEROSCLEROSIS

In comparison to monocytes/macrophages, the number of T cells present in atherosclerotic plaques is far fewer. In human plaques, the ratio of macrophage to T cell is between a range of 4:1 to 10:1 [16]. Despite this, findings suggest that T cells are critical in the development, aggravation, and progression of atherosclerosis. $CD4^+$ Th play essential roles in regulating inflammation and immune responses via differentiation into various Th functional subtypes. Upon activation, $CD4^+$ T

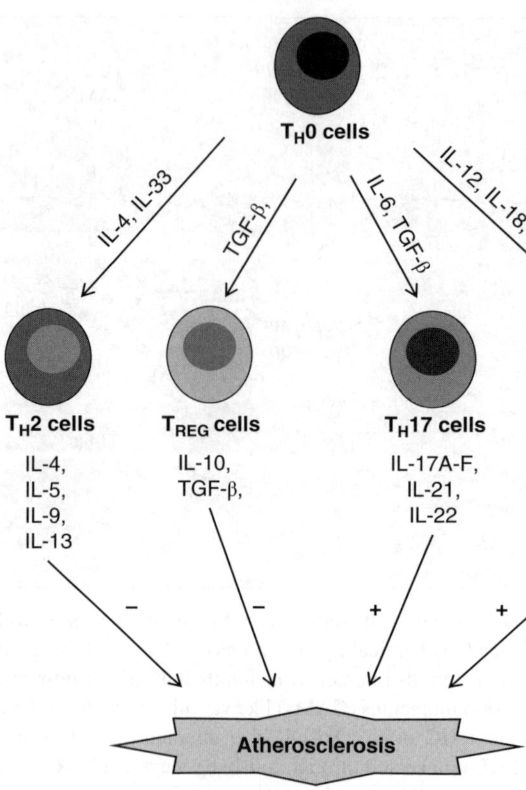

Figure 13.4 Naïve T lymphocytes can differentiate into several subsets including: T helper 1 cells (Th1), T helper 2 cells (Th2), T helper 17 cells (Th17), and regulatory T cells (Treg). Differentiation into a particular subset is dependent on the presence of specific cytokines that are indicated. Upon differentiation, these subsets produce unique cytokines which can either have a proatherogenic or antiatherogenic effect as indicated.

cells differentiate into different Th subsets, which are defined by their unique cytokine profiles (see Figure 13.4). Classically, based on their cytokine generation *in vitro*, $CD4^+$ Th cells were classified into two effector subsets: Th1 and Th2. However, when Mossman and Coffman first proposed their Th1/Th2 paradigm, they also noted the possibility that other subsets of Th cells exist *in vivo* [17]. For two decades, the Th1/Th2 hypothesis had dominated and governed the field of immune responses. However, existence of other Th cell subsets was raised when studies found that organ-specific autoimmune diseases could not be simply explained by the existing paradigm. The characterization of additional Th subsets including Th1, Th2, Th17, Th9, Th22, follicular Th, and Tregs were established [18]. These subsets are primarily defined by the cytokines they secrete since cytokines are responsible for carrying out the majority of Th cell functions. A Th1-driven immune response predominately ensues throughout all stages of atherosclerosis. In early lesion formation, Th1 cells perpetuate the proinflammatory environment with the production of proinflammatory cytokines such as IFN-γ and tumor necrosis factor-α (TNF-α) (see Figure 13.5) [19]. The cytokine IFN-γ promotes atherosclerosis plaque formation and progression by activating macrophages, which results in the production of prothrombotic mediators, other proinflammatory cytokines, and vasoactivators. In advanced plaques, IFN-γ is also believed to cause destabilization by mitigating endothelial proliferation as well as reducing the collagen production,

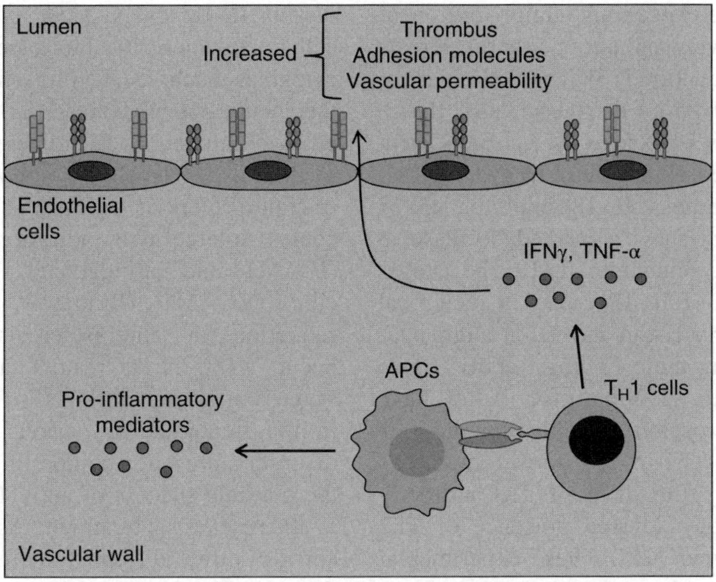

Figure 13.5 Th1 cells within the vessel wall perpetuate a proinflammatory environment through the release of cytokines such as IFN-γ and TNF-α that induce further endothelial activation, the upregulation of cell adhesion molecules, vascular permeability, and thrombus formation. Th1 cells also physically interact with other cell types such as antigen-presenting cells (APCs) leading to the production of proinflammatory mediators by these cells.

proliferation, and differentiation of SMCs [6]. Less circumstantial evidence of IFN-γ contributions to atherosclerosis has been elucidated in several mouse models. Mice genetically devoid of either IFN-γ or its receptor clearly demonstrate a significant reduction in lesion formation, while mature plaques contain elevated levels of collagen [20]. Conversely, the exogenous administration of IFN-γ into apoE$^{-/-}$ mice results in the augmentation of atherosclerosis [20]. In a similar study, the administration of IFN-γ induces atherosclerosis in human arterial segments transplanted into SCID mice [21]. These results strongly support the specific importance of IFN-γ in atherosclerosis. Meanwhile, TNF-α is known to activate the NF-KB pathway leading to the production of prothrombotic tissue factor, reactive oxygen species, and proteolytic enzymes in endothelial cells. It can also lead to the elevation of lipoproteins within the blood by reducing the activity of lipoprotein lipase [6]. Taken together, these findings strongly suggest that IFN-γ and TNF-α are critical in atherosclerosis.

The proatherogenic role of Th1 cells has been indirectly and directly assessed with the use of several animal models. In a model designed for the specific examination of T-cell contributions, atherosclerosis prone Ldlr$^{-/-}$ mice also deficient for T-box expressed in T cells (T-bet), a Th1-specific transcription factor led to reduced atherosclerotic lesion formation. Other mouse models evaluate specific cytokines closely associated with Th1 cells. Produced by monocytes, macrophages, and dendritic cells, IL-12 is a critical cytokine in spurring Th1 cell differentiation. IL-12 activates signal transducer and activator of transcription (STAT)-4 and T-box, which subsequently enhances the production of IFN-γ while simultaneously mitigating IL-4 and IL-5 expression [22]. Mice within an apoE$^{-/-}$ background also deficient for IL-12 subunit p40 demonstrate significant reductions in atherosclerosis development compared with apoE$^{-/-}$ controls [23].

Conversely, the exogenous administration of IL-12 definitively enhances atherosclerosis and IFN-γ production [16]. Th1 cell polarization is also driven by the presence of IL-18, which enhances T-box activity. ApoE-deficient mice transfected with a plasmid overexpressing the endogenous IL-18 inhibitor, IL-18 binding protein, show a marked decrease in atherosclerosis. Similar to IL-12, the exogenous injection of IL-18 causes a reciprocal effect with early lesion formation augmentation. Supporting these findings, $apoE^{-/-}$ mice also genetically deficient for IL-18 have mitigated atherosclerosis. This phenotype is attributed to a Th1 to Th2 subset switch as determined by the drop of Th1 antibody isotypes and the elevated presence of Th2 antibody isotypes [22]. These experimental findings clearly demonstrate that cytokines such as IL-12 and IL-18, which polarize Th1 differentiation, are proatherogenic through the instigation of Th1 cell activity.

Therapeutically, an approach that limits Th1 activity is promising. This strategy can be implemented by either blocking Th1 polarization or the effect of the Th1 cytokines. As aforementioned, IL-12 and IL-18 drive Th1 differentiation. Therefore, inhibition of either cytokine would diminish the proatherogenic effects of Th1 cells. In fact, two human antibodies designed to inhibit the p40 subunit of IL-12 are currently in stage III clinical trials for use in the treatment of psoriasis [24]. If validated, these therapeutics could potentially prove useful in the treatment of other Th1 cell-driven illnesses. It should be noted that the p40 subunit of IL-12 is also shared with IL-23 that promotes Th17 survival and expansion; thus before using such antibodies, the possible effects of mitigating Th17 cell activity must also be taken into account. Meanwhile, the suppression of IL-18, with the treatment of the endogenous IL-18-binding protein, results in reduced atherosclerosis as mentioned earlier. Therefore, the generation of other anti-IL-18 antibodies could prove beneficial in preventing atherosclerosis plaque progression as well as other cardiac events that are correlated with plasma IL-18 levels. Aside of blocking Th1 cell polarization, the blockade of associated cytokines such as TNF-α and IFN-γ hold therapeutic potential. In fact, TNF-α-deficiency crossed into an $apoE^{-/-}$ background results in diminished levels of monocyte chemoattractant protein as well as the adhesion molecules intercellular adhesion molecule 1 (ICAM1) and vascular cell adhesion molecule 1 (VCAM1). Ultimately, in this model, a reduction in atherosclerosis lesion size occurs [22]. These results, along with the success of current anti-TNF-α therapies used in the treatment of various inflammatory disorders, present a compelling argument for the possible success of anti-TNF-α therapies in atherosclerosis treatment. As with any therapeutics directed against the immune system, they must be administered with caution. Inhibition of such powerful proinflammatory cytokines, such as TNF-α and IFN-γ, can have dramatic effects on immune activity, leading to infection susceptibility.

Recent publications suggest that microRNAs (miRNAs), a newly characterized class of short (18–24 nucleotide long), endogenous, noncoding RNAs, contribute to the development of particular disease states through the regulation of diverse biological processes such as cell growth, differentiation, proliferation, and apoptosis. This biological control is accomplished by post-transcriptional gene silencing through Watson and Crick base-pairing predominately at the 3′-untranslatedregion (3′UTR) of messenger RNAs (mRNAs). This pairing can be further characterized as "perfect" or "near perfect," leading to target mRNA cleavage and degradation, or "imperfect," causing the inhibition of mRNA translation. Our recent report showed that 19 miRNAs may inhibit proatherogenic T-cell functions although we have not mapped the functions of these antiatherogenic miRNAs to the Th cell subsets [25]. Another report showed that miRNA-146a in peripheral blood mononuclear cells modulates Th1 function in patients with acute coronary syndrome [26].

THE CONTRIBUTIONS OF TYPE 2 Th CELLS TO ATHEROSCLEROSIS

The discriminating feature between Th1 and Th2 cells is the cytokines that they secrete. Th2 differentiation is instigated by the presence of IL-4, which subsequently activates STAT6 leading to the activation of GATA3, a transcription factor. GATA3 activity inhibits IFN-γ production while augmenting IL-4 and IL-5 secretion, providing Th2 cell phenotype [22]. Functionally, Th2 cells support B-cell antibody production through the secretion of particular cytokines as well as partake in allergic diseases such as asthma. With the production of both proatherosclerotic and antiatherosclerotic cytokines such as IL-4 and IL-5 or IL-10 and IL-33, respectively, the contributions of Th2 cells in atherosclerosis have been hotly contested. The presence of Th2 cells within atherosclerotic lesions is rare, and the participation of Th2 cells in atherosclerosis appears to be stage- and site-specific due to conflicting reports [27]. Promotion of Th2 cell differentiation in atherosclerosis-prone mice results in diminished early lesion formation, while the administration of IL-33, a cytokine associated with Th2-cell polarization, into apoE$^{-/-}$ mice led to a significant reduction in atherosclerosis development. Following IL-33 administration, elevated levels of Th2 cytokines were found within the lymph nodes and serum along with increased production of atheroprotective IgM-type oxidized LDL antibodies. In fact, the atheroprotective effects of vaccination with modified LDL particles are attributed to the induction of IL-5 secretion from Th2 cells, which augment the secretion of IgM-type oxidized LDL antibodies [22]. In contention with these atheroprotective findings, some reports have indicated Th2 cells in a proatherogenic role. Atherosclerotic-prone mice also deficient in the Th2 hallmark cytokine, IL-4, show diminished development of atherosclerosis when compared to littermate controls [4]. Specifically, this phenotype is particularly notable in late-stage lesions located in the aortic arch and thoracic aorta [23]. It should be noted that when this paradigm was carried out again by the same authors, this phenotype was not witnessed. Despite conflicting reports on the proatherogenic or atheroprotective effects of Th2 cells in atherosclerosis, a shift from a Th1 to a Th2 paradigm would appear to passively mitigate atherosclerosis development at the very least. At this point, the therapeutic opportunity of Th2 cells seems rather limited outside of offsetting Th1 cell differentiation.

THE CONTRIBUTIONS OF Tregs TO ATHEROSCLEROSIS

There are two mechanisms attributed to establishing self-tolerance and immune homeostasis. The first involves the deletion of self-reactive T cells through apoptosis, receptor complex editing, or rendering those T cells inactive. The second involves a supervisory T-cell subset, referred to as Tregs, which can be distinguished from other T-cell subsets by the expression of CD4, the forkhead transcription factor Foxp3, and high levels of CD25. The majority of Tregs develop and mature within the thymus under normal physiological conditions and are referred to as natural Tregs (nTregs). Those Tregs that develop in the periphery during active immune responses can be differentiated from naïve T cells by the presence of transforming growth factor-β (TGF-β) or IL-10 and are thus referred to as induced Treg cells (iTregs) [22]. Despite the origin, the maintenance and activation of all Tregs are dependent on the expression of IL-2. In fact, in IL-2-deficient mice, the quantity of Tregs is significantly reduced in the thymus and periphery. This phenotype can be rescued with injection of IL-2 containing spleen cells or thymocytes. Interestingly, the primary source of IL-2 is from effector T cells [28]. This suggests a codependent relationship between Tregs and self-reactive T cells in the maintenance of immune homeostasis. Responder T cells secrete IL-2 that activates and maintains Tregs, which then consequentially inhibit IL-2 production by mitigating responder T cells, thus maintaining proper immune balance.

Following the maturation of nTregs, the cells hone to the periphery where they perpetuate self-tolerance by curbing proinflammatory lymphocytes. This regulation of other T-cell subsets occurs through four basic cellular functions: (i) suppression of cytolysis, (ii) suppression through modulation of dendritic cell maturity and function, (iii) secretion of inhibitory cytokines, and (iv) suppression of metabolic disruption. An in-depth review of these mechanisms has already been written and therefore will not be covered here [29].

The inherit ability of Tregs to suppress the activity of other effector T cells logically led researchers to evaluate them in atherosclerosis. The exploration for the presence of Tregs within atherosclerotic lesions yields meager Treg numbers, representing 1–5% of the T cells present. This number is low when compared to other chronic inflammation models where Tregs can account for 25% of localized T cells [30]. Noteworthy is the fact that apoE$^{-/-}$ mice exhibit significantly fewer Tregs within the spleen than wild-type counterparts. Our recent reports suggest that reduced Tregs in apoE$^{-/-}$ mice may result from increased expression of proapoptotic protein Bax and decreased expression of antiapoptotic Bax-antagonist protein, translationally control tumor protein (TCTP) [31–33]. Taken together, these results suggest that the Tregs apoptosis pathway is a novel therapeutic target for suppressing atheroprogression [34].

The two cytokines released by Tregs that are most accountable for Tregs' inhibitory function are IL-10 and TGF-β, which both have antiatherogenic effects. In fact, the blockade of either IL-10 or TGF-β with antibodies or by genetic deletion results in the exacerbation of Th1 and Th2 cell activity. When the functionality of either of these cytokines is mitigated within atherosclerosis-prone mouse models, accelerated development of atherosclerotic lesions is witnessed in both apoE$^{-/-}$ and Ldlr$^{-/-}$ mice. Mechanistically, the restriction of the TGF-β pathway within T cells results in uncontrolled T-cell proliferation leading to inflammation and thereby accelerated atherosclerosis. These reports indicate that the anti-inflammatory cytokines secreted by Tregs can dramatically reduce atherosclerosis development. In a more cellular-dependent study, the adoptive transfer of Treg cells into apoE-deficient mice led to a significant reduction in the development of inflammation and atherosclerotic plaques in comparison with littermate controls [35–37]. Further, reciprocal results are obtained in Ldlr$^{-/-}$ mice that receive aC D4^{+}25^{+} Tregs-deficient bone marrow transplant [37]. An alternative methodology where Treg apoptosis is induced through the administration of dendritic cells exposed to FOXP3-directed antisense mRNA leads to the augmentation of atherosclerosis in Ldlr$^{-/-}$ mice [38]. Taken together, these reports clearly demonstrate that Tregs act in an atheroprotective fashion, and therefore possess a strong therapeutic potential.

The ability to augment peripheral tolerance prior to exposure to specific antigens is a sought after therapeutic approach in the treatment of allergies, inflammation, autoimmunity, and atherosclerosis. Thus far, several studies have been conducted with Tregs garnering a degree of success. The mucosal administration of proatherogenic antigens apo B-100, heat shock protein 60, and oxidized LDL into atherosclerosis-prone mice results in the reduction of atherosclerosis development. In all three instances, the augmentation of Treg population, TGF-β production, and/or IL-10 production was attributed with conferring tolerance [22]. In comparable studies, the oral or nasal administration of heat shock protein 65 to Ldlr-deficient mice lead to a degree of atherosclerosis tolerance. Although not directly attributed to Tregs, the reduction of lesion size was reportedly due to a Th2-cell shift and the associated production of IL-4 [39, 40]. This shift in cell-type response, along with the elevated levels of IL-10, is quite possibly the result of augmented Treg activity. In fact, multiple subcutaneous injections of influenza peptide have been shown to transform mature effector T cells into Treg [22]. Moreover, when low doses of adjuvant-free apo B-100-derived peptides are administered to apoE$^{-/-}$, a 40% reduction in atherosclerosis lesion development was witnessed. Similar results are also seen with infusions into elderly mice with well-established lesions. Also witnessed is a reduction in T-cell proliferation, infiltration into atherosclerotic lesions, and cytokine production. These atheroprotective effects are attributed to the significant elevation of Tregs within the lymph nodes of the treated mice. Conversely, the atheroprotective effects bestowed upon the treated mice are nullified with the depletion of Tregs from these mice [41]. These studies provide conclusive evidence that the Treg subset plays an indispensable role instilling proatherogenic antigen tolerance.

Aside from bestowing tolerance against proatherogenic antigens, Tregs have also been attributed as the basis for the atheroprotective effects seen with other potential atherosclerotic treatments. The measles virus is capable of attenuating effector T-cell proliferation, dendritic cell activity, and the production of IL-12. Therefore expectantly, apoE$^{-/-}$ mice treated with a measles virus nucleoprotein have demonstrated a reduction in atherosclerosis. Notable is the fact that this protection was accompanied with an increase in Treg numbers. Meanwhile, the antibody blockade of the T-cell marker CD3 reduces autoimmunity by the re-establishment of self-tolerance. Within an atherosclerosis-prone mouse model, the administration of CD3-directed antibodies results in diminished atherosclerosis lesion size. This protection occurs in conjunction with an increase in Tregs as well as elevated levels of TGF-β and FOXP3 within the lymphoid organs [22]. Moreover, expression of chemokine CCL17 by DCs limits the expansion of Tregs by restricting their maintenance leading to increased atherosclerosis development conferred by a T-cell mechanism. Conversely, a blocking antibody specific for CCL17 expands Tregs and reduces atherosclerosis progression [42]. This data clearly implies that augmentation to Treg populations may be a foundation for these therapeutic benefits and therefore be useful in novel therapies designed to combat inflammation and autoimmunity.

CONTRIBUTIONS OF Th17 CELLS TO ATHEROSCLEROSIS

In 2005, a new subset of Th cells was characterized and named "Th17." This subset was defined by its signature cytokine IL-17, although it also secreted IL-17F, IL-21, and IL-22 [43]. IL-17 was initially thought to be a Th1 cytokine, but it has been reclassified with the reorganization of the Th17 subset. IL-17 is a proinflammatory cytokine that was first cloned from a murine T-cell hydridoma and identified as a cytotoxic T-lymphocyte antigen 8 CTLA-8 [44]. High levels of IL-17 from human peripheral blood CD4^{+} was observed upon stimulation indicating its inducible nature [45]. Since its characterization, Th17 and its cytokines have been linked to host defense against extracellular bacterial and fungal infections. In addition, Th17 and IL-17 are implicated in many inflammatory and autoimmune diseases including multiple sclerosis, psoriasis, inflammatory bowel disease, and rheumatoid arthritis [46]. The role of Th17 cells in promoting inflammatory diseases suggests the potential involvement of Th17 cells in atherosclerosis. In fact, IL-17 cytokines and IL 17-producing T cells can be found in atherosclerotic plaques [47]. However, the role of Th17 cells and whether they are proatherogenic or antiatherogenic remains a debate. Using neutralizing approaches, several studies indicate that IL-17 has a proatherogenic role. Blockade of IL-17, by an adenovirus-produced soluble IL-17 receptor or inhibition by monoclonal antibody against IL-17, attenuates lesion formation in apoE$^{-/-}$ mice [48]. In addition, a LDLR$^{-/-}$ mouse bone marrow transfer model resulting in the disruption of IL-17R signaling leads to a reduction in atherosclerosis [49]. Meanwhile, other experimental evidence indicates an antiatherogenic role for IL-17. Suppressor of cytokine signaling (SOCS) 3 protein is a negative regulator of Th17 cell formation since its genetic deficiency leads to enhanced Th17 generation and increased IL-17 production. In regard to atherosclerosis, a loss of SOCS3 expression in T cells of LDLR$^{-/-}$ mice via bone marrow reconstitution results in an atheroprotective phenotype. Specifically, this model enhances IL-17 and IL-10 production, induces an anti-inflammatory phenotype in macrophages, and reduces atherosclerosis formation. Moreover, neutralization of IL-17 with monoclonal antibody in the mouse chimera increases lesion formation [48]. Based on these studies, utilization of neutralization and bone marrow transfer strategies has yielded conflicting results in the role of Th17 and IL-17 in atherosclerosis.

Citing that previous methods used in depleting IL-17 may lack specificity or fail to completely eliminate the cytokines, several laboratories have generated double knockout apoE$^{-/-}$/IL-17$^{-/-}$ and apoE$^{-/-}$/IL-17R$^{-/-}$ mice in an attempt to clarify previous inconsistency. Madhur et al. found that IL-17 deficiency had no effect on atherosclerotic plaque area in apoE$^{-/-}$ mice after 3 months of high fat feeding. However, they found that vascular inflammation was decreased in apoE/IL-17-deficient mice when compared with apoE$^{-/-}$. Specifically, leukocytes and dendritic cells were decreased in the aorta while macrophage within atherosclerotic plaques was reduced. The authors of this study concluded that IL-17 may play a role in some aspect of vascular inflammation and function, but IL-17 inhibition was not adequate to reduce plaque burden [50]. In another study by Butcher et al., IL-17 deficiency in apoE$^{-/-}$ background reduced atherosclerotic lesion size in the whole aorta after 15 weeks of high-fat diet by 35% when compared with controls. Similarly, IL-17R deficiency in apoE$^{-/-}$ mice yielded a 25% reduction in lesion size. In addition, CD45^{+} leukocytes and CD3^{+} T cells were also found to be reduced in the double knockout mice [51]. Data from both laboratories suggest that IL-17 has a proatherogenic role. Although there was no consensus of the effect of IL-17 in lesion formation, both studies found that vascular inflammation was diminished by IL-17 deficiency as judged by leukocyte, T cell, and macrophage numbers in the atherosclerotic aorta [50, 51]. It should be noted that in another study using apoE/IL-17 double knockout mice, Danzaki et al. observed results that were contradictory to those found earlier. IL-17 deficiency was found to accelerate atherosclerotic lesion formation in apoE$^{-/-}$ mice after high fat feeding for 8 or 16 weeks. In addition, increased MOMA-2-positive macrophages were found to infiltrate into atherosclerotic lesions. Moreover, increased CD4^{+}IFN-γ^{+} Th1 cells and IFN-γ production were found in apoE$^{-/-}$/IL-17$^{-/-}$ mice [52]. The authors of these findings suggested that IL-17 deficiency may induce a proatherogenic response by modulating cytokine production and altering Th1 responses. In fairness, Butcher et al. did not find any effect of IL-17 or IL-17R deficiency on the generation and maintenance of Th1 cells in their models [51]. These conflicting reports demonstrate that despite significant progress, the roles of Th17 and IL-17 in atherosclerosis remain undefined. Further investigation is warranted to elucidate these conflicting results.

CONTRIBUTIONS OF γδ T CELLS TO ATHEROSCLEROSIS

Less than 5% of the peripheral T-cell population expresses the γδ TCR instead of the conventional αβ TCR. Limited TCR variability in comparison with typical T cells and the ability to respond to certain antigens despite sensitization suggests that γδ T cells uniquely bridge the gap between the innate and adaptive immune systems. Interestingly, the capability of γδ T cells to produce both Th1 and Th2 cytokines such as IL-2, IL-4, IL-5, IL-10, and IFN-γ indicates an ability to function as a regulator of other T-cell subsets [53]. These characteristics make γδ T cells an attractive research topic. Thus far, γδ T cells have been detected in the intima of human atherosclerotic plaques, particularly during the early stages of lesion development [54]. This early participation is sensible due to γδ T cells innate characteristics. Despite this, the contributions of γδ T cells have yet to be thoroughly elucidated. This is primarily an attribution of the few studies completed with γδ T cells to this point. In one study, 18-week-old apoE$^{-/-}$ mice with a TCRδ-deficiency demonstrated no significant change in atherosclerosis development [55]. However, no other major studies have been conducted to verify or dispute this finding. With innate and adaptive immune characteristics, the presence in early lesions, and immunological modulatory capabilities, γδ T cells represent an intriguing area for future research.

CONTRIBUTIONS OF NKT CELLS TO ATHEROSCLEROSIS

NKT cells are similar to γδ T cells, as they represent a T-cell subset where innate and adaptive immunity intersect. NKT cells are characterized by markers associated with NK cells of the innate immune system, to be discussed later, yet maintain a functional TCR complex. These hybrid T cells are of particular interest to atherosclerosis researchers due to the ability of this cell subset to recognize lipid and glycolipid antigens presented on CD1 complexes of certain antigen presenting cells [4]. This capability possibly suggests a critical role of NKT cells in the initiation of atherosclerosis development through the recognition of endogenously altered lipid products that may be displayed by antigen-presenting cells. Once activated, NKT cells secrete IFN-γ and other proinflammatory cytokines known to drive atherosclerosis development. Further, it has been demonstrated *in vivo* that specific activation of NKT cells results in the subsequent activation of NK cells, CD8$^+$ T cells, and B cells indicating that NKT cells can instigate a strong immunological response through the activation of several cell types [54]. Direct evidence that NKT cells play a role in atherosclerosis development includes the discovery of NKT cell-specific TCR mRNA within atherosclerotic plaques [5]. In addition, a 25% reduction in atherosclerotic lesion size is observed in apoE$^{-/-}$ mice with a CD1-deficiency. To verify that this effect is the result of NKT cell inactivation and not an alternative effect of the CD1-deficiency, an NKT cell activator, α-galactosylceramide, was administered to the apoE/CD1 double knockout mice. In these mice, a 50% increase in atherosclerosis was witnessed implying that the phenotype was the result of NKT inactivation and that augmentation of NKT cell activity will lead to increased atherosclerosis development [55].

THE CONTRIBUTIONS OF B CELLS TO ATHEROSCLEROSIS

The primary role of B cells is to produce antibodies and act as professional antigen-presenting cells. The majority of B cells, referred to as B-2 cells, are produced in the bone marrow and function within the adaptive immune system by secreting IgG antibodies following activation by specific antigens. Functionally, B-2 cells participate in antigen presentation and cytokine production. On the contrary, a small percentage of B lymphocytes, called B-1 cells, participate in the innate immune system and produce IgM antibodies in response to several antigens. B-1 cells carry out similar duties to B-2 cells although they do not demonstrate "memory." Throughout the stages of atherosclerosis development, IgM and IgG antibodies can be found within plaques [22]. As discussed earlier, the effect of T cells in atherosclerosis development is subset dependent. Recent findings suggest that this principle also holds true for B cells. There are several reports which indicate that B cells mainly play an atheroprotective role. Bone marrow adoptive transfer from B cell-deficient mice into atherosclerosis-susceptible recipient mice resulted in increased early and late-stage atherosclerosis development [56]. In further support, atherosclerosis development is aggravated in apoE$^{-/-}$ mice, which undergo a splenectomy suggesting a protective nature for B cells. In addition, splenectomized mice receiving a splenic B cell transfer from late-stage atherosclerotic $apoE^{-/-}$ mice demonstrate a significant degree of protection from atherosclerosis [55]. In contrast to these atheroprotective findings, more recent reports indicate that certain subsets of B cells may be proatherogenic. In particular, the transfer of B-2 cells into recipient mice resulted in exacerbation of atherosclerosis development that was not observed with B-1 transfer [22]. Moreover, treatment with B cell-depleting antibody in both $Ldlr^{-/-}$ and $apoE^{-/-}$ atherosclerosis mouse models resulted in dramatic reductions in atherosclerosis. It is important to note that under this model, B-cell depletion drove T-cell differentiation from a proatherogenic Th1 subset to a Th17 phenotype. In fact, abrogation of the IL-17 production with a blocking peptide resulted in the loss of atheroprotection associated with CD20 antibody administration. This suggests that the protective effects of B-cell depletion are the direct result of alterations to the T-cell subset indicating a critical role of B cells in driving pathogenic T-cell differentiation and supporting the atheroprotective nature of Th17 cells discussed earlier [57].

Future therapies involving B cells should focus on the augmentation of the seemly atheroprotective B-1 subset population while marginalizing the activity of the proatherogenic B-2 subset. Potential therapeutic routes include the blockade of B-cell-activating factor of the TNF family (BAFF). Critical to the survival and proliferation of peripheral B cells, antibody-directed suppression of BAFF leads to the loss of follicular B cells in the spleen of wild-type mice. It should be noted that the function and population of B-1 cells are unaffected with this treatment [22]. Currently, clinical trials are being conducted examining the use of a humanized monoclonal antibody for BAFF in individuals with systemic lupus erythematosus (SLE). It is reasonable to extrapolate that if this treatment shows therapeutic promise in SLE patients, then it may be useful in other chronic inflammatory diseases such as atherosclerosis.

CONTRIBUTIONS OF NK CELLS TO ATHEROSCLEROSIS

Consisting of approximately 15% of lymphocytes in humans, NK cells are a unique, bone marrow-derived subset that partakes in the innate immune system. Functionally, NK cells act as effectors, releasing cytokines and mediating the cytotoxicity of aberrant cells. The ability to act without previous sensitization, in conjunction with the production of or sensitivity to proatherogenic cytokines such as IFN-γ, IL-12, and IL-18 suggests that NK cells participate in early atherosclerosis development [58]. In fact, the presence of NK cells within human atherosclerotic plaques has been detected [55]. However, evaluating the contributions of NK cells in atherosclerosis has proven difficult due to the absence of an adequate mouse model. NK cells develop normally in immunologically compromised mouse models such as recombination-activating gene (RAG)-deficient mice [54]. The only mouse models that do exist retain the presence of NK cells that are not fully functioning. Therefore, the results gathered from these models should be interrupted with caution. The beige mouse model contains a defect in NK cell cytolysis stemming from an inability of NK cells to secrete the contents of their cytoplasmic granules. Crossing these mice into an $Ldlr^{-/-}$ background results in a reduction of atherosclerosis, implying a protective quality of NK cells [59]. However, this result is in contradiction to logic and must be interrupted carefully since the beige mouse model contains a very complex phenotype which opens the possibility to unforeseen contributions. In addition, a similar study carried out by an independent group did not yield the same significant results [60]. Perhaps the most promising data to link NK cells to atherosclerosis has come from a mouse model where bone marrow from a transgenic mouse model expressing Ly49A under the control of the granzyme A promoter is transplanted into irradiated Ldlr null mice. Ly49A is an inhibitory major histocompatibility complex class I-specific receptor that is present on all NK cells. Once this receptor is engaged, activation of NK cells is inhibited [61]. In this model, the activation of NK cells can be completely abrogated leading to a 70 and 38% reduction in atherosclerotic lesion formation within the aortic root and arch, respectively [62]. It is important to note that the original scientists to characterize the Ly49A transgenic mouse model reported a 50% reduction in NK cells within the spleen, which was not reported in this study. Further, the original group noted that even though the populations of B and T cells remained unchanged, approximately 50% of T cells expressed the Ly49A transgene [61]. If these T cells were also suppressed, then the witnessed phenotype cannot be solely attributed to NK cells. Despite the consensus that NK cells play a role in atherosclerosis, the exact contributions remain murky. The development of a more sufficient mouse model would greatly benefit the research examining NK cells' contributions to atherosclerosis.

CONCLUSION

In this chapter, we have clearly outlined how lymphocytes can contribute to or protect against atherosclerosis formation, progression, and aggravation through the production of various cytokines and cellular interactions. Many of the cell types discussed have well-characterized subsets that have been shown to uniquely contribute to atherosclerosis. Continued evaluation of these subsets remains critical to fully elucidate the function these cells perform in the complex immunological process that leads to atherosclerosis development. Future research will hopefully further illuminate this pathological process and yield new findings that will lead to the development of effective cytokine and cell type-based therapeutics.

REFERENCES

1. Frostegard J. Atherosclerosis in patients with autoimmune disorders. *Arteriosclerosis, Thrombosis, and Vascular Biology*. 2005; 25(9):1776–1785.
2. Mor A, Luboshits G, Planer D, Keren G, George J. Altered status of CD4(+)CD25(+) regulatory T cells in patients with acute coronary syndromes. *European Heart Journal*. 2006;27(21): 2530–2537.
3. Zhou X, Nicoletti A, Elhage R, Hansson GK. Transfer of CD4(+) T cells aggravates atherosclerosis in immunodeficient apolipoprotein E knockout mice. *Circulation*. 2000;102(24): 2919–2922.
4. Hansson GK, Robertson AK, Soderberg-Naucler C. Inflammation and atherosclerosis. *Annual Review of Pathology*. 2006;1: 297–329.
5. Paulsson G, Zhou X, Tornquist E, Hansson GK. Oligoclonal T cell expansions in atherosclerotic lesions of apolipoprotein E-deficient mice. *Arteriosclerosis, Thrombosis, and Vascular Biology*. 2000;20(1):10–17.
6. Hansson GK, Libby P. The immune response in atherosclerosis: a double-edged sword. *Nature Reviews*. 2006;6(7):508–519.
7. Luchtefeld M, Grothusen C, Gagalick A, et al. Chemokine receptor 7 knockout attenuates atherosclerotic plaque development. *Circulation*. 2010;122(16):1621–1628.
8. Grivel JC, Ivanova O, Pinegina N, et al. Activation of T lymphocytes in atherosclerotic plaques. *Arteriosclerosis, Thrombosis, and Vascular Biology*. 2011;31(12):2929–2937.
9. Choi JH, Do Y, Cheong C, et al. Identification of antigen-presenting dendritic cells in mouse aorta and cardiac valves. *The Journal of Experimental Medicine*. 2009;206(3):497–505.
10. Choi J, Enis DR, Koh KP, Shiao SL, Pober JS. T lymphocyte-endothelial cell interactions. *Annual Review of Immunology*. 2004;22: 683–709.
11. Gerdes N, Zirlik A. Co-stimulatory molecules in and beyond co-stimulation – tipping the

balance in atherosclerosis? *Thrombosis and Haemostasis*. 2011;106(5):804–813.

12. Cole JE, Georgiou E, Monaco C. The expression and functions of toll-like receptors in atherosclerosis. *Mediators of Inflammation*. 2010;2010: 393946.
13. Lundberg AM, Hansson GK. Innate immune signals in atherosclerosis. *Clinical Immunology*. 2010;134(1):5–24.
14. Pryshchep S, Sato K, Goronzy JJ, Weyand CM. T cell recognition and killing of vascular smooth muscle cells in acute coronary syndrome. *Circulation Research*. 2006;98(9):1168–1176.
15. Ammirati E, Vecchio V, Banfi M, Vermi AC, De Metrio M, Grigore L, Pirillo A, Garlaschelli K, Manfredi AA, Catapano AL, Maseri A, Norata GD. Effector memory T cells are associated with atherosclerosis in humans and animal models. *Journal of the American Heart Association*. 2012;1(1):15.
16. Hansson GK, Hermansson A. The immune system in atherosclerosis. *Nature Immunology*. 12(3):204–212.
17. Mosmann TR, Cherwinski H, Bond MW, Giedlin MA, Coffman RL. Two types of murine helper T cell clone. I. Definition according to profiles of lymphokine activities and secreted proteins. *Journal of Immunology*. 1986;136(7): 2348–2357.
18. Mai J, Wang H, Yang XF. Th 17 cells interplay with Foxp3+ Tregs in regulation of inflammation and autoimmunity. *Frontiers in Bioscience : A Journal and Virtual Library*. 2010;15: 986–1006.
19. Tedgui A, Mallat Z. Cytokines in atherosclerosis: pathogenic and regulatory pathways. *Physiological Reviews*. 2006;86(2):515–581.
20. Whitman SC, Ravisankar P, Elam H, Daugherty A. Exogenous interferon-gamma enhances atherosclerosis in apolipoprotein E–/– mice. *The American Journal of Pathology*. 2000;157(6):1819–1824.
21. Tellides G, Tereb DA, Kirkiles-Smith NC, et al. Interferon-gamma elicits arteriosclerosis in the absence of leukocytes. *Nature*. 2000;403(6766):207–211.
22. Lahoute C, Herbin O, Mallat Z, Tedgui A. Adaptive immunity in atherosclerosis: mechanisms and future therapeutic targets. *Nature Reviews Cardiology*. 2011;8(6):348–358.
23. Davenport P, Tipping PG. The role of interleukin-4 and interleukin-12 in the progression of atherosclerosis in apolipoprotein E-deficient mice. *The American Journal of Pathology*. 2003;163(3):1117–1125.
24. Gandhi M, Alwawi E, Gordon KB. Anti-p40 antibodies ustekinumab and briakinumab: blockade of interleukin-12 and interleukin-23 in the treatment of psoriasis. *Seminars in Cutaneous Medicine and Surgery*. 2010;29(1): 48–52.
25. Virtue A, Mai J, Yin Y, et al. Structural evidence of anti-atherogenic microRNAs. *Frontiers in Bioscience : A Journal and Virtual Library*. 2012;17:3133–3145.
26. Guo M, Mao X, Ji Q, et al. miR-146a in PBMCs modulates Th1 function in patients with acute coronary syndrome. *Immunology and Cell Biology*. 2010;88(5):555–564.
27. Frostegard J, Ulfgren AK, Nyberg P, et al. Cytokine expression in advanced human atherosclerotic plaques: dominance of pro-inflammatory (Th1) and macrophage-stimulating cytokines. *Atherosclerosis*. 1999;145(1):33–43.
28. Sakaguchi S. Naturally arising Foxp3-expressing CD25+CD4+ regulatory T cells in immunological tolerance to self and non-self. *Nature Immunology*. 2005;6(4):345–352.
29. Vignali DA, Collison LW, Workman CJ. How regulatory T cells work. *Nature Reviews. Immunology*. 2008;8(7):523–532.
30. de Boer OJ, van der Meer JJ, Teeling P, van der Loos CM, van der Wal AC. Low numbers of FOXP3 positive regulatory T cells are present in all developmental stages of human atherosclerotic lesions. *PloS One*. 2007;2(8):e779.
31. Xiong Z, Song J, Yan Y, et al. Higher expression of Bax in regulatory T cells increases vascular inflammation. *Frontiers in Bioscience : A Journal and Virtual Library*. 2008;13:7143–7155.
32. Xiong Z, Yan Y, Song J, et al. Expression of TCTP antisense in CD25(high) regulatory T cells aggravates cuff-injured vascular inflammation. *Atherosclerosis*. 2009;203(2):401–408.
33. Yan Y, Xiong Z, Zhang S, et al. CD25high T cells with a prolonged survival inhibit development of diabetes. *International Journal of Immunopathology and Pharmacology*. 2008; 21(4):767–780.
34. Yang XF. Factors regulating apoptosis and homeostasis of CD4+ CD25(high) FOXP3+ regulatory T cells are new therapeutic targets. *Frontiers in Bioscience : A Journal and Virtual Library*. 2008;13:1472–1499.

35. Mallat Z, Gojova A, Brun V, et al. Induction of a regulatory T cell type 1 response reduces the development of atherosclerosis in apolipoprotein E-knockout mice. *Circulation*. 2003;108(10):1232–1237.
36. Mor A, Planer D, Luboshits G, et al. Role of naturally occurring CD4+ CD25+ regulatory T cells in experimental atherosclerosis. *Arteriosclerosis, Thrombosis, and Vascular Biology*. 2007;27(4):893–900.
37. Ait-Oufella H, Salomon BL, Potteaux S, et al. Natural regulatory T cells control the development of atherosclerosis in mice. *Nature Medicine*. 2006;12(2):178–180.
38. van Es T, van Puijvelde GH, Foks AC, et al. Vaccination against Foxp3(+) regulatory T cells aggravates atherosclerosis. *Atherosclerosis*. 2010;209(1):74–80.
39. Harats D, Yacov N, Gilburd B, Shoenfeld Y, George J. Oral tolerance with heat shock protein 65 attenuates Mycobacterium tuberculosis-induced and high-fat-diet-driven atherosclerotic lesions. *Journal of the American College of Cardiology*. 2002;40(7):1333–1338.
40. Maron R, Sukhova G, Faria AM, et al. Mucosal administration of heat shock protein-65 decreases atherosclerosis and inflammation in aortic arch of low-density lipoprotein receptor-deficient mice. *Circulation*. 2002;106(13):1708–1715.
41. Herbin O, Ait-Oufella H, Yu W, et al. Regulatory T-cell response to apolipoprotein b100-derived peptides reduces the development and progression of atherosclerosis in mice. *Arteriosclerosis, Thrombosis, and Vascular Biology*. 2012;32(3):605–612.
42. Weber C, Meiler S, Doring Y, et al. CCL17-expressing dendritic cells drive atherosclerosis by restraining regulatory T cell homeostasis in mice. *The Journal of Clinical Investigation*. 2011;121(7):2898–2910.
43. Bettelli E, Korn T, Oukka M, Kuchroo VK. Induction and effector functions of T(H)17 cells. *Nature*. 2008;453(7198):1051–1057.
44. Rouvier E, Luciani MF, Mattei MG, Denizot F, Golstein P. CTLA-8, cloned from an activated T cell, bearing AU-rich messenger RNA instability sequences, and homologous to a herpesvirus saimiri gene. *Journal of Immunology*. 1993;150(12):5445–5456.
45. Yao Z, Painter SL, Fanslow WC, et al. Human IL-17: a novel cytokine derived from T cells. *Journal of Immunology*. 1995;155(12):5483–5486.
46. Korn T, Bettelli E, Oukka M, Kuchroo VK. IL-17 and Th17 Cells. *Annual Review of Immunology*. 2009;27:485–517.
47. Eid RE, Rao DA, Zhou J, et al. Interleukin-17 and interferon-gamma are produced concomitantly by human coronary artery-infiltrating T cells and act synergistically on vascular smooth muscle cells. *Circulation*. 2009;119(10):1424–1432.
48. Butcher M, Galkina E. Current views on the functions of interleukin-17A-producing cells in atherosclerosis. *Thrombosis and Haemostasis*. 2009;106(5):787–795.
49. van Es T, van Puijvelde GH, Ramos OH, et al. Attenuated atherosclerosis upon IL-17R signaling disruption in LDLr deficient mice. *Biochemical and Biophysical Research Communications*. 2009;388(2):261–265.
50. Madhur MS, Funt SA, Li L, et al. Role of interleukin 17 in inflammation, atherosclerosis, and vascular function in apolipoprotein e-deficient mice. *Arteriosclerosis, Thrombosis, and Vascular Biology*. 2011;31(7):1565–1572.
51. Butcher MJ, Gjurich BN, Phillips T, Galkina EV. The IL-17A/IL-17RA axis plays a proatherogenic role via the regulation of aortic myeloid cell recruitment. *Circulation Research*. 2012;110(5):675–687.
52. Danzaki K, Matsui Y, Ikesue M, et al. Interleukin-17A deficiency accelerates unstable atherosclerotic plaque formation in apolipoprotein E-deficient mice. *Arteriosclerosis, Thrombosis, and Vascular Biology*. 2012;32(2):273–280.
53. Aljurf M, Ezzat A, Musa OM. Emerging role of gammadelta T-cells in health and disease. *Blood Reviews*. 2002;16(4):203–206.
54. Vanderlaan PA, Reardon CA. Thematic review series: the immune system and atherogenesis. The unusual suspects: an overview of the minor leukocyte populations in atherosclerosis. *Journal of Lipid Research*. 2005;46(5):829–838.
55. Galkina E, Ley K. Immune and inflammatory mechanisms of atherosclerosis (*). *Annual Review of Immunology*. 2009;27:165–197.
56. Major AS, Fazio S, Linton MF. B-lymphocyte deficiency increases atherosclerosis in LDL receptor-null mice. *Arteriosclerosis, Thrombosis, and Vascular Biology*. 2002;22(11):1892–1898.
57. Ait-Oufella H, Herbin O, Bouaziz JD, et al. B cell depletion reduces the development

of atherosclerosis in mice. *The Journal of Experimental Medicine*. 2010;207(8): 1579–1587.

58. Yokoyama WM, Kim S, French AR. The dynamic life of natural killer cells. *Annual Review of Immunology*. 2004;22:405–429.
59. Schiller NK, Boisvert WA, Curtiss LK. Inflammation in atherosclerosis: lesion formation in LDL receptor-deficient mice with perforin and Lyst(beige) mutations. *Arteriosclerosis, Thrombosis, and Vascular Biology*. 2002;22(8): 1341–1346.
60. Paigen B, Holmes PA, Novak EK, Swank RT. Analysis of atherosclerosis susceptibility in mice with genetic defects in platelet function. *Arteriosclerosis*. 1990;10(4):648–652.
61. Kim S, Iizuka K, Aguila HL, Weissman IL, Yokoyama WM. In vivo natural killer cell activities revealed by natural killer cell-deficient mice. *Proceedings of the National Academy of Sciences of the United States of America*. 2000;97(6):2731–2736.
62. Whitman SC, Rateri DL, Szilvassy SJ, Yokoyama W, Daugherty A. Depletion of natural killer cell function decreases atherosclerosis in low-density lipoprotein receptor null mice. *Arteriosclerosis, Thrombosis, and Vascular Biology*. 2004;24(6):1049–1054.

14

STEM CELLS

ARJUN DEB
Department of Medicine and Molecular, Cell and Developmental Biology, Broad Stem Cell Research Center, University of California, Los Angeles, CA, USA

"One is as old as one's endothelium" [1]

INTRODUCTION

Stem cells can be defined as specialized cells that are able to renew themselves (self-renewal) as well as differentiate into cells of specific lineage; these two properties of self-renewal and differentiation are considered characteristic hallmarks of stem cells. Stem cells are usually classified according to their potency (Table 14.1). For instance, a totipotent stem cell, such as fertilized zygote can generate an entire organism, while a pluripotent stem cell can generate cells belonging to all three germ layers (ectoderm, mesoderm, and endoderm), but are unable to generate an entire organism. Embryonic stem cells are examples of pluripotent cells. By contrast, multipotent stem cells are more restricted in their potency, are unable to generate cells belonging to all three germ layers but can differentiate into different types of cells usually arising from the same germ layer. Mesenchymal stem cells are multipotent stem cells that can differentiate into cells of mesodermal origin such as fat, bone, and cartilage. A unipotent stem cell or a stem cell with further restriction of its differentiation capacity is called a progenitor and can usually differentiate into a specific type of cell or cells in a given tissue. For example, an endothelial progenitor cell (EPC) gives rise to endothelial cells and an erythroid progenitor differentiates into red blood cells. However, more recently, such traditional classification schemes based on the potency of progenitors have been questioned with the generation of induced pluripotent (iPS) cells [2, 3]. Yamanaka and colleagues demonstrated that the introduction of specific transcription factors can reprogram a terminally differentiated cell such as a fibroblast to a pluripotent embryonic stem cell [2, 3]. Since this pioneering observation, many groups have reprogrammed different types of somatic cells as well as progenitors into pluripotent cells with different cocktails of transcription factors [4–6].

Atherosclerosis: Risks, Mechanisms, and Therapies, First Edition. Edited by Hong Wang and Cam Patterson.

TABLE 14.1 Stem cell glossary

Totipotent	Stem cells that are capable of forming the entire organism including formation of extra-embryonic tissues
Pluripotent	Stem cells that are not capable of forming the entire organism but can form cells belonging to all three germ layers
Multipotent	Potency more restricted than pluripotent stem cells, multipotent stem cells can give rise to different types of cells but not to cells derived from all three germ layers
Progenitor	Also known as unipotent, progenitors can give rise to cells belonging to a specific tissue
Induced pluripotent	Pluripotent stem cell that has been created by reprogramming of a somatic cell via introduction of specific transcription factors
Vasculogenesis	Process of de novo formation of blood vessels from progenitors
Angiogenesis	Process of new blood vessel formation from preexisting blood vessels

These observations suggest that different somatic cells and stem cells irrespective of their potency can undergo nuclear reprogramming and become pluripotent. In this chapter, we broadly discuss the role of stem cells and vascular progenitors in the development and maintenance of blood vessels, their role in vascular disease, and therapeutic possibilities of stem cell-based therapy for new blood vessel formation.

VASCULAR PROGENITORS IN DEVELOPMENT

The cardiovascular system is one of the earliest organ systems to form during embryonic development, and blood vessels initially form outside of the embryo in the yolk sac [7]. Vascular progenitors during development are thought to arise from bipotential precursors called hemangioblasts. Hemangioblasts can differentiate into hematopoietic stem cells (HSCs) that form cells of the hematopoietic system or angioblasts that give rise to endothelial cells and form new vessels [8, 9]. Angioblasts in the yolk sac are vascular progenitors that form the earliest vascular network during development and ultimately lead to formation of the placental vasculature [10]. Shortly after formation of blood vessels in extraembryonic tissue, intraembryonic angioblasts appear and start forming intraembryonic blood vessels of the developing embryo. This process of *de novo* formation of new blood vessels from vascular precursors such as angioblasts is known as vasculogenesis and is to be differentiated from angiogenesis that refers to the formation of new blood vessels from preexisting ones (Table 14.1). Formation of blood vessels in the developing embryo particularly during early development occurs almost entirely via vasculogenesis, where clumps of endothelial precursor cells coalesce together and subsequently form a lumen. It is not clear how angioblasts form new vessels in specific patterns, but a combination of patterning instructions and inductive cues from nonendothelial cells play an important role in specific patterning of the arteries and veins that are seen in postnatal life. By contrast, angiogenesis is primarily driven by a combination of patterning instructions and metabolic requirements, hypoxia being one of the most potent inducers of angiogenesis both during development and postnatal life [11].

Considerable work over the past decade has elucidated the molecular signals regulating the function of angioblasts in early development. Vascular endothelial growth factor (VEGF) signaling is critical for formation, proliferation, and migration of angioblasts as well as proliferation of endothelial and hematopoietic cells. VEGF receptor 2, (VEGFR2) also known as Flk-1 is expressed in blood islands of the mouse yolk sac and deletion of the Flk-1 receptor leads to early embryonic lethality in mice from poor formation of blood islands in the yolk sac [12]. Moreover, Flk-1-deficient ES cells are unable to form endothelium or contribute to hematopoiesis in wild-type embryos [12]. These studies highlight the importance of VEGF/Flk-1 signaling for hemangioblast development and differentiation. Various other genes and signaling pathways have been implicated in the regulation of angioblast formation and function such as the stem cell leukemia (SCL) and bone morphogenetic protein (BMP) pathways. Scl-deficient ES cells are unable to differentiate into endothelial cells even in the presence of intact Flk-1 expression underscoring the importance of Scl in hemangioblast function and differentiation of angioblasts [13]. Scl is also known to specify hemangioblast formation in zebrafish. BMP4 promotes Flk-1 expression in hemangioblasts [14] and endothelial regulator protein of BMP, known as Bmper, is expressed in hemangioblasts and inhibits BMP4 signaling [15]. Bmper thus antagonizes endothelial precursor formation from hemangioblasts and provides a negative feedback mechanism for tightly regulating the formation of endothelial precursors. Similarly, the TGFβ1 pathway is necessary for vascular development and inhibits the growth and differentiation of hematopoietic progenitors [14, 16] and perhaps serves as a checkpoint to direct hemangioblast differentiation to form angioblasts rather than HSCs (Figure 14.1).

Although angioblasts form endothelial cells that ultimately form blood vessels, mature vessels contain a smooth muscle layer. Smooth muscle cells are thought to arise from different types of progenitors in the developing embryo [17]. Smooth muscle cells of the great vessels are thought to originate from neural crest cells, a migrating population of cells that leave the dorsal surface of neural tube and adopt a variety of fates in different tissues of the body. Smooth muscle cells of the coronary arteries arise from epithelial-mesenchymal transition of mesothelial cells residing in the proepicardial organ, a collection of villous-like projections at the venous pole of the heart, that form the majority of the epicardium. Finally, smooth muscle cells of the rest of the aorta and visceral arteries arise from smooth muscle precursors residing in the lateral or splanchnic mesoderm. As smooth muscle cells can arise from progenitors present in different regions of the embryo, this implies that smooth muscle cells in a local region of a great vessel may have different embryonic origins. Defects in smooth muscle progenitors in one region usually cannot be rescued by smooth muscle progenitors arising from a different embryonic region suggesting that smooth muscle progenitors adopt prespecified fates that cannot be easily altered by the local niche [18]. Similar to endothelial differentiation of angioblasts, several signaling pathways such as endothelin, Wnt, retinoic acid, and FGF8 have been

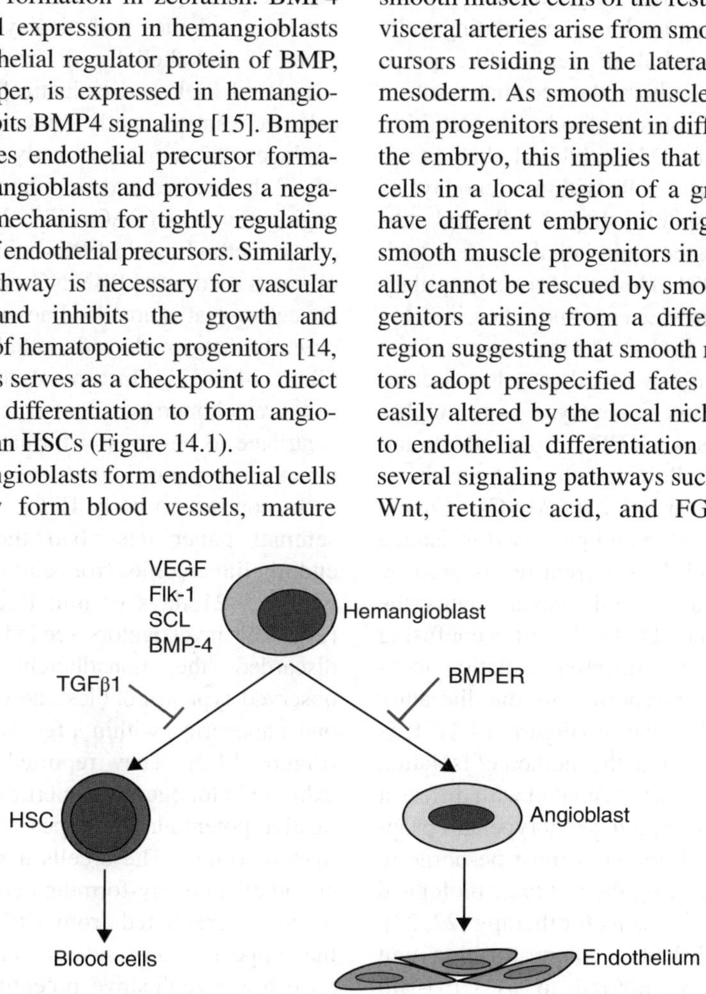

Figure 14.1 Differentiation of hemangioblast into HSCs or endothelial cells and critical molecules that regulate this process.

implicated in regulating the function of smooth muscle precursors during development [19].

VASCULAR PROGENITORS IN POSTNATAL LIFE

Circulating EPCs

Despite a large body of evidence demonstrating the role of vascular progenitors in the development of the vasculature in the embryo, the identity and role of vascular progenitors in postnatal life remained elusive. More than a decade ago, Isner et al., in a pioneering piece of work, demonstrated that circulating cells can have endothelial progenitor-like properties [20]. The authors showed that a subset of circulating peripheral blood cells identified by expression of CD34 exhibited characteristics of EPCs. These cells within a few days of culture formed spindle-shaped cells and subsequently expressed endothelial markers such as Flk-1 and CD31. The cells formed capillary tube-like structures *ex vivo* and when injected into the hind limbs of ischemic rabbits appeared to form new blood vessels (vasculogenesis). Since this discovery, a large number of groups have isolated EPCs from rodents and human subjects following culture of peripheral blood mononuclear cells (PBMNCs). Owing to variations in the techniques used to isolate human EPCs and the different terms used by investigators to describe the isolated cells, the terminology in the EPC field is often confusing [21]. Broadly, three different isolation techniques have been reported in the literature for human EPC isolation (Figure 14.2). It is important to appreciate the method of isolation of human EPCs as EPCs isolated with different techniques differ in their phenotype and properties, and such differences must be borne in mind while comparing the utility or biological applications of these cells for therapy [22, 23]. Hill et al. published a colony-forming unit assay (commonly referred to as CFU-Hill assay) for isolation of EPCs from a sample of peripheral venous blood [24]. PBMNCs isolated from peripheral blood were seeded on fibronectin-coated dishes, and the nonadherent cells were subsequently transferred to another dish and cultured. Hill et al. reported colony formation from culture of the nonadherent cells within 4–9 days of initial seeding, with a central aggregate of round cells with spindle-shaped cells extending toward the periphery displaying endothelial cell-like characteristics (Figure 14.2) [24]. In a variant of this method, several other groups of investigators seeded PBMNCs on fibronectin dishes as reported by Hill but did not separate adherent and nonadherent cells. At 4 days following initial seeding, the investigators noted round cells that did not form colonies and displayed some properties of endothelial cells such as uptake of acetylated LDL and isolectin [25, 26]. These cells were subsequently shown to contribute to angiogenesis without directly forming blood vessel themselves and were termed circulating angiogenic cells (CACs) [27, 28]. Both of the above methods yield EPCs within a few days of initial seeding of PBMNCs, and collectively the cells isolated are also known as early EPCs or early outgrowth endothelial cells (OECs) (Table 14.2) [21]. In general, they are thought to secrete important angiogenic cytokines that contribute to new vessel formation, but do not directly incorporate into new vessels themselves.

In contrast to early EPCs, Lin et al. in a seminal paper described the isolation of endothelial colonies from culture of PBMNCs within 7–21 days of initial seeding of cells [29]. The investigators seeded the PBMNCs, discarded the nonadherent fraction and observed typical cobblestone endothelial colonies appearing within a few weeks of culture (Figure 14.2). They reported that the cells exhibited clonogenic properties, had high replicative potential, and appeared to be of bone marrow origin. These cells have been termed endothelial colony-forming cells (ECFCs) and to be differentiated from CFU-Hill colonies that appear within several days of seeding, have lower replicative potential, and are not clonogenic [21]. As Lin et al. observed, these colonies in a few weeks rather than days of

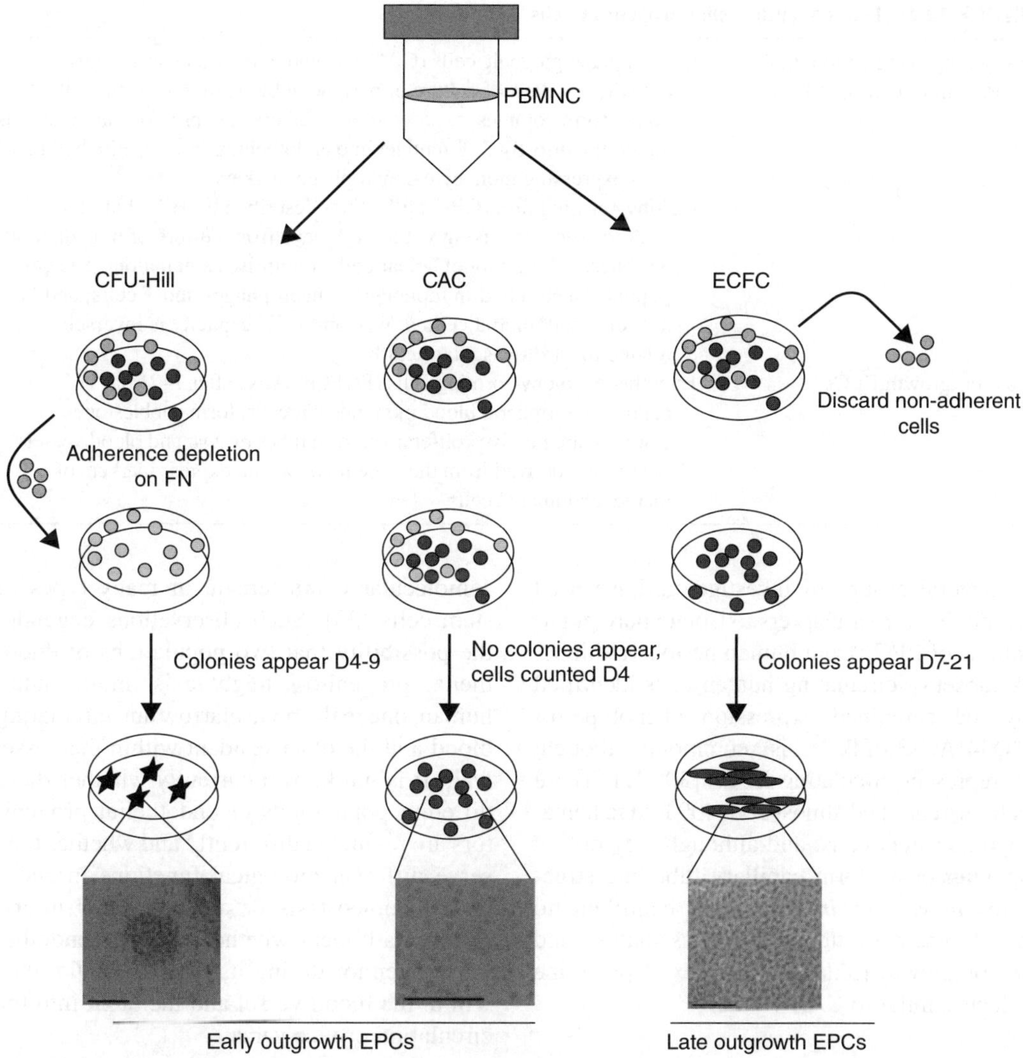

Figure 14.2 Peripheral blood mononuclear cell (PBMNC) isolated from human peripheral blood are seed onto culture dishes, and subsequently nonadherent cells are replated (CFU-Hill) or discarded (ECFC) or kept in culture with the adherent cells (CACs). Each of these methods yields endothelial progenitor cells (EPCs) with different properties and time of appearance from culture (Adapted from Hirschi et al. [21]. © 2008, Lippincott Williams & Wilkins).

culture, these cells have also been termed late EPCs or late outgrowth endothelial cells (Table 14.2). ECFCs formed capillary tube-like structures *ex vivo* and when injected into immunosuppressed rodents formed blood vessels with human endothelial cells demonstrating the ability of these cells to form blood vessels *in vivo*. Given the properties of ECFCs in directly contributing to vessel formation (vasculogenesis), they appear to have more endothelial progenitor characteristics than early EPCs. However, no combination of surface markers has as yet identified the characteristics of the putative progenitor giving rise to ECFCs and intense investigation continues to find the ECFC progenitor.

TABLE 14.2 Types of endothelial progenitor cells

Early outgrowth endothelial progenitor cells or EPCs	Circulating angiogenic cells (CACs). Spindle-shaped cells that grow out within 4 days of culture of peripheral blood mononuclear cells, do not form colonies, express growth factors, and promote angiogenesis but do not directly differentiate into endothelium *in vivo*, enriched in cells expressing monocyte–macrophage markers
	Colony-forming unit (CFU-Hill). First described by Hill et al., they refer to colonies arising within 4–9 days from culture of nonadherent peripheral blood mononuclear cells; comprises a hematopoietic cell population enriched in monocytes–macrophages and T cells; and do not form endothelial cells *in vivo* and CFU capacity is inversely related to cardiovascular health
Late outgrowth EPCs	Endothelial colony-forming cells (ECFCs). Arise after 7–21 days of culture of peripheral blood mononuclear cells, form cobblestone colonies, are highly proliferative, form tubes *ex vivo* and blood vessels *in vivo*, are derived from the bone marrow, and express markers of mature endothelial cells

Several groups of investigators have used a combination of markers to isolate pure populations of EPCs from human peripheral blood. A subset of circulating human cells identified by the combined expression phenotype of $CD34^{+}AC133^{+}Flk\text{-}1^{+}$ is commonly thought to represent circulating EPCs [30, 31]. These cells are enriched almost 300–400 fold in hematopoietic stem cells; and although they do not spontaneously form capillary tube-like structures *in vivo* or *in vitro*, they contribute to development of tumor microvasculature and tumor growth [32] and likely exert paracrine effects similar to CACs.

Vessel Wall EPCs

More recently, Ingram et al. have proposed that the vessel wall itself comprises a hierarchy of EPCs [33]. Vessel wall EPCs possess endothelial characteristics and the authors demonstrated using clonogenic assays that a subset of endothelial cells isolated from the human aorta as well as human umbilical vein are highly proliferative; a single cell could rapidly give rise to clonal populations of endothelial cells similar to endothelial cells derived from human cord blood. These cells also exhibited high telomerase activity, a molecular characteristic of many types of stem cells [33]. Such observations engender the possibility that two populations of endothelial progenitors might exist in the adult human, one in the bone marrow and circulating blood and the other resident within the vessel wall. It is not known currently whether these different populations of endothelial progenitors are regulated differently and whether they serve different biological functions in endothelial homeostasis or after vascular injury. Neither is it clear whether the same endothelial progenitor during its lifetime could exist within the blood vessel and then exit into the circulation or vice versa.

Other Sources of EPCs

While the above sources of EPCs have been extensively investigated, there are several other types of progenitor cells that can be differentiated into endothelial cells. Resident cardiac progenitor cells refer to a population of progenitor cells that reside within specialized niches with the heart [34, 35]. These cells are multipotent, can be isolated based on expression of cell surface markers such as C-kit, and can differentiate not only into cardiac muscle cell-like cells but generate new blood vessels as well. Umbilical cord blood also serves as a rich source of endothelial progenitors [36]. EPCs have been isolated from cord blood using a panel of typical surface markers and appear to be able to form blood vessel-like structures both *in vivo* and *in vitro*. More recently, the ability to reprogram any terminally differentiated cell to a more primitive cell or a specific cell type has opened up new possibilities regarding the field of stem cell-based angiogenesis. The ability of reprogramming technologies to generate an autologous source of endothelial cells potentially represents an important source for future stem cell-based angiogenic therapies [37].

Smooth Muscle Progenitor Cells

Despite immense advances in our understanding of the biology of endothelial progenitors in the past decade, little is understood of the biology of smooth muscle progenitors. Simper et al. reported that human PBMNCs grown in the presence of growth medium enriched with platelet-derived growth factor (PDGF BB) grew out into smooth muscle cells expressing smooth muscle-specific proteins and expressed adhesion to extracellular matrices similar to mature smooth muscle cells [38]. More recent work in this regard has demonstrated that a population of bone marrow cells expressing the chemokine receptor CX3CR1 retains the ability to differentiate into smooth muscle cells and adopt a smooth muscle-like phenotype in the vessel wall after arterial injury [39]. The arterial adventitia also possesses Sca-1 expressing smooth muscle progenitor cells and local sonic hedgehog signaling appears to regulate fate of adventitial smooth muscle progenitors [40]. Despite their diverse origins, it is not known whether a single vascular progenitor in the adult mammal can give rise to both smooth muscle and endothelial progenitors or whether smooth muscle progenitors have a completely different lineage tree than endothelial progenitors.

VASCULAR PROGENITORS IN VASCULAR DISEASE AND HOMEOSTASIS

Hill et al. reported that number and function of early EPCs (CFU-Hill assay) were indices of cardiovascular health [24]. The authors noted that following seeding of PBMNCs, the number of ECF units was inversely related to the presence of cardiovascular disease or risk factors for cardiovascular disease such as diabetes and hypertension. Not just colony-forming units, endothelial cells isolated from diseased subjects were found to inversely correlate with the degree of brachial reactivity (an index of vascular health), and there was a significant increase in endothelial cell senescence in culture as well [24]. Vasa et al. similarly demonstrated that the number of early EPCs (CACs) and the number of circulating $CD34^+Flk\text{-}1^+$ cells were reduced by 40 and 48%, respectively, in patients with coronary artery disease [41]. Smokers had significantly decreased EPCs as well as numbers of circulating CD34+Flk-1-positive cells. Moreover the authors noted significant functional impairment in EPCs isolated from patients with coronary artery disease. Since then several other groups of investigators have explored the association between endothelial progenitor numbers and cardiovascular risk factors both in rodent models and human subjects. Peripheral endothelial progenitor counts increased by threefold in mice that had been subjected to 28 weeks of running wheels and progenitor counts increased within 7 days of subjecting the animals to an exercise program [42]. In humans with coronary artery disease, CD34+Flk-1-positive cells increased in the circulation within 7 days of starting exercise and proliferative capacity of early EPCs isolated from these patients was restored within a month of starting exercise [43]. Studies have suggested that high blood glucose impairs proliferative capacity and functional capacity of circulating EPCs isolated form healthy subjects, likely in NO-dependent manner, providing a potential mechanism of EPC dysfunction observed in diabetic patients [44]. Smoking, both active and passive, has also been shown to deleteriously affect EPC number and function and in animal models of hind limb ischemia, smoking inhibited angiogenic abilities of early EPCs [41]. Dimmeler's group has demonstrated that the beneficial effects of statins may be partly related to their beneficial effects on EPC function. Statins increased differentiation of peripheral blood mononuclear cells to EPCs and appeared to exert its effects through the PI3kinase/Akt pathway [45]. Taken together, these studies suggest that number and function of EPCs in are in some measure a reflection of cardiovascular health and have led to the notion of using EPCs as "biomarkers" of vascular health.

The question of how EPCs contributed to endothelial homeostasis was addressed similarly in human subjects as well as in rodent models. Simper et al. examined hearts of patients who had undergone gender mismatched cardiac transplantation and noted the presence of endothelial cells of recipient origin in coronary arteries as well as in microvessels that had transplant arteriosclerosis (Figure 14.3) [47]. A study by Minami et al. demonstrated that in a cohort of male patients who had received female hearts during cardiac transplantation, approximately 24% of the endothelial cells in microvessels and coronary arteries were of male origin [48]. The investigators noted that the degree of endothelial chimerism was established in a month following transplantation with no significant increases in chimerism even 10 years following transplantation. Rauscher et al. in a provocative study demonstrated that chronic infusion of bone marrow-derived progenitor cells from young nonatherosclerotic mice to aged $ApoE^{-/-}$ mice prevented progression of atherosclerosis in the presence of hypercholesterolemia [49]. EPCs were decreased in peripheral blood of aging $ApoE^{-/-}$ mice and infused bone marrow-derived progenitors from young mice seeded on atherosclerosis-prone areas of the vasculature and prevented atherosclerosis progression. Several other studies demonstrated that in animal models of endothelial injury, infusion of circulating endothelial progenitors or mobilization of circulating endothelial progenitors by cytokines decreased neointima formation and improved vessel wall structure [49]. Taken together, the above animal and human studies suggested that endothelial progenitors contributed to endothelial turnover and could be potentially used therapeutically to enhance vascular function.

The contribution of smooth muscle progenitors to vascular homeostasis and atherosclerosis was less clear. Several studies in rodent models suggested that smooth muscle progenitors of bone marrow origin contributed to smooth muscle cells in plaques [46, 50]. Caplice et al. demonstrated that in patients who had undergone gender mismatched bone marrow transplantation, smooth muscle cells of bone marrow origin populated atherosclerotic regions of vessels (Figure 14.3) [51]. Interestingly, vessels that were normal and atherosclerotic did not exhibit enrichment of bone marrow derived smooth muscle cells in the vessel wall. The field of smooth muscle progenitors and its role in atherosclerosis is still debated, and more rigorous studies need to be performed with better characterization of the circulating smooth muscle progenitor and its fate.

HUMAN EPC THERAPY

Cell Therapy for Ischemic Disease

Animal and human studies demonstrate that number and function of endothelial progenitors are inversely related to cardiovascular function [41, 52]. Further, numerous animal studies suggest that EPCs or PBMNCs exert beneficial effects when injected into ischemic tissues, increase blood vessel formation, and blood flow to ischemic tissues [53]. For instance, several animal studies show that transendocardial or intramyocardial delivery of EPCs improves capillary density, increases ejection fraction, and leads to better preservation of myocardial

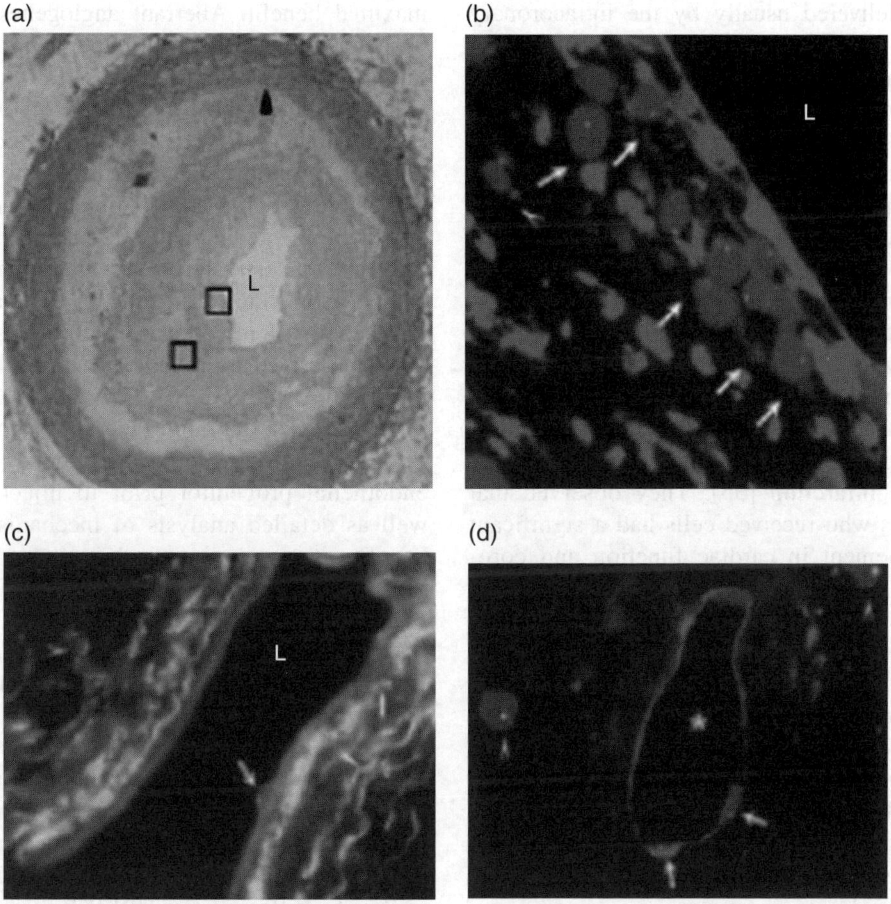

Figure 14.3 (a) Atherosclerotic coronary artery of female patient who underwent gender mismatched bone marrow transplantation; (b) the *arrow* points to internal elastic lamella Y-chromosome-positive (bone marrow origin) smooth muscle cells (*arrows*) in area of plaque highlighted in (a): α-smooth muscle actin shown in *red*, nuclei in *blue*, and Y-chromosome represented by the *green dot*; (c) Y-chromosome-positive endothelial cells (*arrow*) lining a coronary vessel of male patient who received a female heart during cardiac transplantation; and (d) similar Y-chromosome-positive endothelial cells lining the microvessels in the same patient (Adapted from Walter et al. [45] © 2004, Lippincott Williams & Wilkins and Sata et al. [46]. © Nature). (*See insert for color representation of the figure.*)

function after acute myocardial infarction [53, 54]. Further, EPCs injected into atherosclerotic mice appear to retard the development of atherosclerotic lesions. These and other similar observations gave rise to the premise that EPCs could be used as therapeutic agents to increase new blood vessel formation in ischemic tissues.

Thus a large number of clinical trials were initiated almost a decade ago to determine the benefit of injected human EPCs for the treatment of ischemic insults such as acute myocardial infarction and peripheral vascular disease. All these trials involved delivery of autologous cells and usually the injected cells comprised unfractionated bone marrow cells or mononuclear cells isolated from the bone marrow or peripheral circulation or specific subsets of cells expressing a specific EPC surface marker such as CD34 and AC133. Cells were delivered usually by the intracoronary route or via transendocardially with mapping techniques to determine areas of hypocontractile myocardium. Most of these early human studies demonstrated improvement in ejection fraction following delivery of EPC, decrease in infarct size, and improvement in regional wall motion abnormalities of the heart commonly seen after acute myocardial infarction [53]. In another randomized study, investigators administered the cytokine G-CSF to mobilize stem cells from the bone marrow to the circulation, isolated CD34-expressing cells, and administered the cells via the intracoronary route to patients after acute myocardial infarction [55]. They observed that patients who received cells had a significant improvement in cardiac function and coronary flow at a 6-month follow-up compared with subjects who received G-CSF alone without any cells [55]. Similarly, trials of human EPC for hind limb ischemia also demonstrated improved ankle-brachial indices, increased distances of pain free walking, and improved oxygenation of the affected extremity. Overall, the results of these trials are promising and demonstrate the safety and feasibility of administering EPCs for treatment of ischemic injury [53].

Pitfalls of EPC Therapy

Notwithstanding the encouraging results, the majority of trials of EPC therapy were not blinded or randomized and rigorous mechanistic studies to determine cellular mechanisms of benefit were not performed [56]. Initial benefits on cardiac function were not persistent on long-term follow-up, and mechanistically it was not clear whether the cells were contributing to formation of new blood vessels. The trials also injected various types of EPCs ranging from bone marrow-derived mononuclear cells to cells expressing a specific hematopoietic marker such as CD34, and it is still currently unclear as to what is the best type of EPC to be administered for maximal benefit. Aberrant angiogenesis can have deleterious effects in patients harboring tumors and can lead to increased tumor vasculature and tumor growth. Moreover in rodent models, there is a strong association between mononuclear cell infiltration and plaque growth [22]. Such potential adverse effects of injected human EPCs have not been rigorously investigated. A recent double-blind placebo-controlled trial did not demonstrate any benefit of injection of autologous bone marrow-derived mononuclear cells after myocardial infarction [57]. These recent results suggest the importance of performing rigorous studies to define the identity of the endothelial progenitor prior to injection as well as detailed analysis of mechanisms of benefit.

MECHANISMS OF ACTION OF ENDOGENOUS EPCS IN VASCULAR REPAIR

The uninjured endothelium is supposed to turn over slowly [31]. When the endothelium gets injured, endothelial reconstitution occurs from proliferation of the remaining endothelium as well as by differentiation of EPCs recruited to the injured site. The contribution of endogenous progenitors toward replenishment of the injured endothelium is not clear. Hypoxia is known to induce expression of the chemokine CXCL12 on endothelial cells and CXCL12, a potent chemoattractant that helps in recruitment of CXCR4 (receptor for CXCL12) expressing progenitor cells to the ischemic tissue [58]. It is likely that endothelial repair involves close coordination of hematopoietic and endothelial cells to recruit endothelial progenitors for vasculogenesis [31]. We illustrate a hypothetical model of vasculogenesis in endothelial repair, where hematopoietic cells and circulating angiogenic cells are recruited to the site of endothelial injury or denudation and release angiogenic cytokines that recruit circulating or vessel wall ECFC (Figure 14.4). ECFCs

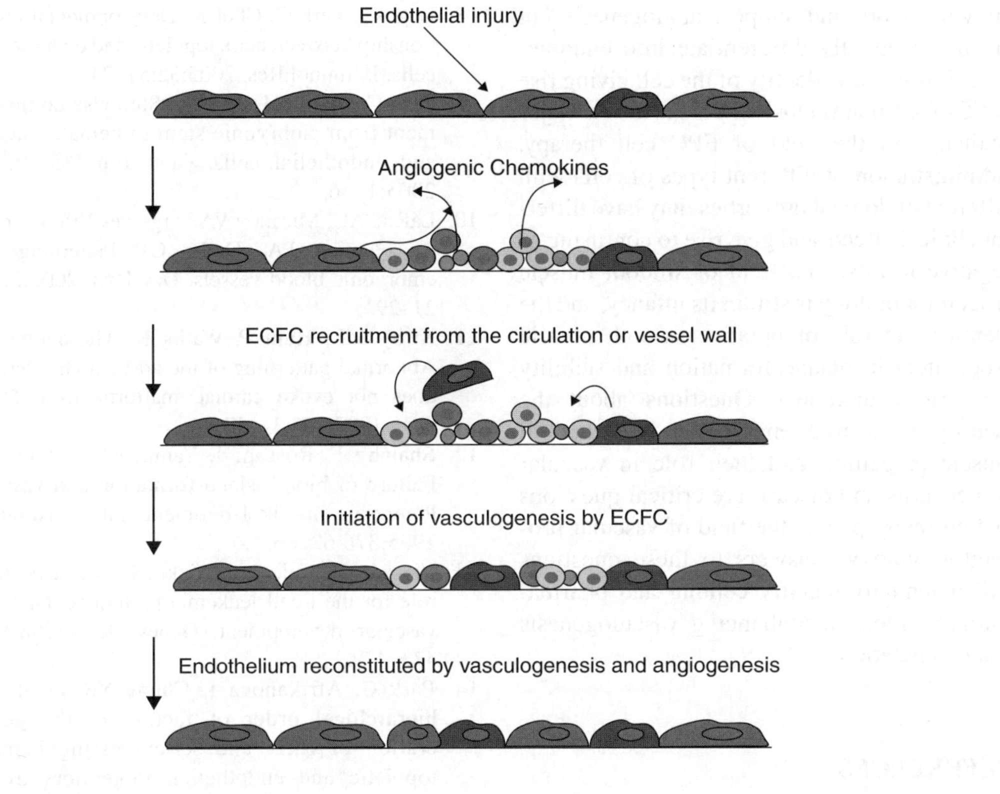

Figure 14.4 In a model of ECFC-mediated vasculogenesis for endothelial repair after injury, hematopoietic cells including platelets, monocytes/macrophages, and CACs attach to areas of endothelial injury. The cellular milieu including neighboring endothelial cells express angiogenic chemokines that attract ECFC from the circulation or vessel wall. ECFCs attach to areas of endothelial injury and form new endothelium (vasculogenesis). The endothelium finally is restored by a combination of ECFC mediated vasculogenesis and angiogenesis. (*See insert for color representation of the figure.*)

recruited to the area of injury then initiate a process of vasculogenesis, and the endothelium is finally replenished with a combination of new endothelial cells formed from vasculogenesis and angiogenesis. In this regard, EPCs administered into ischemic tissues may also exert significant paracrine effects that facilitate angiogenesis or vasculogenesis via recruitment of endogenous endothelial progenitors.

VASCULAR PROGENITORS AND THE FUTURE

Several daunting challenges lie ahead in the field of vascular progenitor biology [59]. The identity of a true endothelial progenitor has not been established. Current markers used to isolate endothelial progenitors identify a cell with hematopoietic properties that can express growth factors and support angiogenesis but unable to directly differentiate into endothelium. Further, the identity of the cell giving rise to ECFCs remains elusive. This is also a major challenge in the field of EPC cell therapy. Administration of different types of cells with different biological properties may have different clinical effects and give rise to confusing or negative results. The field of smooth muscle progenitor biology is still in its infancy, and the identity and role of putative smooth muscle progenitors in plaque formation and stability are largely unknown. Questions about the identity of a true endothelial and smooth muscle progenitor and their role in vascular homeostasis and disease are critical questions that currently plague the field of vascular progenitor biology. Answers to these questions will potentially identify cellular and pharmacologic targets for enhancing vasculogenesis and regeneration.

REFERENCES

1. Libby P. Bone marrow: A fountain of vascular youth? Circulation. 2003;108:378–379.
2. Takahashi K, Yamanaka S. Induction of pluripotent stem cells from mouse embryonic and adult fibroblast cultures by defined factors. Cell. 2006;126:663–676.
3. Takahashi K, Tanabe K, Ohnuki M, et al. Induction of pluripotent stem cells from adult human fibroblasts by defined factors. Cell. 2007;131:861–872.
4. Sridharan R, Plath K. Illuminating the black box of reprogramming. Cell Stem Cell. 2008; 2:295–297.
5. Maherali N, Hochedlinger K. Guidelines and techniques for the generation of induced pluripotent stem cells. Cell Stem Cell. 2008; 3:595–605.
6. Stadtfeld M, Maherali N, Breault DT, Hochedlinger K. Defining molecular cornerstones during fibroblast to ips cell reprogramming in mouse. Cell Stem Cell. 2008;2: 230–240.
7. Keller G, Lacaud G, Robertson S. Development of the hematopoietic system in the mouse. Exp Hematol. 1999;27:777–787.
8. Lugus J, Park C, Choi K. Developmental relationship between hematopoietic and endothelial cells. Immunol Res. 2005;32:57–74.
9. Park C, Lugus JJ, Choi K. Stepwise commitment from embryonic stem to hematopoietic and endothelial cells. Curr Top Dev Biol. 2005:1–36.
10. LaRue AC, Mironov VA, Argraves WS, Czirok A, Fleming PA, Drake CJ. Patterning of embryonic blood vessels. Dev Dyn. 2003;228: 21–29.
11. Kirby ML, Hunt P, Wallis K, Thorogood P. Abnormal patterning of the aortic arch arteries does not evoke cardiac malformations. Dev Dyn. 1997;208:34–47.
12. Shalaby F, Rossant J, Yamaguchi TP, et al. Failure of blood-island formation and vasculogenesis in flk-1-deficient mice. Nature. 1995;376:62–66.
13. Visvader JE, Fujiwara Y, Orkin SH. Unsuspected role for the t-cell leukemia protein scl/tal-1 in vascular development. Genes Dev. 1998;12: 473–479.
14. Park C, Afrikanova I, Chung YS, et al. A hierarchical order of factors in the generation of flk1- and scl-expressing hematopoietic and endothelial progenitors from embryonic stem cells. Development. 2004;131: 2749–2762.
15. Moser M, Binder O, Wu Y, et al. Bmper, a novel endothelial cell precursor-derived protein, antagonizes bone morphogenetic protein signaling and endothelial cell differentiation. Mol Cell Biol. 2003;23:5664–5679.
16. Larsson J, Goumans MJ, Sjostrand LJ, et al. Abnormal angiogenesis but intact hematopoietic potential in tgf-beta type i receptor-deficient mice. EMBO J. 2001;20:1663–1673.
17. Hirschi KK, Majesky MW. Smooth muscle stem cells. Anat Rec A: Discov Mol Cell Evol Biol. 2004;276A:22–33.
18. Kirby ML. Plasticity and predetermination of mesencephalic and trunk neural crest transplanted into the region of the cardiac neural crest. Dev Biol. 1989;134:402–412.
19. Wu J, Saint-Jeannet JP, Klein PS. Wnt-frizzled signaling in neural crest formation. Trends Neurosci. 2003;26:40–45.
20. Asahara T, Murohara T, Sullivan A, et al. Isolation of putative progenitor endothelial cells for angiogenesis. Science. 1997;275: 964–967.

21. Hirschi KK, Ingram DA, Yoder MC. Assessing identity, phenotype, and fate of endothelial progenitor cells. Arterioscler Thromb Vasc Biol. 2008;28:1584–1595.
22. Caplice NM, Deb A. Myocardial-cell replacement: The science, the clinic and the future. Nat Clin Pract Cardiovasc Med. 2004;1:90–95.
23. Ingram DA, Caplice NM, Yoder MC. Unresolved questions, changing definitions, and novel paradigms for defining endothelial progenitor cells. Blood. 2005;106:1525–1531.
24. Hill JM, Zalos G, Halcox JP, et al. Circulating endothelial progenitor cells, vascular function, and cardiovascular risk. N Engl J Med. 2003;348: 593–600.
25. Urbich C, Dimmeler S. Endothelial progenitor cells functional characterization. Trends Cardiovasc Med. 2004;14:318–322.
26. Urbich C, Dimmeler S. Endothelial progenitor cells: Characterization and role in vascular biology. Circ Res. 2004;95:343–353.
27. Walter DH, Dimmeler S. Endothelial progenitor cells: Regulation and contribution to adult neovascularization. Herz. 2002;27:579–588.
28. Dimmeler S, Zeiher AM. Vascular repair by circulating endothelial progenitor cells: The missing link in atherosclerosis? J Mol Med. 2004;82:671–677.
29. Lin Y, Weisdorf DJ, Solovey A, Hebbel RP. Origins of circulating endothelial cells and endothelial outgrowth from blood. J Clin Invest. 2000;105:71–77.
30. Peichev M, Naiyer AJ, Pereira D, et al. Expression of vegfr-2 and ac133 by circulating human cd34(+) cells identifies a population of functional endothelial precursors. Blood. 2000;95:952–958.
31. Richardson MR, Yoder MC. Endothelial progenitor cells: Quo vadis? J Mol Cell Cardiol. 2011;50:266–272.
32. Bertolini F, Mancuso P, Kerbel RS. Circulating endothelial progenitor cells. N Engl J Med. 2005;353:2613–2616; author reply 2613–2616.
33. Ingram DA, Mead LE, Moore DB, Woodard W, Fenoglio A, Yoder MC. Vessel wall-derived endothelial cells rapidly proliferate because they contain a complete hierarchy of endothelial progenitor cells. Blood. 2005;105:2783–2786.
34. Beltrami AP, Barlucchi L, Torella D, et al. Adult cardiac stem cells are multipotent and support myocardial regeneration. Cell. 2003; 114:763–776.
35. Bearzi C, Rota M, Hosoda T, et al. Human cardiac stem cells. Proc Natl Acad Sci USA. 2007;104:14068–14073.
36. Broxmeyer HE, Srour E, Orschell C, et al. Cord blood stem and progenitor cells. Methods Enzymol. 2006;419:439–473.
37. Deb A, Patterson C. Closer to fine: Fewer steps to endothelial stemness. Arterioscler Thromb Vasc Biol. 30:1880–1881.
38. Simper D, Stalboerger PG, Panetta CJ, Wang S, Caplice NM. Smooth muscle progenitor cells in human blood. Circulation. 2002;106: 1199–1204.
39. Metharom P, Kumar AH, Weiss S, Caplice NM. A specific subset of mouse bone marrow cells has smooth muscle cell differentiation capacity-brief report. Arterioscler Thromb Vasc Biol. 2010;30:533–535.
40. Passman JN, Dong XR, Wu SP, et al. A sonic hedgehog signaling domain in the arterial adventitia supports resident sca1+ smooth muscle progenitor cells. Proc Natl Acad Sci USA. 2008;105:9349–9354.
41. Vasa M, Fichtlscherer S, Aicher A, et al. Number and migratory activity of circulating endothelial progenitor cells inversely correlate with risk factors for coronary artery disease. Circ Res. 2001;89:E1–E7.
42. Laufs U, Werner N, Link A, et al. Physical training increases endothelial progenitor cells, inhibits neointima formation, and enhances angiogenesis. Circulation. 2004;109:220–226.
43. Van Craenenbroeck EM, Conraads VM. Endothelial progenitor cells in vascular health: Focus on lifestyle. Microvasc Res. 2010;79: 184–192.
44. Chen YH, Lin SJ, Lin FY, et al. High glucose impairs early and late endothelial progenitor cells by modifying nitric oxide-related but not oxidative stress-mediated mechanisms. Diabetes. 2007;56:1559–1568.
45. Walter DH, Zeiher AM, Dimmeler S. Effects of statins on endothelium and their contribution to neovascularization by mobilization of endothelial progenitor cells. Coron Artery Dis. 2004;15:235–242.
46. Sata M, Saiura A, Kunisato A, et al. Hematopoietic stem cells differentiate into vascular cells that participate in the pathogenesis of atherosclerosis. Nat Med. 2002;8:403–409.
47. Simper D, Wang S, Deb A, et al. Endothelial progenitor cells are decreased in blood of

cardiac allograft patients with vasculopathy and endothelial cells of noncardiac origin are enriched in transplant atherosclerosis. Circulation. 2003;108:143–149.
48. Minami E, Laflamme MA, Saffitz JE, Murry CE. Extracardiac progenitor cells repopulate most major cell types in the transplanted human heart. Circulation. 2005;112:2951–2958.
49. Rauscher FM, Goldschmidt-Clermont PJ, Davis BH, et al. Aging, progenitor cell exhaustion, and atherosclerosis. Circulation. 2003;108:457–463.
50. Shimizu K, Sugiyama S, Aikawa M, et al. Host bone-marrow cells are a source of donor intimal smooth- muscle-like cells in murine aortic transplant arteriopathy. Nat Med. 2001;7:738–741.
51. Caplice NM, Bunch TJ, Stalboerger PG, et al. Smooth muscle cells in human coronary atherosclerosis can originate from cells administered at marrow transplantation. Proc Natl Acad Sci USA. 2003;100:4754–4759.
52. Schmidt-Lucke C, Rossig L, Fichtlscherer S, et al. Reduced number of circulating endothelial progenitor cells predicts future cardiovascular events: Proof of concept for the clinical importance of endogenous vascular repair. Circulation. 2005;111:2981–2987.
53. Dzau VJ, Gnecchi M, Pachori AS, Morello F, Melo LG. Therapeutic potential of endothelial progenitor cells in cardiovascular diseases. Hypertension. 2005;46:7–18.
54. Melo LG, Pachori AS, Kong D, et al. Gene and cell-based therapies for heart disease. FASEB J. 2004;18:648–663.
55. Kang HJ, Kim HS, Zhang SY, et al. Effects of intracoronary infusion of peripheral blood stem-cells mobilised with granulocyte-colony stimulating factor on left ventricular systolic function and restenosis after coronary stenting in myocardial infarction: The magic cell randomised clinical trial. Lancet. 2004;363: 751–756.
56. Melo LG, Pachori AS, Kong D, et al. Molecular and cell-based therapies for protection, rescue, and repair of ischemic myocardium: Reasons for cautious optimism. Circulation. 2004;109: 2386–2393.
57. Perin EC, Willerson JT, Pepine CJ, et al. Effect of transendocardial delivery of autologous bone marrow mononuclear cells on functional capacity, left ventricular function, and perfusion in chronic heart failure: The Focus-cctrn trial. J Am Med Assoc. 2012;307: 1717–1726.
58. Ceradini DJ, Kulkarni AR, Callaghan MJ, et al. Progenitor cell trafficking is regulated by hypoxic gradients through hif-1 induction of sdf-1. Nat Med. 2004;10:858–864.
59. Deb A, Patterson C. Hard luck stories: The reality of endothelial progenitor cells continues to fall short of the promise. Circulation. 2010; 121:850–852.

15

PLATELETS

Jerry Ware and Adam Corken
Department of Physiology and Biophysics, University of Arkansas for Medical Sciences, Little Rock, AR, USA

OVERVIEW

Circulating platelets have a repertoire of surface membrane receptors that support adhesion to a wide range of substrates and ligands. Their well-characterized role is that of recognizing a damaged vascular surface and, via a series of molecular events, preventing excessive blood loss (normal hemostasis). By contrast, the same series of events occurs following rupture of an atherosclerotic plaque and may lead to complete vessel occlusion (thrombosis). Indeed, arterial thrombosis is the most common cause of mortality in developed countries. These acute events have been precisely defined at the molecular level in exquisite detail. As such, the platelet paradigm in hemostasis and thrombosis describes a series of events supported by rapid changes to a circulating platelet. A platelet with limited adhesive properties becomes an activated platelet that aggressively recruits more platelets for aggregation and releases a plethora of ligands and agonists that further support activation and subsequent wound healing. Similar processes aided by a less-pronounced adhesive state can effectively contribute to the pathways associated with inflammation. Here, the role of the platelet is viewed as a chronic effecter interfacing with many of the molecular pathways known to support the development of atherosclerosis. That leads to an interface of the platelet with inflammation. Thus, we can consider the platelet's dual role in acute atherothrombosis and its role in the chronic progression of atherosclerosis. The potential of antiplatelet therapy to alter these disease processes has emerged, and the platelet is now recognized as participatory in a variety of seemingly unrelated disease processes.

Atherosclerosis: Risks, Mechanisms, and Therapies, First Edition. Edited by Hong Wang and Cam Patterson.

THE PLATELET PARADIGM IN HEMOSTASIS AND THROMBOSIS

Platelets circulate as anucleate blood cells originating from a differentiated megakaryocyte in the bone marrow [1]. During human fetal development, the liver is the major hematopoietic organ and, consequently, the source of megakaryocytes—a process that shifts to the bone marrow in the neonate. In mice, the situation is slightly different in the adult with megakaryocytes found in the both the spleen and liver where larger number of megakaryocytes can be found in certain disease settings [2]. A striking attribute of a normal circulating platelet is its low-affinity interaction for a several adhesive ligands that circulate in blood with the platelet, such as von Willebrand factor (VWF) and fibrinogen [3]. The initiating event in the formation of a platelet-rich thrombus is the disruption of the endothelial cell barrier that separates the extravascular matrix and flowing blood. In the event that this barrier is altered, a rapid platelet response occurs characterized by three sequential events: adhesion, activation, and aggregation (Figure 15.1) [3].

A series of recent reviews detail the molecular participants in these events [4, 5]. An emerging theme in these studies in the participation of specific receptor/ligand interactions that are unique to individual vascular beds, or more precisely the relevant rheological, or flow, conditions. For example, arterial thrombus formation under conditions of high flow and high shear stress requires the exclusive adhesion of platelets via their glycoprotein (GP) Ib-IX receptor for VWF. This interaction is characterized by a rapid on and off rate requiring further engagement of other receptors to achieve irreversible adhesion [6, 7]. Thus, if the initial tethering leads to platelet activation, irreversible adhesion occurs and subsequent platelet–platelet interactions can occur mediated by the major platelet ligand of plasma, fibrinogen. The net effect is the growth of a platelet thrombus coinciding with the generation of insoluble fibrin from the coagulation cascade, and the formation of stable thrombus. Again, these processes are rapid, occurring in response to vascular injury, and represents the acute response in a process termed "atherothrombosis" [3].

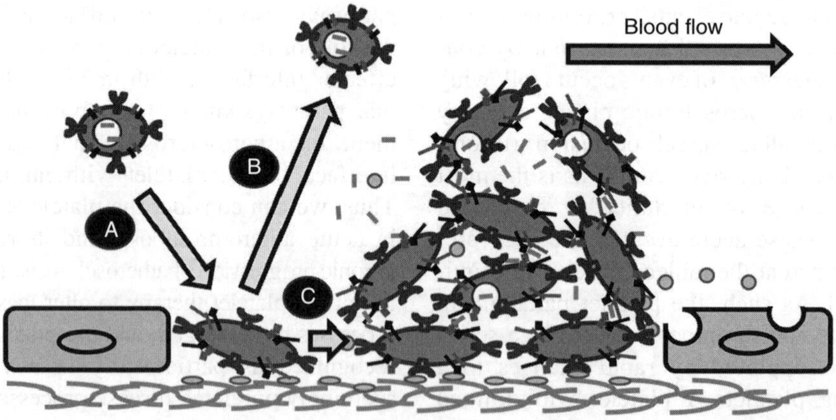

Figure 15.1 The temporal sequence of events that lead to a platelet-rich thrombus. Typically, platelets circulate sequestered near the vascular wall with a low affinity to other blood cells and a normal endothelium. Upon damage to the vascular surface, platelets adhere to a damaged vascular surface (A) with one of two possible outcomes; a reversible detachment (B), or an irreversible attachment (C) that coincides with platelet activation. The latter results in recruitment of more platelets generating a platelet-rich thrombus. Upon activation, stored adhesive ligands such as VWF (endothelial cells and platelets) and fibrinogen (platelets) are released that further support platelet aggregation.

THE PLATELET/INFLAMMATION AXIS

The acute response of the platelet in atherothrombosis has been intensively studied for years. Many of the key insights described earlier were provided by the study and characterization of congenital bleeding disorders that highlight the relevance of specific receptor/ligand interactions and corresponding mouse models that allow more detailed studies [8]. The platelet's contribution to the chronic response in atherothrombosis is more difficult to study but is increasingly recognized to be an ongoing process that contributes to atherosclerosis. Although the acute event is rapid and at some point becomes irreversible, the concept of platelet interactions with an intact endothelium has the potential to lead to participation of the platelet in inflammation. In this case, it is the interaction of platelets with an inflamed endothelium that provides a chronic phase of platelet participation that could be occurring for many years. This has led some to propose the platelet as a "proinflammatory" cell in the circulation [9]. When considered in the context of inflammatory pathways linking dyslipidemia and atherosclerosis, the role of the platelet in inflammation becomes a focus as we try to understand the multifaceted mechanisms of atherosclerosis [10].

Indeed, the interactions between platelets and an intact endothelium are incompletely understood. In the normal situation, the endothelium provides inhibitory mechanisms that limit platelet interaction with the endothelium. Such inhibitory pathways can involve COX-2, PGI_2, and prostanoid synthetic systems [9]. However, *in vitro* cultures of endothelial cells when activated present a proadhesive surface for the platelet involving ligands associated with hemostasis or thrombosis [11, 12]. Herein lies the chronic potential for a platelet-mediated proinflammatory state via two different pathways (Figure 15.2).

First, the platelet is inherently loaded with mediators that stimulate inflammation if platelet activation occurs [13]. Whereas adhesion itself does not cause release of stored platelet granule components, platelet adhesion is a prerequisite to irreversible adhesion that does coincide with

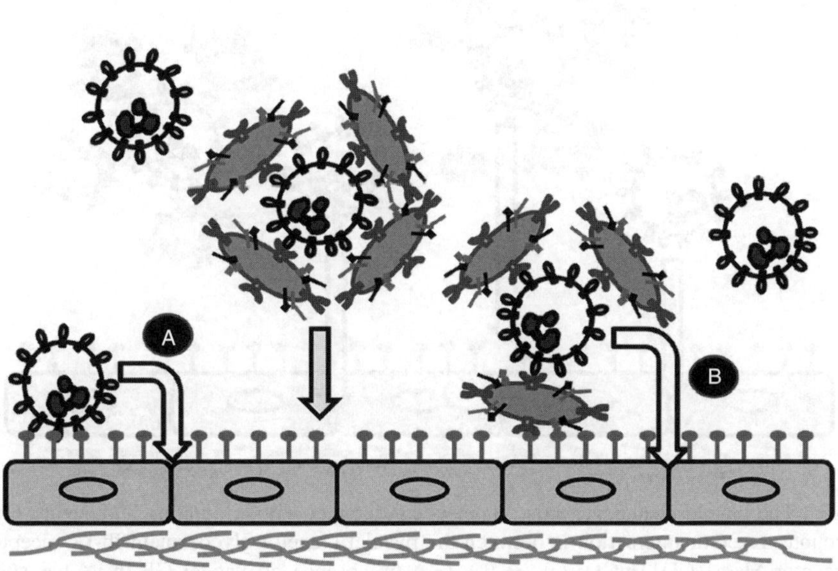

Figure 15.2 The proinflammtory platelet–leukocyte axis. The most extensive set of studies describe a proinflammtory role for platelets. Leukocytes may interact with the endothelium directly (A) or the interaction of platelets with an activated endothelium bridges the leukocyte to a close proximity to the inflamed endothelium (B). The net effect supports migration of the leukocyte to the extravascular space.

platelet activation and the release of a plethora of proinflammatory and secondary activators of platelet activation [14]. Thus, a vascular endothelium sequestering platelets generates a microenvironment with proinflammatory potential.

A second outcome for a platelet endothelium interaction relates to the platelet–leukocyte axis (Figure 15.2) [15]. Platelets have receptors that mediate attachment to circulating neutrophils and monocytes. First is the integrin receptor of monocytes and neutrophils, Mac-1 (CD11b/CD17 or $\alpha_M\beta_2$), that interacts with the platelet adhesion receptor, GPIb-IX [16, 17]. Indeed, if platelet adhesion to monocytes via GPIb-IX is similar to the role of GPIb-IX in hemostasis, it might be predicted that this adhesive event would lead to platelet activation and induction of inflammatory pathways in monocytes [18]. The *in vivo* consequences of this interaction have not been fully elucidated. In fact, some preliminary studies using mice genetically missing the platelet GPIb-IX receptor have increased susceptibility to atherosclerosis [19]. However, others have suggested an opposite effect just examining the relevance of the GPIb-IX receptor with antibody inhibition of the complex [20]. It should be pointed out that the two models are very different ways of examining the inflammatory pathways involved in atherosclerosis. The genetic model is a long-term chronic change, whereas the reported antibody inhibition studies were short-term studies with the potential of activating platelets. Chronic studies using mouse models of hyperlipidemia suggest the major platelet adhesion receptor, GPIb-IX, is inhibitory in promoting atherosclerosis (Figure 15.3). At first glance, this appears contradictory to a range of publications suggesting the proinflammatory role for platelets in inflammation [9, 21]. However, using the temporal sequence of events so well-described for the relevance of platelets in hemostasis, the interaction of platelets with inflammatory cells may follow a temporal sequence of events. For example,

Figure 15.3 The platelet–leukocyte axis. Platelets can interact with circulating leukocytes. One defined receptor–counter-receptor interaction is between the platelet receptor glycoprotein Ib-IX receptor and the integrin receptor Mac-1(CD11b/CD17 or $\alpha_M\beta_2$). Genetic absence of platelet GP Ib-IX has suggested an inhibitory role for this axis in models of atherosclerosis, but further work is needed to understand the anti- versus proinflammatory platelet role in inflammation. This would most likely depend on the activation state of the platelet with an unactivated platelet having a low affinity (A) and an activated platelet having a high affinity for the endothelium (Figure 15.2).

monocyte–platelet adhesion might be a reversible event analogous to platelet binding to a damaged vascular surface. As such, platelet adhesion receptors might not have the same functional consequence for neutrophil migration or other aspects of inflammation depending on the activation state of the platelet.

When platelet activation occurs, a series of dramatic changes take place. One hallmark feature of platelet activation is the increased expression of the membrane receptor P-selectin [22]. Monocytes express a P-selectin ligand, PSGL-1, allowing the platelet to be an intermediary bringing the monocyte in close proximity to an abnormal endothelium [23]. The interaction of P-selectin with PSGL-1 can initiate translational signals to a monocyte causing upregulation of various proteins, including urokinase plasminogen-activated receptor (uPAR) a regulator of *in vivo* leukocyte adhesion [24]. P-selectin is also found on activated endothelial cells where it can serve as a counter-receptor for platelet GPIb-IX and further support a platelet–endothelium or monocyte–platelet–endothelium interaction [25]. Thus, the relevance of platelets to a changing endothelial surface and the possibility of bridging monocytes to the activated endothelium are all mechanisms whereby the platelet is interfacing with aspects of vessel wall inflammation. In addition, the transient interaction of unactivated platelets with circulating neutrophils and monocytes in the circulation could be a negative regulator of inflammation and represents a direction of study that requires more investigation.

PLATELETS AND THE PROGRESSION OF ATHEROSCLEROSIS

With adhesive potential of a leukocyte–platelet–endothelium scenario established, we can consider microenvironmental changes leading to vascular wall changes. As mentioned earlier, the healthy vascular lining is an antiplatelet surface with intrinsic pathways, such as ecto-ADPase, prostaglandin I_2 ($PGEI_2$) release, thrombomodulin-activated protein C, and nitric oxide all inhibiting a platelet–endothelium interaction [5]. However, the hyperlipidemic state alters the vascular wall leading to the expression of the receptors and ligands described earlier and generating a local microenvironment that differs significantly from the unactivated vascular wall.

Platelets contain a plethora of stored proteins that upon activation are released. Beyond the obvious inflammatory mediators, such as interleukin-1β (IL-1β), the release of various chemokines, such as platelet factor 4 (PF4) and chemokine ligand 5 (RANTES), synergize to promote further sequestration of inflammatory cells. In addition, PF4 is a known stimulant of smooth muscle proliferation that contributes to local blood hemodynamics and more changes within the vascular wall [5, 9].

As a major ligand participating in platelet function, the genetic absence of VWF is the cause of the most common hereditary bleeding disorder in humans [26]. VWF is expressed by both endothelial cells and within megakaryocytes, the platelet precursor [27]. It is stored in the Weibel–Palade bodies of endothelial cells and α-granules of platelets released from both as a part of an activation response. VWF also circulates in plasma as a constitutive secretory product of endothelial cells. Whether VW disease (VWD) patients are less prone to atherosclerosis is not known and difficult to discern because of the many other compounding factors of heart disease that preclude a definitive study. Most platelet abnormalities are autosomal recessive in nature and represent such a limited cohort the ability to make any definitive conclusion relevant to atherosclerosis is near impossible [28]. However, large animal models of VWF deficiency do exist, and they have been informative for the relevance of VWF in atherosclerosis [29]. Models of porcine VWF deficiency have been particularly appealing because of the similar VWF levels between pigs and humans and the development of atherosclerosis in the absence of a high cholesterol diet [30]. These models have shown an atheroprotective effect in the absence of VWF [29]. In addition, the use of a mouse model of VWF deficiency when bred into a hyperlipidemic background also displayed less susceptibility to the fatty streak lesion typical of the murine model [31].

While the large animal models have some distinct advantages in these studies, the use of smaller genetic models, such as mice, has other advantages and disadvantages. From the advantage standpoint, the congenic background strains available with low-density lipoprotein receptor (LDLR) or apoliprotein E (ApoE) deficiencies make the ability to examine relevance in a highly controlled genetic background appealing. Absence of P-selectin is atheroprotective in mouse models as might be expected [32]. The surprising result, mentioned earlier, is the absence of platelet GPIb-IX leading to increased atherosclerosis in ApoE-deficient animals, and this suggests the dynamic nature of fatty streak formation is not always proatherosclerotic with respect to platelet adhesion [19].

Overall, both large and small animal models have supported the hypothesis that platelets are a participatory cell in the progression of atherosclerosis. Perhaps, the temporal sequence of events that so well characterizing in hemostasis and thrombosis are in play during chronic interactions with the abnormal vessel wall. There is a need to better understand the similarities and dissimilarities among the various animal models. How the results translate to human studies remains to be determined. Eventually, some insights may be gained from antiplatelet therapy for human thrombosis since it is so widely prescribed.

ANTIPLATELET THERAPY IN ATHEROSCLEROSIS

Targeting platelets for modulating the development of atherosclerosis has several levels of potential efficacy. First is the obvious acute relevance of platelets in the atherothrombotic event. Here, the platelet is a key player in the development of an occlusive thrombus and antiplatelet therapy can be beneficial. Indeed, the targets are widely defined from adhesion and aggregation receptors to inhibitors of platelet activation. The challenge with all of these targets is maintaining the hemostatic balance. To date, every antiplatelet medication comes with a risk of bleeding. This has led to an ongoing debate as to whether it is possible to inhibit thrombosis, the pathological event, and maintain hemostasis, the normal physiologic response of platelets to vascular injury [33]. Currently, a thrombosis-specific inhibition has not been achieved, but this does not mean that such a specific inhibition is impossible.

The most widely prescribed antiplatelet medication is the common aspirin. It is a significant inhibitor of thromboxane A2 (TXA2) formation through the acetylation of cyclo-oxygenase. Thus, aspirin inhibits a major cascade of platelet activation required for platelet aggregation and growth of a thrombus. The Antithrombotic Trialists' Collaboration reported in 2002 provided clinical evidence that aspirin can prevent vascular events in patients with high-risk cardiovascular disease [34]. However, low-dose aspirin in patients with peripheral artery disease has not been beneficial and may have led to some of the confusion surrounding platelet resistance to aspirin therapy [35, 36].

Recent studies on platelet activation may provide an explanation for some of the ineffectiveness of aspirin therapy. *In vitro* studies have documented platelet activation via agonist-independent pathways [37]. Shear stress alone can upregulate platelet P-selectin expression, the hallmark surface marker of platelet activation, despite the use of aspirin and other antiplatelet medications [38, 39]. While rupture of the vulnerable plaque is considered, the major consequence of atherothrombosis, it is clear occlusive thrombi are not always associated with the ruptured plaque [40]. Therefore, understanding how dynamic flow conditions in the vasculature regulate the activation state of the platelet is clearly an area that requires more investigation. Considering the diseased arterial vessel and the changes to rheological conditions as platelets pass through a stenosed vessel, these mechanistic insights provide some of the clearest insights into how current treatments could be improved and the possibility of targeting specific antithrombotic pathways [33].

The second most common antiplatelet medication are inhibitors of the P2Y12 platelet receptor, which bind adenosine diphosphate (ADP) and initiate a secondary wave of platelet activation. Platelets store ADP in their granule compartment and upon activation release ADP that subsequently creates a proactivating microenvironment for a growing thrombus. Starting with ticlopidine and now the more commonly prescribed clopidogrel, these medications have been tested in a wide range of trials alone or in combination with aspirin [41]. The combination of clopidogrel and aspirin has provided the best results for conditions of high thrombotic risk, such as following a thrombotic event, but less effective in long-term administration and in preventing the first major event [41]. Again, the complicating factor in shear causing the upregulation of platelet activation markers and the ineffectiveness of aspirin or clopidogrel inhibiting shear-dependent activation may explain some of the ineffectiveness of long-term treatment with aspirin and/or clopidogrel [38, 39]. Thus, newer therapeutic options are likely to be highly beneficial [41].

Beyond the acute platelet response in atherothrombosis is the chronic inflammatory state discussed earlier. Here, the relevance of antiplatelet medication is very poorly defined but is likely to be important. The widely used mouse models of primary atherosclerosis have a lesion progression that appears to mimic human atherogenesis, and here the relevance of platelets is clear [42]. However, more data on platelet inhibition via the common medications, such as aspirin and clopidogrel, are needed. Some limited preclinical data exist for short-term inhibition of lesion development, but how this translates to the long-term chronic treatment is not known [20].

CONCLUSIONS

The past few decades has seen remarkable progress in understanding normal platelet physiology and pathophysiology. A plethora of receptors and signaling pathways have been identified leading to a platelet paradigm that involves dramatic changes in response to vascular injury or vascular disease. Indeed, the "state of the platelet" dramatically influences its relevance in a wide range of disease events [43]. In the events associated with atherosclerosis, the future studies will focus on how the platelet reacts to the localized vessel environment, flow conditions, endothelial cell properties, and abilities to interact with leukocytes and monocytes. Understanding these *in vivo* events will guide development of new therapeutics that could potentially lead to dramatic reductions in the inflammation associated with atherosclerosis and the acute event of atherothrombosis. It is a clear that a wider selection of new antiplatelet medications might be safer and more widely acceptable in chronic treatments. One thing is clear. Heart disease and the management of heart disease is the major health problem in developing countries and growing at an epidemic pace. Thus, the pursuit of knowledge about cardiovascular disease must be at the forefront to bring awareness and research on this important topic.

REFERENCES

1. Italiano J, Jr., Hartwig J. Megakaryocyte development and platelet formation. In: Michelson AD, editor. Platelets. Academic Press, 2007: 23–44.
2. Suva LJ, Hartman E, Dilley JD et al. (2008) Platelet dysfunction and a high bone mass phenotype in a murine model of platelet-type von Willebrand disease. *Am J Pathol* 172, 430–439.
3. Ruggeri ZM. (2002) Platelets in atherothrombosis. *Nat Med* 8, 1227–1234.
4. Ruggeri ZM, Mendolicchio GL. (2007) Adhesion mechanisms in platelet function. *Circ Res* 100, 1673–1685.

5. Jackson SP. (2011) Arterial thrombosis-insidious, unpredictable and deadly. *Nat Med* 17, 1423–1436.
6. Ruggeri ZM, Dent JA, Saldivar E. (1999) Contribution of distinct adhesive interactions to platelet aggregation in flowing blood. *Blood* 94, 172–178.
7. Savage B, Almus-Jacobs F, Ruggeri ZM. (1998) Specific synergy of multiple substrate-receptor interactions in platelet thrombus formation under flow. *Cell* 94, 657–666.
8. Ware J. (2004) Dysfunctional platelet membrane receptors: from humans to mice. *Thromb Haemost* 92, 478–485.
9. Gawaz M, Langer H, May AE. (2005) Platelets in inflammation and atherogenesis. *J Clin Invest* 115, 3378–3384.
10. Libby P. (2002) Inflammation in atherosclerosis. *Nature* 420, 868–874.
11. Gawaz M, Neumann FJ, Dickfeld T et al. (1997) Vitronectin receptor (alpha(v)beta3) mediates platelet adhesion to the luminal aspect of endothelial cells: implications for reperfusion in acute myocardial infarction. *Circulation* 96, 1809–1818.
12. Bombeli T, Schwartz BR, Harlan JM. (1998) Adhesion of activated platelets to endothelial cells: evidence for a GPIIbIIIa-dependent bridging mechanism and novel roles for endothelial intercellular adhesion molecule 1 (ICAM-1), alphavbeta3 integrin, and GPIbalpha. *J Exp Med* 187, 329–339.
13. Wagner DD, Burger PC. (2003) Platelets in inflammation and thrombosis. *Arterioscler Thromb Vasc Biol* 23, 2131–2137.
14. Maynard DM, Heijnen HF, Horne MK, White JG, Gahl WA. (2007) Proteomic analysis of platelet alpha-granules using mass spectrometry. *J Thromb Haemost* 5, 1945–1955.
15. McEver RP. (2001) Adhesive interactions of leukocytes, platelets, and the vessel wall during hemostasis and inflammation. *Thromb Haemost* 86, 746–756.
16. Simon DI, Chen Z, Xu H et al. (2000) Platelet glycoprotein ibalpha is a counterreceptor for the leukocyte integrin Mac-1 (CD11b/CD18). *J Exp Med* 192, 193–204.
17. Wang Y, Sakuma M, Chen Z et al. (2005) Leukocyte engagement of platelet glycoprotein Iba via the integrin Mac-1 is critical for the biological response to vascular injury. *Circulation* 112, 2993–3000.
18. Weyrich AS, Elstad MR, McEver RP et al. (1996) Activated platelets signal chemokine synthesis by human monocytes. *J Clin Invest* 97, 1525–1534.
19. Strassel C, Hechler B, Bull A, Gachet C, Lanza F. (2009) Studies of mice lacking the GPIb-V-IX complex question the role of this receptor in atherosclerosis. *J Thromb Haemost* 7, 1935–1938.
20. Massberg S, Brand K, Gruner S et al. (2002) A critical role of platelet adhesion in the initiation of atherosclerotic lesion formation. *J Exp Med* 196, 887–896.
21. Seizer P, Gawaz M, May AE. (2008) Platelet-monocyte interactions-a dangerous liaison linking thrombosis, inflammation and atherosclerosis. *Curr Med Chem* 15, 1976–1980.
22. McEver RP. (1994) Role of selectins in leukocyte adhesion to platelets and endothelium. *Ann N Y Acad Sci* 714, 185–189.
23. Zarbock A, Ley K, McEver RP, Hidalgo A. (2011) Leukocyte ligands for endothelial selectins: specialized glycoconjugates that mediate rolling and signaling under flow. *Blood* 118, 6743–6751.
24. Preissner KT, Kanse SM, May AE. (2000) Urokinase receptor: a molecular organizer in cellular communication. *Curr Opin Cell Biol* 12, 621–628.
25. Romo GM, Dong JF, Schade AJ et al. (1999) The glycoprotein Ib-IX-V complex is a platelet counterreceptor for P-selectin. *J Exp Med* 190, 803–809.
26. Ruggeri ZM. (2007) The role of von Willebrand factor in thrombus formation. *Thromb Res* 120 Suppl 1, S5–S9.
27. Ruggeri ZM. (2003) Von Willebrand factor. *Curr Opin Hematol* 10, 142–149.
28. Nurden AT, Nurden P, George JN. (2001) Are patients with Glanzmann thrombasthenia and the Bernard-Soulier syndrome protected against atherosclerosis? *Adv Exp Med Biol* 489, 13–29.
29. Fuster W, Bowie EJ, Lewis JC, Fass DN, Owen CAJr, Brown AL. (1978) Resistance to arteriosclerosis in pigs with von Willebrand's disease. Spontaneous and high cholesterol diet-induced arteriosclerosis. *J Clin Invest* 61, 722–730.
30. Vilahur G, Padro T, Badimon L. (2011) Atherosclerosis and thrombosis: insights from large animal models. *J Biomed Biotechnol* 2011, 907575.
31. Methia N, Andre P, Denis CV, Economopoulos M, Wagner DD. (2001) Localized reduction of

atherosclerosis in von Willebrand factor- deficient mice. *Blood* 98, 1424–1428.

32. Burger PC, Wagner DD. (2003) Platelet P-selectin facilitates atherosclerotic lesion development. *Blood* 101, 2661–2666.
33. Kroll MH, Feng S. (2005) Targeting shear stress-induced platelet activation: is lesion-specific antiplatelet therapy a realistic clinical goal? *Expert Rev Cardiovasc Ther* 3, 941–951.
34. Antithrombotic Trialists' Collaboration. (2002) Collaborative meta-analysis of randomised trials of antiplatelet therapy for prevention of death, myocardial infarction, and stroke in high risk patients. *BMJ* 324, 71–86.
35. Berger JS, Krantz MJ, Kittelson JM, Hiatt WR. (2009) Aspirin for the prevention of cardiovascular events in patients with peripheral artery disease: a meta-analysis of randomized trials. *JAMA* 301, 1909–1919.
36. Drouet L, Bal dit SC, Henry P. (2010) The basis of platelets: platelets and atherothrombosis: an understanding of the lack of efficacy of aspirin in peripheral arterial disease (PAD) and diabetic patients. *Drugs* 70 Suppl 1, 9–14.
37. Nesbitt WS, Westein E, Tovar-Lopez F J et al. (2009) A shear gradient-dependent platelet aggregation mechanism drives thrombus formation. *Nat Med* 15, 665–673.
38. Andrews RK, Berndt MC. (2011) An event of shear importance. *Blood* 117, 5–7.
39. Yong AS, Pennings GJ, Chang M et al. (2011) Intracoronary shear-related up-regulation of platelet P-selectin and platelet-monocyte aggregation despite the use of aspirin and clopidogrel. *Blood* 117, 11–20.
40. Lutgens E, van Suylen RJ, Faber BC et al. (2003) Atherosclerotic plaque rupture: local or systemic process? *Arterioscler Thromb Vasc Biol* 23, 2123–2130.
41. Coccheri S. (2010) Antiplatelet drugs--do we need new options? With a reappraisal of direct thromboxane inhibitors. *Drugs* 70, 887–908.
42. Xiangdong L, Yuanwu L, Hua Z, Liming R, Qiuyan L, Ning L. (2011) Animal models for the atherosclerosis research: a review. *Protein Cell* 2, 189–201.
43. Leslie M. (2010) Cell biology. Beyond clotting: the powers of platelets. *Science* 328, 562–564.

16

DEVELOPMENT AND USE OF MOUSE MODELS OF ATHEROSCLEROSIS

PEGGY ROBINET AND JONATHAN D. SMITH

Department of Cellular and Molecular Medicine, Cleveland Clinic Lerner Research Institute/NC10, Cleveland, OH, USA

INTRODUCTION

Atherosclerosis is a pathology of the artery wall and the primary cause of cardiovascular disease, which is the leading cause of death worldwide [1]. Atherosclerotic lesions are driven by hypercholesterolemia, specifically by excess levels of non-high density lipoprotein (HDL) cholesterol found in the apolipoprotein B (apoB) containing lipoproteins such as very-low-density lipoprotein (VLDL), low-density lipoprotein (LDL), and remnant particles formed from lipolyzed chylomicrons and VLDL. This dyslipidemia allows excess cholesterol to accumulate in the subendothelial intima leading to endothelial dysfunction, the recruitment of monocytes, and the formation of fatty streaks due to the accumulation of cholesterol-loaded monocyte-derived macrophage foam cells. As the lesion progresses, foam cells secrete cytokines attracting more macrophages as well as smooth muscle cells, which then secrete extracellular matrix protein creating a fibrous cap over a lipid core. Advanced lesions often contain necrotic zones where foam cells have died and deposited cholesterol clefts. Advanced lesions can rupture allowing blood to come into contact with the plaque core, leading to thrombus formation that is responsible for acute ischemic events such as myocardial infarction or stroke [2]. In this context, animal models have proven very useful to increase our knowledge about the molecular and cellular mechanisms of lipoprotein metabolism and atherosclerosis. In addition, these models can be used to evaluate therapeutic strategies and interventional procedures.

Ideally, animal models should show similarities with humans in regard to their lipoprotein metabolism, atherosclerosis pathogenesis, and lesion localization. Other considerations include the time frame of lesion development, the time and ease of breeding, the cost of acquisition and maintenance, and the ability to perform *in vivo* imaging and interventions. In this chapter, we will focus on mouse models, as

Atherosclerosis: Risks, Mechanisms, and Therapies, First Edition. Edited by Hong Wang and Cam Patterson.

another recent publication amply reviews atherosclerosis studies in pigs, nonhuman primates, and rabbits [3].

Hundreds of inbred strains and substrains of mice are available. The short generation time, the ability to perform sophisticated genetic engineering, the relatively low cost to acquire and maintain, and the relatively short time until lesions develop make mice the most widely used species for atherosclerosis studies. For example, JAX offers about 400 mouse strains for cardiovascular studies, over 200 of them specifically developed for atherosclerosis research. However, wild-type mice are very different from humans in their lipoprotein metabolism and arterial dimensions. For example, mice do not express cholesteryl ester transfer protein (CETP)—a key component of lipoprotein remodeling and reverse cholesterol transport pathway in humans. The small size of mice does not make them an ideal model for human surgical interventions or imaging, and can also present an issue in terms of tissue availability and repeated blood sampling. While humans carry most of their plasma cholesterol on LDL, mice, which have total plasma cholesterol values of about 100 mg/dl, carry most of their plasma cholesterol in HDL, and are naturally resistant to atherosclerosis; thus, human intervention is required for the development of lesions. This can be achieved through dietary or genetic modification, or a combination of both. Here, we give a sampling of some of the mouse systems used to study atherosclerosis and lipoprotein metabolism, although, this chapter is not comprehensive as an entire book would be needed to report on all of the mouse models generated in this field.

DIET-INDUCED MOUSE MODELS

Paigen et al. developed a widely used diet to induce atherosclerosis, which contains 15% fat, 1.25% cholesterol, and 0.5% cholic acid. A large variation in atherosclerosis susceptibility was observed upon feeding this diet to 10 inbred mouse strains, which was not correlated with their plasma cholesterol levels, implying that genetic background effects cholesterol absorption or atherogenesis [4]. While the Paigen diet model has been widely used for atherosclerosis studies, the lesions in mice fed this diet for several months are present primarily in the aortic root (not a common site for lesions in human) and at most time points are restricted to fatty streaks [4]. In addition, this diet also induces hepatic inflammation and an acute phase response [5]. Using this model, recombinant inbred strains were generated from parental strains with different lesion propensity (C57BL6xC3H) and used to map the localization of atherosclerosis susceptibility genes [6]. Further, this diet-induced model has been successfully used for testing candidate atherosclerosis modifier genes and for environment and drug studies. Since testing an atherosclerosis-modifier candidate gene in this diet-induced model does not require crossing onto an atherosclerosis prone models discussed later, it is still the quickest way to get a preliminary answer. For example, apoAI overexpression decreases atherosclerosis in this model [7], while tumor necrosis factor receptor knockout (KO) increases lesion area [8].

ApoE GENE KO

In an attempt to create better mouse models of atherosclerosis, new strains have be engineered that lack or overexpress genes involved in lipid transport and metabolism. Two laboratories simultaneously reported the development of apoE-deficient (apoE KO) mice by homologous recombination in embryonic stem cells [9, 10]. ApoE is a ligand for remnant clearance and its KO induces severe hypercholesterolemia (~500 mg/dl), with accumulation of cholesterol ester-enriched VLDL and chylomicron remnants. These mice develop spontaneous atherosclerosis even when fed a chow diet, and upon feeding a Western-type diet (21% milk fat and 0.15% cholesterol), these mice develop extreme hypercholesterolemia (~1200 mg/dl) and accelerated lesion progression [10]. The time course and progression from fatty streak to intermediate and necrotic core lesions on both the chow and Western-type diet has been described, with lesions found in the aortic root, arch, and branch points, as well as in the brachiocephalic, carotid, and pulmonary arteries, but not in the coronary arteries [11]. Thus, many features of lesion progression in this model are similar to those seen in humans, although the medial layer in mice has many fewer cell layers than observed in larger animals, and plaque rupture is not observed without additional genetic or environmental interventions. Although vaso vasorum are not observed in the medial and intimal regions of the aorta in wild-type mice, angiogenesis leading to vaso vasorum in the media and intima are found in the advanced lesions in the thoracic aorta of Western-diet-fed apoE KO mice [12].

ApoE KO mice have been used extensively in hundreds of labs around the world to test the effects of candidate genes on lesion development, to map and identify naturally occurring atherosclerosis modifier genes, and to test environmental and drug effects on lesion development and progression. In an early example of the candidate approach, apoE KO mice were bred to osteopetrotic (op) mice that have a mutagen-induced defect in the macrophage colony-stimulating factor gene, which leads to decreased number and function of blood monocytes and tissue macrophages. The apoE KO op/op mice confirmed the important role of macrophages in atherogenesis, as these mice have vastly decreased lesion area despite having 2.5-fold higher levels of non-HDL cholesterol [13]. ApoE and scavenger receptor-BI double KO mice were bred and, even on a chow diet, these mice develop accelerated and severe obstructive lesions in the coronary arteries leading to death by 8 weeks of age [14]. By now, over 100 candidate genes have been tested by breeding to apoE KO mice. An example of identification of potential natural modifier genes comes from the genetic and genomic studies from the Lusis lab. ApoE KO mice on the C57Bl/6 (susceptible) and C3H (resistant) backgrounds were crossed, and nine atherosclerosis modifier loci were identified using classical quantitative trait locus (QTL) methods in an F_2 cohort fed the Western-type diet. Using transcriptome data from the liver and adipose tissue of these mice, 16 candidate genes at these 9 loci were identified as potentially causal atherosclerosis modifiers [15]. There are many studies of dietary, environmental, and even psychosocial effects on atherosclerosis in apoE KO mice; for example, second-hand cigarette smoke markedly increases lesion area in female apoE KO mice that are fed a Western-type diet, without altering plasma cholesterol levels [16]. ApoE KO mice have been used to test many different therapeutics; a recent example demonstrated that chronic daily administration of an inhibitor of bone morphogenic protein signaling led to decreased macrophage iron levels, increased ABCA1 and ABCG1 expression, and decreased lesion formation [17].

The advanced lesions in older apoE KO mice have been utilized to study lesion regression. Fisher and colleagues used surgical transplantation of aortic arches from hyperlipidemic apoE KO mice that are fed a Western-type diet into wild-type mice with their normal atheroprotective lipoprotein profile [18]. They observed a rapid and dramatic decrease in lesion area and lesion macrophage content, as early as 3 days post transplantation, accompanied by an increase in the smooth muscle cell cap. The loss of foam cells was largely due to cellular emigration dependent on the cytokine receptor CCR7 on the foam cells and its circulating ligands CCL19 and CCL21 [18]. The HDL-C levels of the recipient mice have a profound effect on regression, as transplantation into human apoAI-transgenic (Tg) apoE KO hosts that still have high levels of both HDL and non-HDL-C led to more regression than transplantation into apoAI-deficient hosts wild type for apoE with low levels of both HDL and non-HDL-C [19].

Adenoviral gene transfer of apoE can also correct the dyslipidemia in apoE KO mice [20], as can bone marrow transplantation with wild-type bone marrow, since apoE is produced by bone marrow-derived macrophages [21, 22].

In fact, this is a limitation of this model, as leukocyte expressed atherosclerosis modifier candidate genes cannot be tested by transplantation of marrow from the candidate gene KO donor into apoE KO hosts, unless that candidate gene is first crossed onto the apoE KO background. In this regard, the LDLr KO discussed in the following is superior.

ApoE also plays a role in the response to vascular injury. Guidewire-induced endothelial denudation does not lead to much neointima formation in wild-type mice, but large neointimas develop over 18 days in apoE KO mice, which can be reduced by increased HDL levels obtained by apoAI adenoviral gene transfer [23]. In addition, chronic infusion of angiotensin II into apoE KO mice via osmotic minipumps leads to the formation of aortic aneurysms [24], thus, creating a small animal model in which aneurysm formation and therapeutics can be studied.

Although the apoE KO model is a recessive model, since both alleles need to be knocked out to develop hypercholesterolemia and atherosclerosis, Maeda and colleagues have described that hemizygous apoE$^{-/+}$ mice are hyper-responsive to the Paigen diet, thus allowing apoE-deficiency to be used as a dominant trait [25]. In conclusion, the apoE KO model has been successfully used to learn much about the molecular and cellular events associated with atherogenesis and lesion regression. The major limitations of this model are the absence of spontaneous coronary artery lesions and plaque rupture.

LDLr GENE KO

The LDLr KO mouse serves as a model for familial hypercholesterolemia (FH) due to LDLr mutations [26]. The original LDLr KO on an outbred background and fed a chow diet have high total cholesterol levels of about 230 mg/dl, compared with about 100 mg/dl in wild-type mice, but these cholesterol levels are not high enough to promote atherosclerosis. However, when LDLr KO mice were fed a 0.2% cholesterol supplemented diet, their plasma total cholesterol levels rose to about 500 mg/dl with elevation of both LDL and remnants. Subsequently, these mice were fed the Paigen diet that raised their plasma cholesterols to about 1500 mg/dl and resulted in xanthomas and extensive atherosclerosis in the root, arch and descending aorta [27]. After partial backcrossing onto the C57BL/6 background, LDLr KO mice fed the Western-type diet have total cholesterol levels of about 900 mg/dl and develop significant atherosclerosis in the root, arch, and descending aorta [28]. Interestingly, hypercholesterolemia and atherosclerosis can occur when these mice are fed a semisynthetic casein-based diet, even without the addition of cholesterol or excess fat, which can be exacerbated by adding dietary cholesterol [29]. Similar to the apoE KO model, LDLr KO mice have been used extensively for candidate gene studies, mapping atherosclerosis modifier genes, and environmental and drug studies. One of the advantages of this model over the apoE KO model is that the LDLr expression on macrophages does not alter atherogenesis, such that bone marrow transplantation of candidate gene KO donor into LDLr KO hosts can be accomplished without specific back crossing of the donor onto the LDLr KO background.

ADDITIONAL apoE MODELS

Although apoE KO mice are widely used, absence of apoE in humans is extremely rare. There are three common human apoE isoforms due to single nucleotide substitutions: apoE2, apoE3, and apoE4. Homozygosity for apoE2, which is deficient in binding to the LDLr, is associated with type III hyperlipidemia. Maeda and colleagues, in an elegant series of studies, replaced the mouse apoE locus with the genes encoding the three human apoE isoforms, thus creating the three apoE knock in (KI) lines. The apoE2 KI have a two- to threefold increases in plasma levels of cholesterol and triglycerides mainly located on β-VLDL. When fed a chow diet, this model accumulates foam cells in the aortic root with more advance lesions developing on feeding the Western-type diet [30]. ApoE3 KI mice have a normal plasma lipid profile when fed a chow diet; however, they are more susceptible to diet-induced atherosclerosis than the wild-type animals. ApoE4 KI mice present a similar lipid profile and atherosclerosis susceptibility as the apoE3 KI mice with the exception of having slower VLDL clearance [31]. Since apoE4 is the strongly associated with Alzheimer disease in humans, apoE KO and KI mice models have been bred to mouse models of Alzheimer disease-like pathology, which overexpress a mutant amyloid precursor protein. Holtzman and colleagues found apoE allele-specific effects on cerebral amyloid plaque burden, with the apoE4 KI having more, and apoE2 KI having less dense plaques. They discovered an apoE isoform specific effect on the clearance of soluble amyloid-β peptide that can account for this phenotype [32].

ApoE3-Leiden is a rare human variant associated type III hyperlipidemia. Transgenic mice expressing this variant (apoE3L Tg) develop hyperlipidemia and are very susceptible to diet-induced atherosclerosis [33], making them an excellent dominant model for atherosclerosis studies and candidate gene testing. ApoE3L Tg mice fed a Western-type diet present a human-like lipid profile and are very responsive to lipid-lowering treatments. To further humanize lipoprotein metabolism in these mice, they were crossed with human CETP transgenic mice, which shifted the cholesterol distribution from the HDL toward the LDL/VLDL fraction. Upon feeding a high cholesterol diet, the apoE3L/CETP double transgenic mice had sevenfold larger lesions than the apoE3L Tg mice [34]. Niacin, a commonly prescribed HDL raising medication, does not raise HDL levels in mice in general, and specifically in the apoE3L Tg mice. However, niacin treatment of the apoE3L/CETP double transgenic mice increases HDL-C in dose-dependent manner. This effect is associated with downregulation of the transgenic CETP expression in the liver and decreased plasma CETP levels and activity, suggesting that CETP downregulation is the mechanism by which niacin increases HDL-C [35].

ADDITIONAL LDL MODELS

ApoB is the only apolipoprotein associated with mature LDL. In humans, apoB made in the intestine is found in two isoforms—the full-length apoB100 isoform and in its RNA-edited shorter apoB48 isoform—while the liver secretes exclusively the apoB100 isoform. However, mice produce both apoB100 and apoB48 in both the intestine and liver. The first human apoB transgenic mice made from an 80 kb genomic fragment expressed high levels of human apoB, but most of the liver apoB transcript was edited to produce apoB48. These mice have a mild increase in total cholesterol to about 150 mg/dl with increased LDL, but this increase is not a sufficient to lead to atherosclerosis. When fed a high-fat diet, these mice have total cholesterol levels of about 310 mg/dl with elevated VLDL and LDL and develop robust aortic root lesions that progress into intermediate lesions after 6 months on this diet, thereby creating another dominant model of atherosclerosis [36]. When these mice were crossed onto the LDLr KO background, they had elevated VLDL and LDL and developed lesions even on a chow diet [37].

To create mice that can make only mouse apoB100, a knock in approach was used to target codon 2153 (the site of editing) from CCA to CTA, such that the editing would not produce a stop codon. Similarly, this codon was replaced with a TGA stop codon to make the apoB48 only mouse [36]. Upon breeding to the LDLr KO background, the apoB100 only mice had elevated total and LDL cholesterol levels, but lower levels were seen with the apoB48 only LDLr KO mice. Another strategy was also used to make apoB100 only mice, by knocking out the RNA-editing enzyme apoBEC-1. This model was also bred onto the LDLr KO background leading to severe hypercholesterolemia and atherosclerosis even on a chow diet [38].

A regression model of atherosclerosis has also been prepared on the LDLr KO apoB100 only background. This model, called the Reversa mouse, is also homozygous for a conditional loxP-flanked allele of the microsomal triglyceride transfer protein (MTTP) gene, and has an Mx1-Cre transgene, which upon activation by injection of a drug leads to deletion of the MTTP gene in the liver. On a high-fat diet, these mice have basal cholesterol levels of about 530 mg/dl, but after hepatic MTTP expression is abolished plasma cholesterol falls to approximately 100 mg/dl. When MTTP expression is deleted prior to placing the mice on a high-fat diet, atherosclerosis is prevented [39]. But even more exciting was the finding of atherosclerosis lesion regression when MTTP gene expression is abolished after lesion formation [40].

TECHNOLOGICAL ADVANCES AND THE CREATION OF NEW ANIMAL MODELS

Recently, new technologies have emerged for germline gene editing using sleeping beauty transposons, zinc finger nucleases, and TAL nucleases, which can be applied to create new mouse models [41, 42]. Zinc finger nuclease technology has also been successfully applied to make rat KOs, and apoE KO and LDLr KO rats are commercially available through SAGE Labs. Further, the discovery that micro-RNAs can control gene expression posttranscriptionally is yielding new mouse models. For example, KO or knockdown of miR-33, a micro-RNA that controls a subset of genes involved in lipid metabolism (i.e., ABCA1), leads to increased HDL and reverse transport cholesterol. Further, the inhibition of miR-33 in mice with established atherosclerosis leads to a marked regression of atherosclerosis with reduced plaque area as well as decreased lipid and macrophages content [43].

ASSESSING MOUSE ATHEROSCLEROSIS

One of the key issues for evaluating a model of atherosclerosis is the method used for lesion quantification. The majority of mouse studies use lesion quantification in the aortic root or in the entire aortic tree, but other locations such as the brachiocephalic artery may also be assessed. A variety of techniques have been described for *ex vivo* analysis of atherosclerosis, and much progress is being made toward the detection of atherosclerosis *in vivo*. This latter technique would allow repeated measures in the same animal to follow lesion progression and/or regression over time.

AORTIC ROOT LESION ANALYSIS

This technique, first described by Paigen and slightly modified since, is the most commonly used method for evaluation of atherosclerosis in mice. Briefly, it analyses the cross-sectional lesion area in the aortic root after neutral lipid staining with Oil Red O (Figure 16.1). The site of lesion quantification is anatomically defined, starting from the aortic sinus to the region where the valve bases regress, but are still present, a regions comprising about 600 μm in length. Lesion areas are expressed as average lesion per section through this region [44]. This assay is probably the most sensitive assessment of early lesions as this is the earliest site of lesion formation in mice; however, this is not a typical site of human lesion formation. Care must be taken in sectioning consistently in the target region, and the plane of sectioning has to be adjusted to make sure the sections are perpendicular to the axis of the aortic root. Oil Red O staining helps to visualize lesions; however, the plaque is not made exclusively of neutral lipid laden cells but also contains smooth muscle cells, extracellular matrix, and free cholesterol. Thus, the analysis of lesion size requires manual drawing using image software to precisely outline the entire lesion area not only the area of red staining [45]. For

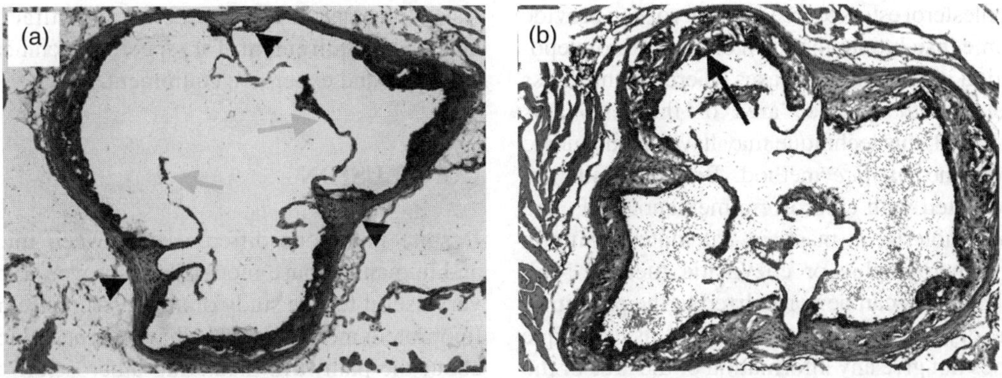

Figure 16.1 Oil Red O and hematoxylin stained aortic root lesions from apoE KO mice. (a) Typical lesion in 16-week-old chow-fed apoE KO mice on the C57BL/6 background. The *black arrow* heads point to the three valve bases that are visible throughout the zone of sectioning. The *green arrows* point to two of the six valve leaflets. This section shows the medial layer stained only with hematoxylin and dark red-stained fatty streak lesions in the intima over most, but not all of the lumen surface. (b) Advanced lesions in 20-week-old apoE KO mice fed a Western-type diet. The entire lumen area is covered with extensive lesions, although the lesion is not completely stained with Oil Red O, indicating the presence of cells, necrotic area, and matrix without cholesterol ester deposits. The *black arrow* points to a lesion containing many cholesterol clefts, the crystalline cholesterol deposited after the death of free cholesterol enriched foam cells. (*See insert for color representation of the figure.*)

testing candidate genes or treatments in apoE KO mice, we recommend starting with a 16-week time point on the chow diet and assaying early lesions consisting mostly of fatty streaks by this method (Figure 16.1a). Care must be taken to sacrifice mice at the specified time points, as aortic root lesions can double in area in apoE KO mice in as short as a 2-week period [44].

EN FACE ANALYSIS OF AORTIC INTIMA

Although mouse lesions usually appear first in the aortic root, lesion progression there may not follow the same dynamics as in other locations. In fact, there are examples where lesion area is affected in the aortic tree but not in the root strongly suggesting that atherosclerosis should be assessed at more than one site [3]. The *en face* technique assesses the overall aortic surface covered with lesions, starting at the aorta as it leaves the heart until reaching the ileac bifurcation. In this method, the aorta is opened longitudinally and pinned or fixed flat and stained with a lipophilic dye. The results are expressed as the percentage lesion coverage and can be analyzed separately in the arch, thoracic, and abdominal regions [46]. A major limitation of the *en face* quantification is that the measurement is two-dimensional and does not take in account lesion thickness [45]. In addition, lesions develop more slowly at this location, therefore a later time point or use of an atherogenic diet may be required. A study comparing *en face* and aortic root lesion quantification in apoE KO and LDLr KO mice has shown a good correlation between the two methods [47]. For follow-up testing candidate genes or treatments in apoE KO mice, we recommend feeding the Western-type diet from 4 to 20 weeks of age and assaying advanced lesions by both the aortic root (Figure 16.1b) and *en face* methods.

AORTIC CHOLESTEROL MASS

The determination of the cholesterol mass in the aortic tree can also be performed as an estimate of atherosclerosis, and the ratio of cholesterol esters to free cholesterol can provide some insight into the form of cholesterol deposition in the lesion. In cases where the thickness but not the surface area of the lesions is affected, this technique may have an advantage over the *en face* method. Briefly, lipids are extracted with chloroform:methanol (2:1) and the cholesterol mass is calculated using a chromatographic or enzymatic method [46]. The results can be normalized to the protein or the dry weight of the sample. As this technique does not give any information of the area or the volume of the lesions, it is best used to complement other techniques. In addition, this method only measures the lesion cholesterol content and not other lesion components [45].

IMAGING TECHNIQUES

All the techniques described earlier involve sacrificing the mice such that studies on the progression or regression of atherosclerosis have to be done with separate mice for each time point. The development of *in vivo* imaging techniques such as ultrasound, magnetic resonance imaging (MRI), or computed tomography (CT), can overcome this issue. For example, CT has been adapted for mouse models of atherosclerosis to detect lesions *ex vivo* as well as *in vivo* [48, 49]. This method also allows the collection of information on plaque composition, particularly when specific contrast agents are used. For example, monocytes recruitment or activated matrix metalloproteinase secretion have been successfully observed *in vivo* [50, 51]. Similarly, MRI has been used for *ex vivo* and *in vivo* detection of mouse lesions in the abdominal aorta and also to determine cellular or molecular plaque characteristics [52]. These techniques have many advantages compared with the methods described earlier, including the possibility to image the same animal repeatedly, to retain *in vivo* structural characteristics of the observed tissue, and to reconstruct three-dimensional (3D) images. However, these techniques have many barriers to success due to the small size of mouse arteries and their lesions, the rapid heartbeat and motion artifacts, and the requirement for specific technical expertise and expensive equipment.

CONCLUSION

Despite numerous differences between mice and humans, mouse models have proven to be very useful for the study of atherosclerosis and lipoprotein metabolism. We have learned much about the pathways mediating atherosclerosis progression and regression by crossing candidate genes into mouse models of atherosclerosis. Further, the development of more advanced genetic tools will give rise to new specific models. For example, drug-induced gene expression/silencing systems will allow the study of the role of a gene at a specific time frame during the development of atherosclerosis. Similarly, conditional KOs with different promoters are useful tools to study the role of a gene product in particular cell types. Finally, progress in *in vivo* imaging will make it possible to study the natural history of atherosclerosis over the course of lesion progression/regression at several sites within one mouse.

ACKNOWLEDGMENT

This work is supported by the NIH grant R01 HL091893.

REFERENCES

1. WHO, *Global atlas on cardiovascular disease: prevention and control*, ed. P.P.a.B.N. Shanthi Mendis. 2011: WHO, World Heart Federation, World Stroke Organization. 164.
2. Glass, C.K. and J.L. Witztum, *Atherosclerosis. the road ahead*. Cell, 2001. **104**(4): p. 503–16.
3. Getz, G.S. and C.A. Reardon, *Animal models of atherosclerosis*. Arterioscler Thromb Vasc Biol, 2012. **32**(5): p. 1104–15.
4. Paigen, B., et al., *Variation in susceptibility to atherosclerosis among inbred strains of mice*. Atherosclerosis, 1985. **57**(1): p. 65–73.

5. Liao, F., et al., *Genetic control of inflammatory gene induction and NF-kappa B-like transcription factor activation in response to an atherogenic diet in mice*. J Clin Invest, 1993. **91**(6): p. 2572–9.
6. LeBoeuf, R.C., et al., *Phenotypic characterization of the Ath-1 gene controlling high density lipoprotein levels and susceptibility to atherosclerosis*. J Lipid Res, 1990. **31**(1): p. 91–101.
7. Rubin, E.M., et al., *Inhibition of early atherogenesis in transgenic mice by human apolipoprotein AI*. Nature, 1991. **353**(6341): p. 265–7.
8. Schreyer, S.A., J.J. Peschon, and R.C. LeBoeuf, *Accelerated atherosclerosis in mice lacking tumor necrosis factor receptor p55*. J Biol Chem, 1996. **271**(42): p. 26174–8.
9. Zhang, S.H., et al., *Spontaneous hypercholesterolemia and arterial lesions in mice lacking apolipoprotein E*. Science, 1992. **258**(5081): p. 468–71.
10. Plump, A.S., et al., *Severe hypercholesterolemia and atherosclerosis in apolipoprotein E-deficient mice created by homologous recombination in ES cells*. Cell, 1992. **71**(2): p. 343–53.
11. Nakashima, Y., et al., *ApoE-deficient mice develop lesions of all phases of atherosclerosis throughout the arterial tree*. Arterioscler Thromb, 1994. **14**(1): p. 133–40.
12. Moulton, K.S., et al., *Inhibition of plaque neovascularization reduces macrophage accumulation and progression of advanced atherosclerosis*. Proc Natl Acad Sci U S A, 2003. **100**(8): p. 4736–41.
13. Smith, J.D., *Mouse models of atherosclerosis*. Lab Anim Sci, 1998. **48**(6): p. 573–9.
14. Braun, A., et al., *Loss of SR-BI expression leads to the early onset of occlusive atherosclerotic coronary artery disease, spontaneous myocardial infarctions, severe cardiac dysfunction, and premature death in apolipoprotein E-deficient mice*. Circ Res, 2002. **90**(3): p. 270–6.
15. Wang, S.S., et al., *Identification of pathways for atherosclerosis in mice: integration of quantitative trait locus analysis and global gene expression data*. Circ Res, 2007. **101**(3): p. e11–30.
16. Gairola, C.G., et al., *Sidestream cigarette smoke accelerates atherogenesis in apolipoprotein E-/- mice*. Atherosclerosis, 2001. **156**(1): p. 49–55.
17. Saeed, O., et al., *Pharmacological suppression of hepcidin increases macrophage cholesterol efflux and reduces foam cell formation and atherosclerosis*. Arterioscler Thromb Vasc Biol, 2012. **32**(2): p. 299–307.
18. Williams, K.J., J.E. Feig, and E.A. Fisher, *Rapid regression of atherosclerosis: insights from the clinical and experimental literature*. Nat Clin Pract Cardiovasc Med, 2008. **5**(2): p. 91–102.
19. Feig, J.E., et al., *HDL promotes rapid atherosclerosis regression in mice and alters inflammatory properties of plaque monocyte-derived cells*. Proc Natl Acad Sci U S A, 2011. **108**(17): p. 7166–71.
20. Stevenson, S.C., et al., *Phenotypic correction of hypercholesterolemia in apoE-deficient mice by adenovirus-mediated in vivo gene transfer*. Arterioscler Thromb Vasc Biol, 1995. **15**(4): p. 479–84.
21. Linton, M.F., J.B. Atkinson, and S. Fazio, *Prevention of atherosclerosis in apolipoprotein E-deficient mice by bone marrow transplantation*. Science, 1995. **267**(5200): p. 1034–7.
22. Boisvert, W.A., J. Spangenberg, and L.K. Curtiss, *Treatment of severe hypercholesterolemia in apolipoprotein E-deficient mice by bone marrow transplantation*. J Clin Invest, 1995. **96**(2): p. 1118–24.
23. De Geest, B., et al., *Effects of adenovirus-mediated human apo A-I gene transfer on neointima formation after endothelial denudation in apo E-deficient mice*. Circulation, 1997. **96**(12): p. 4349–56.
24. Daugherty, A., M.W. Manning, and L.A. Cassis, *Angiotensin II promotes atherosclerotic lesions and aneurysms in apolipoprotein E-deficient mice*. J Clin Invest, 2000. **105**(11): p. 1605–12.
25. Zhang, S.H., et al., *Diet-induced atherosclerosis in mice heterozygous and homozygous for apolipoprotein E gene disruption*. J Clin Invest, 1994. **94**(3): p. 937–45.
26. Ishibashi, S., et al., *Hypercholesterolemia in low density lipoprotein receptor knockout mice and its reversal by adenovirus-mediated gene delivery*. J Clin Invest, 1993. **92**(2): p. 883–93.
27. Ishibashi, S., et al., *Massive xanthomatosis and atherosclerosis in cholesterol-fed low density lipoprotein receptor-negative mice*. J Clin Invest, 1994. **93**(5): p. 1885–93.
28. Merat, S., et al., *Western-type diets induce insulin resistance and hyperinsulinemia in LDL receptor-deficient mice but do not increase aortic atherosclerosis compared with normoinsulinemic mice in which similar plasma cholesterol levels are achieved by a fructose-rich diet*. Arterioscler Thromb Vasc Biol, 1999. **19**(5): p. 1223–30.

29. Teupser, D., A.D. Persky, and J.L. Breslow, *Induction of atherosclerosis by low-fat, semisynthetic diets in LDL receptor-deficient C57BL/6J and FVB/NJ mice: comparison of lesions of the aortic root, brachiocephalic artery, and whole aorta (en face measurement).* Arterioscler Thromb Vasc Biol, 2003. **23**(10): p. 1907–13.
30. Sullivan, P.M., et al., *Type III hyperlipoproteinemia and spontaneous atherosclerosis in mice resulting from gene replacement of mouse Apoe with human Apoe*2*. J Clin Invest, 1998. **102**(1): p. 130–5.
31. Knouff, C., et al., *Apo E structure determines VLDL clearance and atherosclerosis risk in mice*. J Clin Invest, 1999. **103**(11): p. 1579–86.
32. Castellano, J.M., et al., *Human apoE isoforms differentially regulate brain amyloid-beta peptide clearance*. Sci Transl Med, 2011. **3**(89): p. 89ra57.
33. van Vlijmen, B.J., et al., *Apolipoprotein E*3-Leiden transgenic mice as a test model for hypolipidaemic drugs*. Arzneimittelforschung, 1998. **48**(4): p. 396–402.
34. Westerterp, M., et al., *Cholesteryl ester transfer protein decreases high-density lipoprotein and severely aggravates atherosclerosis in APOE*3-Leiden mice*. Arterioscler Thromb Vasc Biol, 2006. **26**(11): p. 2552–9.
35. van der Hoorn, J.W., et al., *Niacin increases HDL by reducing hepatic expression and plasma levels of cholesteryl ester transfer protein in APOE*3Leiden.CETP mice*. Arterioscler Thromb Vasc Biol, 2008. **28**(11): p. 2016–22.
36. Kim, E. and S.G. Young, *Genetically modified mice for the study of apolipoprotein B*. J Lipid Res, 1998. **39**(4): p. 703–23.
37. Sanan, D.A., et al., *Low density lipoprotein receptor-negative mice expressing human apolipoprotein B-100 develop complex atherosclerotic lesions on a chow diet: no accentuation by apolipoprotein(a)*. Proc Natl Acad Sci U S A, 1998. **95**(8): p. 4544–9.
38. Powell-Braxton, L., et al., *A mouse model of human familial hypercholesterolemia: markedly elevated low density lipoprotein cholesterol levels and severe atherosclerosis on a low-fat chow diet*. Nat Med, 1998. **4**(8): p. 934–8.
39. Lieu, H.D., et al., *Eliminating atherogenesis in mice by switching off hepatic lipoprotein secretion*. Circulation, 2003. **107**(9): p. 1315–21.
40. Feig, J.E., et al., *Reversal of hyperlipidemia with a genetic switch favorably affects the content and inflammatory state of macrophages in atherosclerotic plaques*. Circulation, 2011. **123**(9): p. 989–98.
41. Takeda, J., V.W. Keng, and K. Horie, *Germline mutagenesis mediated by Sleeping Beauty transposon system in mice*. Genome Biol, 2007. **8 Suppl 1**: p. S14.
42. Osiak, A., et al., *Selection-independent generation of gene knockout mouse embryonic stem cells using zinc-finger nucleases*. PLoS One, 2011. **6**(12): p. e28911.
43. Rayner, K.J., C. Fernandez-Hernando, and K.J. Moore, *MicroRNAs regulating lipid metabolism in atherogenesis*. Thromb Haemost, 2012. **107**(4): p. 642–7.
44. Baglione, J. and J.D. Smith, *Quantitative assay for mouse atherosclerosis in the aortic root*. Methods Mol Med, 2006. **129**: p. 83–95.
45. Daugherty, A., et al., *Modes of defining atherosclerosis in mouse models: relative merits and evolving standards*. Methods Mol Biol, 2009. **573**: p. 1–15.
46. Daugherty, A. and S.C. Whitman, *Quantification of atherosclerosis in mice*. Methods Mol Biol, 2003. **209**: p. 293–309.
47. Tangirala, R.K., E.M. Rubin, and W. Palinski, *Quantitation of atherosclerosis in murine models: correlation between lesions in the aortic origin and in the entire aorta, and differences in the extent of lesions between sexes in LDL receptor-deficient and apolipoprotein E-deficient mice*. J Lipid Res, 1995. **36**(11): p. 2320–8.
48. Martinez, H.G., et al., *Images in cardiovascular medicine: microscopic computed tomography-based virtual histology for visualization and morphometry of atherosclerosis in diabetic apolipoprotein e mutant mice*. Circulation, 2009. **120**(9): p. 821–2.
49. Lloyd, D.J., et al., *A volumetric method for quantifying atherosclerosis in mice by using microCT: comparison to en face*. PLoS One, 2011. **6**(4): p. e18800.
50. Kircher, M.F., et al., *Noninvasive in vivo imaging of monocyte trafficking to atherosclerotic lesions*. Circulation, 2008. **117**(3): p. 388–95.
51. Razavian, M., et al., *Atherosclerosis plaque heterogeneity and response to therapy detected by in vivo molecular imaging of matrix metalloproteinase activation*. J Nucl Med, 2011. **52**(11): p. 1795–802.
52. Weinreb, D.B., et al., *Non-invasive MRI of mouse models of atherosclerosis*. NMR Biomed, 2007. **20**(3): p. 256–64.

SECTION III

MOLECULAR MECHANISMS

17

MECHANISMS OF THE RENIN ANGIOTENSIN SYSTEM INFLUENCING ATHEROSCLEROSIS

Hong Lu and Alan Daugherty
Saha Cardiovascular Research Center, University of Kentucky, Lexington, KY, USA

INTRODUCTION

The existence of the renin angiotensin system (RAS) was initially discovered more than a century ago, and the major bioactive octapeptide, angiotensin II (AngII), has been recognized for over seven decades. During the past decade, studies on this angiotensin peptide and other classical components have led to new insights into the RAS. While AngII is a critical regulator for blood pressure and water/sodium homeostasis, many other important roles of AngII have also been determined. In addition, it is now acknowledged that the RAS is a complex system that involves a spectrum of bioactive angiotensin peptides, multiple receptors that use these angiotensin peptides as ligands, and diverse enzymes that have the ability to generate these peptides. In addition to AngII, three other angiotensin peptides, that is, AngIII, AngIV, and Ang(1–7) are also biologically active. While AT1 and AT2 receptors are well-recognized AngII-binding receptors, AngIII, a downstream peptide generated from AngII, also binds to these two receptors with lower affinity than AngII. Angiotensin-converting enzyme 2 (ACE2) converts AngII into Ang(1–7) as well as cleaves AngI into Ang(1–9). Although the biological effect of Ang(1–9) is unclear, a recent study has provided evidence that Ang(1–9) can bind AT2 receptors [1].

In addition to the growing array of bioactive peptides of the RAS, many components of this system have been detected in different organs, tissues, and cell types, thereby adding another layer of complexity to this intricate system. The elaborate link among components in the RAS also leads to an impediment in understanding the "Yin" and the "Yang" of this hormonal system in pathophysiological conditions such as atherosclerosis.

There is compelling evidence from animal studies that the RAS influences atherosclerosis. Chronic infusion of AngII into hypercholesterolemic mice accelerates atherosclerosis

Atherosclerosis: Risks, Mechanisms, and Therapies, First Edition. Edited by Hong Wang and Cam Patterson.

[2–4]. Conversely, blockade of AngII synthesis or action through pharmacological inhibition or genetic disruption of some of the RAS components prevents the development of atherosclerosis [5–12]. Despite lack of direct evidence from human studies, large-scale clinical trials have demonstrated that pharmacological inhibition of either ACE [13] or AT1 receptor [14] has beneficial effects on atherosclerosis-related morbidity and mortality. In the Heart Outcomes Prevention Evaluation (HOPE) population, high plasma renin activity has also been identified as an independent predictor of major vascular events and mortality, inferring this rate-limiting enzyme for AngII synthesis is important in the development of atherosclerosis [15].

We have recently summarized the literature regarding the effects of modulation of the RAS on development of atherosclerotic lesions [16]. In this chapter, we will focus on recent mechanistic studies in animal models.

THE RENIN ANGIOTENSIN COMPONENTS INFLUENCE ATHEROSCLEROSIS

Accumulating evidence that the RAS influences atherosclerosis is predominately from mouse studies. Genetic manipulation and pharmacological approaches have been the most powerful approaches to studying the contribution of the RAS to atherosclerosis. However, there are some limitations of both approaches. The use of genetically engineered mice with deletions of critical components of the RAS is confounded by the presence of severe renal impairment. This occurs in mice with deficiency of angiotensinogen, renin, and angiotnesin-converting enzyme (ACE) with a less-severe phenotype present in mice with deletions of AT1 receptor subtypes. The potential confounding feature of pharmacological approaches is that many of the drugs have properties beyond influences on RAS. Irrespective of these disadvantages, the combination of these two approaches has strongly enhanced our understanding on how the RAS influences atherosclerosis.

Bioactive Angiotensin Peptides

To study the effects of bioactive angiotensin peptides on atherosclerosis, one direct approach has been to infuse an angiotensin peptide into hypercholesterolemic mice. Among the four recognized bioactive angiotensin peptides, AngII has been studied extensively in mouse models, while there is scant information regarding roles of AngIII, AngIV, and Ang(1–7) in the development of atherosclerosis. Many research groups have consistently demonstrated that infusion of AngII accelerates atherosclerosis in hypercholesterolemic mice [2–4]. As discussed in our previous review [17], evidence from AngII-infusion studies implicates that the RAS contributes to atherosclerosis beyond its well-known effect on blood pressure regulation. One example is that low-dietary sodium (Na 0.01% weight/weight in diet) intake reduces systolic blood pressure, but activates the RAS and augments atherosclerosis compared with high-dietary sodium intake (Na 2%) [18].

In contrast to the contribution of AngII to atherosclerosis, infusion of Ang(1–7) reduces atherosclerosis in hypercholesterolemic mice without influencing blood pressure [19, 20]. There is a single study reported that AngIV infusion has no effect on the progression of atherosclerosis in mice, but leads to improved endothelial function of atherosclerotic lesions [21].

Angiotensin Peptide Receptors

Two receptor types have been the focus of many mechanistic studies on AngII effects. Of the two AngII receptor subtypes (AT1 and AT2), AT1 receptor is the predominant receptor for AngII biological and pathophysiological actions. In rodents, there are two subtypes of the AT1 receptor: AT1a and AT1b. Pharmacological inhibition that targets both AT1a and AT1b receptors results in profound reductions of experimental atherosclerosis in multiple animal models. There is a single report that AT1b receptor has no effect on AngII-induced atherosclerosis [22], while there are consistent findings that AT1a receptor deficiency reduces atherosclerosis in hypercholesterolemic mice [5, 23–25]. This AT1a receptor-specific role has also been demonstrated by that coadministration of an AT1 receptor antagonist, telmisartan, in *Apoe*$^{-/-}$ mice with AT1a receptor deficiency does not further reduce atherosclerotic lesion size [25].

Reports on the role of AT2 receptor have been highly inconsistent, and include demonstrations that AT2 receptor deficiency attenuates [26], has no effects [5], augments [27], or changes cellularity without changing plaque size [28]. One study has shown that AT2 receptor overexpression through adenoassociated viral system decreases atherosclerosis in mice [29]. Comparably, pharmacological inhibition of AT2 receptor using PD123319 compound has also generated conflicting results as discussed earlier in Lu et al. [16]. It remains undefined what has caused these discrepant results regarding the contribution of AT2 receptor to atherosclerosis. Our previous study demonstrated that AT2 antagonism with PD123319 augmented atherosclerosis in hypercholesterolemic mice infused with AngII 1 μg/kg/min [30], inferring that AngII–AT2 receptor interaction may somewhat antagonize AngII–AT1 receptor interaction. A recent study reported that blockade of AT2 with PD123319 or inhibition of ACE with ramipril augmented atherosclerosis in hypercholesterolemic mice with AT1a receptor deficiency [25]. By contrast, infusion of AngII 0.4 μg/kg/min resulted in attenuation of atherosclerosis in AT1a receptor-deficient mice administered ramipril [25]. The findings from this article also imply that AngII–AT2 receptor interaction antagonizes atherosclerotic formation under certain conditions. However, it is unclear what receptors AngII interacts to augment atherosclerosis in both depletion of AT1a receptor and inhibition of AT2 receptor, since AT1b receptor appears to be not relevant to AngII interaction in the development of atherosclerosis [22, 25].

Among several identified functions of AT4 receptor, one role is to serve as the binding receptor for AngIV. No study has determined roles of AT4 receptor deficiency in the development of atherosclerosis. As determined in a single study, AT4 receptor antagonism through administration of a compound divalinal-AngIV does not affect the progression of atherosclerosis in mice fed a high-fat diet and infused with AngIV [21]. Ang(1–7)-binding receptor is Mas. While genetic manipulation of Mas influencing atherosclerosis has not been studied, two recent studies have reported that AVE0991, a Mas agonist, inhibits atherosclerotic formation in *Apoe*$^{-/-}$ mice [31, 32].

Enzymes Involved in Angiotensin Peptide Generation

Renin is the rate-limiting enzyme in the synthesis of all angiotensin peptides. No study has directly determined the role of whole body deficiency of renin in atherosclerosis. Mice with overexpression of both human angiotensinogen and renin have augmented atherosclerosis when fed a high-fat diet [33]. Since aliskiren was approved to be used in patients with hypertension, many research laboratories [6, 10–12, 34] have reported that this human renin inhibitor attenuates experimental atherosclerosis through suppressing bioactive angiotensin peptide generation [6].

Although ACE can cleave multiple substrates, it is well-known as the major converting enzyme for the synthesis of AngII. Heterozygous deficiency of ACE does not change atherosclerotic lesion size in *Apoe*$^{+/-}$ mice [35]. In mice, lack of ability (homozygous) or having diminished ability (heterozygous) to convert AngI into AngII due to truncation of ACE protein (termed ACE.2), atherosclerotic lesion size is profoundly reduced, compared to their wild-type *Apoe*$^{-/-}$ littermates [8, 36]. In contrast to the conflicting findings regarding genetic disruption of ACE, it is impressively consistent that ACE inhibition reduces atherosclerosis in multiple animal models as discussed recently [16].

Since ACE2 was reported by two independent groups in 2000 [37, 38], this protein has been studied in multiple research areas, including renal disorders, cardiac dysfunctions, and vascular diseases. While ACE2 has been identified as a multifunctional protein, one primary role of this monocarboxypeptidase is to convert AngII into Ang(1–7), which has been proposed as the major atheroprotective effect of ACE2. Deficiency of ACE2 augments atherosclerosis in mouse models [20, 39], and overexpression of ACE2 through an adenoviral approach attenuates atherosclerosis in rabbits [40, 41]. Although ACE2 protein expression has also been detected in early and advanced stages of atherosclerotic lesions in humans [42], it is unclear whether the presence of ACE2 is an automatic protective process in the development of atherosclerosis in humans.

As a summary, despite the complexity of the RAS, the three well-established modes of pharmacological inhibition of the RAS have been demonstrated to equivalently ameliorate the development of atherosclerosis [7, 11]. Despite lack of data whether concentrations of the four bioactive angiotensin peptides are changed upon inhibition of renin, ACE, or AT1 receptor, findings from these studies infer that the major beneficial effects of the RAS inhibition is ascribed to inhibition of either AngII synthesis (inhibition of renin or ACE) or AngII action (blockade of AT1 receptor).

CELLULAR AND MOLECULAR MECHANISMS OF RAS INFLUENCING ATHEROSCLEROSIS

The initiation of atherosclerosis involves endothelial dysfunction and monocyte adhesion, followed by monocyte infiltration into subendothelial area and becoming lipid-laden macrophages. In advanced stages, leukocyte-related inflammatory responses and proliferation and migration of smooth muscle cells become active. There is no direct evidence on how the RAS influences these cellular events in the initiation and progression of atherosclerosis. Many components of the RAS have been detected in multiple cell types of atherosclerotic lesions from both human studies and animal models [5, 16]. In addition, there is *in vitro* evidence that both macrophages [6] and T lymphocytes [43] contain the classic components of the RAS and have the ability to synthesize bioactive angiotensin peptides. This section will update the current knowledge regarding the roles of cell-specific manipulation of the RAS components on atherosclerosis and molecules that influence atherosclerosis via activation of AngII.

Cell-Specific Manipulations of RAS Components on Atherosclerosis

Macrophages are the predominant infiltrating cell type in atherosclerotic lesions. Many of the RAS components are present in this inflammatory cell type [5, 6, 20]. One widely used approach to determining whether the RAS components in macrophages play a critical role in atherosclerotic formation is bone marrow transplantation. Despite the important role of AT1a receptor in atherosclerosis, there are conflicting findings regarding the contribution of AT1a receptor deficiency in macrophages to atherosclerosis. Studies show that deficiency of renin in bone marrow-derived cells reduces atherosclerosis [6], and deficiency of ACE2 augments atherosclerosis in hypercholesterolemic mice [20], respectively. The findings from these two studies are consistent with the proatherogenic role of AngII and the potential atheroprotective role of Ang(1–7). Currently, no studies have reported contributions of angiotensinogen, ACE, AT1b, and AT2 receptors in macrophages to atherosclerosis.

Although the RAS components have also been detected in T lymphocytes, there is no literature evidence whether the RAS components in this leukocyte type contribute to atherosclerosis. In addition to infiltrating leukocytes, two resident cell types, endothelium and smooth muscle cells, also play critical roles in the development of atherosclerosis. ACE, ACE2, and AT1a receptors are ubiquitously present in these two cell types. Our recent study has demonstrated that deficiency of AT1a receptor in a single cell type (either endothelial or smooth muscle cells) has no effect on the development of atherosclerosis [44].

Molecular Mechanisms of RAS Influence on Atherosclerosis

A spectrum of molecules that might contribute to the development of atherosclerosis have been proposed through *in vitro* experiments. Well-known pathways can be stimulated by AngII–AT1 receptor activation including components in reactive oxygen species (ROS) cascade and nuclear factor-kappaB (NF-κB) signaling pathway, both leading to inflammatory responses. Conversely, blockade of AT1 receptor (action of AngII) or inhibition of ACE (synthesis of AngII) suppresses ROS and NF-κB activations. Using hypercholesterolemic mouse models with chronic AngII infusion (typically 0.5 or 1 μg/kg/min for 4 weeks), many potential molecular mechanisms by which this bioactive angiotensin peptide promotes atherosclerosis have been studied [45].

Molecules that affect macrophage number or functions Although studies of macrophage RAS components in atherosclerosis development have yielded contradictory outcomes, several studies have reported molecules that affect the quantitative and functional importance of macrophages in AngII-accelerated atherosclerosis. Macrophage-colony stimulating factor (M-CSF) plays an essential role in maintaining macrophage number and activity. M-CSF-deficient mice have diminished monocyte number in circulation and macrophage number in tissues. AngII infusion into this mouse model in a hypercholesterolemic background leads to attenuated atherosclerotic lesion formation [46].

Multiple molecules are important for functions of macrophages. Monocyte chemoattractant protein-1 (MCP-1), an important monocyte infiltration molecule, can be activated by AngII stimulation. Transferring a mutated MCP-1 into *Apoe*$^{-/-}$ mice leads to a profound reduction of AngII-induced atherosclerosis [47]. CCR2, the receptor for MCP-1, plays a crucial role in AngII-induced atherosclerosis as demonstrated using CCR2 deficient mice [48, 49] or pharmacological inhibition of CCR2 in *Apoe*$^{-/-}$ mice expressing human CCR2 [50]. There is further evidence that CCR2 in macrophages is sufficient to promote AngII-induced atherosclerosis [49].

Cell signaling pathways that relate to macrophage-mediated immune responses in the development of atherosclerosis have also been studied comprehensively. Myeloid differentiation factor 88 (MyD88) is a universal adapter protein that is used by multiple toll-like receptors (TLRs) in the activation of NF-κB and its consequent inflammatory responses. Hypercholesterolemic mice deficient with MyD88 or its related TLRs (TLR2 and TLR4) show profoundly reductions of AngII-induced atherosclerosis [51]. In addition, deficiency of MyD88 in macrophages is adequate to attenuate AngII-accelerated atherosclerosis [51].

Molecules affecting lymphocyte number or functions In addition to macrophages, T lymphocytes are present in atherosclerotic lesions. Recombinant activating gene-1 (Rag-1) encodes proteins that are essential for B and T cell maturation. Consequently, Rag-1-deficient mice do not have mature B and T lymphocytes. Although there is no apparent gender propensity of atherosclerosis in AngII-infused mice, deficiency of Rag-1 results in reductions of atherosclerosis in male, but not female, hypercholesterolemic mice infused with AngII [52].

Interferon γ plays a critical role in T-helper type 1 immune response. Deficiency of this cytokine reduces atherosclerosis in *Apoe*$^{-/-}$ mice infused with AngII [53]. In agreement with this finding, deficiency of CXCL10, the interferon γ inducible T-cell chemoattractant, also attenuates atherosclerosis in *Apoe*$^{-/-}$ mice infused with AngII [53].

In contrast to roles of interferon γ, interferon β, a comprehensive immune modulator, attenuates AngII-accelerated atherosclerosis in *Apoe*$^{-/-}$ mice achieved via subcutaneous injection of interferon β [54].

Molecules affecting endothelium CD31, the transhomophilic inhibitory immunoreceptor, is expressed constitutively by endothelium and is important to prevent blood leukocyte–endothelium adhesion. Reinforcement of CD31 through subcutaneous infusion of a murine synthetic CD31 peptide reduces atherosclerosis in *Apoe*$^{-/-}$ mice infused with AngII [55] and the reduction of atherosclerosis is associated with suppression of both T lymphocyte and macrophage activation by administration of a mouse CD31 peptide.

Molecules affecting lipid homeostasis Hypercholesterolemia increases plasma AngII concentrations in mice [5]. Conversely, there is *in vitro* evidence that AngII regulates macrophage cholesterol synthesis, the lipid transporter ABCA1, and lectin-like oxidized LDL receptor-1 (LOX-1) as discussed by Lu et al. [16]. However, there is no direct evidence in animal models that activation of the RAS promotes atherosclerosis through interruption of these molecules.

Receptor-associated protein (RAP) regulates many proteins in the LDL receptor family, thereby contributing to lipid metabolism [56]. Deficiency of RAP in *Ldlr*$^{-/-}$ mice fed a saturated fat-enriched diet leads to increased plasma cholesterol concentrations that are attributed to increased VLDL and LDL cholesterol concentrations [56, 57]. Irrespective of the increased cholesterol concentrations, deficiency of RAP attenuates AngII-induced atherosclerosis [56].

Molecules affecting arachidonic acid pathway Molecules that contribute to arachidonic acid metabolism are implicated in atherosclerotic development. Secretory phospholipase A2 catalyzes phospholipids of the arachidonic acid cascade. Deficiency of group X secretory phospholipase A2 attenuates AngII-induced atherosclerosis without changing plasma cholesterol concentrations and lipoprotein distributions [58]. Prostaglandin E2 (PGE2), an end product of the arachidonic acid pathway, inhibits inflammatory cytokines through binding its receptor EP4. Depletion of EP4 in bone marrow-derived cells augments AngII-induced atherosclerosis in hypercholesterolemic *Ldlr*$^{-/-}$ mice, accompanied by pronounced increases of plasma cholesterol concentrations [59].

Molecules affecting oxidative stress There are well-recognized roles of AngII stimulation on activation of ROS. However, as demonstrated in mouse models, deficiency of p47 subunit of NADPH or inhibition of oxidative stress via administration of vitamin E in hypercholesterolemic mice does not affect AngII-induced atherosclerosis [60, 61].

Apelin and its cognate G protein-coupled receptor APJ constitute a signaling pathway that regulates cardiovascular function. Deficiency of apelin augments AngII-induced atherosclerosis in *Apoe*$^{-}$ mice through impairment of nitric oxide production and enhancement of ROS [62].

Proteases contributing to AngII-induced atherosclerosis Proteases, particularly matrix metalloproteinases (MMPs), have been detected in atherosclerotic lesions. Roles of many proteases and extracellular matrix proteins have been studied in hypercholesterolemic mice infused with AngII. Calpain is a calcium-activated neutral cysteine protease. Pharmacological inhibition of this protease attenuates AngII-induced atherosclerosis in hypercholesterolemic mice [63]. Acetylcholinesterase, a serine protease, also plays a critical role in atherosclerosis by demonstrations that donepezil and physostigmine, reversible acetylcholinesterase inhibitors, reduces AngII-induced atherosclerosis in *Apoe*$^{-/-}$ mice [64]. Despite the potential role of MMPs in the development of atherosclerosis, broad inhibition of MMPs by administration of doxycycline does not attenuate AngII-induced atherosclerosis [65, 66].

Osteopontin, an extracellular matrix protein that has pleiotropic functions, is ubiquitously expressed and secreted by many cell types. Its role in atherosclerosis development through activation of inflammation was demonstrated in AngII-infused hypercholesterolemic mice.

Notably, macrophage-specific deficiency of osteopontin is able to reduce AngII-induced atherosclerosis [67].

PPARs influence on AngII-induced atherosclerosis Infusion of AngII into hypercholesterolemic mice activates NF-κB-mediated proinflammatory genes but suppresses peroxisome proliferator-activated receptors (PPARs) α and γ [68]. Absence of PPARα attenuates atherosclerosis in high-fat diet-fed mice that express both human angiotensinogen and renin transgenes [69]. Pioglitazone, a PPARγ agonist, reduces atherosclerosis in hypercholesterolemic mice infused with AngII [70, 71]. Besides PPAR α and γ, activation of PPARδ via administration of its agonist GW0742 attenuates AngII-induced atherosclerosis in hypercholesterolemic mice [72]. These data indicate that PPARs may have differential effects on AngII-induced atherosclerosis.

SUMMARY AND PERSPECTIVE

There are impressively consistent demonstrations that activation of AngII augments atherosclerosis, while inhibition of renin, ACE, or AT1 receptor attenuates atherosclerosis in animal models. There are also promising outcomes that pharmacological inhibition of ACE or AT1 receptor has beneficial effects on human atherosclerotic diseases. Although the proposed mechanisms in the current literature have provided insights into better understanding this chronic disease, targeting an identified molecule such as an inflammatory regulator or a component that contributes to oxidative stress has led to either no beneficial effects or oppositely harmful effects. Conflicting findings between animal models and clinical investigations are a common issue in the research field of atherosclerosis, which may be attributable to both an incomplete understanding of the intricate molecular mechanisms and the differences between well-controlled animal experiments versus the byzantine nature of human diseases. While understanding what have caused the contradicting outcomes is a prerequisite in our comprehension of atherosclerosis, the feedback loop formed between animal experiments and clinical trials will continuously be a powerful approach to enhancing our knowledge of this complex disease and providing mechanistic base in improving current therapeutic strategies.

ACKNOWLEDGMENTS

The authors' research work is supported by the grants from the NIH (PPG and R01 HL80100, HL062846, and HL107319).

REFERENCES

1. Flores-Munoz M, Work LM, Douglas K, et al.: Angiotensin-(1-9) attenuates cardiac fibrosis in the stroke-prone spontaneously hypertensive rat via the angiotensin type 2 receptor. *Hypertension* 2012;59:300–307.
2. Daugherty A, Cassis L: Chronic angiotensin II infusion promotes atherogenesis in low density lipoprotein receptor-/- mice. *Ann NY Acad Sci* 1999;892:108–118.
3. Daugherty A, Manning MW, Cassis LA: Angiotensin II promotes atherosclerotic lesions and aneurysms in apolipoprotein E-deficient mice. *J Clin Invest* 2000;105:1605–1612.
4. Weiss D, Kools JJ, Taylor WR: Angiotensin II-induced hypertension accelerates the development of atherosclerosis in apoE-deficient mice. *Circulation* 2001;103:448–454.
5. Daugherty A, Rateri DL, Lu H, Inagami T, Cassis LA: Hypercholesterolemia stimulates angiotensin peptide synthesis and contributes to atherosclerosis through the AT1A receptor. *Circulation* 2004;110:3849–3857.
6. Lu H, Rateri DL, Feldman DL, et al.: Renin inhibition reduces hypercholesterolemia-induced atherosclerosis in mice. *J Clin Invest* 2008; 118:984–993.
7. Lu H, Balakrishnan A, Howatt DA, et al.: Comparative effects of different modes of renin angiotensin system inhibition on hypercholesterolaemia-induced atherosclerosis. *Br J Pharmacol* 2012;165:2000–2008.

8. Weiss D, Bernstein KE, Fuchs S, Adams J, Synetos A, Taylor WR: Vascular wall ACE is not required for atherogenesis in ApoE(-/-) mice. *Atherosclerosis* 2009;209:352–358.
9. Imanishi T, Tsujioka H, Ikejima H, et al.: Renin inhibitor aliskiren improves impaired nitric oxide bioavailability and protects against atherosclerotic changes. *Hypertension* 2008;52: 563–572.
10. Nussberger J, Aubert JF, Bouzourene K, Pellegrin M, Hayoz D, Mazzolai L: Renin inhibition by aliskiren prevents atherosclerosis progression: comparison with irbesartan, atenolol, and amlodipine. *Hypertension* 2008;51: 1306–1311.
11. Weiss D, Taylor WR: Deoxycorticosterone acetate salt hypertension in apolipoprotein E-/- mice results in accelerated atherosclerosis: the role of angiotensin II. *Hypertension* 2008;51:218–224.
12. Poss J, Werner C, Lorenz D, Gensch C, Bohm M, Laufs U: The renin inhibitor aliskiren upregulates pro-angiogenic cells and reduces atherogenesis in mice. *Basic Res Cardiol* 2010; 105: 725–735.
13. Yusuf S, Sleight P, Pogue J, Bosch J, Davies R, Dagenais G: Effects of an angiotensin-converting-enzyme inhibitor, ramipril, on cardiovascular events in high-risk patients. The Heart Outcomes Prevention Evaluation Study Investigators. *N Engl J Med* 2000;342:145–153.
14. ONTARGET Investigators, Yusuf S, Teo KK, et al.: Telmisartan, ramipril, or both in patients at high risk for vascular events. *N Engl J Med* 2008;358:1547–1559.
15. Verma S, Gupta M, Holmes DT, et al.: Plasma renin activity predicts cardiovascular mortality in the Heart Outcomes Prevention Evaluation (HOPE) study. *Eur Heart J* 2011;32:2135–2142.
16. Lu H, Cassis LA, Daugherty A: The Renin Angiotensin System in Atherosclerosis. *Advances in Atherosclerosis: Treatment and Prevention* 2012; Chapter 8: 215–252. Edited by Christian Weber and Oliver Soëhnlein. Pan Stanford Publishing Pte. Ltd, New York.
17. Lu H, Cassis LA, Daugherty A: Atherosclerosis and arterial blood pressure in mice. *Curr Drug Targets* 2007;8:1181–1189.
18. Lu H, Wu C, Howatt DA, et al.: Differential effects of dietary sodium intake on blood pressure and atherosclerosis in hypercholesterolemic mice. *J Nutr Biochem* 2012; 24(1):49–53.
19. Tesanovic S, Vinh A, Gaspari TA, Casley D, Widdop RE: Vasoprotective and atheroprotective effects of angiotensin (1-7) in apolipoprotein E-deficient mice. *Arterioscler Thromb Vasc Biol* 2010;30:1606–1613.
20. Thatcher SE, Zhang X, Howatt DA, et al.: Angiotensin-converting enzyme 2 deficiency in whole body or bone marrow-derived cells increases atherosclerosis in low-density lipoprotein receptor-/- mice. *Arterioscler Thromb Vasc Biol* 2011;31:758–765.
21. Vinh A, Widdop RE, Drummond GR, Gaspari TA: Chronic angiotensin IV treatment reverses endothelial dysfunction in ApoE-deficient mice. *Cardiovasc Res* 2008;77:178–187.
22. Poduri A, Owens III AP, Howatt DA, et al.: Regional variation in aortic AT1b receptor mRNA abundance is associated with contractility but unrelated to atherosclerosis and aortic aneurysms. *PLoS One* 2012;7(10):e48462.
23. Wassmann S, Czech T, Van Eickels M, Fleming I, Bohm M, Nickenig G: Inhibition of diet-induced atherosclerosis and endothelial dysfunction in apolipoprotein E/Angiotensin II type 1A receptor double-knockout mice. *Circulation* 2004;110:3062–3067.
24. Cassis LA , Rateri DL, Lu H, Daugherty A: Bone marrow transplantation reveals that recipient AT1a receptors are required to initiate angiotensin II-induced atherosclerosis and aneurysms. *Arterioscler Thromb Vasc Biol* 2007;27:380–386.
25. Tiyerili V, Mueller CF, Becher UM, et al.: Stimulation of the AT2 receptor reduced atherogenesis in ApoE(-/-)/AT1A(-/-) double knockout mice. *J Mol Cell Cardiol* 2012;52:630–637.
26. Koitka A, Cao Z, Koh P, et al.: Angiotensin II subtype 2 receptor blockade and deficiency attenuate the development of atherosclerosis in an apolipoprotein E-deficient mouse model of diabetes. *Diabetologia* 2010;53:584–592.
27. Iwai M, Chen R, Li Z, et al.: Deletion of angiotensin II type 2 receptor exaggerated atherosclerosis in apolipoprotein E-null mice. *Circulation* 2005;112:1636–1643.
28. Sales VL, Sukhova GK, Lopez-Ilasaca MA, Libby P, Dzau VJ, Pratt RE: Angiotensin type 2 receptor is expressed in murine atherosclerotic lesions and modulates lesion evolution. *Circulation* 2005;112:3328–3336.
29. Hu C, Dandapat A, Chen J, et al.: Over-expression of angiotensin II type 2 receptor

(agtr2) reduces atherogenesis and modulates LOX-1, endothelial nitric oxide synthase and heme-oxygenase-1 expression. *Atherosclerosis* 2008;199:288–294.

30. Daugherty A, Manning MW, Cassis LA: Antagonism of AT2 receptors augments angiotensin II-induced abdominal aortic aneurysms and atherosclerosis. *Br J Pharmacol* 2001;134:865–870.
31. Toton-Zuranska J, Gajda M, Pyka-Fosciak G, et al.: AVE 0991-angiotensin-(1-7) receptor agonist, inhibits atherogenesis in apoE-knockout mice. *J Physiol Pharmacol* 2010;61:181–183.
32. Jawien J, Toton-Zuranska J, Gajda M, et al.: Angiotensin-(1-7) receptor Mas agonist ameliorates progress of atherosclerosis in apoE-knockout mice. *J Physiol Pharmacol* 2012;63:77–85.
33. Sugiyama F, Haraoka S, Watanabe T, et al. Acceleration of atherosclerotic lesions in transgenic mice with hypertension by the activated renin-angiotensin system. *Lab Invest* 1997; 76:835–842.
34. Kuhnast S , van der Hoorn JW, van den Hoek AM, et al.: Aliskiren inhibits atherosclerosis development and improves plaque stability in APOE*3Leiden.CETP transgenic mice with or without treatment with atorvastatin. *J Hypertens* 2012;30:107–116.
35. Krege JH, Moyer JS, Langenbach LL, et al.: Angiotensin-converting enzyme gene and atherosclerosis. *Arterioscler Thromb Vasc Biol* 1997;17:1245–1250.
36. Hayek T, Pavlotzky E, Hamoud S, et al.: Tissue angiotensin-converting-enzyme (ACE) deficiency leads to a reduction in oxidative stress and in atherosclerosis. Studies in ACE-knockout mice type 2. *Arterioscler Thromb Vasc Biol* 2003;23:2090–2096.
37. Tipnis SR , Hooper NM, Hyde R, Karran E, Christie G, Turner AJ: A human homolog of angiotensin-converting enzyme. Cloning and functional expression as a captopril-insensitive carboxypeptidase. *J Biol Chem* 2000;275: 33238–33243.
38. Donoghue M, Hsieh F, Baronas E, et al.: A novel angiotensin-converting enzyme-related carboxypeptidase (ACE2) converts angiotensin I to angiotensin 1-9. *Circ Res* 2000;87:E1–E9.
39. Thomas MC, Pickering RJ, Tsorotes D, et al.: Genetic Ace2 deficiency accentuates vascular inflammation and atherosclerosis in the apoE knockout mouse. *Circ Res* 2010;107:888–897.
40. Zhang C, Zhao YX, Zhang YH, et al.: Angiotensin-converting enzyme 2 attenuates atherosclerotic lesions by targeting vascular cells. *Proc Natl Acad Sci U S A* 2010;107: 15886–15891.
41. Dong B, Zhang C, Feng JB, et al.: Overexpression of ACE2 enhances plaque stability in a rabbit model of atherosclerosis. *Arterioscler Thromb Vasc Biol* 2008;28: 1270–1276.
42. Sluimer JC, Gasc JM, Hamming I, et al.: Angiotensin-converting enzyme 2 (ACE2) expression and activity in human carotid atherosclerotic lesions. *J Pathol* 2008;215: 273–279.
43. Hoch NE, Guzik TJ, Chen W, et al.: Regulation of T-cell function by endogenously produced angiotensin II. *Am J Physiol Regul Integr Comp Physiol* 2009;296:R208–R216.
44. Rateri DL, Moorleghen JJ, Knight V, et al.: Depletion of endothelial or smooth muscle cell-specific angiotensin II type 1a receptors does not influence aortic aneurysms and atherosclerosis in LDL receptor deficient mice. *PLoS One* 2012;7:e51483.
45. Daugherty A, Lu H, Rateri DL, Cassis LA: Augmentation of the renin-angiotensin system by hypercholesterolemia promotes vascular diseases. *Future Lipidol* 2008;3: 625–636.
46. Babamusta F, Rateri DL, Moorleghen JJ, Howatt DA, Li XA, Daugherty A: Angiotensin II infusion induces site-specific intra-laminar hemorrhage in macrophage colony-stimulating factor deficient mice. *Atherosclerosis* 2005; 186:282–290.
47. Ni W, Kitamoto S, Ishibashi M, et al.: Monocyte chemoattractant protein-1 Is an essential inflammatory mediator in angiotensin II-induced progression of established atherosclerosis in hypercholesterolemic mice. *Arterioscler Thromb Vasc Biol* 2004;24:534–539.
48. Daugherty A, Rateri DL, Charo IF, Owens AP, Howatt DA, Cassis LA: Angiotensin II infusion promotes ascending aortic aneurysms: attenuation by CCR2 deficiency in apoE-/- mice. *Clin Sci (Lond)* 2010;118:681–689.
49. Ishibashi M, Egashira K, Zhao Q, et al.: Bone marrow-derived monocyte chemoattractant protein-1 receptor CCR2 is critical in angiotensin II-induced acceleration of atherosclerosis and aneurysm formation in hypercholesterolemic

mice. *Arterioscler Thromb Vasc Biol* 2004; 24:174–178.

50. Olzinski AR, Turner GH, Bernard RE, et al.: Pharmacological inhibition of C-C chemokine receptor 2 decreases macrophage infiltration in the aortic root of the human C-C chemokine receptor 2/apolipoprotein E-/- mouse: magnetic resonance imaging assessment. *Arterioscler Thromb Vasc Biol* 2010;30:253–259.
51. Owens AP 3rd, Rateri DL, Howatt DA, et al.: MyD88 deficiency attenuates angiotensin II-induced abdominal aortic aneurysm formation independent of signaling through toll-like receptors 2 and 4. *Arterioscler Thromb Vasc Biol* 2011;31:2813–2819.
52. Uchida HA, Kristo F, Rateri DL, et al.: Total lymphocyte deficiency attenuates AngII-induced atherosclerosis in males but not abdominal aortic aneurysms in apoE deficient mice. *Atherosclerosis* 2010;211:399–403.
53. King VL, Lin AY, Kristo F, et al.: Interferon-gamma and the interferon-inducible chemokine CXCL10 protect against aneurysm formation and rupture. *Circulation* 2009;119:426–435.
54. Zhang LN, Velichko S, Vincelette J, et al.: Interferon-beta attenuates angiotensin II-accelerated atherosclerosis and vascular remodeling in apolipoprotein E deficient mice. *Atherosclerosis* 2007;197:204–211.
55. Fornasa G, Clement M, Groyer E, et al.: A CD31-derived peptide prevents angiotensin II-induced atherosclerosis progression and aneurysm formation. *Cardiovasc Res* 2012;94: 30–37.
56. Wang S, Subramanian V, Lu H, et al.: Deficiency of receptor-associated protein attenuates angiotensin II-induced atherosclerosis in hypercholesterolemic mice without influencing abdominal aortic aneurysms. *Atherosclerosis* 2012;220:375–380.
57. Willnow TE, Armstrong SA, Hammer RE, Herz J: Functional expression of low density lipoprotein receptor-related protein is controlled by receptor-associated protein in vivo. *Proc Natl Acad Sci USA* 1995;92:4537–4541.
58. Zack M, Boyanovsky BB, Shridas P, et al.: Group X secretory phospholipase A(2) augments angiotensin II-induced inflammatory responses and abdominal aortic aneurysm formation in apoE-deficient mice. *Atherosclerosis* 2011;214:58–64.
59. Tang EH, Shvartz E, Shimizu K, et al.: Deletion of EP4 on bone marrow-derived cells enhances inflammation and angiotensin II-induced abdominal aortic aneurysm formation. *Arterioscler Thromb Vasc Biol* 2011;31:261–269.
60. Thomas M, Gavrila D, McCormick ML, et al.: Deletion of p47phox attenuates angiotensin II-induced abdominal aortic aneurysm formation in apolipoprotein E-deficient mice. *Circulation* 2006;114:404–413.
61. Gavrila D, Li WG, McCormick ML, et al.: Vitamin E inhibits abdominal aortic aneurysm formation in angiotensin II-infused apolipoprotein E-deficient mice. *Arterioscler Thromb Vasc Biol* 2005;25:1671–1677.
62. Chun HJ, Ali ZA, Kojima Y, et al.: Apelin signaling antagonizes Ang II effects in mouse models of atherosclerosis. *J Clin Invest* 2008; 118:3343–3354.
63. Subramanian V, Uchida HA, Ijaz T, Moorleghen JJ, Howatt DA, Balakrishnan A: Calpain inhibition attenuates angiotensin II-induced abdominal aortic aneurysms and atherosclerosis in low-density lipoprotein receptor-deficient mice. *J Cardiovasc Pharmacol* 2012;59:66–76.
64. Inanaga K , Ichiki T, Miyazaki R, et al.: Acetylcholinesterase inhibitors attenuate atherogenesis in apolipoprotein E-knockout mice. *Atherosclerosis* 2010;213:52–58.
65. Manning MW, Cassis LA, Daugherty A: Differential effects of doxycycline, a broad-spectrum matrix metalloproteinase inhibitor, on angiotensin II-induced atherosclerosis and abdominal aortic aneurysms. *Arterioscler Thromb Vasc Biol* 2003;23:483-488.
66. Xie X, Lu H, Moorleghen JJ, et al.: Doxycycline does not influence established abdominal aortic aneurysms in angiotensin ii-infused mice. *PLoS One* 2012;7:e46411.
67. Bruemmer D, Collins AR, Noh G, et al.: Angiotensin II-accelerated atherosclerosis and aneurysm formation is attenuated in osteopontin-deficient mice. *J Clin Invest* 2003;112:1318–1331.
68. Tham DM, Martin McNulty B, Wang YX, et al.: Angiotensin II is associated with activation of NF-kappa B-mediated genes and downregulation of PPARs. *Physiol Genomics* 2002;11:21–30.

69. Tordjman KM, Semenkovich CF, Coleman T, et al.: Absence of peroxisome proliferator-activated receptor-alpha abolishes hypertension and attenuates atherosclerosis in the Tsukuba hypertensive mouse. *Hypertension* 2007;50: 945–951.
70. Subramanian V, Golledge J, Ijaz T, Bruemmer D, Daugherty A: Pioglitazone-induced reductions in atherosclerosis occur via smooth muscle cell-specific interaction with PPAR{gamma}. *Circ Res* 2010;107:953–958.
71. Golledge J, Cullen B, Rush C, et al.: Peroxisome proliferator-activated receptor ligands reduce aortic dilatation in a mouse model of aortic aneurysm. *Atherosclerosis* 2010;210:51–56.
72. Takata Y, Liu J, Yin F, et al.: PPARdelta-mediated antiinflammatory mechanisms inhibit angiotensin II-accelerated atherosclerosis. *Proc Natl Acad Sci U S A* 2008;105: 4277–4282.

18

THE ROLE OF CAVEOLAE AND CAVEOLINS IN ATHEROGENESIS

VICTOR RIZZO

Department of Anatomy and Cell Biology, Cardiovascular Research Center, Temple University School of Medicine, Philadelphia, PA, USA

INTRODUCTION

Atherosclerosis is characterized by chronic damage to the endothelial lining with the formation of lipid deposits and smooth muscle proliferation within the vessel wall. While significant work has been done to understand the role of atherosclerosis in cardiovascular disease, the process for initiation and development of atherosclerotic plaques remains unclear. Several risk factors show strong correlation with the development of atherosclerotic lesions including increased plasma lipids, hypertension, smoking, diabetes, and obesity. Interestingly, the localization of plaques to regions of arterial bifurcations where fluid shear stress is lower on average and turbulent indicates that local hemodynamics also affects disease progression [1]. Recent evidence suggests that risk factors and shear stress influence endothelial cell phenotype by activating second-messenger signaling events in caveolae organelles. In this chapter, we discuss the role of endothelial cell caveolae in atherogenesis. In addition, we briefly consider the function of caveolae in other cell types that contribute to atheroma formation.

CAVEOLAE AND CAVEOLINS

Caveolae were first described as morphologically distinct subcellular organelles of the plasma membrane, characteristically observed as omega-shaped invaginations of 50–100 nm in size [2]. Caveolae comprise proteins and lipids that are organized in a cohesive structure. The lipid composition consists of cholesterol and sphingolipids, which together provide the liquid-ordered phase critical for caveolae shape [3]. The overall structural stability of caveolae is also dependent on the presence of coat proteins termed "caveolins" [4]. Caveolins bind directly to cholesterol and, in turn, cholesterol binding stabilizes the formation of caveolinoligomeric complexes that drive formation

Atherosclerosis: Risks, Mechanisms, and Therapies, First Edition. Edited by Hong Wang and Cam Patterson.

of the invaginated shape [5]. More recently, an additional set of caveolae regulatory proteins have been described—the cavins [6]. They play a role in organizing caveoale morphology as well as in signaling and vesicle budding from the plasma membrane.

The caveolin gene family is a small group of integral membrane proteins 21–25kDa in size. The family consists of three distinct isoforms: caveolin-1, -2, and -3, with caveolin-1 displaying an alternate initiation site during translation to generate splice variants caveolin-1 [α (full length) and β (amino acid 1–31 truncation)]. Caveolin-1 and -2 are expressed ubiquitously, while caveolin-3 appears to be muscle specific [7]. Structurally, caveolins have several key domains critical for their localization and function. The caveolin proteins have a central 33 amino acid hydrophobic region that interdigitates as a hairpin loop into lipid membranes, while the N- and C-terminal regions remain cytosolic where they mediate homo- and hetero-oligomeric complex formation. These polymeric complexes comprise the assembly units of the striated coat observed on the cytoplasmic face of the caveolae structure [8]. Both caveolin-1 and -3 are palmitylated on three cystienes in the C-terminal region. This post-translational modification stabilizes the caveolin oligomers and promotes caveolin membrane association [9]. Caveolin-1 and -3 contain a critical regulatory domain termed the "scaffolding domain" ranging from amino acids 82–101 adjacent to the hydrophobic domain [10]. This region binds an array of signaling molecules that contain the scaffolding-domain-binding sequences ΦXΦXXXXΦ, ΦXXXXΦXXΦ or ΦXΦXXXXΦXXΦ where Φ represents a phenolic amino acid residue. Finally, caveolin-1 contains a phosphorylation site at tyrosine (Y) 14 hypothesized to function as an SH2-binding site [11].

CAVEOLAE/CAVEOLIN-1 IN ATHEROSCLEROSIS

The first indications that caveolae/caveolin-1 are involved in atherosclerosis came from studies that examined the distribution and expression of caveolin-1 in diseased human and rabbit blood vessels. These studies showed that caveolin-1 expression levels were significantly elevated in arteries obtained from patients undergoing coronary artery bypass surgery and rabbits fed a high-cholesterol diet [12]. While cause and effect could not be established in these studies, it was speculated that enhanced caveolin-1 levels inhibited the function of endothelial nitric oxide synthase (eNOS), thereby limiting the antiatherogenic effects of nitric oxide (NO).

Direct evidence that caveolae/caveolin-1 regulate processes that lead to plaque formation came from studies performed in genetically modified mice. In work reported by Lisanti and coworkers, apoE-deficient mice failed to develop significant plaques when cross bred with caveolin-1 knockout mice despite having a proatherogeneic lipid profile [13]. The reduction in plaque burden was attributed to a decrease in the caveolae-mediated transcytosis of oxidized lipids. These observations were later confirmed and by Sessa's group using an endothelial specific re-expression system for caveolin-1/caveolae [14]. With this model, they were able to re-establish lesion formation validating the importance of the endothelium and their caveolae in atherogenesis. In a follow-up study, plaques were evaluated in transgenic mice that overexpressed caveolin-1, specifically in endothelial cells. They found enhanced formation and progression of atherosclerosis in these mice [15], again supporting the concept that caveolin-1 regulates athroma development. As we discuss further, regulation of NO production, leukocyte adhesion molecule expression, mechanotransduction, oxidative stress, and LDL transport provide some mechanistic insight into the role of caveolae/caveolin-1 in atherosclerosis.

eNOS

In the vasculature, NO produced by eNOS is best recognized for its role in vasorelaxation. Moreover, many studies demonstrate that NO is generally protective against pathology. This is primarily achieved through its anti-inflammatory, antithrombogenic and anti-proliferative actions on various vascular cell types. Thus, the loss of NO bioavailability has become synonymous with endothelial, and by extension, vascular dysfunction.

In relationship to atherosclerosis, impaired NO-mediated vasotone regulation is present prior to frank disease. Perhaps, the most compelling evidence for the protective nature of NO against atherosclerosis comes from genetic approaches where ApoE/eNOS double knockout mice display markedly enhanced atherosclerosis compared with ApoE knockout mice [16]. While eNOS is regulated by a large cadre of cofactors, allosteric modulators and post-translation modifications, the association of caveolae and caveolin-1 with eNOS is considered to be fundamental in regulating enzyme activity and NO production.

When appropriately lipid acylated, eNOS targets to the cholesterol- and sphingolipid-rich membrane of caveolae [17]. Once compartmentalized in caveolae, eNOS binds to caveolin-1 via a consensus binding site (AA350-358) directed against the caveolin-1 scaffolding domain (CDS; AA82-101) [18]. Similar to other molecules that interact with caveolin-1 through the CDS, eNOS is rendered functionally inactive by this association and must dissociate from caveolin-1 to properly produce NO [19]. Given that receptors, calcium entry channels, and several upstream modulators important for eNOS activation also reside in caveolae, these microdomains allow for rapid and efficient production of NO. Thus, any alteration in eNOS homing to caveolae or interaction with caveolin-1 would be expected to have profound effects on enzyme function. Indeed, eNOS mistargeting studies demonstrate that the enzymes subcellular localization is crucial for achieving desirable NO effects [20].

Several studies indicate that atherogenic risk factors can impact vascular function by altering eNOS localization in caveolae or eNOS/caveolin-1 interactions. Following exposure to oxLDL, eNOS translocates from plasmalemmal caveolae to internal membrane compartments, resulting in impaired NO release from the endothelium [21]. Mechanistically, oxLDL has the capacity to sequester membrane cholesterol, thereby disrupting caveolae domains and dislodging molecules housed within [22]. In addition, LDL can enhance the expression of caveolin-1 providing a greater opportunity for caveolin-1 to bind and inhibit eNOS [23]. Likewise, endothelial cells exposed to serum from hypercholesterolemic patients promotes caveolin-1/eNOS association and limits NO release [24]. Conversely, statins [25] and HDL [23] have been reported to improve eNOS function by reducing caveolin-1 levels in the endothelium. Last, eNOS activity and NO production is significantly elevated in Cav-$1^{-/-}$ mice [14], while the caveolin-1 transgenic overexpressing mouse generates very little NO [15]. These findings support the concept that caveolae/caveolin-1 regulation of eNOS/NO function provides a mechanistic basis for the observed atheroprotective and atherogenic tendencies of the respective caveolin-1 mouse models.

TRANSCYTOSIS/PERMEABILITY

Caveolae play a primary role in transcytosis of macromolecules from lumen to the interstitial space and in regulation of endothelial barrier functions. Both processes are integral to plaque formation. It is well established that caveolae serve as transport vesicles to shuttle albumin, insulin, LDL, and other elements across the endothelium [26]. For LDL, the current data suggest that both native and oxidized lipoprotein can enter caveolae via fluid-phase uptake or by binding to cognate receptors present in caveolae, such as CD36, SR-BI, or the LDL receptor [27]. Following entry into caveolae, signaling mechanisms, most likely involving dynamin2 activation and Src-mediated phosphorylation events, induce caveolae budding and trafficking across the endothelium. LDL is then released into the subendothelial cell space where they can be engulfed by macrophages initiating the development of foam cells and ensuing atherosclerotic lesion formation. The observation that lipids do not accumulate in the arterial wall of caveolin-1 knockout mice partially validates this concept [13].

An alternative pathway for LDL transit is via the paracellular route. While the mechanism for caveolae/caveolin-1 regulation of LDL flux through endothelial cell junctions is not completely clear, studies in brain microvascular endothelium indicate that tight junctional complex proteins, such as occludin and claudin-5, rapidly internalized via a caveolae-mediated endocytosis mechanism in response to inflammatory mediators. The translocation of these proteins away from the cell–cell junction results in a loss of junctional stability and enhanced vascular permeability [28]. In addition, loss of junctional integrity has also been observed in lung microvascular endothelial cells depleted of caveolin-1 [29]. Interestingly, junctional morphology and enhanced permeability could be reversed by the nitric oxide synthase inhibitor, L-NAME, suggesting the elevated levels of NO exerts proinflammatory effects in the microcirculation. It seems unlikely however, that similar mechanisms are at play in the macrocirculation given the high amount of NO and resistance to plaque formation observed in the caveolin-1 knockout mouse. Moreover, vessels form the endothelial specific caveolin-1 overexpressing mouse show enhanced lipid deposition with a concomitant decrease in NO production [15]. Overall, the existing evidence strongly suggests that key athrogenic factors such as oxLDL require caveolae/caveolin-1 for access to the subendothelial intima where they activate disease-inducing processes.

MECHANOTRANSDUCTION

While risk factors for atherosclerosis circulate through the entire vascular network, plaque commonly forms at arterial curvatures, branches, and bifurcations. Within these regions, blood flow transitions from a uniform pattern to one that is turbulent. These types of flow generate specific fluid mechanical forces. Of particular importance is the shear stress component of hemodynamic blood flow. Within straight vascular segments, normal blood flow produces an average shear stress of 20 dynes/cm^2. However, at sites prone to atheroma development, average shear stress values are very low due to an oscillating pattern of blood in the forward and then reverse direction. As a result, steep spatiotemporal gradients of shear stress are established in these regions.

We now know that shear stress plays a significant role in regulating endothelial cell structure and function. Normal, unidirectional laminar shear stress actively induces and maintains the shape of the endothelium with the cells long axis oriented parallel to the direction of flow. In contrast, endothelial cells take on a polygonal shaped when subjected to oscillatory flow. In addition, transcriptome analysis of endothelial cells isolated from undisturbed and disturbed flow regions of the aorta revealed a differential pattern in gene expression [30]. While cells from both regions expressed antioxidant and anticoagulative genes, proinflammatory and procoagulant genes were largely expressed in cells residing in the area exposed to disturbed flow. These findings suggest that disturbed flow induces an athero-susceptible genotype and phenotype.

A key question in regards to flow effects on the endothelium is how the cells detect particular patterns of shear stress and then transduce a mechanical force into biologically relevant pathways. To date, several mechanosensitive proteins have been described in the endothelium. They include structural components such as the glycocalyx [31], primary cilia [32], the cytoskeleton [33], receptors such as VEGFR2 [34], Tie-2 [35], P2X4 [36], and Bradykinin B2 [37] and adhesive proteins including integrins [38] and PECAM-1 [39]. Our earlier work has identified caveolae as an important site for mechanotransduction [40, 41]. Many second-messenger signaling molecules that reside within caveolae have been implicated in rapid endothelial cell response to flow [40–43]. For instance, Src-family tyrosine kinases have been shown to be enriched in caveolae and copurify with caveolin-1 [10].

Increasing flow and pressure *in situ* stimulated protein tyrosine phosphorylation at the luminal endothelial plasma membrane, specifically within caveolae implying a role for these signaling molecules in mechanotransduction [41]. In addition, Ca^{2+}-ATPase, IP_3-activated channels, which may mediate flow-induced calcium fluxes and TRVP channels are enriched in endothelial caveolae [44]. G-proteins, which modulate shear-sensitive K^+ channels and PKC generation of IP_3 and DAG in mechanically stimulated endothelial cells, are also present in caveolae [45, 46]. As mentioned earlier, caveolae are enriched in eNOS, an enzyme strongly stimulated by shear stress to produce NO resulting in vasoregulatory and cellular signaling events [42]. In addition, several of the purported mechanosensory elements listed earlier associate with caveolae. Of these, we recently described a novel mechanosignaling complex comprising caveolae, its structural protein caveolin-1, and the β1 integrin subunit in regulation of endothelial morphology induced by shear stress [47]. Finally, the general concept that caveolae play a direct role in mechanotransduction has been validated in the caveolin-1 knockout mouse [48]. Here, vessels lacking caveolin-1 were unable to mount a vasodilatory response to increases in flow. Reconstituting caveolin-1 specifically in the endothelium re-established appropriated flow-mediated responses.

While caveolae are important for mechanotransduction, it is currently unclear how this function relates to atherosclerosis. This is primarily due to a paucity of *in vivo* data that examine the expression level and subcellular positioning of caveolae in endothelium exposed to atheroprotective and atheroprone types of flow. In cultured endothelial cells, caveolae localize to several cellular regions such as the plasma membrane and Golgi apparatus. Upon exposure to laminar shear stress, much of the Golgi localized caveolae translocate to the plasma membrane [40, 49]. This results in an enrichment of caveolae residing molecules such as eNOS on the cell surface where it is better positioned to react to further changes in flow. If similar events occur *in vivo* then flow-induced alteration of caveolae/caveolin-1 expression and localization would be expected to profoundly affect endothelial cell function.

INFLAMMATION AND SIGNALING

There exists a large body of literature that demonstrates that inflammatory mediators such as TNFα, IL1B, and AngII contribute to endothelial dysfunction and associated atherogenesis [50]. These compounds initiate signaling pathways that induce and endothelial inflammatory phenotype. Mechanistically, they activate the NFkB transcription factor resulting in enhanced expression of adhesion molecules on the endothelial cell surface. Consequently, immune cells can bind to and eventually migrate across the endothelial barrier.

In addition to mechanotransduction, caveolae function in a broad range of cell signaling including receptor-mediated events [51]. Thus, caveolae are viewed as platforms that coordinate signaling transduction by compartmentalizing receptors and their downstream second messenger. This concept has been extensively validated in experiments where disruption or loss of caveolae microdomains leads to impairment of normal receptor activation, signal propagation, and functional responses in the cell. Besides receptors for LDL and elements involved in mechanotransduction, TNFα [52], IL-1β [53], and AngII [54] receptors are enriched in caveolae. Upon activation by ligand, cofactors necessary for signal enhancement are recruited to these plasma membrane domains. In caveolin-1-deficient endothelial cell, signal transduction initiated by these inflammatory mediators is significantly impaired as evidence by a marked attenuation of NFκB activation [53] and downregulation of adhesion molecules (Figure 18.1). What more, caveole-mediated proinflammatory signaling is also a feature of lipopolysaccharide induced sepsis in the lung vasculature [55]. While direct analysis of caveolae in signaling events

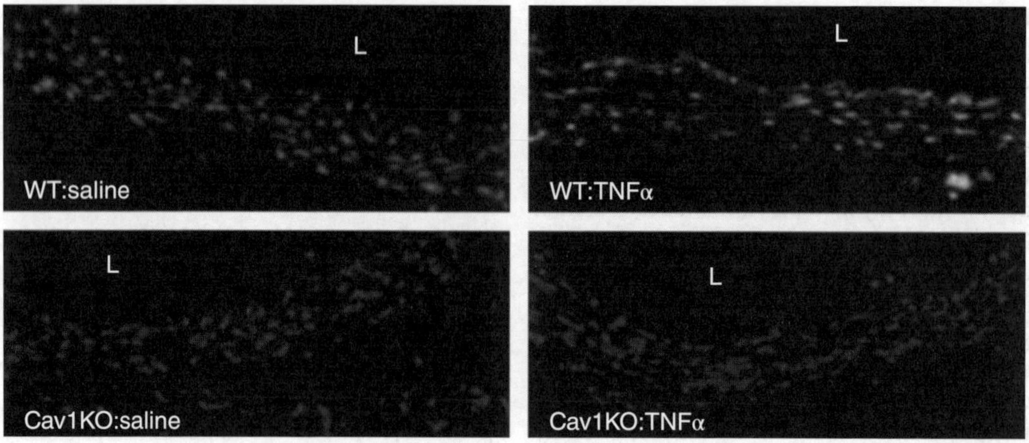

Figure 18.1 Caveolae are necessary for upregulation of adhesion molecules expression in endothelial cells exposed to proinflammatory cytokines. WT and Cav1KO mice were intravenously injected with a bolus of saline or TNFα (20 μg/kg). After 2 hours, aortic tissue was fixed with paraformaldehyde, embedded in paraffin wax and sectioned at 5 μm. Tissue sections were processed for labeling of VCAM (*red*) and cell nuclei (Dapi; *blue*). The data show that TNFα enhanced VCAM expression only in vessels from wild-type mice. L indicates lumen. (*See insert for color representation of the figure.*)

induced by specific inflammatory stimuli have not been examined in a setting of atherogenesis, observations showing a strong correlation between the expression level of adhesion molecules and that of caveolae/caveolin-1 imply that caveolae play a general role in proinflammatory and proatherogenic signaling in the endothelium.

OXIDATIVE STRESS

Many atherogenic risk factors enhance the production of reactive oxygen species (ROS) that damage the vessel wall. Interestingly, experimental evidence also indicates that normal endothelial cell functions may require some degree of ROS production within the cell [56]. This apparent paradox may be explained by the source generator, type, and quantity of the ROS produced in the cell. In addition, the subcellular positioning of oxidant-producing enzyme systems may provide a mechanism that allows products to access local targets and mediate redox signaling events.

A relationship between caveolae and oxidants is slowly emerging. Early reports showed that subjecting endothelial cells to exogenous radicals induce protein tyrosine phosphorylation with caveolin-1 as a phosphorylated target. More recent studies have begun to elucidate the significance of caveolin-1 phosphorylation under oxidative stress [57]. In transformed epithelial cells, oxidant-induced caveolin-1 phosphorylation was responsible for delayed epidermal growth factor receptor (EGF-R) internalization and shifted the internalization pathway from clathrin-coated pits to caveolae that targeted the receptor to a perinuclear location where it was not degraded but remained active.

Prolonged treatment of endothelial cells (24 hours) with superoxide-generating compounds was shown to have profound effects on caveolae number and proper subcellular targeting of eNOS [58], suggesting that chronic oxidative stress can modulates endothelial function by regulating caveolae formation and eNOS–caveolin-1 interactions. Recently, it was reported that endothelial cell cultures respond to acute oxidative stress challenge by activating several signaling pathways [59]. We provided evidence that membrane caveolae are important regions for propagation of compensatory survival pathways activated in response to damaging levels of ROS. Our observations suggest that caveolae expression may be an important mechanism utilized by endothelial cells to adjust their sensitivity to external oxidants.

Indications that oxidant generation by NADPH oxidase may be localized to particular membrane subdomains were first suggested in a study showing that transactivation of the EGFR by angiotensin II (Ang II) stimulation of the angiotensin 1 receptor (AT_1), a process that is dependent on NADPH oxidase-derived ROS, was attenuated in vascular smooth muscle cells (VSMCs) pretreated with cholesterol sequestering compounds to disrupt raft and caveolae compartments [60]. Direct evidence that NADPH oxidase has a heterogeneous distribution within the plasma membrane came from a study showing that NADPH oxidase subunits were localized to lipid raft domains of the neutrophil plasma membrane [61]. In VSMCs, Nox1 and Nox4 were found to have distinct subcellular locations with Nox1 localizing to caveolin-1-enriched plasma membrane domains and Nox4 associating with focal adhesions [62].

Until our studies [43], the few accounts of NADPH oxidase distribution within endothelial cells showed a very different picture than that described in other vascular cell types. In cultured endothelial cells, NADPH oxidase subunits were observed in the vicinity of the endoplasmic reticulum [63]. In a separate study, NADPH oxidase components and superoxide generation was found predominately within a TritonX-100 insoluble cell fraction [64]. Based on this finding, it was concluded that functional NADPH oxidase enzyme complexes are associated with the cytoskeleton in endothelial cell. While a detergent insoluble cell fraction is most often considered to comprise cytoskeletal elements, Triton-X 100 insolubility is also a primary characteristic of plasma membrane lipid rafts and caveolae. With this realization, we tested whether NADPH oxidase was present and functional in membrane rafts in general and caveolae specifically in endothelial cells. We found that the all major components of NADPH oxidase are present, preassembled, and functional in endothelial caveolae [43]. More importantly, both caveolin-1 knockout mice (Figure 18.2) and cultured endothelial cells lacking caveolae (our unpublished observations) lose the capacity to generate ROS in response to proinflammatory stimuli. These initial findings have been confirmed by others and together with more recent studies link caveolae-mediated ROS production to activation of the NFκB pathway and induction of an endothelial inflammatory phenotype [53].

We have come to understand the consequences of oxidant production including protein and lipid oxidation and sequestration of NO. The result is cellular dysfunction characterized by induction of a proinflammatory phenotype. Given the short half-life and limited diffusion distance of primary ROS, such as superoxide, central questions emerge as to how ROS can confer directionality to signaling events and interact with NO, another short-lived compound. The first indication that subcellular compartmentation may provide a mechanism that facilitates these events came in our prior study where we demonstrated that in addition to activating NADPH oxidase, proatherogenic factors acutely enhanced the activity of eNOS present in the same caveolae microdomain [43]. As a result, reactive nitrogen species (RNS) rapidly formed and nitrated several caveolae-localized proteins. In follow-up studies, we found that protein tyrosine nitration is significantly attenuated in cells pretreated with reagents to 1) disrupt caveolae, 2) inhibit NADPH oxidase and eNOS, and 3) scavenge peroxynitrite (our preliminary observations). Further, TNFα-induced protein tyrosine nitration (Figure 18.3) and VCAM-1 upregulation are essentially blocked in caveolin-1 knockout mice (see Figure 18.1). Based on these findings, we propose that caveolae allow for close spatial proximity of ROS/NO-generating enzyme systems, which results in a favorable environment for efficient reduction of NO bioavailability and the formation of the long-live reactive nitrogen compounds. These radicals

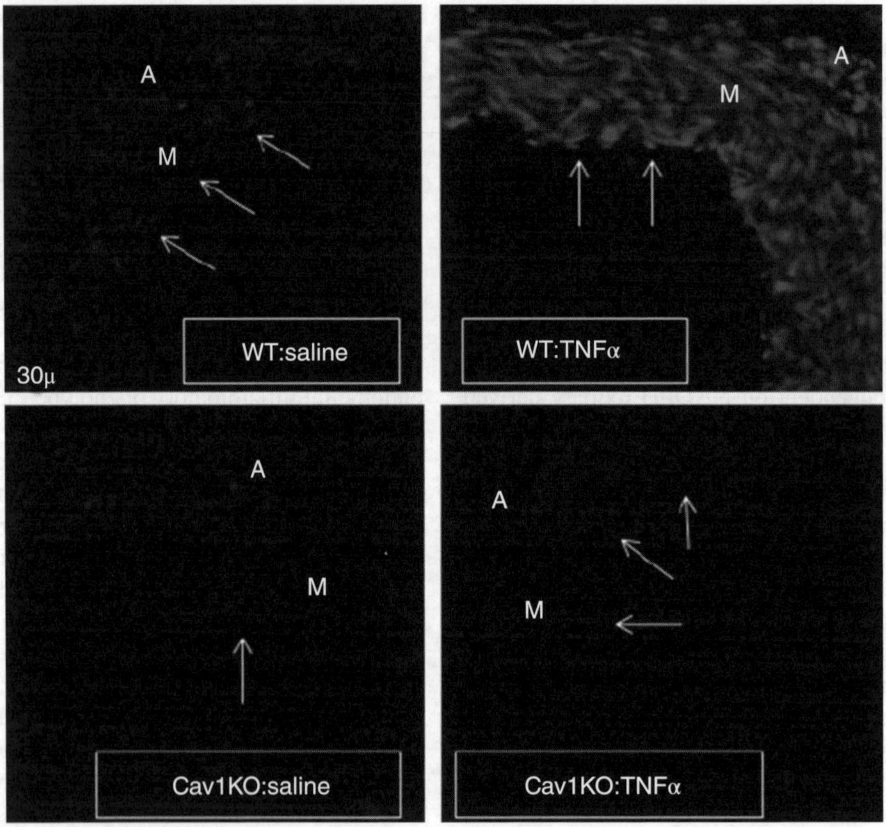

Figure 18.2 Caveolae are necessary for reactive oxygen species production. WT and Cav1KO mice were intravenously injected with a bolus of saline or TNFα. Following 2 hours of exposure, aortic tissue was removed, embedded in OCT, sectioned (30 μm), and incubated with DHE (2 μmol/l). Oxidized DHE was detected by fluorescence microscopy (Nikon TE2000; excitation 488 nm, emission 610 nm). The *arrows* indicate luminal endothelial cell layer, as verified by secondary staining for VE-cadherin (data not shown). There was little evidence of inflammatory leukocyte adherence to the vessel wall (CD 45 staining—data not shown), ruling out leukocytes as a source of ROS at this time point. A, adventitia and M, medial layer. (*See insert for color representation of the figure.)*

initiate redox signaling events that exacerbate ROS production, potentially involving eNOS uncoupling, leading to EC dysfunction. Such events would be expected to play a central role in the atherogenic process (Figure 18.4).

VSMCS, MACS, AND LYMPHOCYTES

While we focused our discussion on endothelial cell caveolae and their potential impact on atherosclerosis, it should be noted that caveolin-1 and caveolae are expressed in other cell types that participate in plaque formation. VSMCs contain a large population of caveolae and express all three isoforms of the caveolin family of proteins. Investigation in this cell type has revealed a variety of important functions regulated by caveolae/caveolins. As in the endothelium, VSMC caveolae contain calcium-regulatory proteins, such as L-type $Ca2^+$ channels, that are involved in initiation of $Ca2^+$ sparks and waves. Depletion of caveolae microdomains, either through pharmacological extraction of membrane cholesterol or knockout of the caveolin-1 protein, perturbs

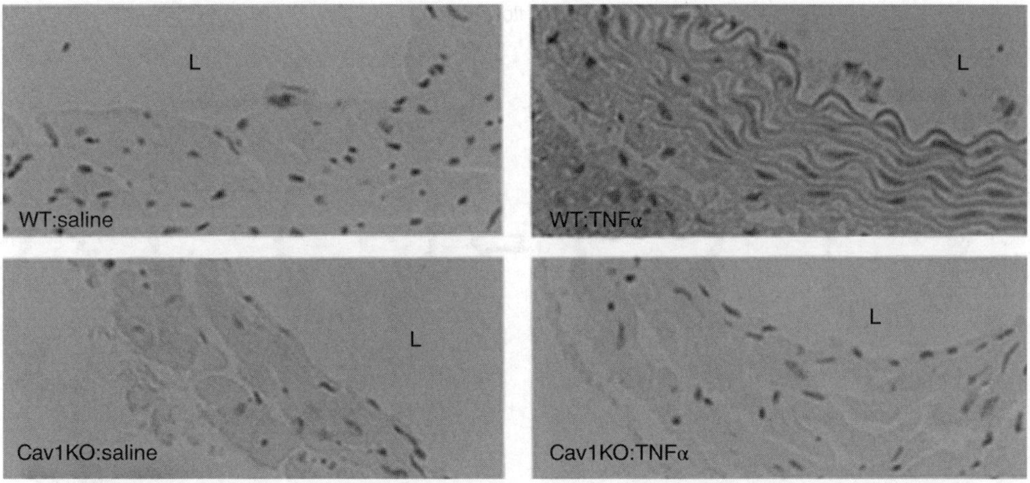

Figure 18.3 Caveolae are necessary for induction of protein tyrosine nitration by proinflammatory cytokines. WT and Cav1KO mice were intravenously injected with a bolus of saline or TNFα (20 mg/kg). After 2 hours, aortic tissue was fixed with paraformaldehyde, embedded in paraffin wax and sectioned at 5 μm. Tissue sections were incubated with primary antibody against nitrotyrosine followed by horseradish peroxidase-tagged secondary antibody. Sections were then incubated with DAB. Results indicate the presence of a brown reaction product only in vessels from TNFα-treated mice. As was the case for ROS, TNFα failed to induce protein tyrosine nitration in vessels from Cav1KO mice. (*See insert for color representation of the figure.*)

normal calcium entry and contractile responses to vasoconstrictors. Hence, caveolin/caveolae play a key role in mediating vascular tone and VSMC contractility. In addition, caveolin/caveolae modulate mitogenic signals and proliferative responses in VSMCs. Initial observations noted that caveolin expression levels were significantly reduced in proliferating VSMCs [65]. Consistent with these findings, reduction in caveolae or caveolin-1 expression levels enhance, while overexpression of caveolin-1 inhibits VSMC proliferation. In these studies, caveolin-1 was linked to ERK1/2 MAP kinases and cyclin D1 pathways [66]. Moreover, *in vivo* studies indicate that caveolin-1 expression levels are substantially reduced in the media of injured arteries and caveolin-1-null mice display VSMC hyperplasia and enhanced neointima formation in response to vascular injury [67].Given these observations, one would have expected that VSMC hyperplasia would be observed in vessels from the ApoE/caveolin-1 double knockout mouse, which is not the case. One can speculate that VSMC proliferation and neointima formation, both hallmarks of atherosclerosis, may be blocked in this model due to endothelial dependent mechanisms such as enhancement of basal NO or perhaps lack of mitogenic signals directed toward the VSMC as a consequence of poor oxLDL transport across the endothelium and uptake by macrophages.

Atheosclerosis has a strong autoimmune background with lesions containing macrophages and activated T cells that secrete many proinflammatory cytokines. Macrophages express caveolin-1 and -2 that appear to modulate a number of key cellular functions. Enhanced levels of caveolin-1 expression correlates with induction of macrophage apoptosis that may influence atheroma development and plaque stability [68]. Caveolin-1 also exerts anti-inflammatory effects by modulating signaling pathways that downregulate macrophage production of TNFα and IL-6 [69]. Further, both cholesterol and lipoprotein uptake is mediated by caveolin-1. These findings imply that macrophage activation and foam

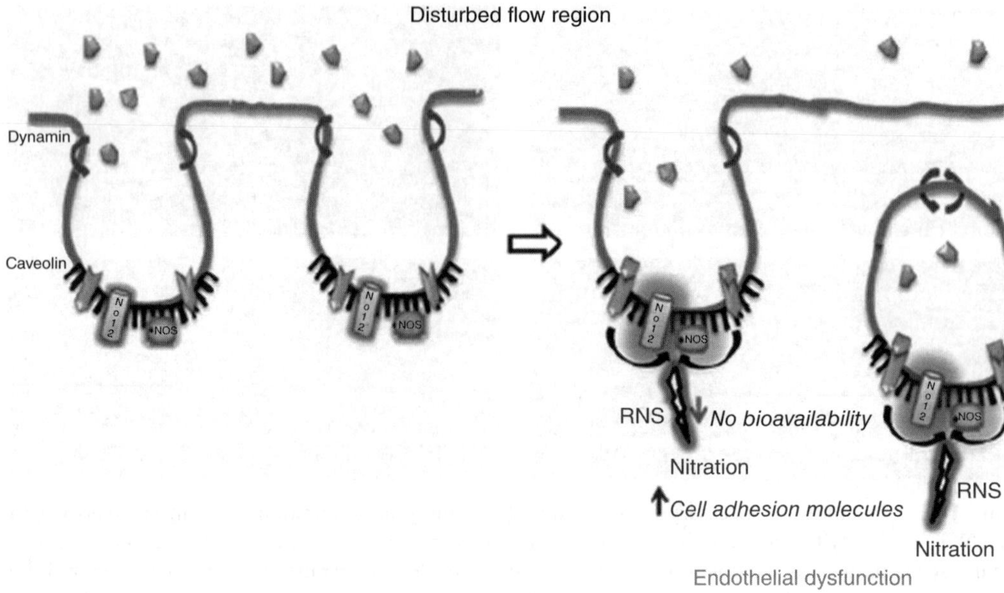

Figure 18.4 Conceptual model of caveolae membrane compartments and its impact on redox signaling and endothelial cell function. Our past and preliminary studies show that Nox2 and eNOS are targeted to caveolae. Vascular disease-inducing factors (*green* and *orange pentagons*) engage receptors found in caveolae (*green* and *orange hexagons*) and stimulate the *acute* production of ROS (*red halo*) and NO (*blue halo*) via simultaneous activation of Nox2 and eNOS, respectively. In addition, disturbed patterns of blood flow activate mechanotransduction pathways that involve caveolae and activation of Nox2 and eNOS. Past reports indicate that these events require internalization of Nox2 and eNOS via a mechanism involving caveolae and dynamin2. Due to the physical proximity of Nox2 and eNOS in caveolae, ROS and NO can rapidly react resulting in loss of NO bioavailability and formation of reactive nitrogen species (RNS). Whether at the plasma membrane or within a dynamic caveolae-carrier (redoxosome), we hypothesize that RNS serve as "kindling radicals" necessary to perpetuate oxidative stress in the vasculature. Based on our preliminary findings and examination of the literature, we suspect that caveolae-mediated nitration events and subsequent activation of an NFκB/adhesion molecule axis serve as a crucial pathway for endothelial/vascular dysfunction that contributes to atherogenesis.

cell development is impaired in caveolin-1 knockout mice. T cells also influence plaque size. Regulatory T cells (Tregs), in particular, are involved in immune suppression, and the expansion of Tregs correlates with significant reduction in plaque formation. A close examination of T-cell subgroups in caveolin-1 knock animals revealed an increase in Tregs by greater than 50% while CD4+ effector T cell number was reduced by nearly 5% [70]. Whether this anti-inflammatory T-cell profile directly contributes to reduction in atheroma in the caveolin-1 knockout mouse, however, requires further testing.

CONCLUSIONS/PERSPECTIVE

Atherosclerosis is a complex disease with multifactoral elements contributing to plaque initiation and progression. A central feature of the disease is endothelial dysfunction that is characterized by a shift to a proinflammatory phenotype. Within the endothelium, caveolae organelles regulate several important processes that maintain endothelial cell homeostasis. However, as in so many other disease settings, normal cell functions are often utilized to induce pathology. Emerging evidence suggest that atherogenic risk factors impart their pathological effects through exploiting the natural function of caveolae.

With the advent of methods to isolate caveolae, approaches to manipulate the expression levels of caveolae/caveolin-1 and the development of caveolae/caveolin-1 knockout and transgenic mice, the extent to which the caveolin proteins and caveolae structures participate in vascular health and disease can be examined. In this chapter, we highlighted that fact that caveolin-1 deficient mice are protected from atherosclerosis and discussed several mechanisms through which caveolae-mediated processes can contribute to plaque formation. Based on our current understanding of the findings in this field, it would appear that strategies to decrease the level of caveolin-1/caveolae would be of clinical benefit. The challenge however, is achieving a level of caveolae reduction that limits plaque formation or progression while maintaining normal physiologic function of the endothelium. We consider that therapeutic reduction of caveolae may be within reach given the following observations. First, HMG-CoA reductase inhibitors have been shown to reduce the expression levels of caveolin-1 by 40–50% in endothelial cells [25], which may explain some of the atheroprotective effects exerted by statins apart from the well-known lipid-lowering ability of the compound. Second, the subcellular positioning of caveolae within the endothelium is sensitive to prevailing hemodynamic flow profiles. Hence, changing flow patterns over "at risk" endothelium, which can be achieved through exercise training, may alter caveolae abundance at the plasma membrane and limit atherogenic potential in these cells. Last, the ability to selectively target ROS/RNS produced in these domains would limit oxidative stress damage and inflammatory processes that contribute to atheroma formation. Thus, further investigation of these and other caveolae/caveolin-1-associated processes may uncover new mechanisms that can be targeted to ameliorate atherosclerosis.

ACKNOWLEDGMENTS

The author thanks Harinder Singh for his efforts in providing the experimental data included in this chapter. Funding was provided by grants awarded to the author from the American Heart Association, National Institutes of Health, and the WW Smith Charitable Trust.

REFERENCES

1. Cunningham, K. and A. Gotlieb, *The role of shear stress in the pathogenesis of atherosclerosis*. Lab Invest, 2005. **85**: p. 9–23.
2. Palade, G.E., *Fine structure of blood capillaries*. J Appl Phys, 1953. **24**: p. 1424.
3. Brown, D. and E. London, *Functions of lipid rafts in biological membranes*. Annu Rev Cell Dev Biol, 1998. **14**: p. 111–136.
4. Kurzchalia, T.V., P. Dupree, R.G. Parton, et al., *VIP21, a 21-kD membrane protein is an integral component of trans-Golgi-network-derived transport vesicles*. J Cell Biol, 1992. **118**: p. 1003–1014.
5. Fra, A.M., E. Williamson, K. Simons, and R.G. Parton, *De novo formation of caveolae in lymphocytes by expression of VIP21-caveolin*. Proc Natl Acad Sci U S A, 1995. **92**(19): p. 8655–8659.
6. Briand, N., I. Dugail, and S. Le Lay, *Cavin proteins: New players in the caveolae field*. Biochimie, 2011. **93**(1): p. 71–77.
7. Okamoto, T., A. Schlegel, P.E. and M.P. Lisanti, *Caveolins, a family of scaffolding proteins for organizing "preassembled signaling complexes" at the plasma membrane*. J Biol Chem, 1998. **273**: p. 5419–5422.
8. Sargiacomo, M., P.E. Scherer, Z. Tang, et al., *Oligomeric structure of caveolin: Implications for caveolae membrane organization*. Proc Natl Acad Sci U S A, 1995. **92**(20): p. 9407–9411.
9. Dietzen, D.J., W.R. Hastings, and D.M. Lublin, *Caveolin is palmitoylated on multiple cysteine residues. Palmitoylation is not necessary for localization of caveolin to caveolae*. J Biol Chem, 1995. **270**(12): p. 6838–6842.
10. Li, S., J. Couet, and M.P. Lisanti, *Src tyrosine kinases, Galpha subunits, and H-Ras share a common membrane-anchored scaffolding protein, caveolin. Caveolin binding negatively regulates the auto-activation of Src tyrosine*

kinases. J Biol Chem, 1996. **271**(46): p. 29182–29190.

11. Li, S., R. Seitz, and M.P. Lisanti, *Phosphorylation of caveolin by src tyrosine kinases. The alpha-isoform of caveolin is selectively phosphorylated by v-Src in vivo*. J Biol Chem, 1996. **271**(7): p. 3863–3868.
12. Zulli, A., B.F. Buxton, M.J. Black, Z. Ming, A. Cameron, and D.L. Hare, *The immunoquantification of caveolin-1 and eNOS in human and rabbit diseased blood vessels*. J Histochem Cytochem, 2006. **54**(2): p. 151–159.
13. Frank, P.G., H. Lee, D.S. Park, N.N. Tandon, P.E. Scherer, and M.P. Lisanti, *Genetic ablation of caveolin-1 confers protection against atherosclerosis*. Arterioscler Thromb Vasc Biol, 2004. **24**(1): p. 98–105.
14. Fernandez-Hernando, C., J. Yu, Y. Suárez, et al., *Genetic evidence supporting a critical role of endothelial caveolin-1 during the progression of atherosclerosis*. Cell Metab, 2009. **10**(1): p. 48–54.
15. Fernandez-Hernando, C., J. Yu, A. Dávalos, J. Prendergast, and W.C. Sessa, *Endothelial-specific overexpression of caveolin-1 accelerates atherosclerosis in apolipoprotein E-deficient mice*. Am J Pathol, 2010. **177**(2): p. 998–1003.
16. Kuhlencordt, P.J., R. Gyurko, F. Han, et al., *Accelerated atherosclerosis, aortic aneurysm formation, and ischemic heart disease in apolipoprotein E/endothelial nitric oxide synthase double-knockout mice*. Circulation, 2001. **104**(4): p. 448–454.
17. Garcia-Cardena, G., P. Oh, J. Liu, J.E. Schnitzer, and W.C. Sessa, *Targeting of nitric oxide synthase to endothelial cell caveolae via palmitoylation: Implications for nitric oxide signaling*. Proc Natl Acad Sci U S A, 1996. **93**(13): p. 6448–6453.
18. Feron, O., L. Belhassen, L. Kobzik, T.W. Smith, R.A. Kelly, and T. Michel, *Endothelial nitric oxide synthase targeting to caveolae. Specific interactions with caveolin isoforms in cardiac myocytes and endothelial cells*. J Biol Chem, 1996. **271**(37): p. 22810–22814.
19. Ju, H., R. Zou, V.J. Venema, and R.C. Venema, *Direct interaction of endothelial nitric oxide synthase and caveolin-1 inhibits synthase activity*. J Biol Chem, 1997. **272**: p. 18522–18525.
20. Church, J.E. and D. Fulton, *Differences in eNOS activity because of subcellular localization are dictated by phosphorylation state rather than the local calcium environment*. J Biol Chem, 2006. **281**(3): p. 1477–1488.
21. Blair, A., P.W. Shaul, I.S. Yuhanna, P.A. Conrad, and E.J. Smart EJ, *Oxidized low density lipoprotein displaces endothelial nitric-oxide synthase (eNOS) from plasmalemmal caveolae and impairs eNOS activation*. J Biol Chem, 1999. **274**(45): p. 32512–32519.
22. Graf, G.A., S.V. Matveev, and E.J. Smart, *Class B scavenger receptors, caveolae and cholesterol homeostasis*. Trends Cardiovasc Med, 1999. **9**(8): p. 221–225.
23. Uittenbogaard, A., P.W. Shaul, I.S. Yuhanna, A. Blair, and E.J. Smart, *High density lipoprotein prevents oxidized low density lipoprotein-induced inhibition of endothelial nitric-oxide synthase localization and activation in caveolae*. J Biol Chem, 2000. **275**(15): p. 11278–11283.
24. Feron, O., C. Dessy, S. Moniotte, J.P. Desager, and J.L. Balligand, *Hypercholesterolemia decreases nitric oxide production by promoting the interaction of caveolin and endothelial nitric oxide synthase*. J Clin Invest, 1999. **103**(6): p. 897–905.
25. Feron, O., C. Dessy, J.P. Desager, and J.L. Balligand, *Hydroxy-methylglutaryl-coenzyme A reductase inhibition promotes endothelial nitric oxide synthase activation through a decrease in caveolin abundance*. Circulation, 2001. **103**(1): p. 113–118.
26. Bendayan, M. and E. Rasio, *Transport of insulin and albumin by the microvascular endothelium of the rete mirabile*. J Cell Sci, 1996. **109**: p. 1857–1856.
27. Frank, P.G., G.S. Hassan, J.A. Rodriguez-Feo, and M.P. Lisanti, *Caveolae and caveolin-1: Novel potential targets for the treatment of cardiovascular disease*. Curr Pharm Des, 2007. **13**(17): p. 1761–1769.
28. Stamatovic, S.M., R.F. Keep, M.M. Wang, I. Jankovic, and A.V. Andjelkovic, *Caveolae-mediated internalization of occludin and claudin-5 during CCL2-induced tight junction remodeling in brain endothelial cells*. J Biol Chem, 2009. **284**(28): p. 19053–19066.
29. Miyawaki-Shimizu, K., D. Predescu, J. Shimizu, M. Broman, S. Predescu, and A.B. Malik, *siRNA-induced caveolin-1 knockdown in mice increases lung vascular permeability via the junctional pathway*. Am J Physiol Lung Cell Mol Physiol, 2006. **290**(2): p. L405–L413.

30. Passerini, A.G., D.C. Polacek, C. Shi, et al., *Coexisting proinflammatory and antioxidative endothelial transcription profiles in a disturbed flow region of the adult porcine aorta*. Proc Natl Acad Sci U S A, 2004. **101**(8): p. 2482–2487.
31. Tarbell, J.M. and M.Y. Pahakis, *Mechanotransduction and the glycocalyx*. J Intern Med, 2006. **259**(4): p. 339–350.
32. Nauli, S.M., Y. Kawanabe, J.J. Kaminski, W.J. Pearce, D.E. Ingber, and J. Zhou, *Endothelial cilia are fluid shear sensors that regulate calcium signaling and nitric oxide production through polycystin-1*. Circulation, 2008. **117**(9): p. 1161–1171.
33. Loufrani, L. and D. Henrion, *Role of the cytoskeleton in flow (shear stress)-induced dilation and remodeling in resistance arteries*. Med Biol Eng Comput, 2008. **46**(5): p. 451–460.
34. Jin, Z.G., H. Ueba, T. Tanimoto, A.O. Lungu, M.D. Frame, and B.C. Berk, *Ligand-independent activation of vascular endothelial growth factor receptor 2 by fluid shear stress regulates activation of endothelial nitric oxide synthase*. Circ Res, 2003. **93**(4): p. 354–363.
35. Lee, H.J. and G.Y. Koh, *Shear stress activates Tie2 receptor tyrosine kinase in human endothelial cells*. Biochem Biophys Res Commun, 2003. **304**(2): p. 399–404.
36. Yamamoto, K., R. Korenaga, A. Kamiya, and J. Ando, *Fluid shear stress activates Ca(2+) influx into human endothelial cells via P2X4 purinoceptors*. Circ Res, 2000. **87**(5): p. 385–391.
37. Chachisvilis, M., Y.L. Zhang, and J.A. Frangos, *G protein-coupled receptors sense fluid shear stress in endothelial cells*. Proc Natl Acad Sci U S A, 2006. **103**(42): p. 15463–15468.
38. Katsumi, A., A.W. Orr, E. Tzima, and M.A. Schwartz, *Integrins in Mechanotransduction*. J Biol Chem, 2004. **279**(13): p. 12001–12004.
39. Tzima, E., M. Irani-Tehrani, and W.B. Kiosses, et al., *A mechanosensory complex that mediates the endothelial cell response to fluid shear stress*. Nature, 2005. **437**(7057): p. 426–431.
40. Rizzo, V., C. Morton, N. DePaola, J.E. Schnitzer, and P.F. Davies, *Recruitment of endothelial caveolae into mechanotransduction pathways by flow conditioning in vitro*. Am J Physiol Heart Circ Physiol, 2003. **285**(4): p. H1720–H1729.
41. Rizzo, V., A. Sung, P. Oh, and J.E. Schnitzer, *Rapid mechanotransduction in situ at the luminal cell surface of vascular endothelium and its caveolae*. J Biol Chem, 1998. **273**(41): p. 26323–26329.
42. Rizzo, V., D.P. McIntosh, P. Oh, and J.E. Schnitzer, *In situ flow activates endothelial nitric oxide synthase in luminal caveolae of endothelium with rapid caveolin dissociation and calmodulin association*. J Biol Chem, 1998. **273**(52): p. 34724–34729.
43. Yang, B. and V. Rizzo, *TNF-{alpha} potentiates protein-tyrosine nitration through activation of NADPH oxidase and eNOS localized in membrane rafts and caveolae of bovine aortic endothelial cells*. Am J Physiol Heart Circ Physiol, 2007. **292**(2): p. H954–H962.
44. Schnitzer, J.E., P. Oh, B.S. Jacobson, and A.M. Dvorak, *Caveolae from luminal plasmalemma of rat lung endothelium: Microdomains enriched in caveolin, Ca^{2+}-ATPase, and inositol trisphosphate receptor*. Proc. Natl. Acad. Sci. USA, 1995. **92**: p. 1759–1763.
45. Oh, P. and J.E. Schnitzer, *Segregation of heterotrimeric G proteins in cell surface microdomains. G(q) binds caveolin to concentrate in caveolae, whereas G(i) and G(s) target lipid rafts by default*. Mol Biol Cell, 2001. **12**(3): p. 685–698.
46. Schnitzer, J.E., J. Liu, and P. Oh, *Endothelial caveolae have the molecular transport machinery for vesicle budding, docking, and fusion including VAMP, NSF, SNAP, annexins, and GTPases*. J Biol Chem, 1995. **270**: p. 14399–14404.
47. Yang, B., C. Radel, D. Hughes, S. Kelemen, and V. Rizzo, *p190 RhoGTPase-activating protein links the beta1 integrin/caveolin-1 mechanosignaling complex to RhoA and actin remodeling*. Arterioscler Thromb Vasc Biol, 2011. **31**(2): p. 376–383.
48. Yu, J., S. Bergaya, T. Murata, et al., *Direct evidence for the role of caveolin-1 and caveolae in mechanotransduction and remodeling of blood vessels*. J Clin Invest, 2006. **116**(5): p. 1284–1291.
49. Boyd, N.L., H. Park, H. Yi, et al., *Chronic shear induces caveolae formation and alters ERK and Akt responses in endothelial cells*. Am J Physiol Heart Circ Physiol, 2003. **285**(3): p. H1113–H1122.
50. Young, J.L., P. Libby, and U. Schonbeck, *Cytokines in the pathogenesis of atherosclerosis*. Thromb Haemost, 2002. **88**(4): p. 554–567.
51. Schwencke, C., R.C. Braun-Dullaeus, C. Wunderlich, and R.H. Strasser, *Caveolae and caveolin in transmembrane signaling: Implications*

for human disease. Cardiovasc Res, 2006. **70**(1): p. 42–49.

52. D'Alessio, A., R.S. Al-Lamki, J.R. Bradley, and J.S. Pober, *Caveolae participate in tumor necrosis factor receptor 1 signaling and internalization in a human endothelial cell line*. Am J Pathol, 2005. **166**(4): p. 1273–1282.
53. Oakley, F.D., R.L. Smith, and J.F. Engelhardt, *Lipid rafts and caveolin-1 coordinate interleukin-1beta (IL-1beta)-dependent activation of NFkappaB by controlling endocytosis of Nox2 and IL-1beta receptor 1 from the plasma membrane*. J Biol Chem, 2009. **284**(48): p. 33255–33264.
54. Ishizaka, N., K.K. Griendling, B. Lassègue, and R.W. Alexander, *Angiotensin II type 1 receptor: Relationship with caveolae and caveolin after initial agonist stimulation*. Hypertension, 1998. **32**(3): p. 459–466.
55. Garrean, S., X.P. Gao, V. Brovkovych, et al., *Caveolin-1 regulates NF-kappaB activation and lung inflammatory response to sepsis induced by lipopolysaccharide*. J Immunol, 2006. **177**(7): p. 4853–4860.
56. Ray, R. and A.M. Shah, *NADPH oxidase and endothelial cell function*. Clin Sci (Lond), 2005. **109**(3): p. 217–226.
57. Khan, E.M., J.M. Heidinger, M. Levy, M.P. Lisanti, T. Ravid, and T. Goldkorn, *Epidermal growth factor receptor exposed to oxidative stress undergoes Src- and caveolin–1-dependent perinuclear trafficking*. J Biol Chem, 2006. **281**(20): p. 14486–14493.
58. Peterson, T.E., V. Poppa, H. Ueba, A. Wu, C. Yan, and B.C. Berk, *Opposing effects of reactive oxygen species and cholesterol on endothelial nitric oxide synthase and endothelial cell caveolae*. Circ Res, 1999. **85**(1): p. 29–37.
59. Yang, B., T.N. Oo, and V. Rizzo, *Lipid rafts mediate H2O2 prosurvival effects in cultured endothelial cells*. Faseb J, 2006. **20**(9): p. 1501–1503.
60. Ushio-Fukai, M., L. Hilenski, N. Santanam et al., *Cholesterol depletion inhibits epidermal growth factor receptor transactivation by angiotensin II in vascular smooth muscle cells: Role of cholesterol-rich microdomains and focal adhesions in angiotensin II signaling*. J Biol Chem, 2001. **276**(51): p. 48269–48275.
61. Shao, D., A.W. Segal, and L.V. Dekker, *Lipid rafts determine efficiency of NADPH oxidase activation in neutrophils*. FEBS Lett, 2003. **550**(1-3): p. 101–106.
62. Hilenski, L.L., R.E. Clempus, M.T. Quinn, J.D. Lambeth, and K.K. Griendling, *Distinct subcellular localizations of Nox1 and Nox4 in vascular smooth muscle cells*. Arterioscler Thromb Vasc Biol, 2004. **24**(4): p. 677–683.
63. Bayraktutan, U., L. Blayney, and A.M. Shah, *Molecular characterization and localization of the NAD(P)H oxidase components gp91-phox and p22-phox in endothelial cells*. Arterioscler Thromb Vasc Biol, 2000. **20**(8): p. 1903–1911.
64. Li, J.M. and A.M. Shah, *Intracellular localization and preassembly of the NADPH oxidase complex in cultured endothelial cells*. J Biol Chem, 2002. **277**(22): p. 19952-19960.
65. Peterson, T.E., L.S. Kleppe, N.M. Caplice, S. Pan, C.S. Mueske, and R.D. Simari, *The regulation of caveolin expression and localization by serum and heparin in vascular smooth muscle cells*. Biochem Biophys Res Commun, 1999. **265**(3): p. 722–727.
66. Peterson, T.E., M.E. Guicciardi, R. Gulati, et al., *Caveolin-1 can regulate vascular smooth muscle cell fate by switching platelet-derived growth factor signaling from a proliferative to an apoptotic pathway*. Arterioscler Thromb Vasc Biol, 2003. **23**(9): p. 1521–1527.
67. Hassan, G.S., J.F. Jasmin, W. Schubert, P.G. Frank, and M.P. Lisanti, *Caveolin-1 deficiency stimulates neointima formation during vascular injury*. Biochemistry, 2004. **43**(26): p. 8312–8321.
68. Gargalovic, P. and L. Dory, *Cellular apoptosis is associated with increased caveolin-1 expression in macrophages*. J Lipid Res, 2003. **44**(9): p. 1622–1632.
69. Wang, X.M., H.P. Kim, R. Song, and A.M. Choi, *Caveolin-1 confers antiinflammatory effects in murine macrophages via the MKK3/p38 MAPK pathway*. Am J Respir Cell Mol Biol, 2006. **34**(4): p. 434–442.
70. Engel, D., L. Beckers, E. Wijnands, et al., *Caveolin-1 deficiency decreases atherosclerosis by hampering leukocyte influx into the arterial wall and generating a regulatory T-cell response*. Faseb J. **25**(11): p. 3838–3848.

19

LYSOSOMAL MOLECULAR DERANGEMENTS IN ATHEROSCLEROSIS

Pin-Lan Li and Yang Zhang

Department of Pharmacology and Toxicology, Medical College of Virginia Campus, Virginia Commonwealth University, Richmond, VA, USA

In 1974, Christian de Duve received the Nobel Prize, which honored his work on lysosome structure and functions. In his studies, the single- or double-membraned vesicles within cells were described as lysosomes, which contain cytoplasm including organelles and digest various exogenous and endogenous particles via a series of enzymatic reactions important for the normal function of cells and their defense and surviving [1]. Over the past half century, numerous studies have been carried out to elucidate the physiological functions of lysosomes and their pathological relevance to human diseases. With recent advancement of molecular biology and genetics, the molecular machinery and regulatory mechanisms of lysosome structural and functional integrity have been intensively studied. These studies have led to many new insights into the role of lysosomal molecular defects in the development of human diseases. In this chapter, we will briefly summarize some current progresses of knowledge on the lysosomal molecular derangement in atherosclerosis, which provides an overview regarding the role of lysosomes as a critical pathogenic organelle in the atherogenesis and related pathophysiologic alterations in arterial wall or cells. Targeting molecules determining or regulating lysosome functions may be a critical therapeutic strategy for treatment or prevention of atherosclerosis.

LYSOSOMES AND THEIR FUNCTIONS

Characteristics of Lysosomes

Lysosomes are membrane-bound organelles that originate as membrane-bound vesicles from Golgi apparatus (called Golgi vesicles) and exist in the cytoplasm of all eukaryotic cells. These cytoplasmic organelles contain several dozens of acid hydrolases that are primarily responsible for intracellular digestion. Based on different functions of lysosomes, they are divided into two types: conventional

Atherosclerosis: Risks, Mechanisms, and Therapies, First Edition. Edited by Hong Wang and Cam Patterson.

lysosome and secretory lysosome. Conventional lysosome is just the common lysosomes what we all are very familiar with. These lysosomes are a digestive organelle in cells, which conducts intracellular digestion for cell defense, contributing to autophagic process. Another type of lysosomes, namely, secretory lysosomes are able to fuse with cell plasma membrane and secret its content outside of the cell. Many cells including vascular endothelial cells (ECs) have been found to have secretory lysosomes that mediate secretion of different substances by the so-called exocytosis. Recently, another type of lysosomes has been reported to mediate membrane repair and transmembrane signaling, which may have the nature of conventional lysosomes, but work like secretory lysosomes to fuse to plasma membrane to repair the damaged membrane [2–5]. This type of lysosomes has been found to play an important role in receptor-mediated endocytosis, and thereby mediates events of receptor recycling [6].

More recently, lysosomal vesicles have been reported to contribute to exocytosis in nonsecretory cells [7], where these vesicles can fuse with the plasma membrane to excrete the contents of the vesicle and to incorporate the vesicle membrane components into the cell membrane [3]. In addition, some studies in our laboratory and by others have demonstrated that lysosomes as a Ca^{2+} store importantly participate in the regulation of cell functions in a variety of tissues or cells [8, 9], where lysosomal Ca^{2+} can be mobilized to mediate a two-phase Ca^{2+} release [10]. These lysosomes have been referred to as signaling lysosomes [11–13].

Function of Lysosomes

It is well known that lysosome functions very much rely on different molecules in lysosomes or on lysosomal membrane. Lysosomes contain several dozens of acid hydrolases that are primarily responsible for intracellular digestion. On the lysosomal membrane, there is an H^+-ATPase residing and operating to acidify a vesicle compartmental environment that facilitates the activity of various acid hydrolases. It has been shown that Ca^{2+} enters lysosomal compartment by H^+/Ca^{2+} exchange under resting condition and releases during different stimulations [14, 15]. With their intracellular digestion function, lysosomes can protect the cell from foreign invaders such as bacteria by breaking them down and enzyme-mediated digestion. Lysosomes may also keep cell robust via destroying various old or aged organelles within the cell such as mitochondria. This process of aged organelles is a process known as autophagy. In addition, lysosomes also play roles in oocyte maturation and fertilization during acrosome reaction, where lysosomal enzymes from the sperm effectively bore a hole through the egg membrane, thereby facilitating the sperm that enters the egg [16]. With respect to its role in the regulation of vascular function and in the development of atherosclerosis, lysosome has been reported to participate in transmembane signaling of different death receptors by activation or promotion of membrane rafts (MRs) redox signaling platforms, which regulates endothelial function or leading to endothelial dysfunction upon different stimuli [17]. It is possible that this lysosome-associated membrane signalosome is of great importance in endothelial injury during atherosclerosis such as abnormal leukocyte adhesion, macrophage invasion or infiltration, and local oxidative stress. In macrophages, lysosome is a center for cholesterol or low-density lipoprotein (LDL) metabolism and transport. It is well known that the oxidized form of cholesterol is taken up by macrophages through scavenger receptors and endocytically delivered to lysosomes for degradation. Under normal condition, the cholesteryl esters are hydrolyzed by lysosomal acid lipase into free cholesterol, which is then actively transported out of lysosomes through lysosomal Niemann–Pick-type C1 (NPC1) protein. Such metabolism and transport of cholesterol in lysosomes are regulated by a number of lysosomal molecules such as mucolipin-1, acid sphingomyelinase (ASMase), and H^+-ATPase. Any defect of these molecular events may result in an insufficiency of free cholesterol transport from lysosomes to cytosol and deficient clearance of cholesterol from macrophages, leading to lipid deposition, foam cell formation, and ultimate atherosclerosis.

Lysosome-centered autophagy has been implicated in the degradation of damaged or aged materials in vascular cells, which may help clean up such wastes and recover cells from the damage in the arterial wall upon atherogenic stimuli. It also interferes with cell apoptosis due to engulfment of defective or damaged mitochondria by autophagosomes (APs) that limit the release of proapoptotic proteins [18]. Upon atherogenic stimulations, lysosomes may be damaged to release its hydrolases due to excessive oxidative stress, which enhances the degree of cellular damage such as endothelial dysfunction or injury. In arterial smooth muscle cells (SMCs), however, lysosome-centered autophagy regulates cell differentiation and maturation, which maintains the contractile phenotype of SMCs and controls their proliferation, growth, and production of extracellular matrix [18–20]. As mentioned earlier, moreover, lysosomes are involved in cellular signaling such as receptor-mediated endocytosis, receptor recycling [6], cell membrane repair, Ca^{2+} uptake, and release [21, 22]. All these signaling events via lysosomes or its enzymes may contribute to the regulation of cell or organ functions.

LYSOSOME-MEMBRANE CROSS-TALK IN ENDOTHELIAL INJURY

MR Redox Signaling Platforms in ECs

Recently, lysosome-membrane cross-talk through the formation and activation of MR signaling platforms have been reported to regulate endothelial function and lead to endothelial injury upon pathological stimuli [17]. The MR redox signaling platform is referred to as a multiple protein complex that uses lipid rafts (LRs) or MRs as a platform to conduct redox signaling. Although individual MRs are too small to be observed on the cell surface by standard light microscopy, clustered MRs form relatively larger macrodomains or platforms on the cell membrane could be visualized by fluorescence or other staining techniques such as confocal microscopy with fluorescent labeled-cholera toxin (CTX), which is used based on its capacity of binding to the raft constituent ganglioside GM1, a glycosphingolipid that consists of a ceramide backbone [23]. Using different methods, various agonists such as FasL, endostatin, TNF-α, TRIAL, visfatin, and other stimuli were found to stimulate the formation of these MR clusters on the membrane of ECs [17, 24, 25], which result in the formation of endothelial redox signalosomes. It has been demonstrated that endothelial redox signalosome is centered on NADPH oxidase (Nox) subunits clustering and its activation, and it works as an enzyme complex. Nox subunits are the major components of such endothelial redox signalosomes [17]. In addition to MR components and Nox subunits, some other molecules were also demonstrated to be present in MR redox signalosomes. These molecules include various receptors that may bind to agonists to stimulate the formation of signalosomes and molecules promoting MR platform formation and functions. In this regard, ASMase has been found to be an important enzyme in MR redox signalosomes, which activates or facilitates the formation of these signalosomes. The function of MR redox signalosomes is to produce O_2^-, and thereby lead to redox regulation of cell and organ functions as shown by O_2^- production in isolated MR fractions from cells stimulated by FasL or other stimuli [11, 17, 26, 27]. This Nox-derived O_2^- has been reported to be importantly implicated in the redox regulation of vascular endothelial and smooth muscle function; and if overly enhanced, it may lead to endothelial injury promoting atherosclerosis [28, 29]. Indeed, Eum SY et al. reported that upon exposure to polychlorinated biphenyl, LR-dependent Nox/JAK/EGFR signaling mechanisms can regulate the expression of cell adhesion molecules (CAMs) in brain ECs and increase adhesion of leukocytes to endothelial monolayers [30]. Another study showed that high-density lipoprotein (HDL) inhibits Nox activation probably by interrupting the assembly of Nox subunits at MRs, which may contribute to the vascular protective actions of HDL against inflammation-mediated oxidative damage [31].

Lysosome Trafficking to Membrane to Form Redox Signalosomes

Since ASMase is mainly present in lysosomes but its substrate sphingomyelin is one of the major components in the outer leaflet of the plasma membrane, an important question is how the lysosomal ASMase is able to approach to its substrate, sphingomyelin in the outer leaflets of the cell membrane, where ceramide is produced to lead to MRs clustering or the formation of MR-Nox signalosomes. Recent work in our laboratory [12, 32] and by others [2, 3] demonstrated that lysosomes are rapidly fused into the cell membrane in response to different stimuli, leading to ASMase secreted locally or attached to the surface of the cell membrane. This lysosomal ASMase hydrolyzes sphingomyelin to generate ceramide, leading to the formation of ceramide-enriched platforms via clustering. Therefore, in MR redox signalosomes, several lysosome-specific molecules such as Lamp-1 and sortilin-1 could be detected [33, 34]. Sortilin-1, a glycoprotein that is responsible for transferring ASMase from the Golgi apparatus to lysosomes, is essential in initiating the movement of lysosomes and promoting their fusion to the cell membrane [33, 35]. The coupled sortilin-1 and ASMase work together to promote the movement of lysosomes toward cell membrane, which leads to LR clustering and Nox activation in ECs. This ASMase-dependent clustering of receptors was also observed for other receptors such

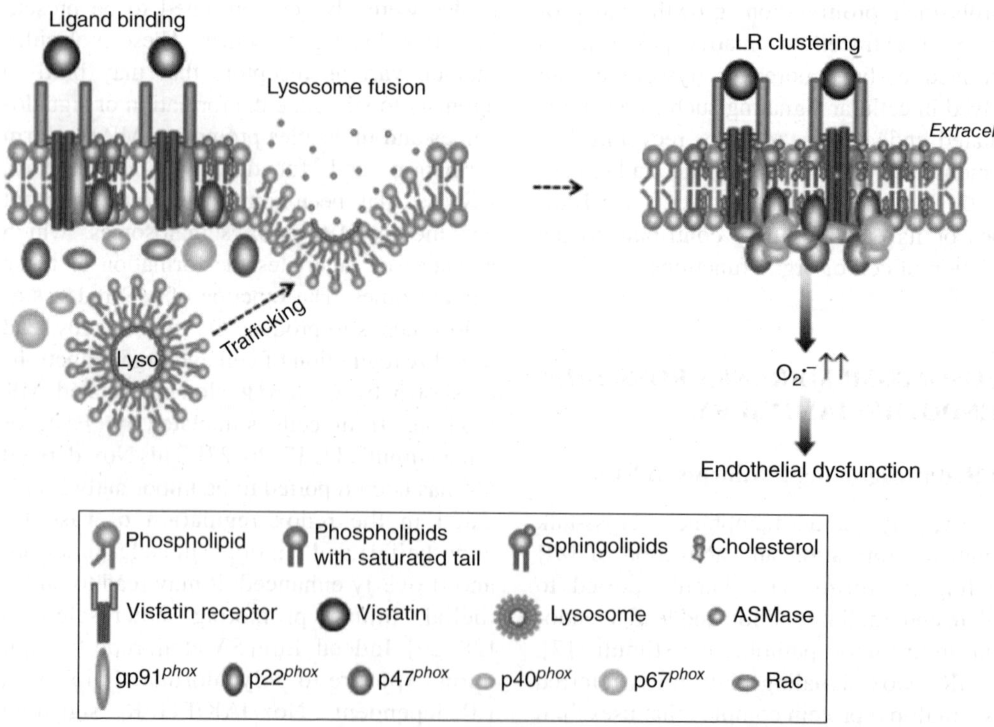

Figure 19.1 Lysosome fusion to plasma membrane to form MRs signaling platforms in ECs. In response to stimuli, lysosomes can be mobilized to traffic and fuse into cell membrane, where acid sphingomyelinase (ASMase) is activated and ceramide produced, resulting in MRs clustering and formation of signaling platforms (Adapted from Xia et al. [13]). (*See insert for color representation of the figure.*)

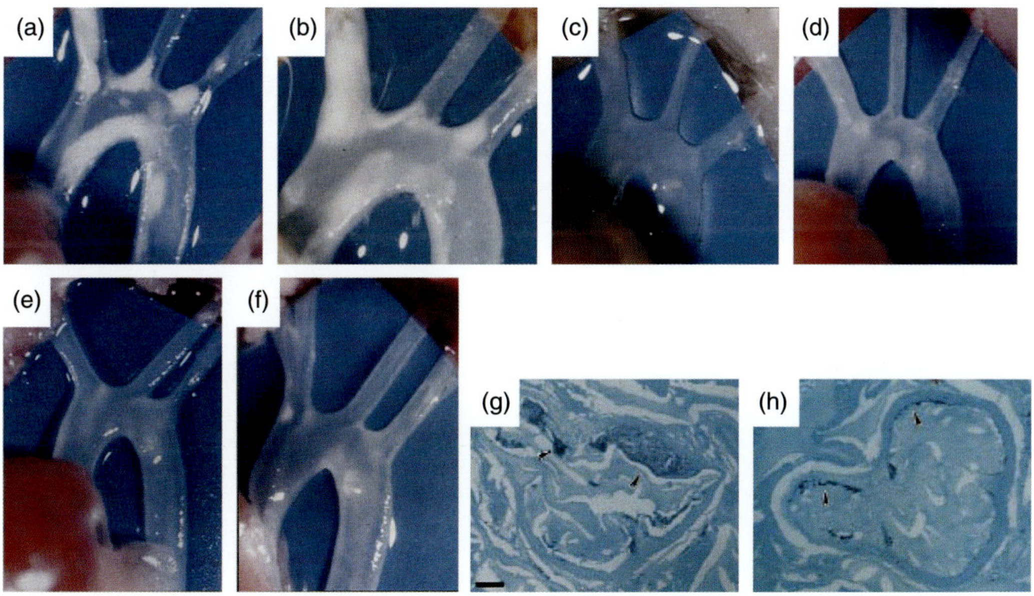

Figure 3.1 Accelerated atherosclerosis in diabetic apoE-null mice. (a–f) Dissection microscopy. ApoE-null mice were either made diabetic with streptotozocin (STZ) or treated with citrate buffer as control. After 6 weeks of diabetes or control treatment, the heart and aorta were dissected under microscopy and photographed. Diabetic mice treated with murine serum albumin (MSA) at 80 μg/day (a) or sRAGE at 3 μg/day (b), 20 μg/day (c) or 40 μg/day (d); control without diabetes treated with MSA at 40 μg/day (e) or sRAGE at 20 μg/day (f). (g, h) Histologic analysis. Typical histologic sections from the aortic sinus are shown from diabetic (g) and nondiabetic (h) apoE-null mice. *Arrows* indicate areas of atherosclerotic lesion as determined by staining with Oil Red O. Scale bar represents 33 μm (Adapted from Park et al. [3]. © Nature).

Atherosclerosis: Risks, Mechanisms, and Therapies, First Edition. Edited by Hong Wang and Cam Patterson.

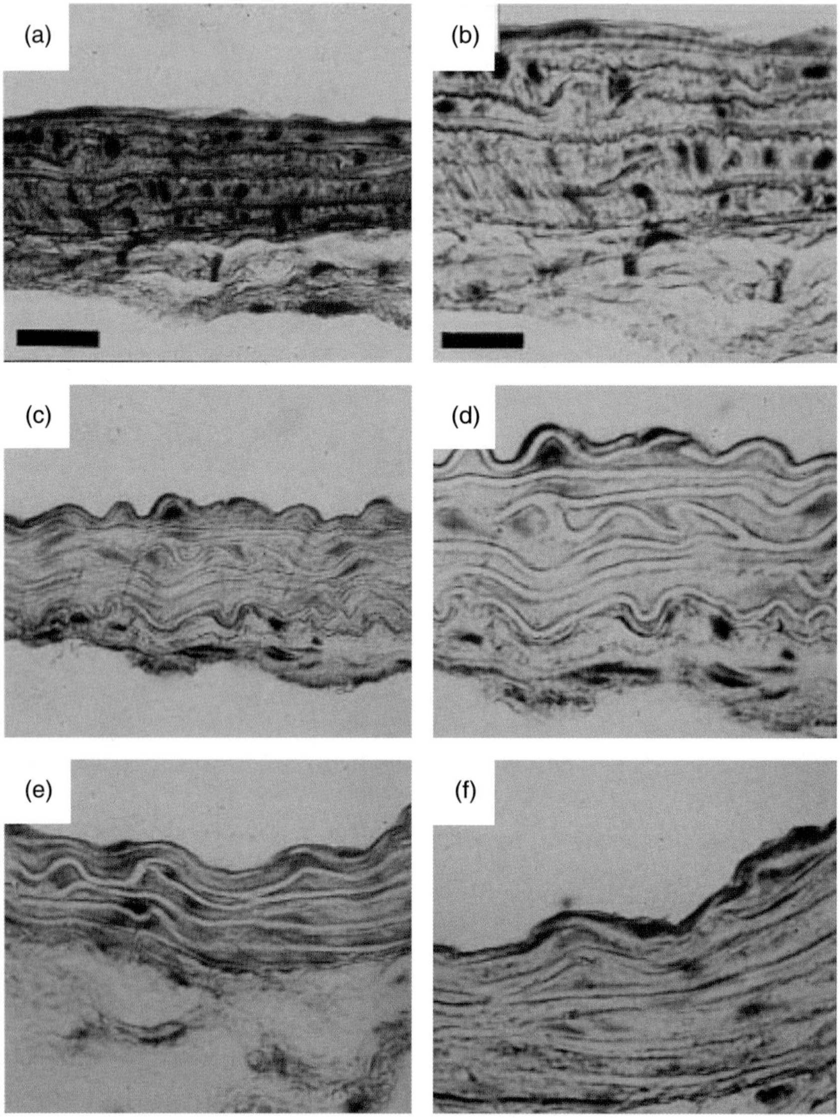

Figure 3.2 Increased AGE-reactive epitopes in diabetic atherosclerosis: studies in ApoE-null mice. (a–f) Vascular sections from the indicated ApoE-null mice were subjected to immunohistochemistry with anti-AGE IgG. a and b, diabetic mice; c and d, control mice; e and f, diabetes treated with sRAGE (20 μg/day). Scale bars represent 33 μm (a, c, and e) and 20 μm (b, d, and f). In all cases, controls with appropriate nonimmune IgG were negative (not shown) (Adapted from Park et al. [3]. © Nature).

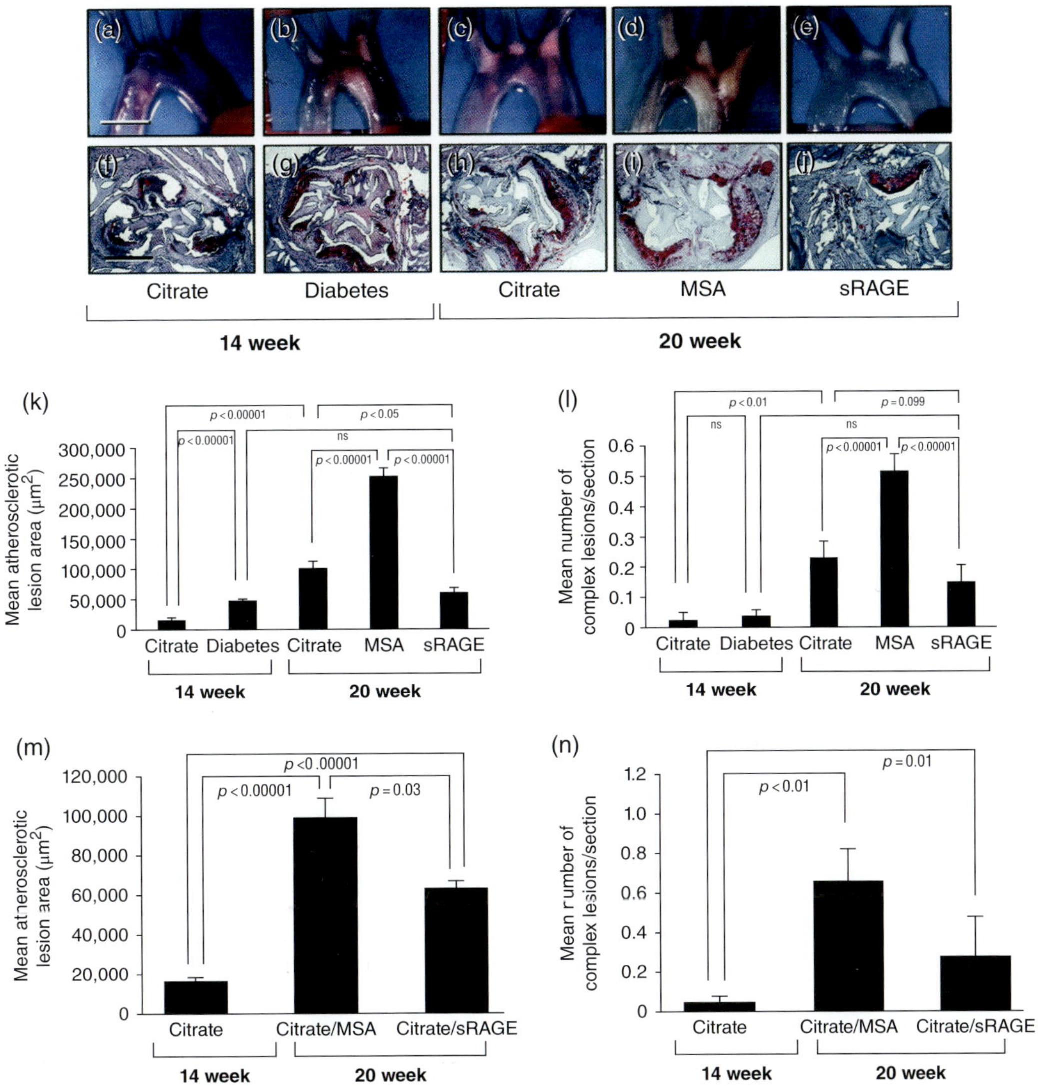

Figure 3.3 Blockade of RAGE suppresses progression of established atherosclerosis in apoE-null mice. Male apoE-null mice were rendered diabetic with streptozotocin or control at age 6 weeks; at age 14 weeks, mice were either sacrificed or treated with sRAGE or vehicle. At age 20 weeks, mice were sacrificed and the aortic tree dissected photographed (a–e) and sections from the aortic arch subjected to staining with Oil Red O (f–j). Quantification of mean atherosclerotic lesion area (k and m) and mean number of complex lesions per section (l, n) was determined. Scale bar: a–e, 0.3 cm; f–j, 125 μm (Adapted from Bucciarelli et al. [36]. © Lippincott Williams & Wilkins).

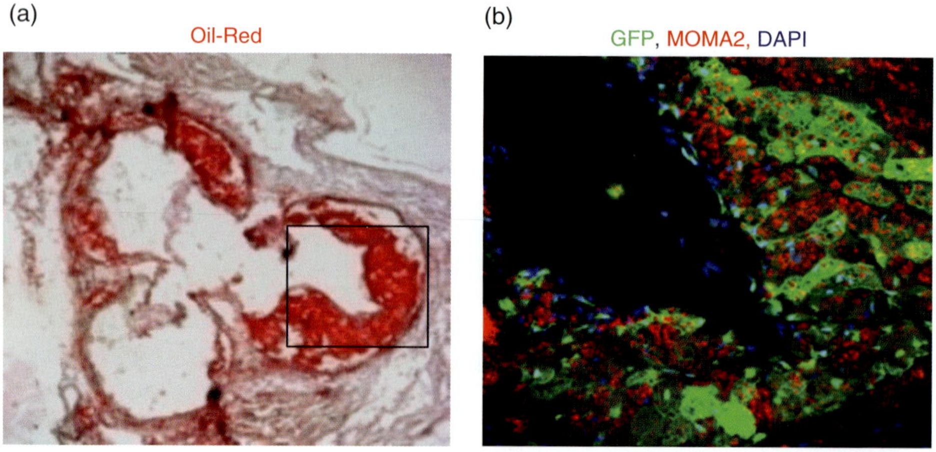

Figure 12.1 Monocyte/macrophage infiltration in atherosclerotic lesion. *Ldlr*$^{-/-}$ mice were lethally irradiated, received bone marrow transplantation from GFP mice at the age of 8 weeks and fed with high-fat diet for 6 weeks. (a) Oil Red O staining of mouse aortic sinus cross-section. Oil Red O staining is a classic staining method to detect triglyceride and lipids. The red staining indicates lipid deposition in the atherosclerotic lesion. (b) Immunostaining of aortic sinus cross-section. Only the section within the black box in (a) was shown. Antibodies against GFP (*green*), MOMA-2 (*red*, monocyte/macrophage marker), and DAPI (*blue*, nuclear marker).

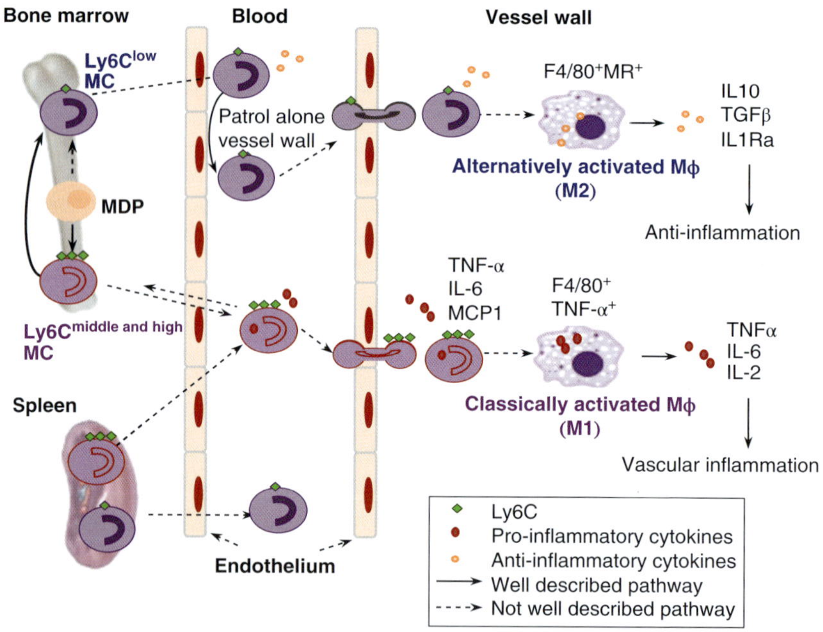

Figure 12.3 Trafficking and differentiation of circulating MC and Mϕ subsets. MDPs in the bone marrow give rise to Ly6C^{high} MC, which can serve as an intermediate for Ly6C^{low} MC. Ly6C^{high} MCs exit the bone marrow in a CC-chemokine receptor 2 (CCR2)-dependent manner. In the absence of inflammation, some circulating Ly6C^{high} MCs could return to the bone marrow. In inflammatory conditions, Ly6C^{hi} MCs migrate through the inflamed endothelium and potentially differentiate into classically activated Mϕ (M1). M1 Mϕ can be distinguished as F4/80^{+}TNF-α^{+}, that secret proinflammatory cytokines such as TNF-α, IL-6, and IL-2 and contribute to vascular inflammation and atherosclerosis formation. Ly6C^{low} MCs patrol along the blood vessel lumen by associating with the vascular endothelium. A small fraction of Ly6C^{low} MCs are also recruited to vessel wall and can differentiate into alternatively activated Mϕ (M2). M2 Mϕ are F4/80^{+}MR^{+} and secret anti-inflammatory cytokines such as IL-10, TGF-β, and IL-1 receptor antagonist (IL-1Rα) and is suggested to play a role in tissue repair. The spleen functions as a reservoir for blood MC. MDP, macrophage and dendritic cell precursor; MC, monocyte; and Mϕ, macrophages.

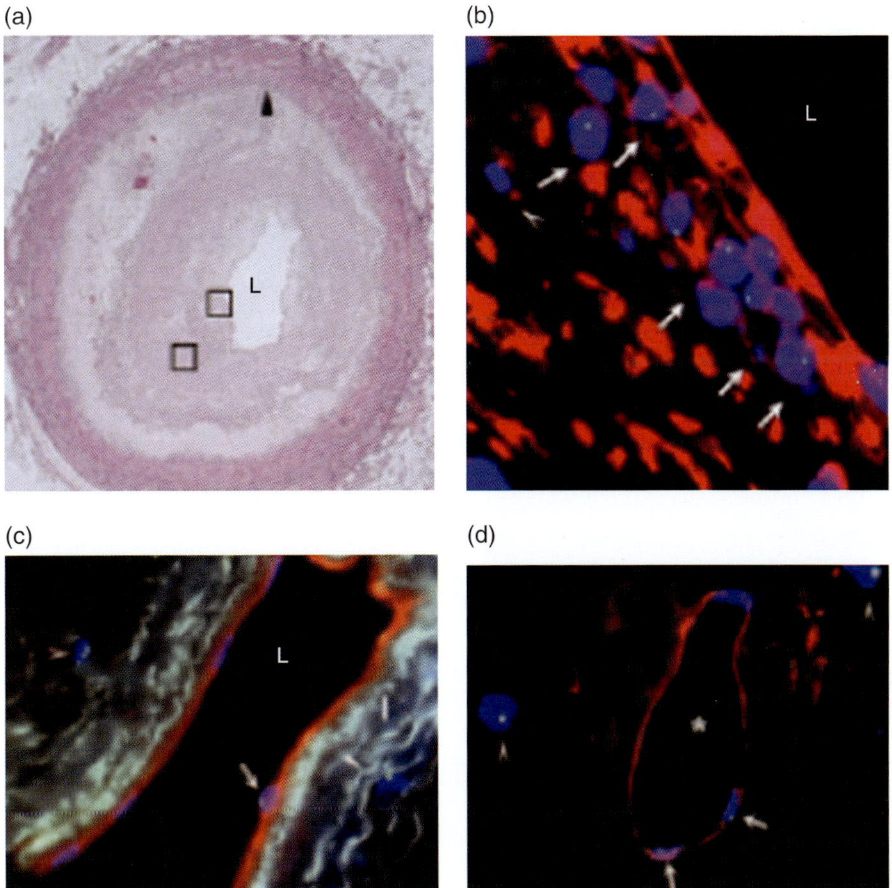

Figure 14.3 (a) Atherosclerotic coronary artery of female patient who underwent gender mismatched bone marrow transplantation; (b) the *arrow* points to internal elastic lamella Y-chromosome-positive (bone marrow origin) smooth muscle cells (*arrows*) in area of plaque highlighted in (a): α-smooth muscle actin shown in *red*, nuclei in *blue*, and Y-chromosome represented by the *green dot*; (c) Y-chromosome-positive endothelial cells (*arrow*) lining a coronary vessel of male patient who received a female heart during cardiac transplantation; and (d) similar Y-chromosome-positive endothelial cells lining the microvessels in the same patient (Adapted from Walter et al. [45] © 2004, Lippincott Williams & Wilkins and Sata et al. [46]. © Nature).

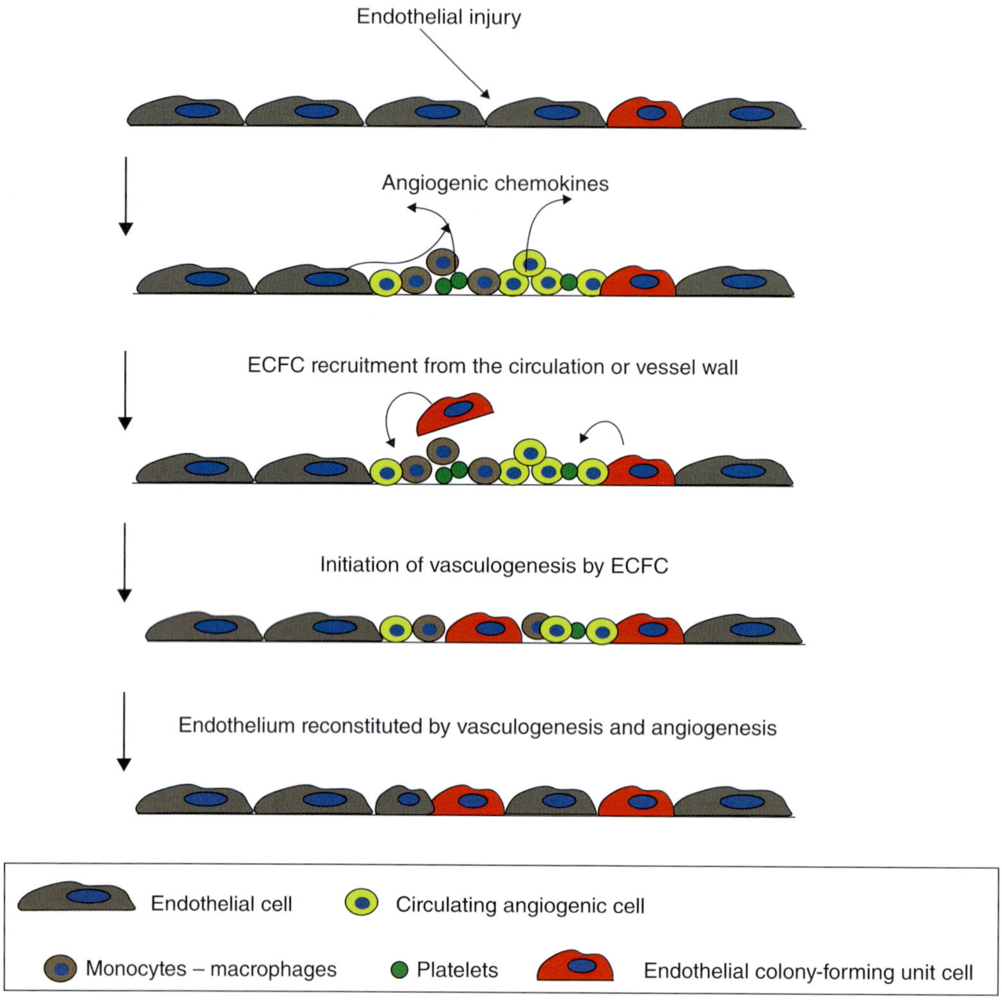

Figure 14.4 In a model of ECFC-mediated vasculogenesis for endothelial repair after injury, hematopoietic cells including platelets, monocytes/macrophages, and CACs attach to areas of endothelial injury. The cellular milieu including neighboring endothelial cells express angiogenic chemokines that attract ECFC from the circulation or vessel wall. ECFCs attach to areas of endothelial injury and form new endothelium (vasculogenesis). The endothelium finally is restored by a combination of ECFC mediated vasculogenesis and angiogenesis.

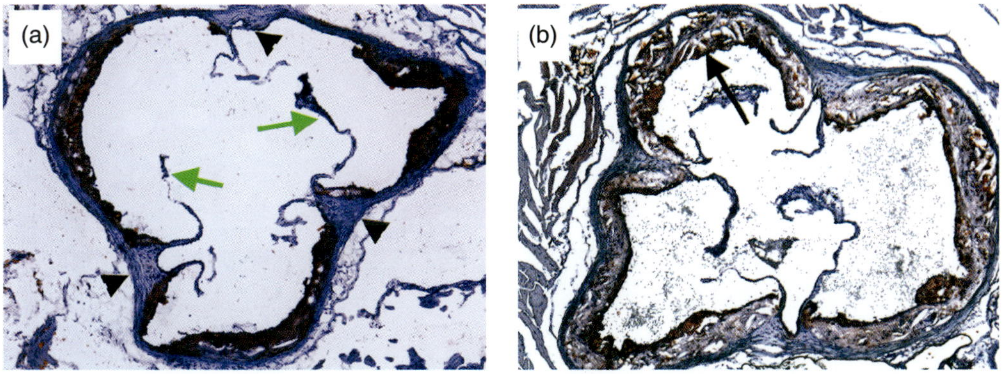

Figure 16.1 Oil Red O and hematoxylin stained aortic root lesions from apoE KO mice. (a) Typical lesion in 16-week-old chow-fed apoE KO mice on the C57BL/6 background. The *black arrow* heads point to the three valve bases that are visible throughout the zone of sectioning. The *green arrows* point to two of the six valve leaflets. This section shows the medial layer stained only with hematoxylin and dark red-stained fatty streak lesions in the intima over most, but not all of the lumen surface. (b) Advanced lesions in 20-week-old apoE KO mice fed a Western-type diet. The entire lumen area is covered with extensive lesions, although the lesion is not completely stained with Oil Red O, indicating the presence of cells, necrotic area, and matrix without cholesterol ester deposits. The *black arrow* points to a lesion containing many cholesterol clefts, the crystalline cholesterol deposited after the death of free cholesterol enriched foam cells.

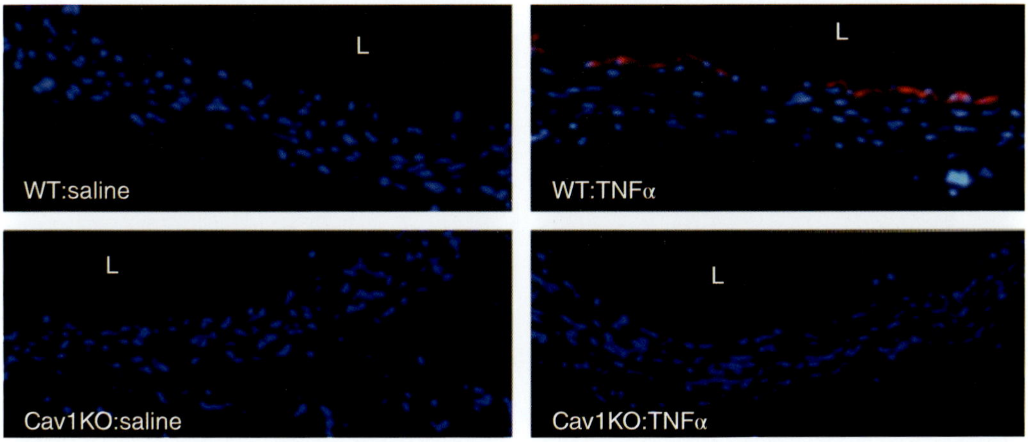

Figure 18.1 Caveolae are necessary for upregulation of adhesion molecules expression in endothelial cells exposed to proinflammatory cytokines. WT and Cav1KO mice were intravenously injected with a bolus of saline or TNFα (20 μg/kg). After 2 hours, aortic tissue was fixed with paraformaldehyde, embedded in paraffin wax and sectioned at 5 μm. Tissue sections were processed for labeling of VCAM (*red*) and cell nuclei (Dapi; *blue*). The data show that TNFα enhanced VCAM expression only in vessels from wild-type mice. L indicates lumen.

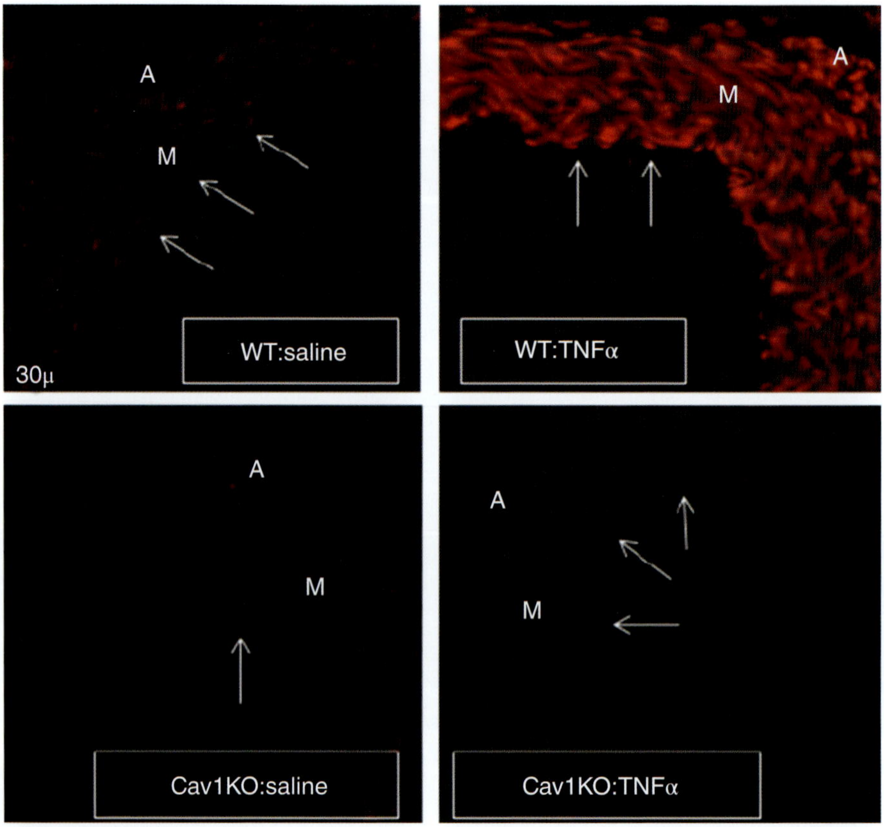

Figure 18.2 Caveolae are necessary for reactive oxygen species production. WT and Cav1KO mice were intravenously injected with a bolus of saline or TNFα. Following 2 hours of exposure, aortic tissue was removed, embedded in OCT, sectioned (30 μm), and incubated with DHE (2 μmol/l). Oxidized DHE was detected by fluorescence microscopy (Nikon TE2000; excitation 488 nm, emission 610 nm). The *arrows* indicate luminal endothelial cell layer, as verified by secondary staining for VE-cadherin (data not shown). There was little evidence of inflammatory leukocyte adherence to the vessel wall (CD 45 staining—data not shown), ruling out leukocytes as a source of ROS at this time point. A, adventitia and M, medial layer.

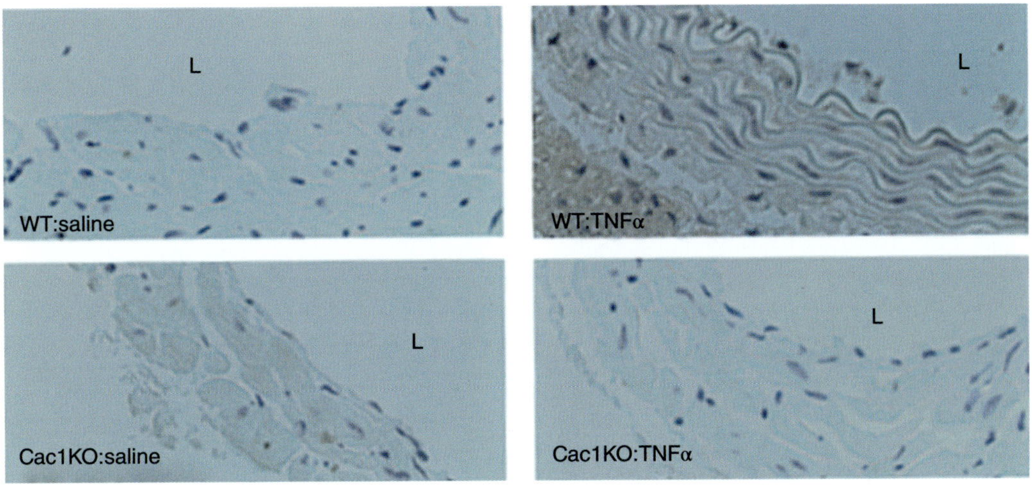

Figure 18.3 Caveolae are necessary for induction of protein tyrosine nitration by proinflammatory cytokines. WT and Cav1KO mice were intravenously injected with a bolus of saline or TNFα (20 mg/kg). After 2 hours, aortic tissue was fixed with paraformaldehyde, embedded in paraffin wax and sectioned at 5 μm. Tissue sections were incubated with primary antibody against nitrotyrosine followed by horseradish peroxidase-tagged secondary antibody. Sections were then incubated with DAB. Results indicate the presence of a brown reaction product only in vessels from TNFα-treated mice. As was the case for ROS, TNFα failed to induce protein tyrosine nitration in vessels from Cav1KO mice.

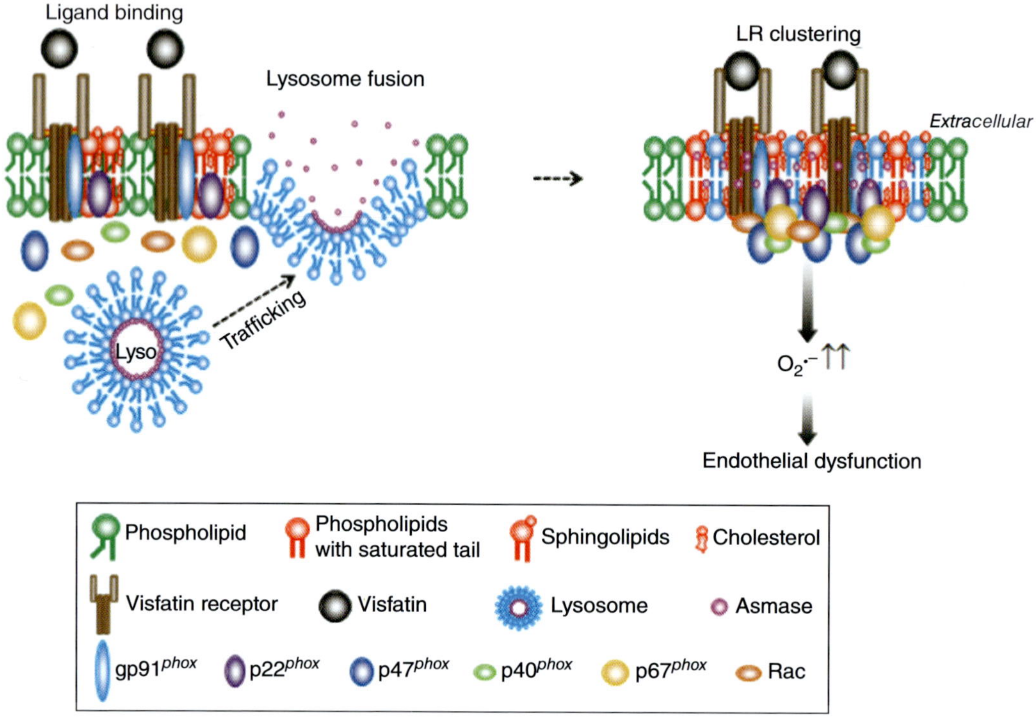

Figure 19.1 Lysosome fusion to plasma membrane to form MRs signaling platforms in ECs. In response to stimuli, lysosomes can be mobilized to traffic and fuse into cell membrane, where acid sphingomyelinase (ASMase) is activated and ceramide produced, resulting in MRs clustering and formation of signaling platforms (Adapted from Xia et al. [13]).

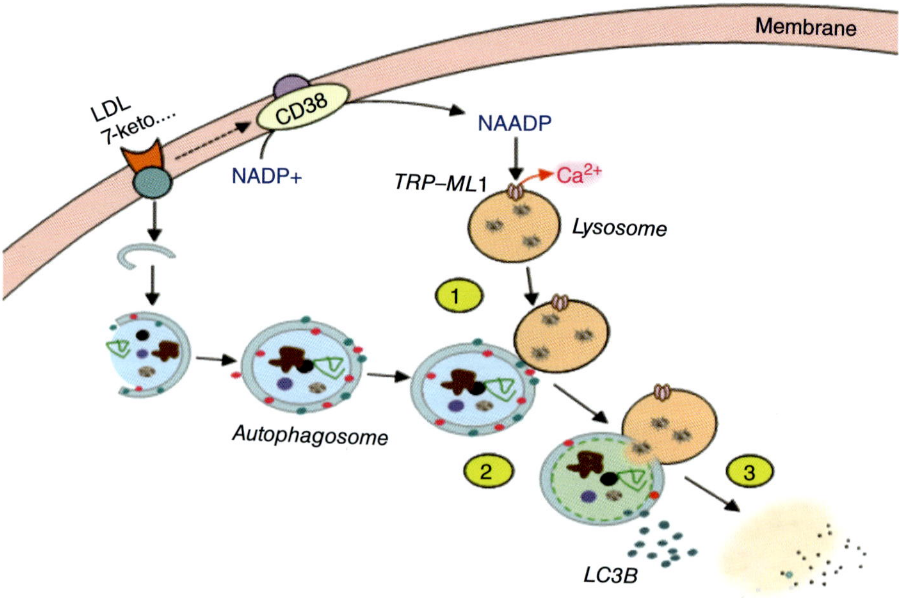

Figure 19.2 Regulation of lysosome trafficking and fusion in autophagy. Lysosome trafficking and fusion with autophagosomes (APs) is not a random process, which is fine controlled by different signaling mechanisms. CD38-NAADP signaling pathway is importantly involved in such lysosome regulation, and therefore the defect of CD38 such as gene knockout resulted in deficient lysosome fusion and autophagic flux, which may result in SMCs proliferation, growth, and production of extracellular matrix, leading to atherosclerosis.

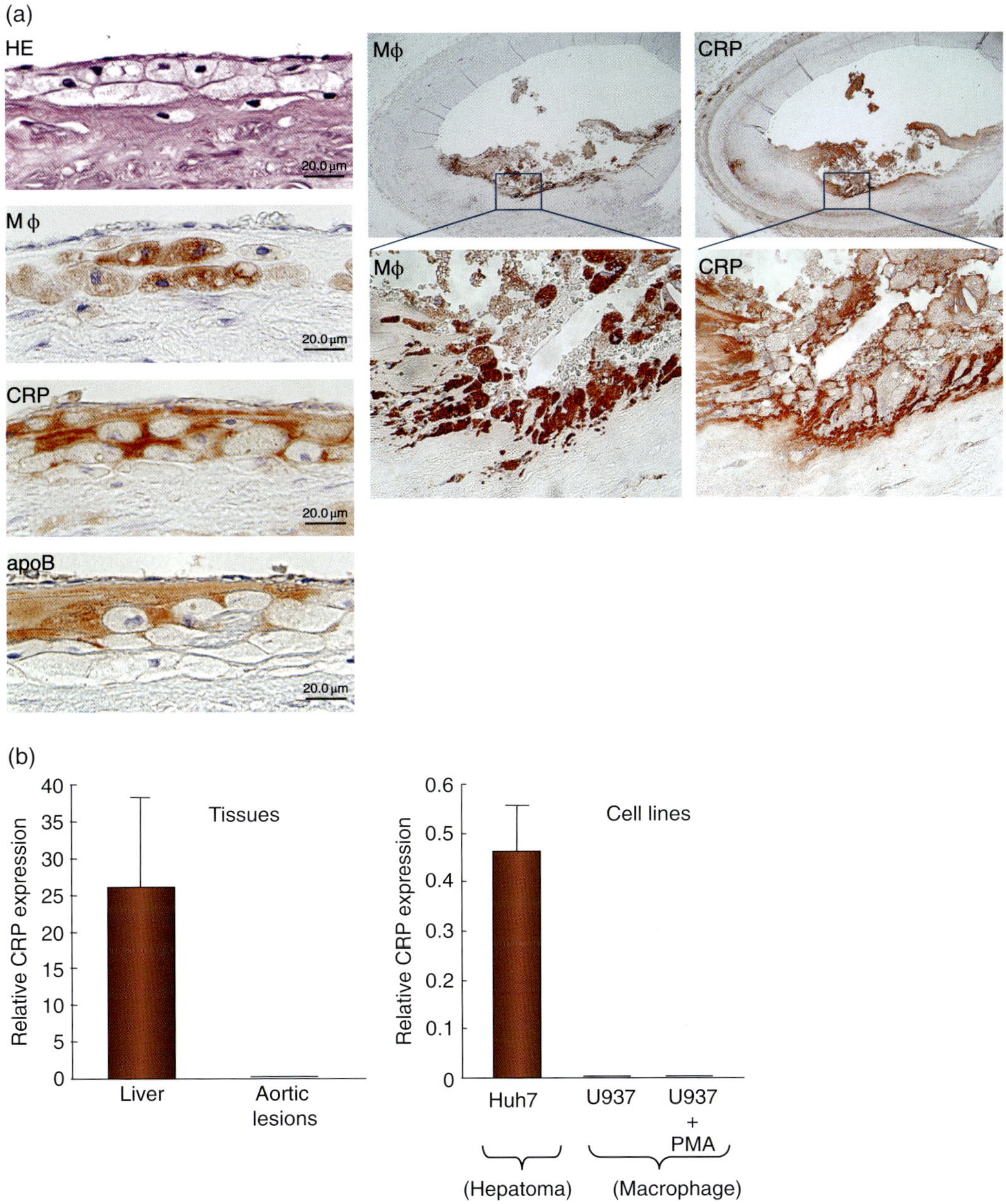

Figure 20.1 (a) Detection of CRP in human aorta and coronary plaque. Immunocytochemistry demonstrating the presence of CRP proteins in atherosclerotic lesions of human aorta (left) and in ruptured plaque in coronary atherosclerosis (right). Note that CRP distribution is associated with that of apoB. Mϕ, macrophages staining. (b) CRP mRNA expression in human tissues and cultured cells. Human liver and aortas were collected from autopsy cases, while the hepatoma (Huh7) and monocytic (U937) cell lines were cultured *in vitro*. CRP mRNA expression was determined by quantitative real-time RT-PCR (Adapted from Sun et al. [18] with modification and permission. © 2005, Elsevier).

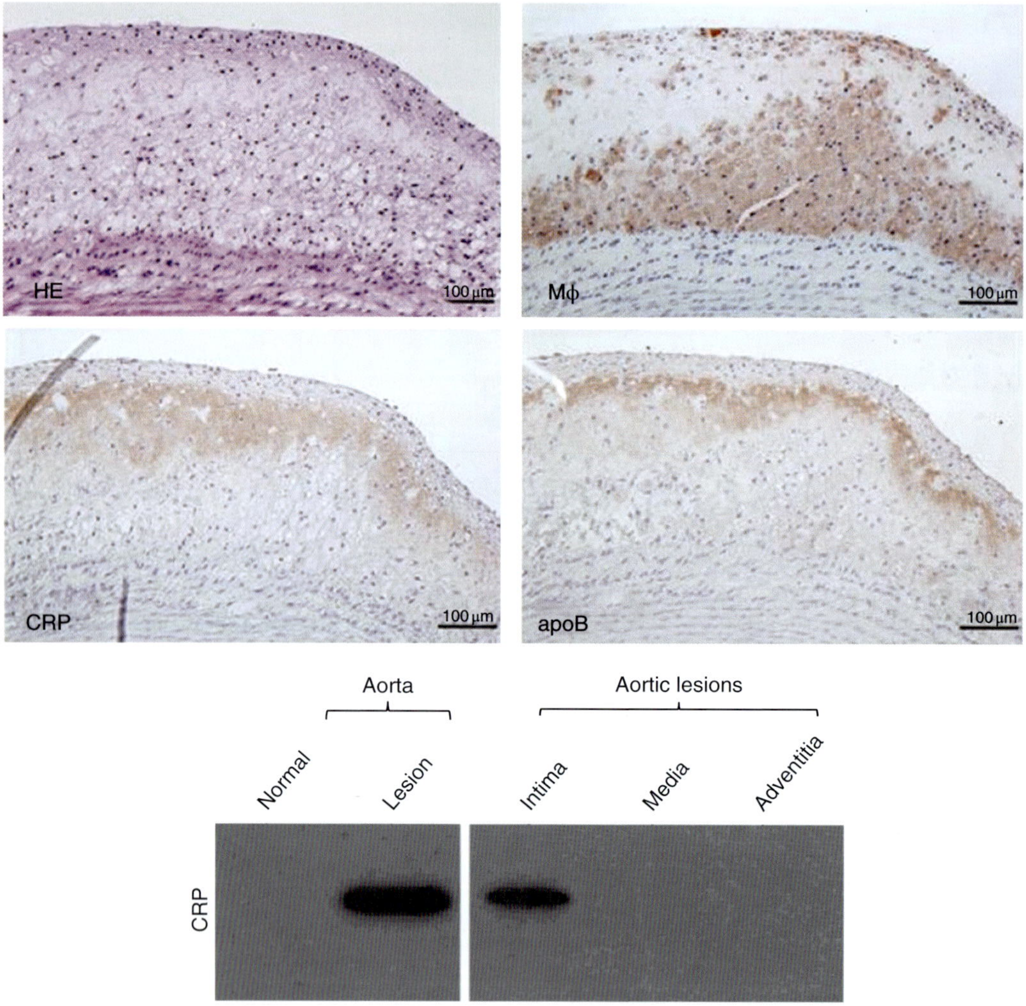

Figure 20.2 Detection of CRP in the lesions of WHHL rabbits. CRP is present in rabbit aortic lesions by immunohistochemical staining (top) and Western blotting (bottom). Compared to human atherosclerotic lesions, positive immunostaining for CRP is frequently observed in rabbit aortic lesions, which is mainly localized to the intima (Adapted from Sun et al. [18] with modification and permission. © 2005, Elsevier).

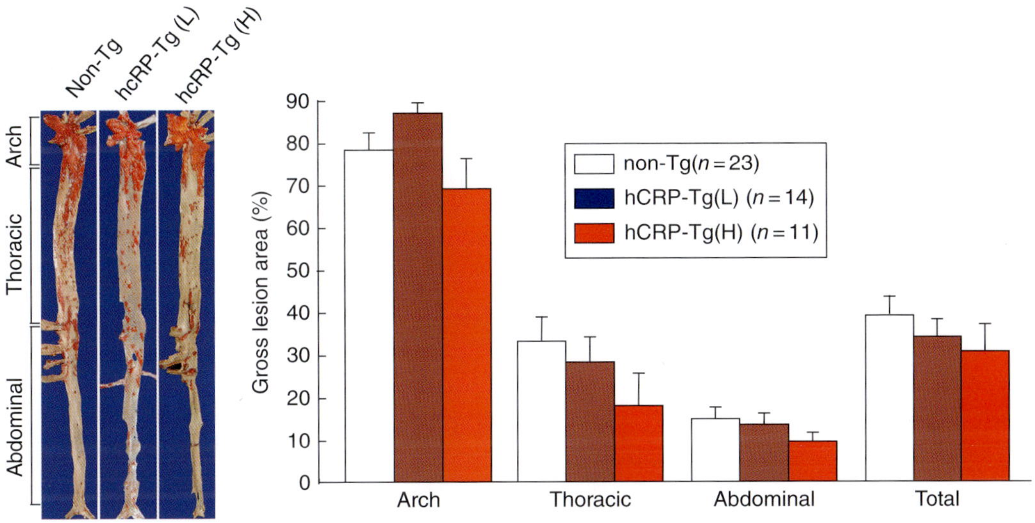

Figure 20.3 Aortic atherosclerotic lesion analysis. Representative photographs of pinned-out aortic trees stained with Sudan IV from hCRP transgenic (Tg) and wild-type (non-Tg) rabbits (left), and aortic atherosclerotic lesions (defined by sudanophilic area) on the surface were quantified by an image analysis system (right). The data show that high levels of hCRP are not proatherogenic in transgenic rabbits (Adapted from Koike et al. [38] with permission. © 2009, Lippincott Williams & Wilkins).

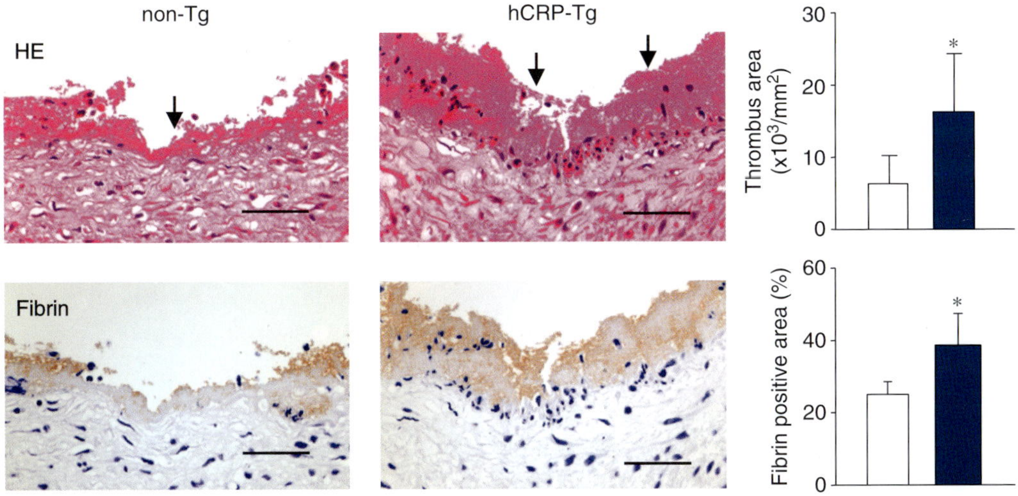

Figure 20.4 Human CRP enhances thrombus formation in transgenic rabbits after balloon injury. Thrombus size (*arrows*) and immunoreactive protein fibrin on neointimas were quantified by image analysis, and results are shown in the bar graph on the right. The scale bar: 100 μm. *: $P < 0.05$ (Adapted from Matsuda et al. [44] with modification and permission. © 2011, John Wiley & Sons).

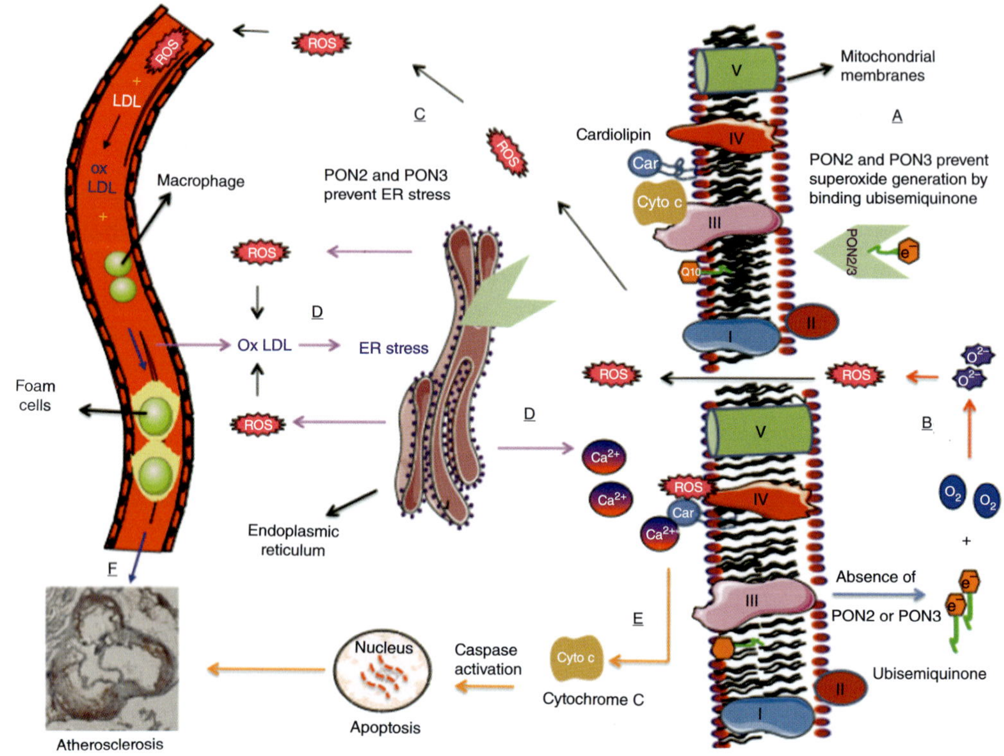

Figure 21.2 Working model for the antiatherogenic properties of PON2 and PON3 proteins. In the inner mitochondrial membrane, ubisemiquinone is released from electron transport chain (ETC) during Q cycle, which is inactivated by PON2/PON3, thereby preventing the formation of superoxides (A). In the absence of PON2 or PON3, ubisemiquinone donates electron to molecular oxygen to form superoxides that generate other reactive oxygen species (ROS) (B). ROS oxidize LDL phospholipids (e.g., PAPC) to form oxLDL phospholipids (e.g., oxPAPC) (C). OxLDL causes ER stress and results in an increased ROS and calcium release to the cytosol (D). Enhanced uptake of calcium by mitochondria causes release of cytochrome c to cytosol (E). Decrease in cytochrome c in the mitochondria causes reduction in the complex IV activity, which results in more ubisemiquinone causing more lipid oxidation as described above in steps (B) and (C) above (E). Increase in cytochrome c levels in the cytosol causes caspase activation and apoptosis (E). Accumulation of oxLDL in macrophages leads to foam cell formation and atherosclerosis (F). Abbreviations: I-NADH, ubiquinone oxidoreductase; II, succinate coenzyme Q reductase; III, ubiquinol cytochrome oxidoreductase; IV, cytochrome c oxidase; V, ATP synthase; Car, cardiolipin; and Q10, coenzyme Q10.

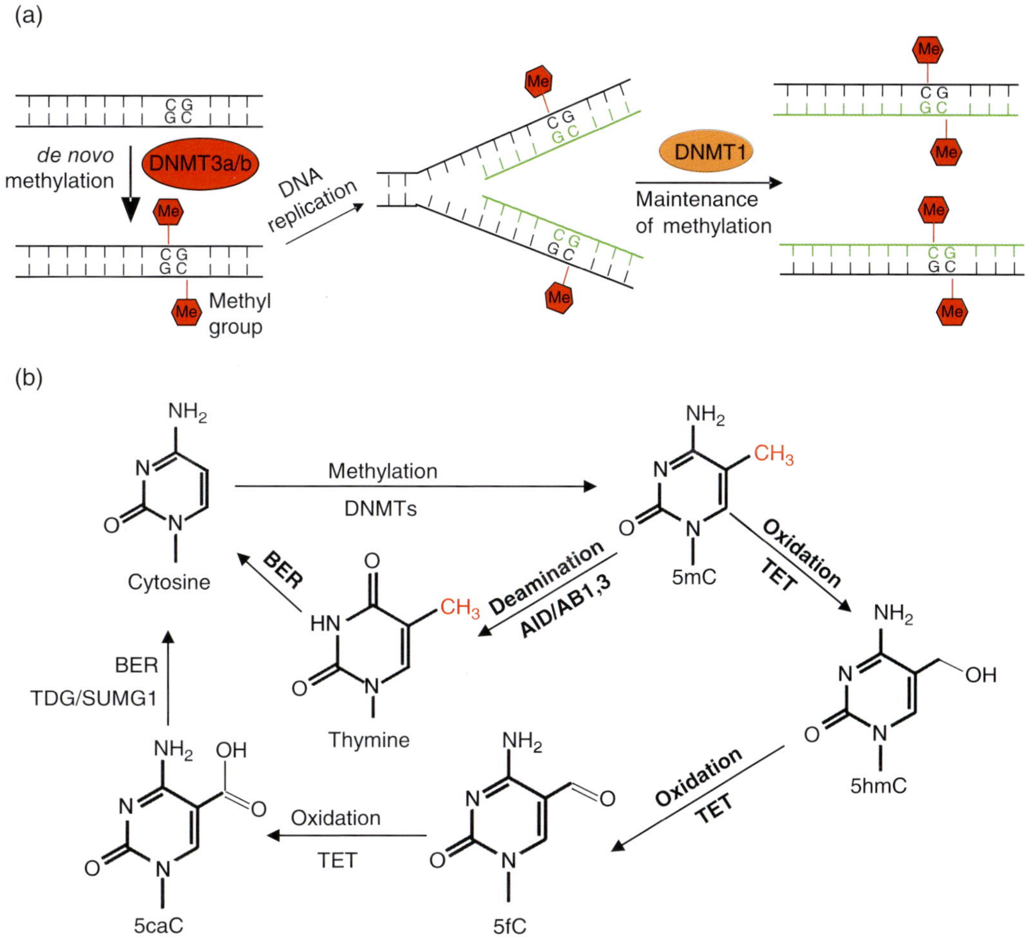

Figure 32.1 DNA methylation and demethylation. (a) Establishment and maintenance of DNA methylation. In mammals, global demethylation occurs immediately after fertilization, erasing all existing methylation patterns. After implantation, DNA methyltransferases (DNMTs) 3a/b de novo transfer methyl group to cytosine residue of CpG dinucleotides and establish new DNA methylation patterns. During cell division, DNMT1 methylates newly synthesized DNA strands to maintain the methylation patterns. (b) DNA demethylation pathway. Three enzyme families have been implicated in active DNA demethylation. (i) 5-methylcytosine (5mC) can be hydroxylated by 10–11 translocation (TET) family of enzymes to form 5-hydroxymethylcytosine (5hmC) or further oxidized to 5-formylcytosine (5fC) and 5-carboxylcytosine (5caC). (ii) 5mC (or 5hmC) can be deaminated by the activation-induced (cytidine) deaminases/apolipoprotein B mRNA editing enzyme, catalytic polypeptide-like (AID/APOBEC) family members to form thymine. (iii) Replacement of these intermediates (i.e., thymine or 5caC) is initiated by the UDG family of base excision repair (BER) glycosylases like TDG or SMUG1, culminating in cytosine replacement and DNA demethylation.

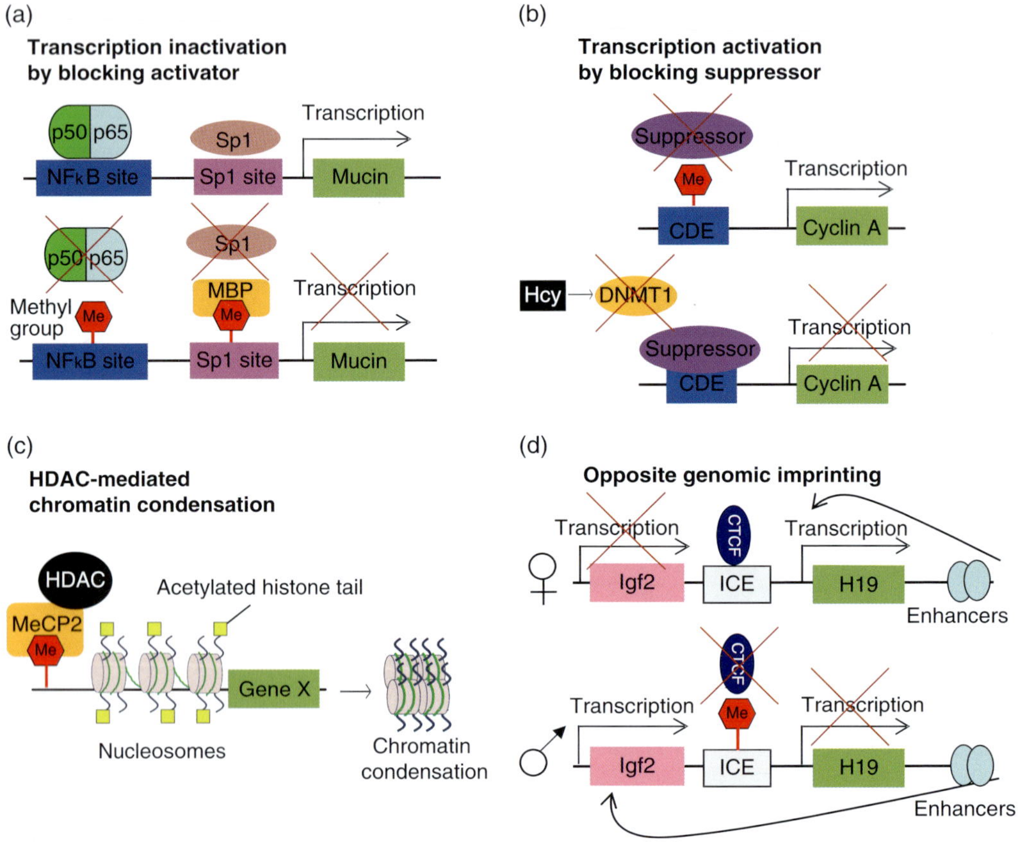

Figure 32.2 DNA methylation regulates gene transcription. (a) DNA methylation blocks transcription activator to repress transcription. The NFκB site and Sp1 site on the promoter of actively transcribed *mucin* gene are not methylated. Methylation of the NFκB site blocks p50–p65 dimer binding. Methylation on Sp1 site recruits methyl-CpG-binding protein (MBP) to block Sp1 binding. These modifications together lead to the inhibition of mucin transcription. (b) DNA methylation blocks transcription suppressor to facilitate transcription. The cell cycle-dependent element (CDE), a transcription suppressor binding site, on the promoter of actively transcribed cyclin A gene is methylated in EC. Homocysteine (Hcy) treatment inhibits DNMT1 activity and induces hypomethylation of CDE on cyclin A promoter in EC. This allows transcription suppressor to bind to the unmethylated CDE, thereby inhibiting cyclin A transcription. (c) DNA methylation induces chromatin condensation to repress transcription. Methyl- CpG-binding protein 2 (MeCP2) binds to methyl CpG and mediates transcription repression by HDACs, which remove acetyl groups from histones and trigger chromatin condensation, leading to transcription repression. (d) DNA methylation causes opposite genomic imprinting. The H19 and Igf2 genes are on the same mouse chromosome and are oppositely "imprinted." On the maternal chromosome, the protein CTCF binds to the unmethylated ICE, blocking the access of enhancers to Igf2. Thus, H19 is expressed instead. On the paternal chromosome, the ICE is methylated, so CTCF cannot bind; therefore, the enhancers have access to Igf2. Igf2 is expressed, whereas H19 is silenced.

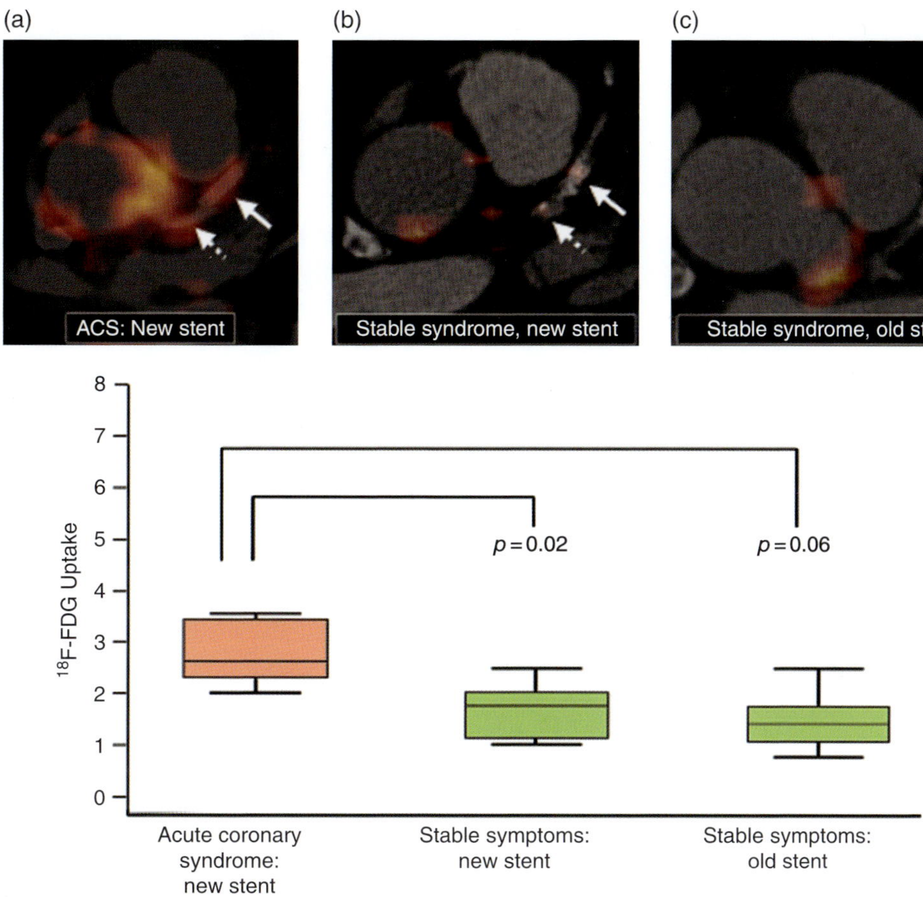

Figure 39.2 ^{18}F-FDG PET/CT coronary imaging (top panels) and FDG target-to-background ratios (bottom graph) in patients with acute coronary syndrome (panel a) and stable angina (panels b and c), respectively, that underwent percutaneous intervention. (a) ^{18}F-FDG PET/CT imaging demonstrated ^{18}F-FDG uptake in the left main (hatched arrows) and recently stented left anterior descending artery culprit lesion (solid arrows). (b and c) In patients with stable chest pain syndromes, significantly lower levels of ^{18}F-FDG uptake were associated with culprit lesions recently or previously stented, respectively. ^{18}F-FDG PET, ^{18}F-fluorodeoxyglucose positron emission tomography (Modified from Rogers et al. [38]. With permission. © Elsevier).

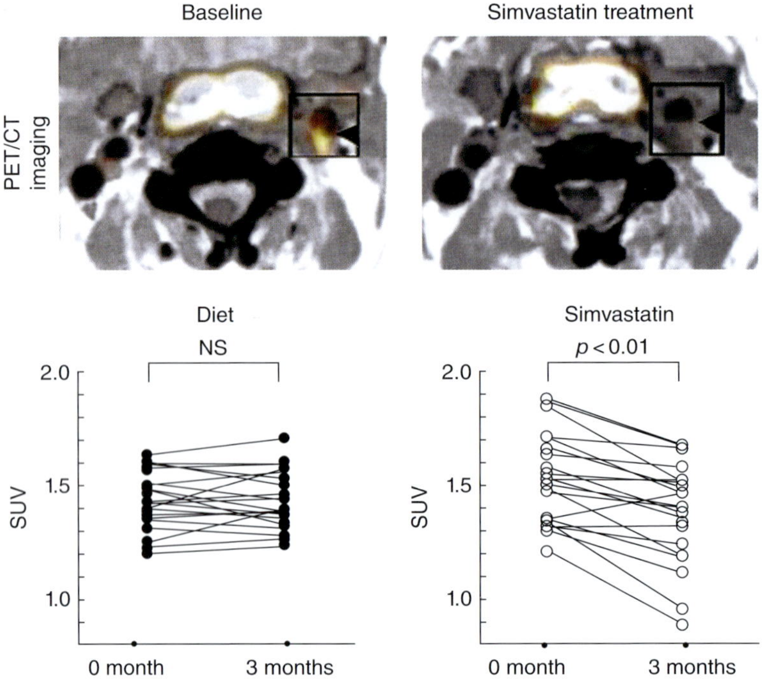

Figure 39.3 Simvastatin therapy decreases ^{18}F-FDG accumulation in carotid atherosclerosis. (Top panels) Serial ^{18}F-FDG PET/CT imaging at baseline (right panel) and after 3 months of simvastatin therapy (left panel) demonstrates a reduction in ^{18}F-FDG carotid uptake (see insets). (Bottom panels) Quantitative PET analysis showed that ^{18}F-FDG carotid uptake after 3 months was unchanged in those undergoing dietary modification (left panel) but was significantly decreased in subjects treated with simvastatin (right panel). ^{18}F-FDG PET, ^{18}F-fluorodeoxyglucose positron emission tomography; NS, not significant; SUV, standardized uptake value (Modified from Tahara et al. [39]. With permission. © Elsevier).

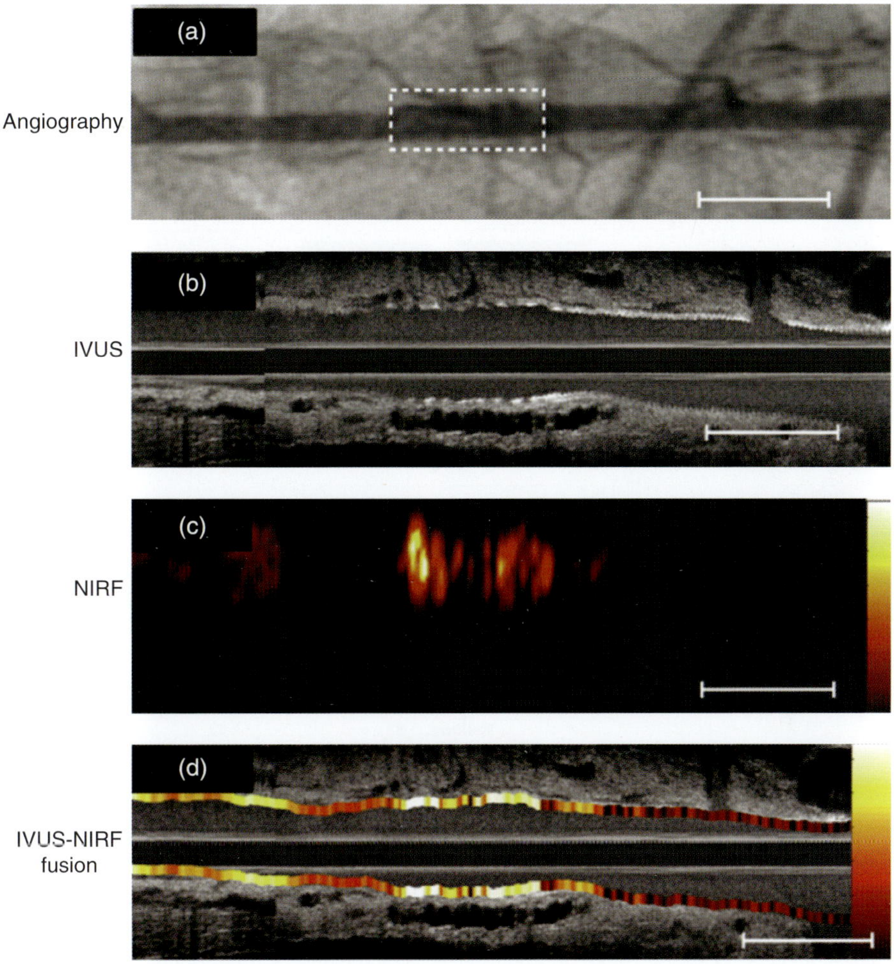

Figure 39.4 Intravascular two-dimensional NIRF imaging of plaque inflammation in coronary-sized arteries through blood. The NIRF molecular imaging catheter detects vascular inflammation (cathepsin protease activity) 7 days following stent implantation in rabbits. (a) The dashed box denotes the position of a stent deployed in the abdominal aorta. (b and c) Longitudinal and rotational IVUS (panel b) and NIRF (panel c) interrogation facilitated by automated pullback demonstrates increased proteolytic activity localized to the stented region. (d) Longitudinal NIRF–IVUS fusion image (yellow/white color denotes strongest intensity NIRF signal; red/black color denotes lowest intensity NIRF signal). IVUS, intravascular ultrasound; NIRF, near-infrared fluorescence (Modified from Jaffer et al. [48]. With permission. © Elsevier).

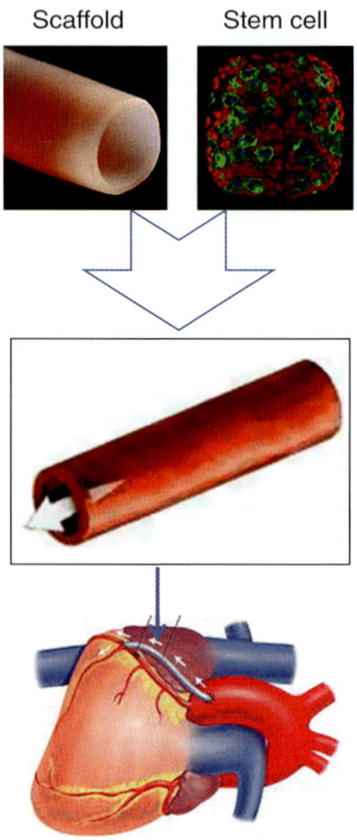

Figure 43.3 Schematic representation for vascular tissue engineering. Scaffold and stem cells could interact in a bioreactor where they form a blood vessel, which can be grafted to coronary artery to replace severely diseased vessel.

as CD20, CD40, TNFR, and epidermal growth factor receptor (EGFR) [36, 37]. As shown in Figure 19.1, the LR redox signaling mechanism associated with LRs clustering may mediate the detrimental action of some injurious stimuli such as FasL, visfatin, or atherogenic stimuli to induce endothelial dysfunction. When these danger factors act on ECs, lysosomes proximal to plasma membrane are stimulated to traffic and fuse into cell membrane, resulting in activation of ASMase, production of ceramide, and formation of MR-Nox redox signaling platforms. This redox signaling platforms contain a variety of signaling molecules that interact to form signalosomes, which produce O_2^- and induce endothelial dysfunction, resulting in atherogenesis when the injurious stimuli are sustained [13].

Redox Signaling or Injury and Amplification by Lysosomes

As mentioned earlier, the major function of MR redox signalosomes is to transmit and amplify signals received by corresponding receptors through aggregation of signaling molecules to work in a platform manner. However, there is also evidence that a feedforward amplifying mechanism exists in the formation of such MR redox signaling platforms, which is associated with lysosomal molecular activation or lysosome dysfunction. It has been proposed that the production of ROS within MR redox signaling platforms can further enhance the formation of larger platforms, thus amplifying the whole process. In such enhancement of platform formation, increased activation of lysosomal ASMase by ROS is found to be a key contributor. The formation of ASMase dimer by modification of the free C-terminal cysteine is mainly responsible for enhancement of ASMase activity [38, 39], which produces more ceramide, further promoting MR platform formation. It may also be related to the enhanced lysosome trafficking since selective activation of lysosomal ASMase by pharmacological intervention leads to lysosome trafficking and fusion to the MR area on the plasma membrane in ECs [34]. This lysosome-mediated amplifying loop for MR redox signaling drives an efficient signaling; and if excessively enhanced, this feedforward signaling amplification may result in the progress and development of severe oxidative stress, which has been considered as a hallmark of advanced atherosclerotic lesions. Another amplification role of lysosomes in vascular regulation relates to the damage of lysosome membrane. It has been demonstrated that largely enhanced production or accumulation of ROS may damage the lysosomal membrane, which results in alteration of the lysosomal compartments. The latter in turns prevents lysosome fusion with autophagic vacuoles containing damaged components and, thereby, results in release of potent hydrolases. This will further enhance cellular damage even though the initial damaging stimuli may disappear. This augmented intracellular oxidative stress and lysosome amplification by release of lysosomal enzymes will cause cell death, which may lead to the destabilization of atherosclerotic plaques if occurring in SMCs. This is due to the reduced synthesis of collagen and thinning of the fibrous cap. If such death occurs in ECs, it promotes lesional thrombosis in the involved arteries.

LYSOSOMAL CHOLESTEROL METABOLISM AND TRANSPORT IN FOAM CELL FORMATION

Cholesterol Metabolism and Transport Via Lysosomes

It has been well known that under normal conditions, there are distinct trafficking pathways for de novo synthesized cholesterol and cholesterol obtained via the receptor-mediated endocytosis of LDLs. Exogenously obtained cholesterol cannot be directly used by the cells, and it must pass through the endosomal/lysosomal compartment prior to its trafficking to the endoplasmic reticulum, the Golgi apparatus, or the plasma membrane. LDL usually binds to LDL receptor (LDLR) resulting in subsequent endocytosis of the receptor-ligand complex at clathrin-coated pits [40]. After endocytosis, acidic conditions of the endosomes catalyze disassociation of LDL from the LDLR. After release of the lipoprotein, the LDLR peptide is recycled back to the membrane. Meanwhile, LDL particles released from LDLR fuse into lysosomes, and they are degraded into lipid components and amino acids by enzymes of the vesicle. The lipid components released are mainly cholesteryl esters that are hydrolyzed by lysosomal acid lipase (LIPA) into free cholesterol [41]. Free cholesterol is then actively transported out of lysosomes to the cytosol, presumably through lysosomal NPC1 protein. This lysosome-originated free cholesterol usually has three distinct destinations including its incorporation into cell membrane, re-esterification with fatty acids by acyl-CoA:cholesterol acyltransferase-1 followed by storage in cytoplasmic inclusions, and removal from plasma membrane by high HDL. It has been reported that the effective transport of free cholesterol from lysosomes into cytosol plays a critical role in clearance of cholesterol from macrophages, which may prevent the foam cell formation and ultimate atherosclerosis.

Foam Cell Formation Due to Lysosomal Molecular Derangement

It is known that a pathological hallmark of atherosclerosis is the excessive accumulation of lipids in macrophages, which is featured by the lipid-swollen lysosomes. Cholesterol in both free and esterified form can be detected in lysosomes, but free cholesterol is dominant. There are evidences that the deposition of cholesterol in lysosomes is associated with a derangement of a number of molecules such as insufficient LIPA activity, defect of NPC1 or mucolipin-1, ASMase deficiency, CD38 loss, and so on [42] (and unpublished results). It has been proposed that the development of macrophage lysosomal lipids accumulation during foam cell formation has two distinct consecutive phases, that is, the primary accumulation of free cholesterol in the initial phase followed by the late phase of cholesteryl ester buildup. In addition, the accumulated cholesterol in lysosomes also have inhibitory effects on lysosomal V-H^+-ATPase activity, which in turn results in decreased in lysosomal acidity and causes dysfunction in the conversion of esterified cholesterol into free cholesterol, further preventing cholesterol transport out of lysosomes. This forms a vicious cycle leading to a failure of cholesterol metabolism and efflux from lysosomes that ultimately result in the formation of foam cells. In recent studies, we demonstrated that a lysosome function regulator, nicotinic acid adenine dinucleotide phosphate (NAADP) importantly participates in cholesterol transport from lysosomes. This nucleotide as a CD38-derived Ca^{2+} messenger activates lysosomal transient receptor potential-mucolipin-1 (TRP-ML1) and functionally promotes free cholesterol efflux from lysosomes in macrophages. It was found that the deficiency of CD38 resulted in a defect of lysosome-free cholesterol efflux and consequent lipid deposition in macrophages leading to foam cell formation and atherosclerosis. It is assumed that not only lysosomal molecular defect or derangement, but also dysregulation of lysosomal molecules or functions by related signaling pathways may contribute to the formation of foam cells and atherogenesis [42].

Atherosclerosis as a Lysosomal Storage Disorder in the Arterial Wall

In 1960s, biochemical and histochemical studies already demonstrated that the abnormal lysosomal activity was associated with the development of atherosclerosis [43, 44]. The accumulated lipids in macrophage foam cells mainly that occur within large, lipid-engorged lysosomes were shown in both human and animal atherosclerotic lesions [45, 46]. These pathological changes in the foam cells during atherosclerosis are similar to tissue damage during inherited lysosomal storage disorders, in particular, mucolipidosis type IV [47, 48] and NPC1 diseases [49], both of which are characterized by accumulation of lipids in lysosomes or cytosol of macrophages or arterial wall. Therefore, the same mechanisms of lysosomal storage diseases, namely, defects of some lysosomal enzymes or transporters, were considered as the molecular basis to form macrophage foam cells and to induce atherosclerosis [44]. Despite extensive studies of a half century, many conflicting results were reported, some of which did not support the assumption of atherosclerosis as lysosomal storage disease in the arterial wall because there was no evidence of lysosomal molecular defects in atherosclerosis despite several of lysosomal enzymes were tested such as lysosomal acid cholesterol esterase (LACE), cysteine proteases (CPs), acylcholesterol acyl transferase (ACAT), and TRP-ML1. Although no molecular identity was confirmed that is responsible for the lipid deposition in microphages, the central role for lysosomes in the formation of foam cells has been still continuously shown in many recent studies [44, 50]. It is possible that some dysregulation of lysosome function rather than any defects of lysosomal molecules may be implicated in the formation of foam cells. Our recent studies have indicated that the regulation of lysosomal TRPML activity by a CD38-NAADP signaling pathway plays a significant role in the transport of cholesterol from lysosomes [42].

LYSOSOME-CENTERED AUTOPHAGY IN ATHEROSCLEROSIS

Role of Lysosomes in Autophagy

Three forms of autophagy have been reported since 1960s, which include chaperone-mediated autophagy, microautophagy, and macroautophagy. Thus far, the macroautophagy has been well characterized, which is a major regulated catabolic mechanism that eukaryotic cells use to degrade long-lived proteins and organelles [18, 51]. Unless there is special definition or description in a study about other two types of autophagy, macroautophagy is often referred to as autophagy in literature. In this chapter, the term "autophagy" also means macroautophagy, which will be discussed to explore its role in atherogenesis or development of atherosclerotic plaques in the arterial wall. Although autophagy was described in 1960s, the knowledge of autophagy was still limited to understanding of its morphological and biochemical characteristics until the early 1990s. Over the past decade, due to the discovery of yeast autophagy genes (Atg genes) followed by the identification of their mammalian homologues, numerous studies have been performed to elucidate the molecular machinery of this main cellular homeostatic process and its regulatory mechanisms [52–55]. It has now been known that autophagy occurs through four stages including induction, autophagosome (APL) formation, docking, and fusion with lysosomes (namely, the formation of autophagolysosome [APLs] or autolysosomes), and breakdown of autophagic vesicles [54]. Although the formation of APs has been well studied, the molecular mechanisms behind their fusion with lysosomes and subsequent breakdown of the autophagic vesicles in APLs are still poorly understood. Recently, some studies in our laboratories focus on such mechanisms revealing that the normal regulation of lysosome trafficking and fusion through NAADP or ceramide as well as lysosomal and cytosolic Ca^{2+} levels [56].

Autophagy in Atherosclerosis

Given different cell types in the artery wall, the role of autophagy in the development of atherosclerosis is complex due to its action on different vascular cell functions. It has been recently assumed that autophagy may have both protective and detrimental roles during atherosclerosis, depending on the status of autophagy or stages of atherosclerosis [19, 57]. Since autophagy is important in degradation of damaged materials, it is possible that autophagy in the arterial wall helps clean up damaged components and recover cells from the damage upon atherosclerotic stimuli. In addition, autophagy activation interferes with cell apoptosis due to engulfment of defective or damaged mitochondria by APs, which limits the release of proapoptotic proteins [20, 58, 59]. This autophagic process may protect arterial cells from atherogenic injury. If acute or persistent oxidative stress occurs in atherosclerosis, however, lysosomes may be damaged to release its hydrolases; and in this way, autophagy is engaged as part of oxidative stress in enhancement of cellular damages [19, 60–62].

Enhanced or reduced autophagy may play different roles in the development of atherosclerosis upon its actions on different involved cells. In macrophages, autophagy increases cholesterol transport out of these cells, which may prevent lipid droplet formation reducing foam cell formation. Similarly, enhanced autophagic death of macrophages also possibly attenuates the foam cell formation, reducing atherosclerotic injury. However, excessive activation of autophagy in ECs may lead to damage of the endothelium enhancing atherogenic injury. In arterial SMCs, enhanced autophagy induces their modulation to a differentiated, quiescent, and contractile phenotype decreasing cell proliferation and preventing fibrosis. Nevertheless, excessive autophagy in arterial SMCs may result in their death increasing the instability of atherosclerotic plaques [60, 62–65]. Although the role of augmented autophagy in atherosclerosis has been extensively studied, there is no evidence that a defective or reduced autophagy is involved in the pathogenesis of atherosclerosis. More recently, we demonstrated that the defective autophagy is also importantly involved in atherogenesis, which is associated with molecular dysregulation of lysosome function. As shown in Figure 19.2, proatherogenic stimuli such as 7-keto-cholesterol (7-keto) or oxy-LDL activates autophagy in arterial SMCs, leading to the formation of APs. Under normal condition, lysosome trafficking and fusion to

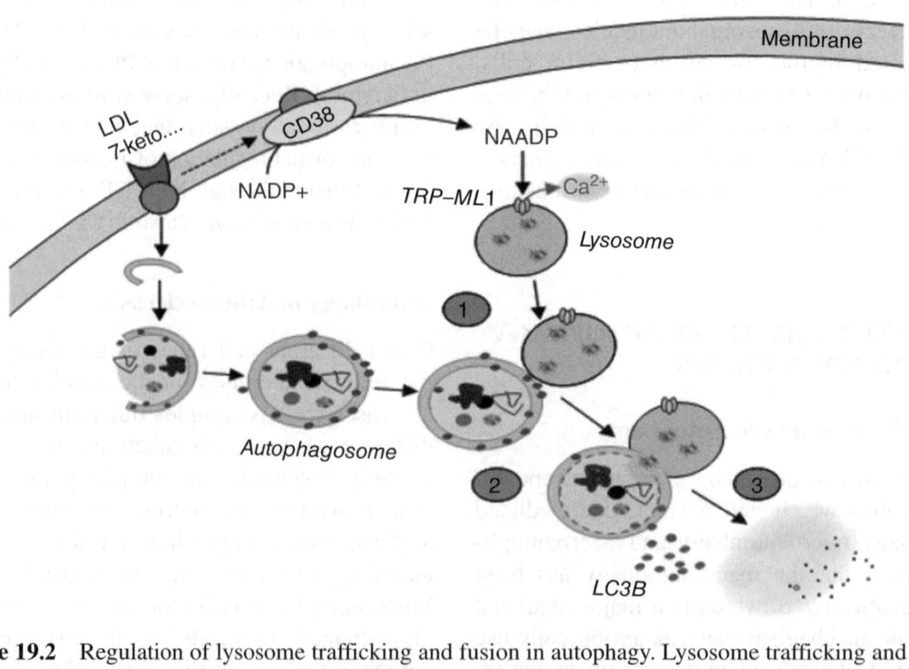

Figure 19.2 Regulation of lysosome trafficking and fusion in autophagy. Lysosome trafficking and fusion with autophagosomes (APs) is not a random process, which is fine controlled by different signaling mechanisms. CD38-NAADP signaling pathway is importantly involved in such lysosome regulation, and therefore the defect of CD38 such as gene knockout resulted in deficient lysosome fusion and autophagic flux, which may result in SMCs proliferation, growth, and production of extracellular matrix, leading to atherosclerosis. (*See insert for color representation of the figure.*)

APs are controlled by CD38-ADP-ribosylcyclase-mediated regulation or other signaling mechanisms, which leads to the formation APLs and subsequent breakdown of the autophagic vesicles within the cells. This regulated autophagic process protects SMCs from atherosclerotic injury upon atherogenic stimulations. When the controlling mechanism of lysosome function is insufficient such as impaired CD38-ADP-ribosyl cyclase activity, the formation of APLs and breakdown of autophagic vesicles are impaired, which activates cell dedifferentiation, proliferation, and growth and, thereby, stimulates production of extracellular matrix, ultimately inducing or accelerating atherosclerosis. Indeed, CD38 gene knockout mice developed atherosclerosis when they were exposed to the atherogenic diet [56]. Although there is no lysosomal molecular derangement under such condition, dysregulation of lysosome function by the nonlysosomal molecular mechanism also causes a deficient autophagy, in particular, failed formation of APLs, which contributes to the development of atherosclerosis.

CONCLUDING REMARKS

Lysosomal molecular derangements and dysregulation of lysosome functions have been implicated in the development of atherosclerosis. This pathogenic role of lysosome dysfunction relates to endothelial injury, macrophage lipid deposition, and arterial SMC growth and production of ECM. It is proposed that both enhancement and reduction of lysosome function may be involved in the pathogenesis. Depending on the stages of atherosclerosis development, lysosome-associated mechanisms may produce different pathological actions such as endothelial injury, foam cell formation, SMC dedifferentiation, growth and production of ECM, and destabilizing atherosclerotic plaques. In conclusion, the role of lysosomes in the development of atherosclerosis is complicated, and it is such complex role of lysosome molecular derangement or function dysregulation in different atherogenic cells in combination with other pathogenic factors that determines the chronic progression of atherosclerosis.

REFERENCES

1. De Duve, C. The lysosome. *Sci Am* **208**, 64–72 (1963).
2. Huynh, C., Roth, D., Ward, D.M., Kaplan, J. & Andrews, N.W. Defective lysosomal exocytosis and plasma membrane repair in Chediak-Higashi/beige cells. *Proc Natl Acad Sci U S A* **101**, 16795–16800 (2004).
3. Jaiswal, J.K., Andrews, N.W. & Simon, S.M. Membrane proximal lysosomes are the major vesicles responsible for calcium-dependent exocytosis in nonsecretory cells. *J Cell Biol* **159**, 625–635 (2002).
4. McNeil, P.L. & Kirchhausen, T. An emergency response team for membrane repair. *Nat Rev Mol Cell Biol* **6**, 499–505 (2005).
5. Reddy, A., Caler, E.V. & Andrews, N.W. Plasma membrane repair is mediated by Ca(2+)-regulated exocytosis of lysosomes. *Cell* **106**, 157–169 (2001).
6. Bennett, E.M., Lin, S.X., Towler, M.C., Maxfield, F.R. & Brodsky, F.M. Clathrin hub expression affects early endosome distribution with minimal impact on receptor sorting and recycling. *Mol Biol Cell* **12**, 2790–2799 (2001).
7. Haimovitz-Friedman, A., Kan, C.C., Ehleiter, D., et al. Ionizing radiation acts on cellular membranes to generate ceramide and initiate apoptosis. *J Exp Med* **180**, 525–535 (1994).
8. Galione, A. NAADP, a new intracellular messenger that mobilizes Ca2+ from acidic stores. *Biochem Soc Trans* **34**, 922–926 (2006).
9. Lee, H.C. Nicotinic acid adenine dinucleotide phosphate (NAADP)-mediated calcium signaling. *J Biol Chem* **280**, 33693–33696 (2005).
10. Churchill, G.C. & Galione, A. NAADP induces Ca2+ oscillations via a two-pool mechanism by priming IP3- and cADPR-sensitive Ca2+ stores. *EMBO J* **20**, 2666–2671 (2001).
11. Jin, S., Yi, F. & Li, P.L. Contribution of lysosomal vesicles to the formation of lipid raft redox signaling platforms in endothelial cells. *Antioxid Redox Signal* **9**, 1417–1426 (2007).
12. Jin, S., Yi, F., Zhang, F., Poklis, J.L. & Li, P.L. Lysosomal targeting and trafficking of acid sphingomyelinase to lipid raft platforms in

coronary endothelial cells. *Arterioscler Thromb Vasc Biol* **28**, 2056–2062 (2008).
13. Xia, M., Zhang, C., Boini, K.M., Thacker, A.M. & Li, P.L. Membrane raft-lysosome redox signalling platforms in coronary endothelial dysfunction induced by adipokine visfatin. *Cardiovasc Res* **89**, 401–409 (2011).
14. Christensen, K.A., Myers, J.T. & Swanson, J.A. pH-dependent regulation of lysosomal calcium in macrophages. *J Cell Sci* **115**, 599–607 (2002).
15. Churchill, G.C., Okada, Y., Thomas, J.M., Genazzani, A.A., Patel, S. & Galione, A. NAADP mobilizes Ca(2+) from reserve granules, lysosome-related organelles, in sea urchin eggs. *Cell* **111**, 703–708 (2002).
16. Jha, K.N., Kameshwari, D.B. & Shivaji, S. Role of signaling pathways in regulating the capacitation of mammalian spermatozoa. *Cell Mol Biol (Noisy-le-grand)* **49**, 329–340 (2003).
17. Zhang, A.Y., Yi, F., Zhang, G., Gulbins, E. & Li, P.L. Lipid raft clustering and redox signaling platform formation in coronary arterial endothelial cells. *Hypertension* **47**, 74–80 (2006).
18. Ravikumar, B., Sarkar, S., Davies, J.E., et al. Regulation of mammalian autophagy in physiology and pathophysiology. *Physiol Rev* **90**, 1383–1435 (2010).
19. Martinet, W. & De Meyer, G.R. Autophagy in atherosclerosis: a cell survival and death phenomenon with therapeutic potential. *Circ Res* **104**, 304–317 (2009).
20. Zhu, H., Tannous, P., Johnstone, J.L., et al. Cardiac autophagy is a maladaptive response to hemodynamic stress. *J Clin Invest* **117**, 1782–1793 (2007).
21. Kinnear, N.P., Boittin, F.X., Thomas, J.M., Galione, A. & Evans, A.M. Lysosome-sarcoplasmic reticulum junctions. A trigger zone for calcium signaling by nicotinic acid adenine dinucleotide phosphate and endothelin-1. *J Biol Chem* **279**, 54319–54326 (2004).
22. Yamasaki, M., Masgrau, R., Morgan, A.J., et al. Organelle selection determines agonist-specific Ca2+ signals in pancreatic acinar and beta cells. *J Biol Chem* **279**, 7234–7240 (2004).
23. Grassme, H., Jekle, A., Riehle, A., et al. CD95 signaling via ceramide-rich membrane rafts. *J Biol Chem* **276**, 20589–20596 (2001).
24. Ikeda, S., Ushio-Fukai, M., Zuo, L., et al. Novel role of ARF6 in vascular endothelial growth factor-induced signaling and angiogenesis. *Circ Res* **96**, 467–475 (2005).
25. Legler, D.F., Micheau, O., Doucey, M.A., Tschopp, J. & Bron, C. Recruitment of TNF receptor 1 to lipid rafts is essential for TNFalpha-mediated NF-kappaB activation. *Immunity* **18**, 655–664 (2003).
26. Zuo, L., Ushio-Fukai, M., Ikeda, S., Hilenski, L., Patrushev, N. & Alexander, R.W. Caveolin-1 is essential for activation of Rac1 and NAD(P)H oxidase after angiotensin II type 1 receptor stimulation in vascular smooth muscle cells: role in redox signaling and vascular hypertrophy. *Arterioscler Thromb Vasc Biol* **25**, 1824–1830 (2005).
27. Yang, B. & Rizzo, V. TNF-alpha potentiates protein-tyrosine nitration through activation of NADPH oxidase and eNOS localized in membrane rafts and caveolae of bovine aortic endothelial cells. *Am J Physiol Heart Circ Physiol* **292**, H954–H962 (2007).
28. Cai, H. NAD(P)H oxidase-dependent self-propagation of hydrogen peroxide and vascular disease. *Circ Res* **96**, 818–822 (2005).
29. Suzuki, Y.J. & Ford, G.D. Redox regulation of signal transduction in cardiac and smooth muscle. *J Mol Cell Cardiol* **31**, 345–353 (1999).
30. Eum, S.Y., Andras, I., Hennig, B. & Toborek, M. NADPH oxidase and lipid raft-associated redox signaling are required for PCB153-induced upregulation of cell adhesion molecules in human brain endothelial cells. *Toxicol Appl Pharmacol* **240**, 299–305 (2009).
31. Peshavariya, H., Dusting, G.J., Di Bartolo, B., Rye, K.A., Barter, P.J. & Jiang, F. Reconstituted high-density lipoprotein suppresses leukocyte NADPH oxidase activation by disrupting lipid rafts. *Free Radic Res* **43**, 772–782 (2009).
32. Jin, S., Zhang, Y., Yi, F. & Li, P.L. Critical role of lipid raft redox signaling platforms in endostatin-induced coronary endothelial dysfunction. *Arterioscler Thromb Vasc Biol* **28**, 485–490 (2008).
33. Bao, J.X., Jin, S., Zhang, F., Wang, Z.C., Li, N. & Li, P.L. Activation of membrane NADPH oxidase associated with lysosome-targeted acid sphingomyelinase in coronary endothelial cells. *Antioxid Redox Signal* **12**, 703–712 (2010).
34. Bao, J.X., Xia, M., Poklis, J.L., Han, W.Q., Brimson, C. & Li, P.L. Triggering role of acid sphingomyelinase in endothelial lysosome-membrane fusion and dysfunction in coronary arteries. *Am J Physiol Heart Circ Physiol* **298**, H992–H1002 (2010).

35. Ni, X. & Morales, C.R. The lysosomal trafficking of acid sphingomyelinase is mediated by sortilin and mannose 6-phosphate receptor. *Traffic* **7**, 889–902 (2006).
36. Rodighiero, S., De Simoni, A. & Formenti, A. The voltage-dependent nonselective cation current in human red blood cells studied by means of whole-cell and nystatin-perforated patch-clamp techniques. *Biochim Biophys Acta* **1660**, 164–170 (2004).
37. Tsunoda, S., Mazda, O., Oda, Y., et al. Sonoporation using microbubble BR14 promotes pDNA/siRNA transduction to murine heart. *Biochem Biophys Res Commun* **336**, 118–127 (2005).
38. Qiu, H., Edmunds, T., Baker-Malcolm, J., et al. Activation of human acid sphingomyelinase through modification or deletion of C-terminal cysteine. *J Biol Chem* **278**, 32744–32752 (2003).
39. Zhang, A.Y., Yi, F., Jin, S., et al. Acid sphingomyelinase and its redox amplification in formation of lipid raft redox signaling platforms in endothelial cells. *Antioxid Redox Signal* **9**, 817–828 (2007).
40. Anderson, R.G., Vasile, E., Mello, R.J., Brown, M.S. & Goldstein, J.L. Immunocytochemical visualization of coated pits and vesicles in human fibroblasts: relation to low density lipoprotein receptor distribution. *Cell* **15**, 919–933 (1978).
41. Daniels, T.F., Killinger, K.M., Michal, J.J., Wright, R.W., Jr. & Jiang, Z. Lipoproteins, cholesterol homeostasis and cardiac health. *Int J Biol Sci* **5**, 474–488 (2009).
42. Zhang, F., Xu, M., Boini, K.M., Xia, M. & Li, P.L. Enhanced lipid deposition and foam cell formation in coronary artery wall by deletion of CD38 gene. *Hypertens Council Meet Proceed* **56**, E57 (2010).
43. Antonius, J. & Hill, L.D. Lysosomes in atherosclerosis. *Bull Mason Clinic* **20**, 137–154 (1966).
44. Jerome, W.G. & Yancey, P.G. The role of microscopy in understanding atherosclerotic lysosomal lipid metabolism. *Microsc Microanal* **9**, 54–67 (2003).
45. Miller, B.F. & Kothari, H.V. Increased activity of lysosomal enzymes in human atherosclerotic aortas. *Exp Mol Pathol* **10**, 288–294 (1969).
46. Fowler, S., Berberian, P.A., Shio, H., Goldfischer, S. & Wolinsky, H. Characterization of cell populations isolated from aortas of rhesus monkeys with experimental atherosclerosis. *Circ Res* **46**, 520–530 (1980).
47. Bach, G. Mucolipidosis type IV. *Mol Genet Metab* **73**, 197–203 (2001).
48. Slaugenhaupt, S.A. The molecular basis of mucolipidosis type IV. *Curr Mol Med* **2**, 445–450 (2002).
49. Loftus, S.K., Morris, J.A., Carstea, E.D., et al. Murine model of Niemann-Pick C disease: mutation in a cholesterol homeostasis gene. *Science* **277**, 232–235 (1997).
50. Ouimet, M., Franklin, V., Mak, E., Liao, X., Tabas, I. & Marcel, Y.L. Autophagy regulates cholesterol efflux from macrophage foam cells via lysosomal acid lipase. *Cell Metab* **13**, 655–667 (2011).
51. Levine, B. & Kroemer, G. Autophagy in the pathogenesis of disease. *Cell* **132**, 27–42 (2008).
52. Klionsky, D.J., Cregg, J.M., Dunn, W.A., Jr, et al. A unified nomenclature for yeast autophagy-related genes. *Dev Cell* **5**, 539–545 (2003).
53. Levine, B. & Klionsky, D.J. Development by self-digestion: molecular mechanisms and biological functions of autophagy. *Dev Cell* **6**, 463–477 (2004).
54. Reggiori, F. & Klionsky, D.J. Autophagy in the eukaryotic cell. *Eukaryot Cell* **1**, 11–21 (2002).
55. Wang, C.W. & Klionsky, D.J. The molecular mechanism of autophagy. *Mol Med* **9**, 65–76 (2003).
56. Xu, M., Zhang, F., Xia, M. & Li, P.L. Lysosomal regulation of autophagy efflux via CD38-mediated signaling in mouse coronary arterial myocytes. *Hyptertens Coucil Meet Proceed* **58**, E178 (2011).
57. Bampton, E.T., Goemans, C.G., Niranjan, D., Mizushima, N. & Tolkovsky, A.M. The dynamics of autophagy visualized in live cells: from autophagosome formation to fusion with endo/lysosomes. *Autophagy* **1**, 23–36 (2005).
58. Gutierrez, M.G., Master, S.S., Singh, S.B., Taylor, G.A., Colombo, M.I. & Deretic, V. Autophagy is a defense mechanism inhibiting BCG and Mycobacterium tuberculosis survival in infected macrophages. *Cell* **119**, 753–766 (2004).
59. Kim, J.S., Nitta, T., Mohuczy, D., et al. Impaired autophagy: a mechanism of mitochondrial dysfunction in anoxic rat hepatocytes. *Hepatology* **47**, 1725–1736 (2008).
60. Jia, G., Cheng, G., Gangahar, D.M. & Agrawal, D.K. Insulin-like growth factor-1 and TNF-alpha regulate autophagy through c-jun N-terminal kinase and Akt pathways in

human atherosclerotic vascular smooth cells. *Immunol Cell Biol* **84**, 448–454 (2006).

61. Martinet, W. & De Meyer, G.R. Autophagy in atherosclerosis. *Curr Atheroscler Rep* **10**, 216–223 (2008).
62. Xu, K., Yang, Y., Yan, M., Zhan, J., Fu, X. & Zheng, X. Autophagy plays a protective role in free cholesterol overload-induced death of smooth muscle cells. *J Lipid Res* **51**, 2581–2590 (2010).
63. Jia, G., Cheng, G. & Agrawal, D.K. Autophagy of vascular smooth muscle cells in atherosclerotic lesions. *Autophagy* **3**, 63–64 (2007).
64. Schrijvers, D.M., De Meyer, G.R., Herman, A.G. & Martinet, W. Phagocytosis in atherosclerosis: molecular mechanisms and implications for plaque progression and stability. *Cardiovasc Res* **73**, 470–480 (2007).
65. Verheye, S., Martinet, W., Kockx, M.M., et al. Selective clearance of macrophages in atherosclerotic plaques by autophagy. *J Am Coll Cardiol* **49**, 706–715 (2007).

20

C-REACTIVE PROTEIN AND ITS PATHOPHYSIOLOGICAL ROLES IN ATHEROSCLEROSIS

JIANGLIN FAN[1], JIFENG ZHANG[2], AND YUQING EUGENE CHEN[2]

[1]*Department of Molecular Pathology, Interdisciplinary Graduate School of Medicine and Engineering, University of Yamanashi, Chuo-City, Yamanashi, Japan*

[2]*Center for Advanced Models for Translational Sciences and Therapeutics, University of Michigan Medical Center, Ann Arbor, MI, USA*

INTRODUCTION

C-reactive protein (CRP) is a classical plasma protein marker that is significantly elevated in the acute phase of inflammation, infection, and tissue damage and, thus, has been broadly used for monitoring and differential diagnosis [1]. The major physiological functions of CRP include its ability to bind to various ligands exposed on damaged tissues or bacteria (opsonization) for the enhancement of phagocytosis and activation of the complement pathway, thereby enabling CRP to exert both anti- and proinflammatory functions [1]. In the past decade, the broad use of the high-sensitive CRP (hsCRP) assay method enabled measurement of low levels of CRP values in patients for assessing cardiovascular risk [2]. Ample evidence from epidemiological and clinical studies has shown that higher levels of plasma CRP are associated with or predict cardiovascular events; therefore, CRP is a useful biomarker for cardiovascular disease. On the contrary, CRP has been inadequately considered as a biomediator or maker for many inflammatory diseases, especially atherosclerosis [3]. Whether increased plasma CRP is only a marker or also acts as a maker has fueled tremendous research in this field; but thus far, the extensive controversy remains to be settled. While this issue has been widely investigated through both *in vitro* and *in vivo* studies and both in humans and experimental animals, many questions remain unanswered, and some of them even puzzle both researchers and clinicians. This chapter will address four broad topics regarding CRP. First, it will provide a brief description of CRP biology, essential for understanding CRP pathophysiological significance. Second, it will review the connection between CRP and cardiovascular disease. Third, it will focus on the functional roles of CRP in atherosclerosis, which is the major cause for myocardial infarction. Fourth, it will discuss questions and quandaries regarding animal models for the study of CRP.

Atherosclerosis: Risks, Mechanisms, and Therapies, First Edition. Edited by Hong Wang and Cam Patterson.

CRP BIOLOGY

CRP was first discovered as an acute phase reactant more than 80 years ago [4]. Tillet and Francis observed a substance in the serum of people with pneumococcal infections that formed a precipitate when mixed with the C-polysaccharide coat of *Streptococcus pneumonia* [4]. CRP is named for its capacity to precipitate the somatic C-polysaccharide. MacLeod and Avery subsequently found this substance to be a protein in the serum of patients with various acute infections [5]. Later on, it was found that phosphocholine is a specific ligand for CRP [6]. The human CRP (hCRP) molecule (115 kDa) comprises five identical nonglycosylated polypeptide subunits or protomers or monomers (23 kDa), each containing 206 amino acid residues, and they are noncovalently associated to form a pentamer. The hCRP gene maps to chromosome [1] and the CRP protein is synthesized mainly in the liver where CRP expression is increased in response to the cytokines, especially interleukin (IL)-6, IL-1β, and tumor necrosis factor-α (TNF-α). CRP mRNA expression was also reported in adipose tissue and vascular cells [7, 8]. Serum levels of CRP in the healthy population are less than 1 mg/L but rise dramatically (up to 10,000-fold) in response to cytokines in conditions of tissue injury, inflammation, and infection.

Therefore, CRP values can be used as a suitable marker for the assessment of many pathological states in the clinical practice. CRP values, just as measuring patient body temperature, however, are nonspecific and can never be diagnostic on their own. In addition to being a systemic inflammatory marker, through binding to phosphocholine residues present on the surface of dead or apoptotic cells or pathogens, hCRP can activate classical complement system via the C1Q complex, enhance phagocytic activity of macrophages (known as opsonization) and bind to immunoglobulin receptors (FcγR). Thus, CRP may contribute to host defense against infection as a part of the innate immunity. Serum CRP can be measured easily and inexpensively using both standard and high-sensitivity CRP (hsCRP) assays in the clinical practice.

Standard assays can only report CRP levels greater than 3 mg/L, while shCRP assay can report CRP values as low as 0.1 mg/L, for which the latter is often used for evaluating cardiovascular disease (CVD) as will be discussed later.

PLASMA CRP LEVEL IS A MARKER OF CARDIOVASCULAR DISEASE

In the past decade, CRP has received much attention for its role in CVD. Three reasons can be considered to account for this "CRP booming." First, CRP has been recognized as a new cardiovascular risk marker. Second, CRP is not only a marker but also a maker (mediator) for CVD. In fact, it should be correctly described as a maker for atherosclerosis rather than CVD because CVD is a broad clinical definition and is not equal to atherosclerosis. Third, both extensive and intensive attempts at developing therapeutic strategies for lowering CRP to treat and prevent CVD have been undertaken.

Epidemiological and clinical studies have revealed many associations between modestly raised baseline CRP values and CVD patients although these associations cannot clearly answer whether CRP is a causal factor for CVD. For example, increased CRP is considered as a predictor, prognostic factor, or even risk factor of future cardiovascular events [9–11]. Based on the current available data, the Centers for Disease Control and Prevention and the American Heart Association Scientific Statement on Markers of Inflammation and Cardiovascular Disease has recommended that CRP may be measured at the physician's discretion in asymptomatic people with an intermediate risk of coronary heart disease (Class IIa recommendation) [12–14]. In general, patients with CRP levels less than 1 mg/l are categorized as having lower relative risk for cardiovascular events. Those with levels of

1–3 mg/l are at intermediate risk and those with levels greater than 3 mg/l are at higher relative risk [12]. In addition, the recent Justification for the Use of Statins in Primary Prevention: an Intervention Trial Evaluating Rosuvastatin (JUPITER) trial showed that a lipid-lowering drug, rosuvastatin (Crestor) can significantly reduce the incidence of major cardiovascular events even in apparently healthy subjects not exhibiting established risk factors such as hyperlipidemia but with relatively high baseline plasma CRP levels (≥2 mg/l) [15]. In general, all these associations suggest that increased CRP levels (as a marker) reflect the CVD progression but cannot answer whether CRP itself has a causal role in CVD.

IS CRP A MAKER OF ATHEROSCLEROSIS DEVELOPMENT?

Atherosclerosis is the major cause of CVD and inflammatory reactions play a central role in the development of atherosclerosis. In addition to its role as marker in CVD, CRP is an inflammatory factor, leading to the hypothesis that CRP may actively participate in the pathogenesis of atherosclerosis, which implies that CRP acts not only as a marker but also as a maker (mediator). The first line of evidence that CRP may be atherogenic is based on the observations that CRP can bind to atherogenic lipoproteins [16, 17], and positive immunostaining for CRP is shown in the lesions of atherosclerosis [18]. As shown in Figures 20.1a and Figure 20.2, CRP protein is present in the lesions of both human and rabbit atherosclerosis as evidenced by immunohistochemical staining and Western blotting analysis using CRP specific antibodies [18]. Analysis of pathological samples revealed that CRP is frequently seen in atherosclerotic lesions regardless of lesion stage (early and advanced) but basically, CRP is mainly associated with extracellular matrix and seldom with macrophages. CRP is also closely colocalized with apolipoprotein B (apoB) and the terminal complement complex, suggesting that possible interactions amongst CRP-apoB-complement are present in the lesions [18, 19]. Origin of the CRP protein in the lesion has been studied by different molecular biology methods, but it is generally believed that lesional CRP is derived from circulation rather than produced by vascular cells and other cells present in the lesion (such as macrophages) [20]. As shown in Figure 20.1b, we could not detect CRP mRNA expression in the aortic lesion samples. The consensus is that CRP synthesized in the liver, which determines plasma levels of CRP [20]. Despite this, the presence of CRP protein in atherosclerotic lesions has engendered a sustained effort in the cardiovascular field during the past decade in an attempt to elucidate whether CRP truly constitutes an independent risk factor, as hypercholesterolemia, and whether it is indeed causally involved in the initiation and progression of atherosclerotic disease. If CRP is a proinflammatory factor in the development of CVD, can we target CRP for the prevention and treatment of atherosclerosis and other inflammatory process such as metabolic syndrome and obesity [21]?

To elucidate the possible roles of CRP in terms of its proinflammatory activity, numerous articles reported using a variety of cultured cells including endothelial cells, vascular smooth muscle cells, and macrophages, all of which are closely involved in atherosclerosis. Treatment with recombinant CRP results in multiple pro-inflammatory reactions in these cultured cells [3]. For example, CRP modulates endothelial functions *in vitro* resulting in decreased nitric oxide (NO) and prostacyclin (PGI2) and increased endothelin-1 (ET-1), increased expression of adhesion molecules, such as intracellular adhesion molecules (ICAM) and vascular cellular adhesion molecules (VCAM) and E-selectin, increased production of monocyte chemoattractant protein-1 (MCP-1), interleukin-8 (IL-8), and plasminogen activator inhibitor-1 (PAI-1). CRP also induces reactive oxygen species, cytokines, and tissue factor release in monocytes [22–24]. Further, CRP induces monocyte chemotaxis and increases oxidized low-density lipoprotein

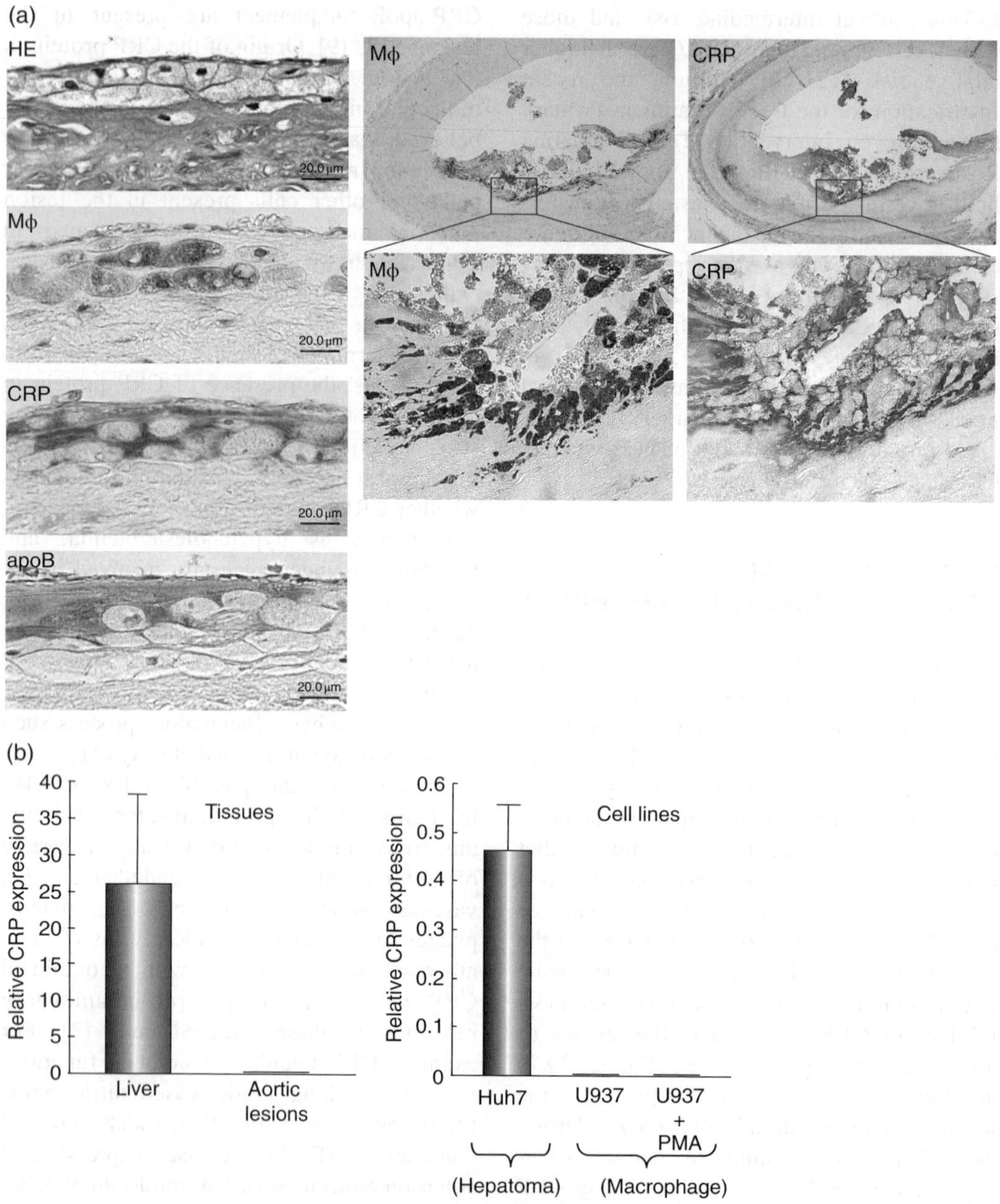

Figure 20.1 (a) Detection of CRP in human aorta and coronary plaque. Immunocytochemistry demonstrating the presence of CRP proteins in atherosclerotic lesions of human aorta (left) and in ruptured plaque in coronary atherosclerosis (right). Note that CRP distribution is associated with that of apoB. Mϕ, macrophages staining. (b) CRP mRNA expression in human tissues and cultured cells. Human liver and aortas were collected from autopsy cases, while the hepatoma (Huh7) and monocytic (U937) cell lines were cultured *in vitro*. CRP mRNA expression was determined by quantitative real-time RT-PCR (Adapted from Sun et al. [18] with modification and permission. © 2005, Elsevier). (*See insert for color representation of the figure.*)

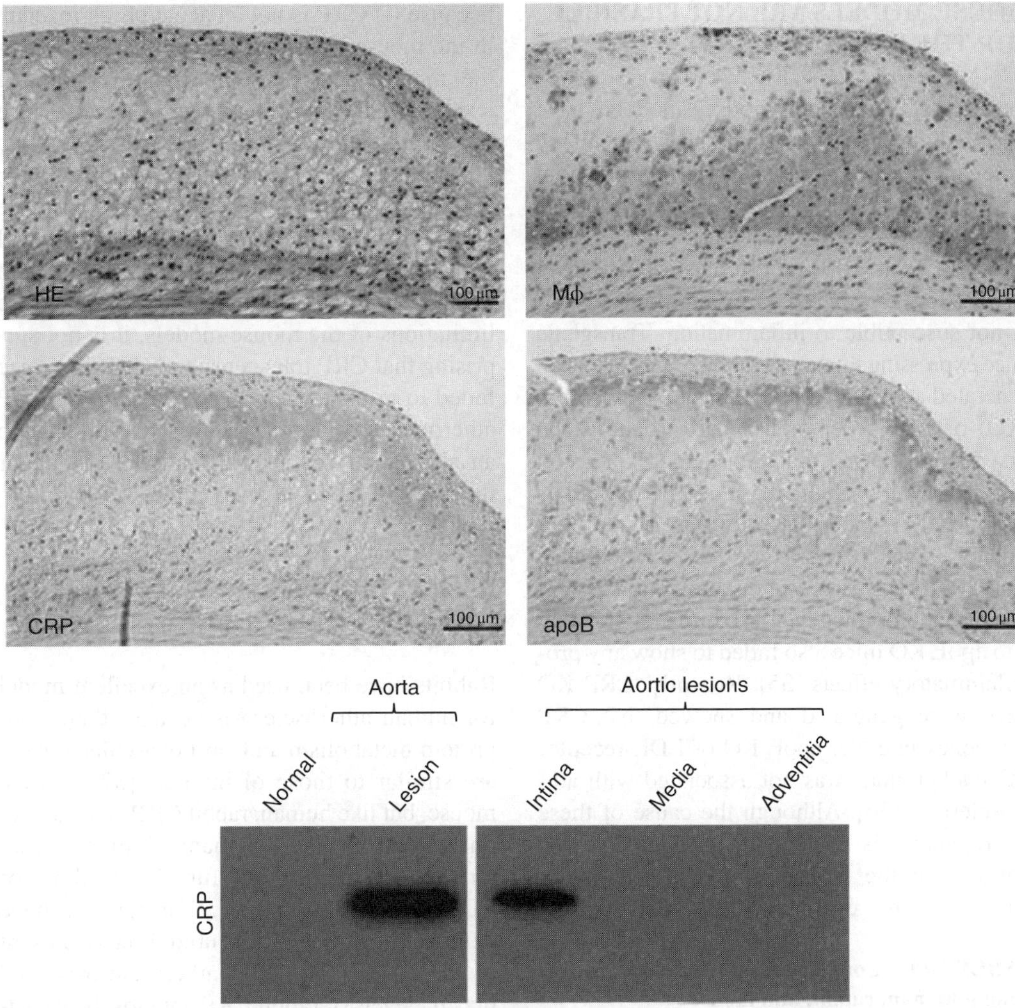

Figure 20.2 Detection of CRP in the lesions of WHHL rabbits. CRP is present in rabbit aortic lesions by immunohistochemical staining (top) and Western blotting (bottom). Compared to human atherosclerotic lesions, positive immunostaining for CRP is frequently observed in rabbit aortic lesions, which is mainly localized to the intima (Adapted from Sun et al. [18] with modification and permission. © 2005, Elsevier). (*See insert for color representation of the figure.*)

(LDL) uptake [25]. In addition, CRP has been shown to increase inducible NO production, NFκb pathway activation and mitogen-activated protein kinase (MAPK) activities; and CRP promotes vascular smooth muscle cells proliferation [26]. Almost all of these *in vitro* findings are in favor of the notion that CRP is potentially atherogenic. Unfortunately, these studies have been fiercely criticized to be artificial because commercial recombinant CRP is often contaminated by LPS and preservatives (azide) [11], and many of these experiments cannot be reproduced when endotoxin-free CRP is used [27]. Therefore, the question regarding the role of CRP as a "maker" in the initiation and progression of atherosclerosis still remains unanswered; and further studies, particularly those using different *in vivo* approaches, may help resolve this lingering issue.

MOUSE MODELS ARE NOT FEASIBLE FOR THE STUDY OF hCRP

Genetically modified mouse models have provided a powerful means to investigate many gene functions *in vivo,* which holds potential for CRP studies. Yet, the study of CRP functions in mouse models has proven challenging. As shown in Table 20.1, mouse plasma CRP level is very low compared to human, and also it is not susceptible to inflammation. Transgenic mice expressing human or rabbit CRP have been generated and studied in either apo-E knock out (KO) or LDL-receptor KO backgrounds. The first report showed that transgenic expression of hCRP led to mild increase of atherosclerosis in apo-E KO mice [28]; however, subsequent studies using different types of mice showed that CRP is either not proatherogenic [29–33] or even atheroprotective [34]. Infusion of hCRP into apoE KO mice also failed to show any pro-inflammatory effects [35]. Recently, CRP KO mice were generated and showed that CRP deficiency in either apoE KO or LDL-receptor KO backgrounds was not associated with atherosclerosis [36]. Although the cause of these discrepancies is unclear, they underscore the notion that the mouse is considered inadequate for the evaluation of CRP biology because (i) CRP is not an acute phase reactant in the mouse, where serum amyloid protein is the major acute phase reactant; (ii) plasma levels of CRP in mice, even in the presence of inflammatory stimuli, are extremely low (<2 mg/l) compared with humans and rabbits; (iii) human and rabbit CRP cannot activate complement in the transgenic mouse model [21, 32]; and (iv) no CRP has been detected in the mouse atherosclerotic lesions. Given the limitations of the mouse models, it is not surprising that CRP transgenic and CRP KO mice failed to provide clear answers regarding CRP atherogenecity, and it is imperative to develop an alternative and adequate animal model for the study of hCRP *in vivo*.

TABLE 20.1 Comparison of C-reactive protein among human, rabbit, and mouse

	Human CRP	Rabbit CRP	Mouse CRP
Plasma base level	<1 mg/l	<1 mg/l	Trace amount
Response to inflammation	Increase >1000 fold	Increase >1000 fold	Remains low
Activation of complement	Yes	Yes	No
Binding to plasma VLDL and LDL	Yes	Yes	No
Deposition in the atheroma lesions	Yes	Yes	No

RABBIT MODELS ARE SUITABLE FOR THE STUDY OF hCRP

Rabbits have been used as an excellent model for human atherosclerosis because their lipoprotein metabolism and cardiovascular system are similar to those of humans [37]. Unlike mouse, but like human, rabbit CRP is the acute-phase reactant and has many similar characteristics in structure and function to those of hCRP (Table 20.1) [38]. Further, we have demonstrated that CRP is ubiquitously present in the lesions of rabbit atherosclerosis, and plasma levels of rabbit CRP are closely associated with lesion severity in cholesterol-fed and Watanabe heritable hyperlipidemic (WHHL) rabbits [18, 39]. We generated two lines of transgenic rabbits expressing different levels of hCRP [38]. The hCRP was specifically expressed in the liver and the plasma levels of hCRP protein in the line 1 (designated as Tg-1) and the line 2 (Tg-2) rabbits were 0.4 ± 0.13 ($n=14$) and 57.8 ± 20.6 mg/l ($n=12$), respectively. To investigate whether the increased hCRP expressed by Tg rabbits was physiologically functional, we isolated hCRP from Tg rabbits and conducted a complement consumption assay using enzymatically modified LDL (E-LDL) as a CRP-ligand. We found that hCRP from Tg rabbits exhibited the same ability to augment activation of rabbit serum complement in the presence of E-LDL as native rabbit CRP and as hCRP does. On a chow diet, hCRP-Tg rabbits did not show either any obvious abnormalities or atherosclerotic lesion formation. To define the role of hCRP in atherosclerosis, we compared the susceptibility of hCRP-Tg rabbits to cholesterol-rich diet-induced aortic and coronary atherosclerosis to that of non-Tg rabbits. After being fed with a cholesterol-rich diet for 16 week, Tg and non-Tg rabbits developed similar hypercholesterolemia and their aortic and coronary lesion sizes were not different from those of control littermates (Figure 20.3).

Immunohistochemical staining along with Western blotting revealed that hCRP was indeed present in the lesions of Tg rabbits but did not affect macrophage accumulation and smooth muscle cell proliferation in the lesions [38]. Taken together, these observations from hCRP-Tg rabbits suggest that hCRP at these levels is not proatherogenic. Our data are also in support of recent studies showing that genetically elevated CRP does not play a causal role in ischemic vascular disease [40, 41]. Nevertheless, we cannot exclude the possibility that hCRP may have a pathophysiological role in other aspects of CVD such as myocardial infarction [42] and thrombosis [43]. Using double balloon-injury models, we found that hCRP enhanced thrombosis formation in hCRP-Tg rabbits (Figure 20.4) [44], but hCRP overexpression did not change the myocardial infarction sizes in coronary artery ligation experiments (unpublished data). Further, CRP seems to be involved in the metabolic syndrome because overexpression of hCRP in SHR rats is detrimental to metabolism [45]. Therefore, just as other inflammatory factors, the function of CRP may be context-dependent, and it may either enhance or inhibit inflammatory responses depending on the circumstances.

In summary, as a biomarker, plasma CRP values definitely help evaluate the cardiovascular risk, but CRP does not play any causative roles in development of atherosclerosis based on CRP transgenic animal studies and

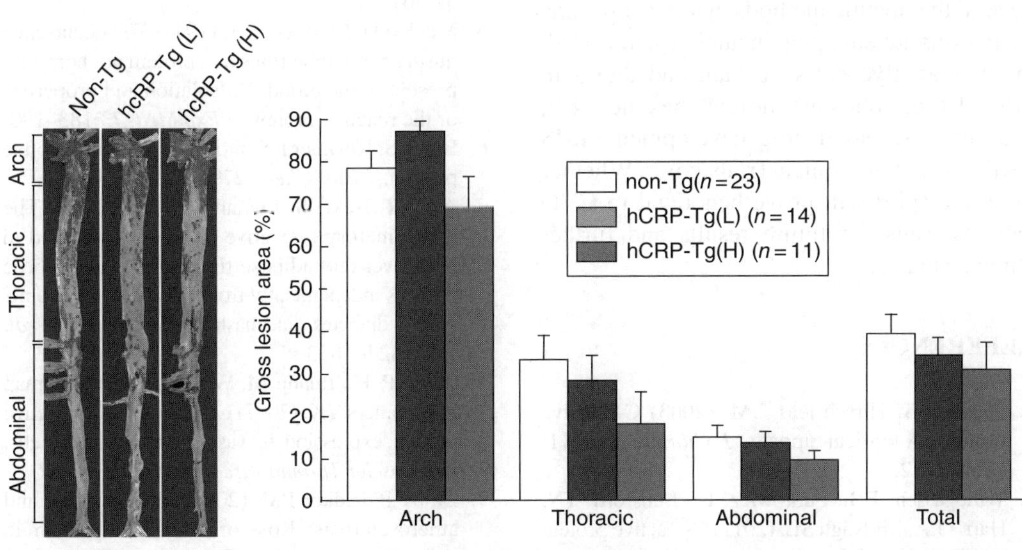

Figure 20.3 Aortic atherosclerotic lesion analysis. Representative photographs of pinned-out aortic trees stained with Sudan IV from hCRP transgenic (Tg) and wild-type (non-Tg) rabbits (left), and aortic atherosclerotic lesions (defined by sudanophilic area) on the surface were quantified by an image analysis system (right). The data show that high levels of hCRP are not proatherogenic in transgenic rabbits (Adapted from Koike et al. [38] with permission. © 2009, Lippincott Williams & Wilkins). (*See insert for color representation of the figure.*)

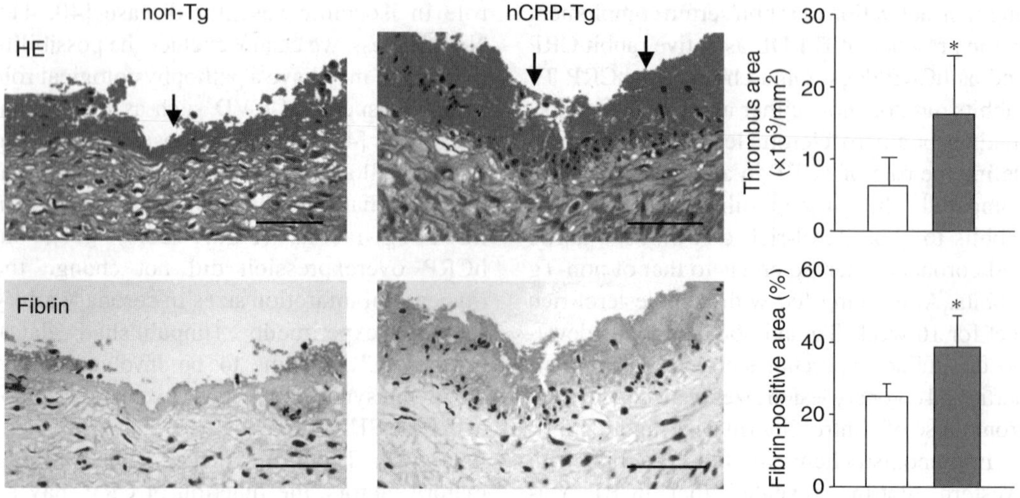

Figure 20.4 Human CRP enhances thrombus formation in transgenic rabbits after balloon injury. Thrombus size (*arrows*) and immunoreactive protein fibrin on neointimas were quantified by image analysis, and results are shown in the bar graph on the right. The scale bar: 100 μm. *: $P < 0.05$ (Adapted from Matsuda et al. [44] with modification and permission. © 2011, John Wiley & Sons). (*See insert for color representation of the figure.*)

human population genetic studies. It remains unclear yet whether CRP is a mediator in other disorders such as the metabolic syndrome. Several therapeutic methods that lower plasma CRP levels are emerging including statins [15], the cardiac glycosides ouabain and digitoxin [46], 1,6-bis(phosphocholine)-hexane [42], and antisense-based drug development (ISIS CRP Rx. www.ClinicalTrials.gov). Whether these therapies will prove beneficial to CVD patients waits for future results and further clinical studies.

REFERENCES

1. Pepys MB, Hirschfield GM. (2003) C-reactive protein: A critical update. *J Clin Invest.* 111: 1805–1812.
2. Windgassen EB, Funtowicz L, Lunsford TN, Harris LA, Mulvagh SL. (2011) C-reactive protein and high-sensitivity c-reactive protein: An update for clinicians. *Postgrad Med.* 123:114–119.
3. Verma S, Devaraj S, Jialal I. (2006) Is c-reactive protein an innocent bystander or proatherogenic culprit? C-reactive protein promotes atherothrombosis. *Circulation.* 113:2135–2150; discussion 2150.
4. Tillett WS, Francis T. (1930) Serological reactions in pneumonia with a non-protein somatic fraction of pneumococcus. *J Exp Med.* 52:561–571.
5. Macleod CM, Avery OT. (1941) The occurrence during acute infections of a protein not normally present in the blood: II. Isolation and properties of the reactive protein. *J Exp Med.* 73:183–190.
6. Black S, Kushner I, Samols D. (2004) C-reactive protein. *J Biol Chem.* 279:48487–48490.
7. Anty R, Bekri S, Luciani N, et al. (2006) The inflammatory c-reactive protein is increased in both liver and adipose tissue in severely obese patients independently from metabolic syndrome, type 2 diabetes, and nash. *Am J Gastroenterol.* 101:1824–1833.
8. Singh P, Hoffmann M, Wolk R, Shamsuzzaman AS, Somers VK. (2007) Leptin induces c-reactive protein expression in vascular endothelial cells. *Arterioscler Thromb Vasc Biol.* 27:e302–e307.
9. Libby P, Ridker PM. (2004) Inflammation and atherosclerosis: Role of c-reactive protein in risk assessment. *Am J Med.* 116 Suppl 6A: 9S–16S.
10. Ridker PM. (2001) High-sensitivity c-reactive protein: Potential adjunct for global risk assessment in the primary prevention of cardiovascular disease. *Circulation.* 103:1813–1818.

11. Scirica BM, Morrow DA. (2006) Is c-reactive protein an innocent bystander or proatherogenic culprit? The verdict is still out. *Circulation*. 113:2128–2134; discussion 2151.
12. Pearson TA, Mensah GA, Alexander RW, et al. (2003) Markers of inflammation and cardiovascular disease: Application to clinical and public health practice: A statement for healthcare professionals from the centers for disease control and prevention and the american heart association. *Circulation*. 107:499–511.
13. Ridker PM. (2003) Clinical application of c-reactive protein for cardiovascular disease detection and prevention. *Circulation*. 107:363–369.
14. Heusch G, Schulz R, Erbel R. (2003) Inflammatory markers in coronary heart disease: Coronary vascular versus myocardial origin? *Circulation*. 108:e4; author reply e4.
15. Ridker PM, Danielson E, Fonseca FA, et al. (2008) Rosuvastatin to prevent vascular events in men and women with elevated c-reactive protein. *N Engl J Med*. 359:2195–2207.
16. Pepys MB. (2008) C-reactive protein is neither a marker nor a mediator of atherosclerosis. *Nat Clin Pract Nephrol*. 4:234–235.
17. Hirschfield GM, Pepys MB. (2003) C-reactive protein and cardiovascular disease: New insights from an old molecule. *QJM*. 96:793–807.
18. Sun H, Koike T, Ichikawa T, et al. (2005) C-reactive protein in atherosclerotic lesions: Its origin and pathophysiological significance. *Am J Pathol*. 167:1139–1148.
19. Torzewski J, Torzewski M, Bowyer DE, et al. (1998) C-reactive protein frequently colocalizes with the terminal complement complex in the intima of early atherosclerotic lesions of human coronary arteries. *Arterioscler Thromb Vasc Biol*. 18:1386–1392.
20. Fan J, Chen YE. (2006) Source of crp in atherosclerotic lesions. *Am J Pathol*. 168: 1055–1056.
21. Torzewski J. (2005) C-reactive protein and atherogenesis: New insights from established animal models. *Am J Pathol*. 167:923–925.
22. Schwartz R, Osborne-Lawrence S, Hahner L, et al. (2007) C-reactive protein downregulates endothelial no synthase and attenuates reendothelialization in vivo in mice. *Circ Res*. 100: 1452–1459.
23. Dasu MR, Devaraj S, Du Clos TW, Jialal I. (2007) The biological effects of crp are not attributable to endotoxin contamination: Evidence from tlr4 knockdown human aortic endothelial cells. *J Lipid Res*. 48:509–512.
24. Wu J, Stevenson MJ, Brown JM, Grunz EA, Strawn TL, Fay WP. (2008) C-reactive protein enhances tissue factor expression by vascular smooth muscle cells: Mechanisms and in vivo significance. *Arterioscler Thromb Vasc Biol*. 28:698–704.
25. Cermak J, Key NS, Bach RR, Balla J, Jacob HS, Vercellotti GM. (1993) C-reactive protein induces human peripheral blood monocytes to synthesize tissue factor. *Blood*. 82:513–520.
26. Cirillo P, Golino P, Calabro P, et al. (2005) C-reactive protein induces tissue factor expression and promotes smooth muscle and endothelial cell proliferation. *Cardiovasc Res*. 68:47–55.
27. Pepys MB, Hawkins PN, Kahan MC, et al. (2005) Proinflammatory effects of bacterial recombinant human c-reactive protein are caused by contamination with bacterial products, not by c-reactive protein itself. *Circ Res*. 97: e97–e103.
28. Paul A, Ko KW, Li L, et al. (2004) C-reactive protein accelerates the progression of atherosclerosis in apolipoprotein e-deficient mice. *Circulation*. 109:647–655.
29. Torzewski M, Reifenberg K, Cheng F, et al. (2008) No effect of c-reactive protein on early atherosclerosis in ldlr-/- / human c-reactive protein transgenic mice. *Thromb Haemost*. 99:196–201.
30. Tennent GA, Hutchinson WL, Kahan MC, et al. (2008) Transgenic human crp is not pro-atherogenic, pro-atherothrombotic or pro-inflammatory in apoe-/- mice. *Atherosclerosis*. 196:248–255.
31. Trion A, de Maat MP, Jukema JW, et al (2005) No effect of c-reactive protein on early atherosclerosis development in apolipoprotein e*3-leiden/human c-reactive protein transgenic mice. *Arterioscler Thromb Vasc Biol*. 25:1635–1640.
32. Reifenberg K, Lehr HA, Baskal D, et al. (2005) Role of c-reactive protein in atherogenesis: Can the apolipoprotein e knockout mouse provide the answer? *Arterioscler Thromb Vasc Biol*. 25:1641–1646.
33. Hirschfield GM, Gallimore JR, Kahan MC, et al. (2005) Dhillon AP, Tennent GA, Pepys MB. Transgenic human c-reactive protein is not pro-atherogenic in apolipoprotein e-deficient mice. *Proc Natl Acad Sci USA*. 102:8309–8314.

34. Kovacs A, Tornvall P, Nilsson R, Tegner J, Hamsten A, Bjorkegren J. (2007) Human c-reactive protein slows atherosclerosis development in a mouse model with human-like hypercholesterolemia. *Proc Natl Acad Sci U S A*. 104:13768–13773.
35. Ortiz MA, Campana GL, Woods JR, et al. (2009) Continuously-infused human c-reactive protein is neither proatherosclerotic nor proinflammatory in apolipoprotein e-deficient mice. *Exp Biol Med (Maywood)*. 234:624–631.
36. Teupser D, Weber O, Rao TN, Sass K, Thiery J, Fehling HJ. (2011) No reduction of atherosclerosis in c-reactive protein (crp)-deficient mice. *J Biol Chem*. 286:6272–6279.
37. Fan J, Watanabe T. (2003) Transgenic rabbits as therapeutic protein bioreactors and human disease models. *Pharmacol Ther*. 99:261–282.
38. Koike T, Kitajima S, Yu Y, et al. (2009) Human c-reactive protein does not promote atherosclerosis in transgenic rabbits. *Circulation*. 120: 2088–2094.
39. Yu Q, Li Y, Wang Y, et al. (2012) C-reactive protein levels are associated with the progression of atherosclerotic lesions in rabbits. *Histol Histopathol*. 27:529–535.
40. Elliott P, Chambers JC, Zhang W, et al. (2009) Genetic loci associated with c-reactive protein levels and risk of coronary heart disease. *JAMA*. 302:37–48.
41. Zacho J, Tybjaerg-Hansen A, Jensen JS, Grande P, Sillesen H, Nordestgaard BG. (2008) Genetically elevated c-reactive protein and ischemic vascular disease. *N Engl J Med*. 359:1897–1908.
42. Pepys MB, Hirschfield GM, Tennent GA, et al. (2006) Targeting c-reactive protein for the treatment of cardiovascular disease. *Nature*. 440:1217–1221.
43. Danenberg HD, Szalai AJ, Swaminathan RV, et al. (2003) Increased thrombosis after arterial injury in human c-reactive protein-transgenic mice. *Circulation*. 108:512–515.
44. Matsuda S, Yamashita A, Sato Y, et al. (2011) Human c-reactive protein enhances thrombus formation after neointimal balloon injury in transgenic rabbits. *J Thromb Haemost*. 9:201–208.
45. Pravenec M, Kajiya T, Zidek V, et al. (2011) Effects of human c-reactive protein on pathogenesis of features of the metabolic syndrome. *Hypertension*. 57:731–737.
46. Kolkhof P, Geerts A, Schafer S, Torzewski J. (2010) Cardiac glycosides potently inhibit c-reactive protein synthesis in human hepatocytes. *Biochem Biophys Res Commun*. 394: 233–239.

21

PARAOXONASES

SRINIVASA T. REDDY
Department of Medicine/Cardiology, University of California Los Angeles, Los Angeles, CA, USA

INTRODUCTION

Infection, inflammation, and oxidative stress correlate with mechanisms underlying a variety of diseases, including atherosclerosis, diabetes mellitus, systemic lupus erythematosus, rheumatoid arthritis, and several ailments associated with age. A number of enzymes have evolved with pro-oxidant, antioxidant, proinflammatory, and anti-inflammatory properties for maintaining oxidative homeostasis in cells and tissues. Understanding the biology and function of such enzymes will pave way for the discovery of novel therapeutic agents in the fight against inflammatory diseases including atherosclerosis.

Atherosclerosis is a disease of chronic inflammation and lipid accumulation. Increasing evidence suggests that oxidative stress plays a key role in the pathogenesis of this disease [1]. The earliest lesion in the development of an atherosclerotic plaque is an inflammatory reaction known as the fatty streak. This lesion is characterized by the accumulation of monocytes in the subendothelial space where they eventually undergo conversion to lipid-laden macrophage foam cells. Oxidation of low-density lipoproteins (LDLs) is a major factor in human atherosclerosis [2, 3]. Oxidation of LDL phospholipids makes the arterial wall cells react as if there were a bacterial infection. Entrapment and oxidation of LDL in the subendothelial space, and the subsequent interactions between endothelial cells and monocytes, is a key process in the initiation of atherosclerotic lesion development [4]. Several biologically active oxidized phospholipids associated with LDL have been identified in atherosclerotic lesions of animal models and humans, and in several inflammatory diseases [5, 6]. High-density lipoproteins (HDLs) play an antiatherogenic role, and its protective capacities have been ascribed primarily to its ability to remove excess cholesterol from peripheral tissues in the cholesterol transport pathway and also to its ability to protect against LDL

Atherosclerosis: Risks, Mechanisms, and Therapies, First Edition. Edited by Hong Wang and Cam Patterson.

oxidation [7]. The inverse relationship between serum HDL and the risk of atherosclerosis is well established. Although there does not appear to be a single explanation for the antiatherogenic role of HDL, it has become clear that the functional status of HDL, which is largely dependent on its apolipoprotein components, is an important determinant of coronary heart disease [8]. PON1, lecithin–cholesterol acyl transferase, platelet-activating factor acetyl hydrolase proteinase (elastase-like), phospholipase D, albumin, ApoJ, and ApoA-I are proteins in HDL, and some of these proteins have been demonstrated to possess antiatherogenic properties.

PON GENES

The PON gene family consists of three members: PON1, PON2, and PON3, located adjacent to each other on the long arm of chromosome 7 in human and chromosome 6 in mice [9]. The three human PON genes share about 65% similarity at the amino acid level, and about 70% similarity at the nucleotide level. From an evolutionary standpoint, PON2 appears to be the oldest member, followed by PON3 and PON1 [10]. PON1 was the first one of the family to be identified in a screen for plasma hydrolases of paraoxon, the active metabolite of insecticide parathion [11, 12]. PON1 is the predominant member of the three members of the PON family in the circulation, and thus the most studied of the three members in toxicological research. Following the discovery that PON1 is associated with HDLs in the circulation and that PON1 plays an important role in the protective effects of HDL [12], a new research focus for PON1 emerged in both its disease association and physiological function. Epidemiological studies suggest that expression of PON genes negatively correlates with a number of inflammatory diseases including atherosclerosis [13]. A definitive proof for the antiatherogenic role for PON genes came first from PON1-deficient mice, which are susceptible to organophosphate toxicity and atherosclerosis [14] and, more recently, from overexpression of the human paraoxonase gene cluster in transgenic mice, which are resistant to atherosclerosis [15].

LOCALIZATION

In humans, PON1 and PON3 messenger RNA (mRNA) expression is high in the liver [16], whereas PON2 is widely expressed and is found in a variety of tissues including the heart, kidney, liver, lung, placenta, small intestine, spleen, stomach, and testis [9, 17, 18]. Further, unlike PON1 and PON3, PON2 message is detected in the cells of the artery wall, including endothelial cells, smooth muscle cells, and macrophages [17]. In addition, while PON1 and PON3 associate with HDL in the circulation [18], PON2 protein is undetectable in HDL, LDL, or cell supernatants, but appears to remain intracellular, associated with membrane fractions [17, 18]. Marsillach et al. recently demonstrated the presence of both PON1 and PON3 in human atheromatous plaques [19]. Interestingly, although PON3 is present in human and rabbit HDL, PON3 along with PON2 are not detectable in mouse HDL. Moreover, mouse PON3 is cell-associated and is expressed in a wide range of tissues such as liver, adipose, macrophage, and the artery wall. Recently, *in situ* hybridization analysis showed that mouse PON3 gene is expressed in sections of the atherosclerotic lesion areas at the aortic root of apoE$^{-/-}$ hearts [20]. Taken together, the localization studies to date suggest that PON protein expression is spatially and temporally controlled.

ACTIVITIES

PON1 is a calcium-dependent esterase that hydrolyzes metabolites of a number of insecticides in addition to paraoxon and also certain nerve agents [21]. PON1 has also been reported to have a number of other enzymatic activities including arylesterase, lactonase, and relatively low levels of peroxidase and phospholipase

A2-like activities [22–25]. One mechanism by which PON1 is believed to exert its antiatherogenic effects is by hydrolyzing biologically active oxidized phospholipids and destroying lipid hydroperoxides [22, 23]. Harel and colleagues reported the partial crystal structure of PON1 based on a recombinant variant of PON1 obtained by directed evolution [26] and postulated that PON1 is a six-bladed β-propeller, which structurally resembles the Loligo vulgaris diisopropyl fluorophosphatase (DFPase) [26]. In addition, PON1's active site and catalytic mechanism is reminiscent of secreted phospholipase A2, supporting previous reports that PON1 has phospholipase A2-like activities. However, Teiber et al. reported that purified human serum PON1 does not protect LDL against oxidation *in vitro* [27], and Connelly et al. demonstrated that PON1 does not reduce or modify oxidation of phospholipids by peroxynitrite [28]. Moreover, evidence that PON3 lacks any significant paraoxonase activity suggested that the name PON might be a misnomer [16, 29]. In 2005, Draganov et al. reported that PON1, PON2, and PON3 are lactonases with overlapping and distinct substrate specificities [30], thus, for the first time, describing a common activity for all the three members of the PON family [31].

PON PROTEINS AND ATHEROSCLEROSIS

PON1 has been demonstrated to protect LDL against oxidation, to reverse the biological effects of oxidized LDL (oxLDL), and to preserve the function of HDL by inhibiting its oxidation [22, 23]. PON1 null mice by gene targeting [14] and transgenic mice [15] corroborate the hypothesis that PON1 protects against atherogenesis and is an important contributor to HDL's antioxidant capacity. The *in vivo* studies combined with the antioxidant and antiatherogenic nature of PON1 underscore the potential of PON1 as a therapeutic agent to prevent atheroma [32, 33]. Recent mechanistic studies on PON1 and its association with HDL [34, 35] further support at least, in part, a causal relationship between PON1 and cardiovascular and its associated pathologies [36–38].

PON2 is a widely expressed intracellular protein with a molecular mass of approximately 44 kDa [9, 17]. Polymorphisms in PON2 are associated with a number of pathophysiological conditions including coronary artery disease, ischemic stroke in patients with type 2 diabetes mellitus, late-onset Alzheimer's disease, and reduced bone mass in postmenopausal women [39–41]. *In vitro* studies show that PON2 can protect against LDL oxidation [17]. Using PON2-deficient mice, Ng et al. and Devarajan et al. demonstrated that PON2 protects against the development of atherosclerosis in mice [42, 43]. Devarajan et al. further demonstrated that PON2 regulates calcium homeostasis and cell survival under endoplasmic reticulum (ER) stress conditions and is sufficient to prevent the development of aggravated atherosclerosis in apoE$^{-/-}$ mice on a Western diet [44].

PON3 is an approximately 40 kDa protein that associates with HDL in the circulation, albeit in much lower levels than PON1 [16, 29]. Similar to PON1 and PON2, PON3 also possesses antioxidant properties. As mentioned earlier, human PON3 and mouse PON3 exhibit different expression patterns. It is not known how these differences in tissue expression pattern and association with HDL will affect the roles of mouse PON3 and human PON3 on atherogenesis and other biological processes. *In vitro* studies have shown that PON3 can prevent LDL oxidation [16]. Previous studies also showed that human PON3 transgenic mice were protected from obesity and atherosclerosis in both the C57BL/6J wild type and LDL receptor (LDLR) knockout genetic background [45].

NEW ANTIATHEROGENIC ROLES FOR PON PROTEINS-BASED ON INFECTION MODELS

Lactonase activity is common to all PON proteins [31]. PON proteins can hydrolyze a number of acylhomoserine lactones (AHLs)

[30], molecules that mediate bacterial quorum-sensing (QS) signals that are important in regulating expression of virulence factors and in inducing a host inflammatory response [46]. Many gram-negative bacteria utilize AHLs that either diffuse or are transported across cell membranes and function as ligands for a family of inducible transcription factors [47]. The mechanism of QS has been extensively characterized in several pathogenic bacteria including *Pseudomonas aeruginosa*, which causes catastrophic infections in immunocompromised hosts, such as individuals with cystic fibrosis, cancer, and severe burns [48]. In the rational pursuit of systems that disarm the QS systems of microbes, a range of quorum-quenching enzymes and inhibitors have been identified from both prokaryotic and eukaryotic organisms [48]. These novel enzymes and inhibitors are the key molecules that may have evolved for establishing the concept of quorum quenching, antipathogenic, and signal interference [49].

Greenberg's laboratory was the first to show AHL-inactivation activity from the cell membrane of differentiated human airway epithelia [50]. The enzyme activity was able to inactivate 3OC12-HSL and was subsequently shown to be a PON [50]. These findings laid the foundation for a new function for PON proteins namely inactivation of QS signals. Indeed, recent studies demonstrate that PON proteins are related to PON-like quorum-quenching bacterial lactonases [51]. Moreover, overexpression of PON1 inhibited oxidative stress and prevented bacterial colonization in Drosophila model [52]. Further, Devarajan et al. demonstrated in a mouse model of PON2 deficiency that PON2 plays an important role in innate immune defense against *P. aeruginosa* infection [53]. Infections are recognized as likely risk factors for atherosclerosis and play a significant role in the development of the atherosclerotic plaque [54]. Accumulating evidence suggests that chronic inflammation associated with low-level endotoxemia, such as that resulting from chronic subclinical gram-negative bacterial infections, may be an important risk factor for cardiovascular disease. Thus, the quorum-quenching lactonase activity of paraoxonases may be contributing to their antiatherogenic properties.

NEW ANTIATHEROGENIC ROLES FOR PON PROTEINS-BASED ON INTRACELLULAR STRESS MODELS

ER association and function. PON1 is found associated almost entirely on circulating HDL. PON2 and PON3, as discussed earlier, are found in many organs intacellularly. Cells overexpressing PON2 exhibit significantly lower levels of intracellular oxidative stress when exposed to oxidized phospholipids [17]. PON2 deficiency in mice increases susceptibility to atherosclerosis, and this has primarily been associated with increased liver and macrophage oxidative stress, which leads to LDL oxidation [42]. Further, LDL from PON2-def mice induced significantly more monocyte chemotaxis than LDL from their control littermates, and it was associated with higher levels of macrophage immunoreactivity in the aortic sections of PON2-deficient mice compared with their wild-type controls [42]. PON2 also reduces oxidative stress in vascular cells and decreases ER-stress-induced caspase activation [55] and apoptosis, which is linked to calcium homeostasis [56]. In mouse macrophages, PON2 regulates calcium homeostasis and cell survival under ER stress conditions [44]. Given the fact that oxidized lipids are implicated in ER stress, these reports suggest a link between PON2 and ER physiology. More recently, it was demonstrated that HDLs protect pancreatic β-cells against ER stress by restoring protein folding and trafficking [57], suggesting that PON1 present on HDL may also participate as an ER-protective molecule similar to PON2. It remains to be seen whether calcium homeostasis is a common mechanism downstream of PON action.

Mitochondrial association and function. Devarajan et al. were the first to report that PON2 protein is associated with mitochondria following a yeast two-hybrid study, which showed an interaction between PON2 and complex III of mitochondria [58]. Altenhöfer et al. demonstrated that PON2 prevents mitochondrial superoxide formation and apoptosis independent from its lactonase activity [59], and Devarajan et al. demonstrated that PON2 deficiency alters mitochondrial function and exacerbates the development of atherosclerosis. More recently, PON3 has been shown to

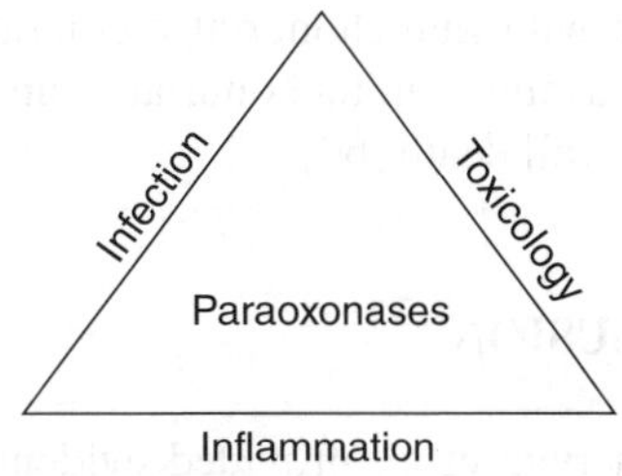

Figure 21.1 Paraoxonases are implicated in inflammation, toxicology, and infection.

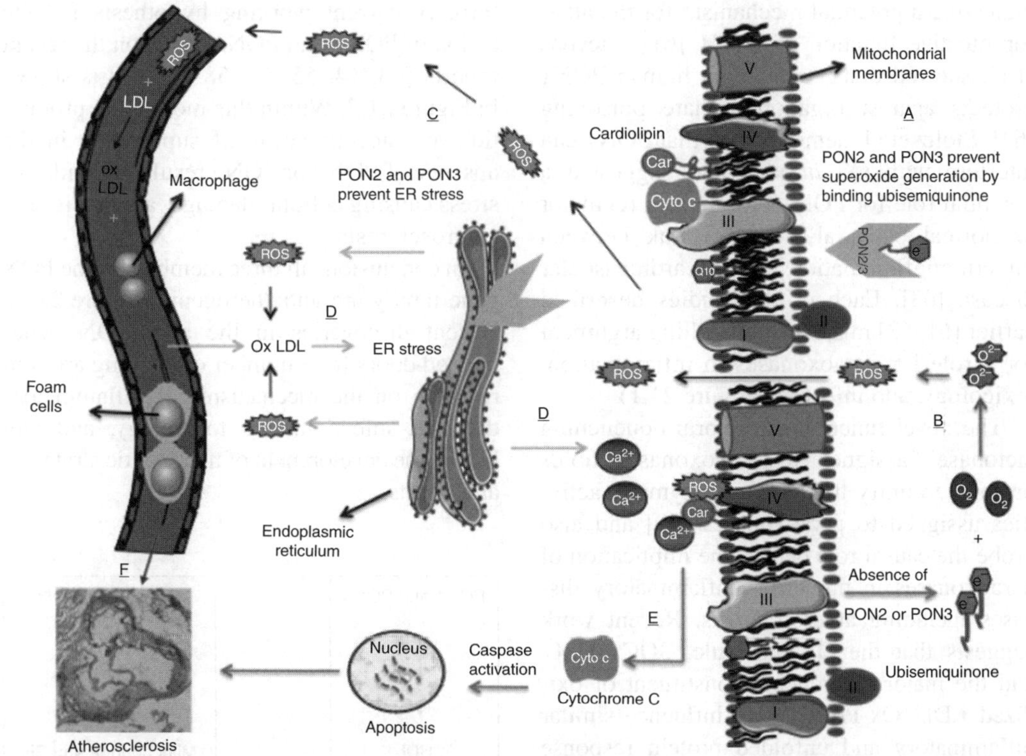

Figure 21.2 Working model for the antiatherogenic properties of PON2 and PON3 proteins. In the inner mitochondrial membrane, ubisemiquinone is released from electron transport chain (ETC) during Q cycle, which is inactivated by PON2/PON3, thereby preventing the formation of superoxides (A). In the absence of PON2 or PON3, ubisemiquinone donates electron to molecular oxygen to form superoxides that generate other reactive oxygen species (ROS) (B). ROS oxidize LDL phospholipids (e.g., PAPC) to form oxLDL phospholipids (e.g., oxPAPC) (C). OxLDL causes ER stress and results in an increased ROS and calcium release to the cytosol (D). Enhanced uptake of calcium by mitochondria causes release of cytochrome c to cytosol (E). Decrease in cytochrome c in the mitochondria causes reduction in the complex IV activity, which results in more ubisemiquinone causing more lipid oxidation as described above in steps (B) and (C) above (E). Increase in cytochrome c levels in the cytosol causes caspase activation and apoptosis (E). Accumulation of oxLDL in macrophages leads to foam cell formation and atherosclerosis (F). Abbreviations: I-NADH, ubiquinone oxidoreductase; II, succinate coenzyme Q reductase; III, ubiquinol cytochrome oxidoreductase; IV, cytochrome c oxidase; V, ATP synthase; Car, cardiolipin; and Q10, coenzyme Q10. (*See insert for color representation of the figure.*)

associate with mitochondrial membranes and protect against mitochondrial superoxide-mediated cell death [60].

CONCLUSION

Bhattacharyya et al. provided evidence for a mechanistic link between genetic determinants and activity of PON1 with systemic oxidative stress and prospective cardiovascular risk, indicating a potential mechanism for the atheroprotective function of PON1 [61]. Stevens et al. showed that recombinant human PON1 protects against organophosphate poisoning [62]. Stoltz et al. demonstrated that PON1 can interfere with QS *in vivo* and suggested a potential role for PON1 protein as a regulator of normal bacterial florae, a link between infection/inflammation and cardiovascular disease [63]. Each of the studies described earlier [61–63] make the compelling argument for a role for paraoxonases in inflammation, toxicology, and infection (Figure 21.1).

The novel function of "quorum-quenching lactonase" assigned to paraoxonases holds promise to unify the various enzymatic activities assigned to paraoxonases [49] and also probe the causal role behind the implication of paraoxonases in numerous inflammatory diseases including atherosclerosis. Recent work suggests that the QS molecule, $3OC_{12}$-HSL, and the major atherogenic constituent of oxidized LDL, Ox-PAPC, both influence similar inflammatory and unfolded protein response (UPR) pathways [64]. The authors of the work suggest that the antiatherogenic effects of PON2 include destruction of $3OC_{12}$-HSL [64]. Future research will need to focus on understanding the mechanism of quorum sensors in the host and the role of various PON gene polymorphisms in susceptibility and resistance to infection.

PON2 deficiency results in enhanced oxidative stress accompanied by increased mitochondrial oxidative stress with reduced ETC complex I+III activities, oxygen consumption, and ATP levels and exacerbated atherosclerosis [43, 58]. PON2 overexpression prevents mitochondrial superoxide formation and apoptosis independent from its lactonase activity [59]. In addition to the mitochondria, PON2 overexpression reduces ER stressor-mediated UPR and apoptosis [55, 56]. These and recent studies in PON3 [60] implicate that there is a novel PON2/PON3-dependent cell survival pathway. PON2 has ben suggested to bind coenzyme Q10 [43], and PON3 has recently been shown to bind coenzyme Q10 [60]. A current working hypothesis for the action of PON2 and PON3 based on the recent reports [43, 44, 55, 56, 58–60, 64] is shown in Figure 21.2. Within this model, the production and accumulation of superoxide in the absence of PON2 or PON3 results in oxidative stress causing cellular damage, apoptosis, and atherosclerosis.

In conclusion, all three members of the PON gene family are antiatherogenic (Figure 21.3). Recent discoveries in the area PON genes opened doors for a number of exciting areas of research on the mechanisms of inflammatory diseases, infection, and toxicology, and will lead to the development of therapeutic strategies and agents.

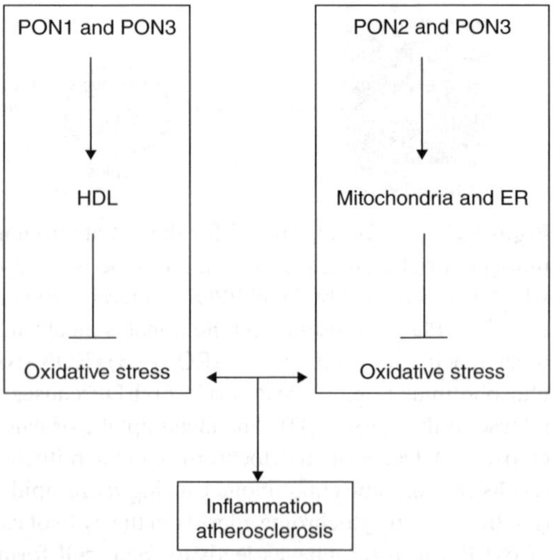

Figure 21.3 PON proteins protect against the development of atherosclerosis by preventing oxidative stress-mediated inflammation.

REFERENCES

1. Navab, M., Hama, S.Y., Reddy, S.T., et al. (2002) Oxidized lipids as mediators of coronary heart disease. *Curr Opin Lipidol*. 13, 363–372.
2. Witztum, J.L. and Steinberg, D. (2001) The oxidative modification hypothesis of atherosclerosis: does it hold for humans? *Trends Cardiovasc Med*. 11, 93–102.
3. Witztum, J.L. and Steinberg, D. (1991) Role of oxidized low density lipoprotein in atherogenesis. *J Clin Invest*. 88, 1785–1792.
4. Berliner, J.A., Navab, M., Fogelman, A.M., et al. (1995) Atherosclerosis: basic mechanisms. Oxidation, inflammation, and genetics. *Circulation*. 91, 2488–2496.
5. Watson, A.D., Leitinger, N., Navab, M., et al. (1997) Structural identification by mass spectrometry of oxidized phospholipids in minimally oxidized low density lipoprotein that induce monocyte/endothelial interactions and evidence for their presence in vivo. *J Biol Chem*. 272, 13597–13607.
6. Leitinger, N., Tyner, T.R., Oslund, L., et al. (1999) Structurally similar oxidized phospholipids differentially regulate endothelial binding of monocytes and neutrophils. *Proc Natl Acad Sci USA*. 96, 12010–12015.
7. Fielding, C.J. and Fielding, P.E. (1995) Molecular physiology of reverse cholesterol transport. *J Lipid Res*. 36, 211–228.
8. Navab, M., Berliner, J.A., Subbanagounder, G., et al. (2001) HDL and the inflammatory response induced by LDL-derived oxidized phospholipids. *Arterioscler Thromb Vasc Biol*. 21, 481–488.
9. Primo-Parmo, S.L., Sorenson, R.C., Teiber, J., and La Du, B.N. (1996) The human serum paraoxonase/arylesterase gene (PON1) is one member of a multigene family. *Genomics*. 33, 498–507.
10. Draganov, D.I. and La Du, B.N. (2004) Pharmacogenetics of paraoxonases: a brief review. *Naunyn Schmiedebergs Arch Pharmacol*. 369, 78–88.
11. Aldridge, W.N. (1953) Serum esterases I. Two types of esterase (A and B) hydrolysing p-nitrophenyl acetate, propionate and butyrate and a method for their determination. *Biochem J*. 53, 110–117.
12. Aldridge, W.N. (1953) Serum esterases II. An enzyme hydrolyzing diethyl p-nitrophenyl acetate (E600) and its identity with the A-esterase of mammalian sera. *Biochem J*. 53, 117–124.
13. Mackness, M.I., Arrol, S., and Durrington, P.N. (1991) Paraoxonase prevents accumulation of lipoperoxides in low-density lipoprotein. *FEBS Lett*. 286, 152–154.
14. She, Z.G., Chen, H.Z., Yan, Y., Li, H., and Liu, D.P. (2012) The human paraoxonase gene cluster as a target in the treatment of atherosclerosis. *Antioxid Redox Signal*. 16, 597–632.
15. She, Z.G., Zheng, W., Wei, Y.S., et al. (2009) Human paraoxonase gene cluster transgenic overexpression represses atherogenesis and promotes atherosclerotic plaque stability in ApoE-null mice. *Circ Res*. 104, 1160–1168.
16. Reddy, S.T., Wadleigh, D.J., Grijalva, V., et al. (2001) Human paraoxonase-3 is an HDL-associated enzyme with biological activity similar to paraoxonase-1 protein but is not regulated by oxidized lipids. *Arterioscler Thromb Vase Biol*. 21, 542–547.
17. Ng, C.J., Wadleigh, D.J., Gangopadhyay, A., et al. (2001) Paraoxonase-2 is a ubiquitously expressed protein with antioxidant properties and is capable of preventing cell-mediated oxidative modification of low density lipoprotein. *J Biol Chem*. 276, 44444–44449.
18. Ng, C.J., Shih, D.M., Hama, S.Y., Villa, N., Navab, M., and Reddy, S.T. (2005) The paraoxonase gene family and atherosclerosis. *Free Radic Biol Med*. 38, 153–163.
19. Marsillach, J., Camps, J., Beltran-Debón, R., et al. (2011) Immunohistochemical analysis of paraoxonases-1 and 3 in human atheromatous plaques. *Eur J Clin Invest*. 41, 308–314.
20. Shih, D.M., Xia, Y.R., Yu, J.M., and Lusis, A.J. (2010) Temporal and tissue-specific patterns of Pon3 expression in mouse: in situ hybridization analysis. *Adv Exp Med Biol*. 660, 73–87.
21. Costa, L.G., Li, W.F., Richter, R.J., Shih, D.M., Lusis, A., and Furlong, C.E. (1999) The role of paraoxonase (PON1) in the detoxication of organophosphates and its human polymorphism. *Chem Biol Interact*. 119–120, 429–438.
22. Aviram, M., Rosenblat, M., Bisgaier, C.L., Newton, R.S., Primo-Parmo, S.L., and La Du, B.N. (1998) Paraoxonase inhibits high-density lipoprotein oxidation and preserves its functions. A possible peroxidative role for paraoxonase. *J Clin Invest*. 101, 1581–1590.
23. Watson, A.D., Berliner, J.A., Hama, S.Y., et al. (1995) Protective effect of high density lipoprotein associated paraoxonase. Inhibition of the biological activity of minimally oxidized

low density lipoprotein. *J Clin Invest.* 96, 2882–2891.

24. Teiber, J.F., Draganov, D.I., and La Du, B.N. (2003) Lactonase and lactonizing activities of human serum paraoxonase (PON1) and rabbit serum PON3. *Biochem Pharmacol.* 66, 887–896.
25. Rozenberg, O., Shih, D.M., and Aviram, M. (2003) Human serum paraoxonase 1 decreases macrophage cholesterol biosynthesis: possible role for its phospholipase-A2-like activity and lysophosphatidylcholine formation. *Arterioscler Thromb Vasc Biol.* 23, 461–467.
26. Harel, M., Aharoni, A., Gaidukov, L., et al. (2004) Structure and evolution of the serum paraoxonase family of detoxifying and anti-atherosclerotic enzymes. *Nat Struct Mol Biol.* 11, 412–419.
27. Teiber, J.F., Draganov, D.I., and La Du, B.N. (2004) Purified human serum PON1 does not protect LDL against oxidation in the in vitro assays initiated with copper or AAPH. *J Lipid Res.*45, 2260–2268.
28. Connelly, P.W., Draganov, D., and Maguire, G.F. (2005) Paraoxonase-1 does not reduce or modify oxidation of phospholipids by peroxynitrite. *Free Radic Biol Med.* 38, 164–174.
29. Draganov, D.I., Stetson, P.L., Watson, C.E., Billecke, S.S., and La Du, B.N. (2000) Rabbit serum paraoxonase 3 (PON3) is a high density lipoprotein-associated lactonase and protects low density lipoprotein against oxidation. *J Biol Chem.* 275, 33435–33442.
30. Draganov, D.I., Teiber, J.F., Speelman, A., Osawa, Y., Sunahara, R., and La Du, B.N. (2005) Human paraoxonases (PON1, PON2, and PON3) are lactonases with overlapping and distinct substrate specificities. *J Lipid Res.* 46, 1239–1247.
31. Draganov, D.I. (2010) Lactonases with organophosphatase activity: structural and evolutionary perspectives. *Chem Biol Interact.* 187, 370–372.
32. Deakin, S.P., Bioletto, S., Bochaton-Piallat, M.L., and James, R.W. (2011) HDL-associated PON1 can redistribute to cell membranes and influence sensitivity to oxidative stress. *Free Radic Biol Med.* 50, 102–109.
33. Aviram, M. (2012) Atherosclerosis: cell biology and lipoproteins – paraoxonases protect against atherosclerosis and diabetes development. *Curr Opin Lipidol.* 23, 169–171.
34. Kim, D.S., Marsillach, J., Furlong, C.E., and Jarvik, G.P. (2013) Pharmacogenetics of paraoxonase activity: elucidating the role of high-density lipoprotein in disease. *Pharmacogenomics.* 14, 1495–1515.
35. Huang, Y., Wu, Z., Riwanto, M., et al. (2013) Myeloperoxidase, paraoxonase-1, and HDL form a functional ternary complex. *J Clin Invest.* 123, 3815–3828.
36. Mackness, M. and Mackness, B. (2013) Targeting paraoxonase-1 in atherosclerosis. *Expert Opin Ther Targets.* 17, 829–837.
37. Tang, W.H., Hartiala, J., Fan, Y., et al. (2012) Clinical and genetic association of serum paraoxonase and arylesterase activities with cardiovascular risk. *Arterioscler Thromb Vasc Biol.* 32, 2803–2812.
38. Charles-Schoeman, C., Lee, Y.Y., Shahbazian, A., et al. (2013) Association of paraoxonase 1 gene polymorphisms and enzyme activity with carotid plaque in rheumatoid arthritis. *Arthritis Rheum.* 65, 2765–2772.
39. Chen, Q., Reis, S.E., Kammerer, C.M., et al. (2003) Association between the severity of angiographic coronary artery disease and paraoxonase gene polymorphisms in the National Heart, Lung, and Blood Institute-sponsored Women's Ischemia Syndrome Evaluation (WISE) study. *Am J Hum Genet.* 72, 13–22.
40. Shi, J., Zhang, S., Tang, M., et al. (2004) Possible association between Cys311 Ser polymorphism of paraoxonase 2 gene and late-onset Alzheimer's disease in Chinese. *Brain Res Mol Brain Res.* 120, 201–204.
41. Martinelli, N., Girelli, D., Olivieri, O., et al. (2004) Interaction between smoking and PON2 SerSHCys polymorphism as a determinant of the risk of myocardial infarction. *Eur J Clin Invest.* 34, 14–20.
42. Ng, C.J., Bourquard, N., Grijalva, V., et al. (2006) Paraoxonase-2 deficiency aggravates atherosclerosis in mice despite lower apolipoprotein-B-containing lipoproteins: anti-atherogenic role for paraoxonase-2. *J Biol Chem.* 281, 29491–29500.
43. Devarajan, A., Bourquard, N., Hama, S., et al. (2011) Paraoxonase 2 deficiency alters mitochondrial function and exacerbates the development of atherosclerosis. *Antioxid Redox Signal.* 14, 341–351.
44. Devarajan, A., Grijalva, V.R., Bourquard, N., et al. (2012) Macrophage paraoxonase 2 regulates calcium homeostasis and cell survival under endoplasmic reticulum stress conditions and is

sufficient to prevent the development of aggravated atherosclerosis in paraoxonase 2 deficiency/apoE-/- mice on a Western diet. *Mol Genet Metab*. 107, 416–427.

45. Shih, D.M., Xia, Y.R., Wang, X.P., et al. (2007) Decreased obesity and atherosclerosis in human paraoxonase 3 transgenic mice. *Circ Res*. 100, 1200–1207.
46. Yang, F., Wang, L.H., Wang, J., Dong, Y.H., Hu, J.Y., and Zhang, L.H. (2005) Quorum quenching enzyme activity is widely conserved in the sera of mammalian species. *FEBS Lett*. 579, 3713–3717.
47. Fuqua, C. and Greenberg, E.P. (2002) Listening in on bacteria: acyl-homoserine lactone signaling. *Nat Rev Mol Cell Biol*. 3, 685–695.
48. Shiner, E.K., Rumbaugh, K.P., and Williams, S.C. (2005) Interkingdom signaling: deciphering the language of acyl homoserine lactones. *FEMS Microbiol Rev*. 29, 935–947.
49. Elias, M. and Tawfik, D.S. (2012) Divergence and convergence in enzyme evolution: parallel evolution of paraoxonases from quorum-quenching lactonases. *J Biol Chem*. 287, 11–20.
50. Chun, C.K., Ozer, E.A., Welsh, M.J., Zabner, M., and Greenberg, E.P. (2004) Inactivation of a pseudomonas aeruginosa quorum-sensing signal by human airway epithelia. *Proc Natl Acad Sci USA*. 101, 3587–3590.
51. Bar-Rogovsky, H., Hugenmatter, A., and Tawfik, D.S. (2013) The evolutionary origins of detoxifying enzymes: the mammalian serum paraoxonases (PONs) relate to bacterial homoserine lactonases. *J Biol Chem*. 288, 23914–23927.
52. Pezzulo, A.A., Hornick, E.E., Rector, M.V., et al. Expression of human paraoxonase 1 decreases superoxide levels and alters bacterial colonization in the gut of Drosophila melanogaster. *PLoS One*. 7, e43777.
53. Devarajan, A., Bourquard, N., Grijalva, V.R., et al. (2013) Role of PON2 in innate immune response in an acute infection model. *Mol Genet Metab*. 110, 362–370.
54. Rosenfeld, M.E. and Campbell, L.A. (2011) Pathogens and atherosclerosis: update on the potential contribution of multiple infectious organisms to the pathogenesis of atherosclerosis. *Thromb Haemost*. 106, 858–867.
55. Horke, S., Witte, I., Wilgenbus, P., Krüger, M., Strand, D., and Förstermann (2007) Paraoxonase-2 reduces oxidative stress in vascular cells and decreases endoplasmic reticulum stress-induced caspase activation. *Circulation*. 115, 2055–2064.
56. Horke, S., Witte, I., Wilgenbus, P., et al. (2008) Protective effect of paraoxonase-2 against endoplasmic reticulum stress-induced apoptosis is lost upon disturbance of calcium homoeostasis. *Biochem J*. 416, 395–405.
57. Pétremand, J., Puyal, J., Chatton, J.Y., et al. (2012) HDLs protect pancreatic β-cells against ER stress by restoring protein folding and trafficking. *Diabetes*. 61, 1100–1111.
58. Devarajan, A. and Reddy, S.T. (2008) Paraoxonase 2 (PON2) localizes to and modulates function of mitochondria in response to oxidative stress. 3rd International Conference on Paraoxonases. University of California, Los Angeles. http://www.cardiology.med.ucla.edu/paraoxonases. Accessed on August 13, 2014.
59. Altenhöfer, S., Witte, I., Teiber, J.F., et al. (2010) One enzyme, two functions: PON2 prevents mitochondrial superoxide formation and apoptosis independent from its lactonase activity. *J Biol Chem*. 285, 24398–24403.
60. Schweikert, E.M., Devarajan, A., Witte, I., et al. (2012) PON3 is upregulated in cancer tissues and protects against mitochondrial superoxide-mediated cell death. *Cell Death Differ*. 19, 1549–1560.
61. Bhattacharyya, T., Nicholls, S.J., Topol, E.J., et al. (2008) Relationship of paraoxonase 1 (PON1) gene polymorphisms and functional activity with systemic oxidative stress and cardiovascular risk. *JAMA*. 299, 1265–1276.
62. Stevens, R.C., Suzuki, S.M., Cole, T.B., Park, S.S., Richter, R.J., and Furlong, C.E. (2008) Engineered recombinant human paraoxonase 1 purified from Escherichia coli protects against organophosphate poisoning. *Proc Natl Acad Sci USA*. 105, 12780–12784.
63. Stoltz, D.A., Ozer, E.A., Taft, P.J., et al. (2008) Drosophila are protected from Pseudomonas aeruginosa lethality by transgenic expression of paraoxonase-1. *J Clin Invest*. 118, 3123–3131.
64. Kim, J.B., Xia, Y.R., Romanoski, C.E., et al. (2011) Paraoxonase-2 modulates stress response of endothelial cells to oxidized phospholipids and a bacterial quorum-sensing molecule. *Arterioscler Thromb Vasc Biol*. 31, 2624–2633.

22

PROSTANOIDS, THEIR RECEPTORS, AND ATHEROGENESIS

DOMENICO PRATICÒ

Center for Translational Medicine and Department of Pharmacology, Temple University School of Medicine, Philadelphia, PA, USA

INTRODUCTION

Atherosclerosis is chronic, complex, and multifactorial vascular disease characterized by progressive and chronic vascular inflammatory reactions [1–4]. While acute inflammation is *per se* a beneficial event that typically leads to the removal of damaging factors and restoration of structure and physiological functions, the failure of this results in chronic inflammation that can then predispose to excessive chronic tissue damage.

Prostanoids and isoprostanoids are lipid autacoids that can have an impact on its phenotype development since they are active modulators of inflammatory processes in the vasculature [5]. In this chapter, we will review our current knowledge on the biological roles that these lipid mediators play in atherosclerosis. We will discuss their biosynthesis, their biologic effects on the vasculature, and the pharmacology of their blockade in modulating the atherogenetic process. This information could help develop therapeutic strategies aimed at preventing or reducing the mortality resulting from atherosclerosis.

PROSTANOIDS

Prostaglandins and thromboxanes A_2 (TxA_2) belong to a large family of lipid mediators called prostanoids. They are metabolic derivatives of arachidonic acid (AA), and the various subclasses are generated by distinct metabolic pathways. AA undergoes oxidation by one of three types of enzyme: (1) lipoxygenases (LOX), to form leukotrienes and lipoxins; (2) cyclooxygenases (COX), to form prostanoids (i.e., prostaglandins and thromboxanes); (3) and cytochrome P-450 monooxygenases, to form epoxides (Figure 22.1).

In this chapter, will focus on TxA_2 and PGI_2. From a biochemical viewpoint, both prostanoids are produced in two steps from AA via the COX pathway and the subsequent action

Atherosclerosis: Risks, Mechanisms, and Therapies, First Edition. Edited by Hong Wang and Cam Patterson.

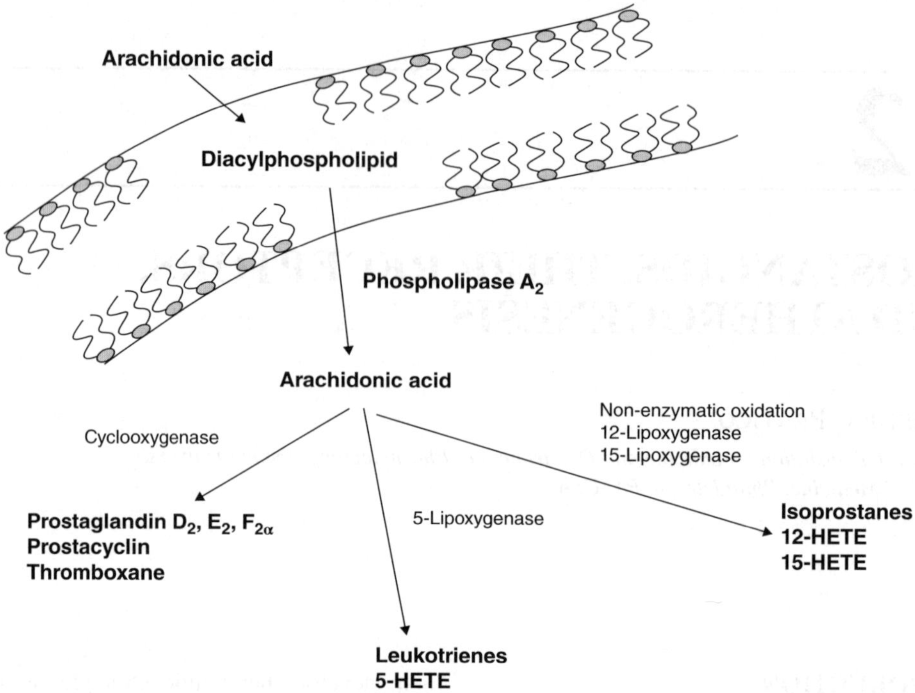

Figure 22.1 Arachidonic acid metabolism via two main enzymatic pathways: the cyclooxygenase (COX) and the lipoxygenase (LOX). COX catalyses the production of prostaglandins and TxA_2. 5-, 12-, and 15-LOX generate leukotrienes, hepoxylins, and lipoxins, respectively. The isoprostanes are prostaglandin-like compounds formed from the free radical-catalyzed oxidation of the arachidonic acid without the action of COX.

of two specific enzymes: the thromboxane and prostacyclin synthase, respectively. There are two main COX isoforms—COX-1 and COX-2—that are encoded by different genes [6]. COX-1 is expressed constitutively in most tissues and generally does not undergo changes in its gene expression and mediates "housekeeping" functions. By contrast, COX-2 isoform expression is typically induced at sites of inflammation by many factors including tumor promoters, cytokines, and inflammatory stimuli [7]. For this reason, COX-2 has also been defined as the "inducible" version of COX isoform. However, it is now clear that there are exceptions to this "rule." For instance, COX-1 expression levels can be regulated, whereas COX-2 is also expressed constitutively in several tissues such as vascular endothelium, kidney, and brain [8, 9].

In vivo, TXA_2 synthesis involves coordinated action of a series of enzymes, including phospholipases that release arachidonate from phospholipids, COX-1 or -2 that convert arachidonate into PGH_2, and, finally, thromboxane synthase (TXS). Despite the fact that from a chemical viewpoint TxA_2 is a very unstable compound with a half life of 20–30 s, it possesses a wide range of activities toward different members of the vascular system. Thus, it is a strong direct platelet activator, and can also amplify the platelet response to other aggregating agents (i.e., collagen and ADP). Further, it has potent biological activities directly on blood vessels by acting as a potent vasoconstrictor and a smooth-muscle cell mitogen. The thromboxane metabolic pathway has important implication in vascular homeostasis, as demonstrated by the observation that by blocking platelet TxA_2-dependent formation is beneficial in the primary and secondary prevention of the atherosclerotic vascular complication [10]. PGI_2, which is produced mainly by vascular endothelial cells via the prostacyclin synthase enzymatic pathway, is the physiological antagonist of TxA_2 and has potent platelet-inhibitory and vasodilating properties [11].

PROSTANOIDS RECEPTORS

Prostanoids exert their biologic effects via the activation of rhodopsin-like 7-transmembrane-spanning G protein-coupled receptors (GPCRs) (Table 22.1). These receptors form a superfamily comprising eight members: E prostanoid (EP) receptors, E1, EP2, EP3, EP4; PGD receptor (DP); PGF receptor (FP); PGI receptor (IP); and thromboxane receptor (TP). The effects of TXA_2 are mediated through its specific GPCR, the TP of which two isoforms are known in humans: TP-α and TP-β (Table 22.1). These two isoforms are generated by alternative splicing. Thus, they are identical with regard to their N-terminal 328 amino acid residues, but differ exclusively in their C-terminal cytoplasmic domains. Prostacyclin biologic effects on the contrary are mediated by a specific GPCR called prostacyclin receptor (IP) (Table 22.1).

These receptors couple to a variety of intracellular signaling pathways that ultimately mediate the biologic effect of its activation in the cell biology context. EP2, EP4, IP, and DP receptors activate adenylyl cyclase via Gs, increasing cAMP. EP1 and FP activate phosphatidylinositol metabolism via Gq, leading to the synthesis of inositol triphosphate with intracellular calcium mobilization. TP couples mainly to two types of G-protein, the Gq and the G13 families, resulting in the activation of phospholipase C and guanine nucleotide exchange factor of the small G protein Rho, respectively. In addition, TP can also couple via Gh to phospholipase C, as well as via Gi and Gs to adenylate cyclase. Both TP isoforms couple to phospholipase C activation, but TPα stimulates adenylyl cyclase, whereas TPβ prevents it (Table 22.1).

PROSTANOIDS AND ATHEROGENESIS

COX-1 and TXS are responsible for the synthesis of TxA_2 in platelets where both enzymes are highly expressed. By contrast, the synthesis of PGI_2 depends primarily on COX-2 activity and prostacyclin synthase in the vascular wall.

TABLE 22.1 The major types and subtypes of receptors for prostanoids

Prostanoid	Prostanoid receptors	G protein-coupled receptor	Prostanoid receptor cellular expression
PGD_2	DP	Gs	Platelets and VSMCs
PGE_2	EP1	?	Monocytes and VSMCs
PGE_2	EP2	Gs	Monocytes and VSMCs
PGE_2	EP3	Gi (Gs,Gq)	Monocytes, VSMCs, and platelets
PGE_2	EP4	Gs	Monocytes and VSMCs
PGF_{2_α}	FP	Gq	VSMCs
PGI_2	IP	Gs (Gq)	Platelets, monocytes, endothelial cells, and VSMCs
TXA_2	TP-α and TP-β	Gq (Gi,Gs)	Platelets, monocytes, endothelial cells, and VSMCs

The different effects of prostanoids are dependent on the activation of specific receptors. In the nomenclature for the classical prostanoid receptors, the first letter indicates the prostanoid with the greatest affinity for this receptor. The prostanoid receptors are membrane seven transmembrane domain G protein-coupled receptors.

PGD_2, prostaglandin D_2; PGE_2, prostaglandin E_2; PGF_{2_α}, prostaglandin F_{2_α}; PGI_2, prostacyclin; TP, thromboxane receptor; TXA_2, thromboxane A_2; and VSMC, vascular smooth muscle cell.

Because they are biological opposites, the balance between these two bioactive lipids is crucial in both healthy and diseased vasculature. Once formed *in vivo*, PGI_2 and TxA_2 act locally in a paracrine fashion during vascular wall/platelets interactions, their biological importance is based on the fact that TxA_2 amplifies the aggregatory responses of platelets to all the known agonists, whereas PGI_2 prevents it.

Recent studies have shown that the ratio TxA_2:PGI_2 is more important than the absolute amount of these two mediators produced *in vivo* [12]. Thus, preferential inhibition of COX-1 with low-dose aspirin (ASA) or selective inhibitors of COX-1, reduces the levels of prothrombotic TxA_2 but has little effect on PGI_2, which maintain the antiplatelet and antithrombotic properties *in vivo*. Conversely, the use of selective COX-2 inhibitors (coxibs) results in a potent and selective depletion of PGI_2 compared with TxA_2, which is not influenced [13].

As mentioned earlier, the biological activity of prostanoids is mediated by their binding to specific receptors on target cells (Table 22.1). TP receptor is present in vascular tissue, and highly expressed in endothelial cells and smooth muscle cells in the medial layer of the vasculature. A significant increase in TP receptor expression has been reported in both the aorta and epicardial coronary arteries from ischemic atherosclerotic vessels [14]. Activation of the TP receptor on endothelial cells increases the expression of adhesion molecules such as intercellular adhesion molecule 1 (ICAM-1). This molecule, which promotes monocyte adherence, is expressed in human atherosclerosis lesions, and its circulating levels are associated with atherosclerosis progression [15]. TP stimulation of both rat and human VSMCs potentiates cytokine-induced VCAM-1 expression, supporting the concept that the activation of TP may potentiate the development of vascular disease [16]. Finally, cigarette smoking, a well-known risk factor of cardiovascular diseases, is associated with early onset of atherosclerosis. Interestingly, TP receptor expression is enhanced by lipid-soluble smoking particles in smooth muscle cells [17]. Thus, activation of TP receptor can also directly accelerate atherogenesis and plaque growth by these mechanisms.

CYCLOXYGENASE INHIBITORS, THROMBOXANE MODULATORS, AND ATHEROSCLEROSIS

Consistent evidence supports the importance of COX-derived prostanoids in experimental atherogenesis. Traditional nonsteroidal anti-inflammatory drugs (NSAIDs), which inhibit COX activity, decrease atherosclerosis in apolipoprotein E (apoE) knockout mice. On the contrary, pharmacological inhibition of COX1-dependent TxA_2 formation and genetic deletion of the gene encoding for COX-1, the main isoform responsible for TxA_2 formation, delay atherogenesis also in both the apoE knockout and low-density lipoprotein (LDL) receptor knockout mice [18, 19].

Low-dose ASA, by significantly blocking the production of TxA_2, reduces atherogenesis, modulates the vascular inflammatory phenotype typical of the disease, and induces plaque stability in LDL receptor knockout mice [13]. However, while ASA inhibits TxA_2 formation, this therapeutic regimen does not affect the production of other TP receptor endogenous ligands such as hydroxyeicosaenoic acids, derivatives of the lipoxygenase, and F_2 isoprostanes, derivative of the free radical-mediated oxidation of AA (Figure 22.2). Both products can be formed by endothelial cells or circulating monocytes and macrophages in the same milieu of the atherosclerotic plaque. Thus, despite the lack of circulating TxA_2, the availability of TP receptor to those endogenous ligands and activators leaves open a possible way for preserving an atherosclerotic vascular phenotype. This line of evidence has led to a new area of research addressing the functional role that the TP receptor system has in vascular biology and atherogenesis.

The beneficial effect of TP receptor pharmacologic antagonism in preventing early atherogenesis has been demonstrated in different studies and animal models [20, 21]. The same therapeutic regimen significantly reduces the

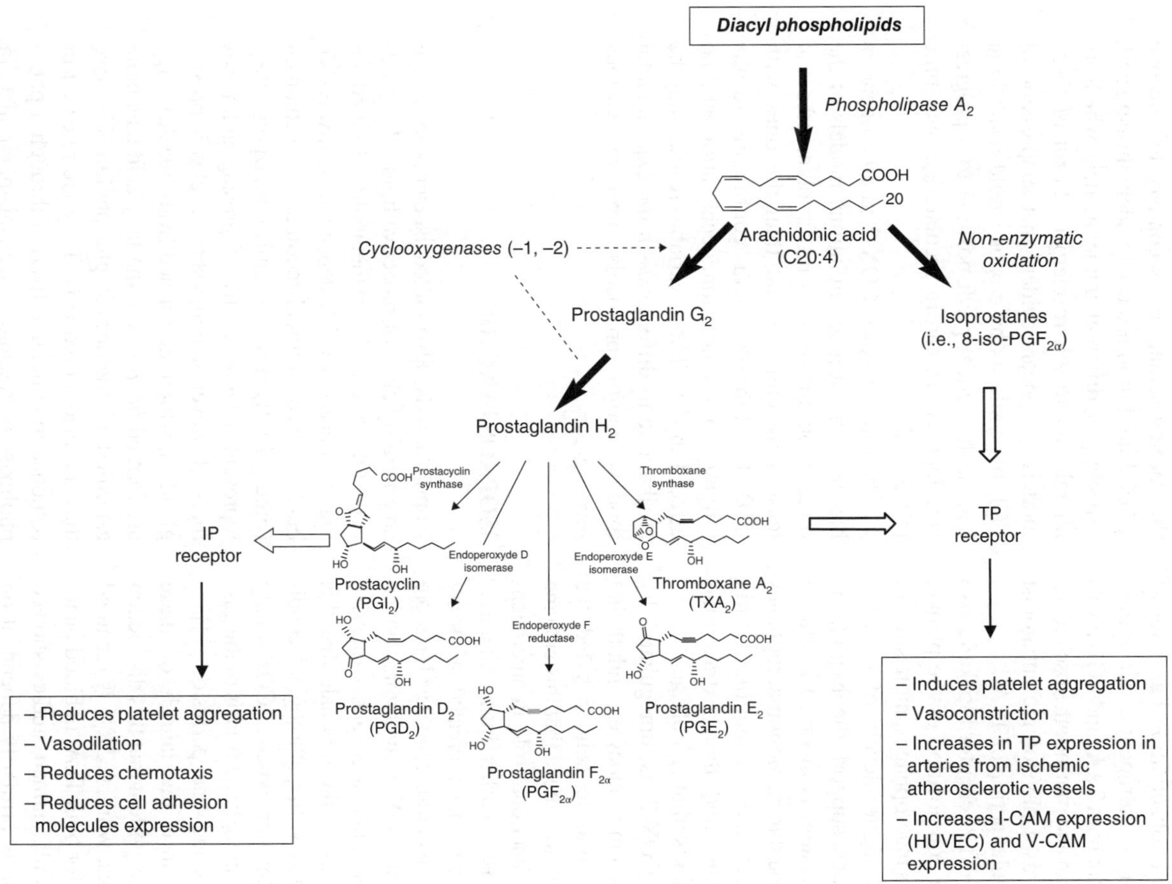

Figure 22.2 The biosynthesis of prostanoids and isoprostanoids is typically in response to cell activation, when arachidonic acid is released from the diacylphospholipids within the cell membrane via the action of a phospholipase A_2. Next, it is enzymatically oxidized to form unstable intermediate cyclic endoperoxides prostaglandin G_2, and then PGH_2. The latter is transformed enzymatically by different prostaglandin synthases into prostaglandins D_2 (PGD_2), E_2 (PGE_2), F_{2_α} (PGF_{2_α}), prostacyclin (PGI_2), and TXA_2. Arachidonic acid can also undergo nonenzymatic oxidation and form isoprostanes, which can activate the same receptor used by TxA_2, called TP, to evoke potent biological actions in the vasculature. TxA_2 amplifies the aggregatory responses of platelets, whereas PGI_2 prevents it. Activation of the TP receptor on endothelial cells increases the expression of adhesion molecules such as intercellular adhesion molecule 1 (ICAM-1). TP stimulation of VSMCs potentiates cytokine-induced VCAM-1 expression, supporting the concept that the activation of TP may potentiate the development of vascular disease. Activation of IP receptor reduces atherogenesis and plaque growth by suppressing chemotaxis, cell adhesion expression, and platelet/endothelial reactivity.

progression of established atherosclerotic vascular lesions in the LDL receptor knockout mice, which manifested a complete inhibition of *ex vivo* platelet aggregation, partial reduction of TxA_2 biosynthesis but fully preserved PGI_2 levels [22, 23]. This observation has a lot of implications from a therapeutic viewpoint, since it is known that both TxA_2 and PGI_2 levels are elevated in human and murine atherosclerosis, and that the latter as a physiologic antagonist modulates the vascular action of TxA_2.

Pharmacological TP receptor blockade also affects the cellular composition of the plaque, which results in less macrophages, more smooth muscle cells, and higher levels of collagen [23]. These findings support the hypothesis that this receptor system is involved in plaque development by modulating monomacrophages and smooth muscle cell interactions within the vascular lesions, and their extracellular derivative components, that is, collagen.

The fact that COX-2 is upregulated by inflammatory stimuli in monocytes, endothelial cells, and smooth muscle cells has led the scientific community to believe that this enzyme should also play a functional role in atherogenesis, where, as discussed earlier, inflammation is an active player. This idea created a strong impulse for the pharma industry to develop more selective and potent COX-2 inhibitors, which were called coxibs. This large and expensive effort was backed up by the rationale that this enzyme is associated with pathological conditions such as inflammation, whereas COX-1 only with platelet aggregation, homeostasis of the gastrointestinal mucosa, and renal perfusion [24].

Unfortunately, since the launch of these drugs, it has become evident that this distinction is not so strict. Thus, while COX-1 is regarded as the main source for both TxA_2 in blood platelets and vascular wall, human studies showed that PGI_2 formation is primarily dependent on COX-2 activation. Based on this evidence, it is clear that coxibs are theoretically harmful in the cardiovascular system since they block PGI_2 biosynthesis but have no effect on the TxA_2 formation; and as a result, they would favor a prothrombotic vascular phenotype. By contrast, nonspecific NSAIDs, by blocking both COX isoforms, would have a balanced effect of reducing at the same time both the prothrombotic effect of TxA_2 and the antithrombotic effect of PGI_2. The assessment of the cardiovascular risk secondary to selective COX-2 inhibition has been widely investigated, especially in different animal models, which are all relevant to atherosclerosis. In all of these models, it appears evident that suppression of PGI_2 biosynthesis by coxibs would result in at least a theoretical predisposition to a progressive increase in cardiovascular risk over time [12, 24, 25].

The direct role of COX-2 in atherogenesis has been investigated in animal models of the disease, but the results thus far obtained are at least more controversial than the ones with COX-1 inhibitors, with conflicting results ranging from retardation, acceleration, and no effect [26–30]. These variable results may be attributable to differences in the experimental models and/or protocols used by various investigators.

ISOPROSTANOIDS

Isoprostanoids, also called isoeicosanoids, are an emerging class of bioactive lipids also products of the AA oxidative metabolism formed not via the classical COX-dependent pathways(s), but a free radical-mediated mechanism (Figure 22.2). From a chemical viewpoint, they represent a large family of prostaglandin isomers. However, in contrast to the classic prostaglandins, which are formed from free AA, they are formed *in situ* from the fatty acid backbone esterified in membrane phospholipids, where they are also stored [31]. They are released in response to cellular activation through a phospholipase A_2-mediated mechanism, circulate in plasma, and are excreted in the urine. Considering the wide distribution of their main precursor, that is, AA, isoprostanoid formation can occur in all of the cellular players of the atherogenic process, including endothelial cells, monocytes, and smooth muscle cells.

ISOPROSTANOIDS AND ATHEROGENESIS

Since their discovery, these chemically stable prostaglandin isomers, generally called F_2-isoprostanes (F_2-iPs), have emerged as one of the most reliable biomarkers of *in vivo* lipid peroxidation and oxidative stress, a mechanism widely involved in atherogenesis [32, 33]. Consistent with this concept, there is an increased accumulation of F_2-iPs in atherosclerotic plaques in humans and animals [34]. In hypercholesterolemic patients, elevated concentration of F_2-iPs correlates with cholesterol levels and falls on statin therapy. It is known that diabetic subjects are predisposed to accelerated atherogenesis and F_2-iPs are elevated in these patients, even without overt atherosclerosis. Further, isoprostanes are elevated in several cardiovascular diseases such as reperfusion syndromes, coronary angioplasty, coronary artery bypass graft surgery, unstable angina and in symptomatic plaques, which indicates that they are associated with plaque instability. They are increased also in different mouse models of genetic hypercholesterolemia and atherogenesis, where exogenous antioxidants, such as vitamin E, reduce both their levels and the development of the disease [35]. Interestingly, in coronary artery disease, urinary levels of F_2-iPs are an independent risk marker to develop future cardiovascular events [36]. This study is very important because it stresses the fact that since these lipids have potent vascular biological activity, they may be also involved in the pathogenesis of the disease. For instance, F_2-iPs promote platelet activation and formation of TxA_2, induce mitogenesis of vascular smooth muscle cells and proliferation of fibroblast and endothelial cells, and increase the expression of endothelin 1 in aortic endothelial cells. All of these effects are prevented by pharmacologic blockade of the TP receptor, which indicates that this receptor is involved in their vascular effects [37] (Figure 22.2). Previously, we showed that the isoprostane 8-iso-$PGF_2\alpha$ (also known as $iPF_2\alpha$-III) (Figure 22.3), directly promotes atherogenesis in two different mouse

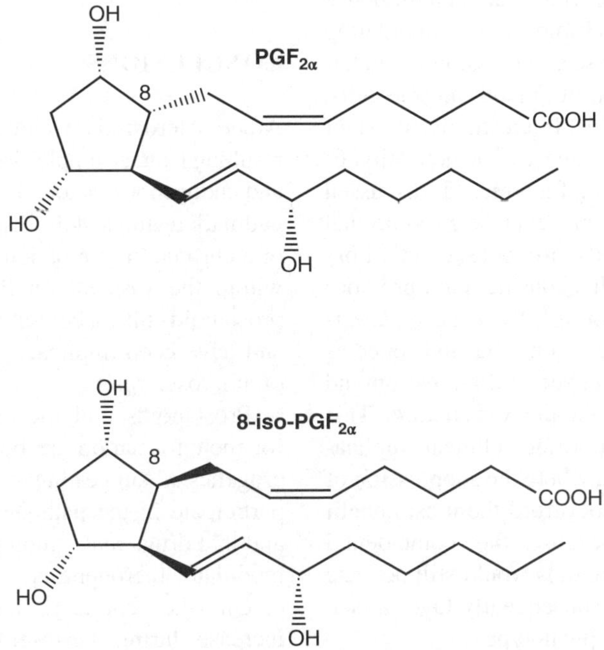

Figure 22.3 An example of the chemical structure of the enzyme-derived prostanoid prostaglandin F_{2_α} (PGF_{2_α}) compared with the nonenzymatically derived isomer F_2-isoprostane, 8-iso-prostaglandin F_{2_α} (8-iso-PGF_{2_α}).

models—the apoE- and LDL receptor knockout mice—by activating the TP receptor [38]. This effect was mediated by the induction of potent proinflammatory vascular reactions, but independent from TxA_2 levels, changes in blood pressure or lipid profile. Thus, pharmacologic blockade or genetic absence of the TP receptor suppressed the vascular proatherogenic effects of 8-iso-$PGF_2\alpha$, but not of other oxidized lipids [38]. These results demonstrated that in atherosclerosis, F_2-iPs are not only biomarkers of oxidative stress but also active mediators of its vascular phenotype whereby modulating the disease development and progression.

TP RECEPTOR SYSTEM AS BIOLOGICAL LINK BETWEEN OXIDATIVE STRESS AND INFLAMMATION IN ATHEROGENESIS

We have extensively discussed the notion that human and experimental atherosclerosis is characterized by an altered biosynthesis of both these mediators, which can then influence atherogenesis in multiple ways: modulating inflammatory responses, expression of metalloproteinases, growth of smooth muscle cells, and platelet reactivity. Interestingly, most of these effects are mediated via the activation of the TP receptor system, for which, as discussed earlier, while TxA_2 represents the conventional ligand isoprostanoids are potent activators too (Figure 22.2). This consideration provides support to the novel concept that in atherogenesis the TP receptor system could be envisioned as the biological link between inflammation and oxidative stress within the vasculature. This hypothesis has an enormous clinical implication in atherogenesis, where the suppression of TxA_2 alone would not afford the most benefit for the patients because the coincidental presence of isoprostanoids would still activate the TP receptor and subsequently favor a proatherogenic vascular phenotype.

In line with this idea, administration of a TP receptor antagonist improves endothelial dysfunction in patients with coronary artery disease and already treated with ASA whose levels of TxA_2 are suppressed [39]. This report was recently confirmed by our study where we showed that in an experimental setting where TxA_2 is suppressed, the coincidental blockade of TP receptor increases the beneficial anti-inflammatory action of COX-1 inhibition and results in an additional antiatherogenic effect [23]. Under the same experimental conditions, there was a decrease in monocyte/macrophages and an increase in collagen content in the atherosclerotic lesions, suggesting a more stable plaque phenotype [23].

The same hypothesis was recently tested in an opposite experimental setting, where the genetic absence the 12/15-lipoxygenase enzyme, which is an important source of oxidative stress and isoprostanoids, is known to result in less atherogenesis in both apoE- and LDL receptor knockout mice. In this scenario, the addition of a selective TP receptor antagonism resulted in an additive and more potent anti-inflammatory and antiatherogenic effect [40].

CONCLUSIONS

Atherosclerosis is a chronic vascular disease resulting from a complex interplay of exogenous and endogenous factors. It is well accepted that hallmark features of the disease are the presence of a chronic inflammation and oxidative stress within the vasculature. Prostanoids and isoprostanoids play a central role in inflammation and have been implicated in the pathogenesis of atherosclerosis.

Prostanoids and the enzymes responsible for their formation are both increased in atherogenesis, suggesting that these mediators participate in the pathogenesis of the disease and that drugs that inhibit their activity should modulate atherogenesis.

On the contrary, since isoprostanoids increase during the disease development and represent an independent risk marker for coronary artery disease, they are also considered active players in atherogenesis. This concept is supported by the evidence that these lipids, by directly activating the TP receptor system, can mediate the vascular effects of inflammation and oxidative stress in atherogenesis.

From a translational viewpoint, this recent observation has an important clinical implication, since suppression of TxA_2 production alone will not afford the best protection because activation of TP receptor by isoprostanoids will still promote an atherosclerotic vascular phenotype during the evolution of the disease.

In summary, pharmacologic modulation of the TP receptor system represents an important additional therapeutic target by modulating both inflammation as well as oxidative stress responses of the vasculature. We conclude that this class of drugs deserves to be further investigated and explored in either preventing or retarding human atherogenesis.

REFERENCES

1. Libby, P., Ridker, P.M., Maseri, A. (2002) Inflammation and atherosclerosis. *Circulation* 105, 1135–1143.
2. Ingelsson, E., Pencina, M.J., Tofler, G.H., et al. (2007) Multimarker approach to evaluate the incidence of the metabolic syndrome and longitudinal changes in metabolic risk factors. The Framingham Offspring Study. *Circulation* 116, 984–992.
3. Weber, C., Noels, H. (2011) Atherosclerosis: current pathogenesis and therapeutic options. *Nat. Med.* 17, 1410–1422.
4. Libby, P., Ridker, P.M., Hansson, G.K. (2011) Progress and challenges in translating the biology of atherosclerosis. *Nature* 473, 317–325.
5. Dogne, J.M., Hanson, J., Pratico, D. (2005) Thromboxane, prostacyclin and isoprostanes: therapeutic targets in atherogenesis. *Trends Pharmacol. Sci.* 26, 639–644.
6. Simmons, D.L., Botting, R.M., Hla, T. (2004) Cycloxygenase isozymes: the biology of prostaglandin synthesis and inhibition. *Pharmacol. Rev.* 56, 387–437.
7. Inoue, I., Taba, Y., Miwa, C., Yokota, C., Miyagi, M., Sasaguri, T. (2002) Transcriptional and posttranslational regulation of cyclooxygenase-2 expression by fluid shear stress in vascular endothelial cells. *Arterioscler. Thromb. Vasc. Biol.* 22,1415–1420.
8. Khan, K.N., Stanfield, K.M., Harris, R.K., Baron, D.A. (2001) Expression of cyclooxygenase-2 in the macula densa of human kidney in hypertension, congestive heart failure, and diabetic nephropathy. *Ren. Fail.* 23, 321–330.
9. Yaksh, T.L., Dirig, D.M., Conway, C.M. (2001) The acute antihyperalgesic action of nonsteroidal, anti-inflammatory drugs and release of spinal prostaglandin E2 is mediated by the inhibition of constitutive spinal cyclooxygenase-2 (COX-2) but not COX-1. *J. Neurosci.* 21, 5847–5853.
10. Dogne, J.M., de Leval, X., Hanson, J. et al. (2004) New developments on thromboxane and prostacyclin modulators part I: thromboxane modulators. *Curr. Med. Chem.* 11, 1233–1241.
11. de Leval, X., Hanson, J., David, L.J., Masereel, B., Pirotte, B., Dogné, J.M. (2004) New developments on thromboxane and prostacyclin modulators part II: prostacyclin modulators. *Curr. Med. Chem.* 11, 1243–1252.
12. Praticò, D., Dogne, J. (2005) Selective cyclooxygenase-2 inhibitors development in cardiovascular medicine. *Circulation* 112, 1073–1079.
13. Cyrus, T., Sung, S., Zhao, L., Funk, C.D., Tang, S., Praticò, D. (2002) Effect of low-dose aspirin on vascular inflammation, plaque stability, and atherogenesis in low-density lipoprotein receptor deficient mice. *Circulation* 106, 1282–1287.
14. Katugampola, S.D., Davenport, A.P. (2001) Thromboxane receptor density is increased in human cardiovascular disease with evidence for inhibition at therapeutic concentrations by the AT1 receptor antagonist losartan. *Br. J. Pharmacol.* 134, 1385–1392.
15. Tzoulaki, I., Murray, G.D., Lee, A.J., Rumley, A., Lowe, G.D., Fowkes, F.G. (2005) C-reactive protein, interleukin-6, and soluble adhesion molecules predictors of progressive peripheral atherosclerosis in the general population: Edinburg Artery Study. *Circulation* 112, 976–983.
16. Bayat, H., Xu, S., Pimentel, D., Cohen, R.A., Jiang, B. (2008) Activation of thromboxane receptor upregulates interleukin (IL)-1β-induced VCAM-1 expression through JNK signal. *Arterioscler. Thromb. Vasc. Biol.* 28, 127–134.

17. Zhang, W., Zhang, Y., Edvinsson, L., Xu, C. (2008) Up-regulation of thromboxane A_2 receptor expression by lipid soluble smoking particles through post-transcriptional mechanisms. *Atherosclerosis* 196, 608–616.
18. Praticò, D., Tillmann, C., Zhang, Z.B., Li, H., FitzGerald, G.A. (2001) Acceleration of atherogenesis by COX-1-dependent prostanoid formation in low density lipoprotein receptor knockout mice. *Proc. Natl. Acad. Sci. USA* 98, 3358–3363.
19. Belton, O., Duffy, A., Toomey, S., Fitzgerald, D. (2003) Cyclooxygenase isoforms and platelet vessel wall interactions in the apolipoprotein E knockout mouse model of atherosclerosis. *Circulation* 108, 3017–3023.
20. Viles-Gonzales, J.F., Fuster, V., Corti, R., et al. (2005) Atherosclerosis regression and TP receptor inhibition: effect of S18886 on plaque size and composition—A magnetic resonance imaging study. *Eur. Heart J.* 26, 1557–1561.
21. Worth, N.F., Berry, C.L., Thomas, A.C., Campbell, J.H. (2005) S18886, a selective TP receptor antagonist, inhibits development of atherosclerosis in rabbits. *Atherosclerosis* 183, 65–73.
22. Cyrus, T., Yao, Y., Ding, T., Dogne, J., Pratico, D. (2007) A novel thromboxane receptor antagonist and synthase inhibitor, BM-573, reduces development and progression of atherosclerosis in LDL receptor deficient mice. *Eur. J. Pharmacol.* 561, 105–111.
23. Cyrus, T., Yao, Y., Ding, T., Dogne, J.M., Pratico, D. (2007) Thromboxane receptor blockade improves the antiatherogenic effect of thromboxane A_2 suppression in LDLR KO mice. *Blood* 109, 3291–3296.
24. Dogne, J.M., Hanson, J., Supuran, C., Praticó, D. (2006) Coxibs and cardiovascular side-effects: from light to shadow. *Curr. Pharm. Des.* 12, 971–975.
25. Pidgeon, G.P., Tamosiuniene, R., Chen, G., et al. (2004) Intravascular thrombosis after hypoxia-induced pulmonary hypertension: regulation by cyclooxygenase-2. *Circulation* 110, 2701–2707.
26. Burleigh, M.E., Babaev, V.R., Oates, J.A., et al. (2002) Cycloxygenase-2 promotes early atherosclerotic lesion formation in LDL receptor-deficient mice. *Circulation* 105, 1816–1823.
27. Rott, D., Zhu, J., Burnett, M.S., et al. (2003) Effects of MF-tricyclic, a selective cycloxygenase-2 inhibitor, on atherogenesis progression and susceptibility to cytomegalovirus replication in apolipoprotein-E knockout mice. *J. Am. Coll. Cardiol.* 41, 1812–1819.
28. Bea, F., Blessing, E., Bennett, B.J., et al. (2003) Chronic inhibition of cycloxygenase-2 does not alter plaque composition in a mouse model of advanced unstable atherosclerosis. *Cardiovasc. Res.* 60, 198–204.
29. Olensen, M., Kwong, E., Meztli, A., et al. (2002) No effect of cyclooxygenase inhibition on plaque size in atherosclerosis-prone mice. *Scand. Cardiovasc. J.* 36, 362–367.
30. Metzner, J., Popp, L., Marian, C., et al. (2007) The effects of COX-2 selective and non-selective NSAIDs on the initiation and progression of atherosclerosis in apoE-/- mice. *J. Mol. Med.* 85, 623–633.
31. Pratico, D. (2010) The neurobiology of isoprostanes. *Biochim. Biophys. Acta* 1801, 930–933.
32. Hansson, G.K. (2005) Inflammation, atherosclerosis and coronary artery disease. *N. Engl. J. Med.* 352, 1685–1695.
33. Hulsmans, M., Dekeyser, D., Holvoet, P. (2011) MicroRNAs regulating oxidative stress and inflammation in realtion to obesity and atherosclerosis. *FASEB J.* 25, 2515–2527.
34. Steinberg, D., Witztum, J.L. (2002) Is the oxidative modification hypothesis relevant to human atherosclerosis? *Circulation* 105, 2107–2111.
35. Cyrus, T., Yao, Y., Rokach, J., et al. (2003) Vitamin E reduces progression of atherosclerosis in low-density lipoprotein receptor-deficient mice with established vascular lesions. *Circulation* 107, 521–523.
36. Schwedhelm, E., Bartling, A., Lenzen, H., et al. (2004) Urinary 8-iso-prostaglandin $F_{2\alpha}$ as a risk marker in patients with coronary heart disease. *Circulation* 109, 843–848.
37. Comporti, M., Signorini, C., Arezzini, B., Vecchio, D., Monaco, B., Gardi, C. (2008) F_2-isoprostanes are not just markers of oxidative stress. *Free Radic. Biol. Med.* 44, 247–256.
38. Tang, M., Cyrus, T., Yao, Y., Vocun, L., Praticò, D. (2005) Involvement of thromboxane receptor in the proatherogenic affect of isoprostane

$F_{2\alpha}$-III. Evidence from apolipoprotein E- and LDL receptor-deficient mice. *Circulation* 112, 2867–2874.

39. Belhassen, L., Pelle, G., Dubois-Rande, J.L., Adnot, S. (2003) Improved endothelial function by the thromboxane A_2 receptor antagonist S18886 in patients with coronary artery disease treated with aspirin. *J. Am. Coll. Cardiol.* 41, 1198–1204.
40. Tang, L., Ding, T., Pratico, D. (2008) Additive anti-atherogenic effect of thromboxane receptor antagonism with 12/15lipoxygenase gene disruption in apolipoprotein E deficient mice. *Atherosclerosis* 199, 265–270.

23

EXTRACELLULAR PHOSPHOLIPASES: ROLE IN INFLAMMATION AND ATHEROSCLEROTIC CARDIOVASCULAR DISEASE

Uwe J.F. Tietge

Department of Pediatrics, University of Groningen, University Medical Center Groningen, Groningen, The Netherlands

INTRODUCTION

Two main driving forces have been established to mechanistically contribute to the development of atherosclerotic cardiovascular disease, inflammation, and dyslipidemia [1]. Atherogenic dyslipidemia thereby comprises low plasma levels of high-density lipoprotein (HDL) cholesterol and increased levels of triglycerides as well as cholesterol within apolipoprotein (apo)-B containing lipoproteins [2]. In addition to changes in mass levels, the functionality of the different lipoprotein subclasses can be altered [2]. Extracellular, that is, secretory phospholipases are linked to all three of these processes: (1) inflammation, (2) atherogenic dyslipidemia, and (3) dysfunctionality of lipoproteins [3–8]. In general, phospholipases cleave phospholipids either at the sn-1 position (sn-1 phospholipases such as endothelial lipase, EL) [7, 8] or the sn-2 position [sn-2 phospholipases such as the secretory phospholipases ($sPLA_2$) group IIA, V, or X] [3, 4, 6]. As reaction products, a lysophospholipid and a liberated fatty acid are formed (Figure 23.1) [4]. In case of sn-2 phospholipases, this reaction product mainly consists of arachidonic acid (AA), since AA is enriched at this position in cellular membrane and lipoprotein phospholipids [4]. AA can then be fueled into subsequent cyclooxygenase (COX) or lipoxygenase (LOX) pathways and can be converted into important mediators of inflammation [4, 6]. However, also a variety of other fatty acids can be released including saturated, unsaturated, and polyunsaturated fatty acids, from which anti-inflammatory mediators can subsequently be synthesized [4]. In addition, the $sPLA_2$s have been suggested to modulate inflammation via signaling through cell surface receptors such as the M-type $sPLA_2$ receptor [4]. However, the *in vivo* impact of signaling events of $sPLA_2$s independent of their catalytic activity is still rather unclear.

Atherosclerosis: Risks, Mechanisms, and Therapies, First Edition. Edited by Hong Wang and Cam Patterson.

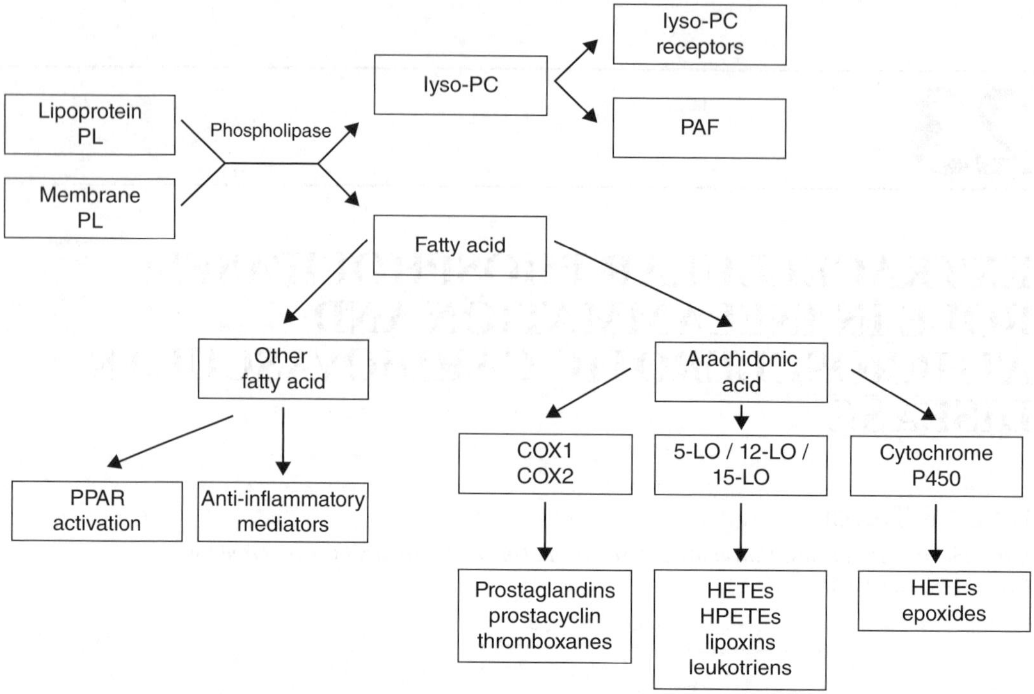

Figure 23.1 General overview over potential metabolic pathways on the generation of bioactive mediators by the activity of secreted phospholipases. Please note that this figure is intended to place phospholipases in the context of bioactive lipid mediator formation and that the details of these pathways, the respective contribution of specific phospholipases on bioactive mediator generation, and the (patho)physiological consequences *in vivo* are still areas of active investigation. PL, phospholipid; PC, phosphatidylcholine; COX, cyclooxygenase; LO, lipoxygenase; HETE, hydroxyeicosatetraenoic acid; HPETE, hydroxyperoxyeicosatetraenoic acid; PPAR, peroxisome proliferator-activated receptor.

Essentially, the M-type receptor represents a mammalian homologue of the snake-type α-$sPLA_2$ inhibitory protein, which serves to bind $sPLA_2$ and inhibit its biological activity [4]. Consistently, the mammalian M-type receptor internalizes upon ligand binding and is directed to phagolysosomes [4], thus making true downstream signaling events somewhat less likely. However, a cooperative action with toll-like receptors might exist [4]. Due to the current unclarity regarding the potential relevance of $sPLA_2$ signaling via the M-type receptor, this will not be discussed further in detail in the following text.

Using lipoprotein phospholipids as substrate extracellular phospholipases can exert substantial effects on the composition and, thereby, metabolism of lipoproteins resulting in altered plasma levels of these [3, 4, 6]. In the following, the main sn-1 and sn-2 phospholipases that have been mechanistically linked to atherosclerotic disease in humans and animal models are discussed, including a summary of their respective specific impact on the process of atherogenesis.

THE SN-1 PHOSPHOLIPASE ENDOTHELIAL LIPASE

Endothelial lipase (EL) is encoded by the LIPG gene and is a 68-kD glycoprotein cloned in 1999, independently by two different research groups [7, 8]. EL is a member of the triacylglycerol lipase gene family, but in contrast to the other members of this family almost completely lacks TG hydrolyzing activity and possesses strong phospholipase activity [7]. In addition to being expressed in endothelial cells, EL expression has also been detected in macrophages, hepatocytes, kidney, thyroid gland, placenta, ovary, and testes [7].

Impact of EL on HDL Metabolism and Reverse Cholesterol Transport

All lipoprotein subclasses have been shown to be a potential substrate for EL; however, by far the strongest affinity of EL is for HDL [7]. Thereby, EL is a strong negative regulator of plasma HDL cholesterol levels, which has been shown *in vivo* in gain of function (adenoviral overexpression) and lack of function studies (specific knockout mice, antibody-mediated inhibition) [7, 8]. Kinetic studies established that hydrolysis of HDL phospholipids by EL results in a significantly enhanced catabolic rate of the HDL particle [9]. As working model integrating available experimental data [8], hydrolysis of HDL phospholipids by EL results in destabilization of the particle with the subsequent shedding of poorly lipidated apoA-I molecules. Cholesterol within the remainder of the particle then becomes more susceptible toward scavenger receptor BI (SR-BI)-mediated selective uptake, a process that also constitutes a prerequisite for HDL remodeling by EL to continue, as evidenced by experiments using EL overexpression in SR-BI-deficient mice [10]. Increased SR-BI-mediated selective uptake results in a higher hepatic cholesterol content in EL overexpressing mice [11]; however, no subsequent impact on biliary cholesterol secretion could be detected [11]. Although EL can also have some metabolic effects by mediating bridging between lipoproteins and cell surface proteoglycans, by far the majority of its effect on HDL is dependent on the catalytic activity of the enzyme [7].

In addition to mechanistic data from experimental animals, human studies point to a similar role of EL in regulating plasma levels of HDL cholesterol. Plasma EL mass was correlated positively with small HDL and negatively with large HDL particles in 510 individuals with a family history of premature coronary heart disease [7, 8]. Further, rare variants of EL resulting in a substantial reduction in plasma EL activity were linked to significantly increased plasma HDL cholesterol levels [7, 8]. In addition, in genome-wide association studies variation at the LIPG locus has been associated with HDL phenotypes [7]. In addition, a recent study using deep resequencing of the LIPG promoter and the 5′UTR together with functional promoter studies and phenotype data in subjects with extreme HDL phenotypes could substantiate a causative negative association between EL and plasma levels of HDL-C [12].

In contrast to the effects of EL on HDL metabolism, the impact of EL on the function of HDL in reverse cholesterol transport (RCT), the major established functionality of HDL, is not so clear-cut. EL has been reported to have variable effects on macrophage cholesterol efflux, the first step in RCT [8]. While EL modification of HDL enhances SR-BI-mediated selective uptake and subsequently the hepatic cholesterol content [11], there was, however, no impact of EL overexpression on either the biliary secretion of cholesterol or the fecal mass output of either neutral sterols or bile acids [11], important hepatic metabolites of cholesterol. Consequently, given the dependence of functional *in vivo* RCT on the biliary secretion process of sterols [13], EL knockout mice did not show altered *in vivo* macrophage-to-feces RCT [8]. On the contrary, however, the same group reported that EL overexpression resulted in a significant decrease in *in vivo* RCT [8]. The impact of EL on RCT in humans remains to be shown.

Association of EL with Inflammation

EL has been demonstrated both as a target and as a modulator of inflammatory responses. Proinflammatory stimuli transcriptionally upregulate EL expression in several cell types with high relevance for atherosclerotic lesion development such as endothelial cells and macrophages [7, 8]. Further, in humans plasma EL levels not only correlate positively with markers of inflammation [7, 8] but are also induced by experimental LPS administration [14]. The impact of EL on inflammatory responses is, however, less well established. While two studies showed that EL increases monocyte adhesion to the vessel wall [15, 16], another indicated a downregulation of adhesion molecules in endothelial cells expressing EL [17]. In a lung inflammation model and in THP-1 macrophages, a reduced or absent expression of EL resulted in a reduction in the inflammatory response. Although not unequivocally decided, taken together available data argue rather in favor of EL promoting inflammatory responses than against. The pathways involved, however, remain to be characterized in detail.

Impact of EL on Atherosclerosis

Studies aiming at providing an integrative view on the impact of EL on atherosclerosis have not generated conclusive results. The first publication addressing this question reported data on EL knockout mice crossed to the pro-atherogenic apoE knockout mouse genetic background and revealed a 70% reduction in atherosclerotic lesion formation in the group lacking EL on chow and a lesser, but still significant, effect in mice fed a Western diet [15]. On the contrary, a later study failed to see any effect of a lack of EL on atherosclerosis phenotypes either in LDLR-deficient or in apoE-deficient mice [18]. The reason for these discrepant results is currently unclear. Further, also genetic association studies in humans did not yield conclusive results [8]. The only prospective data investigating the predictive value of EL as biomarker of future cardiovascular disease (CVD) events are from a relatively small group of hemodialysis patients that were followed for 2 years; there, patients experiencing events had higher baseline plasma levels of EL compared with the event-free group [19].

Conclusion on the Impact of EL on Plasma Lipids and Atherosclerosis

Summing up the available data on EL, it appears well established that decreasing EL activity results in increased plasma HDL cholesterol levels indicating that for such a purpose, EL represents a potential target for pharmacological inhibition. On the contrary, however, the association between EL and atherosclerosis appears less clear and straightforward.

THE SN-2 PHOSPHOLIPASE GROUP IIA SECRETORY PHOSPHOLIPASE A_2

Secretory phospholipase (sPLA$_2$-IIA) is a 14-kD acute phase protein that is potentially expressed by a variety of tissues, mostly of mesenchymal origin [4–6]. With special relevance to atherosclerosis is the expression by hepatocytes, vascular smooth muscle cells, platelets, and monocytes/macrophages [4–6]. Plasma levels can increase substantially by up to a 1000-fold during an acute phase response [6]. Interestingly, while phospholipids within lipoproteins are a good substrate for sPLA$_2$-IIA, normal cell membranes are not. This has been ascribed to the lack of tryptophane residues in sPLA$_2$-IIA, which are thought to be required for efficient interfacial binding to neutral phospholipid surfaces such as the outer leaflet of mammalian cell membranes [4]. Although through various arginine and lysine residues sPLA$_2$-IIA can bind to heparane sulfate proteoglycans [4], substantial amounts of sPLA$_2$-IIA circulate in plasma, where the bulk of the enzyme has been detected in association with the HDL fraction [20], although binding appears to be rather loose so that sPLA$_2$-IIA is easily spun off the particle, when HDL is isolated by ultracentrifugation.

Impact of sPLA$_2$-IIA on HDL Metabolism and RCT

Not surprisingly, HDL particles represent a good substrate for sPLA$_2$-IIA. Similar to the observation made with the sn-1 phospholipase

EL, also the sn-2 phospholipase $sPLA_2$-IIA is a negative regulator of plasma HDL cholesterol levels [20–23]. Hereby, hydrolysis of HDL phospholipids also leads to significantly increased catabolic rates for HDL apolipoproteins and HDL cholesteryl ester resulting in a net effect of substantially decreased plasma HDL cholesterol in transgenic mice overexpressing the human $sPLA_2$-IIA enzyme [21–23]. Also upon $sPLA_2$-IIA overexpression, the SR-BI-mediated hepatic uptake of cholesterol from HDL is higher consequently leading to increased hepatic cholesterol levels [24]. However, also in this case, increased flux of cholesterol via SR-BI is not associated with altered biliary cholesterol secretion rates in the face of unchanged hepatic SR-BI expression [24]. In addition, fecal neutral sterol and bile acid output is not altered in mice overexpressing $sPLA_2$-IIA compared with wild-type C57BL/6 controls [24], which lack the endogenous murine $sPLA_2$-IIA enzyme because of a frame shift mutation in the corresponding murine gene [25]. Finally, *in vivo* macrophage-to-feces RCT experiments revealed that also RCT remains unchanged even when $sPLA_2$-IIA is massively expressed [26]. However, recent studies demonstrated that $sPLA_2$-IIA modification of HDL decreases the cholesterol efflux properties and the antioxidative function of these particles, indicating that $sPLA_2$-IIA activity generates dysfunctional HDL [27].

Association of $sPLA_2$-IIA with Inflammation

As an acute phase enzyme, a variety of proinflammatory stimuli have been characterized to increase transcriptional expression of $sPLA_2$-IIA [4, 5]. Besides being a target of inflammation, $sPLA_2$-IIA has a number of interesting features that impact on an inflammatory response. $sPLA_2$-IIA belongs to the innate immune system and possesses potent antibacterial activity, especially against Gram-positive bacteria [4]. These properties are closely related to the presence of anionic phospholipids in the cell membrane of such bacteria. Regarding gram-negative bacteria, $sPLA_2$-IIA has been shown to act in concert with bactericidal permeability increasing (BPI) protein [4]. As exemplified in a mouse arthritis model, $sPLA_2$-IIA transgenic mice exhibit increased inflammation, whereas balb/c mice specifically deficient in murine $sPLA_2$-IIA show a reduction in the inflammatory response [4].

Impact of $sPLA_2$-IIA on Atherosclerosis

Transgenic mice expressing human $sPLA_2$-IIA on the genetic background of a C57BL/6 mouse lacking the endogenous murine $sPLA_2$-IIA enzyme had a remarkable increase in atherosclerotic lesion formation and developed aortic atherosclerosis even on a chow diet, indicating that $sPLA_2$-IIA is causally linked to the process of atherogenesis [28]. In addition, three independent studies established using bone marrow transplanation experiments that macrophage-specific expression of $sPLA_2$-IIA is sufficient to result in increased atherosclerotic lesion formation [5]. Mechanistically, several factors might contribute to increased atherogenesis upon expression of $sPLA_2$-IIA. First, $sPLA_2$-IIA results in increased plasma levels of oxidized phospholipids and increased formation of isoprostanes indicating increased *in vivo* oxidative stress [29]. Second, $sPLA_2$-IIA causes a severe endothelial dysfunction phenotype that is dependent on the functional expression of COX2 [30]; endothelial dysfunction has been noted to be a precursor of atherosclerosis and an independent predictor of later CVD events. Third, $sPLA_2$-IIA results in decreased plasma levels of HDL and in the generation of dysfunctional HDL particles [20, 21, 27]. Fourth, macrophage $sPLA_2$-IIA causes increased oxidation of LDL in a process largely dependent on the expression of 12/15-LOX [29].

$sPLA_2$-IIA has been detected in human atherosclerotic lesions, specifically within smooth muscle cells and macrophages and extracellularly associated with collagen and [6]. Also in human studies, plasma $sPLA_2$-IIA mass levels and activity are linked to increased *in vivo* oxidative stress determined as plasma isoprostane levels [30]. Further, cross-sectional studies demonstrated that circulating levels of sPLA$_2$-IIA are significantly increased in patients with atherosclerotic CVD, and plasma sPLA$_2$-IIA levels and activity predict the occurrence of future atherosclerotic events in several larger studies [6]. While these data together with the above-mentioned mechanistic animal studies strongly suggest that sPLA$_2$-IIA is a causative risk factor for CVD, the genetic evidence generated thus far is less strong.

Relatively specific sPLA$_2$-IIA inhibitors have been developed, initially designed for intervention in the acute phase response. First, these were tested in patients with sepsis, but there were no discernible differences in outcome between the treated and untreated groups [31]; therefore, this indication was subsequently dropped. Later, especially one compound, varespladib, was further developed for use in patients with atherosclerotic CVD [6]. Before discussing the results of available clinical trials, it might be interesting to note that varespladib was effective in reducing atherosclerosis in apoE knockout mice on the C57BL/6 background that lack the murine sPL$_2$-IIA [32], the isoform this inhibitor supposedly specifically targets. Thus far, the results of four human intervention trials using sPLA$_2$ inhibition with varespladib have been published. The phase 2 Phospholipase Levels and Serological Markers of Atherosclerosis (PLASMA) study found that varespladib administration dose-dependently reduced plasma sPLA$_2$-IIA levels in 396 patients with stable CHD, from which 66% received statin treatment [33]. This result is *per se* surprising, since it remains unexplained how an inhibitor of enzymatic activity is able to reduce circulating mass levels of an enzyme; sPLA$_2$-IIA activity was not determined in this trial. While LDL-C levels were overall reduced by 8% and small dense LDL by 7.3%, there was not effect of the sPLA$_2$-IIA inhibitor on HDL-C levels, which is in contrast to data generated in transgenic mice. A follow-up trial intended to administer varespladib on top of statin treatment confirmed a dose-dependent reduction in circulating sPLA$_2$-IIA levels by 73% in patients receiving 250 mg daily and 85% in the group receiving 500 mg daily of the inhibitor [34]. Again, additional favorable changes in the levels of apoB-containing lipoproteins were noted, however, also a decrease in the HDL particle number and here specifically in smaller HDL that might be more active in atherosclerosis protection, for example, in promoting cholesterol efflux and RCT. The FRANCIS-ACS trial assessed the impact of varespladib on biomarkers in 625 patients with an acute coronary syndrome [35], confirming that the sPLA$_2$ inhibitor had the potential to reduce LDL-C and suggesting that varespladib might have a greater anti-inflammatory potential over placebo, as judged by a rather moderate but significant difference in plasma CRP levels. The sPLA$_2$ Inhibition to Decrease Enzyme Release after Percutaneous Coronary Intervention (SPIDER-PCI) trial tested the impact of varespladib 500 mg daily given around elective PCI on biomarkers of myonecrosis in 144 stable patients with coronary disease [36]. Treatment with varespladib had hereby no impact on plasma levels of troponin I or creatine kinase-MB. However, in the placebo group, there was no increase in plasma sPLA$_2$-IIA levels indicating that the patient population was not adequate to test the initial hypothesis. In summary, thus far available trials with varespladib mainly established that the drug is reasonably safe and has the potential of lowering plasma levels of apoB-containing lipoproteins even when given on top of statins. More definitive data regarding the clinical potential of sPLA$_2$ inhibition to reduce CVD events were expected from the VISTA-16 trial (www.clinicaltrials.gov), an outcome trial initially designed to include a total of 6500 patients with an acute coronary syndrome. However, just recently, this trial has been stopped prematurely because of lack of efficacy (press release Anthera Pharmaceuticals, Inc, March 9, 2012). Interestingly and worth noting, even currently used pharmacological agents such as the angiotensin II type 1 receptor blocker irbesartan and the HMGCoA reductase inhibitor pravastatin are effective in lowering plasma sPLA$_2$-IIA levels and activity in CVD patients [37].

Conclusion on the Impact of sPLA$_2$-IIA on Plasma Lipids and Atherosclerosis

While available data from animal models and from human cohorts strongly suggest that sPLA$_2$-IIA is a causative CVD risk factor, confirmation from clinical trials assessing the impact of specifically inhibiting sPLA$_2$-IIA on cardiovascular outcomes is still to be awaited.

THE SN-2 PHOSPHOLIPASE GROUP V SECRETORY PHOSPHOLIPASE A$_2$

The group V secretory phopholipase A$_2$ (sPLA$_2$-V) is a small 14-kD enzyme expressed highest in heart, but also detected in liver, kidney, macrophages, intestinal tract, placenta, prostate, ovary, and testis [4, 5]. Interestingly, the gene for sPLA$_2$-V is located close to the sPLA$_2$-IIA gene on homologous regions of mouse chromosome 4 and human chromosome 1, but they are transcribed in reverse orientation [5]. However, sPLA$_2$-V is not responsive to LPS but more to chronic types of inflammation such as high-fat diet feeding [5]. sPLA$_2$-V has the ability to hydrolyze PC (i) in the outer leaflet of the plasma membrane of cells, a property distinct from sPLA$_2$-IIA, and (ii) in lipoproteins, where both LDL and HDL serve as good substrates *in vitro* [3–5].

In macrophages, sPLA$_2$-V has been localized to the Golgi apparatus and the recycling endosome and is subsequently recruited to the phagosome upon stimulation of the cell [4]. Interestingly, sPLA$_2$-V-deficient macrophages display a 50% decrease in the phagocytic activity toward zymosan particles, indicating that sPLA$_2$-V contributes significantly to macrophage function [4, 38].

Impact of sPLA$_2$-V on HDL Metabolism and RCT

Macrophage-derived sPLA$_2$-V did not impact plasma HDL cholesterol levels in LDLR-deficient mice [39]. On the contrary, hydrolysis of HDL phospholipids by sPLA$_2$-V has been shown to decrease the cholesterol efflux properties of the particle [40]. Thus far, no studies addressing the impact of sPLA$_2$-V on RCT are available.

Association of sPLA$_2$-V with Inflammation

Expression of sPLA$_2$-V does not increase in mice upon LPS administration and sPLA$_2$-V itself has limited bactericidal activity [4, 5]. Transgenic overexpression of sPLA$_2$-V has been shown to induce severe lung inflammation resulting in death soon after birth due to enhanced hydrolysis of PL within surfactant [41]. Conversely, sPLA$_2$-V knockout mice are protected from antigen-induced asthma and lung inflammation [42]. In addition, sPLA$_2$-V knockouts show a decreased inflammatory response in a zymosan-induced peritonitis model as well as in the air pouch model, and these mice had a substantial decrease in the extent of myocardial infarction in ischemia/reperfusion experiments [4]. In contrast to these results, sPLA$_2$-V knockout mice responded with increased inflammation in a murine arthritis model, while administration of a recombinant sPLA$_2$-V ameliorated disease [43]. Taken together, these findings suggest that depending on the respective model sPLA$_2$-V might have pro- and anti-inflammatory effects: (i) making it difficult to devise a rationale for the development of specific sPLA$_2$-V inhibitors and (ii) adding a cautionary note to the potential *in vivo* effects of pan-sPLA$_2$ inhibitors.

Impact of sPLA$_2$-V on Atherosclerosis

Bone marrow transplantation studies in mice on the LDLR knockout background revealed that sPLA$_2$-V expression by macrophages increases atherosclerotic lesion formation without affecting plasma total cholesterol levels and with only minor effects on cholesterol distribution over the different lipoprotein subclasses [39]. Mechanistically, LDL modification by sPLA$_2$-V was demonstrated to enhance LDL aggregation and to increase subsequent macrophage foam cell formation [44]. By contrast, no difference in atherosclerotic lesion formation was seen when $sPLA_2$-V/apoE double knockout mice were compared with apoE single knockouts [45].

$sPLA_2$-V has been detected in murine and human atherosclerotic lesions associated with vascular smooth muscle cells and around macrophage foam cells [6]. Circulating plasma levels of $sPLA_2$-V are apparently quite low; and currently, neither a mass assay nor a specific activity assay is available for $sPLA_2$-V. Therefore, clinical studies either cross-sectional or prospective on an association between $sPLA_2$-V and CVD and its complications are lacking. However, genetic studies demonstrated that certain tagging SNPs for $sPLA_2$-V were significantly associated with plasma total cholesterol, LDL cholesterol, and circulating oxLDL levels [46]. These data indicate that the established activities of $sPLA_2$-V derived from *in vitro* studies might also translate into a pathophysiological impact in the human *in vivo* situation.

Conclusion on the Impact of $sPLA_2$-V on Plasma Lipids and Atherosclerosis

Although $sPLA_2$-V is associated with proatherogenic changes with respect to lipoprotein modifications, evidence for an overall impact of $sPLA_2$-V on atherogenesis is currently too weak to consider $sPLA_2$-V, a promising candidate for specific inhibition to treat atherosclerotic CVD.

THE SN-2 PHOSPHOLIPASE GROUP X SECRETORY PHOSPHOLIPASE A_2

The group X secretory phospholipase A ($sPLA_2$-X) also is a small 14-kD enzyme that in contrast to $sPLA_2$-IIA and $sPLA_2$-V, however, does not bind to proteoglycans and is expressed in brain, skin, heart, lung, liver, pancreas, vascular smooth muscle cells, leukocytes, macrophages, and the gastrointestinal tract [4]. Apparently, $sPLA_2$-X protein is stored intracellularly as an inactive zymogen that is activated proteolytically among, for example, inflammatory stimuli. $sPLA_2$-X contains a tryptophan residue and consequently exhibits the highest catalytic activity toward phosphatidylcholine among the secreted phospholipases [4].

Impact of $sPLA_2$-X on HDL Metabolism and RCT

It has been shown that PC within HDL is a good substrate for $sPLA_2$-X resulting in a decreased capacity of these HDL to mediate cellular cholesterol efflux [3], but no changes in the plasma lipid profile have thus far been reported in mouse models with altered expression of $sPLA_2$-X. In addition, no studies addressing the question, if $sPLA_2$-X expression impacts *in vivo* RCT have been carried out thus far.

Association of $sPLA_2$-X with Inflammation

The murine $sPLA_2$-X has only very limited direct antibacterial activity, while the human enzyme appears to be more potent [4], which is in contrast to the $sPLA_2$-IIA enzyme that displays the highest potency.

Overall, available data indicate that depending on the model $sPLA_2$-X might possess pro- and anti-inflammatory activities. It has recently been shown that $sPLA_2$-X is able to hydrolyze and thereby inactivate platelet-activating factor (PAF), a potential anti-inflammatory activity with relevance for inflammation and particularly atherosclerosis [3]. $sPLA_2$-X knockout mice display an attenuated response to myocardial ischemia-reperfusion injury and also allergic airway responses were dampened in $sPLA_2$-X knockout mice [4]. Consistently, macrophage-specific overexpression of $sPLA_2$-X resulted in a severe form of early lung inflammation with infiltrates consisting of granulocytes and surfactant-laden macrophages, while in contrast in the same study macrophages overexpressing $sPLA_2$-X also showed reduced activation and an anti-inflammatory phenotype [47]. The generation of PGE2 from AA liberated by the action of $sPLA_2$-X occurs also in concert with COX2, similar to findings regarding $sPLA_2$-IIA [3].

Impact of $sPLA_2$-X on Atherosclerosis

To study a potential causative involvement of $sPLA_2$-X in the development of vascular disease, mostly $sPLA_2$-X-deficient mice have been employed. Mice lacking $sPLA_2$-X on the apoE knockout background have been shown to develop less atherosclerosis in response to angiotensin II infusion in the aortic arch, while lesion formation was not different in the aortic root compared with apoE knockout mice expressing $sPLA_2$-X [48]. $sPLA_2$-X knockouts are less susceptible toward the development of aortic aneurysms [48]. These limited studies indicate that deficiency in $sPLA_2$-X might confer protection against vascular disease. Mechanistically, PC within LDL particles represents a good substrate for $sPLA_2$-X, whereby this enzyme is thought to contribute to the generation of smaller denser LDL particles that are more atherogenic [3, 4]. Consistently, LDL modification by $sPLA_2$-X promotes cellular uptake of these LDL and subsequent foam cell formation as compared with control-treated LDL [3, 4]. However, in contrast to $sPLA_2$-IIA and $sPLA_2$-V, the activity of $sPLA_2$-X does not lead to aggregation of LDL [3, 4]. $sPLA_2$-X can also not only modify the efflux properties of HDL [4] but also decrease the expression of the two major cellular cholesterol efflux transporters in macrophages, ABCA1 and ABCG1 [49]. In addition, it has been demonstrated that $sPLA_2$-X-modified LDL significantly stimulates the expression of adhesion molecules in endothelial cells in vitro suggesting that $sPLA_2$-X also contributes to the generation of at least certain features of endothelial dysfunction [3].

Similar to the other $sPLA_2$s also $sPLA_2$-X has been shown to be present in human atherosclerotic lesions, where protein expression is associated with macrophage foam cells and vascular smooth muscle cells, especially those resembling myofibroblasts [6]. Circulating plasma levels of $sPLA_2$-X are apparently low, however, also for $sPLA_2$-X neither a specific mass nor an activity assay is available, which complicates cross-sectional and prospective clinical studies. On a genetic level, several polymorphisms within the PLA2G10 gene were identified; among these is an interesting Arg38Cys substitution in the coding region of the gene that was shown to result in a misfolded catalytically inactive protein, which is rapidly degraded [50]. However, no association could be established between the presence of this mutation and either the presence of CVD or the prospective risk of CVD events [50].

Conclusion on the Impact of $sPLA_2$-X on Plasma Lipids and Atherosclerosis

Available data indicate that the activity of $sPLA_2$-X is proatherogenic, although hard experimental evidence that would identify this specific phospholipase as a promising therapeutic target for the treatment of atherosclerotic CVD is still lacking.

OVERALL CONCLUSION

Available data provide good evidence that the extracellular phospholipases EL and $sPLA_2$-IIA represent valid targets for inhibition with the intention to raise plasma HDL cholesterol levels and potentially also to improve HDL function. With respect to atherosclerosis, only $sPLA_2$-IIA appears to be unequivocally causally linked to the process of atherogenesis, while mixed results have been obtained thus far for EL, $sPLA_2$-V, and $sPLA_2$-X.

REFERENCES

1. Steinberg D. Atherogenesis in perspective: Hypercholesterolemia and inflammation as partners in crime. *Nat Med*. 2002;8: 1211–1217.
2. Rader DJ, Daugherty A. Translating molecular discoveries into new therapies for atherosclerosis. *Nature*. 2008;451:904–913.
3. Karabina SA, Gora S, Atout R, Ninio E. Extracellular phospholipases in atherosclerosis. *Biochimie*. 2010;92:594–600.
4. Murakami M, Taketomi Y, Sato H, Yamamoto K. Secreted phospholipase A2 revisited. *J Biochem*. 2011;150:233–255.

5. Rosengren B, Jonsson-Rylander AC, Peilot H, Camejo G, Hurt-Camejo E. Distinctiveness of secretory phospholipase A2 group IIA and V suggesting unique roles in atherosclerosis. *Biochim Biophys Acta*. 2006;1761:1301–1308.
6. Rosenson RS, Gelb MH. Secretory phospholipase A2: A multifaceted family of proatherogenic enzymes. *Curr Cardiol Rep*. 2009;11: 445–451.
7. Yasuda T, Ishida T, Rader DJ. Update on the role of endothelial lipase in high-density lipoprotein metabolism, reverse cholesterol transport, and atherosclerosis. *Circ J*. 2010;74:2263–2270.
8. Annema W, Tietge UJ. Role of hepatic lipase and endothelial lipase in high-density lipoprotein-mediated reverse cholesterol transport. *Curr Atheroscler Rep*. 2011;13:257–265.
9. Maugeais C, Tietge UJ, Broedl UC, et al. Dose-dependent acceleration of high-density lipoprotein catabolism by endothelial lipase. *Circulation*. 2003;108:2121–2126.
10. Nijstad N, Wiersma H, Gautier T, van der Giet M, Maugeais C, Tietge UJ. Scavenger receptor bi-mediated selective uptake is required for the remodeling of high density lipoprotein by endothelial lipase. *J Biol Chem*. 2009;284: 6093–6100.
11. Wiersma H, Gatti A, Nijstad N, Kuipers F, Tietge UJ. Hepatic SR-BI, not endothelial lipase, expression determines biliary cholesterol secretion in mice. *J Lipid Res*. 2009;50: 1571–1580.
12. Khetarpal SA, Edmondson AC, Raghavan A, et al. Mining the LIPG allelic spectrum reveals the contribution of rare and common regulatory variants to HDL cholesterol. *PLoS Genet*. 2011;7:e1002393.
13. Nijstad N, Gautier T, Briand F, Rader DJ, Tietge UJ. Biliary sterol secretion is required for functional in vivo reverse cholesterol transport in mice. *Gastroenterology*. 2011;140:1043–1051.
14. Badellino KO, Wolfe ML, Reilly MP, Rader DJ. Endothelial lipase is increased in vivo by inflammation in humans. *Circulation*. 2008; 117:678–685.
15. Ishida T, Choi SY, Kundu RK, et al. Endothelial lipase modulates susceptibility to atherosclerosis in apolipoprotein-E-deficient mice. *J Biol Chem*. 2004;279:45085–45092.
16. Kojma Y, Hirata K, Ishida T, et al. Endothelial lipase modulates monocyte adhesion to the vessel wall. A potential role in inflammation. *J Biol Chem*. 2004;279:54032–54038.
17. Ahmed W, Orasanu G, Nehra V, et al. High-density lipoprotein hydrolysis by endothelial lipase activates pparalpha: A candidate mechanism for high-density lipoprotein-mediated repression of leukocyte adhesion. *Circ Res*. 2006;98:490–498.
18. Ko KW, Paul A, Ma K, Li L, Chan L. Endothelial lipase modulates HDL but has no effect on atherosclerosis development in apoE-/- and LDLR-/- mice. *J Lipid Res*. 2005;46:2586–2594.
19. Fujii H, Fukuda A, Tanaka M, et al. Putative role of endothelial lipase in dialysis patients with hypoalbuminemia and inflammation. *Am J Nephrol*. 2008;28:974–981.
20. Tietge UJ, Maugeais C, Lund-Katz S, Grass D, deBeer FC, Rader DJ. Human secretory phospholipase A2 mediates decreased plasma levels of HDL cholesterol and apoA-I in response to inflammation in human apoA-I transgenic mice. *Arterioscler Thromb Vasc Biol*. 2002;22: 1213–1218.
21. Tietge UJ, Maugeais C, Cain W, et al. Overexpression of secretory phospholipase A(2) causes rapid catabolism and altered tissue uptake of high density lipoprotein cholesteryl ester and apolipoprotein A-I. *J Biol Chem*. 2000;275:10077–10084.
22. Tietge UJ, Maugeais C, Cain W, Rader DJ. Acute inflammation increases selective uptake of HDL cholesteryl esters into adrenals of mice overexpressing human sPLA2. *Am J Physiol Endocrinol Metab*. 2003;285:E403–E411.
23. de Beer FC, Connell PM, Yu J, de Beer MC, Webb NR, van der Westhuyzen DR. HDL modification by secretory phospholipase A(2) promotes scavenger receptor class B type I interaction and accelerates HDL catabolism. *J Lipid Res*. 2000;41:1849–1857.
24. Tietge UJ, Nijstad N, Havinga R, et al. Secretory phospholipase A2 increases SR-BI-mediated selective uptake from HDL but not biliary cholesterol secretion. *J Lipid Res*. 2008;49: 563–571.
25. Kennedy BP, Payette P, Mudgett J, et al. A natural disruption of the secretory group II phospholipase A2 gene in inbred mouse strains. *J Biol Chem*. 1995;270:22378–22385.
26. Annema W, Nijstad N, Tolle M, et al. Myeloperoxidase and serum amyloid a contribute to impaired in vivo reverse cholesterol transport during the acute phase response but not group IIA secretory phospholipase A(2). *J Lipid Res*. 2010;51:743–754.

27. Blache D, Gautier T, Tietge UJ, Lagrost L. Activated platelets contribute to oxidized low-density lipoproteins and dysfunctional high-density lipoproteins through a phospholipase A2-dependent mechanism. *FASEB J*. 2012;26:927–937.
28. Ivandic B, Castellani LW, Wang XP, et al. Role of group II secretory phospholipase A2 in atherosclerosis: 1. Increased atherogenesis and altered lipoproteins in transgenic mice expressing group IIA phospholipase A2. *Arterioscler Thromb Vasc Biol*. 1999;19:1284–1290.
29. Tietge UJ, Pratico D, Ding T, et al. Macrophage-specific expression of group IIA sPLA2 results in accelerated atherogenesis by increasing oxidative stress. *J Lipid Res*. 2005;46:1604–1614.
30. van der Giet M, Tolle M, Pratico D, et al. Increased type IIA secretory phospholipase A(2) expression contributes to oxidative stress in end-stage renal disease. *J Mol Med (Berl)*. 2010;88:75–83.
31. Zeiher BG, Steingrub J, Laterre PF, Dmitrienko A, Fukiishi Y, Abraham E. LY315920NA/S-5920, a selective inhibitor of group IIA secretory phospholipase A2, fails to improve clinical outcome for patients with severe sepsis. *Crit Care Med*. 2005;33:1741–1748.
32. Shaposhnik Z, Wang X, Trias J, Fraser H, Lusis AJ. The synergistic inhibition of atherogenesis in apoE-/- mice between pravastatin and the sPLA2 inhibitor varespladib (A-002). *J Lipid Res*. 2009;50:623–629.
33. Rosenson RS, Hislop C, McConnell D, et al. Effects of 1-H-indole-3-glyoxamide (A-002) on concentration of secretory phospholipase A2 (plasma study): A phase II double-blind, randomised, placebo-controlled trial. *Lancet*. 2009;373:649–658.
34. Rosenson RS, Elliott M, Stasiv Y, Hislop C. Randomized trial of an inhibitor of secretory phospholipase A2 on atherogenic lipoprotein subclasses in statin-treated patients with coronary heart disease. *Eur Heart J*. 2011;32: 999–1005.
35. Rosenson RS, Hislop C, Elliott M, Stasiv Y, Goulder M, Waters D. Effects of varespladib methyl on biomarkers and major cardiovascular events in acute coronary syndrome patients. *J Am Coll Cardiol*. 2010;56:1079–1088.
36. Dzavik V, Lavi S, Thorpe K, et al. The sPLA2 inhibition to decrease enzyme release after percutaneous coronary intervention (spider-pci) trial. *Circulation*. 2010;122:2411–2418.
37. Divchev D, Grothusen C, Luchtefeld M, et al. Impact of a combined treatment of angiotensin II type 1 receptor blockade and 3-hydroxy-3-methyl-glutaryl-CoA-reductase inhibition on secretory phospholipase A2-type IIA and low density lipoprotein oxidation in patients with coronary artery disease. *Eur Heart J*. 2008; 29:1956–1965.
38. Balestrieri B, Hsu VW, Gilbert H, et al. Group V secretory phospholipase A2 translocates to the phagosome after zymosan stimulation of mouse peritoneal macrophages and regulates phagocytosis. *J Biol Chem*. 2006;281:6691–6698.
39. Bostrom MA, Boyanovsky BB, Jordan CT, et al. Group V secretory phospholipase A2 promotes atherosclerosis: Evidence from genetically altered mice. *Arterioscler Thromb Vasc Biol*. 2007;27:600–606.
40. Ishimoto Y, Yamada K, Yamamoto S, Ono T, Notoya M, Hanasaki K. Group V and X secretory phospholipase A(2)s-induced modification of high-density lipoprotein linked to the reduction of its antiatherogenic functions. *Biochim Biophys Acta*. 2003;1642:129–138.
41. Ohtsuki M, Taketomi Y, Arata S, et al. Transgenic expression of group V, but not group X, secreted phospholipase A2 in mice leads to neonatal lethality because of lung dysfunction. *J Biol Chem*. 2006;281:36420–36433.
42. Munoz NM, Meliton AY, Arm JP, Bonventre JV, Cho W, Leff AR. Deletion of secretory group V phospholipase A2 attenuates cell migration and airway hyperresponsiveness in immunosensitized mice. *J Immunol*. 2007;179:4800–4807.
43. Boilard E, Lai Y, Larabee K, et al. A novel anti-inflammatory role for secretory phospholipase A2 in immune complex-mediated arthritis. *EMBO Mol Med*. 2010;2:172–187.
44. Boyanovsky BB, van der Westhuyzen DR, Webb NR. Group V secretory phospholipase A2-modified low density lipoprotein promotes foam cell formation by a SR-A- and CD36-independent process that involves cellular proteoglycans. *J Biol Chem*. 2005;280: 32746–32752.
45. Boyanovsky B, Zack M, Forrest K, Webb NR. The capacity of group V sPLA2 to increase atherogenicity of ApoE-/- and LDLR-/- mouse LDL in vitro predicts its atherogenic role in vivo. *Arterioscler Thromb Vasc Biol*. 2009;29: 532–538.
46. Wootton PT, Arora NL, Drenos F, et al. Tagging SNP haplotype analysis of the secretory

PLA2-V gene, PLA2G5, shows strong association with LDL and oxLDL levels, suggesting functional distinction from sPLA2-IIA: Results from the UDACS study. *Hum Mol Genet*. 2007;16:1437–1444.

47. Curfs DM, Ghesquiere SA, Vergouwe MN, et al. Macrophage secretory phospholipase A2 group X enhances anti-inflammatory responses, promotes lipid accumulation, and contributes to aberrant lung pathology. *J Biol Chem*. 2008; 283:21640–21648.
48. Zack M, Boyanovsky BB, Shridas P, et al. Group X secretory phospholipase A(2) augments angiotensin II-induced inflammatory responses and abdominal aortic aneurysm formation in apoE-deficient mice. *Atherosclerosis*. 2011;214:58–64.
49. Shridas P, Bailey WM, Gizard F, et al. Group X secretory phospholipase A2 negatively regulates ABCA1 and ABCG1 expression and cholesterol efflux in macrophages. *Arterioscler Thromb Vasc Biol*. 2010;30: 2014–2021.
50. Gora S, Perret C, Jemel I, et al. Molecular and functional characterization of polymorphisms in the secreted phospholipase A2 group X gene: Relevance to coronary artery disease. *J Mol Med (Berl)*. 2009;87:723–733.

24

APOLIPOPROTEIN-B: THE CRUCIAL PROTEIN OF ATHEROGENIC LIPOPROTEINS

KEVIN JON WILLIAMS[1,2] AND EDWARD A. FISHER[3,4]

[1] *Section of Endocrinology, Diabetes, & Metabolism, Departments of Medicine and Physiology, Cardiovascular Research Center, Temple University School of Medicine, Philadelphia, PA, USA*

[2] *Department of Molecular and Clinical Medicine, Sahlgrenska Center for Cardiovascular and Metabolic Research, University of Gothenburg, Göteborg, Sweden*

[3] *Department of Medicine, Division of Cardiology, and the Marc and Ruti Bell Program in Vascular Biology, New York University School of Medicine, New York, NY, USA*

[4] *The Center for the Prevention of Cardiovascular Disease, NYU School of Medicine, New York, NY, USA*

INTRODUCTION

Atherosclerosis kills roughly half of all Westerners and almost a third of people across the world, with this latter figure likely to rise considerably in the coming years [1, 2]. For a single disease to become such a widespread killer, with the same basic risk factors everywhere it has been studied [3], the underlying pathophysiology must be robust and straightforward. Basic and clinical findings now support this conclusion, while providing clear therapeutic targets where we can interrupt and perhaps reverse the disease.

A single key process initiates atherogenesis—namely, the subendothelial retention of plasma-derived apolipoprotein-B (apoB) lipoproteins, particularly low-density lipoprotein (LDL) [4, 5]. These retained lipoproteins become modified by arterial wall enzymes and other processes to form a uniquely dangerous accumulation (Figure 24.1). The resulting material provokes a series of local biological responses that eventually include a strikingly maladaptive reaction dominated by persistent macrophages and T-cells that accelerate lipoprotein retention and other features of plaque development. The most effective therapy to date against atherothrombotic cardiovascular disease—plasma LDL lowering—works by decreasing circulating apoB-lipoprotein levels, thereby reducing the likelihood that these particles will enter and become retained within the arterial wall.

As a point of clarification, cholesterol *per se* is not the cause of this disease. Owing to its largely hydrophobic nature, unesterified cholesterol can exist at only miniscule concentrations as a free molecule in any body fluid, and esterified cholesterol, which is completely hydrophobic, never does. Both unesterified and esterified cholesterol are transported by carriers.

Atherosclerosis: Risks, Mechanisms, and Therapies, First Edition. Edited by Hong Wang and Cam Patterson.

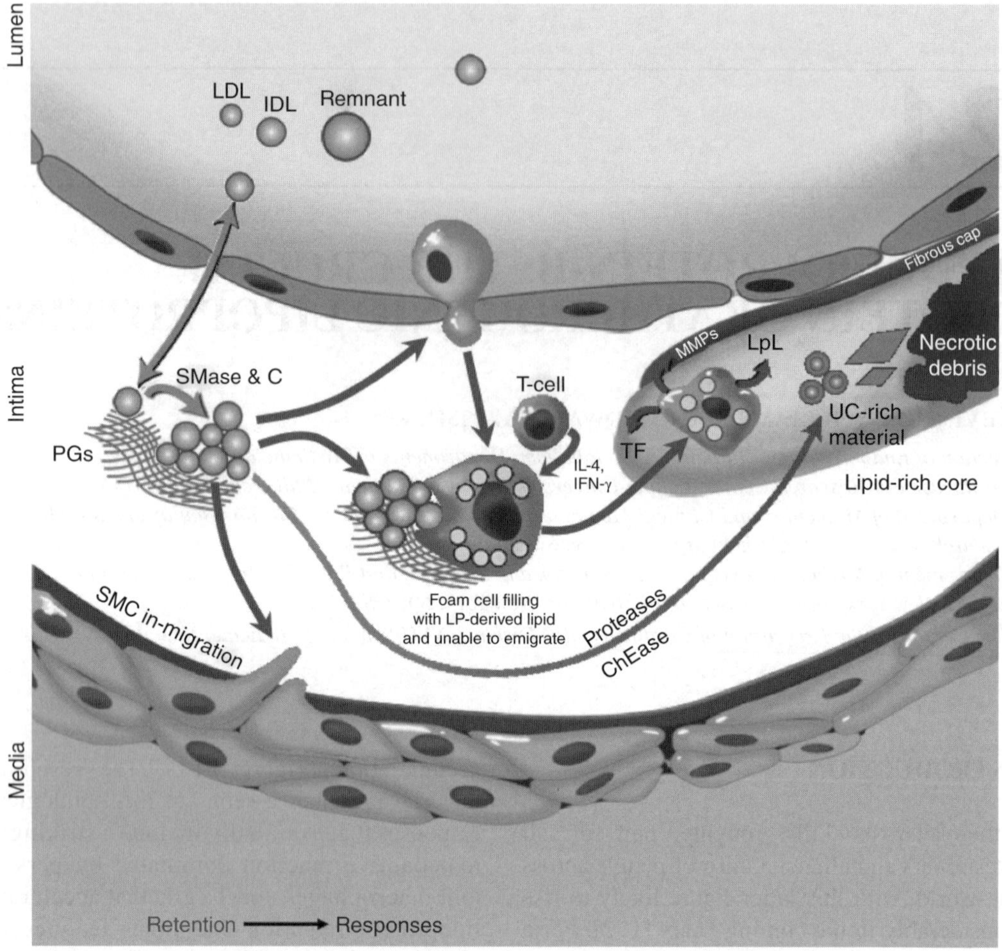

Figure 24.1 The response-to-retention model of early atherogenesis and then later progression. Arrows indicate crucial mechanisms in the retention of atherogenic apolipoprotein-B-lipoproteins within the arterial wall, which is the key initiating step in atherogenesis; and then local responses to the retained and modified lipoproteins that lead to plaque growth and evolution. The main text describes mechanisms for early retention of apoB-lipoproteins within the arterial wall and then the acceleration of lipoprotein retention after plaque initiation and other subsequent maladaptive responses. Abbreviations: ChEase, cholesteryl esterase; IDL, intermediate-density lipoprotein; IFN-γ, interferon γ; IL-4, interleukin 4; LP, lipoprotein; LpL, lipoprotein lipase; MMPs, matrix metalloproteinases; PGs, proteoglylcans; Remnant, remnant lipoprotein; SMase, sphingomyelinase; SMC, smooth muscle cell; TF, tissue factor; UC, unesterified cholesterol. Adapted from Ref. [5].

The nature of these carriers drives physiology and pathophysiology. Red blood cells are major carriers of cholesterol in blood, but they have no known epidemiologic or causal link to atherosclerosis. Moreover, cholesterol carried by normal high-density lipoprotein (HDL) particles is not pathogenic and has been proposed to provide cardiovascular protection (see section "Conclusion: the evolution of epidemiologic risk factors into causative factors, exacerbating factors, and bystander phenomenon in atherosclerotic cardiovascular disease").

Hence, we focus on cholesterol that is transported by apoB-containing lipoproteins, which become pathogenic only after their retention within the arterial wall. The four most commonly used methods to assess the concentrations of these particles in human plasma are total plasma cholesterol (which in most humans is predominantly LDL), assays of plasma LDL itself, the calculation of non-HDL cholesterol (which equals total cholesterol minus HDL cholesterol, thereby including LDL and other apoB-lipoproteins), and specific immunoassays for the apoB protein.

This chapter covers several aspects of the key protein component, apoB, of atherogenic lipoproteins—namely, the regulation of its secretion from liver and intestine (Section "Regulation of the assembly and secretion of apoB-lipoproteins by the liver and intestine"), pathways for its healthy removal from the circulation by the liver (Section "Pathways for the healthy removal of apoB-lipoproteins from the circulation by the liver"), and mechanisms for the retention of apoB-lipoproteins within the vessel wall (Section "Mechanisms for the retention of apoB-lipoproteins within the vessel wall"). In the final section, we use this extensive knowledge to reclassify epidemiologic risk factors for atherosclerotic cardiovascular events into causative factors, exacerbating factors, and mere bystander phenomena. Causative and exacerbating factors are targets for therapy; bystander phenomena are not.

REGULATION OF THE ASSEMBLY AND SECRETION OF apoB-LIPOPROTEINS BY THE LIVER AND INTESTINE

The primary apoB-containing lipoproteins secreted by the liver and intestine are VLDL and chylomicrons, respectively. There is only one apoB molecule on either a VLDL or a chylomicron. In both organs, apoB becomes associated with lipids in the ER through a process dependent on microsomal triglyceride (TG) transfer protein (MTP), which, despite its name, mediates the transfer of all of the major lipid classes found on lipoproteins (besides triglycerides, also cholesterol, cholesteryl esters, and phospholipids). The major lipid class carried on VLDL and chylomicrons is TG, whose fatty acid content is derived from endogenously synthesized or dietary sources, respectively [6]. In both organs, the nascent lipoprotein that exits the ER does so in specialized vesicles in a process dependent on the COPII protein SAR1A [7–9]. In the case of chylomicrons, most of the lipidation appears to be completed prior to their exit from the ER (reviewed in Refs. [10, 11]), whereas VLDL maturation and lipidation continues in a post-ER compartment, most likely the Golgi [12–14].

There are a number of differences between liver and intestinal apoB-lipoprotein biology. For example, in humans, the major form of apoB in liver and on VLDL is full-length $apoB_{100}$, whereas chylomicrons contain a trucated protein, $apoB_{48}$. The relationship between these two forms was revealed by the discovery of an RNA editing activity that converts the C in a glutamine codon into a U by deamination, thereby forming a nongenomic UAA translational stop codon approximately 48% from the amino terminal end of the mature protein (reviewed in Ref. [6]). Thus, in humans, an $apoB_{100}$-particle can be assumed to have a hepatic origin, while $apoB_{48}$-particles are intestinal. In rodents, however, both the liver and the intestine contain RNA editing activity, and therefore $apoB_{100}$ or $apoB_{48}$ can be found on their VLDL particles. Another key difference between the intestine and liver is that the assembly and secretion of chylomicrons increases from the augmented synthesis of apoB and lipids in the prandial and postprandial state [10]. By contrast, the liver primarily regulates apoB secretion not at the level of synthesis, but through post-translational degradation of the protein (reviewed in Refs. [15, 16]).

The recognition that degradation was a key determinant of hepatic secretion of $apoB_{100}$ came from the pioneering studies of Sven Olof-Olofsson [17] and Roger Davis [18] in human hepatocarcinoma HepG2 cells and primary rat hepatocytes, respectively. In both hepatic cell systems, these investigators performed pulse-chase analyses and found that under basal conditions, a substantial amount of newly synthesized $apoB_{100}$ was degraded. Subsequent studies by many laboratories have parsed the degradation of newly synthesized hepatic $apoB_{100}$

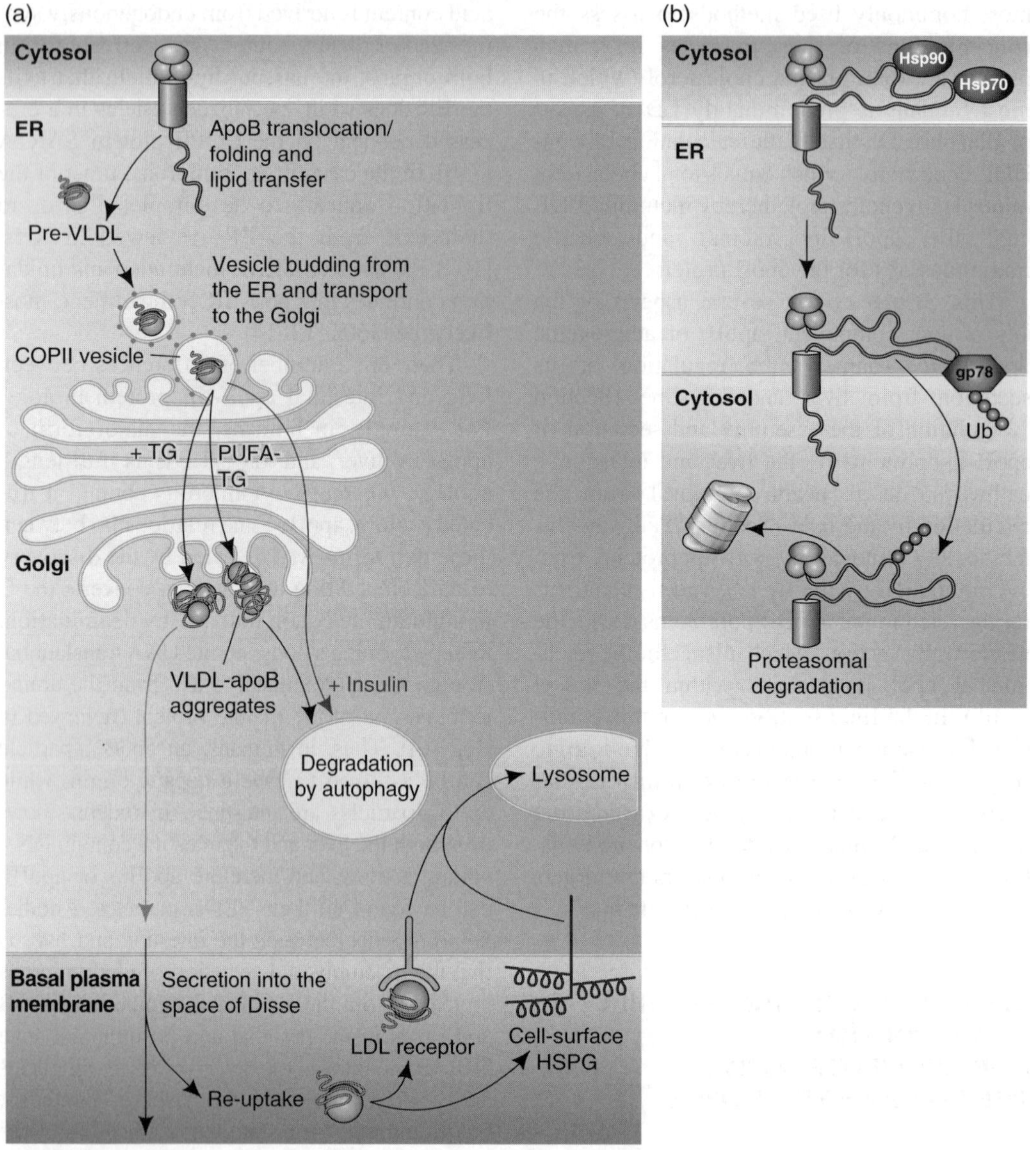

Figure 24.2 (a) Adequate triglyceride for particle assembly and (b) inadequate triglyceride availability or transfer.

into three general pathways (the "triple threat"): endoplasmic reticulum-associated degradation via the proteasome (ERAD/proteasome pathway), post-ER presecretory proteolysis via autophagy (PERPP/autophagy), and reuptake (Figure 24.2; also reviewed in Refs. [15, 16, 19, 20]).

Reuptake was the first pathway to be explicitly identified [21]. Essentially, it represents the opportunity for $apoB_{100}$-lipoproteins reaching the cell surface to encounter a binding partner, such as the LDL-receptor or heparan sulfate proteoglycans (HSPGs), before diffusing away from the unstirred water layer or out of the space of Disse. Both the LDL receptor and certain HSPGs mediate endocytosis and lysosomal delivery, discussed in section "Pathways for the healthy removal of apoB-lipoproteins from the circulation by the liver," and therefore these newly assembled lipoproteins become reinternalized and degraded. This process has been demonstrated in both transformed [21] and primary hepatocyes [19] and was shown to be regulated by statins through their upregulation of the LDL receptor. The role of reuptake in the regulation of hepatic $apoB_{100}$ secretion has also been confirmed *in vivo* in humans [22].

The next degradation pathway to be identified was ERAD. ERAD was first defined as a process in yeast [23] in which a secretory protein that did not fold correctly in the ER was dislocated to the cytosol, where it was tagged with ubiquitin and degraded by the proteasome as part of a quality control system. It was known that degradation of newly synthesized $apoB_{100}$ in liver cells could be induced by either inhibiting MTP [24] or by severely limiting lipid synthesis [25, 26]. It came to be appreciated that these conditions block the conformational stabilization of nascent $apoB_{100}$ polypeptides that normally occurs from co- and post-translational association with small amounts of lipids. Thus, these $apoB_{100}$ polypeptides, which are extremely hydrophobic, would be prone to intra- and intermolecular aggregation unless promptly assembled with lipids or promptly degraded. Indeed, it was shown that apoB in liver cells under lipid "deprivation" conditions was ubiqutinylated and degraded by the proteasome in a process involving a number of cytosolic and ER chaperones [26–29]. Patients with loss-of-function mutations in the *MTP* gene are unable to perform these early steps in the assembly of either $apoB_{48}$ or $apoB_{100}$-lipoproteins and exhibit the disease abetalipoproteinemia, that is, a complete absence of intestinal or hepatic apoB-lipoproteins in their plasma [6]. A small-molecule inhibitor of MTP is under evaluation by the Food and Drug Administration (USA) for lowering plasma LDL concentrations in patients with homozygous familial hypercholesterolemia, in whom other medications work poorly. The MTP inhibitor lowers plasma apoB and LDL concentrations by about 50% in these patients, although fatty liver and intestine occur as side effects that may limit widespread clinical utility [30].

ERAD was initially thought to be the major degradation pathway for $apoB_{100}$, and, therefore, the key regulator of its secretion. We and others subsequently showed, however, that under conditions in which lipid synthesis or transfer were not limiting (which is the usual physiological state in liver), there was still an opportunity for the hepatocyte to regulate apoB secretion by degradation. Our characterization of this third pathway for destruction newly synthesized apoB identified it as post-ER, pre-secretory proteolysis (PERPP) and nonproteasomal [19, 31]. The most studied example has been the cellular effects of highly polyunsaturated Ω-3 fish oil fatty acids (EPA, DHA), which are used clinically as VLDL lowering agents. In a series of studies (e.g., [19, 31, 32]), we have shown that these fatty acids support the initial MTP-dependent lipidation/assembly steps in the ER, but increase the formation of intracellular lipid peroxides that promote oxidative modifications and aggregation of apoB after it leaves the ER. The altered $apoB_{100}$ molecules become substrates for autophagy, known to target damaged proteins and organelles, causing their ultimate degradation in the lysosome.

Because polyunsaturated fats are an essential part of the diets of herbivores and omnivores, yet readily become converted into lipid peroxides inside cells, the PERPP/autophagy pathway appears to be a normal adaptation. Besides protecting the liver from potentially toxic aggregates, autophagic destruction of these particles may also prevent the export of lipid peroxides on VLDL to triglyceride-consuming, but oxidant-sensitive, tissues in the periphery, such as cardiac muscle. Thus, lipid peroxidation is kept within the liver, where it is safe, even beneficial, and away from potentially vulnerable peripheral organs [31]. On the contrary, the intrinsic tendency of damaged or modified $apoB_{100}$ (or $apoB_{48}$) to aggregate may also contribute to the retention and aggregation of apoB-lipoproteins within the arterial wall, discussed in Refs. [31, 33] and in section "Acceleration of lipoprotein retention after plaque initiation."

In the course of carrying out these experiments, we noted that a fraction of $apoB_{100}$

under *basal* conditions (i.e., in cells not incubated with DHA) also appeared to be targeted for autophagy [31], a phenomenon noted by others as well (e.g., [34]), further supporting the concept that PERPP/autophagy is an essential component of the normal, physiologic regulation of hepatic apoB secretion. Under basal conditions, aggregated forms of $apoB_{100}$ can still be detected (though they are not as large or as abundant as in DHA-treated cells [31]), suggesting that PERPP/autophagy, like ERAD, serves as a quality control pathway. Autophagy mediates quality control for a special class of proteins, such as $apoB_{100}$, in which the proper final folding does not occur until after exit from the ER. This model would be consistent with studies showing that a naturally occurring mutant of human $apoB_{100}$, which is predicted to misfold and is associated with hypolipidemia, exits the ER, but is degraded by autophagy in a model system [35]. Moreover, insulin-stimulated degradation of newly synthesized hepatic apoB is also a post-ER [36] autophagic process [37] that preferentially targets apoB-lipoprotein particles before they fully mature [38]. A quantitatively important role for insulin-stimulated PERPP/autophagy of nascent intracellular apoB-lipoproteins may contribute to the overproduction of these particles that occurs in states of defective insulin signaling, such as type 2 diabetes and the metabolic syndrome (recently reviewed in Refs. [39, 40]).

PATHWAYS FOR THE HEALTHY REMOVAL OF apoB-LIPOPROTEINS FROM THE CIRCULATION BY THE LIVER

In humans, the majority of circulating apoB-containing lipoproteins are taken up and disposed of by the liver [41, 42]. Three major receptors mediate this uptake: the LDL receptor, which is high affinity but low capacity; the syndecan-1 heparan sulfate proteoglycan HSPG, which is medium affinity but high

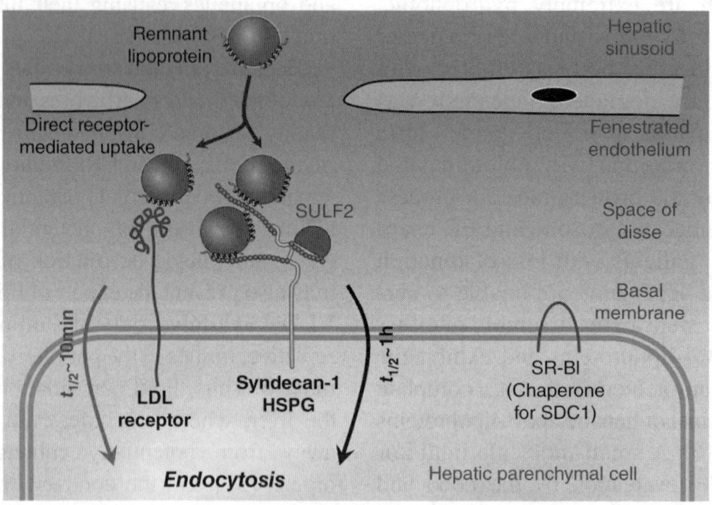

Figure 24.3 Efficient hepatic uptake of atherogenic, apoB-containing remnant lipoproteins. The helices on the surface of the remnant lipoproteins represent amphipathic apolipoproteins, such as apoE and lipoprotein lipase, that serve as ligands for the two major endocytic receptors on the basal membrane of hepatic parenchymal cells for these particles – namely, the LDL receptor and the syndecan-1 HSPG. Syndecan-1 HSPGs have far greater capacity for remnant lipoproteins than do LDL receptors, as indicated in the text and on this schematic. Sulfatase-2 (SULF2) is the endogenous inhibitor of syndecan-1 that is overexpressed and oversecreted from T2DM liver. HSPG, heparan sulfate proteoglycan; SDC1, syndecan-1; SR-BI, scavenger receptor class B type I. Adapted from Refs. [42, 163].

capacity; and the scavenger receptor class B type I (SR-BI), which apparently serves only as a docking receptor for remnant lipoproteins or as a syndecan chaperone, without mediating endocytosis [42] (Figure 24.3). Other molecules have been proposed as hepatic receptors for apoB-lipoproteins, but appear to have detectable effects mainly in extreme circumstances (reviewed in Ref. [42]). Here, we will focus on the LDL receptor and the syndecan-1 HSPG, because they exhibit abnormal function in disease.

Hepatic Receptors for apoB-Lipoproteins: The LDL Receptor

The LDL receptor mediates the clearance of LDL by binding a specific domain in the C-terminal region apoB_{100} that is absent from the truncated intestinal form of apoB, apoB_{48}. To allow clearance of apoB_{48}-lipoproteins, these particles can carry lipoprotein lipase [43], which can act as a ligand for the syndecan-1 HSPG [42, 44], and the particles often have apoE, which is a second ligand for the LDL receptor [45] and also binds HSPGs. In humans, loss-of-function mutations in the LDL receptor cause familial hypercholesterolemia, which is characterized by severe elevations of plasma LDL concentrations and hence accelerated atherosclerosis [46]. Similar monogenic syndromes are produced by loss-of-function mutations affecting the receptor-binding domain of apoB_{100} [47] or in the *LDLRAP* gene, which encodes an adaptor protein required for the internalization of LDL receptors via coated pits. High LDL levels also result from gain-of-function mutations in PCSK9, a proprotein convertase that accelerates degradation of the LDL receptor protein (reviewed in Ref. [45]). An additional protein, the inducible degrader of the LDL receptor (IDOL), has been identified as mediating ubiquitination of the LDLR on its cytoplasmic domain, targeting it for degradation [48, 49]. Human variants in and near the *IDOL* gene were recently associated with plasma LDL levels and responsiveness to statin therapy [49]. Owing to the dual function of apoE as a ligand for the LDL receptor and for HSPGs, loss-of-function mutations in this protein do not produce pure elevations in plasma LDL concentrations, but instead a mixed hypercholesterolemia with hypertriglyceridemia [50, 51].

In humans, over two-thirds of normal LDL clearance occurs through the LDL receptor [52, 53]. Pathways for clearance of LDL that are independent of LDL receptors have not been molecularly characterized, except to note that they are inefficient and of low affinity. Consistent with these findings, heterozygous familial hypercholesterolemia (one defective LDL receptor gene) elevates untreated plasma LDL concentrations to about 250–400 mg/dl, and homozygous familial hypercholesterolemia can elevate LDL levels to over 1000 mg/dl.

Because the LDL receptor binds apoB_{100} with relatively high affinity, it acts at or near capacity. Thus, interventions to increase hepatic expression of LDL receptors will lower plasma levels of LDL. Although the principal medicines to lower plasma cholesterol concentrations were not designed with LDL receptors in mind, we now know that HMG-CoA reductase inhibitors (statins), cholesterol absorption inhibitors, and bile acid sequestrants indirectly induce the LDL receptor gene, via sterol regulatory element-binding proteins (SREBPs), thereby lowering plasma LDL concentrations. After endocytosis, LDL and the LDL receptor separate: LDL is delivered to lysosomes for degradation, and the majority of LDL receptors recycle back to the cell surface for another round of particle binding and internalization.

Because PCSK9 expression is also stimulated by SREBPs, induction of this protein blunts the LDL-lowering effects of statins, cholesterol absorption inhibitors, and bile acid sequestrants. Thus, as described in Section "Conclusion: the evolution of epidemiologic risk factors into causative factors, exacerbating factors, and bystander phenomenon in atherosclerotic cardiovascular disease," PCSK9 has become an attractive pharmacologic target for inhibition. Regarding IDOL, no complete loss-of-function or dominant-negative alleles have been reported to date [49], perhaps because the protein also plays a role in neural outgrowth. Its suitability as a therapeutic target for inhibition remains unknown.

Hepatic Receptors for apoB-Lipoproteins: The Syndecan-1 HSPG

The importance of additional hepatic receptors for apoB-lipoproteins is illustrated by the differences between familial hypercholesterolemia and the postprandial dyslipoproteinemia seen in states of overnutrituion, underexertion, obesity, impaired insulin signaling, and type 2 diabetes (T2DM). Familial hypercholesterolemia exhibits elevations in plasma concentrations of LDL alone, while postprandial remnant lipoproteins are cleared from the plasma of these patients at normal or nearly normal rates. Postprandial dyslipoproteinemia is the mirror image: it does not necessarily affect LDL levels, while the clearance of postprandial remnant lipoproteins is substantially impaired [42, 54–56]. Thus, the remnant lipoproteins from one meal or snack may not be cleared from plasma before the next caloric ingestion; and therefore, these particles stay in plasma throughout nearly the entire 24-hour daily cycle. Several key studies have linked persistent postprandial lipoproteins to the development of human atherosclerosis [57], heart attacks [58, 59], and strokes [60].

The realization that hepatic clearance of remnant lipoproteins does not require LDL receptors launched a long and difficult search for the molecules responsible for this crucial pathway. Unfortunately, the search was hobbled by the conviction that the molecular mediators had to be proteins [56, 61]. Instead, we [62, 63] and others [50, 51, 64, 65] implicated heparan sulfate proteoglycans (HSPGs) as receptors for remnants and other apoB-lipoproteins, leading to hepatic catabolism of these particles (reviewed in Refs. [42, 44, 56, 66]). Each HSPG molecule consists of a polypeptide strand, called the core protein, onto which the cell assembles highly negatively charged sugar polymers, called heparan sulfate (HS), that we showed could capture lipoproteins. Thus, the ligand-binding domains of these molecules are carbohydrate, not polypeptide. Under normal circumstances, the carbohydrate side chains of hepatic HSPGs exhibit extreme structures of high flexibility and negative charge [67] that enhance ligand affinity and direct remnants into the liver [42, 56, 61].

The liver contains several different species of HSPGs, from which our laboratory identified syndecan-1 as a strong candidate for remnant lipoprotein clearance [44, 56] (Figure 24.3). The syndecan-1 HSPG is a transmembrane molecule that is abundantly expressed on the sinusoidal surface of hepatic parenchymal cells [68–70], along microvilli facing the space of Disse [71], where remnant lipoproteins are cleared [44, 61]. Moreover, the syndecan-1 HSPG directly mediates endocytosis of LpL-enriched apoB-lipoproteins through an unusual pathway independent of coated pits [44, 72, 163]. Accordingly, mice lacking the syndecan-1 core protein exhibit substantially impaired remnant lipoprotein clearance [71], similar to the postprandial dyslipoproteinemia of T2DM. In an elegant experiment, no additional lipoprotein defect was found in mice dually deficient for the syndecan-1 core protein and the HS *N*-deacetylase/*N*-sulfotransferase-1 (NDST1) [71], an enzyme that is required for proper assembly of heparan sulfate side chains on all species of hepatic HSPGs. This result implies that the side chains of syndecan-1, but no other hepatic HSPG, mediate remnant clearance under normal circumstances. Thus, the syndecan-1 HSPG is a major hepatic receptor for remnant lipoproteins [44, 71, 72, 163].

These studies indicate that the LDL receptor and the syndecan-1 HSPG provide complementary pathways for apoB-lipoprotein uptake that cannot entirely replace each other (Figure 24.3). The syndecan-1 pathway constitutes high capacity ($>10^6$ lipoprotein binding sites per hepatocyte) [71], a medium affinity [63, 71], and leads to slower endocytosis [44, 72, 163]. It is therefore suitable for handling the flood of postprandial particles created after each meal. The LDL receptor is of low capacity ($<5\times10^4$ binding sites per hepatocyte) [71], high affinity, and leads to rapid endocytosis

[46]. It is suitable for controlling steady-state levels of plasma LDL.

To molecularly characterize the defects in T2DM liver that impair syndecan-1 function, as well as the correction that occurs upon caloric restriction, we used microarrays to compare expression of proteoglycan-related genes in livers from lean *db/m* mice; obese, hyperphagic, T2DM *db/db* littermates fed ad libitum (AL); and *db/db* mice pair-fed to match the intake of *db/m* mice [73]. Surprisingly, the arrays identified only one gene whose dysregulation by T2DM would disrupt HSPG structure: the heparan sulfate glucosamine-6-*O*-endosulfatase-2 (*Sulf2*). The SULF2 enzyme degrades HSPGs by removing 6-*O* sulfate groups, but had no previously known role in diabetes or lipoprotein biology (Figure 24.3). Immunoblots demonstrated induction of SULF2 in AL livers, with restoration toward normal in livers from pair-fed *db/db* mice. Knockdown of SULF2 in cultured hepatocytes doubled HSPG-mediated catabolism of model remnant lipoproteins. Notably, co-immunoprecipitations revealed a persistent physical association of SULF2 with syndecan-1. To identify mechanisms of SULF2 dysregulation in T2DM, our laboratory found that advanced glycation end products provoked a 10-fold induction in SULF2 expression by cultured hepatocytes and an approximately 50% impairment in their catabolism of remnants and very low-density lipoprotein, an effect that was entirely reversed by SULF2 knockdown. Adiponectin and insulin each suppressed SULF2 protein in cultured liver cells and in murine livers *in vivo*, consistent with a role in energy flux. Likewise, both hormones enhanced remnant lipoprotein catabolism *in vitro* [73].

Most importantly, *Sulf2* antisense oligonucleotide treatment flattened postprandial plasma triglyceride excursions in *db/db* mice after corn oil gavage [74]. Thus, despite extensive metabolic derangements in T2DM mice, inhibition of a single dysregulated molecule, SULF2, abolishes postprandial hypertriglyceridemia. These findings provide a key proof of concept *in vivo* to support *Sulf2* inhibition as an attractive strategy to improve metabolic dyslipidemia [74]. Moreover, despite the existence of roughly 50 genes involved in hepatic HSPG assembly and disassembly, our results suggest dysregulation of remarkably few of them in diabetes [42].

MECHANISMS FOR THE RETENTION OF apoB-LIPOPROTEINS WITHIN THE VESSEL WALL

A considerable flux of LDL can enter but then leave the normal arterial wall *in vivo* [75], apparently without harm. Only a small fraction becomes retained. The mechanism for arterial wall retention of apoB-lipoproteins within a pristine, prelesional region of the arterial wall is relatively inefficient; but unfortunately, once a lesion develops, the dominant mechanism for trapping these particles switches to a more avid molecular process. Thus, lifetime plasma concentrations of LDL around 70–100 mg/dl will largely avoid atherosclerotic events [76], whereas the same concentrations of LDL will allow existing plaques to grow [77].

Molecular and Cellular Mechanisms of Early Lipoprotein Retention

In earliest atherogenesis, the molecular mechanism for retention depends on specific, positively charged domains near the C-terminus of $apoB_{100}$ that bind directly to negatively charged elements within the arterial matrix, particularly the sulfated polysaccharide side chains of certain arterial wall chondroitin sulfate proteoglycans (CSPGs). Site-directed mutagenesis of these C-terminal sequences in human $apoB_{100}$, followed by expression in experimental animals, produces a striking phenotype: high plasma concentrations of LDL particles that are poorly retained within the prelesional arterial wall, and therefore poorly initiate atherosclerotic lesion development [78]. Subsequent work showed that molecular manipulations to decrease arterial matrix substantially lessen lipoprotein retention [79] and atherogenesis [79, 80], as predicted [4]. Likewise, an intervention to increase arterial matrix accelerated both lipoprotein retention and atherosclerosis

[81]. These results experimentally established lipoprotein retention *in vivo* as the key causal step in early atherogenesis, while also identifying apoB-CSPG interactions as the dominant molecular mechanism for the early trapping of LDL within the arterial wall.

Unlike murine arteries, healthy human coronary arteries contain a normal structure called diffuse intimal thickening (DIT) [82, 83]. In humans, DIT is considered to be normal intima, because it is invariably present in human coronary arteries and shows well-organized structures mainly comprising SMCs, elastin, and proteoglycans [82]. Moreover, the thickened regions of healthy human intima exhibit no lipid deposition and no macrophage infiltration, and hence are not atherosclerosis. Low levels of cells with dendritic features have been found in normal murine aortic intima, particularly at curvatures and branches. These cells appear to be part of healthy immune surveillance and also do not constitute atherosclerosis [84].

In the presence of high plasma concentrations of apoB-lipoproteins, however, proretentive CSPGs and other matrix elements in DIT will trap these particles, thereby initiating this deadly disease [82, 83]. Several groups documented the extracellular deposition of apoB-lipoproteins and lipoprotein-derived lipids in the intima of very early human lesions [85, 86], well before the infiltration of macrophages [83, 87]. An elegant autopsy study of children and young adults who died of noncardiac causes showed a spectrum of changes in the subendothelium of susceptible areas within the right coronary artery, beginning with no lipid, to small amounts of apoB-lipoproteins and lipoprotein-derived lipids with no immune cells, to larger amounts apoB-lipoproteins and lipoprotein-derived lipids associated with the first signs of macrophage infiltration, and then finally the conversion of these macrophages into frank foam cells [83]. Key roles for the biglycan and versican CSPGs were identified in the trapping of apoB-lipoproteins within these early human lesions [83], similar to findings in more advanced human lesions [88]. Thus, the molecular mediators of early atherogenesis include arterial wall CSPGs, primarily biglycan and versican, specific positively charged domains on the C-terminus of human $apoB_{100}$, and abnormally high plasma concentrations of apoB-lipoproteins.

Several factors modulate these early processes. First, a drop in local pH, especially below 6.0, markedly enhances LDL binding to proteoglycans [89] and subsequent modifications of LDL by arterial wall enzymes [33]. These processes may affect the initiation and progression of atherosclerosis in human DIT, a structure that is thick enough to impede the diffusion of oxygen. One might expect that the earliest accumulation of lipids in human DIT should occur superficially, near the flow of plasma LDL. Instead, accumulation of apoB-lipoproteins in human arteries begins within the deepest areas of DIT [83], where the pH is lowest [33]. By contrast, the subsequent accumulation of macrophages begins in the superficial regions of human DIT, and these cells take on a characteristic lipid-laden, foamy appearance only after they have colocalized with retained apoB-lipoproteins [83].

Second, in addition to LDL, other apoB-lipoproteins can become retained within the arterial wall, most notably Lp(a) and hepatically and intestinally derived postprandial remnant lipoproteins. The unique composition of Lp(a) enhances particle binding to arterial matrix, which likely contributes to its atherogenicity [90–92]. Intestinally derived postprandial remnant lipoproteins, which contain $apoB_{48}$, are retained in lesion-prone areas of the arterial wall, via specific positively charged domains on that protein that become exposed when the C-terminal 52% of $apoB_{100}$ is missing [93]. Patients with high plasma levels of remnant lipoproteins are at markedly increased risk for future atherosclerotic cardiovascular events [58, 59]. Currently, the major causes of persistent postprandial remnant lipoproteins are type 2 diabetes mellitus, obesity, and related syndromes of positive caloric imbalance (recently reviewed in Refs. [40, 56, 66]).

Third, LDL samples from patients with coronary artery disease show enhanced binding affinity to purified arterial wall proteoglycans *in vitro* [94]. Moreover, the fatty-acyl composition of cholesteryl esters within the LDL core has been reported to affect the affinity for arterial proteoglycans [95]. Particles of LDL containing cholesteryl esters of polyunsaturated fatty acids bind protoglycans the least well and are associated with less atherosclerosis in subhuman primates [95]. Fourth, cytokines known to be present under certain conditions within the arterial wall, particularly TGF-ß, stimulate the production of proteoglycans with longer side chains and hence enhanced affinity for apoB-lipoproteins [96]. Likewise, lesion-susceptible arterial branch points exhibit a matrix structure that binds LDL more avidly than does matrix from lesion-resistant regions [97–99]. Fifth, other major conventional risk factors, particularly diabetes and hypertension, produce arterial wall changes likely to enhance apoB-lipoprotein retention [81, 100, 101] and accelerate the subsequent stages of this disease.

Finally, as noted earlier, life-long low plasma levels of LDL essentially block the development of clinically evident vascular disease in humans, even in the presence of other major risk factors [76, 102]. These findings indicate that apoB-lipoproteins are causative, while these other major risk factors become important modulators of lipoprotein retention and atherosclerosis, once sufficient plasma apoB-lipoprotein levels have been reached (see section "Conclusion: the evolution of epidemiologic risk factors into causative factors, exacerbating factors, and bystander phenomenon in atherosclerotic cardiovascular disease").

Acceleration of Lipoprotein Retention after Plaque Initiation

Key accessory molecules that accelerate lipoprotein retention and modification have also been identified (Figure 24.1). Three types of enzymes within the arterial wall bear particular emphasis: the secretory sphingomyelinase (S-SMase), lipoprotein lipase (LpL), and the secretory phospholipase A_2 ($sPLA_2$) enzymes. Each of these has been shown *in vitro* to promote apoB-lipoprotein aggregation, a process that is considerably accelerated when the enzymes are present in combination, owing to synergistic effects, and also when arterial matrix is present [103, 104]. The effects of S-SMase and the $sPLA_2$ enzymes appear to be lipolytic, and S-SMase has been shown to induce extensive conformational changes in $apoB_{100}$, which lead to the formation of large aggregates of LDL, particularly in regions of low pH [33]. By contrast, LpL plays a nonenzymatic, structural role as a "bridging" molecule between apoB-lipoproteins and proteoglycans [63], which in this context accelerates further lipoprotein retention [4, 5, 103, 105–107].

Once LDL and other apoB-lipoproteins in the arterial wall have aggregated, diffusion back into plasma becomes essentially impossible, owing to the large size of the aggregates and their greatly enhanced affinity as multivalent ligands for arterial matrix [103], as well as conformational changes that expose additional positively charged domains on apoB [108, 109]. Lipid hydrolysis and other modifications [110] of the retained, aggregated lipoproteins release biologically active byproducts that recruit macrophages and other cells into the developing lesion. Aggregated LDL that we isolated from advanced human plaques contains ceramide, the signature lipid product of S-SMase [111], and lipoproteins extracted from atherosclerotic lesions show evidence of PLA_2-mediated hydrolysis as well [112].

Aggregation of apoB-lipoproteins by S-SMase, LpL, and PLA_2 substantially increases particle uptake by macrophages, leading to frank foam cell formation *in vitro* (Figure 24.1) [103, 113]. Importantly, manipulations to suppress arterial wall expression of S-SMase [114] or macrophage-derived LpL [106] *in vivo* inhibit lipoprotein retention and atherogenesis. Manipulations to increase the local expression of LpL [115, 116] or PLA_2 [117] accelerate lipid or lipoprotein deposition and atherosclerotic lesion development. Consistent with cell culture data [63, 103], overexpression of catalytically inactive LpL also promotes these effects *in vivo* [116]. These data support a model in which vessel wall LpL is proatherogenic, through a bridging function that enhances lipoprotein retention.

Note that if the phagocytic system of the human body functioned helpfully in this circumstance, the macrophages that enter the arterial wall and consume this retained and modified material would then simply leave [107]. Recent work indicates that downstream products of retained and modified apoB-lipoproteins are particularly dangerous, because they eventually block the normal emigration of macrophages and dendritic cells from the plaque, thereby accelerating disease progression [118, 119]. These cells then become abnormally persistent within the lesion and exhibit a series of strikingly maladaptive responses that include the secretion of lipases, which, as noted earlier, greatly accelerate further lipoprotein retention and modification [105, 106]; proteases, which weaken the overlying fibrous cap [120]; and tissue factor, which ensures vigorous clot formation upon plaque rupture [121] (Figure 24.1). As reviewed elsewhere, many of these features can be quickly corrected in experimental atherosclerosis by large decreases in plasma concentrations of apoB-lipoproteins and increases in cholesterol acceptors [107].

CONCLUSION: THE EVOLUTION OF EPIDEMIOLOGIC RISK FACTORS INTO CAUSATIVE FACTORS, EXACERBATING FACTORS, AND BYSTANDER PHENOMENON IN ATHEROSCLEROTIC CARDIOVASCULAR DISEASE

The modern understanding of atherosclerosis began with cholesterol-feeding studies in animals [122, 123], followed by human epidemiology. The Framingham Study and others established "risk factors" for cardiovascular events in human cohorts [3, 124, 125]. Before the Framingham Study, it was not widely understood that plasma LDL cholesterol levels, hypertension, or smoking were linked to heart attacks, strokes, and peripheral vascular disease [126].

Epidemiology can identify risk factors that associate with disease and can compare the statistical strengths of these associations. Nevertheless, epidemiology cannot establish causal relationships—and from its earliest use, epidemiology has often missed the key causal factors altogether. Following Koch's postulates, the causal role for arterial retention of cholesterol-rich, apoB-containing lipoproteins in atherosclerosis came from extensive animal studies using a variety of methods to elevate plasma concentrations of these particles [107, 122, 127–130]; studies to alter the subsequent retention of apoB-lipoproteins in arteries [78, 79, 81, 106, 114]; research that linked human genetic variants that affect LDL levels with atherosclerotic events [46, 76, 102, 131]; and an extensive series of randomized, controlled clinical trials of different interventions to lower plasma concentrations of apoB-lipoproteins in human subjects (reviewed in Ref. [132] and elsewhere). By contrast, genetic variants or pharmacologic interventions that alter several other strong epidemiologic risk factors, such as plasma levels of C-reactive protein [133] or homocysteine [134–136], do not affect human cardiovascular risk.

Using this extensive knowledge, we can now reclassify nearly all epidemiologic risk factors for atherosclerotic cardiovascular events into causative factors, exacerbating factors, and mere bystander phenomena (Table 24.1). Causative and exacerbating factors are targets for therapy; bystander phenomena are not. Thus, new strategies to lower plasma concentrations of apoB-lipoproteins—the causative agent in atherosclerosis—have become a high priority [30, 74, 137–139]. We have come a long way since the deep, persistent skepticism over LDL lowering as a strategy to control this disease [140–142]. Instead, it now appears that tools for preventing, arresting, and possibly reversing the number 1 killer in the world have come within our grasp.

Unfortunately, much of these elegant scientific and clinical advancements will become eroded if we do not correct our high-caloric, chair-enticing environment [158]. An epidemiologic study to account for the dramatic decrease in US deaths from coronary disease from 1980 to 2000 attributed about half of the decrease to improvements in conventional cardiovascular risk factors,

TABLE 24.1 Reclassification of epidemiologic risk factors for atherosclerotic cardiovascular events into causative factors, exacerbating factors, and mere bystander phenomena

Epidemiologic risk factor	Pathophysiologic reclassification	Comments
Elevated plasma concentrations of total cholesterol (most of which is LDL), LDL-cholesterol, non-HDL-cholesterol, and apoB	*Causative*, via the retention of these lipoproteins within the subendothelium and then the failure of the body to properly deal with the deposited material [4]	1 Atherosclerosis is rare in people with lifelong levels of LDL below 80 mg/dl, regardless of other risk factors [76, 102]. 2 Human atherosclerotic lesions contain lipid and protein derived from apoB-lipoproteins—and these are the first pathologic signs of atherosclerosis to appear in human arteries, long before macrophage infiltration [83]. 3 Genetic, dietary, and other interventions that raise plasma levels of apoB-lipoproteins in animals produce atheromata. By contrast, immune derangements in the absence of elevated plasma levels of apoB-lipoproteins do not. 4 Genetic and other interventions to alter LDL retention in the arteries of experimental animals correspondingly change the progression of disease [78, 79, 81, 106, 114]. 5 Human genetic mutations and polymorphisms that raise LDL levels increase cardiovascular risk, and those that lower LDL levels reduce cardiovascular endpoints [46, 76, 102, 131]. 6 Interventions that lower plasma LDL concentrations decrease human cardiovascular events [132].
Tobacco smoking	*Exacerbating*	Unusual populations that smoke but have low plasma apoB-lipoprotein levels have many health problems, although not many atherosclerotic cardiovascular events [76]. Typical populations that smoke have greatly increased risk of atherosclerotic cardiovascular events, as well as several other diseases for which smoking is exacerbating, not causal, such as tuberculosis. All smokers have increased risk for emphysema, lung cancer, and other malignancies, for which smoke exposure is causal.
Diabetes mellitus	*Exacerbating*	Diabetic patients with low plasma apoB-lipoprotein levels have few atherosclerotic cardiovascular events [76]. The more common situation is that diabetes causes atherogenic dyslipoproteinemia, hypertension, and a greatly increased risk of cardiovascular events.
Hypertension	*Exacerbating*	See Refs. [76, 102].
Male sex, age	*Exacerbating*	Men who cross the age of 50 years with optimal values of just four basic cardiovascular risk factors (total cholesterol below 180 mg/dl, blood pressure below 120/80 mm Hg, no smoking, and no diabetes) have a residual lifetime risk of a cardiovascular event of only 5% and an extended life expectance of healthy years. 50-year old men with two or more suboptimal values have a residual lifetime risk of a cardiovascular event of 70%, that is, 14 times as high [143].

(*Continued*)

TABLE 24.1 (*Continued*)

Epidemiologic risk factor	Pathophysiologic reclassification	Comments
Low plasma HDL concentrations	*Under investigation: exacerbating or bystander*	Low plasma concentrations of HDL associate with increased cardiovascular risk, but also associate with other risk factors, such as hypertriglyceridemia, obesity, and the metabolic syndrome. Raising plasma concentrations of HDL or other cholesterol acceptors in animals by injection or by genetic manipulations to increase HDL production slow or reverse atherosclerosis (reviewed in Ref. [107]). Nevertheless, many human genetic polymorphisms associated with increased HDL levels do not confer cardiovascular protection [131], and several recent clinical trials to raise HDL failed to show benefit [144, 145]. Additional strategies to raise human HDL concentrations or efficacy remain under exploration. Their success or failure in reducing cardiovascular events will demonstrate if low HDL directly exacerbates human atherosclerotic vascular disease or if it is just a biomarker.
Elevated plasma homocysteine concentrations	*Bystander*	Elevated plasma homocysteine levels correlated with future cardiovascular events, but several randomized controlled trials to lower homocysteine levels by administration of B-vitamins failed to show any clinical benefit [134–136]. Plasma levels of homocysteine in humans may be a marker for early renal impairment rather than an independent cardiovascular risk factor [146], in which case cosmetically lowering plasma homocysteine would not fix the underlying problem.
Chlamydia pneumoniae infection; periodontal disease (gingivitis)	*Bystander*	Prospective, randomized clinical trials of antibiotics failed to show cardiovascular benefit [147, 148], consistent with earlier work that a germ-free environment provides no protection against atherosclerosis in hypercholesterolemic animals [149]. While periodontal disease should be treated, a recent position paper from the American Heart Association concludes, "Although periodontal interventions result in a reduction in systemic inflammation and endothelial dysfunction in short-term studies, there is no evidence that they prevent ASVD [atherosclerotic vascular disease] or modify its outcomes" [150].
Elevated plasma C-reactive protein	*Bystander*	Elevated plasma C-reactive protein levels correlated with future cardiovascular events, but also correlate with several conventional cardiovascular risk factors and therefore add little predictive power. Regarding causality, genetic polymorphisms that markedly elevate plasma C-reactive protein levels do not increase the risk of human cardiovascular disease [133].

TABLE 24.1 (*Continued*)

Epidemiologic risk factor	Pathophysiologic reclassification	Comments
Oxidized LDL	*Bystander*	Peroxidized material been found in human plaques, and circulating levels of oxidized LDL have been reported to correlate with cardiovascular risk [151–154]. Nevertheless, numerous large-scale prospective randomized clinical trials of antioxidant administration, including under conditions that protect LDL from oxidative modifications *ex vivo*, failed to show cardiovascular benefit [155]. Moreover, several components of the heart-healthy lifestyle, such as aerobic exercise and consumption of polyunsaturated fat and moderate amounts of alcohol, are pro-oxidant [156]. "But isn't it past the time for the scientific and public health communities to loosen their ties to a theory that lacks predictive ability for human disease?" [157].

but concluded that "increases in the body-mass index accounted overall for about 26,000 additional deaths from coronary heart disease in 2000 and increases in the prevalence of diabetes for about 33,500 additional deaths" [159]. In other words, worsening weight gain and diabetes undid 8% and 10%, respectively, of the overall decrease in coronary deaths in the United States during that period. The data may be worse for young adults [160]. Since 2000, obesity and diabetes have continued to increase, particularly worldwide [1, 161, 162]. Remaining challenges will include improving lifestyle, promoting access and use of proven medical therapies, and the development of new treatments for a major killer that we now understand on a fundamental level.

REFERENCES

1. Gaziano TA, Bitton A, Anand S, Abrahams-Gessel S, Murphy A. Growing epidemic of coronary heart disease in low- and middle-income countries. *Curr Probl Cardiol*. 2010;35:72–115.
2. Roger VL, Go AS, Lloyd-Jones DM, et al. Heart disease and stroke statistics—2012 update: a report from the American Heart Association. *Circulation*. 2012;125:e2–e220.
3. Yusuf S, Hawken S, Ounpuu S, et al. Effect of potentially modifiable risk factors associated with myocardial infarction in 52 countries (the INTERHEART study): case-control study. *Lancet*. 2004;364:937–952.
4. Williams KJ, Tabas I. The response-to-retention hypothesis of early atherogenesis. *Arterioscler Thromb Vasc Biol*. 1995;15:551–561.
5. Williams KJ, Tabas I. Lipoprotein retention – and clues for atheroma regression. *Arterioscler Thromb Vasc Biol*. 2005;25:1536–1540.
6. Kane J, Havel R. Disorders of the biogenesis and secretion of lipoproteins containing the B apolipoproteins. In: Valle D, ed. *Scriver's Online Metabolic and Molecular Bases of Inherited Disease*. McGraw Hill: New York; 2011.
7. Gusarova V, Brodsky JL, Fisher EA. Apolipoprotein B100 exit from the endoplasmic reticulum (ER) is COPII-dependent, and its lipidation to very low density lipoprotein occurs post-ER. *J Biol Chem*. 2003;278:48051–48058.
8. Shoulders CC, Stephens DJ, Jones B. The intracellular transport of chylomicrons requires the small GTPase, Sar1b. *Curr Opin Lipidol*. 2004;15:191–197.
9. Tiwari S, Siddiqi SA. Intracellular trafficking and secretion of VLDL. *Arterioscler Thromb Vasc Biol*. 2012;32:1079–1086.
10. Hussain MM, Fatma S, Pan X, Iqbal J. Intestinal lipoprotein assembly. *Curr Opin Lipidol*. 2005;16:281–285.
11. Abumrad NA, Davidson NO. Role of the gut in lipid homeostasis. *Physiol Rev*. 2012;92: 1061–1085.
12. Tran K, Thorne-Tjomsland G, DeLong CJ, et al. Intracellular assembly of very low density

lipoproteins containing apolipoprotein B100 in rat hepatoma McA-RH7777 cells. *J Biol Chem*. 2002;277:31187–31200.
13. Olofsson SO, Boren J. Apolipoprotein B: a clinically important apolipoprotein which assembles atherogenic lipoproteins and promotes the development of atherosclerosis. *J Intern Med*. 2005;258:395–410.
14. Gusarova V, Seo J, Sullivan ML, Watkins SC, Brodsky JL, Fisher EA. Golgi-associated maturation of very low density lipoproteins involves conformational changes in apolipoprotein B, but is not dependent on apolipoprotein E. *J Biol Chem*. 2007;282:19453–19462.
15. Brodsky JL, Fisher EA. The many intersecting pathways underlying apolipoprotein B secretion and degradation. *Trends Endocrinol Metab*. 2008;19:254–259.
16. Ginsberg HN, Fisher EA. The ever-expanding role of degradation in the regulation of apolipoprotein B metabolism. *J Lipid Res*. 2009;50 (Suppl):S162–S166.
17. Boström K, Wettesten M, Borén J, Bondjers G, Wiklund O, Olofsson S-O. Pulse-chase studies of the synthesis and intracellular transport of apolipoprotein B-100 in HepG2 cells. *J Biol Chem*. 1986;261:13800–13806.
18. Borchardt RA, Davis RA. Intrahepatic assembly of very low density lipoproteins. *J Biol Chem*. 1987;262:16394–16402.
19. Fisher EA, Pan M, Chen X, et al. The triple threat to nascent apolipoprotein B. Evidence for multiple, distinct degradative pathways. *J Biol Chem*. 2001;276:27855–27863.
20. Fisher EA. The degradation of apolipoprotein B100: multiple opportunities to regulate VLDL triglyceride production by different proteolytic pathways. *Biochim Biophys Acta*. 2012;1821:778–781.
21. Williams KJ, Brocia RW, Fisher EA. The unstirred water layer as a site of control of apolipoprotein B secretion. *J Biol Chem*. 1990;265:16741–16744.
22. Millar JS, Maugeais C, Ikewaki K, et al. Complete deficiency of the low-density lipoprotein receptor is associated with increased apolipoprotein B-100 production. *Arterioscler Thromb Vasc Biol*. 2005;25:560–565.
23. McCracken AA, Brodsky JL. Assembly of ER-associated protein degradation in vitro: dependence on cytosol, calnexin, and ATP. *J Cell Biol*. 1996;132:291–298.
24. Benoist F, Grand-Perret T. Co-translational degradation of apolipoprotein B100 by the proteasome is prevented by microsomal triglyceride transfer protein. Synchronized translation studies on HepG2 cells treated with an inhibitor of microsomal triglyceride transfer protein. *J Biol Chem*. 1997;272: 20435–20442.
25. Yeung SJ, Chen SH, Chan L. Ubiquitin-proteasome pathway mediates intracellular degradation of apolipoprotein B. *Biochemistry*. 1996;35:13843–13848.
26. Fisher EA, Zhou M, Mitchell DM, et al. The degradation of apolipoprotein B100 is mediated by the ubiquitin-proteasome pathway and involves heat shock protein 70. *J Biol Chem*. 1997;272:20427–20434.
27. Gusarova V, Caplan AJ, Brodsky JL, Fisher EA. Apoprotein B degradation is promoted by the molecular chaperones hsp90 and hsp70. *J Biol Chem*. 2001;276:24891–24900.
28. Oyadomari S, Yun C, Fisher EA, et al. Cotranslocational degradation protects the stressed endoplasmic reticulum from protein overload. *Cell*. 2006;126:727–739.
29. Hrizo SL, Gusarova V, Habiel DM, Goeckeler JL, Fisher EA, Brodsky JL. The Hsp110 molecular chaperone stabilizes apolipoprotein B from endoplasmic reticulum-associated degradation (ERAD). *J Biol Chem*. 2007;282: 32665–32675.
30. Cuchel M, Meagher EA, du Toit Theron H, et al. Efficacy and safety of a microsomal triglyceride transfer protein inhibitor in patients with homozygous familial hypercholesterolaemia: a single-arm, open-label, phase 3 study. *Lancet*.2013;381:40–46.
31. Pan M, Maitin V, Parathath S, et al. Presecretory oxidation, aggregation, and autophagic destruction of apoprotein-B: a pathway for late-stage quality control. *Proc Natl Acad Sci U S A*. 2008;105:5862–5867.
32. Pan M, Cederbaum AI, Zhang YL, Ginsberg HN, Williams KJ, Fisher EA. Lipid peroxidation and oxidant stress regulate hepatic apolipoprotein B degradation and VLDL production. *J Clin Invest*. 2004;113:1277–1287.
33. Sneck M, Nguyen SD, Pihlajamaa T, et al. Conformational changes of apoB-100 in SMase-modified LDL mediate formation of large aggregates at acidic pH. *J Lipid Res*. 2012;53:1832–1839.

34. Qiu W, Zhang J, Dekker MJ, et al. Hepatic autophagy mediates endoplasmic reticulum stress-induced degradation of misfolded apolipoprotein B. *Hepatology*. 2011;53: 1515–1525.
35. Zhong S, Magnolo AL, Sundaram M, et al. Nonsynonymous mutations within APOB in human familial hypobetalipoproteinemia: evidence for feedback inhibition of lipogenesis and postendoplasmic reticulum degradation of apolipoprotein B. *J Biol Chem*. 2010;285: 6453–6464.
36. Sparks JD, Phung TL, Bolognino M, Sparks CE. Insulin-mediated inhibition of apolipoprotein B secretion requires an intracellular trafficking event and phosphatidylinositol 3-kinase activation: studies with brefeldin A and wortmannin in primary cultures of rat hepatocytes. *Biochem J*. 1996;313:567–574.
37. Andreo U, Guo L, Chieriac DV, et al. Insulin-stimulated degradation of apolipoprotein B100: roles of class II phosphatidylinositol-3-kinase and autophagy. *PLoS One*. 2013;8:e57590.
38. Brown AM, Gibbons GF. Insulin inhibits the maturation phase of VLDL assembly via a phosphoinositide 3-kinase-mediated event. *Arterioscler Thromb Vasc Biol*. 2001;21: 1656–1661.
39. Sparks JD, Sparks CE, Adeli K. Selective hepatic insulin resistance, VLDL overproduction, and hypertriglyceridemia. *Arterioscler Thromb Vasc Biol*. 2012;32:2104–2112.
40. Wu X, Williams KJ. NOX4 pathway as a source of selective insulin resistance and responsiveness. *Arterioscler Thromb Vasc Biol*. 2012;32:1236–1245.
41. Bilheimer DW, Goldstein JL, Grundy SM, Starzl TE, Brown MS. Liver transplantation to provide low-density lipoprotein receptors and lower plasma cholesterol in a child with homozygous familial hypercholesterolemia. *N Engl J Med*. 1984;311:1658–1664.
42. Williams KJ, Chen K. Recent insights into factors affecting remnant lipoprotein uptake. *Curr Opin Lipidol*. 2010;21:218–228.
43. Zheng C, Murdoch SJ, Brunzell JD, Sacks FM. Lipoprotein lipase bound to apolipoprotein B lipoproteins accelerates clearance of postprandial lipoproteins in humans. *Arterioscler Thromb Vasc Biol*. 2006;26:891–896.
44. Fuki IV, Kuhn KM, Lomazov IR, et al. The syndecan family of proteoglycans: novel receptors mediating internalization of atherogenic lipoproteins *in vitro*. *J Clin Invest*. 1997;100:1611–1622.
45. Rios J, Stein E, Shendure J, Hobbs HH, Cohen JC. Identification by whole-genome resequencing of gene defect responsible for severe hypercholesterolemia. *Hum Mol Genet*. 2010; 19:4313–4318.
46. Brown MS, Goldstein JL. A receptor-mediated pathway for cholesterol homeostasis. *Science*. 1986;232:34–47.
47. Borén J, Ekström U, Agren B, Nilsson-Ehle P, Innerarity TL. The molecular mechanism for the genetic disorder familial defective apolipoprotein B100. *J Biol Chem*. 2001;276: 9214–9218.
48. Zelcer N, Hong C, Boyadjian R, Tontonoz P. LXR regulates cholesterol uptake through idol-dependent ubiquitination of the LDL receptor. *Science*. 2009;325:100–104.
49. Zhang L, Reue K, Fong LG, Young SG, Tontonoz P. Feedback regulation of cholesterol uptake by the LXR-IDOL-LDLR axis. *Arterioscler Thromb Vasc Biol*. 2012;32: 2541–2546.
50. Ji ZS, Fazio S, Mahley RW. Variable heparan sulfate proteoglycan binding of apolipoprotein E variants may modulate the expression of type III hyperlipoproteinemia. *J Biol Chem*. 1994;269:13421–13428.
51. Mann WA, Meyer N, Weber W, Meyer S, Greten H, Beisiegel U. Apolipoprotein E isoforms and rare mutations: parallel reduction in binding to cells and to heparin reflects severity of associated type III hyperlipoproteinemia. *J Lipid Res*. 1995;36:517–525.
52. Goldstein JL, Brown MS. Atherosclerosis: the low-density lipoprotein receptor hypothesis. *Metabolism*. 1977;26:1257–1275.
53. Kesaniemi YA, Witztum JL, Steinbrecher UP. Receptor-mediated catabolism of low density lipoprotein in man. Quantitation using glucosylated low density lipoprotein. *J Clin Invest*. 1983;71:950–959.
54. Dane-Stewart CA, Watts GF, Barrett PH, et al. Chylomicron remnant metabolism studied with a new breath test in postmenopausal women with and without type 2 diabetes mellitus. *Clin Endocrinol (Oxf)*. 2003;58:415–420.
55. Hogue JC, Lamarche B, Tremblay AJ, Bergeron J, Gagne C, Couture P. Evidence of increased secretion of apolipoprotein B-48-containing

lipoproteins in subjects with type 2 diabetes. *J Lipid Res*. 2007;48:1336–1342.
56. Williams KJ. Molecular processes that handle – and mishandle – dietary lipids. *J Clin Invest*. 2008;118:3247–3259.
57. Karpe F, Boquist S, Tang R, Bond GM, de Faire U, Hamsten A. Remnant lipoproteins are related to intima-media thickness of the carotid artery independently of LDL cholesterol and plasma triglycerides. *J Lipid Res*. 2001;42:17–21.
58. Nordestgaard BG, Benn M, Schnohr P, Tybjaerg-Hansen A. Nonfasting triglycerides and risk of myocardial infarction, ischemic heart disease, and death in men and women. *JAMA*. 2007;298:299–308.
59. Bansal S, Buring JE, Rifai N, Mora S, Sacks FM, Ridker PM. Fasting compared with non-fasting triglycerides and risk of cardiovascular events in women. *JAMA*. 2007;298: 309–316.
60. Freiberg JJ, Tybjaerg-Hansen A, Jensen JS, Nordestgaard BG. Nonfasting triglycerides and risk of ischemic stroke in the general population. *JAMA*. 2008;300:2142–2152.
61. Williams KJ, Fuki IV. Cell-surface heparan sulfate proteoglycans: dynamic molecules mediating ligand catabolism. *Curr Opin Lipidol*. 1997;8:253–262.
62. Williams KJ, Fless GM, Petrie K, Snyder ML, Brocia RW, Swenson TL. Lipoprotein lipase enhances cellular catabolism of lipoprotein(a). *Circulation*. 1991;84 (Suppl II):566.
63. Williams KJ, Fless GM, Petrie KA, Snyder ML, Brocia RW, Swenson TL. Mechanisms by which lipoprotein lipase alters cellular metabolism of lipoprotein(a), low density lipoprotein, and nascent lipoproteins. Roles for low density lipoprotein receptors and heparan sulfate proteoglycans. *J Biol Chem*. 1992;267: 13284–13292.
64. Ebara T, Conde K, Kako Y, et al. Delayed catabolism of apoB-48 lipoproteins due to decreased heparan sulfate proteoglycan production in diabetic mice. *J Clin Invest*. 2000;105: 1807–1818.
65. MacArthur JM, Bishop JR, Stanford KI, et al. Liver heparan sulfate proteoglycans mediate clearance of triglyceride-rich lipoproteins independently of LDL receptor family members. *J Clin Invest*. 2007;117:153–164.
66. Wu X, Chen K, Williams KJ. The role of pathway-selective insulin resistance and responsiveness in diabetic dyslipoproteinemia. *Curr Opin Lipidol*. 2012;23:334–344.
67. Lyon M, Deakin JA, Gallagher JT. Liver heparan sulfate structure. A novel molecular design. J Biol Chem. 1994;269:11208–11215.
68. Hayashi K, Hayashi M, Jalkanen M, Firestone JH, Trestad RL, Bernfield M. Immunocytochemistry of cell surface heparan sulfate proteoglycan in mouse tissue. A light and electron microscopic study. *J Histochem Cytochem*. 1987;35:1079–1088.
69. Roskams T, Moshage H, De Vos R, Guido D, Yap P, Desmet V. Heparan sulfate proteoglycan expression in normal human liver. *Hepatology*. 1995;21:950–958.
70. Bernfield M, Kokenyesi R, Kato M, et al. Biology of the syndecans: a family of transmembrane heparan sulfate proteoglycans. *Annu Rev Cell Biol*. 1992;8:365–393.
71. Stanford KI, Bishop JR, Foley EM, et al. Syndecan-1 is the primary heparan sulfate proteoglycan mediating hepatic clearance of triglyceride-rich lipoproteins in mice. *J Clin Invest*. 2009;119:3236–3245.
72. Fuki IV, Meyer ME, Williams KJ. Transmembrane and cytoplasmic domains of syndecan mediate a multi-step endocytic pathway involving detergent-insoluble membrane rafts. *Biochem J*. 2000;351:607–612.
73. Chen K, Liu M-L, Schaffer L, et al. Type 2 diabetes in mice induces hepatic overexpression of sulfatase 2, a novel factor that suppresses uptake of remnant lipoproteins. *Hepatology*. 2010;52:1957–1967.
74. Hassing HC, Mooij H, Guo S, et al. Inhibition of hepatic sulfatase-2 *in vivo*: a novel strategy to correct diabetic dyslipidemia. *Hepatology*. 2012;55:1746–1753.
75. Schwenke DC, Carew TE. Initiation of atherosclerotic lesions in cholesterol-fed rabbits. II. Selective retention of LDL vs. selective increases in LDL permeability in susceptible sites of arteries. *Arteriosclerosis*. 1989;9:908–918.
76. Cohen JC, Boerwinkle E, Mosley TH, Jr., Hobbs HH. Sequence variations in PCSK9, low LDL, and protection against coronary heart disease. *N Engl J Med*. 2006;354: 1264–1272.
77. Nissen SE, Tardif J-C, Nicholls SJ, et al. Effect of Torcetrapib on the progression of coronary atherosclerosis. *N Engl J Med*. 2007;356: 1304–1316.

78. Skålén K, Gustafsson M, Rydberg EK, et al. Subendothelial retention of atherogenic lipoproteins in early atherosclerosis. *Nature*. 2002; 417:750–754.
79. Tran-Lundmark K, Tran PK, Paulsson-Berne G, et al. Heparan sulfate in perlecan promotes mouse atherosclerosis: roles in lipid permeability, lipid retention, and smooth muscle cell proliferation. *Circ Res*. 2008;103:43–52.
80. Vikramadithyan RK, Kako Y, Chen G, et al. Atherosclerosis in perlecan heterozygous mice. *J Lipid Res*. 2004;45:1806–1812.
81. Huang F, Thompson JC, Wilson PG, Aung HH, Rutledge JC, Tannock LR. Angiotensin II increases vascular proteoglycan content preceding and contributing to atherosclerosis development. *J Lipid Res*. 2008;49:521–530.
82. Nakashima Y, Chen YX, Kinukawa N, Sueishi K. Distributions of diffuse intimal thickening in human arteries: preferential expression in atherosclerosis-prone arteries from an early age. *Virchows Arch*. 2002;441:279–288.
83. Nakashima Y, Fujii H, Sumiyoshi S, Wight TN, Sueishi K. Early human atherosclerosis: accumulation of lipid and proteoglycans in intimal thickenings followed by macrophage infiltration. *Arterioscler Thromb Vasc Biol*. 2007;27:1159–1165.
84. Cybulsky MI, Jongstra-Bilen J. Resident intimal dendritic cells and the initiation of atherosclerosis. *Curr Opin Lipidol*. 2010;21: 397–403.
85. Torzewski M, Klouche M, Hock J, et al. Immunohistochemical demonstration of enzymatically modified human LDL and its colocalization with the terminal complement complex in the early atherosclerotic lesion. *Arterioscler Thromb Vasc Biol*. 1998;18:369–378.
86. Fukuchi M, Watanabe J, Kumagai K, et al. Normal and oxidized low density lipoproteins accumulate deep in physiologically thickened intima of human coronary arteries. *Lab Invest*. 2002;82:1437–1447.
87. Napoli C, D'Armiento FP, Mancini FP, et al. Fatty streak formation occurs in human fetal aortas and is greatly enhanced by maternal hypercholesterolemia. Intimal accumulation of low density lipoprotein and its oxidation precede monocyte recruitment into early atherosclerotic lesions. *J Clin Invest*. 1997;100:2680–2690.
88. O'Brien KD, Olin KL, Alpers CE, et al. Comparison of apolipoprotein and proteoglycan deposits in human coronary atherosclerotic plaques: colocalization of biglycan with apolipoproteins. *Circulation*. 1998;98: 519–527.
89. Sneck M, Kovanen PT, Öörni K. Decrease in pH strongly enhances binding of native, proteolyzed, lipolyzed, and oxidized low density lipoprotein particles to human aortic proteoglycans. *J Biol Chem*. 2005;280: 37449–37454.
90. Bihari-Varga M, Gruber E, Rotheneder M, Zechner R, Kostner GM. Interaction of lipoprotein Lp(a) and low density lipoprotein with glycosaminoglycans from human aorta. *Arteriosclerosis*. 1988;8:851–857.
91. Klezovitch O, Edelstein C, Zhu L, Scanu AM. Apolipoprotein(a) binds via its C-terminal domain to the protein core of the proteoglycan decorin. Implications for the retention of lipoprotein(a) in atherosclerotic lesions. *J Biol Chem*. 1998;273:23856–23865.
92. Lundstam U, Hurt-Camejo E, Olsson G, Sartipy P, Camejo G, Wiklund O. Proteoglycans contribution to association of Lp(a) and LDL with smooth muscle cell extracellular matrix. *Arterioscler Thromb Vasc Biol*. 1999;19:1162–1167.
93. Flood C, Gustafsson M, Richardson PE, Harvey SC, Segrest JP, Borén J. Identification of the proteoglycan binding site in apolipoprotein B48. *J Biol Chem*. 2002;277:32228–32233.
94. Linden T, Bondjers G, Camejo G, Bergstrand R, Wilhelmsen L, Wiklund O. Affinity of LDL to a human arterial proteoglycan among male survivors of myocardial infarction. *Eur J Clin Invest*. 1989;19:38–44.
95. Degirolamo C, Shelness GS, Rudel LL. LDL cholesteryl oleate as a predictor for atherosclerosis: evidence from human and animal studies on dietary fat. *J Lipid Res*. 2009;50 (Suppl):S434–S439.
96. Little PJ, Tannock L, Olin KL, Chait A, Wight TN. Proteoglycans synthesized by arterial smooth muscle cells in the presence of transforming growth factor-ß1 exhibit increased binding to LDLs. *Arterioscler Thromb Vasc Biol*. 2002;22:55–60.
97. Cardoso LE, Mouraõ PA. Glycosaminoglycan fractions from human arteries presenting diverse susceptibilities to atherosclerosis have different binding affinities to plasma LDL. *Arterioscler Thromb*. 1994;14:115–124.
98. Hurt-Camejo E, Olsson U, Wiklund O, Bondjers G, Camejo G. Cellular consequences of the

association of apoB lipoproteins with proteoglycans. Potential contribution to atherogenesis. *Arterioscler Thromb Vasc Biol.* 1997;17:1011–1017.

99. Kwon GP, Schroeder JL, Amar MJ, Remaley AT, Balaban RS. Contribution of macromolecular structure to the retention of low-density lipoprotein at arterial branch points. *Circulation.* 2008;117:2919–2927.
100. Wasty F, Alavi MZ, Moore S. Distribution of glycosaminoglycans in the intima of human aortas: changes in atherosclerosis and diabetes mellitus. *Diabetologia.* 1993;36:316–322.
101. Figueroa JE, Vijayagopal P. Angiotensin II stimulates synthesis of vascular smooth muscle cell proteoglycans with enhanced low density lipoprotein binding properties. *Atherosclerosis.* 2002;162:261–268.
102. Sankatsing RR, Fouchier SW, de Haan S, et al. Hepatic and cardiovascular consequences of familial hypobetalipoproteinemia. *Arterioscler Thromb Vasc Biol.* 2005;25: 1979–1984.
103. Tabas I, Li Y, Brocia RW, Xu SW, Swenson TL, Williams KJ. Lipoprotein lipase and sphingomyelinase synergistically enhance the association of atherogenic lipoproteins with smooth muscle cells and extracellular matrix. A possible mechanism for low density lipoprotein and lipoprotein(a) retention and macrophage foam cell formation. *J Biol Chem.* 1993;268:20419–20432.
104. Rosengren B, Peilot H, Umaerus M, et al. Secretory phospholipase A2 group V: lesion distribution, activation by arterial proteoglycans, and induction in aorta by a Western diet. *Arterioscler Thromb Vasc Biol.* 2006;26: 1579–1585.
105. Pentikäinen MO, Oksjoki R, Öörni K, Kovanen PT. Lipoprotein lipase in the arterial wall: linking LDL to the arterial extracellular matrix and much more. *Arterioscler Thromb Vasc Biol.* 2002;22:211–217.
106. Gustafsson M, Levin M, Skålén K, et al. Retention of low-density lipoprotein in atherosclerotic lesions of the mouse: evidence for a role of lipoprotein lipase. *Circ Res.* 2007;101:777–783.
107. Williams KJ, Feig JE, Fisher EA. Rapid regression of atherosclerosis: insights from the clinical and experimental literature. *Nat Clin Pract Cardiovasc Med.* 2008;5:91–102.
108. Öörni K, Hakala JK, Annila A, Ala-Korpela M, Kovanen PT. Sphingomyelinase induces aggregation and fusion, but phospholipase A2 only aggregation, of low density lipoprotein (LDL) particles. Two distinct mechanisms leading to increased binding strength of LDL to human aortic proteoglycans. *J Biol Chem.* 1998;273:29127–29134.
109. Flood C, Gustafsson M, Pitas RE, Arnaboldi L, Walzem RL, Boren J. Molecular mechanism for changes in proteoglycan binding on compositional changes of the core and the surface of low-density lipoprotein-containing human apolipoprotein B100. *Arterioscler Thromb Vasc Biol.* 2004;24:564–570.
110. Medeiros LA, Khan T, El Khoury JB, et al. Fibrillar amyloid protein present in atheroma activates CD36 signal transduction. *J Biol Chem.* 2004;279:10643–10648.
111. Schissel SL, Tweedie-Hardman J, Rapp JH, Graham G, Williams KJ, Tabas I. Rabbit aorta and human atherosclerotic lesions hydrolyze the sphingomyelin of retained low-density lipoprotein. Proposed role for arterial-wall sphingomyelinase in subendothelial retention and aggregation of atherogenic lipoproteins. *J Clin Invest.* 1996;98:1455–1464.
112. Rosengren B, Jonsson-Rylander AC, Peilot H, Camejo G, Hurt-Camejo E. Distinctiveness of secretory phospholipase A2 group IIA and V suggesting unique roles in atherosclerosis. *Biochim Biophys Acta.* 2006;1761: 1301–1308.
113. Boyanovsky BB, Shridas P, Simons M, van der Westhuyzen DR, Webb NR. Syndecan-4 mediates macrophage uptake of group V secretory phospholipase A2-modified low density lipoprotein. *J Lipid Res.* 2009;50:641–650.
114. Devlin CM, Leventhal AR, Kuriakose G, Schuchman EH, Williams KJ, Tabas I. Acid sphingomyelinase promotes lipoprotein retention within early atheromata and accelerates lesion progression. *Arterioscler Thromb Vasc Biol.* 2008;28:1723–1730.
115. Babaev VR, Patel MB, Semenkovich CF, Fazio S, Linton MF. Macrophage lipoprotein lipase promotes foam cell formation and atherosclerosis in low density lipoprotein receptor-deficient mice. *J Biol Chem.* 2000; 275:26293–26299.
116. Wang J, Xian X, Huang W, et al. Expression of LPL in endothelial-intact artery results in lipid

deposition and vascular cell adhesion molecule-1 upregulation in both LPL and ApoE-deficient mice. *Arterioscler Thromb Vasc Biol.* 2007;27:197–203.

117. Bostrom MA, Boyanovsky BB, Jordan CT, et al. Group V secretory phospholipase A2 promotes atherosclerosis: evidence from genetically altered mice. *Arterioscler Thromb Vasc Biol*. 2007;27:600–606.

118. Llodrá J, Angeli V, Liu J, Trogan E, Fisher EA, Randolph GJ. Emigration of monocyte-derived cells from atherosclerotic lesions characterizes regressive, but not progressive, plaques. *Proc Natl Acad Sci U S A*. 2004;101: 11779–11784.

119. Trogan E, Fayad ZA, Itskovich VV, et al. Serial studies of mouse atherosclerosis by in vivo magnetic resonance imaging detect lesion regression after correction of dyslipidemia. *Arterioscler Thromb Vasc Biol.* 2004;24: 1714–1719.

120. Watanabe N, Ikeda U. Matrix metalloproteinases and atherosclerosis. *Curr Atheroscler Rep*. 2004;6:112–120.

121. Liu ML, Reilly MP, Casasanto P, McKenzie SE, Williams KJ. Cholesterol enrichment of human monocyte/macrophages induces surface exposure of phosphatidylserine and the release of biologically-active tissue factor-positive microvesicles. *Arterioscler Thromb Vasc Biol*. 2007;27:430–435.

122. Anitschkow N, Chalatow S. Über experimentelle Cholesterinsteatose und ihre Bedeutung für die Entstehung einiger pathologischer Prozesse. *Zentrbl Allg Pathol Pathol Anat.* 1913;24:1–9.

123. Konstantinov IE, Mejevoi N, Anichkov NM. Nikolai N. Anichkov and his theory of atherosclerosis. *Tex Heart Inst J*. 2006;33:417–423.

124. Gofman JW, Young W, Tandy R. Ischemic heart disease, atherosclerosis, and longevity. *Circulation*. 1966;34:679–697.

125. Gordon T, Kannel WB. Multiple risk functions for predicting coronary heart disease: the concept, accuracy, and application. *Am Heart J*. 1982;103:1031–1039.

126. Messerli FH, Messerli AW, Luscher TF. Eisenhower's billion-dollar heart attack–50 years later. *N Engl J Med*. 2005;353:1205–1207.

127. Armstrong ML. Evidence of regression of atherosclerosis in primates and man. *Postgrad Med J*. 1976;52:456–461.

128. Buja LM, Kita T, Goldstein JL, Watanabe Y, Brown MS. Cellular pathology of progressive atherosclerosis in the WHHL rabbit. An animal model of familial hypercholesterolemia. *Arteriosclerosis*. 1983;3:87–101.

129. Paigen B, Morrow A, Holmes PA, Mitchell D, Williams RA. Quantitative assessment of atherosclerotic lesions in mice. *Atherosclerosis*. 1987;68:231–240.

130. Breslow JL. Mouse models of atherosclerosis. *Science*. 1996;272:685–688.

131. Voight BF, Peloso GM, Orho-Melander M, et al. Plasma HDL cholesterol and risk of myocardial infarction: a Mendelian Randomisation Study. *Lancet*. 2012;380:572–580.

132. Robinson JG, Smith B, Maheshwari N, Schrott H. Pleiotropic effects of statins: benefit beyond cholesterol reduction? A meta-regression analysis. *J Am Coll Cardiol*. 2005;46:1855–1862.

133. Zacho J, Tybjaerg-Hansen A, Jensen JS, Grande P, Sillesen H, Nordestgaard BG. Genetically elevated C-reactive protein and ischemic vascular disease. *N Engl J Med*. 2008;359:1897–1908.

134. Toole JF, Malinow MR, Chambless LE, et al. Lowering homocysteine in patients with ischemic stroke to prevent recurrent stroke, myocardial infarction, and death: the Vitamin Intervention for Stroke Prevention (VISP) randomized controlled trial. *JAMA*. 2004;291: 565–575.

135. Bonaa KH, Njolstad I, Ueland PM, et al. Homocysteine lowering and cardiovascular events after acute myocardial infarction. *N Engl J Med*. 2006;354:1578–1588.

136. Lonn E, Yusuf S, Arnold MJ, et al. Homocysteine lowering with folic acid and B vitamins in vascular disease. *N Engl J Med*. 2006;354:1567–1577.

137. Ason B, Tep S, Davis HR, Jr., et al. Improved efficacy for ezetimibe and rosuvastatin by attenuating the induction of PCSK9. *J Lipid Res*. 2011;52:679–687.

138. Visser ME, Witztum JL, Stroes ES, Kastelein JJ. Antisense oligonucleotides for the treatment of dyslipidaemia. *Eur Heart J*. 2012;33:1451–1458.

139. Stein EA, Dufour R, Gagne C, et al. Apolipoprotein B synthesis inhibition with mipomersen in heterozygous familial hypercholesterolemia: results of a randomized, double-blind, placebo-controlled trial to assess efficacy and safety as add-on therapy in patients

with coronary artery disease. *Circulation*. 2012; 126:2283–2292.
140. McMichael J. Fats and atheroma: an inquest. *Br Med J*. 1979;1:173–175.
141. Moore TJ. The cholesterol myth. *Atlantic Monthly*. 1989;264:37.
142. Steinberg D. Thematic review series: the pathogenesis of atherosclerosis. An interpretive history of the cholesterol controversy: part I. *J Lipid Res*. 2004;45:1583–1593.
143. Lloyd-Jones DM, Leip EP, Larson MG, et al. Prediction of lifetime risk for cardiovascular disease by risk factor burden at 50 years of age. *Circulation*. 2006;113:791–798.
144. Nicholls SJ. Is niacin ineffective? Or did AIM-HIGH miss its target? *Cleve Clin J Med*. 2012;79:38–43.
145. Schwartz GG, Olsson AG, Abt M, et al. Effects of dalcetrapib in patients with a recent acute coronary syndrome. *N Engl J Med*. 2012;367: 2089–2099.
146. Potter K, Hankey GJ, Green DJ, Eikelboom JW, Arnolda LF. Homocysteine or renal impairment: which is the real cardiovascular risk factor? *Arterioscler Thromb Vasc Biol*. 2008;28:1158–1164.
147. O'Connor CM, Dunne MW, Pfeffer MA, et al. Azithromycin for the secondary prevention of coronary heart disease events: the WIZARD study: a randomized controlled trial. *JAMA*. 2003;290:1459–1466.
148. Grayston JT, Kronmal RA, Jackson LA, et al. Azithromycin for the secondary prevention of coronary events. *N Engl J Med*. 2005;352: 1637–1645.
149. Wright SD, Burton C, Hernandez M, et al. Infectious agents are not necessary for murine atherogenesis. *J Exp Med*. 2000;191: 1437–1442.
150. Lockhart PB, Bolger AF, Papapanou PN, et al. Periodontal disease and atherosclerotic vascular disease: does the evidence support an independent association? A scientific statement from the American Heart Association. *Circulation*. 2012;125:2520–2544.
151. Glavind J, Hartmann S, Clemmesen J, Jessen KE, Dam H. Studies on the role of lipoperoxides in human pathology. II. The presence of peroxidized lipids in the atherosclerotic aorta. *Acta Pathol Microbiol Scand*. 1952;30:1–6.
152. Tsimikas S, Witztum JL. Measuring circulating oxidized low-density lipoprotein to evaluate coronary risk. *Circulation*. 2001; 103:1930–1932.
153. Holvoet P, Mertens A, Verhamme P, et al. Circulating oxidized LDL is a useful marker for identifying patients with coronary artery disease. *Arterioscler Thromb Vasc Biol*. 2001;21:844–848.
154. Steinberg D, Witztum JL. History of discovery: oxidized low-density lipoprotein and atherosclerosis. *Arterioscler Thromb Vasc Biol*. 2010;30:2311–2316.
155. Stocker R, Keaney JF, Jr. Role of oxidative modifications in atherosclerosis. *Physiol Rev*. 2004;84:1381–1478.
156. Williams KJ, Fisher EA. Oxidation, lipoproteins, and atherosclerosis: which is wrong, the antioxidants or the theory? *Curr Opin Clin Nutr Metab Care*. 2005;8:139–146.
157. Greenberg ER. Vitamin E supplements: good in theory, but is the theory good? *Ann Intern Med*. 2005;142:75–76.
158. Levine JA, Vander Weg MW, Hill JO, Klesges RC. Non-exercise activity thermogenesis: the crouching tiger hidden dragon of societal weight gain. *Arterioscler Thromb Vasc Biol*. 2006;26:729–736.
159. Ford ES, Ajani UA, Croft JB, et al. Explaining the decrease in U.S. deaths from coronary disease, 1980–2000. *N Engl J Med*. 2007;356: 2388–2398.
160. Ford ES, Capewell S. Coronary heart disease mortality among young adults in the U.S. from 1980 through 2002: concealed leveling of mortality rates. *J Am Coll Cardiol*. 2007; 50:2128–2132.
161. Joshi R, Chow CK, Raju PK, et al. Fatal and nonfatal cardiovascular disease and the use of therapies for secondary prevention in a rural region of India. *Circulation*. 2009; 119:1950–1955.
162. Yang W, Lu J, Weng J, et al. Prevalence of diabetes among men and women in China. *N Engl J Med*. 2010;362:1090–1101.
163. Chen K, Williams KJ. Molecular mediators for raft-dependent endocytosis of syndecan-1, a highly conserved, multifunctional receptor. *J Biol Chem*. 2013;288:13988–13999.

25

HDL RECEPTOR SCAVENGER RECEPTOR BI: A MULTIPURPOSE PLAYER IN CHOLESTEROL AND STEROID METABOLISM IN MICE AND MAN

YING ZHAO[1], MENNO HOEKSTRA[1], SUZANNE J. A. KORPORAAL[2], THEO J. C. VAN BERKEL[1], AND MIRANDA VAN ECK[1]

[1] *Division of Biopharmaceutics, Leiden Academic Centre for Drug Research, Gorlaeus Laboratories, Leiden University, Leiden, The Netherlands*

[2] *Department of Clinical Chemistry and Haematology, University Medical Center Utrecht, Utrecht, The Netherlands*

INTRODUCTION

Scavenger receptors, first described by Brown and Goldstein [1, 2], are membrane glycoproteins that are involved in the cellular uptake of a broad range of polyanionic ligands, including modified lipoproteins, collagen, apoptotic cells, and bacterial components. In 1979, Goldstein and Brown were the first to discover that macrophages contain a binding site that mediates the uptake and degradation of acetylated low-density lipoprotein (LDL), leading to massive cholesterol deposition in lipid droplets [1, 2]. Although the physiological importance was unknown at that time, it was hypothesized that uptake of modified LDL might account for the widespread deposition of cholesterol in macrophages of patients with hypercholesterolemia, who lack functional LDL receptors [1]. The scavenger receptor superfamily now encompasses eight different subclasses, labeled A through H, that have been classified based on their structural motifs [3]. Many of the family members are expressed in atherosclerotic plaques and have been implicated in the development of lipid-laden foam cells or play an active role in the regulation of inflammatory reactions during atherogenesis [4].

Scavenger receptor class B type I (SR-BI) belongs to the class B subfamily of scavenger receptors, which also contains it splice variant SR-BII and CD36. SR-BI was cloned first from a variant CHO cell cDNA library that expressed a scavenger receptor activity distinct from the classical SR-A activity [5]. As all members of the class B subfamily of scavenger receptors, SR-BI contains a heavily *N*-glycosylated backbone, consisting of a large membrane spanning extracellular loop and two short intracellular

Atherosclerosis: Risks, Mechanisms, and Therapies, First Edition. Edited by Hong Wang and Cam Patterson.

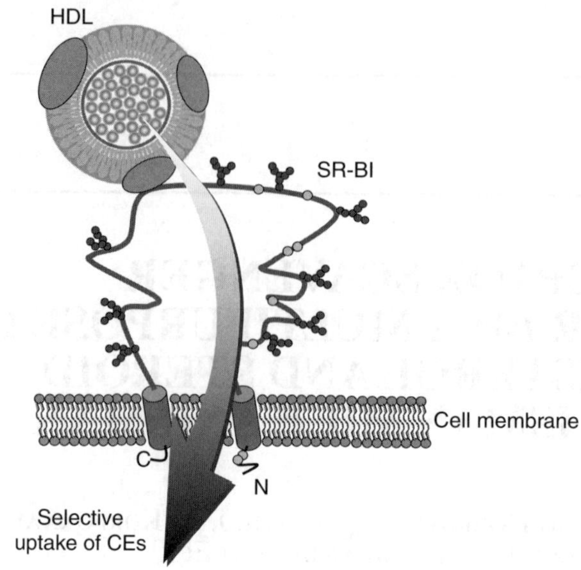

Figure 25.1 Schematic representation of SR-BI-mediating selective uptake of cholesterol esters (CEs) from the core of the HDL particle.

N-terminal and C-terminal domains (Figure 25.1) [6]. SR-BII, representing 40% of total SR-BI/BII mRNA but only 12% of the immunodetectable SR-BI/BII protein, only differs from SR-BI in its C-terminal tail [7].

SR-BI is expressed in various mammalian tissues and cells, including brain, kidney, intestine, heart, placenta, macrophages, endothelial cells, smooth muscle cells, various epithelial cells, and platelets [8]. The highest expression of SR-BI, however, is found in organs with critical roles in cholesterol metabolism (liver) and steroidogenesis (adrenal, ovary, testis) [9]. Further, using *in situ* hybridization, Ji et al. found SR-BI mRNA in the thickened intima of apoliprotein E (apoE) knockout (KO) mice, suggesting that SR-BI could contribute to the development of foam cells [10]. The human homologue of SR-BI, CD36, and LIMP-II analogous-1 (CLA-1) shows a similar tissue expression profile as murine SR-BI [11].

SR-BI was recognized as a scavenger receptor based on its property to bind modified LDL and maleylated BSA (M-BSA) [5]. Currently, SR-BI has been implicated in the binding of a wide array of ligands, including anionic phospholipids [12], advanced glycation end-products [13], apoptotic cells [14], and native and modified lipoproteins [8, 15–17]. Importantly, in the mid 1990s, it was also identified as the first receptor for HDL facilitating the selective uptake of cholesterol esters (CEs) from HDL [8].

REGULATION OF SR-BI EXPRESSION

The transcription of SR-BI can be regulated by sterol regulatory element-binding proteins (SREBPs) [18, 19], steroidogenic factor-1 (SF-1) [18, 20], liver receptor homolog 1 (LRH-1) [21], and the peroxisome proliferator-activated receptor-α (PPAR-α) [22–24]. The stability of the SR-BI protein is modulated by Ras/mitogen-activated protein kinase (MAPK) [25]. Interestingly, SR-BI is regulated differently in liver and in steroidogenic tissues. Overexpression of SREBP1a downregulates the transcription of hepatic SR-BI but induces its expression in ovaries [18, 19]. Moreover, fibrates (PPAR-α agonists) reduce SR-BI in liver while they upregulate SR-BI in macrophages and have no effect on SR-BI in adrenals [22–24]. Of note, the expression of SR-BI is also regulated differently in different types of cells at post-translational level. In hepatocytes as well as epithelial cells of the intestine, the PDZ domain-containing adaptor protein, PDZK1, is essential for the expression of SR-BI on the cell surface. Consequently, PDZK1 KO mice showed about 95% and around 50% reduction in the protein levels of SR-BI in the liver and the small intestine, respectively [26]. In addition, atherogenic diet-induced downregulation of SR-BI protein expression in the liver and intestine is associated with a reduction of PDZK1 [19]. However, the expression of SR-BI in adrenals and macrophages is unchanged in PDZK1 KO mice [26]. Moreover, small PDZK1-associated protein (SAAP) decreases PDZK1 in a liver-specific fashion, thereby resulting in downregulation of hepatic SR-BI, but has no effect on the levels of SR-BI in the adrenals or peritoneal macrophages [27]. SR-BI is thus regulated in a cell-type-specific fashion.

SR-BI FUNCTION IN LIPOPROTEIN METABOLISM

Evidence for the physiological importance of SR-BI in HDL metabolism was obtained from studies in genetically engineered mice. Hepatic overexpression of SR-BI increased the selective uptake of HDL-CE by liver [28, 29], resulting in the virtual absence of plasma HDL cholesterol (HDL-C) [30, 31]. Conversely, impaired hepatic uptake of HDL-CE in mice with attenuated expression of SR-BI or complete SR-BI deficiency led to the accumulation of abnormally large HDL particles and increased plasma HDL-C levels [32–34]. In humans, a clear association between mutations in the coding and promoter regions of human SR-BI and increased plasma HDL-C has been shown in several population studies [35–37]. Recent genome-wide association studies (GWAS) also demonstrated that single nucleotide polymorphism (SNPs) in and near SR-BI are significantly associated with plasma levels of HDL-C in humans [38]. However, only very recently conclusive evidence was provided by Vergeer et al. on the importance of SR-BI in controlling HDL-C levels in humans. They identified a family in which heterozygous carriers of a unique mutation (P297S) in the extracellular domain of SR-BI led to a 37% elevation in plasma HDL-C levels [39]. Importantly, hepatocytes expressing the P297S mutant SR-BI displayed a reduced capacity to take up HDL-CE, thereby explaining the observed increase in HDL-C in the circulation in carriers of the mutant SR-BI [39].

By comparing the uptake of CE from HDL with holoparticle uptake in SR-BI KO mice and wild-type (WT) mice, SR-BI was identified as the sole molecule responsible for the selective uptake of CEs from HDL in mice *in vivo* [40, 41]. The extracellular domain of SR-BI and the proper orientation of apoAI molecules on the HDL particles are crucial for efficient lipid uptake via SR-BI [42–47]. The selective uptake of HDL-CE is considered a two-step process: the first step involves the binding of the lipoprotein to the extracellular domain of SR-BI, followed by internalization of its CEs without net internalization and degradation of the lipoprotein itself [48]. Notably, HDL binding and the subsequent uptake of CEs are independent processes [42, 49]; high-affinity HDL binding to SR-BI is not sufficient for efficient HDL-CE selective uptake [50, 51]. Interestingly, SR-BI reconstituted into liposomes is still capable of avidly binding lipoproteins and selectively taking up CEs, indicating that specific cellular structures and compartments are not required for SR-BI-mediated HDL-CEs uptake [52]. Several recent studies have indicated that the so-called retroendocytosis pathway involving holoparticle uptake of HDL followed by resecretion of CE-poor HDL could also contribute to the selective uptake of HDL-CE via SR-BI [53, 54]. Upon selective uptake via SR-BI, HDL-CEs are delivered into an extralysosomal metabolically active membrane pool and subsequently hydrolyzed by neutral cholesterol ester hydrolase [55–57].

SR-BI has also been implicated in the metabolism of apoB-containing lipoproteins, including LDL and VLDL. Wiersma et al. demonstrated that the expression level of hepatic SR-BI is correlated with the production rate of VLDL by affecting the hepatic cholesterol content and the activity of microsomal triglyceride transfer protein (MTP) [58]. Moreover, *in vitro* studies clearly demonstrated that SR-BI binds apoB lipoproteins and facilitates the subsequent uptake of CEs from these lipoproteins [15–17, 59, 60]. *In vivo*, SR-BI attenuation or deficiency led to increased VLDL and LDL cholesterol levels in WT [59], LDL receptor KO [61], and apoE KO mice [34], while mice overexpressing SR-BI displayed reduced levels of apoB-containing lipoproteins [28, 31, 62]. In addition, SR-BI is also involved in chylomicron metabolism as evidenced by higher postprandial triglyceride (TG) levels in plasma of SR-BI KO mice and reduced association of chylomicron-like emulsion particles to freshly isolated hepatocytes lacking SR-BI [63]. Béaslas et al. recently demonstrated that postprandial micelles supplied to Caco2/TC7 enterocytes induced the clustering of SR-BI at the apical brush border membrane and movement from nonraft to raft domains [64]. Importantly, competition, inhibition, or knockdown of SR-BI impair the trafficking of apoB from apical toward secretory domains, indicating the involvement of SR-BI in the secretion of intestinal TG-rich lipoproteins [64]. Moreover, overexpression of intestinal SR-BI results in accelerated lipid absorption [65]. Several studies on common polymorphisms of CLA-1, the human homologue of SR-BI, also suggested its role in the metabolism of apoB lipoproteins [35, 66–68] and postprandial lipoproteins in humans [69].

SR-BI AND MACROPHAGE CHOLESTEROL EFFLUX

Increased uptake of lipoproteins by macrophages, combined with impaired cholesterol efflux leads to accumulation of cholesterol and transformation of macrophages into foam cells, the first stage of atherosclerotic lesion development. SR-BI is expressed in macrophage foam cells in human and murine atherosclerotic lesions [10, 22, 70]. SR-BI might thus also play an important role locally in the arterial wall. Studies on the role of SR-BI in cellular cholesterol homeostasis have, however, been hampered by the low expression of SR-BI in macrophages in culture. Further, the role of SR-BI in cellular cholesterol homeostasis is complex, and it is the subject of discussion.

Apart from facilitating the selective uptake of CEs from HDL, SR-BI binds both modified and native lipoproteins and mediates their uptake by macrophages [5, 11]. In agreement, macrophages lacking SR-BI show a clearly reduced uptake of β-VLDL, the major lipoprotein that circulates in mouse models of atherosclerosis [71]. Intracellularly, SR-BI localizes in the late endo/lysosomes and might thus be involved in the intracellular cholesterol trafficking. In hepatocytes, SR-BI deficiency is indeed associated with accumulation of lysosomal cholesterol [72]. It is currently unknown if the same accounts for macrophages lacking SR-BI. SR-BI has also been implicated in the efflux of cellular cholesterol. Ji et al. showed by overexpression studies in cultured macrophages that the level of cellular cholesterol efflux to mature HDL correlates with the expression level of SR-BI [10]. Subsequent studies showed that SR-BI stimulates the bidirectional flux of free cholesterol (FC) between cells and mature HDL down the concentration gradient [73]. SR-BI-deficient peritoneal macrophages loaded with cholesterol *in vivo* and subsequently subjected to cholesterol efflux *in vitro* showed 20% lower efflux of cholesterol to HDL as compared to wild-type macrophages [71]. The mechanisms underlying the SR-BI mediated bidirectional flux of cholesterol are largely unclear. Although not proven, it is often assumed that SR-BI accelerates aqueous diffusion of FC between cells and mature HDL. Binding to SR-BI will tether potential cholesterol acceptors in close proximity to the plasma membrane, thereby facilitating aqueous diffusion [74]. The extracellular domain of SR-BI is crucial for mediating the bidirectional flux of FC [75]. However, high-affinity binding alone is not sufficient to stimulate FC flux [76]. Interestingly, SR-BI also facilitates FC efflux by increasing the fraction of membrane cholesterol available for efflux [77]. Cholesterol in this faction of the membrane is not available for lipid-free/poor apoAI, although apoAI can bind to SR-BI. One study, using mutated forms of SR-BI, suggests that low-efficiency export of FC to HDL is related to low cholesterol availability in the plasma membrane rather than impaired binding of acceptors to the mutated forms of SR-BI [75]. In addition, SR-BI-mediated HDL retroendocytosis, that is, the uptake of whole HDL particles followed by resecretion of CE-poor HDL, has been implicated in SR-BI-mediated cholesterol efflux [78, 79]. This process appears independent of cholesterol transport out of the lysosome [78].

The mass of cholesterol derived from macrophage foam cells is only a tiny fraction of the overall flux of cholesterol from peripheral tissues. Mass-based assays can therefore not be used to trace such a small amount of cholesterol derived from macrophage foam cells. For this reason, Rader and colleagues [80] developed a method to trace cholesterol transport, specifically from macrophage foam cells to the feces [81]. Hereto, macrophages are labeled with [^{3}H]cholesterol in the presence of acetylated LDL and injected into the peritoneal cavity of mice. In time the appearance of the [^{3}H]tracer can be detected in the plasma, liver, bile, and faeces. This assay has increased the understanding of the mechanisms involved in the transport of cholesterol from macrophages to the feces. More importantly, the rate of macrophage reverse cholesterol transport (RCT) predicts atherosclerosis susceptibility better than the level of HDL-C in mice [82]. Using this model, however, no effect of macrophage SR-BI deficiency was found on macrophage RCT, while deletion of ATP-binding cassette transporter (ABC) A1 and ABCG1, two other transporters implicated in cellular cholesterol efflux did lead to a substantial decrease [83]. It must be noted that for the Rader *in vivo* RCT model, macrophages are cultured *in vitro* for at least 24 hours prior to injection into the recipient mice, which might have affected the expression of SR-BI on the cells.

Definitive proof for the importance of SR-BI for macrophage foam cell formation has come from *in vivo* studies in which SR-BI was specifically deleted in bone marrow-derived cells of LDL receptor KO mice on a high-fat/high-cholesterol diet. In agreement, with its role in the uptake of modified and native lipoproteins [71], deletion of SR-BI led to a tendency to reduced foam cell formation in the peritoneal cavity [84]. Deletion of SR-BI on top of an ABCA1 deletion, however, led to a massive increase in the amount of heavily lipid-laden macrophages in the peritoneum, and an almost 24-fold increase in the accumulation of CEs as compared to wild-type macrophages and a fivefold increase as compared with ABCA1 KO macrophages [84]. Thus, in line with the dual function of SR-BI in macrophage cholesterol homeostasis, SR-BI can either decrease or increase macrophage foam cell formation.

SR-BI AND ATHEROSCLEROSIS

Atherosclerosis is a chronic inflammatory reaction of the vascular wall in response to dyslipidemia and endothelial distress involving the inflammatory recruitment of leukocytes and the activation of resident vascular cells. Experiments using transgenic and KO mice have established that SR-BI protects against atherosclerosis. Disruption of SR-BI in wild-type as well as LDL receptor KO mice increases atherosclerosis, despite increased levels of circulating HDL [85, 86]. Further, when crossbred onto the apoE KO background, SR-BI-deficiency leads to severe cardiac dysfunction and premature death [34, 87]. Interestingly, subjects with the P297S SR-BI variant were also not protected against the development of atherosclerosis despite elevated HDL levels [39].

Tissue-specific inactivation of SR-BI in mice has dissected the discrete effects of SR-BI in different cell types on atherosclerosis susceptibility. Liver-specific knockdown of SR-BI results in the accumulation of abnormally large HDL particles and accelerated atherosclerosis [85]. Increased oxidative stress in SR-BI KO mice suggests that the abnormally large HDL particles are dysfunctional, which may also potentially promote atherogenesis [88]. Normalization of HDL levels by CETP cannot reverse the oxidative stress and reduce atherosclerosis in SR-BI KO mice [89]. Inhibition of the production of HDL by deletion of ABCA1, however, does reduce the atherosclerosis susceptibily of SR-BI KO mice, although the level of lesion development still remains substantially higher as compared with wild-type mice (Zhao et al., unpublished results). Transgene or adenovirus-mediated hepatic overexpression of SR-BI markedly reduced atherosclerosis [62, 90, 91]. The atheroprotective effect of hepatic SR-BI is largely attributed to its role in HDL-mediated RCT. Of note, hepatic SR-BI is also involved in the clearance of VLDL [59] and chylomicrons [63], which might affect atherosclerotic lesion development.

Extrahepatic SR-BI also protects against atherosclerosis, as evidenced by the finding that deletion of hepatic SR-BI results in less lesion formation as compared to total-body SR-BI deficiency [85]. Bone marrow transplantation studies indicate that macrophage SR-BI plays a unique dual role in the pathogenesis of atherosclerosis, depending on the stage of lesion development [92]. The development of early fatty streak lesions is facilitated by macrophage SR-BI [71], while it reduces the development of advanced atherosclerotic lesions [71, 86]. This unique dual role is likely a direct effect of the multifunctional and multiligand qualities of SR-BI, which binds a wide array of native and modified lipoproteins and mediates the bidirectional flux of cholesterol between HDL and cells [92].

ATHERPROTECTIVE FUNCTIONS OF SR-BI BEYOND LIPID METABOLISM

In addition to its essential role in macrophage lipid homeostasis, evidence is accumulating that SR-BI can also regulate the inflammatory response of macrophages. Macrophage SR-BI supresses TLR-4-mediated nuclear factor-κB (NF-κB) activation, thereby alleviating the inflammatory response to LPS [93]. A very recent study by Cai and colleagues showed a hyperinflammatory response to LPS in macrophages isolated from SR-BI KO mice and a reduced response in macrophages overexpressing SR-BI. Further, reconstitution of WT mice with SR-BI KO bone marrow led to an elevated *in vivo* cytokine response following LPS treatment, while the response was reduced in SR-BI KO mice upon transplantation with WT bone marrow. SR-BI in macrophages thus decreases the proinflammatory response to LPS both in isolated macrophages and *in vivo* [94]. The increased cytokine response in macrophages lacking SR-BI involved both NF-κB and JNK and P38MAPK cell signaling pathways. An increased membrane cholesterol content has been shown to affect the inflammatory response to LPS in ABCA1 KO macrophages [95, 96]. However, the hyperinflammatory response of SR-BI KO macrophages was not associated with alterations in cellular FC content or the level of cholesterol-rich lipid rafts in the plasma membrane. Although the underlying mechanism is not yet fully understood, the results of the study of Cai et al. provide strong evidence that SR-BI is an important modulator of the macrophage inflammatory response. Given the fact that macrophage inflammation is a key feature in the pathogenesis of atherosclerosis, SR-BI in macrophages may thus also protect against atherosclerotic lesion formation through modulating macrophage inflammatory responses.

Evidence is also accumulating indicating atheroprotective functions of SR-BI at sites other than the liver or macrophages. A recent study by Barth and colleagues showed that SR-BI is highly expressed on human monocyte-derived dendritic cells (DCs), the most potent antigen-presenting cells [97]. DCs have been identified in atherosclerotic plaques [98, 99] and are thought to influence atherogenesis by presenting antigens of, for example, oxidized LDL to T cells, leading to T-cell activation and the release of cytokines, which play an important role in the progression of disease. However, DC expansion alleviates atherosclerosis in mice, while depletion of DCs accelerates atherosclerosis, probably due to indirect effects on serum lipids [100]. Importantly, vaccination using mature dendritic cells pulsed with oxidized LDL reduces atherosclerotic lesion development [101]. Considering that SR-BI binds oxidized LDL, it is interesting to speculate that SR-BI on DCs is important for the atheroprotective effects of vaccination with oxidized LDL-pulsed DCs.

SR-BI might also modulate glucocorticoid production by the adrenal in response to inflammation in the arterial wall, and thereby influences the progression of the disease. Recently, van der Sluis and colleagues showed that adrenal-derived steroids indeed protect against the development of initial atherosclerotic lesions in LDL receptor KO mice [102]. SR-BI is highly expressed on adrenocortical cells in the adrenal [8]. In agreement, SR-BI deficiency leads to an impaired uptake of CEs from HDL [40] and a dramatic reduction in the adrenal cholesterol content [33]. Importantly, the impaired uptake of HDL CEs by the adrenal leads to altered steroidogenesis in response to stress triggers that activate the hypothalamus–pituitary–adrenal axis. A 50% lower plasma glucocorticoid level was observed in SR-BI KO mice upon fasting-induced stress [103, 104] and after a challenge with LPS [104, 105]. Importantly, also in humans, impaired SR-BI function leads to primary glucocorticoid insufficiency. Heterozygote carriers of the P297S SR-BI mutant showed a reduced urinary steroid excretion rate, while also the cortisol response to corticotropin stimulation was inhibited, despite unchanged plasma adrenocorticotropic hormone (ACTH) levels [39].

SR-BI is also expressed on endothelial cells (ECs) in the arterial wall and several lines of evidence have indicated that SR-BI on ECs is important for the atheroprotective effects of HDL. Endothelial SR-BI transports HDL through the endothelium by transcytosis to get access to the cells in the intima [106]. Importantly, during the transcytosis, the size of the HDL particle is reduced, but the protein moiety remains intact, which might increase the capacity of HDL as a cholesterol acceptor. Moreover, HDL prevents apoptosis of EC [107], promotes EC migration to maintain the integrity of the endothelium [108], and induces the activation of endothelial nitric oxide synthase (eNOS) via SR-BI [109, 110]. Further, SR-BI expression on endothelial progenitor cells plays an essential role in apoAI-induced accelerated endothelial regeneration [111]. Interestingly, simvastatin stimulates endothelial SR-BI expression, thereby enhancing the HDL-induced activation of NOS, reducing adhesion molecule expression, and inhibiting *ex vivo* monocyte adhesion in aortas from TNF-α-injected WT mice [112].

Platelets, which play an essential role in the development of atherothrombosis after plaque rupture, also express SR-BI [113]. *In vitro*, platelet SR-BI was shown to be essential for oxidized HDL-induced repression of platelet aggregation [114]. Further, SR-BI on platelets is a functional receptor for native HDL_3 that generates an inhibitory signal for platelet activation [115]. The content of negatively charged phospholipids in HDL may be an important determinant of their antithrombotic potential. SR-BI KO mice suffer from markedly reduced platelet counts [116]. However, this was not the consequence of direct effects of deletion of SR-BI in the platelets, but rather due to secondary effects of the high level of HDL-associated FC in the SR-BI KO mice. SR-BI KO platelets circulated in an activated state and showed increased adherence to immobilized fibrinogen. The activated state of the platelets *in vivo* led to desensitization of the receptors for ADP and thromboxane that are essential for positive feedback mechanisms to enhance the platelet response. As a result, *ex vivo* aggregation of platelets from SR-BI KO mice was inhibited in response to agonist stimulation as compared with control platelets. Importantly, the susceptibility of SR-BI KO mice to thrombosis is increased [117]. Platelet-specific SR-BI deficiency also protected mice from a prothrombotic phenotype during hypercholesterolemia [118], indicating that SR-BI on platelets also influences thrombosis susceptibility directly.

Interestingly, platelets from atherosclerotic disease patients express less SR-BI and the levels of SR-BI were inversely correlated with the intracellular cholesterol content of platelets and their ability to aggregate [113], suggesting that SR-BI in platelets is regulated during atherosclerotic lesion development in humans. The recent study by Vergeer et al. on subjects with a defective P297S variant was the first showing that dysfunctional SR-BI on platelets also result in abnormal platelet reactivity in humans [39]. Similar as in SR-BI KO mice, platelets from P297S subjects are enriched in FC, circulate in a more active state, display an augmented ability to spread on immobilized fibrinogen, and show a diminished aggregation response to different agonists.

CONCLUSION

Since its discovery in 1994 and the identification in 1996 as the receptor for HDL, SR-BI has been studied extensively in various *in vivo* and *in vitro* models. In particular, the generation of SR-BI KO mice by the laboratory of Monty

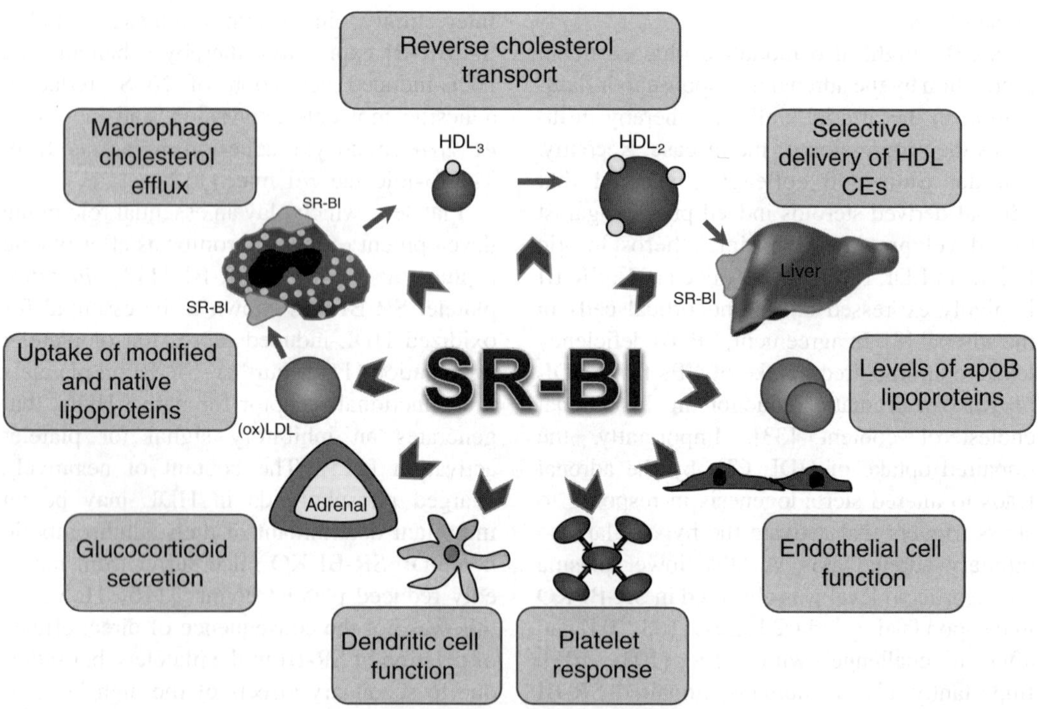

Figure 25.2 SR-BI functions affecting the pathogenesis of atherosclerosis. SR-BI facilitates the reverse cholesterol transport (RCT) pathway by promoting the HDL-mediated removal of excess cholesterol from macrophages in the arterial wall and subsequent delivery to the liver for biliary excretion. SR-BI in macrophages also mediates the uptake of native and modified lipoproteins, thereby promoting foam cell formation. Further, SR-BI directly or indirectly interferes with several other physiological processes that might affect the development of atherosclerosis, including adrenal steroidogenesis, platelet function, endothelial transport and integrity, and possibly dendritic cell function.

Krieger has provided valuable information on the physiological role of SR-BI. Studies using these mice have identified SR-BI as a multipurpose player in atherogenesis (Figure 25.2). It functions as the receptor for delivery of HDL CEs to both the liver and the adrenal. In macrophages, SR-BI not only plays an essential role in lipid homeostasis but also modulates cellular inflammatory functions. Moreover, evidence is accumulating that SR-BI also affects the function of endothelial cells, dendritic cells, and platelets. Most importantly, in the 20 years that have passed since the discovery of SR-BI, it became clear that SR-BI is essentially required for the atheroprotective function of HDL. Elevation of HDL-C is believed to be a promising strategy for treatment of atherosclerosis, as the levels of HDL correlate inversely with cardiovascular risk. However, the studies in SR-BI KO mice and humans with the dysfunctional P297S variant clearly show that the steady-state levels of HDL-C in the circulation poorly reflect HDL function. Clinical approaches to raise HDL-C to battle cardiovascular burden should thus carefully take SR-BI function into account. For example, HDL376, a newly identified HDL raising agent, was shown to inhibit SR-BI-mediated lipid transport [119]. Since at least part of the HDL cholesterol-raising activity of HDL376 may be due to inhibition of SR-BI, the development of this drug for the treatment of atherosclerosis should be continued with caution. Future identification of pharmacological enhancers of SR-BI and extensive testing in animal and clinical studies will hopefully show that targeting SR-BI is a valuable strategy for treating atherosclerosis.

ACKNOWLEDGMENTS

This work was supported by the Netherlands Heart Foundation [grants 2001T4101 (Y.Z.), 2007T056 (M.V.E.), and 2008T070 (M.H.)], the Landsteiner Foundation for Blood Transfusion Research (0912F to S.J.A.K.), and The Netherlands Organization for Scientific Research [VICI Grant 91813603 (M.V.E)].

REFERENCES

1. Goldstein, J.L., Ho, Y.K., Basu, S.K., Brown, M.S. (1979) Binding site on macrophages that mediates uptake and degradation of acetylated low density lipoprotein, producing massive cholesterol deposition. *Proc Natl Acad Sci U S A*. 76, 333–337.
2. Brown, M.S., Goldstein, J.L., Krieger, M., Ho, Y.K., Anderson, R.G. (1979) Reversible accumulation of cholesteryl esters in macrophages incubated with acetylated lipoproteins. *J Cell Biol*. 82, 597–613.
3. Krieger, M. (1997) The other side of scavenger receptors: pattern recognition for host defense. *Curr Opin Lipidol*. 8, 275–280.
4. Kzhyshkowska, J., Neyen, C., Gordon, S. (2012) Role of macrophage scavenger receptors in atherosclerosis. *Immunobiology*. 217, 492–502.
5. Acton, S.L., Scherer, P.E., Lodish, H.F., Krieger, M. (1994) Expression cloning of SR-BI, a CD36-related class B scavenger receptor. *J Biol Chem*. 269, 21003–21009.
6. Babitt, J., Trigatti, B., Rigotti, A., et al. (1997) Murine SR-BI, a high density lipoprotein receptor that mediates selective lipid uptake, is N-glycosylated and fatty acylated and colocalizes with plasma membrane caveolae. *J Biol Chem*. 272, 13242–13249.
7. Webb, N.R., de Villiers, W.J., Connell, P.M., de Beer, F.C., van der Westhuyzen, D.R. (1997) Alternative forms of the scavenger receptor BI (SR-BI). *J Lipid Res*. 38, 1490–1495.
8. Acton, S., Rigotti, A., Landschulz, K.T., Xu, S., Hobbs, H.H., Krieger, M. (1996) Identification of scavenger receptor SR-BI as a high density lipoprotein receptor. *Science*. 271, 518–520.
9. Rigotti, A., Miettinen, H.E., Krieger, M. (2003) The role of the high-density lipoprotein receptor SR-BI in the lipid metabolism of endocrine and other tissues. *Endocr Rev*. 24, 357–387.
10. Ji, Y., Jian, B., Wang, N., et al. (1997) Scavenger receptor BI promotes high density lipoprotein-mediated cellular cholesterol efflux. *J Biol Chem*. 272, 20982–20985.
11. Calvo, D., Vega, M.A. (1993) Identification, primary structure, and distribution of CLA-1, a novel member of the CD36/LIMPII gene family. *J Biol Chem*. 268, 18929–18935.
12. Rigotti, A., Acton, S.L., Krieger, M. (1995) The class B scavenger receptors SR-BI and CD36

are receptors for anionic phospholipids. *J Biol Chem*. 270, 16221–16224.

13. Ohgami, N., Nagai, R., Miyazaki, A., et al. (2001) Scavenger receptor class B type I-mediated reverse cholesterol transport is inhibited by advanced glycation end products. *J Biol Chem*. 276, 13348–13355.
14. Murao, K., Terpstra, V., Green, S.R., Kondratenko, N., Steinberg, D., Quehenberger, O. (1997) Characterization of CLA-1, a human homologue of rodent scavenger receptor BI, as a receptor for high density lipoprotein and apoptotic thymocytes. *J Biol Chem*. 272, 17551–17557.
15. Calvo, D., Gomez-Coronado, D., Lasuncion, M.A., Vega, M.A. (1997) CLA-1 is an 85-kD plasma membrane glycoprotein that acts as a high-affinity receptor for both native (HDL, LDL, and VLDL) and modified (OxLDL and AcLDL) lipoproteins. *Arterioscler Thromb Vasc Biol*. 17, 2341–2349.
16. Stangl, H., Hyatt, M., Hobbs, H.H. (1999) Transport of lipids from high and low density lipoproteins via scavenger receptor-BI. *J Biol Chem*. 274, 32692–32698.
17. Gu, X., Lawrence, R., Krieger, M. (2000) Dissociation of the high density lipoprotein and low density lipoprotein binding activities of murine scavenger receptor class B type I (mSR-BI) using retrovirus library-based activity dissection. *J Biol Chem*. 275, 9120–9130.
18. Lopez, D., McLean, M.P. (1999) Sterol regulatory element-binding protein-1a binds to cis elements in the promoter of the rat high density lipoprotein receptor SR-BI gene. *Endocrinology*. 140, 5669–5681.
19. Niemeier, A., Kovacs, W.J., Strobl, W., Stangl, H. (2009) Atherogenic diet leads to posttranslational down-regulation of murine hepatocyte SR-BI expression. *Atherosclerosis*. 202, 169–175.
20. Martin, G., Pilon, A., Albert, C., et al. (1999) Comparison of expression and regulation of the high-density lipoprotein receptor SR-BI and the low-density lipoprotein receptor in human adrenocortical carcinoma NCI-H295 cells. *Eur J Biochem*. 261, 481–491.
21. Schoonjans, K., Annicotte, J.S., Huby, T., et al. (2002) Liver receptor homolog 1 controls the expression of the scavenger receptor class B type I. *EMBO Rep*. 3, 1181–1187.
22. Chinetti, G., Gbaguidi, F.G., Griglio, S., et al. (2000) CLA-1/SR-BI is expressed in atherosclerotic lesion macrophages and regulated by activators of peroxisome proliferator-activated receptors. *Circulation*. 101, 2411–2417.
23. Mardones, P., Pilon, A., Bouly, M., et al. (2003) Fibrates down-regulate hepatic scavenger receptor class B type I protein expression in mice. *J Biol Chem*. 278, 7884–7890.
24. Fu, T., Kozarsky, K.F., Borensztajn, J. (2003) Overexpression of SR-BI by adenoviral vector reverses the fibrateinduced hypercholesterolemia of apolipoprotein E-deficient mice. *J Biol Chem*. 278, 52559–52563.
25. Wood, P., Mulay, V., Darabi, M., et al. (2011) Ras/Mitogen-activated protein kinase (MAPK) signaling modulates protein stability and cell surface expression of scavenger receptor SR-BI. *J Biol Chem*. 286, 23077–23092.
26. Kocher, O., Yesilaltay, A., Cirovic, C., Pal, R., Rigotti, A., Krieger, M. (2003) Targeted disruption of the PDZK1 gene in mice causes tissue-specific depletion of the high density lipoprotein receptor scavenger receptor class B type I and altered lipoprotein metabolism. *J Biol Chem*. 278, 52820–52825.
27. Silver, D.L., Wang, N., Vogel, S. (2003) Identification of small PDZK1-associated protein, DD96/MAP17, as a regulator of PDZK1 and plasma high density lipoprotein levels. *J Biol Chem*. 278, 28528–28532.
28. Ueda, Y., Royer, L., Gong, E., et al. (1999) Lower plasma levels and accelerated clearance of high density lipoprotein (HDL) and non-HDL cholesterol in scavenger receptor class B type I transgenic mice. *J Biol Chem*. 274, 7165–7171.
29. Ji, Y., Wang, N., Ramakrishnan, R., et al. (1999) Hepatic scavenger receptor BI promotes rapid clearance of high density lipoprotein free cholesterol and its transport into bile. *J Biol Chem*. 274, 33398–33402.
30. Kozarsky, K.F., Donahee, M.H., Rigotti, A., Iqbal, S.N., Edelman, E.R., Krieger, M. (1997) Overexpression of the HDL receptor SR-BI alters plasma HDL and bile cholesterol levels. *Nature*. 387, 414–417.
31. Wang, N., Arai, T., Ji, Y., Rinninger, F., Tall, A.R. (1998) Liver-specific overexpression of scavenger receptor BI decreases levels of very low density lipoprotein ApoB, low density lipoprotein ApoB, and high density lipoprotein in transgenic mice. *J Biol Chem*. 273, 32920–32926.

32. Varban, M.L., Rinninger, F., Wang, N., et al. (1998) Targeted mutation reveals a central role for SR-BI in hepatic selective uptake of high density lipoprotein cholesterol. *Proc Natl Acad Sci U S A*. 95, 4619–4624.
33. Rigotti, A., Trigatti, B.L., Penman, M., Rayburn, H., Herz, J., Krieger, M. (1997) A targeted mutation in the murine gene encoding the high density lipoprotein (HDL) receptor scavenger receptor class B type I reveals its key role in HDL metabolism. *Proc Natl Acad Sci U S A*. 94, 12610–12615.
34. Trigatti, B., Rayburn, H., Vinals, M., et al. (1999) Influence of the high density lipoprotein receptor SR-BI on reproductive and cardiovascular pathophysiology. *Proc Natl Acad Sci U S A*. 96, 9322–9327.
35. Acton, S., Osgood, D., Donoghue, M., et al. (1999) Association of polymorphisms at the SR-BI gene locus with plasma lipid levels and body mass index in a white population. *Arterioscler Thromb Vasc Biol*. 19, 1734–1743.
36. Hsu, L.A., Ko, Y.L., Wu, S., et al. (2003) Association between a novel 11-base pair deletion mutation in the promoter region of the scavenger receptor class B type I gene and plasma HDL cholesterol levels in Taiwanese Chinese. *Arterioscler Thromb Vasc Biol*. 23, 1869–1874.
37. Roberts, C.G., Shen, H., Mitchell, B.D., Damcott, C.M., Shuldiner, A.R., Rodriguez, A. (2007) Variants in scavenger receptor class B type I gene are associated with HDL cholesterol levels in younger women. *Hum Hered*. 64, 107–113.
38. Teslovich, T.M., Musunuru, K., Smith, A.V., et al. (2010) Biological, clinical and population relevance of 95 loci for blood lipids. *Nature*. 466, 707–713.
39. Vergeer, M., Korporaal, S.J., Franssen, R., et al. (2011) Genetic variant of the scavenger receptor BI in humans. *N Engl J Med*. 364, 136–145.
40. Out, R., Hoekstra, M., Spijkers, J.A., et al. (2004) Scavenger receptor class B type I is solely responsible for the selective uptake of cholesteryl esters from HDL by the liver and the adrenals in mice. *J Lipid Res*. 45, 2088–2095.
41. Brundert, M., Ewert, A., Heeren, J., et al. (2005) Scavenger receptor class B type I mediates the selective uptake of high-density lipoprotein-associated cholesteryl ester by the liver in mice. *Arterioscler Thromb Vasc Biol*. 25, 143–148.
42. Gu, X., Trigatti, B., Xu, S., Acton, S., Babitt, J., Krieger, M. (1998) The efficient cellular uptake of high density lipoprotein lipids via scavenger receptor class B type I requires not only receptor-mediated surface binding but also receptor-specific lipid transfer mediated by its extracellular domain. *J Biol Chem*. 273, 26338–26348.
43. Temel, R.E., Walzem, R.L., Banka, C.L., Williams, D.L. (2002) Apolipoprotein A-I is necessary for the in vivo formation of high density lipoprotein competent for scavenger receptor BI-mediated cholesteryl ester-selective uptake. *J Biol Chem*. 277, 26565–26572.
44. Temel, R.E., Parks, J.S., Williams, D.L. (2003) Enhancement of scavenger receptor class B type I-mediated selective cholesteryl ester uptake from apoA-I(-/-) high density lipoprotein (HDL) by apolipoprotein A-I requires HDL reorganization by lecithin cholesterol acyltransferase. *J Biol Chem*. 278, 4792–4799.
45. Thuahnai, S.T., Lund-Katz, S., Dhanasekaran, P., et al. (2004) Scavenger receptor class B type I-mediated cholesteryl ester-selective uptake and efflux of unesterified cholesterol. Influence of high density lipoprotein size and structure. *J Biol Chem*. 279, 12448–12455.
46. Papale, G.A., Nicholson, K., Hanson, P.J., Pavlovic, M., Drover, V.A., Sahoo, D. (2010) Extracellular hydrophobic regions in scavenger receptor BI play a key role in mediating HDL-cholesterol transport. *Arch Biochem Biophys*. 496, 132–139.
47. Papale, G.A., Hanson, P.J., Sahoo, D. (2011) Extracellular disulfide bonds support scavenger receptor class B type I-mediated cholesterol transport. *Biochemistry*. 50, 6245–6254.
48. Rigotti, A., Trigatti, B., Babitt, J., Penman, M., Xu, S., Krieger, M. (1997) Scavenger receptor BI—a cell surface receptor for high density lipoprotein. *Curr Opin Lipidol*. 8, 181–188.
49. Connelly, M.A., Klein, S.M., Azhar, S., Abumrad, N.A., Williams, D.L. (1999) Comparison of class B scavenger receptors, CD36 and scavenger receptor BI (SR-BI), shows that both receptors mediate high density lipoprotein-cholesteryl ester selective uptake but SR-BI exhibits a unique enhancement of cholesteryl ester uptake. *J Biol Chem*. 274, 41–47.

50. Vinals, M., Xu, S., Vasile, E., Krieger, M. (2003) Identification of the N-linked glycosylation sites on the high density lipoprotein (HDL) receptor SR-BI and assessment of their effects on HDL binding and selective lipid uptake. *J Biol Chem*. 278, 5325–5332.
51. Nieland, T.J., Penman, M., Dori, L., Krieger, M., Kirchhausen, T. (2002) Discovery of chemical inhibitors of the selective transfer of lipids mediated by the HDL receptor SR-BI. *Proc Natl Acad Sci U S A*. 99, 15422–15427.
52. Liu, B., Krieger, M. (2002) Highly purified scavenger receptor class B, type I reconstituted into phosphatidylcholine/cholesterol liposomes mediates high affinity high density lipoprotein binding and selective lipid uptake. *J Biol Chem*. 277, 34125–34135.
53. Silver, D.L., Wang, N., Xiao, X., Tall, A.R. (2001) High density lipoprotein (HDL) particle uptake mediated by scavenger receptor class B type 1 results in selective sorting of HDL cholesterol from protein and polarized cholesterol secretion. *J Biol Chem*. 276, 25287–25293.
54. Rhainds, D., Bourgeois, P., Bourret, G., Huard, K., Falstrault, L., Brissette, L. (2004) Localization and regulation of SR-BI in membrane rafts of HepG2 cells. *J Cell Sci*. 117, 3095–3105.
55. Sparrow, C.P., Pittman, R.C. (1990) Cholesterol esters selectively taken up from high-density lipoproteins are hydrolyzed extralysosomally. *Biochim Biophys Acta*. 1043, 203–210.
56. Shimada, A., Tamai, T., Oida, K., et al. (1994) Increase in neutral cholesteryl ester hydrolase activity produced by extralysosomal hydrolysis of high-density lipoprotein cholesteryl esters in rat hepatoma cells (H-35). *Biochim Biophys Acta*. 1215, 126–132.
57. Delamatre, J.G., Carter, R.M., Hornick, C.A. (1993) Evidence that a neutral cholesteryl ester hydrolase is responsible for the extralysosomal hydrolysis of high-density lipoprotein cholesteryl ester in rat hepatoma cells (Fu5AH). *J Cell Physiol*. 157, 164–168.
58. Wiersma, H., Nijstad, N., Gautier, T., et al. (2010) Scavenger receptor BI facilitates hepatic very low density lipoprotein production in mice. *J Lipid Res*. 51, 544–553.
59. Van Eck, M., Hoekstra, M., Out, R., et al. (2008) Scavenger receptor BI facilitates the metabolism of VLDL lipoproteins in vivo. *J Lipid Res*. 49, 136–146.
60. Rohrl, C., Fruhwurth, S., Schreier, S.M., et al. (2010) Scavenger receptor, Class B, Type I provides an alternative means for beta-VLDL uptake independent of the LDL receptor in tissue culture. *Biochim Biophys Acta*. 1801, 198–204.
61. Huszar, D., Varban, M.L., Rinninger, F., et al. (2000) Increased LDL cholesterol and atherosclerosis in LDL receptor-deficient mice with attenuated expression of scavenger receptor B1. *Arterioscler Thromb Vasc Biol*. 20, 1068–1073.
62. Kozarsky, K.F., Donahee, M.H., Glick, J.M., Krieger, M., Rader, D.J. (2000) Gene transfer and hepatic overexpression of the HDL receptor SR-BI reduces atherosclerosis in the cholesterol-fed LDL receptor-deficient mouse. *Arterioscler Thromb Vasc Biol*. 20, 721–727.
63. Out, R., Kruijt, J.K., Rensen, P.C., et al. (2004) Scavenger receptor BI plays a role in facilitating chylomicron metabolism. *J Biol Chem*. 279, 18401–18406.
64. Béaslas, O., Cueille, C., Delers, F., et al. (2009) Sensing of dietary lipids by enterocytes: a new role for SR-BI/CLA-1. *PLoS One*. e4278.
65. Bietrix, F., Yan, D., Nauze, M., et al. (2006) Accelerated lipid absorption in mice overexpressing intestinal SR-BI. *J Biol Chem*. 281, 7214–7219.
66. Perez-Martinez, P., Ordovas, J.M., Lopez-Miranda, J., et al. (2003) Polymorphism exon 1 variant at the locus of the scavenger receptor class B type I gene: influence on plasma LDL cholesterol in healthy subjects during the consumption of diets with different fat contents. *Am J Clin Nutr*. 77, 809–813.
67. Osgood, D., Corella, D., Demissie, S., et al. (2003) Genetic variation at the scavenger receptor class B type I gene locus determines plasma lipoprotein concentrations and particle size and interacts with type 2 diabetes: the Framingham study. *J Clin Endocrinol Metab*. 88, 2869–2879.
68. Morabia, A., Ross, B.M., Costanza, M.C., et al. (2004) Population-based study of SR-BI genetic variation and lipid profile. *Atherosclerosis*. 175, 159–168.
69. Perez-Martinez, P., Lopez-Miranda, J., Ordovas, J.M., et al. (2004) Postprandial lipemia is modified by the presence of the polymorphism present in the exon 1 variant at the SR-BI gene locus. *J Mol Endocrinol*. 32, 237–245.

70. Hirano, K., Yamashita, S., Nakagawa, Y., et al. (1999) Expression of human scavenger receptor class B type I in cultured human monocyte-derived macrophages and atherosclerotic lesions. *Circ Res*. 85, 108–116.
71. Van Eck, M., Bos, I.S., Hildebrand, R.B., Van Rij, B.T., Van Berkel, T.J. (2004) Dual role for scavenger receptor class B, type I on bone marrow-derived cells in atherosclerotic lesion development. *Am J Pathol*. 165, 785–794.
72. Ahras, M., Naing, T., McPherson, R. (2008) Scavenger receptor class B type I localizes to a late endosomal compartment. *J Lipid Res*. 49, 1569–1576.
73. Yancey, P.G., Llera-Moya, M., Swarnakar, S., et al. (2000) High density lipoprotein phospholipid composition is a major determinant of the bi-directional flux and net movement of cellular free cholesterol mediated by scavenger receptor BI. *J Biol Chem*. 275, 36596–36604.
74. Gu, X., Kozarsky, K., Krieger, M. (2000) Scavenger receptor class B, type I-mediated [3H]cholesterol efflux to high and low density lipoproteins is dependent on lipoprotein binding to the receptor. *J Biol Chem*. 275, 29993–30001.
75. Connelly, M.A., Llera-Moya, M., Monzo, P., et al. (2001) Analysis of chimeric receptors shows that multiple distinct functional activities of scavenger receptor, class B, type I (SR-BI), are localized to the extracellular receptor domain. *Biochemistry*. 40, 5249–5259.
76. Favari, E., Calabresi, L., Adorni, M.P., et al. (2009) Small discoidal pre-beta1 HDL particles are efficient acceptors of cell cholesterol via ABCA1 and ABCG1. *Biochemistry*. 48, 11067–11074.
77. Llera-Moya, M., Rothblat, G.H., Connelly, M.A., et al. (1999) Scavenger receptor BI (SR-BI) mediates free cholesterol flux independently of HDL tethering to the cell surface. *J Lipid Res*. 40, 575–580.
78. Pagler, T.A., Neuhofer, A., Laggner, H., Strobl, W., Stangl, H. (2007) Cholesterol efflux via HDL resecretion occurs when cholesterol transport out of the lysosome is impaired. *J Lipid Res*. 48, 2141–2150.
79. Pagler, T.A., Rhode, S., Neuhofer, A., et al. (2006) SR-BI-mediated high density lipoprotein (HDL) endocytosis leads to HDL resecretion facilitating cholesterol efflux. *J Biol Chem*. 281, 11193–11204.
80. Zhang Y, Zanotti I, Reilly MP, Glick JM, Rothblat GH, Rader DJ. (2003) Overexpression of apolipoprotein A-I promotes reverse transport of cholesterol from macrophages to feces in vivo. *Circulation*. 108(6), 661–663.
81. Cuchel, M., Rader, D.J. (2006) Macrophage reverse cholesterol transport: key to the regression of atherosclerosis? *Circulation*. 113, 2548–2555.
82. Rader, D.J., Alexander, E.T., Weibel, G.L., Billheimer, J., Rothblat, G.H. (2009) The role of reverse cholesterol transport in animals and humans and relationship to atherosclerosis. *J Lipid Res*. 50 (Suppl), S189–S194.
83. Wang, X., Collins, H.L., Ranalletta, M., et al. (2007) Macrophage ABCA1 and ABCG1, but not SR-BI, promote macrophage reverse cholesterol transport in vivo. *J Clin Invest*. 117, 2216–2224.
84. Zhao, Y., Pennings, M., Hildebrand, R.B., et al. (2010) Enhanced foam cell formation, atherosclerotic lesion development, and inflammation by combined deletion of ABCA1 and SR-BI in Bone marrow-derived cells in LDL receptor knockout mice on western-type diet. *Circ Res*. 107, e20–e31.
85. Huby, T., Doucet, C., Dachet, C., et al. (2006) Knockdown expression and hepatic deficiency reveal an atheroprotective role for SR-BI in liver and peripheral tissues. *J Clin Invest*. 116, 2767–2776.
86. Covey, S.D., Krieger, M., Wang, W., Penman, M., Trigatti, B.L. (2003) Scavenger receptor class B type I-mediated protection against atherosclerosis in LDL receptor-negative mice involves its expression in bone marrow-derived cells. *Arterioscler Thromb Vasc Biol*. 23, 1589–1594.
87. Braun, A., Trigatti, B.L., Post, M.J., et al. (2002) Loss of SR-BI expression leads to the early onset of occlusive atherosclerotic coronary artery disease, spontaneous myocardial infarctions, severe cardiac dysfunction, and premature death in apolipoprotein E-deficient mice. *Circ Res*. 90, 270–276.
88. Van Eck, M., Hoekstra, M., Hildebrand, R.B., et al. (2007) Increased oxidative stress in scavenger receptor BI knockout mice with dysfunctional HDL. *Arterioscler Thromb Vasc Biol*. 27, 2413–2419.
89. Hildebrand, R.B., Lammers, B., Meurs, I., et al. (2010) Restoration of high-density lipoprotein levels by cholesteryl ester transfer protein

expression in scavenger receptor class B type I (SR-BI) knockout mice does not normalize pathologies associated with SR-BI deficiency. *Arterioscler Thromb Vasc Biol.* 30, 1439–1445.

90. Arai, T., Wang, N., Bezouevski, M., Welch, C., Tall, A.R. (1999) Decreased atherosclerosis in heterozygous low density lipoprotein receptor-deficient mice expressing the scavenger receptor BI transgene. *J Biol Chem.* 274, 2366–2371.
91. Ueda, Y., Gong, E., Royer, L., Cooper, P.N., Francone, O.L., Rubin, E.M. (2000) Relationship between expression levels and atherogenesis in scavenger receptor class B, type I transgenics. *J Biol Chem.* 275, 20368–20373.
92. Van Eck, M., Pennings, M., Hoekstra, M., Out, R., Van Berkel, T.J. (2005) Scavenger receptor BI and ATP-binding cassette transporter A1 in reverse cholesterol transport and atherosclerosis. *Curr Opin Lipidol.* 16, 307–315.
93. Guo, L., Song, Z., Li, M., et al. (2009) Scavenger receptor BI protects against septic death through its role in modulating inflammatory response. *J Biol Chem.* 284, 19826–19834.
94. Cai, L., Wang, Z., Ji, A., Meyer, J.M., van der Westhuyzen, D.R. (2012) Macrophage SR-BI regulates pro-inflammatory signaling in mice and isolated macrophages. *J Lipid Res.* 53, 1472–1481.
95. Zhu, X., Lee, J.Y., Timmins, J.M., et al. (2008) Increased cellular free cholesterol in macrophage-specific Abca1 knock-out mice enhances pro-inflammatory response of macrophages. *J Biol Chem.* 283, 22930–22941.
96. Yvan-Charvet, L., Welch, C., Pagler, T.A., et al. (2008) Increased inflammatory gene expression in ABC transporter-deficient macrophages: free cholesterol accumulation, increased signaling via toll-like receptors, and neutrophil infiltration of atherosclerotic lesions. *Circulation.* 118, 1837–1847.
97. Barth, H., Schnober, E.K., Neumann-Haefelin, C., et al. (2008) Scavenger receptor class B is required for hepatitis C virus uptake and cross-presentation by human dendritic cells. *J Virol.* 82, 3466–3479.
98. Bobryshev, Y.V., Taksir, T., Lord, R.S., Freeman, M.W. (2001) Evidence that dendritic cells infiltrate atherosclerotic lesions in apolipoprotein E-deficient mice. *Histol Histopathol.* 16, 801–808.
99. Ozmen, J., Bobryshev, Y.V., Lord, R.S., Ashwell, K.W. (2002) Identification of dendritic cells in aortic atherosclerotic lesions in rats with diet-induced hypercholesterolaemia. *Histol Histopathol.* 17, 223–237.
100. Gautier, E.L., Huby, T., Saint-Charles, F., et al. (2009) Conventional dendritic cells at the crossroads between immunity and cholesterol homeostasis in atherosclerosis. *Circulation.* 119, 2367–2375.
101. Habets, K.L., van Puijvelde, G.H., van Duivenvoorde, L.M., et al. (2010) Vaccination using oxidized low-density lipoprotein-pulsed dendritic cells reduces atherosclerosis in LDL receptor-deficient mice. *Cardiovasc Res.* 85, 622–630.
102. van der Sluis, R.J., van Puijvelde, G.H., Van Berkel, T.J., Hoekstra, M. (2012) Adrenalectomy stimulates the formation of initial atherosclerotic lesions: reversal by adrenal transplantation. *Atherosclerosis.* 221, 76–83.
103. Hoekstra, M., Meurs, I., Koenders, M., et al. (2008) Absence of HDL cholesteryl ester uptake in mice via SR-BI impairs an adequate adrenal glucocorticoid-mediated stress response to fasting. *J Lipid Res.* 49, 738–745.
104. Hoekstra, M., Ye, D., Hildebrand, R.B., et al. (2009) Scavenger receptor class B type I-mediated uptake of serum cholesterol is essential for optimal adrenal glucocorticoid production. *J Lipid Res.* 50, 1039–1046.
105. Cai, L., Ji, A., de Beer, F.C., Tannock, L.R., van der Westhuyzen, D.R. (2008) SR-BI protects against endotoxemia in mice through its roles in glucocorticoid production and hepatic clearance. *J Clin Invest.* 118, 364–375.
106. Rohrer, L., Ohnsorg, P.M., Lehner, M., Landolt, F., Rinninger, F., von Eckardstein, A. (2009) High-density lipoprotein transport through aortic endothelial cells involves scavenger receptor BI and ATP-binding cassette transporter G1. *Circ Res.* 104, 1142–1150.
107. Li, X.A., Guo, L., Dressman, J.L., Asmis, R., Smart, E.J. (2005) A novel ligand-independent apoptotic pathway induced by scavenger receptor class B, type I and suppressed by endothelial nitric-oxide synthase and high density lipoprotein. *J Biol Chem.* 280, 19087–19096.
108. Seetharam, D., Mineo, C., Gormley, A.K., et al. (2006) High-density lipoprotein promotes endothelial cell migration and reendothelialization

via scavenger receptor-B type I. *Circ Res*. 98, 63–72.

109. Yuhanna, I.S., Zhu, Y., Cox, B.E., et al. (2001) High-density lipoprotein binding to scavenger receptor-BI activates endothelial nitric oxide synthase. *Nat Med*. 7, 853–857.
110. Li, X.A., Titlow, W.B., Jackson, B.A., et al. (2002) High density lipoprotein binding to scavenger receptor, Class B, type I activates endothelial nitric-oxide synthase in a ceramide-dependent manner. *J Biol Chem*. 277, 11058–11063.
111. Feng, Y., van Eck, M., Van Craeyveld, E., et al. (2009) Critical role of scavenger receptor-BI-expressing bone marrow-derived endothelial progenitor cells in the attenuation of allograft vasculopathy after human apo A-I transfer. *Blood*. 113, 755–764.
112. Kimura, T., Tomura, H., Mogi, C., et al. (2006) Role of scavenger receptor class B type I and sphingosine 1-phosphate receptors in high density lipoprotein-induced inhibition of adhesion molecule expression in endothelial cells. *J Biol Chem*. 281, 37457–37467.
113. Imachi, H., Murao, K., Cao, W., et al. (2003) Expression of human scavenger receptor B1 on and in human platelets. *Arterioscler Thromb Vasc Biol*. 23, 898–904.
114. Valiyaveettil, M., Kar, N., Ashraf, M.Z., Byzova, T.V., Febbraio, M., Podrez, E.A. (2008) Oxidized high-density lipoprotein inhibits platelet activation and aggregation via scavenger receptor BI. *Blood*. 111, 1962–1971.
115. Brodde, M.F., Korporaal, S.J., Herminghaus, G., et al. (2011) Native high-density lipoproteins inhibit platelet activation via scavenger receptor BI: role of negatively charged phospholipids. *Atherosclerosis*. 215, 374–382.
116. Dole, V.S., Matuskova, J., Vasile, E., et al. (2008) Thrombocytopenia and platelet abnormalities in high-density lipoprotein receptor-deficient mice. *Arterioscler Thromb Vasc Biol*. 28, 1111–1116.
117. Korporaal, S.J., Meurs, I., Hauer, A.D., et al. (2011) Deletion of the high-density lipoprotein receptor scavenger receptor BI in mice modulates thrombosis susceptibility and indirectly affects platelet function by elevation of plasma free cholesterol. *Arterioscler Thromb Vasc Biol*. 31, 34–42.
118. Ma, Y., Ashraf, M.Z., Podrez, E.A. (2010) Scavenger receptor BI modulates platelet reactivity and thrombosis in dyslipidemia. *Blood*. 116, 1932–1941.
119. Nieland, T.J., Shaw, J.T., Jaipuri, F.A., et al. (2007) Influence of HDL-cholesterol-elevating drugs on the in vitro activity of the HDL receptor SR-BI. *J Lipid Res*. 48, 1832–1845.

26

PRO AND ANTI-INFLAMMATORY CYTOKINE NETWORKS IN ATHEROSCLEROSIS

MICHAEL V. AUTIERI

Department of Physiology, Cardiovascular Research Center, Temple University School of Medicine, Philadelphia, PA, USA

CYTOKINES, INFLAMMATION, AND ATHEROSCLEROSIS

Atherosclerosis is a chronic, progressive, dynamic disease initiated by an excess of serum concentrations of low-density lipoprotein that leads to local inflammation of large- and medium-sized arteries [1]. This inflammation is mediated by numerous cell types: lymphocytes, macrophage, endothelial, and vascular smooth muscle cells, among others that communicate with each other through a series of cytokine-receptor-mediated interactions [2]. This communication allows bidirectional cross-talk between resident vascular cells and inflammatory cells [2]. For effective communication to take place, these disparate cell types evolved a common set of soluble cytokine ligands and specific membrane receptors allowing them to transmit their effects into the cell. It is essential to recognize that most cytokines initiate a complex and varied repertoire of responses on their target cells and as such, can initiate, promote, and potentially resolve atherogenic inflammation. It is also important to appreciate that the microenvironment of the atherosclerotic plaque is a dynamic collection of resident vascular and infiltrating inflammatory cells and their cytokine products. This complex microenvironment results in a complex milieu consisting of multiple cytokines with redundant and opposing effects, and the balance of pro- and anti-inflammatory cytokines often determine plaque severity and stability. Cells present in the lesion need to interrogate and respond to these factors in an appropriate, more often, maladaptive fashion, which results in the development and progression of atherosclerosis (Figure 26.1). This introduces an important caveat that our current understanding of the role of any individual cytokine in this disease should be tempered by the fact that scientists are by necessity experimental reductionists; we need to appropriately control our experiments. Addition of a single cytokine to cultured cells, genetic manipulation of mice

Atherosclerosis: Risks, Mechanisms, and Therapies, First Edition. Edited by Hong Wang and Cam Patterson.

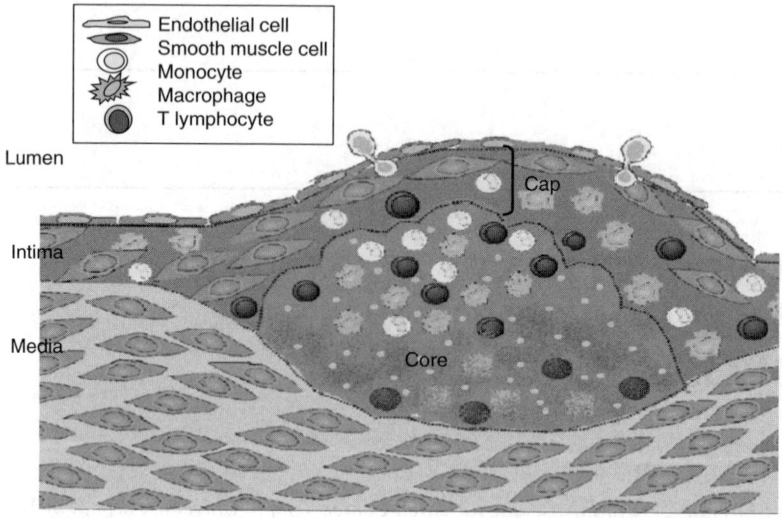

Figure 26.1 Cellular constituents of an atherosclerotic lesion. The microenvironment of the atherosclerotic lesion is a dynamic collection of resident vascular and infiltrating inflammatory cells and their cytokine products. These cells include endothelial, vascular smooth muscle, lymphocyte, monocyte, and macrophage. Other cells that participate here but are not shown include mast and dendritic cells. Driven by a chemotactic gradient, leukocytes exit the circulation and extravasated into the media and developing plaque where they synthesize numerous cytokines. This drives medial to intimal VSMC migration, resulting in an inward expansion of the intima and development of a core, which is sequestered from the lumen by a fibrous cap comprising VSMC and fibrous matrix. The core contains immune cells, esterified cholesterol, and debris from dead and dying cells in a hypoxic environment. This cytokine-rich milieu provides the driving mechanism for continued localized inflammation and progression of the atherosclerotic plaque.

one cytokine at a time does not accurately reflect the *in vivo* scenario where no single cytokine exists or functions in a vacuum. It is more appropriate to think in terms of cytokine networks that are temporally expressed and temporally functional.

CYTOKINE CLASSIFICATIONS

By current consensus, there are six families of cytokines, loosely categorized-based on structure and function. This includes interleukins or ILs (with several subfamilies based on peptide and receptor homology), interferons (IFNs) tumor necrosis factor family (TNF), chemokines, growth factors, and colony-stimulating factors (CSF) [3]. There is abundant pleiotropism and overlap between these factors, and other categorization strategies have been utilized. Alternatively, cytokines can be categorized based on structural homology of their receptors, in which almost all of the above families of cytokines can be grouped into class I or class II [4]. They can further be distinguished by the more broad functional categorization of "pro" or "anti"-inflammatory. Using these classifications, for example, IL-1β can belong to the IL, proinflammatory, and also the class II cytokine family (Table 26.1). Cytokines can also be classified according to their effects on lymphocyte function or maturation as Th1 (T helper), which promote the inflammatory response, and Th2 (T regulatory), which generally resolve inflammation [5]. When we categorize a cytokine as "proatherogenic," we most often are describing a proinflammatory mediator. However, it should be appreciated that a Th1 and Th2 IL may not strictly function directly as a definitive pro- or antiatherosclerotic cytokine, especially considering the complex

TABLE 26.1 Cytokines in atherosclerosis: families, classifications, and signaling

Proatherogenic cytokines	Inflammatory	Class	Cytokine family	Leukocyte phenotype	Detected in plaque	Primary signaling
IL-1β	+	IL-1	IL-1	–	+	p38/NF-κB
IL-2	+	I	γ c chain	Th1	+	JAK1/STAT5
IL-6	+	I	IL-6	–	+	JAK1/STAT3
IL-8	+	CXCR	IL-8	–	+	GPCR/NF-κB
IL-12	+	I	IL-6	Th1	+	JAK2/STAT3
IL-15	+	I	γ c chain	–	+	JAK1/STAT5
IL-17	+/–	IL-17R	IL-17	Th17	+	MAPK/NF-κB
IL-18	+	II	IL-1	Th1	+	p38/NF-κB
IL-20	+/–	II	IL-10	–	+	JAK1/STAT3
IFNγ	+	II	IFN	Th1	+	JAK1/STAT1,2
TNFα	+	TNF	TNF	–	+	p38/NF-κB
Antiatherogenic Cytokines						
IL-10	–	II	IL-10	Th2	+	JAK1/STAT1,3,5
IL-33	+/–	IL-1	IL-1	Th2	+	p38/NF-κB
TGFβ	+/–	TGF	TGF	–	+	SMAD2,3
Predicted anti-atherogenic cytokines						
IL-4	–	I	IL-4	Th2	+	JAK1/STAT5
IL-5	–	I	IL-4	Th2	–	JAK2/STAT5
IL-11	–	I	IL-6	Th2	–	JAK1,2/STAT5
IL-13	–	I	IL-4	Th2	–	JAK1/STAT5
IL-19	–	II	IL-10	Th2	–	JAK1/STAT3

local cytokine milieu of the vascular lesion. When describing a potential role in atherosclerosis, it may be most useful to categorize cytokines into the broad functional categories of pro- and antiatherogenic. Of course, several will be classified as "undetermined."

CYTOKINE SIGNALING AND FUNCTION

Because they are often expressed by cells that also express their receptors, cytokines can signal in an autocrine, paracrine, or juxtacrine fashion, in contrast to soluble peptide hormones that function exclusively in an endocrine manner. Most cytokines often act in synergy with other cytokines [6]. Cytokines can drive multiple, often simultaneous cellular processes, and can be mitogenic, drive development, influence gene expression, chemotactic, and/or fibrotic. They are promiscuous in their use of receptor subunits and frequently share one or more homodimers or heterodimers of receptor subunits with other cytokines. For example, IL-19, IL-20, IL-22, and IL-24 all share receptor subunits, but display distinct cellular effects [7]. Many cytokines and receptors overlap both vascular and inflammatory cells in terms of synthesis and recognition, allowing bidirectional crosstalk between these two organ systems. The fact that these cytokines are synthesized and recognized by resident vascular cells demonstrates that vascular cells can be considered as part of the adaptive immune system and participate in the adaptive immune response (Figure 26.2).

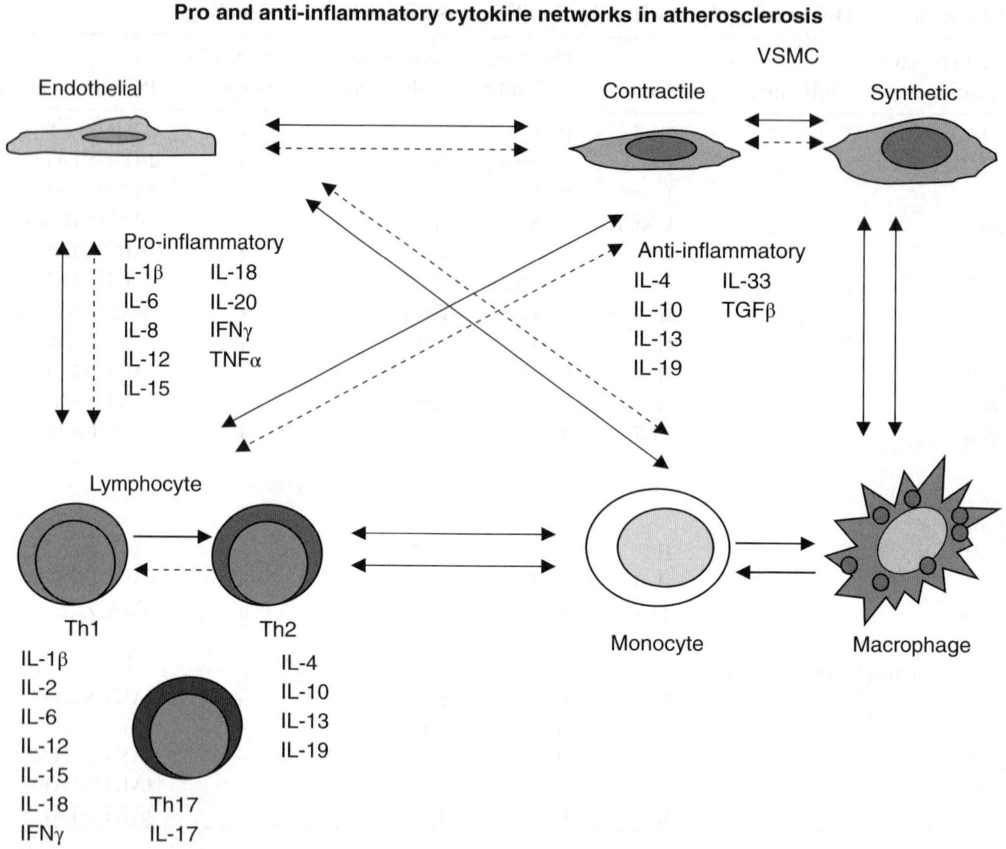

Figure 26.2 Information flow in a dynamic environment. The balance of pro- and anti-inflammatory cytokines determines plaque severity and stability. Cytokines produced by resident vascular and infiltrating immune cells generate a complex milieu consisting of multiple cytokines with often redundant and opposing effects. Resident vascular cells and leukocytes are capable of synthesis of Th1 and Th2 cytokines that enable bidirectional communication among these disparate cell types. Cytokines regulate T-lymphocyte maturation, monocyte-to-macrophage differentiation, and VSMC phenotype modulation. The solid lines indicate proinflammatory communication, while the dashed lines indicate anti-inflammatory.

Cytokine engagement of their cognate receptor initiate a series of intracellular signaling events including activation of mitogen-activated protein kinases (MAPKs), protein kinase C (PKC), and transcription factors including the signal transducer and activator of transcription (STATS) family [8]. Proinflammatory cytokines most often lead to activation of nuclear factor (NF-κB), a transcription factor that acts as a "master switch" for transcription of numerous genes central for inflammatory responses [9]. These genes include matrix and matrix degrading enzymes, cytokines and their receptors, cell cycle proteins, and adhesion molecules that lead to matrix deposition, fibrinolysis, cell proliferation, chemotaxis, and leukocyte–endothelial adhesion; all key processes in initiation and propagation of plaque formation [10]. Anti-inflammatory ILs often dampen the activity of proinflammatory Th1 ILs by a series of negative feedback loops mediated by de novo expression of suppressors of cytokine signaling (SOCS) family of proteins, which bind to the cytoplasmic tail of cytokine receptors and signaling intermediates (raines).

PROATHEROGENIC CYTOKINES

Overall, Th1 ILs are much more prevalent in human atherosclerotic lesions than the Th2 cytokines [5, 11]. Because atherosclerosis is primarily an inflammatory condition, we will categorize our focused presentation into pro- and anti-inflammatory cytokines.

IL-1β

IL-1β is strongly proinflammatory as it initiates expression of other inflammatory cytokines. IL-1 is induced by other proinflammatory stimuli, including TNFα, and expression is detected in multiple cell types in human lesions [5, 11]. IL-1β is produced by macrophage, monocytes, EC, and VSMC. These cells also express IL-1β receptor and thus IL-1β is a potent autocrine proinflammatory mediator. IL-1β signals activate p38 MAPK and converge on NF-κB, thus eliciting expression of cytokines and adhesion molecules, and are mitogenic for VSMC and EC [12]. Direct, proatherosclerotic effects are reported for IL-1β, and subcutaneous administration of IL-1 to apoliprotein E (ApoE)$^{-/-}$ or low-density lipoprotein receptor (LDLR)$^{-/-}$ mice had increased plaque burden [13]. When crossed with atherosclerosis-prone mice, IL-1β$^{-/-}$ had fewer atherosclerotic lesions [13].

IL-2

IL-2 is an established T-lymphocyte-derived autocrine growth and development factor. IL-2 is detected in atherosclerotic plaques, but a direct role in development of atherosclerosis is not yet fully characterized. IL-2 signals through the JAK/STAT pathway in lymphocytes, but the IL-2 receptor is absent on macrophage, EC, and VSMC [6]. Intraperitoneal injection of IL-2 increased atherosclerotic lesions in ApoE$^{-/-}$ mice [14], and neutralizing IL-2 antibody reduced lesion size [5]. Because the IL-2 receptor is limited to T cells, IL-2 is a driver of T-cell development and proliferation, rather than directly proatherogenic.

IL-6

IL-6 was initially thought of as a lymphocyte-specific factor, but it was soon found that IL-6 has potent proinflammatory effects in a variety of cell types. EC, VSMC, and macrophage all synthesize IL-6, and IL-6 is detected in human atherosclerotic plaque and increased in serum in patients with CAD and unstable angina [11]. Soluble IL-6R can activate EC to induce expression of adhesion molecules, and other proinflammatory cytokines, promoting extravasation and local lesion formation. IL-6 signals through the JAK/STAT family of signal transducers, and is mitogenic for VSMC. A direct role for IL-6 in atherogenesis has been established as systemic injection of ApoE$^{-/-}$ with recombinant IL-6 resulted in significantly increased plaque [15]. Surprisingly, atherosclerotic burden was also increased in IL-6$^{-/-}$/ApoE$^{-/-}$ double knockout mice [16]. Potential mechanisms for this counterintuitive result was an unanticipated increased expression of IL-10, increased IL-1Rα and TNFα receptor, which may neutralize bioavailability of these potent proinflammatory mediators. Thus, while a potent proinflammatory cytokine, IL-6 can have complex and somewhat confounding antiatherosclerotic side effects.

IL-8

IL-8 was originally classified as a chemoattractant for leukocytes and more recently for other cell types including VSMC [17]. IL-8 is detected in human atherosclerotic plaque, and serum levels are increased in patients with hypercholesterolemia and unstable angina. IL-8 expression is induced in macrophage by oxidized LDL (ox LDL) and in EC and VSMC when challenged with TNFα and IL-1β. As a potent chemoattractant, IL-8 can be considered to be an important proatherogenic cytokine as it accelerates atherosclerosis by enhancing leukocyte extravasation and EC adhesiveness [18]. IL-8 contributes to plaque instability by inhibition of tissue inhibitor of metalloproteinase-1 (TIMP-1) expression. In animal models, IL-8 proatherogenic effects are likely mediated primarily by leukocytes, as transplantation of IL8$^{-/-}$ bone marrow into LDLR$^{-/-}$ recipients results in less atherosclerosis compared to IL-8$^{+/+}$ bone marrow [12].

IL-12

IL-12 is an atherogenic IL comprising the p35/p40 heterodimeric complex and plays a significant role in lymphocyte Th1 differentiation. IL-12 signals through the JAK/STAT complex and activates the T-bet transcription factor, leading to IFNγ production and polarization of the immune response to a proinflammatory phenotype [11]. IL-12 can be induced in monocytes by adding oxidized LDL, and is expressed by lymphocytes, activated macrophage, and dendritic cells, and is detected in these cells in atherosclerotic lesions. Atherosclerotic lesions in ApoE$^{-/-}$ mice are accelerated by injection of recombinant IL-12 [19]. Concordantly, IL-12$^{-/-}$/ApoE$^{-/-}$ double knockout mice have reduced atherosclerotic lesion formation, and LDLR$^{-/-}$ mice crossed with a mouse with a mutation in which monocytes cannot secrete IL-12 also have reduced plaque formation. With the absence of data reporting IL-12 direct effects on resident vascular cells, the proatherogenic effects of IL-12 are likely a result of Th1-cell polarization.

IL-15

IL-15 is a T-lymphocyte growth factor with no sequence homology to, but a functional profile similar to IL-2. Atherosclerotic plaque in mice and humans contains IL-15 immunoreactive T cells [20], but murine studies on a role for IL-15 participation in atherogenesis are limited. IL-15 is produced by cell types involved in atherogenesis including macrophage, monocytes, and EC [21]. DNA vaccination against IL-15 reduces atherosclerotic lesion size in LDLR$^{-/-}$ mice [22]. IL-15 has potent chemoattractant effects on T lymphocytes and natural killer cells, thus, the major proatherosclerotic effects of IL-15 on plaque formation may be due to promotion of extravasation in addition to T-cell maturation.

IL-17

Th17-positive lymphocytes constitute a third T-lymphocyte phenotype (in addition to Th1 and Th2) and along with natural killer cells, Th17-positive cells are the major source of IL-17. IL-17 is composed of a family of five proinflammatory cytokines currently designated as IL-17 A through F (reviewed in Ref. [23]). Each of these family members signal through subtype-specific receptors, and activate MAP kinases and NF-κB leading to expression of proinflammatory genes including IL-6 and TNFα [23, 24]. Although IL-17 is considered proinflammatory and IL-17 positive cells are present in human atherosclerotic lesions, a direct causative role of this family of cytokines in atherogenesis is currently the subject of debate. Recombinant murine IL-17A reduced atherosclerotic lesions and EC adhesion molecule expression in fat-fed LDLR$^{-/-}$ mice [25], suggesting a protective role for this cytokine. However, neutralizing antibody to IL-17A reduced atherosclerosis in ApoE$^{-/-}$ mice, and further, when IL-17R$^{-/-}$ bone marrow cells are transplanted into LDLR$^{-/-}$ mice, these mice had significantly decreased atherosclerosis [26]. This complex family of ILs requires further study to firmly establish a specific role in an atherosclerosis.

IL-18

IL-18 is potently proinflammatory, which may be expected from a cytokine that has similar sequence homology to IL-1β and signals through a receptor also with a high degree of similarity to the IL-1β receptor. Also similar to IL-1β, IL-18 activates NF-κB, and upregulates IL-6, IL-8, and adhesion molecules on monocytes and EC [27]. IL-18 and its receptor are detected in human atherosclerotic plaque, and serum levels are elevated in patients with unstable angina and CAD [28]. IL-18 is produced primarily by macrophage, but the IL-18 receptor is present on macrophage; EC; and to a lesser degree, VSMC, suggesting bidirectional communication among these cell types. IL-18 has proatherosclerotic effects in several animal models. IL-18 is considered a Th1 IL because it also induces production of IFNγ, promotion of atherosclerosis by exogenous addition of IL-18 is reduced in $IFN^{-/-}/ApoE^{-/-}$ double knockout mice [29, 30]. Systemic administration of IL-18 significantly increased atherosclerotic plaque in $ApoE^{-/-}$ mice, and $IL\text{-}18^{-/-}/ApoE^{-/-}$ double knockout mice had smaller and more stable lesions compared with control mice.

IL-20

IL-20 was originally classified in an IL-10 subfamily, which includes IL-19, IL-22, and IL-24, which share similar receptor complexes [7]. Like these other ILs, IL-20 signals through the JAK-STAT pathway. IL-20 and its receptor subunits are detectible in macrophage and endothelium in human and murine atherosclerotic plaque, and IL-20 is expressed in macrophage and EC when challenged with inflammatory stimuli [31]. Despite its structural similarity to IL-10, IL-20 induces chemokine and VEGF expression in EC, as well as reported proatherosclerotic effects [32]. Intramuscular electroporation of IL-20 cDNA into $ApoE^{-/-}$ mice enhanced atherosclerotic plaque area, but no increase in macrophage infiltrate into lesions in these mice. Because leukocyte infiltrate was not increased in these lesions, it is suggested that IL-20 proatherosclerotic effects are due to increased lesion microvascularization, rather than by a proinflammatory effect.

IFNγ

The majority of T cells present in atherosclerotic plaque are T helper, or Th1-positive, cells that produce abundant amounts of IFNγ. IFNγ polarizes other T cells to the Th1 phenotype and also induces macrophage, EC and VSMC to a pro-inflammatory phenotype (reviewed in Ref. [6]). IFNγ signals through the JAK/STAT pathway and also activates the SMAD signaling complex leading to inflammatory cytokine production. IFNγ is expressed by most cell types present in atherosclerotic plaque in humans and experimental lesions in mice. Several lines of evidence indicate that IFNγ is a potent-pro-atherogenic cytokine. Systemic intraperitoneal IFNγ administration promoted atherosclerosis in $ApoE^{-/-}$ mice. Concordantly, $IFN\gamma R^{-/-}/ApoE^{-/-}$ double knockout mice display a reduced atherosclerotic burden compared with controls. $IFN\gamma^{-/-}/ApoE^{-/-}$ double knockout mice also present reduced atherosclerosis compared with controls. Gene transfer of a secreted IFNγ receptor decoy in $ApoE^{-/-}$ mice prevented progression of atherosclerotic plaque [33, 34]. There are various modes of action that point to IFNγ as a powerful pro-atherosclerotic agent. In addition to Th1 polarization, IFNγ can reduce cap VSMC proliferation and collagen deposition, while also increasing MMP expression in these cells, leading toward plaque instability and potential rupture [33]. IFNγ can increase oxidized lipid uptake and foam cell formation in macrophage and VSMC. Together, IFNγ can influence plaque progression at multiple levels and is a potent proatherosclerotic cytokine.

TNFα

Tumor necrosis factor alpha is a member of the TNF superfamily that on the basis of sequence, functional, and structural similarities currently contains 19 distinct cytokines. Canonical TNFα signaling is mediated by p38 MAPK and NF-κB. TNFα is a powerful proinflammatory cytokine and can induce a cascade of proinflammatory gene expression including IL-1β, IL-8, MCP-1, ICAM1, VCAM1, and MMPs in a variety of cell types including lymphocytes, macrophage, EC, and VSMC. TNFα immunoreactivity is found in all of these cell types in human and experimental atherosclerotic plaque, and serum TNFα levels correlate with atherosclerotic plaque burden assessed by carotid IVUS [35]. Similarly, $TNF\alpha^{-/-}/ApoE^{-/-}$ double knockout mice have significantly reduced atherosclerotic plaque compared with controls and reduced inflammatory serum cytokine levels [36]. Studies in TNFα-receptor-deficient mice surprisingly resulted in increased plaque burden, but likely because of elevated cholesterol levels in these mice. Overall, because of its powerful proinflammatory effects on all cell types present in atherosclerotic lesions, TNFα is a potent protagonist of atherosclerosis.

ANTIATHEROGENIC CYTOKINES

Cytokine abundance in atherosclerotic plaque is overwhelmingly Th1 oriented[5, 11]. Since so many more proatherogenic cytokines and receptors have been identified and characterized, much greater effort has gone into understanding these cytokines and potential for inhibition of their expression or activity. Far fewer studies have pursued characterization of anti-inflammatory cytokines, but perhaps modulation of anti-inflammatory or protective cytokines holds greater therapeutic potential. This section will present both confirmed and potentially antiatherosclerotic cytokines.

EXPERIMENTALLY CONFIRMED ANTIATHEROGENIC CYTOKINES

IL-10

IL-10 is the archetypical Th2 IL, and consequently, the most studied. IL-10 is produced by macrophage, monocytes, and T and B lymphocytes, and is detected in these cells in human atherosclerotic plaque. Multiple approaches demonstrate that IL-10 is considered atheroprotective by the dual mechanisms of Th2 T-cell polarization and attenuation of inflammatory gene expression. IL-10 signals through the IL-10 heterodimeric receptor and activates multiple members of the JAK/STAT pathway, particularly STAT3, and the molecular mechanisms of IL-10 activity are multifaceted. IL-10 blocks MAPK signaling and NF-κB activation in monocytes, macrophage, EC, and VSMC [11, 37]. It has recently been shown in monocytes that IL-10 can reduce mRNA stability of inflammatory transcripts that contain ARE regulatory elements in their 3′ UTR [37]. Thus, IL-10 can dampen the inflammatory response in immune cells at the transcriptional and post-transcriptional levels.

Most *in vivo* studies suggest that IL-10 is strictly an immune modulator; EC and VSMC do not express this cytokine, precluding paracrine/juxtacrine signaling between activated resident vascular and inflammatory cells. In cultured EC, IL-10 has no attenuating effect on expression of inflammatory cytokines, but can reduce ICAM-1 and VCAM-1 expression. In VSMC, IL-10 did not reduce IL-1β-induced expression inflammatory genes of including IL-8, IL-6, or MCP-1, but did reduce proliferation in an NF-κB-dependent manner. In overexpression studies, IL-10 transgenic mice have reduced atherosclerotic plaque burden, and transfer of bone marrow from IL-10 transgenic mice into $LDLR^{-/-}$ mice reduced atherosclerosis [38]. Systemic adenovirus-mediated IL-10 gene expression reduced atherosclerosis by immune cell deactivation and reduction in inflammatory cell infiltrate in plaque. Interestingly, serum cholesterol is reduced in these mice, which also may contribute to decreased vascular inflammation. Concordantly, atherosclerosis is increased in $IL\text{-}10^{-/-}$ mice and $IL\text{-}10^{-/-}/ApoE^{-/-}$ double knockout mice, and transfer of $IL\text{-}10^{-/-}$ BM to $LDLR^{-/-}$ mice shows that it is leukocyte-derived IL-10 that mediates these effects by polarizing the T lymphocyte Th2/Th1 ratio toward a more anti-inflammatory phenotype [39]. The synthesis of these studies demonstrate IL-10 is a potent immune modulator, with pleiotropic, albeit unresolved effects on vascular cells.

IL-33

IL-33 is an IL-1 family member and expressed by Th2 cells and is considered a Th2 IL as it increases synthesis of IL-4, IL-5, and IL-13, while decreasing IFNγ expression. IL-33 is expressed in normal artery, but increased in macrophage in human atherosclerotic plaque [40]. IL-33 is expressed in several cell types, and expression by macrophage expression has direct autocrine effects on macrophage function at several levels. IL-33 signals through the ST2, or orphan IL-1 receptor, and paradoxically, activates the proinflammatory mediators NF-κB and p38 MAPK. Mice injected with neutralizing decoy receptor developed increased atherosclerotic plaque and had reduced IL-5 expression and anti-oxLDL antibody. In ApoE$^{-/-}$ mice-fed high-fat diet, IL-33, reduced macrophage accumulation into atherosclerotic plaque, but plaque area in these mice was not directly assessed [40]. A second level of protection is that IL-33 reduces macrophage foam cell formation and increases cholesterol efflux in these cells by changing the abundance of proteins involved in lipid uptake and efflux [41]. One mechanism for these effects is reduced expression of scavenger receptors such as CD36, and increased expression of genes that promote cholesterol efflux such as ApoE, ABCA-1. Despite this promising data, it remains to be determined if IL-33 can actually reduce atherosclerosis.

TGFβ

Transforming growth factor β is produced by macrophage, T lymphocytes, EC, and VSMC, detected in all of these cells in human atherosclerotic lesions, and serum levels are decreased in patients with severe atherosclerosis (reviewed in Ref. [42]). TGFβ is primarily known as a profibrotic, wound-healing cytokine with VSMC as a major target; however, it also has potent anti-inflammatory and Th2 polarizing effects on leukocytes as well. TGFβ is recognized by a complex family of receptors [42], which signals through the SMAD family of signal transducers. TGFβ heterozygous mice fed an atherogenic diet had increased atherosclerotic plaque, endothelial activation, and leukocyte infiltrate. Administration of neutralizing antibody or soluble TGFβ receptor to ApoE$^{-/-}$ mice increases atherogenesis in those mice [43]. When crossed with ApoE$^{-/-}$ mice, TGFβ$^{-/+}$ mice had accelerated atherosclerosis with increased inflammatory cell content in lesions. Interestingly, T-cell differentiation was accelerated to both Th1 and Th2 phenotypes in these mice. When TGFβ is depleted by receptor decoy, lesions had decreased collagen content and concomitant intraplaque rupture was observed, which is pertinent for human disease where plaque stability is paramount [44]. Mice receiving bone marrow from mice expressing a dominant-negative TGFβ receptor driven by a T-cell-specific promoter demonstrated increased plaque, T-cell infiltration, and decreased VSMC and collagen content, suggesting that T cells, rather than resident vascular cells, are the major effectors for TGFβ activity.

LIKELY ANTIATHEROGENIC

Several presumed anti-inflammatory cytokines have been investigated in nonatherosclerotic disease models or otherwise recently been characterized. It is anticipated that these will also be confirmed as antiatherosclerotic, and merit mention.

IL-4

IL-4 is a confirmed Th2 IL produced by activated T cells that promote autocrine and paracrine immunodampening effects on macrophage and T lymphocytes [6, 11], and some reviews consider IL-4 to be atheroprotective [45]. IL-4 protein and mRNA are detected in human atherosclerotic plaque. IL-4 signals through STAT5, STAT6, and JAK proteins, and can suppress IL-1β and TNFα synthesis in these cells. While conceptually this should be an antiatherogenic cytokine, experimental data do not fully support this assumption. For example, IL-4 can induce VCAM-1 expression on EC, which would presumably expedite leukocyte extravasation and leukocyte retention into the vessel wall. Yet other studies show IL-4 can inhibit macrophage adhesiveness and VSMC proliferation. IL-4$^{-/-}$ mice do not have increased atherosclerosis, and subcutaneous injection of recombinant IL-4 into ApoE$^{-/-}$ mice does not reduce development of atherosclerotic lesions [46]. Further, lesions were actually reduced in area in IL-4$^{-/-}$/ApoE$^{-/-}$ double knockout mice, and reconstitution of LDLR$^{-/-}$ mice with IL-4$^{-/-}$ bone marrow also reduces lesion formation. IL-4 may represent an example where a Th2 IL may not have a clearly direct role in reduction of atherosclerosis, confirming the complexity of the function of the IL and atherosclerosis.

IL-5

IL-5 is synthesized by EC, and its function is primarily associated with stimulation of the eosinophil response and B-lymphocyte immunoglobulin switching and antibody production. It is considered a Th2 cytokine because mice injected with this IL demonstrate a serum cytokine profile shift toward the Th2 cytokine profile. IL-5 mRNA is detected at very low levels in human atherosclerotic plaque. Transfer of IL-5$^{-/-}$ bone marrow into atheroprone mice suggests an atheroprotective role for this IL, apparently by linking adaptive immunity with innate immunity by stimulating the expansion of atheroprotective natural IgM specific for OxLDL [47].

IL-11

IL-11 promotes a Th1 to Th2 phenotypic switch in T cells, and can attenuate macrophage expression of proinflammatory cytokines [6, 11]. IL-11 is expressed by EC, but is not detected in human atherosclerotic plaque. IL-11 can suppress VSMC proliferation and cytokine production by suppression of NF-κB activation [48]. Although we could predict that this cytokine would be antiatherosclerotic, studies directly addressing a role for this IL in atherosclerosis are lacking.

IL-13

IL-13 shares 25% sequence homology and shares a receptor subunit (IL-4Rα) with IL-4. Like IL-4, IL-13 activates STAT proteins, and is considered to be a Th2 cytokine because it is produced by Th2 T cells. IL-13 has anti-inflammatory activity on monocytes in that it can decrease expression of many inflammatory cytokines [49]. As such, while IL-13 can be predicted to be antiatherogenic, IL-13 expression in atherosclerotic plaque, and studies that directly test IL-13 role in atherosclerosis have not been reported.

IL-19

IL-20 was originally classified in an IL-10 subfamily that includes IL-19, IL-22, and IL-24, but signals through different combinations of shared receptor chains complexes, and also activates STAT3 [7]. IL-19 is considered Th2 IL because it can induce expression of Th2 cytokines in leukocytes. While originally assumed to be leukocyte specific, recent reports demonstrate induction by inflammatory stimuli in EC and VSMC [50]. IL-19 is detected in atherosclerotic plaque and expressed in inflammatory and resident vascular cells in chronically inflamed arteries from human transplant patients [50, 51]. IL-19 decreases migration, proliferation, and expression of inflammatory transcripts in VSMC; but unlike either IL-10 or IL-20, IL-19 does not inhibit NF-κB activity. IL-19 decreases inflammatory mRNA transcript stability by inhibition of the mRNA stability factor HuR [52]. IL-19$^{-/-}$ mice have an exaggerated inflammatory response, and adenoviral gene delivery inhibits restenosis in angioplasty injured rat carotid arteries [50]. While little has been reported on IL-19 effects in atherosclerotic models, existing data suggest that IL-19 should be potently antiatherosclerotic.

SUMMARY

Cytokine involvement in atherosclerosis is best thought of as a dynamic, interactive network that not only coordinates the immune response but also directly mediates resident vascular cell response to inflammation. The balance of pro- and anti-inflammatory cytokines can determine plaque severity and stability. Despite this complexity, proinflammatory cytokines, their receptors, or intracellular signaling pathways represent possible targets of antiatherosclerotic therapy. Correspondingly, anti-inflammatory cytokines themselves show promise as therapeutic modalities themselves. More research is necessary to fully understand cytokine expression and mechanisms of activity.

ACKNOWLEDGMENT

This work was funded by NHLBI HL115575 and HL117724. The author acknowledges the original artwork by Matthew Autieri. Because of word limit constraints, it was not possible to cite all original papers and excellent reviews, and the author apologizes to the authors of whose work may not have been included in the references.

REFERENCES

1. Ross R. Atherosclerosis, and inflammatory disease. *N Engl J Med*. 1999;340:115–126.
2. Hansson GK, Libby P. The immune response in atherosclerosis: a double-edged sword. *Nat Rev Immunol*. 2006;6:508–519.
3. Libby P. Inflammation in atherosclerosis. *Nature*. 2002;420:868–874.
4. Langer JA, Cutrone EC, Kotenko S. The Class II cytokine receptor (CRF2) family: overview and patterns of receptor-ligand interactions. *Cytokine Growth Factor Rev*. 2004;15:33–48.
5. Frostegård J, Ulfgren AK, Nyberg P, et al. Cytokine expression in advanced human atherosclerotic plaques: dominance of pro-inflammatory (Th1) and macrophage-stimulating cytokines. *Atherosclerosis*. 1999;145:33–43.
6. Tedgui A, Mallat Z. Cytokines in atherosclerosis: pathogenic and regulatory pathways. *Physiol Rev*. 2006;86:515–581.
7. Parrish-Novak J, Xu W, Brender T, et al. 19, 20, and 24 signal through two distinct receptor complexes. Differences in receptor-ligand interactions mediate unique biological functions. *J Biol Chem*. 2002;277:47517–47523.
8. Paukku K, Silvennoinen O. STATs as critical mediators of signal transduction and transcription: lessons learned from STAT5. *Cytokine Growth Factor Rev*. 2004;15(6):435–455.
9. Brand K, Page S, Walli AK, Neumeier D, Baeuerle PA. Role of nuclear factor-kappa B in atherogenesis. *Exp Physiol*. 1997;82:297–304.
10. Tang J, Raines EW. Are suppressors of cytokine signaling proteins recently identified in atherosclerosis possible therapeutic targets? *Trends Cardiovasc Med*. 2005;15:243–249.
11. von der Thüsen JH, Kuiper J, van Berkel TJ, Biessen EA. Interleukins in atherosclerosis: molecular pathways and therapeutic potential. *Pharmacol Rev*. 2003;55:133–166.
12. Boisvert WA, Santiago R, Curtiss LK, Terkeltaub RA. A leukocyte homologue of the IL-8 receptor CXCR-2 mediates the accumulation of macrophages in atherosclerotic lesions of LDL receptor-deficient mice. *J Clin Invest*. 1998;101: 353–363
13. Kirii H, Niwa T, Yamada Y, et al. Lack of interleukin-1beta decreases the severity of atherosclerosis in ApoE-deficient mice. *Arterioscler Thromb Vasc Biol*. 2003;23:656–660.
14. Upadhya S, Mooteri S, Peckham N, Pai RG. Atherogenic effect of interleukin-2 and antiatherogenic effect of interleukin-2 antibody in apo-E-deficient mice. *Angiology*. 2004;55: 289–294.
15. Huber SA, Sakkinen P, Conze D, Hardin N, Tracy R. Interleukin-6 exacerbates early atherosclerosis in mice. *Arterioscler Thromb Vasc Biol*. 1999;19:2364–2367.
16. Schieffer B, Selle T, Hilfiker A, et al. Impact of interleukin-6 on plaque development and morphology in experimental atherosclerosis. *Circulation*. 2004;110:3493–3500.
17. Reape TJ, Groot PH. Chemokines and atherosclerosis. *Atherosclerosis*. 1999;147:213–225.
18. Gerszten RE, Garcia-Zepeda EA, Lim YC, et al. MCP-1 and IL-8 trigger firm adhesion of monocytes to vascular endothelium under flow conditions. *Nature*. 1999;398:718–723.
19. Lee TS, Yen HC, Pan CC, Chau LY. The role of interleukin 12 in the development of atherosclerosis in ApoE-deficient mice. *Arterioscler Thromb Vasc Biol*. 1999;19:734–742.
20. Wuttge DM, Eriksson P, Sirsjö A, Hansson GK, Stemme S. Expression of interleukin-15 in mouse and human atherosclerotic lesions. *Am J Pathol*. 2001;159:417–423.

21. Oppenheimer-Marks N, Brezinschek RI, Mohamadzadeh M, Vita R, Lipsky PE. Interleukin 15 is produced by endothelial cells and increases the transendothelial migration of T cells in vitro and in the SCID mouse-human rheumatoid arthritis model in vivo. *J Clin Invest*. 1998;101:1261–1272.
22. van Es T, van Puijvelde GH, Michon IM, et al. IL-15 aggravates atherosclerotic lesion development in LDL receptor deficient mice. *Vaccine*. 2011;29:976–983.
23. Miossec P. IL-17 and Th17 cells in human inflammatory diseases. *Microbes Infect*. 2009; 11:625–630.
24. Awane M, Andres PG, Li DJ, Reinecker HC. NF-kappa B-inducing kinase is a common mediator of IL-17-, TNF-alpha-, and IL-1 beta-induced chemokine promoter activation in intestinal epithelial cells. *J Immunol*. 1999;162: 5337–5344.
25. Taleb S, Romain M, Ramkhelawon B, et al. Loss of SOCS3 expression in T cells reveals a regulatory role for interleukin-17 in atherosclerosis. *J Exp Med*. 2009;206:2067–2077.
26. van Es T, van Puijvelde GH, Ramos OH, et al. Attenuated atherosclerosis upon IL-17R signaling disruption in LDLr deficient mice. *Biochem Biophys Res Commun*. 2009;388:261–265.
27. Dinarello CA. IL-18: a TH1-inducing, proinflammatory cytokine and new member of the IL-1 family. *J Allergy Clin Immunol*. 1999;103:11–24.
28. Blankenberg S, Tiret L, Bickel C, et al. AtheroGene investigators. Interleukin-18 is a strong predictor of cardiovascular death in stable and unstable angina. *Circulation*. 2002;106:24–30.
29. Whitman SC, Ravisankar P, Daugherty A. Interleukin-18 enhances atherosclerosis in apolipoprotein E(-/-) mice through release of interferon-gamma. *Circ Res*. 2002;90: E34–E38.
30. Elhage R, Jawien J, Rudling M, et al. Reduced atherosclerosis in interleukin-18 deficient apolipoprotein E-knockout mice. *Cardiovasc Res*. 2003;59:234–240.
31. Chen WY, Cheng BC, Jiang MJ, Hsieh MY, Chang MS. IL-20 is expressed in atherosclerosis plaques and promotes atherosclerosis in apolipoprotein E-deficient mice. *Arterioscler Thromb Vasc Biol*. 2006;26:2090–2095.
32. Hsieh MY, Chen WY, Jiang MJ, Cheng BC, Huang TY, Chang MS. Interleukin-20 promotes angiogenesis in a direct and indirect manner. *Genes Immun*. 2006;7:234–242.
33. Gupta S, Pablo AM, Jiang X, Wang N, Tall AR, Schindler C. IFN-gamma potentiates atherosclerosis in ApoE knock-out mice. *J Clin Invest*. 1997;99:2752–2761.
34. Whitman SC, Ravisankar P, Elam H, Daugherty A. Exogenous interferon-gamma enhances atherosclerosis in apolipoprotein E-/- mice. *Am J Pathol*. 2000;157:1819–1824.
35. Skoog T, Dichtl W, Boquist S, et al. Plasma tumour necrosis factor-alpha and early carotid atherosclerosis in healthy middle-aged men. *Eur Heart J*. 2002;23:376–383.
36. Ohta H, Wada H, Niwa T, et al. Disruption of tumor necrosis factor-alpha gene diminishes the development of atherosclerosis in ApoE-deficient mice. *Atherosclerosis*. 2005;180: 11–17.
37. Rajasingh J, Bord E, Luedemann C, et al. IL-10-induced TNF-alpha mRNA destabilization is mediated via IL-10 suppression of p38 MAP kinase activation and inhibition of HuR expression. *FASEB J*. 2006;20:2112–2114.
38. Pinderski LJ, Fischbein MP, Subbanagounder G, et al. Overexpression of interleukin-10 by activated T lymphocytes inhibits atherosclerosis in LDL receptor-deficient Mice by altering lymphocyte and macrophage phenotypes. *Circ Res*. 2002;90:1064–1071.
39. Mallat Z, Besnard S, Duriez M, et al. Protective role of interleukin-10 in atherosclerosis. *Circ Res*. 1999;85:e17–e24.
40. Miller AM, Xu D, Asquith DL, et al. IL-33 reduces the development of atherosclerosis. *J Exp Med*. 2008;205:339–346.
41. McLaren JE, Michael DR, Salter RC, et al. IL-33 reduces macrophage foam cell formation. *J Immunol*. 2010;185:1222–1229.
42. Grainger DJ. TGF-beta and atherosclerosis in man. *Cardiovasc Res*. 2007;74:213–222.
43. Mallat Z, Gojova A, Marchiol-Fournigault C, et al. Inhibition of transforming growth factor-beta signaling accelerates atherosclerosis and induces an unstable plaque phenotype in mice. *Circ Res*. 2001;89:930–934.
44. Gojova A, Brun V, Esposito B, et al. Specific abrogation of transforming growth factor-beta signaling in T cells alters atherosclerotic lesion

size and composition in mice. *Blood*. 2003;102: 4052–4058.

45. Fisman EZ, Adler Y, Tenenbaum A. Biomarkers in cardiovascular diabetology: interleukins and matrixins. *Adv Cardiol*. 2008;45:44–64.
46. King VL, Cassis LA, Daugherty A. Interleukin-4 does not influence development of hypercholesterolemia or angiotensin II-induced atherosclerotic lesions in mice. *Am J Pathol*. 2007;171: 2040–2047.
47. Binder CJ, Hartvigsen K, Chang MK, et al. IL-5 links adaptive and natural immunity specific for epitopes of oxidized LDL and protects from atherosclerosis. *J Clin Invest*. 2004;114:427–437.
48. Zimmerman MA, Selzman CH, Reznikov LL, et al. Interleukin-11 attenuates human vascular smooth muscle cell proliferation. *Am J Physiol Heart Circ Physiol*. 2002;283: H175–H180.
49. Mijatovic T, Kruys V, Caput D, Defrance P, Huez G. Interleukin-4 and -13 inhibit tumor necrosis factor-alpha mRNA translational activation in lipopolysaccharide-induced mouse macrophages. *J Biol Chem*. 1997;272:14394–14398.
50. Jain S, Gabunia K, Kelemen SE, Panetti TS, Autieri MV. The anti-inflammatory cytokine interleukin 19 is expressed by and angiogenic for human endothelial cells. *Arterioscler Thromb Vasc Biol*. 2011;31:167–175.
51. Tian Y, Sommerville LJ, Cuneo A, Kelemen SE, Autieri MV. Expression and suppressive effects of interleukin-19 on vascular smooth muscle cell pathophysiology and development of intimal hyperplasia. *Am J Pathol*. 2008;173:901–909.
52. Cuneo AA, Herrick D, Autieri MV. Il-19 reduces VSMC activation by regulation of mRNA regulatory factor HuR and reduction of mRNA stability. *J Mol Cell Cardiol*. 2010;49: 647–654.

27

ECM AND ATHEROSCLEROSIS

JIE DU[1,2], YING WANG[1], AND LIXIN JIA[1]

[1] *Beijing An Zhen Hospital, Capital Medical University; The Key Laboratory of Remodeling-Related Cardiovascular Diseases, Ministry of Education, Beijing, China*

[2] *Beijing Institute of Heart Lung and Blood Vessel Diseases, Beijing, China*

INTRODUCTION

The arterial wall comprises a highly ordered structure of cells and extracellular matrix (ECM). ECM constitutes a major portion of blood vessels and atherosclerosis lesions. The ECM in blood vessels and atherosclerosis lesions includes collagens, proteoglycans, and other proteins. In the intimal layer, endothelial cells sit on a basement membrane rich in laminin, fibronectin, and type IV collagen. In the media layer, individual smooth muscle cells (SMCs) are surrounded by a basement membrane and are in turn embedded infibrillar types I, III, V, and VIII collagen, fibronectin, and proteoglycans. In elastic arteries, each layer of SMCs is separated by a well-defined, fenestrated elastic lamina. The outermost layer, known as the adventitia, is rich in fibroblasts, collagen types I and III and elastin.

These extracellular proteins modulate responses of vascular cell (SMCs, endothelial cells (ECs) and inflammatory cells) via specific receptors and signaling pathways. This chapter focuses on the interactions of vascular cells with the ECM in atherosclerosis: (i) how these ECM proteins are regulated and (ii) how ECM proteins play their roles in atherosclerosis, regulating low-density lipoprotein (LDL) metabolism, participating in the infiltration and differentiation of SMCs and macrophages, controlling matrix remodeling through feedback signaling to proteinases, and influencing the development of plaque rupture. Better understanding of the role of ECM will help us understand the pathogenesis of atherosclerosis. The composition of ECM shows pronounced changes after arterial injury and during atherosclerotic plaque development.

STRUCTURE, SYNTHESIS OF ECM

The ECM in blood vessels and atherosclerosis lesions includes collagens, elastin, fibrillin, proteoglycan, laminin, fibronectin, and vitronectin.

Atherosclerosis: Risks, Mechanisms, and Therapies, First Edition. Edited by Hong Wang and Cam Patterson.

Collagen

The collagens include at least 24 different types and 38 distinct polypeptide chains. Collagen types I and III are the most abundant in the vessel wall and are distributed throughout the media and adventitia. The vascular collagens are subdivided into three morphologically distinct groups: fibrillar collagens (collagen I, III, V), nonfibrillar collagens (fibril-associated collagens: collagens XIV, XVI; microfibrillar collagens: collagen VI, XV, XVIII, XIX; basement membrane-associated collagens; membrane-bound collagen XIII; anchoring collagen VII), and network-forming collagen (collagen VIII). SMCs are the main source of vascular collagen. ECs, adventitial fibroblasts, and macrophages can also produce collagens. Collagen biosynthesis includes various transcriptional and posttranslational, intracellular, and extracellular events. Intracellular processes include regulation of transcription and mRNA stability, formation of the α-procollagen chains, hydroxylation, glycosylation, assembly of the procollagen chains, and secretion. Extracellular processes include processing of the procollagen molecules, assembly of the collagen fibrils, and cross-linking. Collagen I modulates the tensile strength of the vessel wall, whereas collagen III accounts for its elasticity. A shift toward a higher proportion of collagen I in atherosclerotic lesions indicates decreased vascular resilience.

Elastin

Elastin is the dominant ECM protein deposited in the arterial wall and makes up to 50% of its dry weight. Elastic fibers comprise a cross-linked elastin core surrounded by a mantle of fibrillin-rich microfibrils. Fibrillin-1 is the major structural component of microfibrils and provides the scaffold for the deposition and crosslinking of elastin. SMCs synthesize and secrete elastin as a monomer. Tropoelastin's lysine residues are post-translationally modified to form covalent cross-links between and within elastin molecules in the extracellular space. The deposition of elastin is limited to the medial layer of the vessel wall, extending from the internal to the external elastic lamina. This biomechanical support provides the compliance that arteries need to absorb and transmit hemodynamic force.

Fibrillin

The fibrillins are a class of extracellular microfibrils that associate with elastic fibers in the arterial wall. Two highly homologous isoforms are fibrillin-1, encoded by a gene on chromosome 15, and fibrillin-2, on chromosome 5. Both fibrillin microfibrils are known to interact with multiple ECM components within the vessel wall, including elastin, vitronectin, fibronectin, and collagen. Fibrillin gene mutations lead to aortic aneurysm.

Proteoglycans

The proteoglycans found in greatest abundance in the vessel wall can be categorized into two classes: (i) large proteoglycans that form large aggregates by interaction with hyaluronan (HA) and (ii) small leucine-rich proteoglycans (SLRPs). The large proteoglycans interact with hyaluronic acid to form an extensive, interconnected polymeric network in the extracellular space. The SLRPs are a family of secreted proteoglycans that include decorin, biglycan, fibromodulin, osteoglycin, and lumican, which do not interact with hyaluronic acid but instead bind to ECM molecules such as collagen, tropoelastin, fibronectin, and fibrillin-containing microfibrils, among others.

Both biglycan and decorin bind to and regulate collagen fibrillogenesis and bind to transforming growth factor β (TGF-β) and sequester it to the ECM [1]. In arterial blood vessels, SMCs synthesize biglycan mediated by TGF-β. Biglycan is present in the intima of normal human blood vessels and is localized in the outer layer of the diffuse intimal thickening in human coronary arteries. The prelesional distribution of biglycan is similar to the distribution of lipids in the early phase of type I lesions, which suggests that biglycan exists in the intima before lipids deposit and plays a pivotal role in initial deposition of lipids in the arterial intima.

Decorin and biglycan share the highest homology between members of SLRP family, with 57% amino acid identity. Decorin has glycosaminoglycan (GAG) chains attached to the core protein that can be chondroitins or dermatan sulfates. Decorin is more prominent in the intima of internal thoracic and radial arteries as compared with the coronary artery and saphenous vein grafts. Decorin is highly expressed in fibrous caps of atherosclerotic lesions, which is consistent with the known collagen-stabilizing function of this proteoglycan. Directional atherectomy samples obtained from both native atherosclerotic segments and segments with restenosis following angioplasty show colocalization of decorin with fibrous regions [2].

Versican is a member of the hyaladherins or proteoglycans, which interact with HA and lectins, providing fixed negative charges that influence the spatial distribution of ions and hydration volumes in pericellular and extracellular gel matrixes. Versican is the largest proteoglycan in the vessel wall and has been localized to the aortic media and endothelial layers. Versican is synthesized by SMCs and ECs, accumulates during human atherosclerosis and restenosis, and has been shown to bind low-density lipoproteins (LDLs). Biochemical analyses of arterial versican of normal vascular and atherosclerotic lesions demonstrated a heterogeneous composition. Versican in advanced lesions is markedly reduced in proportion (~68%) among the total oversulfated disaccharides present in GAG species.

Perlecan is the major heparan sulfate (HS) proteoglycan in the subendothelial matrix. Perlecan is a 450-kDa core protein containing three HS chains of 70 kDa. Perlecan is found in basement membranes underneath the endothelial layer and around medial SMCs. It is downregulated in human carotid atherosclerosis [3]. However, perlecan, together with biglycan, is the predominant proteoglycan in atherosclerotic lesions in mice [4]. Perlecan is synthesized by ECs and SMCs. Vascular endothelial growth factor (VEGF)-165 induces perlecan synthesis in ECs via VEGF receptor 2.

Syndecans are members of proteoglycans and a family of transmembrane core proteins capable of carrying HS and chondroitin sulfate (CS) chains. Four syndecan genes (syndecan 1, 2, 3, and 4) are present in vertebrates. The presence of HS chains allows for interactions with a large number of proteins, including heparin-binding growth factors such as fibroblast growth factors (FGFs), VEGFs, TGF-β, and platelet-derived growth factors (PDGFs). Further, HSs facilitate interactions with various ECMs, including fibronectin and plasma proteins such as antithrombin-1. Each syndecan family member has a distinct temporal and spatial pattern of expression. Syndecan 1 is expressed predominantly in epithelial and mesenchymal tissues, syndecan 2 in cells of mesenchymal origin and neuronal and epithelial cells, and syndecan 3 almost exclusively in neuronal and musculoskeletal tissue, whereas syndecan 4 is found in all cell types. Tumor necrosis factor α (TNF-α) upregulates syndecan 2 and downregulates syndecan 1 in ECs. In SMCs, FGF2 induces the expression of syndecan 4 but not syndecan 1 or 2. Syndecan 4 levels are markedly upregulated in several disease states, including arterial injury and acute myocardial infarction [5, 6].

HA is a member of proteoglycans and a major macromolecular polysaccharide component of ECM that confers structural frameworks for cells. In the arterial wall, HA is mainly produced by fibroblasts and SMCs [7]. The three members of the HA synthase (HAS) gene family are HAS1, HAS2, and HAS3, which share 55–71% amino acid sequence homology. The expression profiles of HAS genes are temporally and spatially regulated during embryogenesis and pathogenesis. Divergence in the transcriptional regulation of HAS genes can be explained to some extent by upstream signaling pathways that are triggered by various growth factors, cytokines, and cellular stress. Several NFκB-binding sites are also present, allowing for cytokine- and growth factor-induced transcription of HAS2. PDGF-BB, TGF-β, and cytokines such as interleukin (IL) 18 induce HAS2 expression.

Vitronectin

Vitronectin circulates as a single-chain (75-kDa) and two-chain (10- and 65-kDa) forms under reducing conditions. Vitronectin contains three glycosylation sites that contribute to approximately 30% of its molecular mass. The complete open reading frame of vitronectin encodes for 459 amino acids preceded by a signal peptide. Vitronectin is present in ECM, plasma, and granules of platelets. The liver is a major site of vitronectin biosynthesis. A second circulatory pool of vitronectin is contained within platelets. This rapidly releasable intracellular form accounts for only 0.8% of the circulating vitronectin pool. Other organs including brain, heart, skeletal muscle, lung, uterus, testis, and thymus are also found to contain vitronectin mRNA.

Fibronectin

Fibronectin is expressed by multiple cell types and involved in cell adhesion and migration. Two variations of fibronectin exist: (i) plasma fibronectin—a dimeric and soluble form secreted by hepatocytes directly into the circulation—lacking the alternatively spliced EDA and EDB sequences and (ii) cellular fibronectin—found in the ECM of tissues as a multimeric form assembled into fibrils—which contain variable proportions of the EDA and EDB domains. The normal arterial wall contains significant amounts of fibronectin that are devoid of the EDA and EDB domains. Atherosclerotic lesions and experimentally induced thickened rat aorta show a marked increase in fibronectin and fibronectin containing the EDA domain. In addition to its cell-binding properties, fibronectin binds other glycoproteins, as well as components of the complement and coagulation systems. Moreover, within the ECM, fibronectin can crosslink collagens, heparansulfate proteoglycans, and itself to provide a stable matrix platform in which cells reside.

Laminin

The basement membrane comprises two independent and distinct networks of collagen IV and laminin. The laminins are approximately cross-shaped heterotrimeric glycoproteins comprising α, β, and γ chains. The vascular endothelium expresses two laminin α chains depending on the EC type and state of growth or activation. Laminin chain isotype composition within atherosclerotic plaque tissue could differ from the chain composition in the media. Laminin 3 is exclusively present in media associated with plaque tissue. Laminin α2 is enriched in medial areas. Plaque tissue is predominantly enriched in laminin α5 chains. This general distribution applies to lesions both with and without a fibrous cap-like structure [8].

Receptors of ECM

The principal matrix receptors on vascular cells include integrins, members of the diverse family of transmembrane heterodimers that lack any inherent kinase activity [9]; CD44, a broadly expressed cell surface protein that is a receptor for hyaluronate; receptor for HA-mediated motility (RHAMM), which has been implicated in the regulation of SMCs migration; and discoidin domain receptors (DDRs), that can be activated by specific types of collagen [10]. The integrin profiles of ECs and SMCs in the normal vascular wall have been characterized. Both the adhesive and signaling properties of the integrins are critical for their biological activities. The adhesion of leukocytes to collagen is mediated by three types of collagen receptors: (1) integrins (α1β1, α2β1), (2) DDRs, and (3) scavenger receptors class A (SRA). Integrin α2β1 is constitutively expressed on macrophages at low levels, whereas α1β1 expression is upregulated during macrophage activation. Clustering β1 integrins with antibodies upregulates inflammatory cytokines expression, including TNF-α and IL-1.

BIOLOGICAL FUNCTIONS OF ECM

SMC Proliferation

SMC proliferation is an important event for the growth of atherosclerotic plaque, with collagen Iα critical regulator of proliferation. Plating SMCs on top of polymerized 3-D collagen I maintains SMC quiescence by upregulating the cyclin-dependent kinase inhibitors sp27^{Kip1} and p21^{Cip1} [11].

Mitogen and PDGF stimulate HA synthesis in SMCs and promote pericellular coat formation as these cells divide and migrate. Breaking the binding of HA to SMCs with competitive oligosaccharides or antibodies to HA receptors such as RHAMM blocks SMC proliferation and migration. In vein SMCs, small interfering RNA against HAS2 significantly suppresses SMC proliferation by attenuating cell cycle progression on the G1/S phase.

Perlecan binds to heparin-binding mitogens, such as FGF2, and prevents them from stimulating SMCs. The underlying mechanism could be related to apolipoprotein E, which was specifically found to induce the expression of perlecan in SMCs. Perlecan, which can be inhibited by PTEN, could also interact with nonheparin-binding mitogens such as thrombin and PDGF. Both thrombin and factor Xa increase SMC proliferation by transcription of FGF receptor-1.

Syndecan 4 is required for thrombin-induced SMC migration and proliferation [12]. Syndecan 4 gene deficiency inhibits injury-induced SMC proliferation and impairs FGF2- or PDGF-BB-induced proliferation, extracellular signal-regulated kinase activation, and cyclin D1 and Bcl-2 expression [13].

SMC Migration

During atherogenesis, SMCs migrate to populate the intima, either from the media or from the circulation via migration of CD34$^+$ hematopoietic progenitor cells, giving rise to smooth muscle progenitor cells. The ECM influences all steps in migration, including adhesion, spreading, and translocation. Plating SMCs on polymerized type I collagen suppresses actin-binding protein production, known to be important in cell spreading. Consistently, suspending SMCs within polymerized 3-D collagen I gels decreases the formation of actin stress fibers and focal adhesion and reduces the phosphorylation of focal adhesion kinase. The effect of HA on SMC migration disappears when CD44 receptors or its signal transmission is blocked. Perlecan interferes with the adhesion of SMCs to fibronectin, in the arterial media and intima. HS chains of proteoglycans also mediate a potent inhibitory signal for migration of SMCs [14]. Shear stress causes syndecan 4 dissociation from the focal adhesions, whereas overexpression of syndecan 4 blocks this dissociation and suppresses mechanical stress-induced cell migration.

EC Proliferation and Migration

Fragments of HA promote new blood vessel formation by stimulating CD44-mediated EC migration and proliferation and ECM synthesis of macromolecules, thus leading to neovascularization, a critical event in determining the severity of the lesions.

FGF2 interacts with the HS chains of proteoglycans on the EC surface [15, 16]. When complexed with HS chains, FGF-2 is protected from proteolytic degradation and maintains a capacity for long-term stimulation of EC proliferation, migration, and increased plasminogen activator activity. Perlecan deficiency in ECs upregulates p27^{kip1} expression and its nuclear localization. Introduction of a dominant-negative mutant of syndecan 4 leads to slow migration and impaired tube formation of ECs on matrigel. Syndecan 4$^{-/-}$ mice show reduced postnatal angiogenesis because of impaired EC migration.

Macrophage Differentiation

Signals transmitted from collagen regulate all aspects of macrophage biology, including differentiation, migration, inflammatory cytokine production, and the elaboration of matrix degrading enzymes. Plating human peripheral blood monocytes on a collagen I matrix results in increased differentiation and activation, as evidenced by increased cell spreading. Furthermore, adhesion to type I collagen stimulates macrophage phagocytic activity and increases the uptake of acetylated LDL. Human monocyte-derived macrophages plated on monomeric collagen I increases matrix metalloproteinase 9 (MMP-9) production, whereas plating cells on polymerized collagen I inhibits it. Macrophages from fibronectin-EIIIA-deficient mice show decreased uptake of lipoproteins, as well as less foam cell formation.

HA degradation products induce cytokine and chemokine expression by macrophages [17]. Also, HA can bind LDL particles and form complexes that are internalized *in vitro* into peritoneal macrophages more intensely than native LDL. Regarding monocytes, low-molecular-weight HA stimulates the incorporation of oxidized LDL (ox-LDL) through a CD44 signaling pathway. Low-molecular-weight HA is a potent inflammatory inducer in tissues, but high-molecular-weight or native HA has no such effect. Moreover, the depolymerization of HA into low-mass polymers occurs after tissue injury and during inflammation. Thus, HA drives the inflammatory response by retaining macrophages and regulating macrophage activation in part.

In macrophages, biglycan acts as proinflammatory protein, activating TNF-α and macrophage inflammatory protein 2 (MIP-2), via p38, ERK, and NF-κB [18]. Biglycan could also activate the inflammasome NLRP3 in macrophages. Soluble biglycan induces the NLRP3/ASC inflammasome, activating caspase-1 and releasing mature IL-1β without additional costimulatory factors.

Syndecan 1 deficiency increases leukocyte–endothelial interactions, and is also involved in chemokine gradient formation for *trans*-endothelial and *trans*-epithelial migration of neutrophils. The former process is facilitated by an IL-8/syndecan 1 complex and regulated by shedding of syndecan 1. *Trans*-epithelial neutrophil migration depends on matrilysin-mediated shedding of syndecan 1 from the mucosal surface of epithelium. Syndecan 4 plays an important role in the process of foam cell formation. Peritoneal macrophages from syndecan 4-deficient mice showed markedly reduced uptake of fluorescently labeled group VLDL as compared with wild-type cells.

Platelet Adhesion

Fibronectin cross-linked to fibrin by FXIII provides a substrate for increased platelet adhesion and thrombus growth. Fibronectin could mediate platelet adhesion by three RGD-recognizing integrins on platelets: (1) αIIbβ3, (2) αvβ3, and (3) α5β1. Antibodies of fibronectin decrease platelet adhesion to the subendothelium, and removing fibronectin from the perfusate decreases platelet adhesion to the subendothelium or fibrillar collagen III. Fibronectin contributes to platelet adhesion under shear conditions. *In vitro*, under static conditions, platelets adhere to absorbed fibronectin and under flow conditions, lysophosphatidic acid-treated platelets adhere to absorbed fibrin, fibronectin, and fibronectin cross-linked to fibrin. Static adhesion is mediated by αvβ3, αIIbβ3, and a β1 integrin.

Lipid Retention

The atherosclerosis development hypothesis of response to retention was first outlined by Williams and Tabas [19] in 1995, which proposes that the underlying mechanism initiating the development of atherosclerotic plaque is the retention of atherogenic (ApoB-containing) lipoproteins in the subendothelial matrix by their interaction with ECM molecules, particularly proteoglycans. This interaction can be direct or mediated by bridging molecules such as lipoprotein lipase (LPL), sphingomyelinase (SMase), or secretory phospholipase A2 (sPLA2). Retention of atherogenic lipoproteins within the subendothelial space leads to increased susceptibility to lipoprotein modification, which can then trigger an inflammatory response with alterations in SMC phenotype and function and infiltration of inflammatory cells [20].

Mice genetically modified to express modified forms of ApoB-100 with defective binding to proteoglycans show greatly reduced rate of development of atherosclerosis [21]. All proteoglycans can bind lipoproteins, and versican has a greater capacity for binding LDL than biglycan *in vitro*, presumably because of its greater number of GAG chains.

Collagen I and III bind to native LDL *in vitro*, and the extent of this binding is related to the net negative charge on LDL. Ox-LDL exhibited even greater collagen binding than did native LDL. Ox-LDL binding to SMC extracellular matrices involves a nonproteoglycan component of the matrix. ApoB-containing lipoproteins can be released from atherosclerotic tissue by elastase treatment. *In vitro*, elastin and LDL have a high-affinity interaction, with increased binding of LDL to elastin isolated from atherosclerotic versus normal arteries.

Role of ECM in Atherosclerosis

α1 Integrin mediates leukocyte invasion during atherogenesis. Mice with double knockout for integrin α1 and ApoE ($\alpha1^{-/-}$; $ApoE^{-/-}$) developed showed plaques with reduced content of macrophages and T cells as compared with their $\alpha1^{+/+}$; $ApoE^{-/-}$ littermates. By contrast, deletion of the α2 integrin in ApoE-null mice does not affect atherosclerotic plaque formation, which suggests that this receptor is not required for atherosclerosis. Overexpression of decorin by administration of an adenoviral vector in ApoE-null mice resulted in reduced development of atherosclerosis with decreased plaque macrophage and collagen content and reduced plasma TGF-β levels. Deficiency in decorin was found to increase atherosclerotic lesion area [22].

Heterozygous deletion of perlecan in ApoE-null mice significantly decreased atherosclerosis lesions in the aorta 12 weeks after chow feeding. Decreased number of atherosclerotic lesions was observed in HS-deficient perlecan mice cross-bred with ApoE-null mice. Binding of labeled mouse triglyceride-rich lipoproteins and human LDL to purified proteoglycans, prepared from SMCs deficient in perlecan HS, is reduced [23]. Atherosclerosis is associated with an increase in total fibronectin within the arterial wall, as well as the appearance of fibronectin containing EIIIA or EIIIB. $ApoE^{-/-}$/FN $EDA^{-/-}$ and $ApoE^{-/-}$/FN $EDA^{+/+}$ mice have smaller atherosclerotic lesions than do $ApoE^{-/-}$ mice with normally spliced fibronectin. Deletion of the EIIIA domain in ApoE ($^{-/-}$) mice also significantly decreased the lesions throughout the aortic tree [24].

In the early stage of atherosclerosis, HA is especially enriched around proliferating and migrating arterial SMCs. The accumulation of HA in early atherosclerotic lesions is often accompanied by an increase in molecules associated with HA, such as versican. *In vivo* investigations with transgenic mice, with HA expression driven by an αSMC-actin promoter, demonstrated that HA overproduction promoted atherosclerosis development in the aorta of ApoE-null mice [25].

The formation of atherosclerotic plaques in blood vessels is associated with an increase in biglycan expression, which may be due to an increase in TGF-β expression or abnormal mechanical stress in the prelesional arteries. Biglycan deposits colocalize with ApoB and E in human atherosclerotic plaques. The increase in biglycan expression in the plaques could be associated with an increase in retention of apolipoproteins by atherosclerotic intima by anchoring the apolipoproteins to the GAG side chains of biglycan. Biglycan also may regulate the binding and localization of other factors that contribute to the progression of the atherosclerotic plaques. The binding of biglycan to the class A scavenger receptor could enhance the adhesion and the retention of macrophages by the atherosclerotic plaques, thus causing the formation of foam cells. Therefore, biglycan may play a role at multiple levels in the pathogenesis of atherosclerosis. Long-term administration of the angiotensin II inhibitor telmisartan significantly reduces biglycan accumulation and inhibits atherosclerosis independently of blood pressure lowering and the collagenous plaque matrix.

Role of ECM in the Atherosclerosis-Related Process

Plaque rupture Atherosclerotic plaques are covered by a cap rich in fibrillar collagens and elastin. The cap confers considerable stability to the plaque, protecting against rupture and the devastating complication of thrombosis. Increased matrix degradation is a mechanism of lesion instability. The three MMP classes—interstitial collagenase, MMP-1; gelatinases, MMP-2 and MMP-9; and stromelysin, MMP-3—are expressed primarily in the shoulder regions of advanced plaques along with their endogenous inhibitors (tissue inhibitors of MMPs (TIMPs) 1 and 2). MMPs secreted by macrophages and SMCs in the shoulder regions of plaque degrade collagens in the fibrous cap, thus rendering the plaque vulnerable to rupture.

Atherosclerotic calcification Atherosclerotic calcification is a feature of advanced atherosclerotic lesions. Intimal and medial calcification proceed via distinct mechanisms, affect unique patient populations, and present different clinical ramifications. Histologically, intimal calcification is observed as hydroxyapatite crystals originating from secreted matrix vesicles near the lipid-rich necrotic core and the cellular debris and inflammatory cell infiltrate. Because hydroxyapatite crystals are propagated along ECM fibers, the composition of the ECM has a profound effect on regulating this process, and collagens, which are abundant in atherosclerotic lesions, are likely to contribute to intima calcification. Vascular calcification requires a permissive matrix. Bone matrix proteins such as osteocalcin, osteopontin, bone sialoprotein, and collagen II and IX can be expressed by SMCs. Roles have been established for several noncollagenous matrix proteins in promoting (e.g., fibronectin, elastin degradation products, elastin precursors, and decorin) or inhibiting (e.g., matrix gla protein-1, and osteopontin) this process. Fibronectin and collagen I matrixes promote calcification of cell lines derived from bovine aortic media *in vitro*, whereas collagen IV matrix is inhibitory. DDR1 deficiency significantly reduced the calcium content of the aortic arch in LDL($^{-/-}$) mice. Deletion of DDR1 in SMCs upregulates the expression nucleotide pyrophosphatase/phosphodiesterase 1, an inhibitor of tissue mineralization.

CONCLUSIONS

Our understanding of the role of the ECM in the initiation and progression of atherosclerosis is expanding at an increasing rate. Once thought to be biologically inert, the ECM is now known to consist of bioactive components that interact with and exert profound effects on many cell types within the vessel wall. In addition to their biological activities, the ECM is indeed a critical structural component of both the healthy and diseased vessel. We have made advancements in understanding how ECMs regulate atherogenesis, but there is still much to be learned, and several avenues of research remain largely unexplored.

REFERENCES

1. Kolb M, Margetts PJ, Sime PJ, Gauldie J. Proteoglycans decorin and biglycan differentially modulate TGF-beta-mediated fibrotic responses in the lung. *Am J Physiol Lung Cell Mol Physiol*. 2001;280:L1327–L1334.
2. Strom A, Ahlqvist E, Franzen A, Heinegard D, Hultgardh-Nilsson A. Extracellular matrix components in atherosclerotic arteries of apo E/LDL receptor deficient mice: An immunohistochemical study. *Histol Histopathol*. 2004;19: 337–347.
3. Tran PK, Agardh HE, Tran-Lundmark K, et al. Reduced perlecan expression and accumulation in human carotid atherosclerotic lesions. *Atherosclerosis*. 2007;190:264–270.

4. Kunjathoor VV, Chiu DS, O'Brien KD, LeBoeuf RC. Accumulation of biglycan and perlecan, but not versican, in lesions of murine models of atherosclerosis. *Arterioscler Thromb Vasc Biol*. 2002;22:462–468.
5. Geary RL, Koyama N, Wang TW, Vergel S, Clowes AW. Failure of heparin to inhibit intimal hyperplasia in injured baboon arteries. The role of heparin-sensitive and -insensitive pathways in the stimulation of smooth muscle cell migration and proliferation. *Circulation*. 1995;91:2972–2981.
6. Li J, Brown LF, Laham RJ, Volk R, Simons M. Macrophage-dependent regulation of syndecan gene expression. *Circ Res*. 1997;81:785–796.
7. Vigetti D, Ori M, Viola M, et al. Molecular cloning and characterization of UDP-glucose dehydrogenase from the amphibian xenopus laevis and its involvement in hyaluronan synthesis. *J Biol Chem*. 2006;281:8254–8263.
8. Rauch U, Saxena A, Lorkowski S, et al. Laminin isoforms in atherosclerotic arteries from mice and man. *Histol Histopathol*. 2011;26:711–724.
9. Shattil SJ, Ginsberg MH. Perspectives series: Cell adhesion in vascular biology. Integrin signaling in vascular biology. *J Clin Invest*. 1997;100:1–5.
10. Schlessinger J. Direct binding and activation of receptor tyrosine kinases by collagen. *Cell*. 1997;91:869–872.
11. Koyama H, Raines EW, Bornfeldt KE, Roberts JM, Ross R. Fibrillar collagen inhibits arterial smooth muscle proliferation through regulation of Cdk2 inhibitors. *Cell*. 1996;87:1069–1078.
12. Rauch BH, Millette E, Kenagy RD, Daum G, Fischer JW, Clowes AW. Syndecan-4 is required for thrombin-induced migration and proliferation in human vascular smooth muscle cells. *J Biol Chem*. 2005;280:17507–17511.
13. Ikesue M, Matsui Y, Ohta D, et al. Syndecan-4 deficiency limits neointimal formation after vascular injury by regulating vascular smooth muscle cell proliferation and vascular progenitor cell mobilization. *Arterioscler Thromb Vasc Biol*. 2011;31:1066–1074.
14. Koyama N, Kinsella MG, Wight TN, Hedin U, Clowes AW. Heparan sulfate proteoglycans mediate a potent inhibitory signal for migration of vascular smooth muscle cells. *Circ Res*. 1998;83:305–313.
15. Yayon A, Klagsbrun M, Esko JD, Leder P, Ornitz DM. Cell surface, heparin-like molecules are required for binding of basic fibroblast growth factor to its high affinity receptor. *Cell*. 1991;64:841–848.
16. Iozzo RV, San Antonio JD. Heparan sulfate proteoglycans: Heavy hitters in the angiogenesis arena. *J Clin Invest*. 2001;108:349–355.
17. McKee CM, Penno MB, Cowman M, et al. Hyaluronan (HA) fragments induce chemokine gene expression in alveolar macrophages. The role of HA size and CD44. *J Clin Invest*. 1996;98:2403–2413.
18. Schaefer L, Babelova A, Kiss E, et al. The matrix component biglycan is proinflammatory and signals through toll-like receptors 4 and 2 in macrophages. *J Clin Invest*. 2005;115: 2223–2233.
19. Williams KJ, Tabas I. The response-to-retention hypothesis of early atherogenesis. *Arterioscler Thromb Vasc Biol*. 1995;15:551–561.
20. Tannock LR, King VL. Proteoglycan mediated lipoprotein retention: A mechanism of diabetic atherosclerosis. *Rev Endocr Metab Disord*. 2008;9:289–300.
21. Skalen K, Gustafsson M, Rydberg EK, et al. Subendothelial retention of atherogenic lipoproteins in early atherosclerosis. *Nature*. 2002; 417:750–754.
22. Williams KJ. Arterial wall chondroitin sulfate proteoglycans: Diverse molecules with distinct roles in lipoprotein retention and atherogenesis. *Curr Opin Lipidol*. 2001;12:477–487.
23. Tran-Lundmark K, Tran PK, Paulsson-Berne G, et al. Heparan sulfate in perlecan promotes mouse atherosclerosis: Roles in lipid permeability, lipid retention, and smooth muscle cell proliferation. *Circ Res*. 2008;103:43–52.
24. Tan MH, Sun Z, Opitz SL, Schmidt TE, Peters JH, George EL. Deletion of the alternatively spliced fibronectin eiiia domain in mice reduces atherosclerosis. *Blood*. 2004;104:11–18.
25. Chai S, Chai Q, Danielsen CC, et al. Overexpression of hyaluronan in the tunica media promotes the development of atherosclerosis. *Circ Res*. 2005;96:583–591.

SECTION IV

BIOCHEMICAL MECHANISMS

28

EFFECTS OF NITRIC OXIDE ON ATHEROSCLEROSIS

Terrence Pong[1] and Paul L. Huang[2]

[1] School of Medicine, University of California, San Francisco, CA, USA
[2] Cardiovascular Research Center and Cardiology Division, Department of Medicine, Massachusetts General Hospital and Harvard Medical School, Boston, MA, USA

INTRODUCTION

The endothelial cell layer plays an important role in the development and pathogenesis of atherosclerosis. Nitric oxide (NO) is produced by the endothelium and serves as an important signaling molecule in the regulation of normal cardiovascular physiology. The endothelium uses NO to signal the relaxation of surrounding vascular smooth muscle, resulting in vasodilation and increased blood flow. NO is generated in blood vessels by the constitutive expression of endothelial nitric oxide synthase (eNOS). In addition, adequate levels of endothelial NO are required to provide various antiatherosclerotic properties including the prevention of leukocyte–endothelium adhesion [1], decreased influx of lipoproteins into the vascular wall through regulation of endothelial permeability [2], and inhibition of vascular smooth muscle cell (SMC) proliferation [3] (Table 28.1).

Atherosclerosis is associated with an impairment of endothelium-dependent relaxation and reduced NO bioavailability. Decreases in NO bioavailability can be caused by inactivation through reactions with reactive oxygen species (ROS) and reductions in the enzymatic synthesis of NO [4]. Approaches that increase the bioactivity of NO or restore eNOS-derived NO represent potential antiatherosclerotic principles. This chapter will highlight the role of NO and eNOS and their roles in vascular physiology and pathophysiological development of atherosclerosis.

IMPAIRED ENDOTHELIAL-DEPENDENT RELAXATION

The endothelial layer of blood vessels is critical to vascular health. Dysfunction of the endothelium is an important factor in the development of atherosclerosis. Endothelial production of NO can be activated through agonist stimulation by bradykinin and acetylcholine and through mechanical activity such

Atherosclerosis: Risks, Mechanisms, and Therapies, First Edition. Edited by Hong Wang and Cam Patterson.

TABLE 28.1 Atherosclerotic potential of functional and dysfunction endothelial nitric oxide systems

Functional endothelial NO system	Endothelial dysfunction
Coupled eNOS	Uncoupled eNOS
Atherosclerotic protection	Proatherosclerotic effects
Antioxidative capacity	Reduced NO availability
ROS scavenging	Increased ROS production
Decreased oxLDL formation	$\bullet O_2^-$ and $ONOO^-$ generation
Inhibition of SMC proliferation	Stimulation of SMC proliferation
Modulation of leukocyte adhesion–endothelial cell interactions	Increased oxLDL formation and recruitment of macrophages
Modulation of platelet aggregation	Platelet adhesion
Antiapoptotic environment	Reduced BH_4 bioavailability
Normal endothelial barrier function	Loss of barrier function
Inhibition of adhesion molecule expression	Increased adhesion molecule expression

as flow-mediated shear stress. Acetylcholine-mediated regulation of vascular tone requires an intact functional endothelium to properly signal the release of NO. The subsequent production of NO causes relaxation of adjacent SMCs. An impairment of endothelial-dependent relaxation is present in atherosclerotic vessels even before structural changes in the vasculature occur. Endothelial dysfunction is defined by the blunted vasodilatory response of the endothelium in response to acetylcholine. Although the underlying mechanisms responsible for causing endothelial dysfunction are multifactorial, the major cause can be attributed to dysfunction in the NO/eNOS pathway. Impairments in this pathway can manifest as reduced production of NO by eNOS, increased degradation of NO through reactions with ROS, and a decreased sensitivity to NO.

GENERATION OF NITRIC OXIDE: NITRIC OXIDE SYNTHASES

There are three identified nitric oxide synthase (NOS) isoforms: (1) endothelial NOS (eNOS), (2) cytokine-inducible NOS (iNOS), and (3) neuronal NOS (nNOS). Despite their names, a variety of cell types express these isoforms. eNOS is predominantly expressed in the endothelium, iNOS is involved in the immune response and produced primarily in cells of macrophage/monocyte lineage, while nNOS is expressed in certain neurons and skeletal muscle. nNOS and eNOS are both constitutively expressed and enzyme activity is primarily regulated by intracellular Ca^{2+} levels. eNOS and nNOS both require Ca^{2+}/calmodulin binding to stimulate NO synthesis. By contrast, iNOS binds Ca^{2+} tightly and remains active even at low levels of intracellular Ca^{2+} concentration. Consequently, NO production in eNOS and nNOS is primarily regulated through intracellular Ca^{2+} concentration, while the main switch for iNOS is at the level of transcription. While eNOS and nNOS are capable of producing NO in the picometer to nanometer range, iNOS in activated macrophages produces large amounts of NO, in the μM range, as a defense mechanism. This difference in NO-generating capacity has implications in the pathogenesis of atherosclerosis. Sustained low concentrations of NO are mainly vasoprotective, but excessive cytotoxic formation of NO may promote atherosclerotic lesion formation [4]. Thus, while endothelial dysfunction is associated with decreased NO bioavailability, excessively high concentrations of NO can also prove to be deleterious.

All three NOS enzymes utilize the amino acid L-arginine as substrate for the generation of NO producing L-citrulline as a product. NOS are NADPH-dependent enzymes that require Ca^{2+}/calmodulin, heat shock protein 90 (hsp90), FAD, FMN, and tetrahydrobiopterin (BH_4) as regulatory cofactors [5]. NOS activity can also be regulated at multiple steps, through transcription, substrate availability, and post-translation modifications.

In blood vessels, NO is produced by the endothelium through the constitutive expression of eNOS. eNOS is post-translationally modified through phosphorylation at several serine and threonine residues. In particular, phosphorylation of eNOS at the human serine^{1177} residue is an important mechanism of regulation of the production of vascular NO [6]. Structural studies indicate that active eNOS requires dimerization that facilitates electron flow through the active site pocket [7]. This interface includes the binding site for BH_4 and L-arginine, which stabilize the active dimeric form of eNOS. eNOS activity is regulated by Ca^{2+}-dependent interaction at the calmodulin binding domain, as well as by phosphorylation. Akt protein kinase phosphorylates serine^{1177} [8, 9], resulting in increased electron flux through the reductase domain and an increase in NO production [10]. eNOS also localizes to the endothelial cell caveolae, highly dynamic invaginations of the plasma membrane. Caveolin-1 is the main structural protein in endothelial caveolae and has been show to directly bind and inhibit eNOS [11]. For optimal eNOS activity, the dissociation of eNOS from caveolin-1 within the caveolae is required.

In blood vessels, the downstream target of NO is soluble guanylate cyclase (sGC). NO binds to the heme moiety of sGC, activating it and resulting in increased cyclic guanosine monophosphate (cGMP) production that mediates vascular SM relaxation. Under normal physiological conditions, eNOS enzymes exist as coupled dimers. However, decreased bioavailability of BH_4 or of the substrate L-arginine can switch eNOS from a coupled to uncoupled state [12]. The uncoupling of eNOS can produce superoxide anion ($\bullet O_2^-$) instead of NO. $\bullet O_2^-$ has a short half-life and is rapidly converted to other ROS including hydrogen peroxide (H_2O_2) and peroxynitrite ($ONOO^-$), adding oxidant stress on the blood vessels [13].

SHEAR STRESS-MEDIATED ROS FORMATION

The flow of blood through vessels constantly exposes the endothelial cell layer to shear stress and cyclic strain. The hemodynamic profile of blood flow, as well as the forces exerted on the endothelium, can differ at varying locations depending on the anatomy of the vasculature. A positive correlation exists between the clinical location of atherosclerotic plaques and regions of the cardiovascular system regularly exposed to low, oscillatory, or disturbed flow [14, 15]. Parts of the vasculature exposed to complex and disturbed flow patterns are more susceptible to atherosclerosis compared with areas with laminar flow where there are fewer discontinuities in flow. Models of blood flow at vascular bifurcations and vessel branches indicate that these areas are characterized by lower shear stress [16, 17]. Shear stress is involved in the modulation of eNOS expression and can induce short-term increases in eNOS transcription and longer term stabilization of eNOS mRNA [18, 19]. Indeed, studies have identified that arterial level shear stress (>15 dyne/cm^2) augments eNOS expression in an atheroprotective fashion while low shear stress (<4 dyne/cm^2) and low oscillatory regions stimulates a pro-atherosclerotic phenotype and a decrease in eNOS expression [20].

The flow-dependent endothelial balance between NO and $\bullet O_2^-$ may modulate the endothelial cell layer's vasoprotective properties and endothelial function. Laminar blood flow provides an antiatherosclerotic potential and increases NO and reduces $\bullet O_2^-$ formation [21]. In addition to shear stress-mediated NO regulation, continuous oscillatory shear stress can induce ROS formation through specific NAD(P)H complexes [22]; and while NO can act as a scavenger of ROS, an increase in endothelial ROS formation can negatively shift the balance between NO and ROS [4]. Therefore, hemodynamic forces can affect the balance between NO and $\bullet O_2^-$ and may modulate the development of atherosclerosis.

eNOS UNCOUPLING IN ATHEROSCLEROSIS

Endothelial dysfunction is considered to be an early marker of atherosclerosis and can develop before clinical angiographic or ultrasound evidence of atherosclerotic plaque formation. Under the pathological conditions of atherosclerosis, the reduction of molecular oxygen by eNOS may no longer be coupled with L-arginine and as a result, superoxide is produced instead of NO. This phenomena is termed "eNOS uncoupling" [23]. In normal vasculature, eNOS-generated NO prevents oxidation of LDL [24]. However, eNOS uncoupling generates increased oxidative stress that can cause the oxidation of LDL (oxLDL) within endothelial walls [25]. Further, oxLDL impairs endothelial progenitor cell survival and function by regulation of eNOS [26]. Progressive accumulation of macrophages and their uptake of oxLDL ultimately lead to the development of fatty streaks and atherosclerotic lesions.

Of the various cofactors involved in the regulation of eNOS activity, uncoupling of eNOS and the development of endothelial dysfunction have been critically linked to the enhanced oxidation of BH_4 and the decrease in BH_4 bioavailability [27]. Adequate BH_4 levels are required to maintain eNOS dimerization; when BH_4 bioavailability decreases, eNOS becomes unstable and uncoupled. Subsequently, less NO is produced and more superoxide generated. Recent reports have drawn interest to the BH_4 :BH_2 ratio, rather than the absolute concentration of BH_4 as the key determinant of eNOS uncoupling [28]. Under pro-oxidant conditions, BH_4 can be oxidized to dihydrobiopterin (BH_2), which can competitively replace eNOS-bound BH_4 [29, 30]. Figure 28.1 summarizes the effects of coupled and uncoupled eNOS in endothelial cell function.

LESSONS FROM GENETIC NOS MOUSE MODELS

NO is a potent oxidant produced by both endothelial cells and macrophages. Depending on the source of production, NO can exert both atherosclerotic and protective effects. Apolipoprotein E (ApoE) knockout mice have been a useful model to study human diet-induced atherosclerosis. Genetic mouse models of endothelial dysfunction have yielded important evidence on the roles of the three NOS isoforms in atherosclerosis. To study the effects of each individual NOS enzyme on atherosclerosis, all three NOS knockout mouse were individually bred to the apoE knockout background to create a double knockout mouse and subjected to a high-fat, Western diet to induce atherosclerosis.

Kuhlencordt et al. [31] demonstrated that the eNOS deficiency promoted atherosclerosis in eNOS/apoE double knockout mice. eNOS/apoE mice on high-fat diets had larger lesion areas and developed atherosclerotic plaques faster than apoE mice on the same diet. The eNOS/apoE model also introduced an array of cardiovascular complications including spontaneous aortic aneurysm and dissection, phenotypes that more closely resemble the spectrum of cardiovascular complications seen in human atherosclerosis. While eNOS/apoE mice are hypertensive, normalization of blood pressure with hydralazine treatment did not account for the accelerated atherosclerosis [32]. These findings indicate that the absence of endogenous eNOS accelerates atherosclerosis. Thus, the restoration of eNOS function in patients with atherosclerosis represents a potential therapeutic goal. Ozaki et al. [33] examined the effects of eNOS over expression in transgenic (eNOS-Tg) mice overexpressing eNOS in the endothelium. Unexpectedly, eNOS-Tg/apoE double knockout mice showed eNOS dysfunction, lower NO production levels relative to eNOS expression, and an acceleration in atherosclerotic lesion formation. Further investigation revealed a marked reduction in BH_4 and increased superoxide production, oral administration of BH_4 was able to reduce the formation of atherosclerotic plaques in these mice. This study indicates that an increase in eNOS expression alone, without complementary increases in BH_4, may be atherosclerotic and promote eNOS uncoupling.

Similar to the eNOS/apoE double knockout mice, nNOS/apoE double knockout mice on a

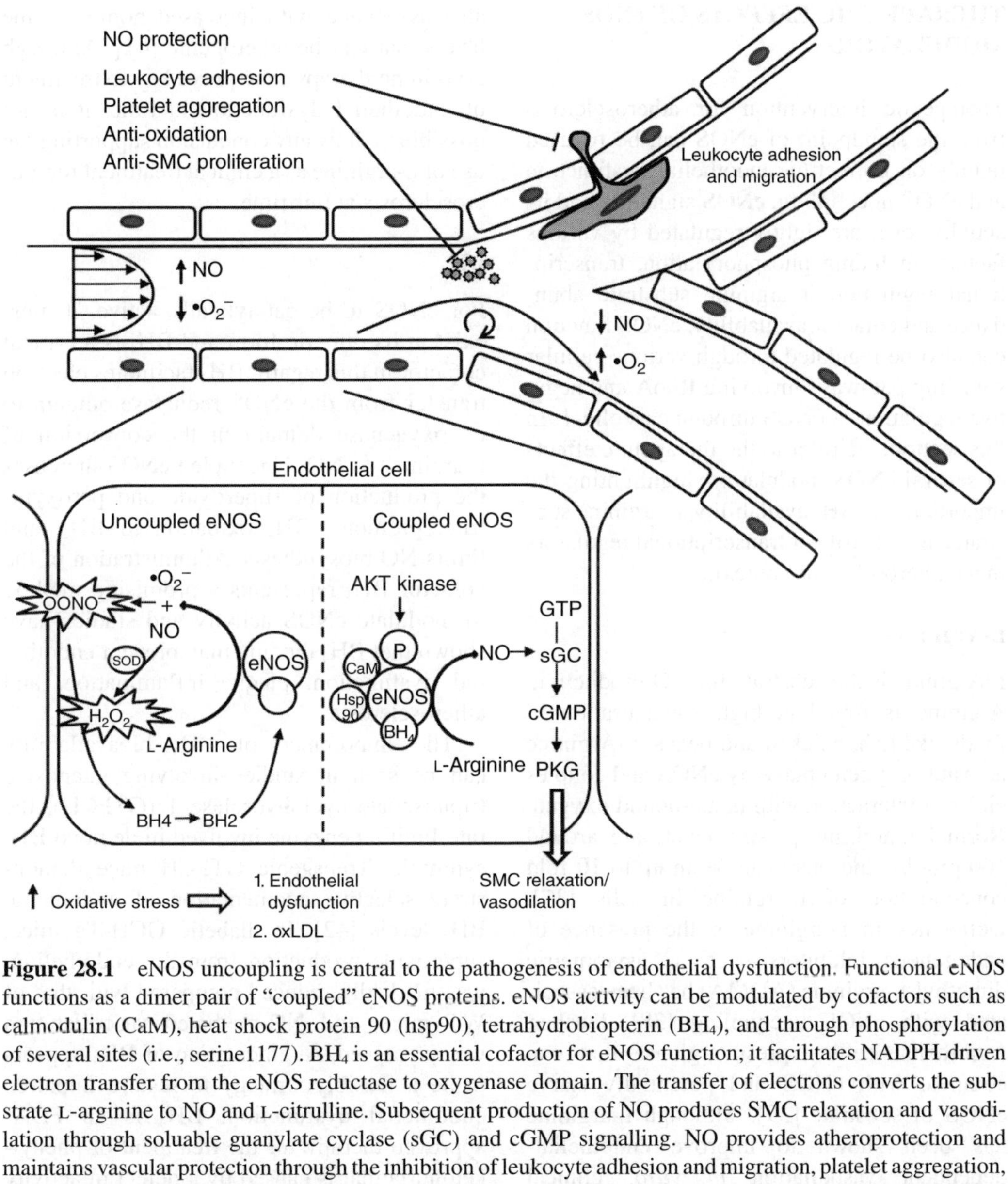

Figure 28.1 eNOS uncoupling is central to the pathogenesis of endothelial dysfunction. Functional eNOS functions as a dimer pair of "coupled" eNOS proteins. eNOS activity can be modulated by cofactors such as calmodulin (CaM), heat shock protein 90 (hsp90), tetrahydrobiopterin (BH_4), and through phosphorylation of several sites (i.e., serine1177). BH_4 is an essential cofactor for eNOS function; it facilitates NADPH-driven electron transfer from the eNOS reductase to oxygenase domain. The transfer of electrons converts the substrate L-arginine to NO and L-citrulline. Subsequent production of NO produces SMC relaxation and vasodilation through soluable guanylate cyclase (sGC) and cGMP signalling. NO provides atheroprotection and maintains vascular protection through the inhibition of leukocyte adhesion and migration, platelet aggregation, and neointimal cell proliferation. When the bioavailability of BH_4 becomes inadequate or the $BH_4 : BH_2$ ratio is upset, eNOS becomes unstable and uncoupling occurs. L-Arginine is instead converted to superoxide, leading to the formation of peroxynitrite and subsequent oxidation of LDL and endothelial dysfunction.

Western diet develop larger atherosclerotic lesions than their apoE counterparts [34]. Thus, there is evidence that nNOS may also play an atheroprotective role. In contrast to eNOS and nNOS, in the increased inflammatory environment of atherosclerosis, iNOS is unregulated in macrophages and produces an increased amount of NO [35]. The deletion of iNOS in apoE knockout mice decreases diet-induced atherosclerosis, indicating that iNOS-mediated oxidant stress has proatherogenic effects [36].

THERAPEUTIC EFFECTS OF eNOS MODULATORS

Therapeutic intervention for atherosclerosis from the standpoint of eNOS can be focused mainly on correcting endothelial dysfunction and eNOS uncoupling. eNOS signaling and its activity level are tightly regulated by various factors, including phosphorylation, transcriptional regulation, L-arginine substrate abundance and cofactor availability. eNOS function can also be regulated through various cellular signaling pathways involving RhoA and negative regulation of eNOS through caveolin-1. In this section, we refer to the therapeutic effects of several eNOS modulators, highlighting the importance of BH_4 availability, L-arginine substrate, and the role of transcriptional regulators in an atherosclerosis context.

L-Arginine

L-Arginine is the substrate for NO production. Arginine is found at high concentration in foods like fish, chicken and beans. L-Arginine activates oxygen uptake by eNOS and couples electron interaction with heme-bound oxygen. Normal L-arginine plasma levels are around 100 μmol/l, and there can be an up to 10-fold concentration of L-arginine in cells [37]. Deficiency in L-arginine or the presence of endogenous inhibitors such as asymmetric dimethyl-L-arginine (ADMA) has been associated with eNOS uncoupling [38]. Further, endothelial cells express arginase that can compete with eNOS and potentially starve eNOS of substrate [39]. Although L-arginine has been shown to improve endothelial-dependent vasodilation *in vitro*, clinical studies involving L-Arginine administration have shown inconsistent effects on endothelial function. The VINTAGE MI trial investigated L-arginine therapy in patients with acute myocardial infarction and found no improvements in vascular stiffness measurements or ejection fraction. Arginine supplementation was also associated with higher postinfarction mortality [40]. Long-term L-arginine supplementation is also associated with increased homocysteine levels that can be atherogenic [41]. Although L-arginine therapy may prove key to treatment of endothelial dysfunction in time, it is not possible to draw any conclusion supporting the use of L-arginine as a clinical treatment for atherosclerosis at this time.

BH_4

For eNOS to be catalytically active, it must exist in its dimeric form and BH_4 is a critical cofactor in this regard. BH_4 facilitates electron transfer from the eNOS reductase domain to its oxygenase domain in the conversion of L-arginine to NO. Uncoupled eNOS increases the production of superoxide and peroxynitrite, promotes BH_4 oxidation to BH_2, and limits NO biosynthesis. Administration of the cofactor BH_4 represents a promising method to modulate eNOS activity and studies have shown that BH_4 therapy may prevent endothelial dysfunction, plaque inflammation, and atherosclerosis.

The importance of BH_4 bioavailability can be seen in studies involving guanosine triphosphate cyclohydrolase 1 (GTPCH), the rate-limiting enzyme involved in de novo BH_4 synthesis. Transgenic GTPCH mice demonstrate selective augmentation of endothelial BH_4 levels [42]. In diabetic GCH-Tg mice, superoxide production from the endothelium was markedly reduced compared with that of WT mice, and NO-mediated vasodilatation was preserved. Augmentation of BH_4 appears to be a feasibly strategy to restore impaired endothelial dysfunction. BH_4 is an FDA-approved therapy for the treatment of phenylketonuria that is caused by a defect in activity of the phenylalanine hydroxylase enzyme. In mice with hypertension-induced heart failure, exogenous BH_4 supplementation increase NOS activity by recoupling eNOS and ameliorates preexisting advanced cardiac hypertrophy and fibrosis [43]. Administration of sepiapterin, the physiological precursor of BH_4, leads to a significant improvement of endothelium-dependent vasodilatation in *ex vivo* human and porcine vessels with atherosclerosis [44]. Treatment of apoE knockout mice (subjected to disturbed flow with partial carotid ligation) with oral BH_4 in drinking water recoupled NOS, decreased superoxide production, improved endothelium-dependent vasodilatation, and virtually eliminated atherosclerosis [45]. Further, chronic studies of orally administered BH_4 have been reported to improve endothelial function. Infusion of BH_4 in humans improves age-related impairment of endothelium-dependent vasodilation through increased in nitric oxide production [46]. Importantly, Moens et al. [47] report a bimodal dose-dependent cardiac response to tetrahydrobiopterin where the efficacy of BH_4 first rises then declines at higher doses. BH_2 rose linearly with dose, and at higher BH_4 dosages, lowered the BH_4/BH_2 ratio back toward control. This study highlights an important ratio that must be considered during the clinical use of BH_4 and the need to maintain suitable BH_4 : BH_2 ratios. Future studies are needed to determine the dose–response relationship of BH_4 and its ability to recouple eNOS in the setting of atherosclerosis and to prevent superoxide production.

RhoA-ROCK Inhibition

RhoA is a member of the Rho protein family. RhoA acts as a molecular switch that cycles between an inactive GDP-bound and an active GTP-bound conformation to elicit signal transduction cascades related to vascular inflammation. RhoA and its first effectors, the Rho-kinases (ROCK), have been implicated in the pathogenesis of cardiovascular disease and have been shown to be upregulated in inflammatory atherosclerotic lesions. In endothelial cells, Rho kinase activation leads to the downregulation of eNOS and activation of proinflammatory pathways [48]. ROCK affects eNOS mRNA stability and reduces eNOS-phyosphorylation and catalytic activity. ROCK inhibitors such as fasudil have been shown to reverse the effects of ROCK on arteriolar SMCs and improve endothelial function [49]. Fasudil administration leads to activation of the Akt/eNOS pathway and cardiovascular protection [50] with decreased leukocyte recruitment and adhesion after ischemia reperfusion injury in wild-type but not eNOS knockout mice. In addition, exercise has been well established as effective prevention against atherosclerotic disease. Matsumoto et al. [51] recently reported that treadmill exercise downregulated gene expression of RhoA in the endothelial cells of rats and administration of fasudil normalized eNOS expression and improved endothelial dysfunction in the aorta of spontaneously hypertensive rats [52]. These studies suggest that fasudil and the use of Rho kinase inhibitors may be effective to limit the development of atherosclerosis.

eNOS Transcriptional Enhancers

Various physiological stimuli, including the shear stress of blood flow, modulate the transcriptional activity of the eNOS gene. Increased transcription of eNOS can result in upregulated expression of eNOS protein and improved eNOS-dependent activity. Recent chemical library screening has yielded two novel small molecules, AVE9488 and AVE3085, that have demonstrated the ability upregulate eNOS gene expression. Both wild-type and apoE knockout mice treated with AVE9488 resulted in enhanced vascular eNOS expression [53]. ApoE knockout mice treated with AVE9488 displayed reduced cuff-induced neointima formation, while these beneficial effects were lost in eNOS knockout mice. Treatment with AVE9488 and AVE3085 was able to reduce atherosclerotic plaque formation in apoE knockout mice, but not in apoE/eNOS double knockout mice. Pharmacological intervention with AVE9488 was also able to protect against myocardial ischemia reperfusion injury [54], while AVE3085 has been shown to restore endothelial function in spontaneously hypertensive model [55]. These studies suggest that upregulation of eNOS expression and a reversal of eNOS uncoupling achieved by these small molecules can decrease endothelial dysfunction and provide and atherosclerotic protection. Further studies will have to be conducted to determine their clinical efficacy.

CONCLUSIONS

NO production through the NO synthases thus represents a double-edged sword. Results from NOS/apoE double knockout models and our understanding of relevant NOS cofactors have demonstrated the importance of functional NOS dimerization and the deleterious effects of eNOS uncoupling. It is important to note that the NO/$\bullet O_2^-$ equilibrium is tightly regulated under normal physiological conditions, and that this balance can be disturbed by endothelial dysfunction. Depending on its source of production (iNOS vs. eNOS/nNOS) and the concentration, NO can exert both atherosclerotic and protective effects.

REFERENCES

1. Kubes P, Suzuki M, Granger DN. Nitric oxide: an endogenous modulator of leukocyte adhesion. *Proceedings of the National Academy of Sciences*. 1991;88(11):4651–5.
2. Draijer R, Atsma DE, van der Laarse A, van Hinsbergh VWM. cGMP and nitric oxide modulate thrombin-induced endothelial permeability. *Circulation Research*. 1995;76(2): 199–208.
3. Garg UC, Hassid A. Nitric oxide-generating vasodilators and 8-bromo-cyclic guanosine monophosphate inhibit mitogenesis and proliferation of cultured rat vascular smooth muscle cells. *The Journal of Clinical Investigation*. 1989; 83(5):1774–7.
4. Muller G, Morawietz H. Nitric oxide, NAD(P)H oxidase, and atherosclerosis. *Antioxidants & Redox Signaling*. 2009;11(7):1711–31.
5. Liu VWT, Huang PL. Cardiovascular roles of nitric oxide: a review of insights from nitric oxide synthase gene disrupted mice. *Cardiovascular Research*. 2008;77(1):19–29.
6. Atochin D, Huang P. Endothelial nitric oxide synthase transgenic models of endothelial dysfunction. *Pflügers Archiv European Journal of Physiology*. 2010;460(6):965–74.
7. Alderton WK, Cooper CE, Knowles RG. Nitric oxide synthases: structure, function and inhibition. *The Biochemical Journal*. 2001;357 (Pt 3):593–615.
8. Fulton D, Gratton J-P, McCabe TJ, et al. Regulation of endothelium-derived nitric oxide production by the protein kinase Akt. *Nature*. 1999;399(6736):597–601.
9. Dimmeler S, Fleming I, Fisslthaler B, Hermann C, Busse R, Zeiher AM. Activation of nitric oxide synthase in endothelial cells by Akt-dependent phosphorylation. *Nature*. 1999; 399(6736):601–5.
10. McCabe TJ, Fulton D, Roman LJ, Sessa WC. Enhanced electron flux and reduced calmodulin dissociation may explain "calcium-independent" eNOS activation by phosphorylation. *Journal of Biological Chemistry*. 2000;275(9): 6123–8.
11. Bernatchez PN, Bauer PM, Yu J, Prendergast JS, He P, Sessa WC. Dissecting the molecular control of endothelial NO synthase by caveolin-1 using cell-permeable peptides. *Proceedings of the National Academy of Sciences of the United States of America*. 2005;102(3):761–6.
12. Landmesser U, Dikalov S, Price SR, et al. Oxidation of tetrahydrobiopterin leads to uncoupling of endothelial cell nitric oxide synthase in hypertension. *The Journal of Clinical Investigation*. 2003;111(8):1201–9.
13. Li J-M, Shah AM. Endothelial cell superoxide generation: regulation and relevance for cardiovascular pathophysiology. *American Journal of Physiology—Regulatory, Integrative and Comparative Physiology*. 2004;287(5): R1014–30.
14. Won D, Zhu S-N, Chen M, et al. Relative reduction of endothelial nitric-oxide synthase expression and transcription in atherosclerosis-prone regions of the mouse aorta and in an in vitro model of disturbed flow. *The American Journal of Pathology*. 2007;171(5):1691–704.
15. Ku DN, Giddens DP, Zarins CK, Glagov S. Pulsatile flow and atherosclerosis in the human carotid bifurcation. Positive correlation between plaque location and low oscillating shear stress. *Arteriosclerosis, Thrombosis, and Vascular Biology*. 1985;5(3):293–302.
16. Malek AM, Alper SL, Izumo S. Hemodynamic shear stress and its role in atherosclerosis. *JAMA: The Journal of the American Medical Association*. 1999;282(21):2035–42.

17. Zhao SZ, Xu XY, Hughes AD, et al. Blood flow and vessel mechanics in a physiologically realistic model of a human carotid arterial bifurcation. *Journal of Biomechanics*. 2000;33(8):975–84.
18. Davis ME, Cai H, Drummond GR, Harrison DG. Shear stress regulates endothelial nitric oxide synthase expression through c-Src by divergent signaling pathways. *Circulation Research*. 2001; 89(11):1073–80.
19. Davis ME, Grumbach IM, Fukai T, Cutchins A, Harrison DG. Shear stress regulates endothelial nitric-oxide synthase promoter activity through nuclear factor κB binding. *Journal of Biological Chemistry*. 2004;279(1):163–8.
20. Cheng C, van Haperen R, de Waard M, et al. Shear stress affects the intracellular distribution of eNOS: direct demonstration by a novel in vivo technique. *Blood*. 2005;106(12):3691–8.
21. Fleming I, Busse R. Molecular mechanisms involved in the regulation of the endothelial nitric oxide synthase. *American Journal of Physiology—Regulatory, Integrative and Comparative Physiology*. 2003;284(1):R1–12.
22. De Keulenaer GW, Chappell DC, Ishizaka N, Nerem RM, Alexander RW, Griendling KK. Oscillatory and steady laminar shear stress differentially affect human endothelial redox state. *Circulation Research*. 1998;82(10):1094–101.
23. Förstermann U, Münzel T. Endothelial nitric oxide synthase in vascular disease. *Circulation*. 2006;113(13):1708–14.
24. Rubbo H, Trostchansky A, Botti H, Batthyány C. Interactions of nitric oxide and peroxynitrite with low-density lipoprotein. *Biological Chemistry*. 2002;383(3–4):547–52.
25. Victor VM, Rocha M, Sola E, Banuls C, Garcia-Malpartida K, Hernandez-Mijares A. Oxidative stress, endothelial dysfunction and atherosclerosis. *Current Pharmaceutical Design*. 2009; 15(26):2988–3002.
26. Ma FX, Zhou B, Chen Z, et al. Oxidized low density lipoprotein impairs endothelial progenitor cells by regulation of endothelial nitric oxide synthase. *Journal of Lipid Research*. 2006;47(6):1227–37.
27. Alp NJ, Channon KM. Regulation of endothelial nitric oxide synthase by tetrahydrobiopterin in vascular disease. *Arteriosclerosis, Thrombosis, and Vascular Biology*. 2004;24(3):413–20.
28. Crabtree MJ, Tatham AL, Al-Wakeel Y, et al. Quantitative regulation of intracellular endothelial nitric-oxide synthase (eNOS) coupling by both tetrahydrobiopterin-eNOS stoichiometry and biopterin redox status. *Journal of Biological Chemistry*. 2009;284(2):1136–44.
29. Bec N, Gorren ACF, Voelker C, Mayer B, Lange R. Reaction of neuronal nitric-oxide synthase with oxygen at low temperature. *Journal of Biological Chemistry*. 1998;273(22):13502–8.
30. Crabtree MJ, Smith CL, Lam G, Goligorsky MS, Gross SS. Ratio of 5,6,7,8-tetrahydrobiopterin to 7,8-dihydrobiopterin in endothelial cells determines glucose-elicited changes in NO vs. superoxide production by eNOS. *American Journal of Physiology—Heart and Circulatory Physiology*. 2008;294(4):H1530–40.
31. Kuhlencordt PJ, Gyurko R, Han F, et al. Accelerated atherosclerosis, aortic aneurysm formation, and ischemic heart disease in apolipoprotein E/endothelial nitric oxide synthase double-knockout mice. *Circulation*. 2001; 104(4):448–54.
32. Chen J, Kuhlencordt PJ, Astern J, Gyurko R, Huang PL. Hypertension does not account for the accelerated atherosclerosis and development of aneurysms in male apolipoprotein E/endothelial nitric oxide synthase double knockout mice. *Circulation*. 2001;104(20):2391–4.
33. Ozaki M, Kawashima S, Yamashita T, et al. Overexpression of endothelial nitric oxide synthase accelerates atherosclerotic lesion formation in apoE-deficient mice. *The Journal of Clinical Investigation*. 2002;110(3):331–40.
34. Kuhlencordt PJ, Hötten S, Schödel J, et al. Atheroprotective effects of neuronal nitric oxide synthase in apolipoprotein E knockout mice. *Arteriosclerosis, Thrombosis, and Vascular Biology*. 2006;26(7):1539–44.
35. Kietadisorn R, Juni RP, Moens AL. Tackling endothelial dysfunction by modulating NOS uncoupling: new insights into its pathogenesis and therapeutic possibilities. *American Journal of Physiology—Endocrinology and Metabolism*. 2012;302(5):E481–95.
36. Kuhlencordt PJ, Chen J, Han F, Astern J, Huang PL. Genetic deficiency of inducible nitric oxide synthase reduces atherosclerosis and lowers plasma lipid peroxides in apolipoprotein E–knockout mice. *Circulation*. 2001;103(25): 3099–104.
37. Förstermann U, Sessa WC. Nitric oxide synthases: regulation and function. *European Heart Journal*. 2012;33(7):829–37.
38. Antoniades C, Shirodaria C, Leeson P, et al. Association of plasma asymmetrical dimethylarginine (ADMA) with elevated vascular

superoxide production and endothelial nitric oxide synthase uncoupling: implications for endothelial function in human atherosclerosis. *European Heart Journal*. 2009;30(9):1142–50.

39. Berkowitz DE, White R, Li D, et al. Arginase reciprocally regulates nitric oxide synthase activity and contributes to endothelial dysfunction in aging blood vessels. *Circulation*. 2003;108(16):2000–6.
40. Schulman SP, Becker LC, Kass DA, et al. L-arginine therapy in acute myocardial infarction: the vascular interaction with age in myocardial infarction (VINTAGE MI) randomized clinical trial. *JAMA: The Journal of the American Medical Association*. 2006;295(1): 58–64.
41. Jahangir E, Vita JA, Handy D, et al. The effect of l-arginine and creatine on vascular function and homocysteine metabolism. *Vascular Medicine*. 2009;14(3):239–48.
42. Alp NJ, Mussa S, Khoo J, et al. Tetrahydrobiopterin-dependent preservation of nitric oxide-mediated endothelial function in diabetes by targeted transgenic GTP-cyclohydrolase I overexpression. *The Journal of Clinical Investigation*. 2003;112(5):725–35.
43. Moens AL, Takimoto E, Tocchetti CG, et al. Reversal of cardiac hypertrophy and fibrosis from pressure overload by tetrahydrobiopterin. *Circulation*. 2008;117(20):2626–36.
44. Tiefenbacher CP, Bleeke T, Vahl C, Amann K, Vogt A, Kübler W. Endothelial dysfunction of coronary resistance arteries is improved by tetrahydrobiopterin in atherosclerosis. *Circulation*. 2000;102(18):2172–9.
45. Li L, Chen W, Rezvan A, Jo H, Harrison DG. Tetrahydrobiopterin deficiency and nitric oxide synthase uncoupling contribute to atherosclerosis induced by disturbed flow. *Arteriosclerosis, Thrombosis, and Vascular Biology*. 2011;31(7): 1547–54.
46. Higashi Y, Sasaki S, Nakagawa K, et al. Tetrahydrobiopterin improves aging-related impairment of endothelium-dependent vasodilation through increase in nitric oxide production. *Atherosclerosis*. 2006;186(2):390–5.
47. Moens AL, Ketner EA, Takimoto E, et al. Bi-modal dose-dependent cardiac response to tetrahydrobiopterin in pressure-overload induced hypertrophy and heart failure. *Journal of Molecular and Cellular Cardiology*. 2011; 51(4):564–9.
48. Satoh K, Fukumoto Y, Shimokawa H. Rho-kinase: important new therapeutic target in cardiovascular diseases. *American Journal of Physiology—Heart and Circulatory Physiology*. 2011;301(2):H287–96.
49. Batchelor TJP, Sadaba JR, Ishola A, Pacaud P, Munsch CM, Beech DJ. Rho-kinase inhibitors prevent agonist-induced vasospasm in human internal mammary artery. *British Journal of Pharmacology*. 2001;132(1):302–8.
50. Wolfrum S, Dendorfer A, Rikitake Y, et al. Inhibition of rho-kinase leads to rapid activation of phosphatidylinositol 3-kinase/protein kinase Akt and cardiovascular protection. *Arteriosclerosis, Thrombosis, and Vascular Biology*. 2004;24(10):1842–7.
51. Matsumoto A, Mason SR, Flatscher-Bader T, et al. Effects of exercise and antioxidant supplementation on endothelial gene expression. *International Journal of Cardiology*. 2012; 158(1):59–65.
52. Tsounapi P, Saito M, Kitatani K, et al. Fasudil improves the endothelial dysfunction in the aorta of spontaneously hypertensive rats. *European Journal of Pharmacology*. 2012; 691(1–3):182–9.
53. Wohlfart P, Xu H, Endlich A, et al. Antiatherosclerotic effects of small-molecular-weight compounds enhancing endothelial nitric-oxide synthase (eNOS) expression and preventing eNOS uncoupling. *Journal of Pharmacology and Experimental Therapeutics*. 2008;325(2):370–9.
54. Frantz S, Adamek A, Fraccarollo D, et al. The eNOS enhancer AVE 9488: a novel cardioprotectant against ischemia reperfusion injury. *Basic Research in Cardiology*. 2009;104(6): 773–9.
55. Yang Q, Xue H-M, Wong W-T, et al. AVE3085, an enhancer of endothelial nitric oxide synthase, restores endothelial function and reduces blood pressure in spontaneously hypertensive rats. *British Journal of Pharmacology*. 2011; 163(5):1078–85.

29

PROTECTIVE ROLE OF HEME OXYGENASE-1 IN ATHEROSCLEROSIS

William Durante

Department of Medical Pharmacology and Physiology, School of Medicine, University of Missouri-Columbia, Columbia, MO, USA

INTRODUCTION

Heme oxygenase (HO) plays a fundamental role in regulating intracellular heme levels by catalyzing the first and rate-limiting step in heme metabolism. HO oxidatively degrades heme into equimolar amounts of biliverdin, free iron, and carbon monoxide (CO) (Figure 29.1). This reaction requires molecular oxygen and reducing equivalents derived from nicotinamide adeninine dinucleotide phosphate (NADPH) that are donated to HO by NADPH-cytochrome P450 reductase [1]. The metabolism of heme by HO is inhibited by various metalloporphyrins, including zinc protoporphyrin-IX (ZnPP) and tin protoporphyrin-IX (SnPP), which compete with heme for binding to HO. Subsequently, biliverdin is converted to bilirubin by biliverdin reductase, while iron is promptly sequestered by ferritin or excreted and recycled for heme synthesis. There are two isoforms of HO that are products of separate genes and exhibit unique tissue distribution and molecular properties. HO-2 is a 36-kDa protein that is constitutively expressed and found in high concentration in specific organs, such as the brain and testes. By contrast, HO-1 is a ubiquitously distributed 32-kDa protein that is strongly induced by its substrate heme and a myriad of biochemical and biophysical factors. HO-1 expression is principally regulated at the transcriptional level by a diverse array of *cis*-acting responsive elements in the promoter region of the gene that preferentially respond to stress-activated transcription factors such as nuclear factor E2-related factor-2 (Nrf2), activator protein-1 (AP-1), and nuclear factor-κB (NF-κB) [2–4]. Interestingly, several known risk factors and stimuli that are associated with the development of cardiovascular disease, including high blood pressure, cigarette smoke, oxidized lipids, homocysteine, inflammatory cytokines, and advanced glycation end products stimulate the induction of HO-1, suggesting a role for HO-1 in regulating the disease process (Figure 29.1).

Atherosclerosis: Risks, Mechanisms, and Therapies, First Edition. Edited by Hong Wang and Cam Patterson.

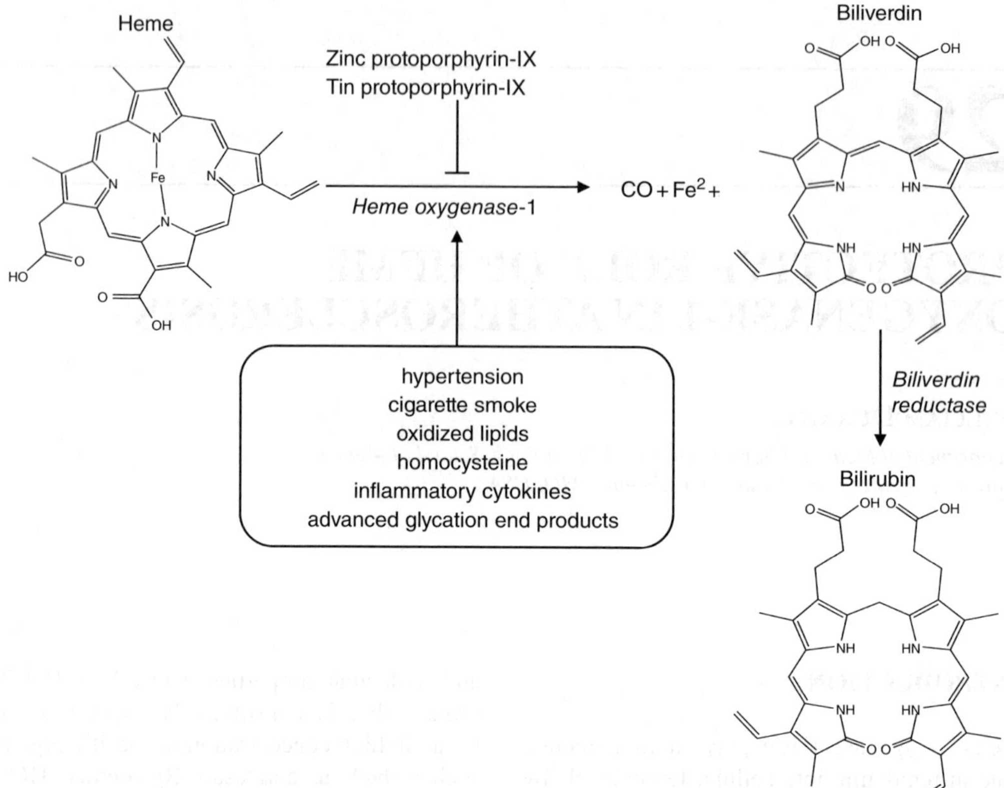

Figure 29.1 Regulation of heme metabolism by heme oxygenase-1 (HO-1). HO-1 catalyzes the degradation of heme into carbon monoxide (CO), ferrous iron (Fe^{2+}), and biliverdin. Biliverdin is promptly metabolized by biliverdin reductase to bilirubin. HO-1 is induced by a number of atherogenic risk factors and molecules, including hypertension, cigarette smoke, oxidized lipids, homocysteine, inflammatory cytokines, and advanced glycation end products. HO-1 activity is inhibited by various metalloporphyrins, such as zinc protoporphyrin-IX and tin protoporphryin-IX.

Original interest in HO-1 centered on its heme degrading properties and for many years CO and bilirubin were regarded as potentially toxic waste products of the reaction. However, research in the past couple of decades discovered that these heme metabolites possess important biological activities, expanding the study of HO-1 well beyond its initial role in heme metabolism. It is now recognized that the induction of HO-1 serves an integral defense mechanism against tissue damage. The cytoprotection conferred by HO-1 arises via several mechanisms, including its ability to decompose pro-oxidant and potentially toxic-free heme, stimulate iron efflux from cells, and induce ferritin expression that sequesters free iron. In addition, the HO-1-derived products biliverdin, bilirubin, and CO possess potent antioxidant, anti-inflammatory, antiapoptotic, antithrombotic, and growth-regulatory properties [3–5]. Emerging evidence indicates that HO-1 protects against vascular disease. Genetic deficiency or reduced expression of HO-1 is associated with oxidative tissue damage, inflammation, thrombosis, hypertension, and atherosclerosis in both mice and humans, whereas increased expression of HO-1 improves vascular dysfunction in several disease states [3, 4]. Furthermore, low serum concentrations of the heme metabolite bilirubin are associated with an increased risk of coronary and peripheral artery disease, stroke, and diabetes [6].

This chapter will specifically review the role of HO-1 in atherosclerosis and examine cellular and molecular mechanisms that contribute to the vasoprotective properties of this enzyme, focusing on the actions of CO and bilirubin. In addition, it will present potential therapeutic options in targeting HO-1 or its reaction products in the treatment and prevention of atherosclerosis.

PROTECTIVE ROLE OF HO-1 IN ATHEROSCLEROSIS

A preponderance of evidence suggests that HO-1 plays a salutary role in atherosclerosis. HO-1 is prominently expressed in atherosclerotic lesions of humans and experimental animals [7, 8]. The enzyme is detected in endothelial cells (ECs) and macrophages in the early stages of atherosclerotic lesion formation and in foam cells and vascular smooth muscle cells (VSMCs) residing in the necrotic core of advanced lesions. The degree of HO-1 expression correlates with plaque burden and phenotype. Increased HO-1 expression is observed in atherosclerotic plaque that contains large amounts of lipids, macrophages, and inflammatory mediators [9, 10]. Furthermore, leukocytes from patients with coronary artery disease express HO-1, and the magnitude of HO-1 expression correlates with the severity of disease with patients suffering from acute myocardial infarction displaying the highest levels of expression [11]. Interestingly, HO-1 expression is more frequent in atherectomy samples derived from asymptomatic compared with symptomatic patients, suggesting a putative role for HO-1 in promoting plaque stability. Although a number of atherogenic stimuli may contribute to the induction of HO-1 in the vasculature, oxidized phospholipids are likely to play a predominant role since they are potent inducers of HO-1 gene expression and colocalize with HO-1 in atherosclerotic lesions [12, 13]. Thus, the induction of HO-1 in atherosclerosis likely represents an adaptive response to the disease process rather than its cause.

Studies in two patients with rare HO-1 deficiency further supports a protective role for HO-1 in the circulation [14, 15]. These individuals exhibited marked intravascular hemolysis, hyperlipidemia, inflammation, and hypertension. Extensive EC damage was also observed and an autopsy report on one of the patients, a 6-year-old boy, detected fatty streaks, and fibrous plaques in the aorta [16]. The notion that HO-1 may protect against vascular disease in humans is also supported by studies analyzing polymorphisms in the 5′-flanking region of the human HO-1 gene. In particular, several studies have reported that a short $(GT)n$ microsatellite polymorphism that is linked to higher HO-1 transcriptional activity and expression is associated with reduced risk of coronary artery disease and coronary events in some patient populations [17]. Chinese diabetic patients possessing long GT repeats have a 4.7-fold increase in the incidence of coronary artery disease. In addition, a reduced frequency of coronary artery disease in carriers of short GT repeats was observed in patients undergoing coronary angiography in high-risk subgroups with hyperlipidemia, diabetes, and in current smokers. However, these findings in the Asian population were not replicated in Caucasian patients. The beneficial effects noted in some studies may reflect the ability of the GT repeat promoter polymorphism to influence EC function. ECs possessing the short GT promoter repeat exhibit higher HO-1 expression and greater survival under conditions of oxidative stress. In addition, they proliferate at a faster rate and generate lower concentrations of proinflammatory cytokines [18].

Animal studies provide strong support for the beneficial effect of HO-1 in atherosclerosis. Pharmacological induction of HO-1 by hemin significantly decreases atherosclerotic lesion formation in low-density lipoprotein (LDL) receptor-deficient mice fed a high-fat diet [13]. Systemic administration of an HO-1 adenovirus also attenuates lesion size in both young and old apolipoprotein E (apoE)-deficient mice relative to littermates injected with a control empty vector [19]. Conversely, inhibition of HO activity by SnPP augments atherosclerotic lesion formation in Watanabe heritable hyperlipidemic rabbits and in LDL-receptor-deficient mice fed a high-fat diet [12, 13]. Interestingly, HO-1 inhibition increases plasma and tissue lipid peroxide levels without affecting plasma lipid composition, suggesting that the antiatherogenic property of HO-1 occurs, in part, through prevention of lipid peroxidation. Consistent with the pharmacological studies, deletion of HO-1 in apoE-knockout mice placed on a Western diet causes larger and more advanced lesions in the absence of any change in total plasma cholesterol levels [20]. Furthermore, bone marrow transplantation experiments conducted in lethally irradiated LDL receptor-deficient mice found that animals reconstituted with bone marrow from HO-1-deficient mice possess atherosclerotic lesions with greater macrophage content relative to animals reconstituted with bone marrow from wild-type animals, indicating that HO-1 regulates plaque composition [21]. Moreover, recent work using a vulnerable plaque mouse model found that pharmacological induction or gene delivery of HO-1 prevents plaque progression into vulnerable lesions by increasing fibrous cap thickness and intimal VSMC accumulation while decreasing intraplaque lipid accretion and necrotic core size [9]. By contrast, inhibition of HO activity by ZnPP augments plaque vulnerability in this animal model. Collectively, clinical and experimental studies provide compelling support for the concept that HO-1 affords protection against atherosclerosis by reducing plaque size and vulnerability.

Interestingly, HO-1 may also ameliorate vascular complications that arise following percutaneous transluminal angioplasty or coronary bypass surgery. These surgical interventions restore blood flow at sites of atherosclerosis, but their long-term success is compromised by restenosis of blood vessels due to aberrant proliferation of VSMCs. Several studies have reported that pharmacological induction of HO-1 or adenoviral gene transfer of HO-1 inhibits VSMC proliferation and neointima formation following arterial injury in animals, while deletion of the HO-1 gene exacerbates VSMC growth and intimal thickening [22–25]. In addition, loss of HO-1 is associated with the development of early, large VSMC-rich intimal lesions in a mouse jugular vein/carotid artery autograft model that mimics the vein graft used in bypass surgery [20]. The antiproliferative effect of HO-1 is mediated by its reaction products biliverdin and CO. Flow cytometry experiments indicate that CO and biliverdin arrests VSMCs in the G_0/G_1 phase of the cell cycle, and this is associated with a marked decrease in DNA synthesis and cyclin A expression and pronounced increase in p21 and p53 expression [26–30]. Significantly, administration of CO, biliverdin, or bilirubin retards intimal thickening of injured carotid rat arteries, while chronic inhalation of CO blocks transplant arteriosclerosis in rats. Moreover, HO-1 restores the endothelial integrity of injured blood vessels by stimulating the proliferation and migration of ECs from adjacent border regions of the injured artery and by mobilizing and recruiting endothelial progenitor cells from the bone marrow to the site of injury [31]. More recently, CO has been implicated in these repair processes, as administration of CO accelerates EC proliferation and the re-endothelialization of injured rodent carotid arteries and enhances the deposition of endothelial progenitor cells to the luminal surface of damaged blood vessels [32].

POTENTIAL MECHANISMS MEDIATING THE ANTIATHEROGENIC ACTIONS OF HO-1

There are several potential mechanisms by which HO-1 inhibits atherosclerosis. Since the accumulation and oxidative modification of lipoproteins and other biologically active lipids within the arterial intima represents the initial event in atherogenesis, the capacity of HO-1 to scavenge reactive oxygen species (ROS) may be especially important. Indeed, induction of HO-1 is paralleled by a decline in circulating and tissue reactive oxygen intermediates in atherosclerotic animals [12, 13]. The bile pigments biliverdin and bilirubin likely mediate the antioxidant action of HO-1 since they are efficient scavengers of various oxidants, including singlet oxygen, hydroxyl radical, hypochlorous acid, peroxynitrite, and peroxide. Of importance, both free and albumin-bound bilirubin scavenges peroxyl radicals and protects LDL against peroxidation [33]. In addition, oxidative metabolites of bilirubin have been detected in atherosclerotic lesions, supporting an antioxidant role for bile pigments in this disease [9]. However, CO may also participate in the antioxidant actions of HO-1 by inhibiting the activity of pro-oxidant enzymes and stimulating the expression of antioxidant genes [34, 35]. Aside from generating CO and bilirubin, HO-1 may evoke an antioxidant effect by diminishing the concentration of pro-oxidant iron in atherosclerotic lesions [19]. The mechanism underlying the reduction in iron overload by HO-1 in atherosclerosis is not fully understood but may involve HO-1-mediated increases in iron efflux from cells [36]. In addition, the rise in ferritin that accompanies the induction of HO-1 may chelate free iron and curb its oxidant activity.

Since inflammation mediates all stages of atherosclerosis from initiation to progression, and eventual plaque rupture, the ability of HO-1 to arrest inflammation is highly significant. As we have recently reviewed [37], HO-1 is a potent inhibitor of vascular inflammation that suppresses both the innate and adaptive arms of the immune response (Figure 29.2). Overexpression of HO-1 using pharmacological or molecular approaches attenuates the expression of vascular cell adhesive molecule-1 (VCAM-1), E-selectin, and CC-chemokine ligand 2 (CCL2) by activated ECs, and the expression of E- and P-selectin in vascular beds of lipopolysaccharide-challenged animals [38, 39]. These anti-inflammatory actions of HO-1 are duplicated by the exogenous administration of bilirubin, and result from inhibition of the transcription factor NF-κB. In addition, overexpression of HO-1 in monocytes attenuates their chemotactic activity by reducing the expression of chemokine receptors [40]. This later response involves both CO and bilirubin since the combined use of both reaction products is needed to mimic the actions of HO-1. The ability of HO-1 to simultaneously block adhesion molecule, chemokine, and chemokine receptor expression provides a potent armament by which HO-1 impedes inflammatory cell recruitment into the vessel wall and initial lesion formation.

HO-1 may further impact the progression of atherosclerosis by directly modulating macrophage function. Peritoneal macrophages collected from HO-1-deficient mice demonstrate increased lipid uptake and foam cell formation compared with macrophages isolated from wild-type littermates [21]. The elevation in lipid uptake by macrophages lacking HO-1 is associated with increased production of reactive oxygen species and is due, in part, to enhanced expression of scavenger receptor A. Absence of HO-1 also promotes the formation of interleukin-6 (IL-6) and CCL2 from activated macrophages, while overexpression of HO-1 diminishes the production of several inflammatory cytokines, including tumor-necrosis factor-α (TNFα) [21]. The suppression of inflammatory cytokine synthesis by HO-1 is usually ascribed to CO and requires the activation of p38 mitogen-activated protein kinase (MAPK) [41]. The anti-inflammatory action of CO on macrophages is further amplified by its ability to stimulate the synthesis of the anti-inflammatory cytokine, IL-10. Interestingly, biliverdin also induces IL-10 production in macrophages, but the underlying mechanism differs between the two HO-1 products: CO upregulates IL-10 formation by activating p38 MAPK, while biliverdin utilizes the phosphatidylinositol-3-kinase-Akt signaling pathway [41, 42]. These findings raise the possibility that CO and biliverdin may

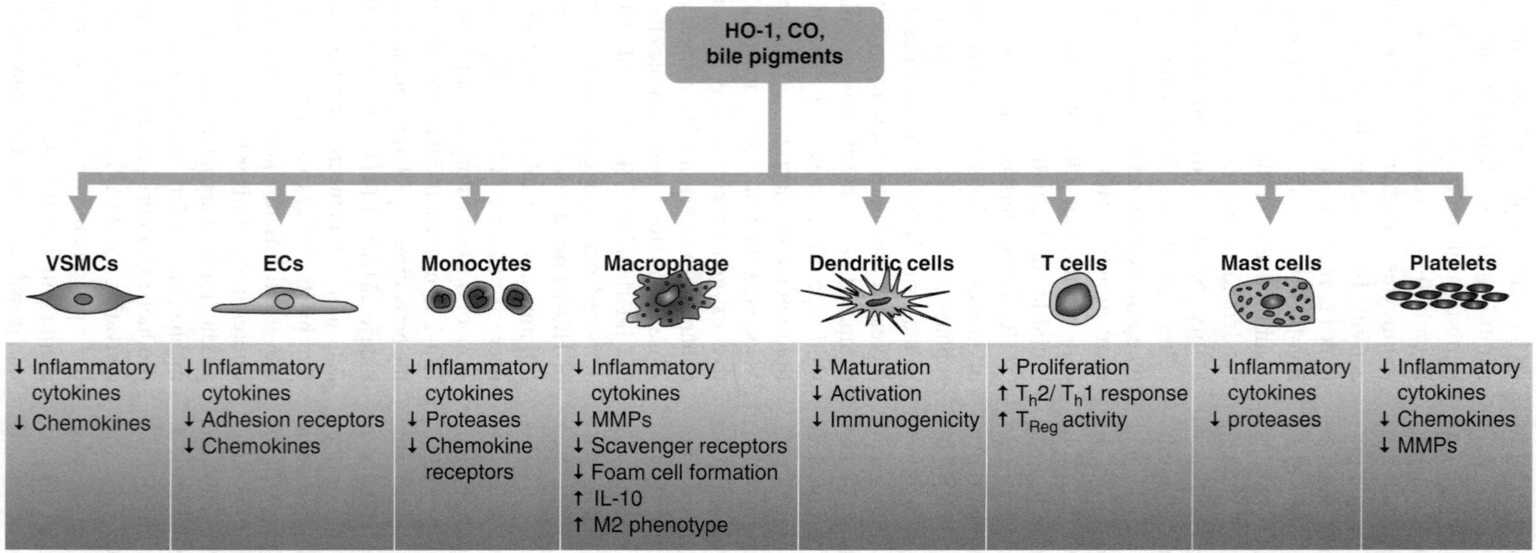

Figure 29.2 Anti-inflammatory actions of HO-1 in atherosclerosis. HO-1 impedes the recruitment and infiltration of immune cells into the vasculature by blocking the expression of adhesion receptors and chemokines by vascular cells and chemokine receptors by monocytes. HO-1 also blocks the synthesis of inflammatory cytokines, proteases, and/or matrix metalloproteinases by endothelial cells (ECs), vascular smooth muscle cells (VSMCs), platelets, mast cells, and macrophages. In addition, HO-1 inhibits the differentiation of macrophages into foam cells by suppressing scavenger receptor expression, polarizes macrophages to the immunomodulatory M2 phenotype, and increases the production of anti-inflammatory interleukin-10 (IL-10). Further, HO-1 may indirectly influence T-lymphocyte (T cell) function by inhibiting the maturation, activation, and immunogenicity of dendritic cells and stimulating the activity of T-regulatory cells (T_{Regs}). Moreover, HO-1 may block T cell proliferation and drive T-cell differentiation to a T helper type 2 (T_h2) reaction that limits the production of proatherogenic cytokines.

function in an additive or synergistic manner to elevate macrophage IL-10 synthesis. The release of CO by HO-1 may also prevent the differentiation of monocytes into macrophages by suppressing the synthesis of granulocyte macrophage colony-stimulating factor [43]. Moreover, HO-1 was recently demonstrated to polarize macrophages toward an anti-inflammatory M2 phenotype [44]. Thus, HO-1 may repress inflammation within atherosclerotic lesions by restricting macrophage activation, differentiation, and polarization.

HO-1 may also prevent vascular inflammation by blocking the activation and degranulation of myeloid cells. Activation of mast cells and platelets results in the release of a large number of inflammatory mediators that alter the adhesive and chemotactic properties of ECs, facilitating the recruitment of monocytes into atherosclerotic lesions. Further, degranulation of these cells leads to the liberation of proteases and matrix metalloproteinases that degrades extracellular matrix and weakens the fibrous cap of plaques. Interestingly, the stabilizing effect of HO-1 on mast cells appears to be mediated by biliverdin and bilirubin, while CO contributes to the anti-aggregatory action of HO-1 by blocking the activation of soluble guanylate cyclase (sGC) in platelets, demonstrating that distinct HO-1 products target the activation of specific myeloid cells [45, 46].

Recent evidence suggests that HO-1 also retards the progression of atherosclerosis by regulating the number and function of antigen-presenting cells and T lymphocytes (T cells) within arterial lesions. Dendritic cells are specialized antigen-presenting cells found in atherosclerotic lesions that present antigens and costimulatory molecules to T cells to initiate the immune response. Several reports indicate that the induction of HO-1 inhibits dendritic cell maturation, activation, and immunogenicity in a manner that is mimicked by both CO and biliverdin/bilirubin [47–50]. Interestingly, deletion of HO-1 in dendritic cells directs the T-cell response preferentially toward a $CD4^+$ T rather than $CD8^+$ T-cell reaction. In a murine model of transplant arteriosclerosis, adoptive transfer of HO-1-null dendritic cells prior to allograft transplantation results in accelerated transplant arteriosclerosis and prominent intragraft $CD4^+$ T-cell infiltration, highlighting the pivotal role of HO-1 in regulating dendritic function and subsequent T-cell priming in arteriosclerosis [51]. Significantly, HO-1 also blocks the proliferation of T cells, and this response is duplicated by both CO and bilirubin [50, 52]. This cytostatic effect is accompanied by a significant decline in activation markers and is likely mediated via a decrease in IL-2 secretion. In addition, transplantation studies suggest that HO-1 favors a T helper 2 (Th2) cell-dominant response that shifts the cytokine balance within grafts toward the production of IL-10 and away from the synthesis of proinflammatory cytokines [53]. Further, HO-1 is able to upregulate the activity of regulatory T cells (T_{Regs}), providing another avenue by which this enzyme dampens the immune response in atherosclerosis [54].

Another important mechanism by which HO-1 arrests atherosclerosis is by promoting cell survival. Apoptosis of macrophages and other cells is a prominent feature of atherosclerotic lesions that aids in the development of a necrotic core. Many atherogenic molecules and processes, including endoplasmic reticulum stress, trigger apoptosis in vascular lesions. However, gene transfer of HO-1 prevents apoptosis in response to many atherogenic factors and provides protection in animal models of transplant arteriosclerosis [3, 5, 55]. The antiapoptotic action of HO-1 is largely driven by CO that blocks multiple steps in the apoptotic cascade [5]. However, the bile pigments also contribute to the antiapoptotic actions of HO-1 in response to oxidative stress [56]. Since apoptosis of VSMCs contributes to plaque rupture, the capacity of HO-1 to block VSMC death in the vulnerable shoulder area of plaques may provide a crucial mechanism by which HO-1 ensures plaque stability [57].

HO-1 may also defend against atherosclerosis by preserving vascular function. EC dysfunction reflected by impaired endothelium-dependent vasodilation, and endothelial nitric oxide (NO) synthesis is one of the earliest manifestations of atherosclerosis. Significantly, overexpression of HO-1 improves endothelium-dependent vasorelaxation and restores endothelial NO synthase (eNOS) expression in LDL receptor-deficient mice [58]. A similar effect is noted after treatment with bilirubin, whereas CO exposure has no effect. Both HO-1- and bilirubin-mediated improvements in endothelial function are paralleled by a decrease in circulating reactive oxygen intermediates, further illustrating the antioxidant action of HO-1. However, the HO-1-mediated release of CO may also modulate vascular tone in atherosclerosis since this gas directly dilates blood vessels by activating sGC and/or calcium-activated potassium channels [3, 4]. Interestingly, we recently reported that HO-1 plays a fundamental role in maintaining sGC in a reduced state that is resistant to degradation and responsive to gaseous monoxides [59]. Thus, the induction of HO-1 may prevent vasospasm in atherosclerosis by preserving eNOS and sGC expression through the production of the antioxidant bile pigments biliverdin and bilirubin while simultaneously activating sGC via the generation of CO.

THERAPEUTIC STRATEGIES TARGETING HO-1 IN ATHEROSCLEROSIS

Atherosclerosis and its cardiovascular complications remain the leading cause of death worldwide. Despite considerable progress during the past few decades, a significant number of patients are not adequately served by current treatment regimens, emphasizing the need to develop alternative therapeutic approaches. Recent clinical and experimental studies have identified HO-1 as promising target in the treatment of atherosclerosis. Multiple mechanisms contribute to the vasoprotective action of HO-1 (Figure 29.3). HO-1 exerts a direct antioxidant effect by catabolizing pro-oxidant heme and blocking intracellular iron accumulation. In addition, the HO-1 reaction products, CO and the bile pigment biliverdin and bilirubin, possess potent antioxidant, anti-inflammatory, antiapoptotic, antithrombotic, and vasodilatory properties. Furthermore, these reaction products regulate the proliferation, migration, activation, maturation, differentiation, immunogenicity, and/or infiltration of vascular cells, platelets, and immune cells within and to atherosclerotic lesions. HO-1 also promotes vascular function by preserving endothelial NO synthase (eNOS) and sGC expression and vascular repair by mobilizing and recruiting endothelial progenitor cells to sites of arterial injury.

Several strategies can be utilized to target HO-1 in atherosclerosis. One promising approach involves the use of pharmacological inducers. There is a growing awareness that many vanguard drugs used to treat atherosclerosis or its complications evoke their therapeutic effect, at least in part, via the induction of HO-1 [3]. Statins are widely prescribed drugs that lower circulating LDL levels and reduce various cardiovascular events. Recent evidence suggests that HO-1 may contribute to the pleiotropic action of statins [60]. Several statins stimulate HO-1 expression, and the induction of HO-1 contributes to the anti-inflammatory and antiproliferative effect of statins. However, it remains to be established whether clinically relevant doses of statins elevates HO-1 expression in humans. Other lipid-modifying agents, including the peroxisome proliferator-activated receptor ligands and niacin, also stimulate HO-1 gene expression, and this may contribute to the anti-inflammatory properties of these compounds [61, 62]. Drugs that are used to combat the acute and chronic complications of atherosclerosis likewise increase HO-1 expression. Aspirin stimulates HO-1 expression and this may promote the drug's antioxidant profile in ECs [63]. In addition, the induction of HO-1 by pentaerythritol tetranitrate prevents the development of tolerance to this drug by suppressing the generation of reactive oxygen species [64]. Moreover, the induction of HO-1 by rapamycin and paclitaxel may enhance the potency of drug-eluting

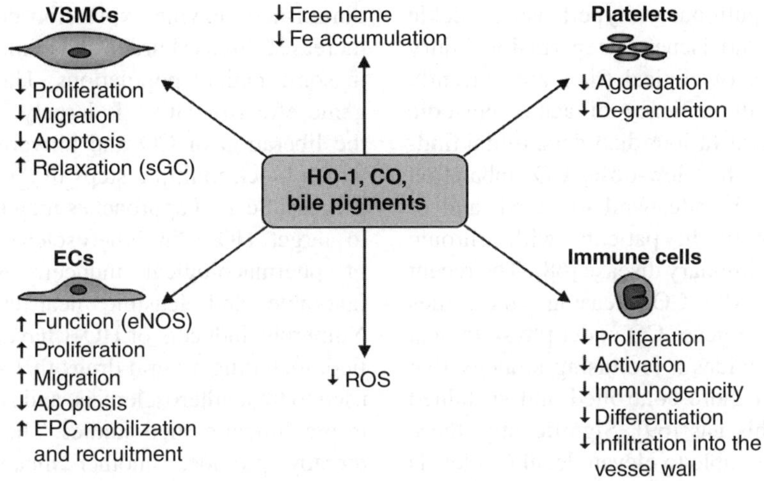

Figure 29.3 Potential mechanisms by which HO-1 inhibits atherosclerosis. HO-1 exerts a direct antioxidant effect by catabolizing pro-oxidant heme and preventing intracellular iron (Fe) accumulation. In addition, the HO-1 reaction products, CO and the bile pigments biliverdin and bilirubin, possess potent antioxidant, anti-inflammatory, antiapoptotic, antiaggregatory, and vasodilatory properties. Further, CO and the bile pigments regulate the proliferation, migration, activation, maturation, differentiation, immunogenicity and/or infiltration of endothelial cells (ECs), vascular smooth muscle cells (VSMCs), platelets, and/or immune cells within or to atherosclerotic lesions. HO-1 also promotes vascular function and repair by preserving endothelial nitric oxide synthase (eNOS) and soluble guanylatecyclase (sGC) expression and the mobilization and recruitment of endothelial progenitor cells to sites of arterial injury.

stents in maintaining blood flow in atherostenotic regions following percutaneous interventions [65, 66].

Another potential approach to upregulating HO-1 expression involves the use of dietary supplements. A large number of dietary antioxidants have been shown to induce HO-1, including quercetin, circumin, resveratrol, sulphoraphane, α-lipoic acid, catechins, and carnosol [3]. Moreover, the induction of HO-1 by some of these common dietary polyphenols is associated with a reduction in atherosclerosis in mice [67]. In addition, heme and its synthetic analogues are both substrates and inducers of HO-1 that protect against atherosclerosis in animal models and are currently approved for the treatment of porphyrin disorders [19]. However, clinical studies are needed to establish the safety and efficacy of any pharmacological or nutritional approach to raise HO-1 levels and ameliorate atherosclerosis.

One potential concern with the use of pharmacological or dietary inducers of HO-1 is the presence of polymorphisms in the promoter that limits the induction of HO-1 in certain patient populations. Thus, genotyping of patients may be necessary to select the appropriate patient population prior to pharmacological intervention. However, increasing HO-1 gene expression via viral-mediated delivery circumvents this problem and allows for a more selective approach in targeting the gene to all patients. Gene therapy approaches with HO-1 have proven highly effective in repressing atherosclerosis and intimal thickening in rodents, and provides a therapeutic option for the future [3, 19].

As an alternative approach, the products of the HO-1 reaction may be directly employed to treat atherosclerosis and its complications. Experimental studies demonstrate that inhalation of CO prevents vascular disease in animal models of transplant arteriosclerosis, intimal hyperplasia, pulmonary hypertension, sickle cell disease, and ischemia-reperfusion injury [3]. A number of clinical trials are currently investigating the efficacy of acute, episodic CO inhalation in various disorders. Initial findings indicate that low-dose CO inhalation for 1–2 hours is safe, well tolerated, and of potential benefit in patients with chronic obstructive pulmonary disease [68]. The recent development of CO-releasing molecules (CORMs) that release CO under physiological conditions provides a promising strategy that may permit a more controlled and sustained delivery of this gas [69]. Significantly, these compounds are able to elevate local CO levels without increasing carboxyhemoglobin levels, which is commonly found with CO inhalation therapy. Several CORMs have been synthesized that possess differing solubility and release kinetics. Moreover, their efficacy in blocking in-stent stenosis has recently been validated in a rodent model [70], but their ability to inhibit atherosclerosis in animals has not yet been established. Finally, bile pigments may also be administered to treat vascular disease. Both epidemiological and experimental studies support the administration of bilirubin against vascular disease [3, 6, 28–30]. While bilirubin is lipophilic and soluble only in organic solvents, biliverdin is soluble in aqueous media providing a vehicle for therapeutic delivery. Alternatively, the development of pharmacological inhibitors that target either the transport or metabolism of bilirubin by the liver affords another mechanism to raise circulating levels of bilirubin.

CONCLUSION

Convincing evidence indicates that HO-1 protects against atherosclerotic vascular disease. Genetic deficiency of HO-1 is associated with increased susceptibility to atherosclerosis in both mice and humans, while HO-1 overexpression retards atherosclerotic lesion formation in multiple animal models. Further, HO-1 polymorphisms in humans resulting in diminished enzyme expression correlate with increased disease burden and clinical outcomes in some patient populations. The antiatherogenic effect of HO-1 is largely mediated via the liberation of CO and biliverdin/bilirubin, which block multiple steps in the atherogenic process. Several approaches may be employed to target HO-1 in atherosclerosis. The use of pharmacological inducers is a highly favorable and feasible near-term strategy. Numerous inducers of HO-1 have been identified, including several drugs that are currently used to treat atherosclerosis, and show promise in preclinical animal studies. The use of gene therapy provides another means to target HO-1. Alternatively, the HO-1 reaction products CO and biliverdin/bilirubin may be directly applied for therapeutic effect. Future translational studies will establish the safety and efficacy of any approach targeting HO-1 or its reaction products in atherosclerosis.

ACKNOWLEDGMENTS

The author acknowledges the support of grants from the National Institutes of Health HL59976 and HL74966, and the American Heart Association Midwest Affiliate 13GRNT 16710017.

REFERENCES

1. Tenhunen, R., Marver, H.S., Schmid, R. (1968) The enzymatic conversion of heme to bilirubin by microsomal heme oxygenase. *Proc Natl Acad Sci U S A* 61, 748–755.
2. Alam, J., Cook, J.L. (2003) Transcriptional regulation of the heme oxygenase-1 gene via the stress response pathway. *Curr Pharm Des* 9, 2499–2511.
3. Durante, W. (2010) Targeting heme oxygenase-1 in vascular disease. *Curr Drug Targets* 11, 1504–1516.
4. Kim, Y.-M., Pae, H.-O., Park, J.E., et al. (2011) Heme oxygenase in the regulation of vascular biology: from molecular mechanisms to therapeutic opportunities. *Antioxid Redox Signal* 14, 137–167.

5. Morse, D., Lin, L., Choi, A.M., Ryter, S.W. (2009) Heme oxygenase-1, a critical arbitrator of cell death pathways in lung injury and disease. *Free Radic Biol Med* 47, 1–12.
6. Ryter, S.W. (2012) Bile pigments in pulmonary and vascular disease. *Front Pharmacol* 3, 39.
7. Wang, L.J., Lee, T.S., Lee, F.Y., Pai, R.C., Chau, L.Y. (1998) Expression of heme oxygenase-1 in atherosclerotic lesions. *Am J Pathol* 152, 711–720.
8. Nakayama, M., Takahashi, K., Komaru, T., et al. (2001) Increased expression of heme oxygenase-1 and bilirubin accumulation in foam cells of rabbit atherosclerotic lesions. *Arterioscler Thromb Vasc Biol* 21, 1373–1377.
9. Cheng, C., Noordeloos, A.M., Jeney, V., et al. (2009) Heme oxygenase-1 determines atherosclerotic lesion progression into a vulnerable plaque. *Circulation* 119, 3017–3027.
10. Li, Y.G., Wang, D.M., Chen, S.M., et al. (2006) Haem oxygenase-1 expression and coronary heart disease—association between levels of haem oxygenase-1 expression and angiographic morphology as well as the quantity of coronary lesions. *Acta Cardiol* 61, 295–300.
11. Chen, S.M., Li, Y.G., Wang, D.M. (2005) Study on the changes of heme oxygenase-1 in patients with coronary artery disease. *Clin Cardiol* 28, 197–201.
12. Ishikawa, K., Sugawara, D., Goto, J., et al. (2001) Heme oxygenase-1 inhibits atherogenesis in Watanabe heritable hyperlipidemic rabbits. *Circulation* 104, 1831–1836.
13. Ishikawa, K., Sugawara, D., Wang, X.P., et al. (2001) Heme oxygenase-1 inhibits atherosclerosis lesion formation in ldl-receptor knockout mice. *Circ Res* 88, 506–512.
14. Yachie, A., Niida, Y., Wada, T., et al. (1999) Oxidative stress causes enhanced endothelial cell injury in human heme oxygenase-1 deficiency. *J Clin Invest* 103, 129–135.
15. Radhakrishnan, N., Yadav, S.P., Sachdeva, A., et al. (2011) Human heme oxygenase-1 deficiency presenting with hemolysis, nephritis, and asplenia. *J Pediatr Hematol Oncol* 33, 74–78.
16. Kawashima, A., Oda, T., Yachie, A., Koizumi, S., Nakanishi, I. (2002) Heme oxygenase-1 deficiency: the first autopsy case. *Hum Pathol* 33, 125–130.
17. Exner, M., Minar, E., Wagner, O., Schillinger, M. (2004) The role of heme oxygenase-1 promoter polymorphisms in human disease. *Free Radic Biol Med* 37, 1097–1104.
18. Taha, H., Skrzypek, K., Guevara, I., et al. (2010) Role of heme oxygenase-1 in human endothelial cells: lesson from the promoter allelic variants. *Arterioscler Thromb Vasc Biol* 30, 1634–1641.
19. Juan, S.H., Lee, T.S., Tseng, K.W., et al. (2001) Adenovirus-mediated heme oxygenase-1 gene transfer inhibits the development of atherosclerosis in apolipoprotein E-deficient mice. *Circulation* 104, 1519–1525.
20. Yet, S.F., Layne, M.D., Liu, X., et al. (2003) Absence of heme oxygenase-1 exacerbates atherosclerosis lesion formation and vascular remodeling. *FASEB J* 17, 1759–1761.
21. Orozco, L.D., Kapturczak, M.H., Barajas, B., et al. (2007) Heme oxygenase-1 expression in macrophages plays a beneficial role in atherosclerosis. *Circ Res* 100, 1703–1711.
22. Togane, Y., Toshisuki, M., Suematsu, M., Ishimura, Y., Yamazaki, J., Katayama, S. (2000). Protective roles of exogenous carbon monoxide in neointimal development elicited by arterial injury. *Am J Physiol Heart Circ Physiol* 278, H623–H632.
23. Tulis, D.A., Durante, W., Peyton, K.J., Evans, A.J., Schafer, A.I. (2001) Heme oxygenase-1 attenuates vascular remodeling following balloon injury in rat carotid arteries. *Atherosclerosis* 155, 113–122.
24. Duckers, H.J., Boehm, M., True, A.L., et al. (2001) Heme oxygenase-1 protects against vascular constriction and proliferation. *Nat Med* 7, 693–698.
25. Tulis, D.A., Durante, W., Liu, X., Evans, A.J., Peyton, K.J., Schafer, A.I. (2001) Adenovirus-mediated heme oxygenase-1 gene delivery inhibits injury-induced vascular neointima formation. *Circulation* 104, 2710–2715.
Peyton, K.J., Reyna, S.V., Chapman, G.B., et al. (2002) Heme oxygenase-1-derived carbon monoxide is an autocrine of inhibitor vascular smooth muscle cell growth. *Blood* 99, 4443–4448.
26. Otterbein, L.E., Zuckerbraun, B.S., Haga, M., et al. (2003) Carbon monoxide suppresses arteriosclerotic lesions associated with chronic graft rejection and with balloon injury. *Nat Med* 9, 183–190.
27. Ollinger, R., Bilban, M., Erat, A., et al. (2005) Bilirubin: a natural inhibitor of vascular smooth muscle cell proliferation. *Circulation* 112, 1030–1039.

28. Nakao, A., Murase, N., Ho, C., Toyokawa, H., Billiar, T.R., Kanno, S. (2005) Biliverdin administration prevents the formation of intimal hyperplasia induced by vascular injury. *Circulation* 112, 587–591.
29. Peyton, K.J., Shebib, A.R., Azam, M.A., Liu, X.M., Tulis, D.A., Durante, W. (2012) Bilirubin inhibits neointima formation and vascular smooth muscle cell proliferation and migration. *Front Pharmacol* 3, 48.
30. Wu, B.J., Midwinter, R.G., Cossano, C., et al. (2009) Heme oxygenase-1 increases endothelial progenitor cells. *Arterioscler Thromb Vasc Biol* 29, 1537–1542.
31. Weigel, B., Gallo, D., Raman, K., et al. (2010) Nitric oxide-dependent bone marrow progenitor mobilization by carbon monoxide enhances endothelial repair after vascular injury. *Circulation* 121, 537–548.
32. Neuzil, J., Stocker, R. (1994) Free and albumin-bound bilirubin are efficient co-antioxidants for α-tocopherol, inhibiting plasma and low density lipoprotein peroxidation. *J Biol Chem* 269, 16712–16719.
33. Piantadosi, C.A., Carraway, M.S., Suliman, H.B. (2006) Carbon monoxide, oxidative stress, and mitochondrial permeability pore transition. *Free Radic Biol Med* 40, 1332–1339.
34. Taille, C., El-Benna, J., Lanone, S., Boczkowski, J., Motterlini, R. (2005) Mitochondrial respiratory chain and NAD(P)H oxidase are targets for the antiproliferative effect of carbon monoxide in human airway smooth muscle. *J Biol Chem* 280, 25350–25360.
35. Ferris, C.D., Jaffrey, S.R., Sawa, S.R., et al. (1999) Haem oxygenase-1 prevents cell death by regulating cellular iron. *Nat Cell Biol* 1, 152–157.
36. Durante, W. (2011) Protective role of heme oxygenase-1 against inflammation in atherosclerosis. *Front Biosci* 16, 2372–2388.
37. Soares, M.P., Seldon, M.P., Gregoire, I.P., et al. (2004) Heme oxygenase-1 modulates the expression of adhesion molecules associated with endothelial activation. *J Immunol* 172, 3553–3563.
38. Vachharajan, T.J., Work, J., Issekutz, A.C., Granger, D.N. (2000) Hemeoxygenase modulates selectin expression in different vascular beds. *Am J Physiol Heart Circ Physiol* 278, H1613–H1617.
39. Morita, T., Imai, T., Yamaguchi, T., Sugiyama, T., Katayama, S., Yoshino, G. (2003) Induction of heme oxygenase-1 suppresses angiotensin II-elicited chemotactic activity through inhibition of CCR2: role of bilirubin and carbon monoxide generated by the enzyme. *Antioxid Redox Signal* 5, 439–447.
40. Otterbein, L.E., Bach, F.H., Alam, J., et al. (2000) Carbon monoxide has anti-inflammatory effects involving the mitogen-activated protein kinase pathway. *Nat Med* 6, 422–428.
41. Weigel, B., Baty, C.J., Gallo, D., et al. (2009) Cell surface biliverdin reductase mediates biliverdin-induced anti-inflammatory effects via phosphatidylinositol-3-kinase and Akt. *J Biol Chem* 284, 21369–21378.
42. Song, R., Ning, W., Liu, F., et al. (2002) Regulation of IL-1β-induced GM-CSF production in human airway smooth muscle cells by carbon monoxide. *Am J Physiol Lung Cell Mol Physiol* 27, L50–L56.
43. Weis, N., Weigert, A., von Knethen, A., Brune, B. (2009) Heme oxygenase-1 contributes to an alternative macrophage activation profile induced by apoptotic cell supernatants. *Mol Biol Cell* 20, 1280–1288.
44. Takamiya, R., Murakami, M., Kajimura, M., et al. (2002) Stabilization of mast cells by heme oxygenase: an anti-inflammatory role. *Am J Physiol Heart Circ Physiol* 283, H861–H870.
45. Wagner, C.T., Durante, W., Christodoulides, N., Hellums, J.D., Schafer, A.I. (1997) Hemodynamic forces induce the expression of heme oxygenase in cultured vascular smooth muscle cells. *J Clin Invest* 100, 589–596.
46. Chauveau, C.S., Remy, S., Royer, P.J., et al. (2005) Heme oxygenase-1 expression inhibits dendritic cell maturation and proinflammatory function but conserves IL-10 expression. *Blood* 106, 1694–1702.
47. Kotsch, K., Martins, P.N., Klemz, R., et al. (2007) Heme oxygenase-1 ameliorates ischemia/reperfusion injury by targeting dendritic cell maturation and migration. *Antioxid Redox Signal* 9, 2049–2063.
48. Remy, S., Blancou, S.P., Tesson, L., et al. (2009) Carbon monoxide inhibits TLR-induced dendritic cell immunogenicity. *J Immunol* 182, 1877–1884.
49. Liu, Y., Li, P., Lu, W., et al. (2008) Bilirubin possesses powerful immunomodulatory activity

and suppresses experimental autoimmune encephalomyelitis. *J Immunol* 181, 1887–1897.
50. Cheng, C., Noorderloos, M., van Deel, E.D., et al. (2010) Dendritic cell function in transplantation arteriosclerosis is regulated by heme oxygenase-1. *Circ Res* 106, 1656–1666.
51. Pae, H.O., Ho, G.S., Choi, B.M., et al. (2004) Carbon monoxide produced by heme oxygenase-1 suppresses T cell proliferation via inhibition of IL-2 production. *J Immunol* 172, 4744–4751.
52. Ke, B., Buelow, R., Shen, X.D., et al. (2002) Heme oxygenase-1 gene transfer prevents CD95/Fas ligand-mediated apoptosis and improves allograft survival via carbon monoxide signaling pathway. *Hum Gene Ther* 13, 1189–1199.
53. Choi, B.M., Pae, H.O., Jeong, Y.R., Kim, Y.M., Chung, H.T. (2005) Critical role of heme oxygenase-1 in Foxp3-mediated immune suppression. *Biochem Biophys Res Commun* 327, 1066–1071.
54. Liu, X.M., Peyton, K.J., Ensenat, D., et al. (2005) Endoplasmic reticulum stress stimulates heme oxygenase-1 gene expression in vascular smooth muscle: role in cell survival. *J Biol Chem* 280, 872–877.
55. Wei, Y., Liu, X.M., Peyton, K.J., et al. (2009) Hypochlorous acid-induced heme oxygenase-1 gene expression promotes human endothelial cell survival. *Am J Physiol Cell Physiol* 297, C907–C915.
56. Liu, X.M., Chapman, G.B., Peyton, K.J., Schafer, A.I., Durante, W. (2002) Carbon monoxide inhibits apoptosis in vascular smooth muscle cells. *Cardiovasc Res* 55, 396–405.
57. Kawamura, K., Ishikawa, K., Wada, Y., et al. (2005) Bilirubin from heme oxygenase-1 attenuates vascular endothelial activation and dysfunction. *Arterioscler Thromb Vasc Biol* 25, 155–160.
58. Jones, A.W., Durante, W., Korthuis, R.J. (2010) Heme oxygenase-1 deficiency leads to alteration of soluble guanylate cyclase redox regulation. *J Pharmacol Exp Ther* 335, 85–91.
59. Lee, T.S., Chang, C.C., Zhu, Y., Shyy, J.Y. (2004) Simvastatin induces heme oxygenase-1: a novel mechanism of vessel protection. *Circulation* 110, 1296–1302.
60. Kronke, G., Kadl, A., Ikonomu, E., et al. (2007) Expression of heme oxygenase-1 in human vascular cells is regulated by peroxisome proliferator-activated receptors. *Arterioscler Thromb Vasc Biol* 27, 1276–1282.
61. Wu, B.J., Chen, K., Barter, P.J., Rye, K.-A. (2012) Niacin inhibits vascular inflammation via the induction of heme oxygenase-1. *Circulation* 125, 150–158.
62. Grosser, N., Abate, A., Oberle, S., et al. (2003) Heme oxygenase-1 induction may explain the antioxidant profile of aspirin. *Biochem Biophys Res Commun* 308, 956–960.
63. Wenzel, P., Oelze, S., Abate, A., et al. (2007) Heme oxygenase-1: a novel player in the development of tolerance in response to organic nitrates. *Arterioscler Thromb Vasc Biol* 27, 1729–1735.
64. Visner, G.A., Lu, F., Zhou, H., Liu, J., Kazemfar, K., Agarwal, A. (2003) Rapamycin induces heme oxygenase-1 in human pulmonary vascular cells: implications in the antiproliferative response to rapamycin. *Circulation* 107, 911–916.
65. Choi, B.M., Kim, Y.M., Jeong, Y.R., et al. (2004) Induction of heme oxygenase-1 is involved in the antiproliferative effect of paclitaxel in rat vascular smooth muscle cells. *Biochem Biophys Res Commun* 25, 39–47.
66. Loke, W.M., Proudfoot, J.M., Hodgson, J.M., et al. (2010) Specific dietary polyphenols attenuate atherosclerosis in apolipoprotein E-knockout mice by alleviating inflammation and endothelial dysfunction. *Arterioscler Thromb Vasc Biol* 30, 749–757.
67. Bathoorn, E., Slebos, D.-J., Postma, D.S., et al. (2007) Anti-inflammatory effects of inhaled carbon monoxide in patients with COPD: a pilot study. *Eur Respir J* 30, 1131–1137.
68. Motterlini, R., Otterbein, L.E. (2010) The therapeutic potential of carbon monoxide. *Nat Rev Drug Discov* 9, 728–743.
69. Hyvelin, J.M., Maurel, B., Uzbekov, R., Motterlini, R., Larmusiaux, P. (2010) Hemin prevents in-stent stenosis in rat and rabbit models by inducing heme oxygenase-1. *J Vasc Res* 51, 417–428.

30

ROLES OF REACTIVE OXYGEN SPECIES IN PHYSIOLOGY AND PATHOLOGY

PING SONG AND MING-HUI ZOU

Section of Molecular Medicine, Department of Internal Medicine and Department of Biochemistry and Molecular Biology, University of Oklahoma Health Sciences Center, Oklahoma City, OK, USA

INTRODUCTION

Reactive oxygen species (ROS) is a collective term for reactive oxygen derivatives. In healthy cells, the continuous and spontaneous production of ROS is balanced by detoxification reactions operated by various antioxidant systems. A moderate level of ROS is physiological and recognized as essential signaling molecules which regulate multiple cellular processes such as proliferation, differentiation, and apoptosis in response to growth factors, cytokines, and vasoactive agents [1]. Nevertheless, excess ROS are pathogenic species which cause tissue injury and/or impair tissue repair, a common mechanism shared by many disease states including cancer, cardiovascular diseases, diabetes, and neurodegenerative disease.

GENERAL CHARACTERISTICS OF ROS

ROS are generally short-lived, highly reactive molecules that are formed by incomplete one-electron reduction of molecular oxygen. ROS include oxygen anions and free radicals, including superoxide anion ($O_2^{\bullet-}$), hydroxyl radicals (•OH), and hydroperoxyl radical ($HO_2\bullet$), as well as peroxides such as hydrogen peroxide (H_2O_2) (Table 30.1). Three ROS, $O_2^{\bullet-}$, H_2O_2, and •OH, are relevant to biological functions. $O_2^{\bullet-}$ and H_2O_2 are two major ROS in cells, and $O_2^{\bullet-}$ is the simplest peroxyl radical. The highly reactive and short-lived $O_2^{\bullet-}$ can be converted to the relatively more stable form of H_2O_2 by superoxide dismutases (SOD). $O_2^{\bullet-}$ inactivates nitric oxide (NO•), one member of

Atherosclerosis: Risks, Mechanisms, and Therapies, First Edition. Edited by Hong Wang and Cam Patterson.

TABLE 30.1 Characteristics of reactive oxygen species (ROS)

ROS/RNS	Half-life (S)	Location	Reactivity	Diffusibility	Detection probes
$O_2^{\bullet-}$	10^{-6}	Mitochondria and cytosol	+++	++	DHE, hydrocyanines
H_2O_2	10^{-3}	Mitochondria and cytosol	++	+++++	DCFH-DA, Amplex Red
•OH	10^{-9}	Mitochondria and cytosol	+++++	+	HPF, EPR spin trapping
NO•	1–10	Mitochondria, cytosol, and plasma membrane	+	+++	Spin trapping
$ONOO^-$	1.9	Mitochondria	++++	+	DHR

Abbreviations: DCFH-DA, dichlorodihydrofluorescein diacetate; DHE, dihydroethidium; DHR, dihydrorhodamine; EPR, electron paramagnetic resonance; HPF, hydroxyphenyl fluorescein.

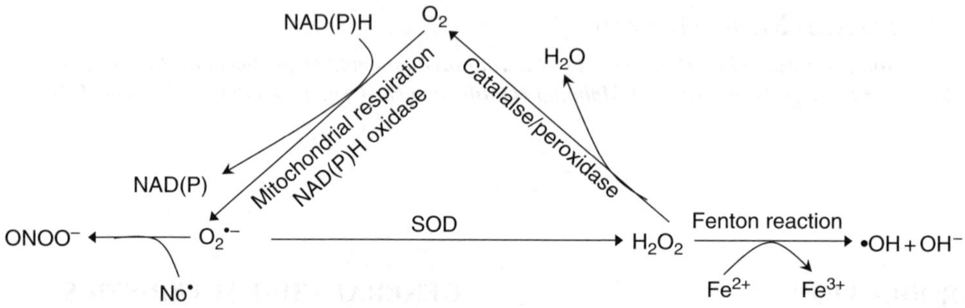

Figure 30.1 Enzyme-mediated ROS conversion. Superoxide ($O_2^{\bullet-}$) is formed from molecular oxygen (O_2) as a by-product of respiration in the mitochondria or by the particular function of NAD(P)H oxidase. $O_2^{\bullet-}$ is converted by superoxide dismutase (SOD) into H_2O_2, which is further detoxified into H_2O and O_2 by catalase or peroxidases, or converted to •OH by Fenton reaction.

reactive nitrogen species (RNS) to yield peroxynitrite ($ONOO^-$), another member of RNS, which can be upregulated by high glucose [2]. H_2O_2 can be further converted, by catalase, glutathione peroxidases, and other peroxidases, to H_2O and O_2 (Figure 30.1). Moreover, H_2O_2 is an attractive signaling molecule in various physiological processes including autophagy [3], since it is relatively stable compared with other ROS and its neutral ionic state enables it to exit the mitochondria or other subcellular organelles freely. In addition, •OH is produced by the Fenton reaction, if a single electron is added to H_2O_2 by a reduced metal ion (e.g., Fe^{2+}) (Figure 30.1). The •OH is extremely reactive with a half-life of less than 1 ns; thus, it reacts very close to its site of formation (Table 30.1). Currently, these ROS/RNS can be measured by several fluorescent probes (Table 30.1) [4].

ROS-GENERATING SYSTEMS

Cellular ROS can arise from both exogenous and endogenous sources. Exogenous ROS sources include UV, ionizing radiation, drugs, and environmental toxins. Endogenous ROS sources include mitochondrial and nonmitochondrial systems. The nonmitochondrial ROS-generating enzymes include membrane-bound nicotinamide adenine dinucleotide (NADH)-dependent oxidase (Nox), aldehyde oxidase (AO), xanthine oxidase (XO), glucose oxidase (GO), uncoupled eNOS [5], cytochrome P-450 enzymes in the endoplasmic reticulum (ER), and peroxisomal flavin oxidases. The major sources of ROS are the mitochondrial respiratory chain in the inner mitochondrial membrane and Nox systems. Emerging data suggest that nonmitochondrial-generated ROS can stimulate mitochondrial ROS production.

Mitochondrial ROS Production

One of the major intracellular sources of ROS production is mitochondria [6], with reports that between 0.2 and 2% of cellular mitochondrial O_2 consumption produces $O_2^{\bullet-}$ [7]. Based on data from isolated mitochondrial samples, organelle concentrations of O_2 and $O_2^{\bullet-}$ range from 3 to 30 μM and 10 to 200 pM [8], respectively. The rate of H_2O_2 generation in isolated mitochondria varies between approximately 0.1 and 0.2 nmol/min/mg protein. Exact tissue-specific $O_2^{\bullet-}$ concentrations and production rates in vivo remain only estimates, as our capacity to make in vivo measurements of mitochondrial ROS remains limited. It is worth noting that mitochondria from different tissues may vary conspicuously in their capacity to produce ROS using different substrates, and this capacity is also related to animal species and age.

There are multiple potential mitochondrial sites for ROS generation. The vast majority of $O_2^{\bullet-}$ from mitochondria is produced by, or modulated by the complexes of the respiratory chain. The mitochondrial respiratory chain, electron transport chain (ETC) consists of five major multisubunit complexes designated as NADH–coenzyme Q (CoQ) reductase (complex I), succinate–CoQ reductase (complex II), ubiquinol–cytochrome c reductase (complex III), cytochrome C oxidase (complex IV), and ATP synthase (complex V). The major ETC sources of ROS from mitochondria are complexes I and III (Figure 30.2), particularly under conditions of high membrane potential. Complex II may also generate ROS, but this may be dependent on complex I. More recent studies confirmed that complex I is a major source of $O_2^{\bullet-}$ production in several types of mitochondria. The $O_2^{\bullet-}$ formation by complex I is higher during the reverse electron transport from succinate to NAD^+ [9], while it is much lower during the forward electron transport. Rotenone, a complex I inhibitor, stimulates ROS formation during forward electron transfer and inhibits it during reverse electron transfer.

Dysfunction of complexes II and IV may result in electron leak and increased ROS formation [10]. In addition, mitochondrial uncoupling, a considerable amount of protons bypass the ATP synthase pathway and leak back to the mitochondrial matrix, also produces ROS. Interestingly, the deficiency of

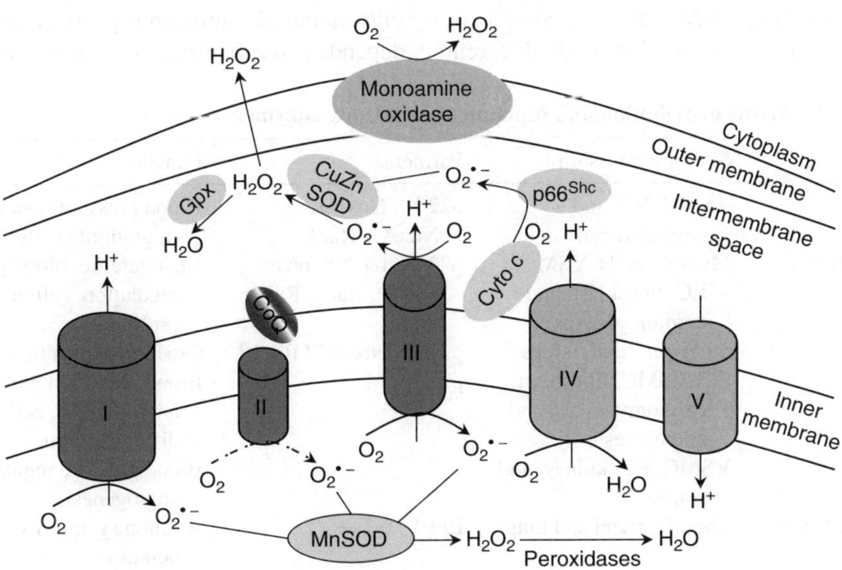

Figure 30.2 ROS production by electron transport chain.

paraoxonase 2 (PON2) localized in respiratory complex III leads to a decreased ETC complex I + III activity and accompanied ROS induction, as well as severe atherosclerosis [11].

Besides ETC, there are several enzymes involved in mitochondrial ROS formation, which include Krebs cycle enzymes implicated in redox reactions such as α-ketoglutarate dehydrogenase and aconitase, glycerol-3-phosphate dehydrogenase from the glycerol phosphate shuttle, monoamine oxidase in outer mitochondrial membrane, and p66Shc (Figure 30.2).

NADPH Oxidases

Recent evidence suggests that the major source of vascular $O_2^{\bullet-}$ and H_2O_2 is Nox systems. Nox is a transmembrane enzyme and generates $O_2^{\bullet-}$ by electron transfer from NADPH to molecular oxygen. Originally designated as **ph**agocytic **ox**idase (phox) involved in defense and innate immunity, the Nox family is now recognized to be widely expressed in a different cell types. Seven Nox isoforms in mammalian cells have been identified: Nox1 to Nox5 and dual oxidases (Duox1 and Duox2) [12]. Nox1, Nox2, Nox4, and Nox5 are reportedly expressed in cardiovascular tissues [13]. Different Nox complexes work as unique combination with different subunits in respective cells (Table 30.2). Nox activation generally requires the association of a membrane-bound catalytic subunit Nox with a variety of regulatory subunits including Rac GTPase; cytosolic phox proteins including $p40^{phox}$, $p47^{phox}$, and $p67^{phox}$; and proteins of the cytoskeleton including $p22^{phox}$ (Table 30.2). Nox activity can be regulated at the level of Nox partner assembly and expression [14].

Nox1 *Nox1* gene is located on the X chromosome. *Nox1* is expressed in both endothelial cells (EC) and vascular smooth muscle cells (VSMC) [14], although it is found predominantly in colon epithelial cells. Nox1 is similar to Nox2 in that its activity requires $p22^{phox}$, $p47^{phox}$ [or the analogous NoxO1 (Nox organizer 1)], $p67^{phox}$ [or the analogous NoxA1 (Nox activator 1)] [15], and Rac1 (Table 30.2).

Nox1 is expressed at low levels under physiological conditions. Moreover, its relative mRNA expression level appears to be remarkably lower than that of the other Nox2 homologues, even under induction. In cultured EC and VSMC, Nox1 is upregulated by shear stress, angiotensin II (AngII), aldosterone, epidermal growth factor (EGF) [16], and platelet-derived growth factor (PDGF). AngII-mediated Nox1 induction in VSMC may be regulated through mitochondrial interaction, possibly via a Ca^{2+}-dependent mechanism. Nox1 is induced in the

TABLE 30.2 Tissue distribution and function of Nox/Duox enzymes

Enzyme	Site of expression	Partners	Function
Nox1	VSMC, EC, and epithelial cell	$p22^{phox}$, NoxO1, NoxA1, Rac1	Blood pressure regulation, cell migration/proliferation
Nox2 ($gp91^{phox}$)	Myeloid cells, VSMC, EC, fibroblast, and cardiomyocytes	$p22^{phox}$, $p47^{phox}$, $p67^{phox}$, $p40^{phox}$, Rac1, Rac2	Host defense, blood pressure regulation, cell migration/ proliferation
Nox3	Inner ear, fetal tissues	$p22^{phox}$, NoxO1? Rac1?	Otoconia biosynthesis
Nox4	EC, VSMC, fibroblast, cardiomyocytes, and adipocytes	$p22^{phox}$	Blood pressure regulation, angiogenesis, cell differentiation
Nox5	VSMC, EC, kidney, and spleen		Blood pressure regulation, angiogenesis
Duox1 and Duox2	Thyroid gland and lung	EFP1	Hormone synthesis, host defense?

Abbreviations: EC, endothelial cells; VSMC, vascular smooth muscle cells.

vasculature in models of hypertension, atherosclerosis, diabetes, and hypercholesterolemia. Nox1 can constitutively produce low levels of $O_2^{\bullet-}$, which increase in a stimulus-dependent manner, involving complex interactions made between NoxO1, NoxA1, and membrane-bound p22^{phox}. Phosphorylation of NoxA1 on Ser282 by MAPK and Ser172 by either protein kinase C (PKC) or protein kinase A (PKA) decreases its binding to Nox1 and Rac1 and consequently inhibits constitutive Nox1 activity [17].

Nox1 has been implicated in VSMC migration, proliferation, and extracellular matrix production, effects mediated by cofilin [18]. Studies on transgenic mice suggest a possible role for Nox1 in acute, but not chronic, forms of AngII-dependent hypertension [19], atherosclerosis, neointima formation post injury [18], endothelial dysfunction, and stroke. Although there are extensive experimental data implicating Nox1 in vascular disease, there is a paucity of clinical data supporting its implication in human vascular diseases, except that the expression of Nox1 and NoxA1 was shown to be induced in human atherosclerotic vessels [15].

Nox2 Nox2, known as gp91^{phox}, originally indentified as the catalytic subunit of the respiratory burst oxidase in phagocytes, is also expressed in EC and VSMC [14] and cardiac [20], renal, and neural cells. In VSMC and EC, Nox2 colocalizes with the cytoskeleton, lipid rafts/caveolac, and perinuclear compartment. However, in neutrophils, Nox2 localizes to intracellular and plasma membranes. Full activation of Nox2 requires membrane-bound subunit p22^{phox} and three or more cytoplasmic subunits (p40^{phox}, p47^{phox}, p67^{phox}, Rac1) (Table 30.2). We recently found Nox2 activation due to p67^{phox} membrane translocation is mediated by PKCζ, which contributes to the elevated $O_2^{\bullet-}$ formation in EC in response to thromboxane A2 receptor (TP) activation [21]. The *Nox2* gene is located on the X chromosome, and its expression is highly regulated by AngII and stretch. Vascular Nox2 is also upregulated in experimental hypertension [22], atherosclerosis, ischemia–reperfusion injury, and neointimal formation models [23]. The physiological functions of Nox2 in the vasculature remain unclear, but pathological upregulation of Nox2 contributes to oxidative injury and vascular damage, including angiogenesis, endothelial dysfunction [24], vascular inflammation, and structural remodeling. Although Nox2 has been shown to be important in models of acute AngII-infused hypertension, it seems not critical in chronic AngII-dependent hypertension and cardiac hypertrophy [25]. Nox2-deficient mice exhibit significantly reduced cerebral infarct size compared to wild-type controls in an ischemia–reperfusion model [26]. In humans, Nox2 has been shown to play a role in endothelial function, as patients with *Nox2* mutation develop chronic granulomatous disease and exhibit a significant increase in forearm-mediated vasodilation with increased NO bioavailability [27].

Nox3 Nox3 shows 58% sequence identity to Nox2, and its mRNA is found in fetal tissues including lung, liver, kidney, and spleen. Nox activation requires p22^{phox} [28]. Nox3 seems to be constitutively active, and can be regulated by NoxO1, p47^{phox}, Rac1, and p67^{phox} [29]. Nox3 is essential for normal vestibular function. Nox3 deficiency causes the absence of otoconia in the inner ear and develops serious balance defects [30].

Nox4 Nox4, initially identified as a renal gp91^{phox} homologue, is expressed in EC, VSMC [31], fibroblasts, osteoclasts [32], and (pre)adipocytes. In the vascular system, Nox4 appears to be the primary Nox isoform in EC and VSMC [14]. Nox4 is localized in the nucleus and endoplasmic reticulum (ER) of both EC and VSMC, and also the mitochondria of VSMC. Additionally, Nox4 was also reported to exist in stress fibers and focal adhesions in VSMC [33]. Nox4 does require p22^{phox} as a regulatory subunit [34], but not p47^{phox}, p67^{phox}, or p40^{phox}. *Nox4* expression appears to be regulated by transforming growth factor-β1 (TGF-β1) in VSMC [35]. Recently, two Nox4-binding proteins, Poldip2 [polymerase

(DNA-directed) δ-interacting protein 2] in VSMC [33] and Tks5 (tyrosine kinase substrate with five Src homology 3 domains) [36], have been shown to be important for Nox4 activation. Unlike Nox1 and Nox2, Nox4 is constitutively active, and produces predominantly H_2O_2 rather than $O_2^{\bullet -}$ [37]. Vascular Nox4 expression is known to be enhanced by several stimuli, including ER stress, mitochondrial dysfunction [38], physiological shear stress, atherosclerosis, arterial injury, and tissue hypoxia [31]. Nox4 constitutively contributes to basal ROS production, but ROS generation by Nox4 is increased in response to AngII, glucose, TNFα, and growth factors.

Nox4 has been implicated in cell differentiation and migration of VSMC [33], proliferation, tube formation, and angiogenesis of EC [31]. Nox4 has been associated with cardiovascular disease (CVD) including hypertension, atherosclerosis and cardiovascular and renal complications of diabetes and in the remodeling of pulmonary arteries in hypoxia-induced pulmonary hypertension [39], although the precise mechanisms are unclear. Additionally, Nox4 is related to cellular senescence of EC [40] and age-related cardiac remodeling [41]. Recent studies suggest that Nox4 may also have protective effects in cardiovasculature. Specifically, Nox4-null mice develop exaggerated contractile dysfunction, hypertrophy, and cardiac dilatation in response to chronic overload, whereas the cardiomyocyte-targeted expression of Nox4 protects transgenic mice from these effects due to enhanced myocardial angiogenesis by Nox4 [42]. Furthermore, Nox4 deletion results in eNOS reduction and impaired recovering from muscle ischemia [43]. In addition, elevated endothelial Nox4 promotes angiogenesis and flow recovery from limb ischemia in an eNOS-dependent manner [31], augments vasodilatation, and decreases blood pressure. These results suggest that Nox4 may be a potential therapeutic target for manipulating angiogenesis and tissue repair.

Nox5 Nox5, the newest member of the Nox isoforms, is expressed in EC from the human microvessels [44], VSMC [45], testes, spleen, lymphoid tissue, and kidney. The *Nox5* gene is not reported in rodents. Nox5 is found in the perinuclear area of human EC which colocalizes with ER [44]. Nox5 is also found in the plasma membrane of human VSMC. Unlike the other vascular Noxes, Nox5 possesses an amino-terminal calmodulin-like domain named EF hands with four binding sites for Ca^{2+} [46]. Nox5 activation is uniquely $p22^{phox}$ or other subunits independent [46]. Nox5 is regulated in a Ca^{2+}-sensitive manner [46]. Moreover, phosphorylation of Nox5 by PKC-dependent pathway [47] increases the sensitivity of Nox5 for Ca^{2+} [48]. Nox5 is also activated by a proto-oncogenic tyrosine kinase c-Abl in a Ca_2^{+}- and H_2O_2-dependent manner [49]. Additionally, Nox5 can be activated by thrombin, PDGF, and ionomycin through PKC and cAMP-response-element-binding protein (CREB) [45]. AngII and endothelin-1 (ET-1) also induces Nox5 mRNA and protein expression [44]. Nox5 plays an important role in EC proliferation and angiogenesis, in PDGF-induced proliferation of VSMC [45], and in oxidative stress in atherosclerosis [50]. It is reported that Nox5 protein and mRNA levels are increased in coronary arteries from patients with coronary artery disease [50]. Nox5 is upregulated in VSMC from hypertensive subjects, implying that Nox5 may be a key source of ROS in hypertension.

Dual oxidases (Duox1 and Duox2) Human Duox1 and Duox2 proteins show 83% sequence similarity. The Nox portion of the protein shows 47% similarity to Nox2. Both Duox1 and Duox2 are highly expressed in the thyroid gland. Duox1 is also expressed in the lung airway epithelium, pancreas, placenta, prostate, and testis. Duox2 mutation results in hypothyroidism due to a lack of peroxidase-dependent hormone synthesis [51].

AO

AO produces carboxylic acids and H_2O_2 from aldehyde. Additionally, AO also use NADH as a substrate to form $O_2^{\bullet -}$ [52]. There is evidence for an increasing relevance of AO in the metabolism and clearance of numerous important drugs including the cardiac stimulant carbazeran, the antimalarial agent quinine, and certain cancer chemotherapeutic agents such as methotrexate. Moreover, the NADH levels have been reported to increase during starvation and other pathophysiological conditions including diabetes, ischemia–reperfusion injury, and alcoholic liver disease. AO was reported to play a critical role in the alcohol-mediated liver injury due to its ability to form ROS from the oxidation of acetaldehyde or NADH. Therefore, AO-derived ROS may also play a role in diabetic cardiovascular complications. However, the pathophysiological importance of this pathway for diabetic vascular disease warrants further study.

XO

XO activity accounts for significant induction of ROS formation in different tissues, as treatment with the XO inhibitor allopurinol reduces the ROS production [53]. An increase of XO expression and activity results in imbalance of ATP/ADP ratio and increased ROS production in skeletal muscle of STZ-induced diabetic mice [54]. These studies demonstrate that $O_2^{\bullet-}$ generated via increased XO seen in experimental diabetic mice is intimately involved in the pathogenesis of diabetic vascular complications.

GO

GO oxidizes glucose, which leads to production of free radicals resulting in oxidative stress during diabetes. There is evidence that glycation of protein depends on the oxidation of glucose since the product of glucose oxidation is attached to the protein to generate advanced glycation end products (AGEs). AGEs cause to induction of Nox2 expression, Nox activity, and reduction of MnSOD levels [55]. This finding indicates that AGEs play key pathophysiological roles in increasing oxidative stress in diabetes along with subsequent endothelial dysfunction through decreasing endothelial nitric oxide synthase (eNOS) phosphorylation.

Nitric Oxide Synthases

Nitric oxide synthases (NOS) are a family of enzymes including endothelial (eNOS), neuronal (nNOS), and inducible NOS (iNOS). NOS requires tetrahydrobiopterin BH_4 to transfer electrons to arginine to release NO•. In the absence of either BH_4 or arginine, NOS uncouples, and reacts with O_2 to produce $O_2^{\bullet-}$ instead of NO•. Peroxynitrite ($ONOO^-$) releases zinc from the zinc–thiolate cluster of eNOS and presumably forms disulfide bonds between the monomers. It is reported that eNOS exposure to oxidants including $ONOO^-$ causes increased enzymatic uncoupling and generation of $O_2^{\bullet-}$ in diabetes, contributing further to endothelial cell oxidative stress [5].

ANTIOXIDANT SYSTEMS

ROS production and ROS-mediated damage in cells are counteracted by different mechanisms including enzymatic antioxidizing and nonenzymatic systems.

Enzymatic Antioxidants

Several groups of enzymes play significant roles in protecting cells from oxidant stress:

SOD catalyzes the dismutation of $O_2^{\bullet-}$ to H_2O_2 and O_2 (Figure 30.1); it accelerates this detoxifying reaction roughly 10,000-fold over the noncatalyzed reaction. Human SOD exists in three isoforms: cytosolic Cu,Zn-SOD (SOD1), mitochondrial MnSOD (SOD2), and extracellular SOD (SOD3). Interestingly, SOD2 is an inducible enzyme that is probably critical for survival upon stress condition including tumor necrosis factor (TNF) stimulation, UV irradiation, and oxidative stress.

Catalase is found in peroxisomes in eukaryotic cells. It degrades H_2O_2 to H_2O and

O_2 and hence finishes the detoxification reaction started by SOD.

Glutathione peroxidase (GPx) is a group of enzymes, the most abundant of which contain selenium. These enzymes, like catalase, degrade H_2O_2. They also reduce organic peroxides to alcohols, providing another route for eliminating toxic oxidants.

Peroxiredoxins (Prdxs) are antioxidant proteins containing essential catalytic cysteine residues that use thioredoxin to scavenge H_2O_2, lipid hydroperoxides, and $ONOO^-$ [56]. Prdxs constitute a potent defense mechanism for maintaining redox balance under normal conditions and oxidative stress. The six mammalian Prdxs (Prdx1 to Prdx6) are distributed at sites of ROS formation, including the cytosol, mitochondria, and peroxisomes [56]. Prdx1 is expressed in the cytosol of many types of cells and tissues and is induced in mouse peritoneal macrophages exposed to oxidative stresses.

In addition to these enzymes, glutathione transferase, ceruloplasmin, hemoxygenase, and possibly several other enzymes may participate in enzymatic control of oxygen radicals and their products.

Nonenzymatic Antioxidants

Several nonenzymatic antioxidants of particular importance are:

Vitamin E is the major lipid-soluble antioxidant and plays a vital role in protecting membranes from oxidative damage. Its primary activity is to trap peroxy radicals in cellular membranes.

Vitamin C or ascorbic acid is a water-soluble antioxidant that can reduce radicals from a variety of sources. It also appears to participate in recycling vitamin E radicals. Interestingly, vitamin C also functions as a pro-oxidant under certain circumstances.

Glutathione, the main redox buffer of the cell, is localized primarily in the cytosol (1–11 mM). It is a tripeptide (glutamyl–cysteinyl–glycine). The cysteine provides an exposed free sulfhydryl group (SH) that is very reactive, providing an abundant target for radical attack. Reaction with radicals oxidizes glutathione, but the reduced form is regenerated in a redox cycle involving glutathione reductase and the electron acceptor NADPH. Thioredoxin directs redox control of specific factors, such as transcription factors [57].

ROS'S PHYSIOLOGICAL AND PATHOLOGICAL FUNCTIONS

Appropriate ROS levels have a significant impact on health and disease.

ROS in Normal Physiological Processes

ROS as signaling molecules It has been widely documented that ROS/RNS act as intracellular signaling molecules to exert their physiological function. ROS are involved in signaling through chemical reactions with specific atoms of target proteins that lead to covalent protein modifications [58]. Thus, ROS molecular recognition occurs at the atomic and not at the macromolecular level. Moreover, ROS exert their particular signaling properties by specifically targeting elements in protein. For example, $O_2^{\bullet-}$ preferably reacts with iron–sulfur (Fe–S) clusters. Both Cysteine residues and Methionine residues are the major targets of H_2O_2. However, •OH has no preferable reactivity toward biological molecules. Furthermore, selenocysteine is more reactive with H_2O_2 than Cysteine, and may operate in H_2O_2-based redox regulation. H_2O_2 can act as a direct modifier of thiol-containing proteins. Because ROS are oxidants, they influence the redox state of signaling proteins, to produce a redox-dependent modification of enzymes. For the same reason not all ROS are equally suitable for signal transduction, for instance the •OH radical is too unspecific to undergo catalyzed reactions.

Many of low grade ROS-mediated acute responses actually protect the cell against oxidative damage and re-establish or maintain the intracellular redox homeostasis. For example, antioxidant hemeoxygenase-1 (HO-1) protein and mRNA are strongly induced by ROS and UVA irradiation. The HO-1 induction in skin fibroblast may serve as an inducible defense pathway to remove heme liberated by oxidants. In addition, cytokines such as TNFα and IL-1β have been shown to increase pro-oxidant generation, which upregulates NFκB-mediated SOD2 expression [59].

ROS are involved in important physiological responses/events, such as defense against environmental pathogens, response to growth factors, cellular proliferation, differentiation, apoptosis, and autophagy. Additionally, Duox expression and H_2O_2 formation are the rate-limiting processes for the synthesis of thyroid hormones [60].

ROS in Pathology

General biochemical mechanisms of ROS in pathology All biomolecules in the cell are targets of ROS and undergo chemical modifications that accumulate with age. However, it is not clearly understood which species are responsible for the damage *in vivo* and how the increased availability of ROS translates into the accumulation of specific oxidative damage. Here are the common biochemical mechanisms for ROS implicated in pathology.

Lipid peroxidation: The polyunsaturated fatty acyl chains of phospholipids are very sensitive to ROS oxidation. Oxidation of ω-6 PUFA yields 4-hydroxy-2-nonenal (4-HNE), while oxidation of ω-3 PUFA produces 4-hydroxy-2-hexanal. 4-HNE is more reactive than malondialdehyde (MDA) produced by proteins oxidation.

Protein oxidation: •OH generated by either ionizing radiation or the Fenton reaction attacks the polypeptide backbone of proteins to form a carbon-centered radical which may react with O_2 to form peroxyl radical. The side chains of all amino acids in proteins are susceptible to oxidation; however, some residues like proline, histidine, arginine, cysteine, and lysine are highly sensitive to oxidation by metal ions. Thus, it is believed that Fe^{2+} binding to the aforementioned residues undergoes the Fenton reaction with H_2O_2 to yield •OH at those specific sites. The amino acid residues that are most vulnerable to ROS attack are arginine, glutamate, histidine, tyrosine, valine, cysteine, proline, threonine, and methionine. Cysteine and methionine residues are particularly susceptible to ROS oxidation. The oxidation of methionine residues by ROS forms methionine sulfoxide (Met-SO). The oxidation of the amino acid residues, as well as the glycation of lysine groups generates carbonyl groups. Protein carbonylation is the most widely used marker of protein oxidation both *in vitro* and in pathological conditions. Peroxynitrite may react with cysteine SH groups, oxidize methionine to methionine sulfoxide, and react with tyrosine forming 3-nitrotyrosine, another important marker of oxidative protein damage [61].

DNA damage: •OH radical reacts with all components of the DNA molecule damaging the purine and pyrimidine bases and the deoxyribose backbone. To date, more than 100 products have been identified from the DNA oxidation, including single- or double-strand DNA breaks; purine, pyrimidine, and deoxyribose modifications; and DNA cross-links. DNA damage can result in either arrest or induction of transcription, induction of signal transduction pathways, replication errors and genome instability.

The formation of 8-hydroxydeoxyguanine (8-OHdG) is the most prominent example among over 20 base modifications. 8-OHdG is the most easily formed base alteration, is mutagenic and carcinogenic, and is a good biomarker of oxidative DNA damage; this modification occurs in one in 10^5 G residues in normal cells but increases in oxidative stress, aging, and a numerous diseases. Mitochondrial DNA is more susceptible than nuclear DNA to oxidative damage.

ROS in atherosclerosis Chronic overproduction of ROS under pathophysiologic conditions is an integral part of the development of CVD, in particular atherosclerosis, a chronic inflammatory disease. There are several molecular mechanisms underlying ROS-accelerated atherosclerosis. First, oxidative stress stimulates EC activation [62], which functions in early atherosclerosis. For example, ROS, including H_2O_2, activate NFκB, which leads to the upregulation of adhesion molecules such as intercellular adhesion molecule (ICAM)-1, vascular cell adhesion molecule (VCAM)-1, monocyte chemotactic protein (MCP)-1, and E-selectin, as well as consequent adhesion and infiltration of inflammatory cells including monocyte and neutrophil. Recently, it has been reported that the deficiency of either Prdx1 [63] or Prdx2 [64], an antioxidant enzyme, enhances endothelial activation and exacerbates atherosclerosis in ApoE$^{-/-}$ mice, due to failure to eliminate ROS. Second, increased ROS formation mediates EC dysfunction, an initial step of atherogenesis. The elevated ROS reduces the bioavailability of NO via the formation of $ONOO^-$, which results in oxidation of the zinc–thiolate complex, consequent uncoupling of eNOS [5], and resultant endothelial dysfunction. Interestingly, energy sensor, AMP-activated protein kinase $\alpha 2$ (AMPK$\alpha 2$) deletion leads to upregulation of Nox2/4 and its partners p67^{phox} and p47^{phox} via NFκB activation. Increased Nox activity results in elevated $O_2^{\bullet-}$ production in EC, which leads to endothelial dysfunction contributing to exacerbated atherosclerosis in low-density lipoprotein receptor knockout (LDLr$^{-/-}$) mice fed a high-fat diet [24]. Moreover, AMPK$\alpha 2$ deletion causes to elevated oxidation of sarco-endoplasmic reticulum calcium ATPase (SERCA) associated with ER stress, which contributes to severe atherosclerosis [65]. Third, low-density lipoproteins (LDL) are oxidation by increased ROS. LDL oxidation generates cytotoxic products, which activate EC and interrupt the normal barrier function of endothelial layer. Oxidized LDL (oxLDL) has higher affinity for scavenger receptors of macrophage than native LDL and is avidly internalized by macrophages through receptor-mediated endocytosis and phagocytosis, resulting in the formation of lipid-laden foam cells, a hallmark of early atherosclerotic lesions. Fourth, ROS activates VSMC. It is well known that ROS enhance aortic VSMC proliferation and migration which play a critical role in the pathophysiology of atherosclerosis. For example, GKT136901, a specific inhibitor of Nox1 and Nox4, attenuated ROS generation and subsequent expression of the adhesion molecule CD44 and its principal ligand, hyaluronan in VSMC, as well as decreased proliferation and migration of VSMC and resultantly attenuated atherogenesis in the aortas of ApoE$^{-/-}$ mice [66]. In addition, oxidized LDL due to increased ROS also promotes the growth and migration of VSMC [67]. Fifth, ROS enhance monocyte-derived macrophage recruitment, differentiation, and activation. For example, thiol oxidative stress enhances macrophage recruitment into vascular lesions by increasing the responsiveness of macrophages to chemoattractants, MCP-1 [68].

Taken together, ROS are involved in many aspects of atherogenesis. However, so far we have not established a specific causal role of ROS in atherogenesis of patients, although there is a large body of evidence connecting the effects of ROS with atherosclerosis in animal models. Emerging evidences indicate that microRNAs may involve in atherogenesis. It is reported that miRNA directly regulates chemotactic factors [69]. For example, miR-10a knockdown in HAEC induces MCP-1, VCAM-1, and E-selectin [70], which mediates monocytes adhesion and transmigration associated with atherogenesis. Therefore, ROS are not the only factors mitigating atherogenesis.

CONCLUSIONS AND PERSPECTIVES

ROS have both beneficial and harmful effects in tissues and organs. Cell has both ROS production and antioxidant defense systems. The balance between these two systems is critical important for normal cell function. Increased ROS formation is usually associated with many diseases including cancer, CVD, diabetes, and aging, but it is not fully defined whether and how ROS actually cause these diseases or whether their removal is beneficial. Moreover, many antioxidant drugs or agents have shown poor outcomes or even unsuccessful. Because the role of the various ROS in disease development is complex, their regulation is likely to be complicated as well, we need to clarify the distinct molecular mechanisms of spatiotemporal production and distribution of different ROS involved in a particular disease, then develop the specific antioxidant strategy for targeted disease.

REFERENCES

1. Finkel, T. Oxidant signals and oxidative stress. *Curr Opin Cell Biol* **15**:247 254; 2003.
2. Song, P.; Wu, Y.; Xu, J. et al. Reactive nitrogen species induced by hyperglycemia suppresses Akt signaling and triggers apoptosis by upregulating phosphatase PTEN (phosphatase and tensin homologue deleted on chromosome 10) in an LKB1-dependent manner. *Circulation* **116**: 1585–1595; 2007.
3. Scherz-Shouval, R.; Shvets, E.; Fass, E.; Shorer, H.; Gil, L.; Elazar, Z. Reactive oxygen species are essential for autophagy and specifically regulate the activity of Atg4. *EMBO J* **26**: 1749–1760; 2007.
4. Kalyanaraman, B.; Darley-Usmar, V.; Davies, K. J. et al. Measuring reactive oxygen and nitrogen species with fluorescent probes: challenges and limitations. *Free Radic Biol Med* **52**: 1–6; 2012.
5. Zou, M. H.; Shi, C.; Cohen, R. A. Oxidation of the zinc-thiolate complex and uncoupling of endothelial nitric oxide synthase by peroxynitrite. *J Clin Invest* **109**:817–826; 2002.
6. Widlansky, M. E; Gutterman, D. D. Regulation of endothelial function by mitochondrial reactive oxygen species. *Antioxid Redox Signal* **15**:1517–1530; 2011.
7. Cadenas, E.; Davies, K. J. Mitochondrial free radical generation, oxidative stress, and aging. *Free Radic Biol Med* **29**:222–230; 2000.
8. Murphy, M. P. How mitochondria produce reactive oxygen species. *Biochem J* **417**: 1–13; 2009.
9. Jezek, P.; Hlavata, L. Mitochondria in homeostasis of reactive oxygen species in cell, tissues, and organism.. *The international journal of biochemistry & cell biology* **37**:2478–2503; 2005.
10. Shen, G. X. Oxidative stress and diabetic cardiovascular disorders: roles of mitochondria and NADPH oxidase. *Canadian journal of physiology and pharmacology* **88**:241–248; 2010.
11. Devarajan, A.; Bourquard, N.; Hama, S. et al. Paraoxonase 2 deficiency alters mitochondrial function and exacerbates the development of atherosclerosis. *Antioxid Redox Signal* **14**: 341–351; 2011.
12. Lambeth, J. D. NOX enzymes and the biology of reactive oxygen. *Nat Rev Immunol* **4**: 181–189; 2004.
13. Leto, T. L.; Morand, S.; Hurt, D.; Ueyama, T. Targeting and regulation of reactive oxygen species generation by Nox family NADPH oxidases. *Antioxid Redox Signal* **11**:2607–2619; 2009.
14. Song, P.; Zou, M. H.. Regulation of NAD(P)H oxidases by AMPK in cardiovascular systems. *Free Radic Biol Med* **52**:1607–1619; 2012.
15. Niu, X. L.; Madamanchi, N. R.; Vendrov, A. E. et al. Nox activator 1: a potential target for modulation of vascular reactive oxygen species in atherosclerotic arteries. *Circulation* **121**: 549–559; 2010.
16. Stanic, B.; Katsuyama, M.; Miller, F. J., Jr. An oxidized extracellular oxidation–reduction state increases Nox1 expression and proliferation in vascular smooth muscle cells via epidermal growth factor receptor activation. *Arterioscler Thromb Vasc Biol* **30**:2234–2241; 2010.

17. Kroviarski, Y.; Debbabi, M.; Bachoual, R. et al. Phosphorylation of NADPH oxidase activator 1 (NOXA1) on serine 282 by MAP kinases and on serine 172 by protein kinase C and protein kinase A prevents NOX1 hyperactivation. *FASEB J* **24**:2077–2092; 2010.
18. Lee, M. Y.; San Martin, A.; Mehta, P. K. et al. Mechanisms of vascular smooth muscle NADPH oxidase 1 (Nox1) contribution to injury-induced neointimal formation. *Arterioscler Thromb Vasc Biol* **29**:480–487; 2009.
19. Yogi, A.; Mercure, C.; Touyz, J. et al. Renal redox-sensitive signaling, but not blood pressure, is attenuated by Nox1 knockout in angiotensin II-dependent chronic hypertension. *Hypertension* **51**:500–506; 2008.
20. Prosser, B. L.; Ward, C. W.; Lederer, W. J. X-ROS signaling: rapid mechano-chemo transduction in heart. *Science* **333**:1440–1445; 2011.
21. Zhang, M.; Song, P.; Xu, J. et al. Activation of NAD(P)H oxidases by thromboxane A2 receptor uncouples endothelial nitric oxide synthase. *Arterioscler Thromb Vasc Biol* **31**:125–132; 2011.
22. Gupte, S. A.; Kaminski, P. M.; George, S. et al. Peroxide generation by p47phox-Src activation of Nox2 has a key role in protein kinase C-induced arterial smooth muscle contraction. *Am J Physiol Heart Circ Physiol* **296**: H1048–H1057; 2009.
23. Chen, Z.; Keaney, J. F., Jr.; Schulz, E. et al. Decreased neointimal formation in Nox2-deficient mice reveals a direct role for NADPH oxidase in the response to arterial injury. *Proc Natl Acad Sci U S A* **101**:13014–13019; 2004.
24. Wang, S.; Zhang, M.; Liang, B. et al. AMPKalpha2 deletion causes aberrant expression and activation of NAD(P)H oxidase and consequent endothelial dysfunction in vivo: role of 26S proteasomes. *Circ Res* **106**: 1117–1128; 2010.
25. Touyz, R. M.; Mercure, C.; He, Y. et al. Angiotensin II-dependent chronic hypertension and cardiac hypertrophy are unaffected by gp91phox-containing NADPH oxidase. *Hypertension* **45**:530–537; 2005.
26. Chen, H.; Song, Y. S.; Chan, P. H. Inhibition of NADPH oxidase is neuroprotective after ischemia-reperfusion. *J Cereb Blood Flow Metab* **29**:1262–1272; 2009.
27. Violi, F.; Sanguigni, V.; Carnevale, R. et al. Hereditary deficiency of gp91(phox) is associated with enhanced arterial dilatation: results of a multicenter study. *Circulation* **120**: 1616–1622; 2009.
28. Ueno, N.; Takeya, R.; Miyano, K. et al. The NADPH oxidase Nox3 constitutively produces superoxide in a p22phox-dependent manner: its regulation by oxidase organizers and activators. *J Biol Chem* **280**:23328–23339; 2005.
29. Cheng, G.; Ritsick, D.; Lambeth, J. D. Nox3 regulation by NOXO1, p47phox, and p67phox. *J Biol Chem* **279**:34250–34255; 2004.
30. Paffenholz, R.; Bergstrom, R. A.; Pasutto, F. et al. Vestibular defects in head-tilt mice result from mutations in Nox3, encoding an NADPH oxidase. *Genes Dev* **18**:486–491; 2004.
31. Craige, S. M.; Chen, K.; Pei, Y. et al. NADPH oxidase 4 promotes endothelial angiogenesis through endothelial nitric oxide synthase activation. *Circulation* **124**:731–740; 2011.
32. Clempus, R. E.; Sorescu, D.; Dikalova, A. E. et al. Nox4 is required for maintenance of the differentiated vascular smooth muscle cell phenotype. *Arterioscler Thromb Vasc Biol* **27**: 42–48; 2007.
33. Lyle, A. N.; Deshpande, N. N.; Taniyama, Y. et al. Poldip2, a novel regulator of Nox4 and cytoskeletal integrity in vascular smooth muscle cells. *Circ Res* **105**:249–259; 2009.
34. Martyn, K. D.; Frederick, L. M.; von Loehneysen, K.; Dinauer, M. C.; Knaus, U. G. Functional analysis of Nox4 reveals unique characteristics compared to other NADPH oxidases. *Cell Signal* **18**:69–82; 2006.
35. Tong, X.; Hou, X.; Jourd'heuil, D. et al. Upregulation of Nox4 by TGFβ1 oxidizes SERCA and inhibits NO in arterial smooth muscle of the prediabetic Zucker rat. *Circ Res* **107**:975–983; 2010.
36. Diaz, B.; Shani, G.; Pass, I.; Anderson, D.; Quintavalle, M.; Courtneidge, S. A. Tks5-dependent, nox-mediated generation of reactive oxygen species is necessary for invadopodia formation. *Sci Signal* **2**:ra53; 2009.
37. Gordillo, G.; Fang, H.; Park, H.; Roy, S. Nox-4-dependent nuclear H_2O_2 drives DNA oxidation resulting in 8-OHdG as urinary biomarker and hemangioendothelioma formation. *Antioxid Redox Signal* **12**:933–943; 2010.
38. Chen, K.; Kirber, M. T.; Xiao, H.; Yang, Y.; Keaney, J. F., Jr. Regulation of ROS signal transduction by NADPH oxidase 4 localization. *J Cell Biol* **181**:1129–1139; 2008.

39. Ismail, S.; Sturrock, A.; Wu, P. et al. NOX4 mediates hypoxia-induced proliferation of human pulmonary artery smooth muscle cells: the role of autocrine production of transforming growth factor-{beta}1 and insulin-like growth factor binding protein-3. *Am J Physiol Lung Cell Mol Physiol* **296**:L489–L499; 2009.
40. Lener, B.; Koziel, R.; Pircher, H. et al. The NADPH oxidase Nox4 restricts the replicative lifespan of human endothelial cells. *Biochem J* **423**:363–374; 2009.
41. Wang, M.; Zhang, J.; Walker, S. J.; Dworakowski, R.; Lakatta, E. G.; Shah, A. M. Involvement of NADPH oxidase in age-associated cardiac remodeling. *J Mol Cell Cardiol* **48**:765–772; 2010.
42. Zhang, M.; Brewer, A. C.; Schroder, K. et al. NADPH oxidase-4 mediates protection against chronic load-induced stress in mouse hearts by enhancing angiogenesis. *Proc Natl Acad Sci U S A* **107**:18121–18126; 2010.
43. Schroder, K.; Zhang, M.; Benkhoff, S. et al. Nox4 is a protective reactive oxygen species generating vascular NADPH oxidase. *Circ Res* **110**:1217–1225; 2012.
44. Montezano, A. C.; Burger, D.; Paravicini, T. M. et al. Nicotinamide adenine dinucleotide phosphate reduced oxidase 5 (Nox5) regulation by angiotensin II and endothelin-1 is mediated via calcium/calmodulin-dependent, rac-1-independent pathways in human endothelial cells. *Circ Res* **106**:1363–1373; 2010.
45. Jay, D. B.; Papaharalambus, C. A.; Seidel-Rogol, B.; Dikalova, A. E.; Lassegue, B.; Griendling, K. K. Nox5 mediates PDGF-induced proliferation in human aortic smooth muscle cells. *Free Radic Biol Med* **45**:329–335; 2008.
46. Banfi, B.; Tirone, F.; Durussel, I. et al. Mechanism of Ca^2+ activation of the NADPH oxidase 5 (NOX5). *J Biol Chem* **279**:18583–18591; 2004.
47. Serrander, L.; Jaquet, V.; Bedard, K. et al. NOX5 is expressed at the plasma membrane and generates superoxide in response to protein kinase C activation. *Biochimie* **89**:1159–1167; 2007.
48. Jagnandan, D.; Church, J.E. ; Banfi, B.; Stuehr, D. J.; Marrero, M. B.; Fulton, D. J. Novel mechanism of activation of NADPH oxidase 5. calcium sensitization via phosphorylation. *J Biol Chem* **282**:6494–6507; 2007.
49. El Jamali, A.; Valente, A. J.; Lechleiter, J. D. et al. Novel redox-dependent regulation of NOX5 by the tyrosine kinase c-Abl. *Free Radic Biol Med* **44**:868–881; 2008.
50. Guzik, T. J.; Chen, W.; Gongora, M. C. et al. Calcium-dependent NOX5 nicotinamide adenine dinucleotide phosphate oxidase contributes to vascular oxidative stress in human coronary artery disease. *J Am Coll Cardiol* **52**:1803–1809; 2008.
51. Moreno, J. C.; Bikker, H.; Kempers, M. J. et al. Inactivating mutations in the gene for thyroid oxidase 2 (THOX2) and congenital hypothyroidism. *N Engl J Med* **347**:95–102; 2002.
52. Kundu, T. K.; Velayutham, M.; Zweier, J. L. Aldehyde oxidase functions as a superoxide generating NADH oxidase: an important redox regulated pathway of cellular oxygen radical formation. *Biochemistry* **51**:2930–2939; 2012.
53. Bravard, A.; Bonnard, C.; Durand, A. et al. Inhibition of xanthine oxidase reduces hyperglycemia-induced oxidative stress and improves mitochondrial alterations in skeletal muscle of diabetic mice. *Am J Physiol Endocrinol Metab* **300**:E581–591; 2011.
54. Bonnard, C.; Durand, A.; Peyrol, S. et al. Mitochondrial dysfunction results from oxidative stress in the skeletal muscle of diet-induced insulin-resistant mice. *J Clin Invest* **118**: 789–800; 2008.
55. Su, J.; Lucchesi, P. A.; Gonzalez-Villalobos, R. A. et al. Role of advanced glycation end products with oxidative stress in resistance artery dysfunction in type 2 diabetic mice. *Arterioscler Thromb Vasc Biol* **28**:1432–1438; 2008.
56. Rhee, S. G.; Chae, H. Z.; Kim, K. Peroxiredoxins: a historical overview and speculative preview of novel mechanisms and emerging concepts in cell signaling. *Free Radic Biol Med* **38**: 1543–1552; 2005.
57. Kern, J. C.; Kehrer, J. P. Free radicals and apoptosis: relationships with glutathione, thioredoxin, and the BCL family of proteins. *Frontiers in bioscience: a journal and virtual library* **10**:1727–1738; 2005.
58. D'Autreaux, B.; Toledano, M. B. ROS as signalling molecules: mechanisms that generate specificity in ROS homeostasis. *Nat Rev Mol Cell Biol* **8**:813–824; 2007.
59. Li, Q.; Engelhardt, J. F. Interleukin-1beta induction of NFkappaB is partially regulated by H_2O_2-mediated activation of NFkappaB-inducing kinase. *J Biol Chem* **281**:1495–1505; 2006.

60. Milenkovic, M.; De Deken, X.; Jin, L. et al. Duox expression and related H_2O_2 measurement in mouse thyroid: onset in embryonic development and regulation by TSH in adult. *The Journal of endocrinology* **192**:615–626; 2007.
61. Zou, M. H.; Cohen, R.; Ullrich, V. Peroxynitrite and vascular endothelial dysfunction in diabetes mellitus. *Endothelium* **11**:89–97; 2004.
62. Alom-Ruiz, S. P.; Anilkumar, N.; Shah, A. M. Reactive oxygen species and endothelial activation. *Antioxid Redox Signal* **10**: 1089–1100; 2008.
63. Kisucka, J.; Chauhan, A. K.; Patten, I. S. et al. Peroxiredoxin1 prevents excessive endothelial activation and early atherosclerosis. *Circ Res* **103**:598–605; 2008.
64. Park, J. G.; Yoo, J. Y.; Jeong, S. J. et al. Peroxiredoxin 2 deficiency exacerbates atherosclerosis in apolipoprotein E-deficient mice. *Circ Res* **109**:739–749; 2011.
65. Dong, Y.; Zhang, M.; Liang, B. et al. Reduction of AMP-activated protein kinase alpha2 increases endoplasmic reticulum stress and atherosclerosis in vivo. *Circulation* **121**:792–803; 2010.
66. Vendrov, A. E.; Madamanchi, N. R.; Niu, X. L. et al. NADPH oxidases regulate CD44 and hyaluronic acid expression in thrombin-treated vascular smooth muscle cells and in atherosclerosis. *J Biol Chem* **285**:26545–26557; 2010.
67. Mitra, S.; Deshmukh, A.; Sachdeva, R. et al. Oxidized low-density lipoprotein and atherosclerosis implications in antioxidant therapy. *The American journal of the medical sciences* **342**:135–142; 2011.
68. Qiao, M.; Zhao, Q.; Lee, C. F. et al. Thiol oxidative stress induced by metabolic disorders amplifies macrophage chemotactic responses and accelerates atherogenesis and kidney injury in LDL receptor-deficient mice. *Arterioscler Thromb Vasc Biol* **29**:1779–1786; 2009.
69. Hulsmans, M.; De Keyzer, D.; Holvoet, P. MicroRNAs regulating oxidative stress and inflammation in relation to obesity and atherosclerosis. *FASEB J* **25**:2515–2527; 2011.
70. Fang, Y.; Shi, C.; Manduchi, E. et al. MicroRNA-10a regulation of proinflammatory phenotype in athero-susceptible endothelium in vivo and in vitro. *Proc Natl Acad Sci U S A* **107**:13450–13455; 2010.

31

HYDROGEN SULFIDE: ROLE IN ATHEROSCLEROSIS, MECHANISMS, AND THERAPEUTIC POTENTIAL

ELVIS A. PETER[1,2], JAI VARMA[1,2], AND CHRISTOPHER G. KEVIL[1,2]

[1]Department of Pathology, School of Medicine in Shreveport, Louisiana State University, Shreveport, Louisiana
[2]Departments of Cardiology, Pathology, and Molecular and Cellular Physiology, LSU Health-Shreveport, Shreveport, LA, USA

INTRODUCTION

The gasotransmitter hydrogen sulfide (H_2S) has been traditionally viewed for the majority of human existence as a poisonous gas capable of widespread annihilation of life forms such as in the Permian–Triassic super anoxic event [1]. However, in 1996, seminal work by Abe and Kimura on the expression of the H_2S-producing enzyme cystathionine β-synthase (CBS) in the hippocampus and the enhancement of long-term memory retention by the potentiation of NMDA receptor activity demonstrated the physiological role of H_2S as a gasotransmitter capable of modulating a myriad of biologic effects [2]. Perhaps not surprisingly then, modern analysis of Stanley Miller's spark discharge experiments demonstrating the creation of amino acids abiotically in a H_2S-rich environment suggests that H_2S may have been indeed the spark of life [3]. In this chapter, we focus on the current state of knowledge about the biochemistry and physiology of H_2S, its role in the pathogenesis of atherogenesis, and its possible therapeutic usefulness for cardiovascular disease.

H_2S BIOCHEMISTRY

H_2S is a colorless lipophilic gas with an odor of rotten eggs and a vapor density of 1.19. It exists in the body as a free form (H_2S) which dissociates into the hydrosulfide anion (HS^-) with a pKa of approximately 6.9 and free sulfide anion (S^{2-}) with a pKa of greater than 13 [4]. In mammalian systems at a temperature of 37 °C and pH of 7.2, the majority of H_2S exists as HS^- (85%) and only 15% as H_2S gas with the contribution from S^{2-} being negligible [5]. In addition to the free form, important bioavailable pools of sulfide include the acid-labile form (Figure 31.1), so called because H_2S is released in response to acidification below pH 5.4 [6]. This pool represents sulfide present in iron sulfur clusters of nonheme proteins including enzymes like aconitase,

Atherosclerosis: Risks, Mechanisms, and Therapies, First Edition. Edited by Hong Wang and Cam Patterson.

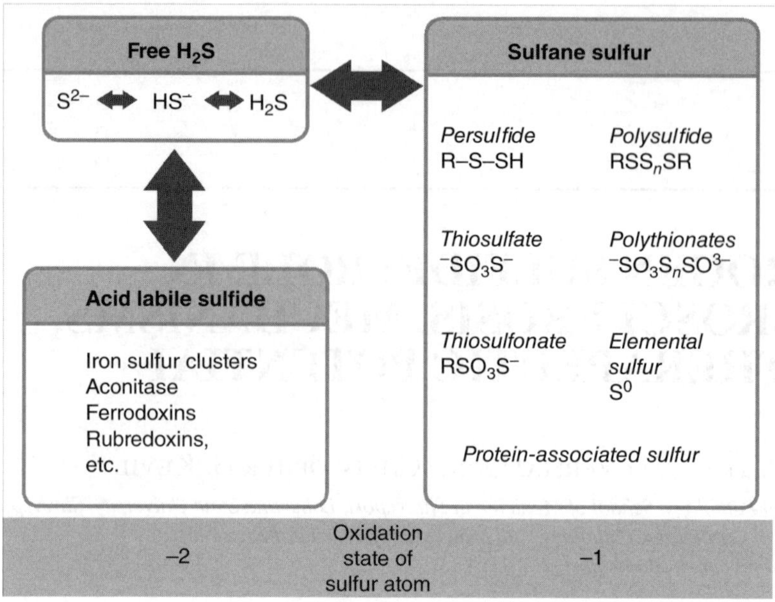

Figure 31.1 Biological pools of H_2S. While sulfur can exist in the human body from a fully reduced divalent state to the fully oxidized hexavalent state, the biologically most relevant states are shown with their constituent molecules. H_2S, hydrogen sulfide; HS^-, hydrosulfide anion; S^{2-}, sulfide anion.

ferredoxins, rubredoxins, and so on. The other major pool of sulfide is present as bound or sulfane sulfur pool (Figure 31.1), where sulfur is bound only to other sulfur atoms. The sulfur present in this pool can be released as H_2S through the action of reducing agents. Ex vivo, this is demonstrated by tissue release of H_2S upon treatment with dithiothreitol (DTT) or more recently tris-2-carboxyethylphospine (TCEP) [7, 8]. The role of these various pools and their relevance to biologic functions continue to be elucidated and remain controversial. However, it is likely that at least one of these two pools represents a storage depot that helps regulate bioavailability of H_2S. The strongest argument for this comes from the fact that the H_2S levels needed to affect biologic function appear to be considerably higher than the free levels measured [9].

Bioavailability and Methods of Detection

A major controversy facing the field is the lack of a true gold standard for detecting H_2S levels. The traditional method employed trapping the H_2S as zinc sulfide and subsequent reaction with *N*,*N*-dimethylphenyl-1,4-diamine in the presence of ferric chloride forming methylene blue which was then spectrophotometrically quantified. However, H_2S levels obtained using this method have been closely examined [10] and found to be invalid [11]. It is now accepted that this method is incapable of measuring bona fide free H_2S at physiological levels and is also subject to several sources of experimental interference. Recently, sensitive and accurate analytical measurement of free H_2S has been accomplished using the fluorescent probe monobromobimane (mBB) which reacts with H_2S forming a sulfide dibimane product which is separated using reverse-phase high performance liquid chromatography (HPLC) and quantified by fluorescence detection [11, 12]. Other analytical methods that are currently utilized include gas chromatography and polarographic electrode detection [13]. Novel H_2S-sensitive fluorescent probes have also been reported that could be used for single cell measurements [14]. Another recent advance in the field is the development of an analytical

HPLC method to measure discrete biologic pools of H_2S (free, acid labile, and sulfane sulfur/persulfides) based on the mBB method using selective acidification and reduction to discretely measure all bioavailable H_2S equivalents in different biochemical compartments [7]. Others have utilized gas chromatography in conjunction with acidification to measure free and acid-labile pools of H_2S [13].

It is now clear that the normal levels of human plasma free H_2S are in the nanomolar range instead of the micromolar range as was previously thought [7]. The question of what constitutes normal levels or levels in various disease states remains unclear. However, it has been reported that plasma free H_2S levels are elevated in patients with cardiovascular disease using the mBB reverse-phase HPLC analytical detection method [15]. It is clear that H_2S levels in different tissues vary with the aorta exhibiting several fold higher levels than other tissues including plasma [13, 16]. Additionally, the acid-labile sulfide pool is several fold higher than the free levels in most tissues, and the presence of microbiota significantly impacts tissue-bound sulfane sulfur levels, supporting the theory of potential tissue H_2S reservoirs to effect local concentrations that could impact multiple biological responses [7].

Production and Metabolism

H_2S production may occur by nonenzymatic reduction reactions as well as be synthesized by the specific enzymes. The enzymatic production occurs by two cytosolic enzymes, cystathionine gamma lyase (CSE) and CBS, which utilize pyridoxine as a cofactor. The other major enzyme is 3-mercaptopyruvate sulfurtransferase (3-MST). 3-MST is predominately a mitochondrial enzyme though a small amount can be found in the cytosol as well. In contrast to earlier reports of the tissue-specific nature of these enzymes, it appears that they are present and operative in various tissues. However, CSE has been shown to have a major role in vascular endothelium and hence is of particular importance to cardiovascular physiology. Indeed, the generation of gene knockout model of CSE has been instrumental in advancing our understanding of this field [17].

Figure 31.2 shows the major biosynthetic pathways and the substrates for the three major enzymes. In brief, in humans, CSE is encoded by *Cth* gene located on chromosome 1p31.1. It is a versatile enzyme that utilizes primarily cysteine but also cystathionine and homocysteine via multiple pathways to generate H_2S. In humans, cystine is not used as a substrate for H_2S production [4, 18]. CSE has been shown to present in humans in the brain, ocular lens, joints, pulmonary artery smooth muscle, and most importantly vascular endothelium [17]. CSE activity can be inhibited by the irreversible inhibitor D,L-propargylglycine or β-cyanoalanine and trifluoroalanine that may be reversible.

CBS is encoded by a gene present on chromosome 21 in humans at locus 21q22.3 and utilizes cysteine or cysteine plus homocysteine to produce lanthionine or cystathionine, respectively. The latter pathway appears to be the predominant one in humans. CBS has been demonstrated in the human brain, liver, colon, ocular lens, joints, as well as uterine tissue and is inhibited by AOAA and hydroxylamine. CBS activity undergoes allosteric upregulation by S-adenosylmethionine. It is of interest that patients with trisomy 21 or Down's syndrome who appear to have some protection against atherosclerosis have an elevated level of H_2S metabolites [19].

The gene for 3-MST is located on chromosome 22 and has it's locus at 22q12.3. It is a sulfurtransferase and catalyzes the transfer of sulfane sulfur from mercaptopyruvate to an active site cysteine forming a persulfide with subsequent release of H_2S via a nonenzymatic reduction reaction. 3-MST has been demonstrated in human brain and vasculature.

The elimination of H_2S occurs via oxidation in the mitochondria. Two sulfide quinone oxidoreductases (SQR) oxidize sulfide to elemental sulfur with simultaneous reduction of cysteine disulfide with the formation of a persulfide group at one of the SQR cysteines. The enzyme sulfur dioxygenase (SDO) then oxidizes one

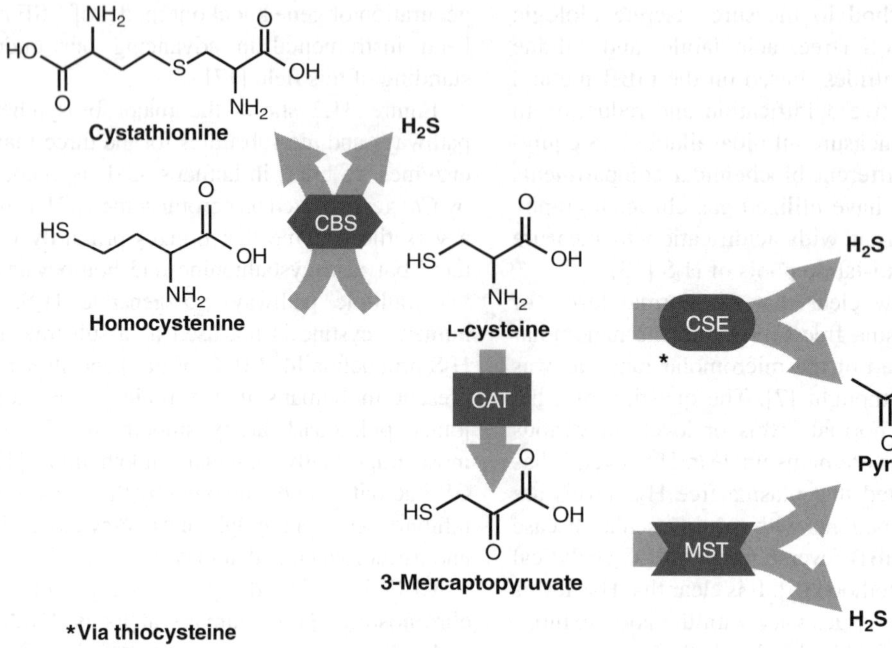

Figure 31.2 Major biosynthetic pathways for production of H_2S. CSE and CBS are cytosolic enzymes and utilize pyridoxal phosphate as a cofactor. MST is a sulfurtransferase and is present in both the cytosol and mitochondria. CSE, γ-cystathionase; CBS, cystathionine β-synthase; MST, mercaptopyruvate sulfur transferase; H_2S, hydrogen sulfide.

persulfide to sulfite (H_2SO_3) utilizing oxygen and water. The second persulfide is transferred to sulfite with formation of thiosulfate ($H_2S_2O_3$). The electrons generated enter the respiratory chain and are ultimately transferred to oxygen [10]. Ethylmalonic encephalopathy (EE) is a rare metabolic disorder caused by dysfunction of ETHE1, a mitochondrial dioxygenase involved in H_2S detoxification. Manifestations include psychomotor retardation, chronic diarrhea, orthostatic acrocyanosis, and relapsing petechiae starting in infancy [20].

Transport, Second Messenger Systems, and Other Interactions

The lipophilic nature of H_2S suggests that it should be diffusible across the lipid bilayer of cellular membranes, and it has been shown that H_2S does not require a specialized membrane transporter [21]. Recent modeling data indicates that upon generation H_2S can diffuse a distance of approximately 300 cells [22]. Conversely, much less is known about hydrosulfide anion transport, which represents the most readily bioavailable free pool of H_2S equivalents.

The earliest second messenger signaling system reported to be affected by H_2S was activation of K_{ATP} [23], which can facilitate an increase in cGMP and modulate vasodilation. The role of H_2S as an endothelium-dependent vasodilator is largely attributed to this and discussed in detail in the following. Since this initial discovery, it is clear that a variety of other intracellular pathways are affected by H_2S including the RISK antiapoptosis pathway, Nrf-2 nuclear translocation, and phosphatase activity. H_2S alters these various signaling pathways through direct protein reaction involving a posttranslational modification process termed sulfhydration. Protein sulfhydration occurs primarily at cysteine residues resulting in persulfide formation (RS–SH) that can alter protein function and/or enzyme activity. For example, sulfhydration of the Kir subunit of the K_{ATP} is responsible for the enhancement of its activity leading to smooth muscle relaxation. Similarly, sulfhydration of the cysteine residues of GADPH increases its catalytic activity 700-fold [24].

Given the cardiovascular pathophysiological importance of H_2S, studies have examined whether H_2S affects NO metabolism in a variety of ways. Several studies using in vitro cell culture and recombinant protein activity assays have shown that H_2S potently inhibits nitric oxide synthase (NOS) activity that would reduce NO synthesis [25–27]. Conversely, other reports have shown that H_2S can augment endothelial NOS phosphorylation in culture and that the beneficial effect of H_2S therapy during global tissue ischemia due to cardiopulmonary arrest is abrogated in *eNOS* gene-deficient mice [28–30]. It has also been suggested that H_2S–NO reactions might occur [31]; however, direct reactions appear biochemically unfavorable with studies indicating that H_2S likely interacts with nitrite (NO_2^-) or nitroxyl anion (HNO^-) [32]. Nonetheless, increasing evidence indicates that H_2S bioavailability can affect NO metabolism, an area that requires further investigation.

H_2S AND CARDIOVASCULAR PATHOPHYSIOLOGY

H_2S and Vasomotor Tone Regulation

Perhaps the most well studied aspect of H_2S on the circulatory system is its role in vasodilation. To appreciate this, one must step back to earlier actions of endothelial derived relaxing factors, the most notable being acetylcholine (Ach) mediated nitric oxide production that activates vascular smooth muscle sGC increasing cGMP formation and subsequent protein kinase G activation resulting in decreased intracellular calcium and vasorelaxation. However, it is well known that the blockade of endothelial nitric oxide as well as prostacyclin, another endothelium-dependent vasorelaxant, does not fully abrogate the vasodilatory response to Ach [33]. The search for this alternative mediator brings us to the role of H_2S in the vasodilation. The first step in this discovery was the recognition of the role of H_2S in relaxing blood vessels and lowering blood pressure by opening ATP-sensitive K^+ channels in vascular smooth muscle. The blockade of CSE, the enzyme that generates H_2S by D,L-propargylglycine, elicits the development of hypertension [34]. The CSE knockout murine model also demonstrates the development of sustained hypertension at 7 weeks of age. Furthermore, the homozygous animals demonstrate higher levels of hypertension relative to the heterozygotes without any change in the levels of eNOS. This effect was reversed by the administration of H_2S and found to be dependent on the Ach-induced increase in calcium calmodulin similar to NO. This study established the role of H_2S as a vasodilator [17].

H_2S has also been shown to cause rise in intracellular calcium via mobilization from intracellular stores such as the endoplasmic reticulum in endothelial cells. The resultant decline in storage levels of calcium can serve as a trigger for influx of extracellular calcium and hence regulate vascular tone [35]. The characteristic response of norepinephrine-precontracted aortic rings to H_2S has been described as triphasic with an initial contraction and then relaxation followed by a second contraction phase. Of interest, this response was enhanced with aging, which is associated with decrease in H_2S and increase in CSE and CBS expression [36].

Angiotensin-converting enzyme is a zinc metalloproteinase that is present on the endothelium and is a part of the renin angiotensin pathway. It catalyzes the conversion of angiotensin I to angiotensin II, a potent vasoconstrictor. The blockade of this enzyme has been demonstrated to be of significant benefit in the treatment of hypertension, heart failure, and coronary artery disease. H_2S perhaps by interaction with the zinc moiety has been shown to decrease the activity of ACE in a dose-dependent fashion [37].

H_2S and Cellular Viability

The demonstration of enhanced longevity of *Caenorhabditis elegans* by exposure of H_2S via modulation of the Sir 2.1, a member of the Sir-2 family of NAD(+)-dependent protein deacetylases, raised intriguing possibilities [38]. While the implication for mammalian survival is less clear, H_2S has been reported to have antiapoptotic activity. In an animal model of cardiac ischemia–reperfusion injury, H_2S has been shown to have direct antiapoptotic activity via phosphorylation/activation of PKCε–STAT3 signaling cascade including increase in expression of HSP90, HSP70, Bcl-2, and Bcl-xl as well as inactivation of the proapoptogen Bad [39].

More recently, H_2S has been demonstrated to have an integral role in the antiapoptotic activity of NFκB. The release of TNF-α has been shown to induce the expression of CSE, and the resultant increase in H_2S causes sulfhydration of the P65 subunit of NfκB augmenting its ability to bind to its coactivator ribosomal protein S3. This complex in turn stimulated the transcription of antiapoptotic genes [40].

H_2S and Redox Regulation

H_2S has been shown to protect neurons from glutamate-induced oxidative damage by increasing activity of γ-glutamylcysteine synthetase, which is the rate-limiting step in the production of the antioxidant glutathione. Glutathione production is also enhanced by attenuation of glutamate inhibition of cystine transport. This cystine is normally converted to cysteine and then onto glutathione [41]. In the murine myocardial ischemia–reperfusion injury model, H_2S has been shown to reduce oxidative stress via rapid increase in Nrf-2 translocation to the nucleus and binding with antioxidant responsive element with subsequent increase in hemeoxygenase-1 (HO-1), thioredoxin (Trx1), and glutathione metabolism [39]. H_2S has also been reported to act as a direct antioxidant that could be beneficial against cardiovascular pathophysiology [42]. However, measurement of H_2S reaction rate constants with various oxidants (e.g., peroxide, peroxynitrite, hypochlorite, etc.) was comparatively slower to that of other low molecular weight thiols (e.g., cysteine and glutathione) that are in greater abundance indicating that H_2S direct scavenging of oxidants most likely doesn't account for all its protective effects [43].

H_2S and Endothelial Cell Function

Diabetes mellitus is a well-known risk factor for atherosclerosis. The hyperglycemic state is well characterized to induce endothelial dysfunction mediated by increase in reactive oxygen species and more recently has been shown to decrease plasma H_2S concentrations. Furthermore the replacement of H_2S via exogenous administration or the by CSE overexpression in transgenic mouse model demonstrated amelioration of the hyperglycemia induced endothelial cell dysfunction by preserving mitochondrial function. Conversely, the use of siRNA to silence CSE worsened ROS production in hyperglycemia [44].

In the normoglycemic state, H_2S impacts endothelial cell function as well. This was demonstrated in the TNF-α-induced endothelial dysfunction model utilizing human umbilical vein endothelial cells (HUVEC). NaHS had an anti-inflammatory effect, specifically decreased expression of intercellular adhesion molecule-1 (ICAM-1) and vascular cell adhesion molecule-1 (VCAM-1). Additionally, mRNA expression of other adhesion molecules, p-selectin and e-selectins, was reduced as was U937 monocyte adhesion to HUVEC. These effects were mediated via upregulation of HO-1 expression and diminished NFκB activation and ROS production [45].

The angiogenic potential of endothelial cells is an important element of the body's response to the consequences of atherosclerosis. H_2S has been demonstrated to enhance the angiogenic potential of endothelial cells via a K_{ATP} channel/MAPK pathway. Moreover,

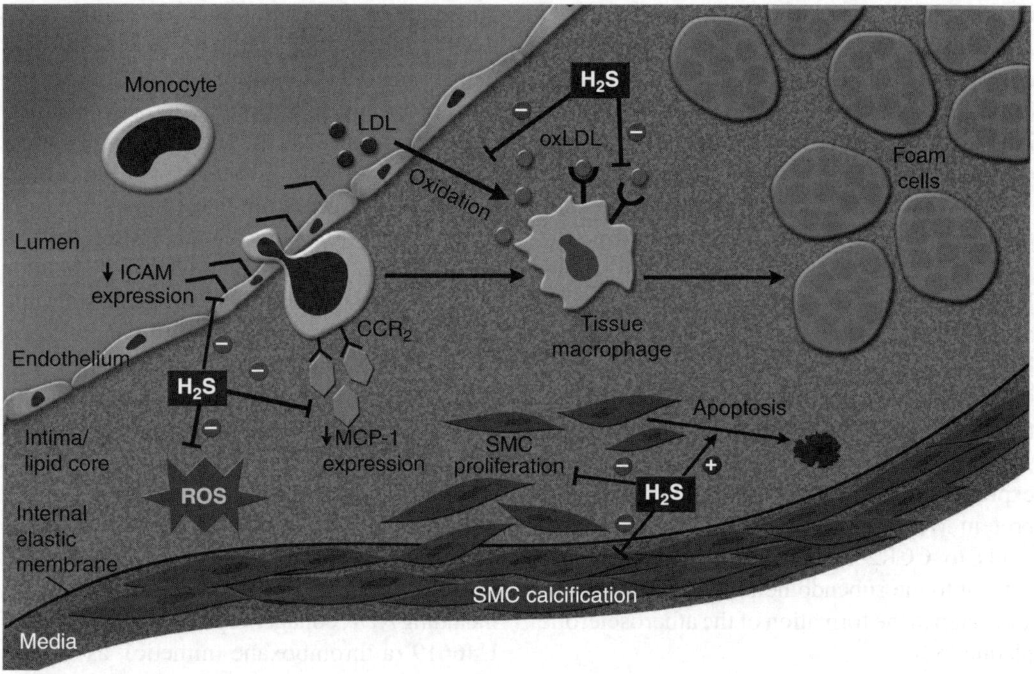

Figure 31.3 The role of H_2S in atherogenesis. This figure illustrates the effect of H_2S on inflammation and resultant decrease in monocyte recruitment and transformation, as well as the net decrease in smooth muscle cell infiltration of the lipid core. ICAM, intercellular adhesion molecule; MCP-1, monocyte chemoattractant protein-1; CCR_2, chemokine receptor 2; LDL, low-density lipoprotein; oxLDL, oxidized LDL; SMC, smooth muscle cell; H_2S, hydrogen sulfide; ROS, reactive oxygen species.

H_2S also facilitates the proangiogenic action of vascular endothelial growth factor (VEGF) on endothelial cells as demonstrated with HUVEC and CSE knockout and siRNA model [46]. It has also been reported that exogenous H_2S therapy stimulates ischemic angiogenesis activity using the femoral artery ligation model that involves increased NO generation through enzymatic and nonenzymatic pathways [47]. Additionally, a deficiency of H_2S bioavailability has been observed in patients with preeclampsia, and experimental inhibition of CSE results in defective placental vascularization [48]. Together, these studies highlight an important role of H_2S for angiogenic activity.

H_2S and Atherosclerosis

There is significant insight on the relationship between H_2S and atherosclerosis. The currently known effects of H_2S on the atherogenesis cascade are illustrated in Figure 31.3. In general, the effects can be characterized as:

1. Decreased oxidative modification of low-density lipoprotein (LDL) and dyslipidemia
2. Effects on adhesion molecule expression, monocyte recruitment, transformation into tissue macrophages, and foam cell formation
3. Decreased smooth muscle cell proliferation, calcification, and migration
4. Decrease in reactive oxygen species

A critical step in the development of the atherosclerotic plaque is the transformation of macrophages into foam cells. This involves the uptake of oxidized LDL by macrophages. Zhao and colleagues showed that treatment with NaHS reduced the binding and uptake of oxLDL by macrophages. Additionally, downregulation of ACAT1, CD-36, and SR-A expression contributes to the inhibition of foam cell formation [49]. Furthermore, H_2S has been shown to inhibit the hypochlorite- and hemin-mediated atherogenic modification of LDL [50].

In the ApoE$^{-/-}$ model of atherosclerosis, plasma H_2S levels as measured by the methylene blue method were shown to be lower, and treatment with NaHS decreased the aortic plaque as well as ICAM-1 expression [51]. ICAM-1 plays an important role in inflammatory cell adhesion and leukocyte transmigration into the lipid core of the atherosclerotic plaque. H_2S also downregulates the expression of the monocyte chemoattractant protein MCP-1, which is a CC chemokine that binds to CCR2 receptor and recruits monocytes into the subendothelial layer, a key initiation step in the formation of the atherosclerotic plaque [52].

Another critical role of H_2S in the pathogenesis of atherosclerosis is the effect on inducing apoptosis of vascular smooth muscle cells, a key player in the generation of the atherosclerotic plaque. Using a model of adenoviral CSE overexpression, increased CSE expression reduced smooth muscle cell proliferation as well as increased apoptosis [53]. Conversely in the CSE knockout model, increase in SMCs in the aortic media was noted. These effects are mediated by phosphorylation of ERK1/2 and altered expression of cyclin D1 and p21 (Cip/WAF-1) [54]. Consistent with the aforementioned findings, *CSE* gene deficiency significantly exacerbates the development of experimental atherosclerosis in the ApoE knockout model involving differential effects on adhesion molecule expression, inflammatory signaling responses, and dyslipidemia [55]. Importantly, H_2S administered as NaHS has also been reported to decrease proliferation of vascular smooth muscle cell via MAPK pathway in a dose-dependent fashion in the rat model [56]. Findings from a rat model of vascular calcification also show that H_2S reduces vascular calcification with reduction in alkaline phosphatase activity and osteopontin gene downregulation [57]. Lastly, exogenous H_2S therapy has been shown to correct CSE knockout enhancement of atherosclerosis, reinforcing the importance of the H_2S/CSE in regulating atherogenesis [55].

Neointima formation plays a critical role in response to percutaneous interventions for atherosclerosis. In animal models of balloon injury to carotid intima, H_2S has been shown to be protective in reducing neointimal proliferation when administered as NaHS, an H_2S donor. CSE expression and H_2S levels were shown to be depressed in the balloon-injured carotid artery [58].

Platelet aggregation is a key pathophysiological event during atherosclerosis and acute coronary syndrome. H_2S decreases platelet aggregation in response to a variety of stimuli including ADP, collagen, epinephrine, thrombin, U46619 (a thromboxane mimetic), as well as arachidonic acid. The mechanism for this has not been clearly delineated but appears to be independent of nitric oxide, cGMP, or cAMP as well as the K_{ATP} channel activity [59].

H_2S AND CARDIOVASCULAR THERAPEUTIC POTENTIAL

Attempts at translating the beneficial effects of H_2S to clinical therapeutics are in its early stages. Several therapeutic attempts have utilized either naturally occurring H_2S-donating drugs such as allicin, the active component of garlic, or synthetic H_2S-donating agents like buffered Na_2S, NaHS, or more complex molecules purported to slowly release GYY4137. Still, other approaches have involved complexing H_2S donors to other therapeutic agents such as combination of anethole trithione with nonsteroidal anti-inflammatory agents to reduce adverse gastrointestinal effects.

The most promising animal studies have been on the effect of H_2S in animal models of ischemia–reperfusion injury; specifically, NaHS has been shown to be cardioprotective in the murine model [60] and Na_2S in the porcine model with the latter showing improvement in global LV systolic function and decrease in infarct size in response to pretreatment prior to reperfusion though cardiac biomarker elevation (CK-MB) levels were not significantly different [61]. Supporting evidence for this is provided by limited infarct size in animals overexpressing CSE and increased infarct size in CSE knockout mice [62]. Finally, garlic oil and its associated organosulfur compounds (e.g., diallyl disulfide and diallyl trisulfide) regulate vascular reactivity in an H_2S-dependent manner and also protect eNOS function against oxidative lipid stress [63, 64]. Together, these and the aforementioned findings suggest that H_2S-based therapies may be useful for cardiovascular disease in the future.

REFERENCES

1. Grice K, Cao C, Love GD, et al. Photic zone euxinia during the permian-triassic superanoxic event. *Science*. 2005; **307**:706–709.
2. Abe K, Kimura H. The possible role of hydrogen sulfide as an endogenous neuromodulator. *J Neurosci*. 1996; **16**:1066–1071.
3. Parker ET, Cleaves HJ, Dworkin JP, et al. Primordial synthesis of amines and amino acids in a 1958 miller H_2S-rich spark discharge experiment. *Proc Natl Acad Sci U S A*. 2011; **108**: 5526–5531.
4. Kabil O, Banerjee R. Redox biochemistry of hydrogen sulfide. *J Biol Chem*. 2010; **285**: 21903–21907.
5. Dombkowski RA, Russell MJ, Olson KR. Hydrogen sulfide as an endogenous regulator of vascular smooth muscle tone in trout. *Am J Physiol Regul Integr Comp Physiol*. 2004; **286**:R678–R685.
6. Ishigami M, Hiraki K, Umemura K, Ogasawara Y, Ishii K, Kimura H. A source of hydrogen sulfide and a mechanism of its release in the brain. *Antioxid Redox Signal*. 2009; **11**:205–214.
7. Shen X, Peter EA, Bir S, Wang R, Kevil CG. Analytical measurement of discrete hydrogen sulfide pools in biological specimens. *Free Radic Biol Med*. 2012; **52**:2276–2283.
8. Ubuka T. Assay methods and biological roles of labile sulfur in animal tissues. *J Chromatogr B Analyt Technol Biomed Life Sci*. 2002; **781**:227–249.
9. Furne J, Saeed A, Levitt MD. Whole tissue hydrogen sulfide concentrations are orders of magnitude lower than presently accepted values. *Am J Physiol Regul Integr Comp Physiol*. 2008; **295**:R1479–R1485.
10. Olson KR. The therapeutic potential of hydrogen sulfide: Separating hype from hope. *Am J Physiol Regul Integr Comp Physiol*. 2011; **301**:R297–R312.
11. Shen X, Pattillo CB, Pardue S, Bir SC, Wang R, Kevil CG. Measurement of plasma hydrogen sulfide *in vivo* and *in vitro*. *Free Radic Biol Med*. 2011; **50**:1021–1031.
12. Wintner EA, Deckwerth TL, Langston W, et al. A monobromobimane-based assay to measure the pharmacokinetic profile of reactive sulphide species in blood. *Br J Pharmacol*. 2010; **160**: 941–957.
13. Levitt MD, Abdel-Rehim MS, Furne J. Free and acid-labile hydrogen sulfide concentrations in mouse tissues: Anomalously high free hydrogen sulfide in aortic tissue. *Antioxid Redox Signal*. 2011; **15**:373–378.
14. Lippert AR, New EJ, Chang CJ. Reaction-based fluorescent probes for selective imaging of hydrogen sulfide in living cells. *J Am Chem Soc*. 2011; **133**:10078–10080.
15. Peter EA, Shen X, Shah SH, et al. Plasma free H_2S levels are elevated in patients with cardiovascular disease. *J Am Heart Assoc*. 2013; **2**:e000387.
16. Shen X, Carlstrom M, Borniquel S, Jadert C, Kevil CG, Lundberg JO. Microbial regulation of host hydrogen sulfide bioavailability and metabolism. *Free Radic Biol Med*. 2013; **60**:195–200.
17. Yang G, Wu L, Jiang B, et al. H_2S as a physiologic vasorelaxant: Hypertension in mice with deletion of cystathionine gamma-lyase. *Science*. 2008; **322**:587–590.
18. Whiteman M, Le Trionnaire S, Chopra M, Fox B, Whatmore J. Emerging role of hydrogen sulfide in health and disease: Critical appraisal of biomarkers and pharmacological tools. *Clin Sci (Lond)*. 2011; **121**:459–488.
19. Kamoun P, Belardinelli MC, Chabli A, Lallouchi K, Chadefaux-Vekemans B. Endogenous hydrogen sulfide overproduction in

down syndrome. *Am J Med Genet A*. 2003; **116A**:310–311.

20. Barth M, Ottolenghi C, Hubert L, et al. Multiple sources of metabolic disturbance in ethe1-related ethylmalonic encephalopathy. *J Inherit Metab Dis*. 2010; **33 Suppl** 3:S443–S453.
21. Mathai JC, Missner A, Kugler P, et al. No facilitator required for membrane transport of hydrogen sulfide. *Proc Natl Acad Sci U S A*. 2009; **106**:16633–16638.
22. Cuevasanta E, Denicola A, Alvarez B, Möller MN. Solubility and permeation of hydrogen sulfide in lipid membranes. *PLoS ONE*. 2012; **7**(4):e34562.
23. Zhao W, Zhang J, Lu Y, Wang R. The vasorelaxant effect of H_2S as a novel endogenous gaseous k(atp) channel opener. *EMBO J*. 2001; **20**:6008–6016.
24. Mustafa AK, Gadalla MM, Sen N, et al. H_2S signals through protein s-sulfhydration. *Sci Signal*. 2009; **2**:ra72.
25. Geng B, Cui Y, Zhao J, et al. Hydrogen sulfide downregulates the aortic L-arginine/nitric oxide pathway in rats. *Am J Physiol Regul Integr Comp Physiol*. 2007; **293**:R1608–R1618.
26. Kubo S, Kurokawa Y, Doe I, Masuko T, Sekiguchi F, Kawabata A. Hydrogen sulfide inhibits activity of three isoforms of recombinant nitric oxide synthase. *Toxicology*. 2007; **241**:92–97.
27. Li XH, Du JB, Bu DF, Tang XY, Tang CS. Sodium hydrosulfide alleviated pulmonary vascular structural remodeling induced by high pulmonary blood flow in rats. *Acta Pharmacol Sin*. 2006; **27**:971–980.
28. Minamishima S, Bougaki M, Sips PY, et al. Hydrogen sulfide improves survival after cardiac arrest and cardiopulmonary resuscitation via a nitric oxide synthase 3-dependent mechanism in mice. *Circulation*. 2009; **120**:888–896.
29. Predmore BL, Julian D, Cardounel AJ. Hydrogen sulfide increases nitric oxide production from endothelial cells by an akt-dependent mechanism. *Front Physiol*. 2011; **2**:104.
30. Yusof M, Kamada K, Kalogeris T, Gaskin FS, Korthuis RJ. Hydrogen sulfide triggers late-phase preconditioning in postischemic small intestine by an no- and p38 MAPK-dependent mechanism. *Am J Physiol Heart Circ Physiol*. 2009; **296**:H868–H876.
31. Whiteman M, Li L, Kostetski I, et al. Evidence for the formation of a novel nitrosothiol from the gaseous mediators nitric oxide and hydrogen sulphide. *Biochem Biophys Res Commun*. 2006; **343**:303–310.
32. Kolluru GK, Shen X, Kevil CG. A tale of two gases: NO and HS, foes or friends for life? *Redox Biol*. 2013; **1**:313–318.
33. Cohen RA. The endothelium-derived hyperpolarizing factor puzzle: A mechanism without a mediator? *Circulation*. 2005; **111**:724–727.
34. Zhao W, Ndisang JF, Wang R. Modulation of endogenous production of H_2S in rat tissues. *Can J Physiol Pharmacol*. 2003; **81**:848–853.
35. Bauer CC, Boyle JP, Porter KE, Peers C. Modulation of Ca^{2+} signalling in human vascular endothelial cells by hydrogen sulfide. *Atherosclerosis*. 2010; **209**:374–380.
36. Predmore BL, Alendy MJ, Ahmed KI, Leeuwenburgh C, Julian D. The hydrogen sulfide signaling system: Changes during aging and the benefits of caloric restriction. *Age (Dordr)*. 2010; **32**:467–481.
37. Laggner H, Hermann M, Esterbauer H, et al. The novel gaseous vasorelaxant hydrogen sulfide inhibits angiotensin-converting enzyme activity of endothelial cells. *J Hypertens*. 2007; **25**:2100–2104.
38. Miller DL, Roth MB. Hydrogen sulfide increases thermotolerance and lifespan in *Caenorhabditis elegans*. *Proc Natl Acad Sci U S A*. 2007; **104**:20618–20622.
39. Calvert JW, Jha S, Gundewar S, et al. Hydrogen sulfide mediates cardioprotection through nrf2 signaling. *Circ Res*. 2009; **105**:365–374.
40. Sen N, Paul BD, Gadalla MM, et al. Hydrogen sulfide-linked sulfhydration of nf-kappab mediates its antiapoptotic actions. *Mol Cell*. 2012; **45**:13–24.
41. Kimura Y, Kimura H. Hydrogen sulfide protects neurons from oxidative stress. *FASEB J*. 2004; **18**:1165–1167.
42. Lavu M, Bhushan S, Lefer DJ. Hydrogen sulfide-mediated cardioprotection: Mechanisms and therapeutic potential. *Clin Sci (Lond)*. 2011; **120**:219–229.
43. Carballal S, Trujillo M, Cuevasanta E, et al. Reactivity of hydrogen sulfide with peroxynitrite and other oxidants of biological interest. *Free Radic Biol Med*. 2011; **50**:196–205.
44. Suzuki K, Olah G, Modis K, et al. Hydrogen sulfide replacement therapy protects the vascular endothelium in hyperglycemia by preserving mitochondrial function. *Proc Natl Acad Sci U S A*. 2011; **108**:13829–13834.

45. Pan LL, Liu XH, Gong QH, Wu D, Zhu YZ. Hydrogen sulfide attenuated tumor necrosis factor-alpha-induced inflammatory signaling and dysfunction in vascular endothelial cells. *PLoS ONE*. 2011; **6**:e19766.
46. Papapetropoulos A, Pyriochou A, Altaany Z, et al. Hydrogen sulfide is an endogenous stimulator of angiogenesis. *Proc Natl Acad Sci U S A*. 2009; **106**:21972–21977.
47. Bir SC, Kolluru GK, McCarthy P, et al. Hydrogen sulfide stimulates ischemic vascular remodeling through nitric oxide synthase and nitrite reduction activity regulating hypoxia-inducible factor-1alpha and vascular endothelial growth factor-dependent angiogenesis. *J Am Heart Assoc*. 2012; **1**:e004093.
48. Wang K, Ahmad S, Cai M, et al. Dysregulation of hydrogen sulfide producing enzyme cystathionine gamma-lyase contributes to maternal hypertension and placental abnormalities in preeclampsia. *Circulation*. 2013; **127**:2514–2522.
49. Zhao ZZ, Wang Z, Li GH, et al. Hydrogen sulfide inhibits macrophage-derived foam cell formation. *Exp Biol Med (Maywood)*. 2011; **236**:169–176.
50. Laggner H, Muellner MK, Schreier S, et al. Hydrogen sulphide: A novel physiological inhibitor of LDL atherogenic modification by HOCL. *Free Radic Res*. 2007; **41**:741–747.
51. Wang Y, Zhao X, Jin H, et al. Role of hydrogen sulfide in the development of atherosclerotic lesions in apolipoprotein E knockout mice. *Arterioscler Thromb Vasc Biol*. 2009; **29**: 173–179.
52. Jin HF, Liang C, Liang JM, Tang CS, Du JB. Effects of hydrogen sulfide on vascular inflammation in pulmonary hypertension induced by high pulmonary blood flow: Experiment with rats. *Zhonghua Yi Xue Za Zhi*. 2008; **88**: 2235–2239.
53. Yang G, Wu L, Wang R. Pro-apoptotic effect of endogenous H_2S on human aorta smooth muscle cells. *FASEB J*. 2006; **20**:553–555.
54. Yang G, Wu L, Bryan S, Khaper N, Mani S, Wang R. Cystathionine gamma-lyase deficiency and overproliferation of smooth muscle cells. *Cardiovasc Res*. 2010; **86**:487–495.
55. Mani S, Li H, Untereiner A, et al. Decreased endogenous production of hydrogen sulfide accelerates atherosclerosis. *Circulation*. 2013; **127**:2523–2534.
56. Du J, Hui Y, Cheung Y, et al. The possible role of hydrogen sulfide as a smooth muscle cell proliferation inhibitor in rat cultured cells. *Heart Vessels*. 2004; **19**:75–80.
57. Wu SY, Pan CS, Geng B, et al. Hydrogen sulfide ameliorates vascular calcification induced by vitamin D3 plus nicotine in rats. *Acta Pharmacol Sin*. 2006; **27**:299–306.
58. Meng QH, Yang G, Yang W, Jiang B, Wu L, Wang R. Protective effect of hydrogen sulfide on balloon injury-induced neointima hyperplasia in rat carotid arteries. *Am J Pathol*. 2007; **170**:1406–1414.
59. Zagli G, Patacchini R, Trevisani M, et al. Hydrogen sulfide inhibits human platelet aggregation. *Eur J Pharmacol*. 2007; **559**:65–68.
60. Elrod JW, Calvert JW, Morrison J, et al. Hydrogen sulfide attenuates myocardial ischemia-reperfusion injury by preservation of mitochondrial function. *Proc Natl Acad Sci U S A*. 2007; **104**:15560–15565.
61. Sodha NR, Clements RT, Feng J, et al. The effects of therapeutic sulfide on myocardial apoptosis in response to ischemia-reperfusion injury. *Eur J Cardiothorac Surg*. 2008; **33**: 906–913.
62. Calvert JW, Coetzee WA, Lefer DJ. Novel insights into hydrogen sulfide—mediated cytoprotection. *Antioxid Redox Signal*. 2010; **12**:1203–1217.
63. Benavides GA, Squadrito GL, Mills RW, et al. Hydrogen sulfide mediates the vasoactivity of garlic. *Proc Natl Acad Sci U S A*. 2007; **104**:17977–17982.
64. Lei YP, Liu CT, Sheen LY, Chen HW, Lii CK. Diallyl disulfide and diallyl trisulfide protect endothelial nitric oxide synthase against damage by oxidized low-density lipoprotein. *Mol Nutr Food Res*. 2010; **54**(Suppl 1):S42–S52.

32

METHYLATION AND ATHEROSCLEROSIS

XIN-YU XIONG[1,4], SHU MENG[2], XIAOFENG YANG[3,4], AND HONG WANG[3,4]

[1]*Thrombosis Research Center, Temple University School of Medicine, Philadelphia, PA, USA*
[2]*Methodist Hospital Research Institute, Cardiovascular Science, Houston, TX, USA*
[3]*Center for Metabolic Disease Research, Temple University School of Medicine, Philadelphia, PA, USA*
[4]*Cardiovascular Research Center, Department of Pharmacology, Temple University School of Medicine, Philadelphia, PA, USA*

DEOXYRIBONUCLEIC ACID METHYLATION

Deoxyribonucleic Acid Methylation and Methyltransferase

Deoxyribonucleic acid (DNA) methylation usually occurs at the 5′ position of the cytosine ring within CpG dinucleotide. The DNA methyltransferases (DNMTs) catalyze the transfer of a methyl group from S-adenosyl methionine (SAM) to DNA. In mammals, although five members of the DNMT family have been found (DNMT1, DNMT2, DNMT3a, DNMT3b, and DNMT3L), only DNMT1, DNMT3a, and DNMT3b possess methyltransferase activity. As shown in Figure 32.1, DNMT3A and DNMT3B are de novo methyltransferases that establish cytosine methylation patterns on unmethylated DNA during embryogenesis [1]. DNMT1 functions to maintain the existing methylation pattern following DNA replication [2]. Studies of knockout mice have shown that DNMT1, DNMT3a, and DNMT3b are essential for normal embryonic development. Loss of DNMT1 and DNMT3b results in embryonic lethality, while absence of DNMT3a leads to death in 4 weeks after birth [1, 3].

DNA Demethylation

DNA demethylation is reversible and occurs either passively in a DNA replication-dependent manner or actively in a cell division-independent process. Passive demethylation happens only during development. Specifically, in mammalian development, the maternal genome is demethylated during preimplantation growth [4] by the inhibition of DNMT1 [5]. On the other hand, three enzymatic families have been implicated in active demethylation mechanism. The 10–11 translocation (TET) family oxidizes methylated cytosine, the activation-induced (cytidine) deaminases/apolipoprotein B mRNA editing enzyme, catalytic polypeptide-like (AID/APOBEC) enzymes deaminate 5-methylcytosine (5mC),

Atherosclerosis: Risks, Mechanisms, and Therapies, First Edition. Edited by Hong Wang and Cam Patterson.

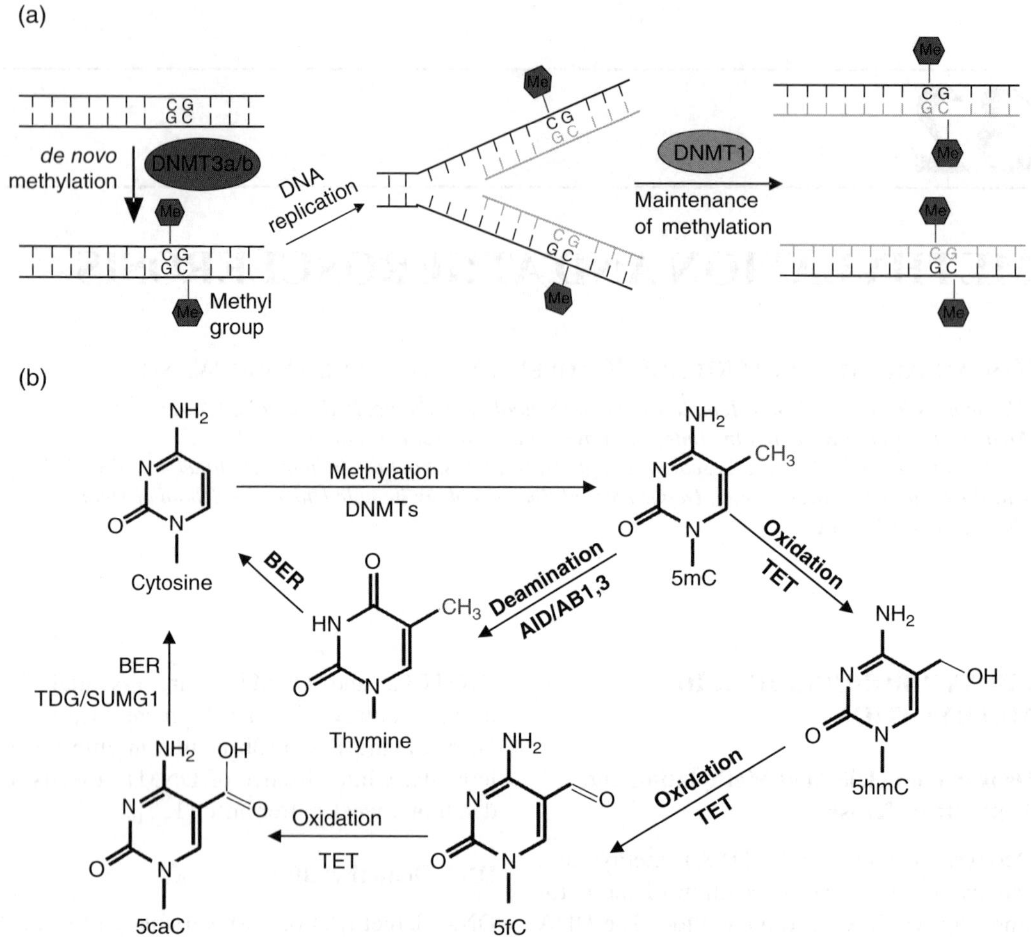

Figure 32.1 DNA methylation and demethylation. (a) Establishment and maintenance of DNA methylation. In mammals, global demethylation occurs immediately after fertilization, erasing all existing methylation patterns. After implantation, DNA methyltransferases (DNMTs) 3a/b de novo transfer methyl group to cytosine residue of CpG dinucleotides and establish new DNA methylation patterns. During cell division, DNMT1 methylates newly synthesized DNA strands to maintain the methylation patterns. (b) DNA demethylation pathway. Three enzyme families have been implicated in active DNA demethylation. (i) 5-methylcytosine (5mC) can be hydroxylated by 10–11 translocation (TET) family of enzymes to form 5-hydroxymethylcytosine (5hmC) or further oxidized to 5-formylcytosine (5fC) and 5-carboxylcytosine (5caC). (ii) 5mC (or 5hmC) can be deaminated by the activation-induced (cytidine) deaminases/apolipoprotein B mRNA editing enzyme, catalytic polypeptide-like (AID/APOBEC) family members to form thymine. (iii) Replacement of these intermediates (i.e., thymine or 5caC) is initiated by the UDG family of base excision repair (BER) glycosylases like TDG or SMUG1, culminating in cytosine replacement and DNA demethylation. (*See insert for color representation of the figure.*)

while the base excision repair (BER) glycosylase mediates DNA mismatch repair [6]. These enzymes modify methylated cytosines, leading to their replacement by unmethylated cytosines by DNA repair (Figure 32.1).

Functions of DNA Methylation Depend on Its Position

CpG sites are concentrated either in CpG islands (CGIs), which are short (~1 kb) CpG-rich DNA regions located in about 60% of human gene promoters, in the gene body, or in regions of large repetitive sequences, such as centromeres and retrotransposon elements [7]. The position of DNA methylation determines its function. Methylation of CGIs in promoter regions mostly blocks transcription initiation, whereas gene body methylation is common in ubiquitously expressed genes and is positively correlated with gene expression [8]. Methylation in the gene body is proposed to prevent spurious transcription initiation and stimulate transcription elongation [9]. Meanwhile, methylation in centromeres is important for chromosomal stability, whereas methylation in retrotransposon elements is likely to suppress the expression of transposable elements to maintain genome stability [10]. In fact, methylation degree of exons is higher than introns. Transitions of methylation degree occurring at exon–intron boundaries suggest that methylation in the gene body may have a role in the regulation of splicing [11].

DNA Methylation Regulates Gene Transcription

DNA methylation regulates gene transcription by interference with transcription activator or suppressor binding, by attracting methyl-CpG-binding proteins (MBPs) to repress transcription via the recruitment of histone deacetylase (HDAC) which leads to chromatin condensation (Figure 32.2). In one mechanism, the presence of a methyl group at specific CpG sites may block the access of transcription activators or suppressors, resulting in either transcription inactivation or activation (Figure 32.2a and b). For example, the NFκB and Sp1 sites on the promoter of actively transcribed *Mucin* gene are not methylated. However, methylation of the NFκB site blocks p50–p65 dimer binding, while methylation of Sp1 site recruits MBP to block Sp1 binding. Together, these modifications lead to the inhibition of mucin transcription [12]. Meanwhile, in proliferative endothelial cells (ECs), the cell cycle-dependent element (CDE) on the cyclin A promoter is methylated to block the binding of a transcription suppressor, thereby facilitating cyclin A transcription and cell cycle progress [13]. During homocysteine (Hcy) treatment, DNMT1 activity is inhibited, leading to hypomethylation of CDE. This allows the transcription suppressor to bind and inhibit cyclin A transcription. As an example of another gene regulation mechanism, methyl-CpG-binding protein 2 (MeCP2) and other methyl-CpG-binding domain (MBD) family members bind to methyl CpG and attract the binding of HDACs [14]. HDACs remove acetyl groups from specific lysine residues on histones 3 and 4 to trigger chromatin condensation and further repress transcription (Figure 32.2c). Another mechanistic example of the consequence of DNA methylation is opposite genomic imprinting (Figure 32.2d). The H19 and Igf2 genes are found on mouse chromosome 7, 100 kb apart. The insulator imprint control element (ICE) is upstream of H19, between H19 and Igf2. CTCF is a transcription factor which binds to ICE. Transcription of the maternal copy of Igf2 gene is blocked by the binding of CTCF to ICE to insulate the promoter from the interaction with a downstream enhancer. However, at the paternal locus, ICE is methylated which prevents CTCF binding, thereby allowing the downstream enhancer to activate Igf2 expression but not H19 [15].

DNA Methylation Plays a Key Role in Development

During embryonic development, the pattern of CpG methylation is established via a dynamic process. As shown in Figure 32.3, methylation on both the paternal and maternal genome is erased shortly after fertilization. After implantation, the embryo undergoes de novo methylation which establishes a new genome methylation pattern [16, 17]. However, CGIs remain hypomethylated until later in development, when some will become methylated.

In regard to cell differentiation, the pluripotency genes are generally hypomethylated in stem cells and gain methylation during cell differentiation [18]. Recent study on hematopoietic stem cells (HSCs) indicates that DNMT3a is highly expressed in HSCs compared to

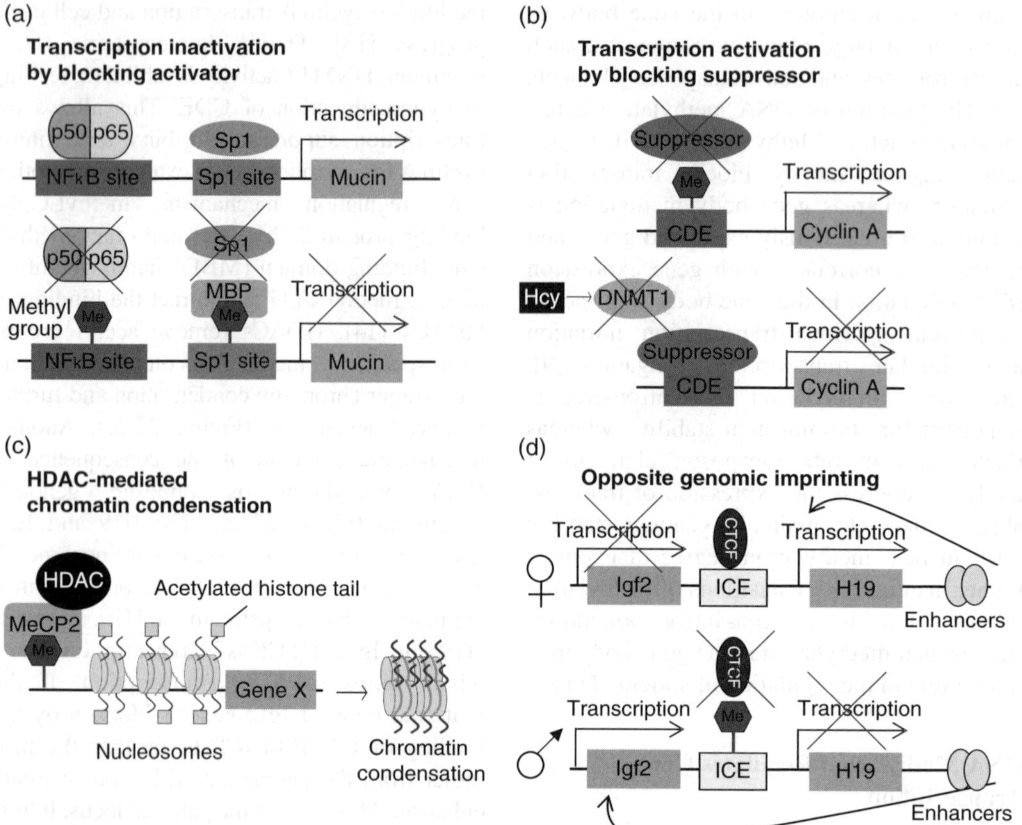

Figure 32.2 DNA methylation regulates gene transcription. (a) DNA methylation blocks transcription activator to repress transcription. The NFκB site and Sp1 site on the promoter of actively transcribed *mucin* gene are not methylated. Methylation of the NFκB site blocks p50–p65 dimer binding. Methylation on Sp1 site recruits methyl-CpG-binding protein (MBP) to block Sp1 binding. These modifications together lead to the inhibition of mucin transcription. (b) DNA methylation blocks transcription suppressor to facilitate transcription. The cell cycle-dependent element (CDE), a transcription suppressor binding site, on the promoter of actively transcribed cyclin A gene is methylated in EC. Homocysteine (Hcy) treatment inhibits DNMT1 activity and induces hypomethylation of CDE on cyclin A promoter in EC. This allows transcription suppressor to bind to the unmethylated CDE, thereby inhibiting cyclin A transcription. (c) DNA methylation induces chromatin condensation to repress transcription. Methyl- CpG-binding protein 2 (MeCP2) binds to methyl CpG and mediates transcription repression by HDACs, which remove acetyl groups from histones and trigger chromatin condensation, leading to transcription repression. (d) DNA methylation causes opposite genomic imprinting. The H19 and Igf2 genes are on the same mouse chromosome and are oppositely "imprinted." On the maternal chromosome, the protein CTCF binds to the unmethylated ICE, blocking the access of enhancers to Igf2. Thus, H19 is expressed instead. On the paternal chromosome, the ICE is methylated, so CTCF cannot bind; therefore, the enhancers have access to Igf2. Igf2 is expressed, whereas H19 is silenced. (*See insert for color representation of the figure.*)

differentiated cells. In fact, loss of DNMT3a leads to impaired HSC differentiation and multipotency gene upregulation via DNA hypomethylation in vivo [19]. Recent studies indicate that methylation occurs at only a small percentage of CpG dinucleotides and inhibits transcription of only a small subset of genes in differentiated cells. Many of these repressed genes are

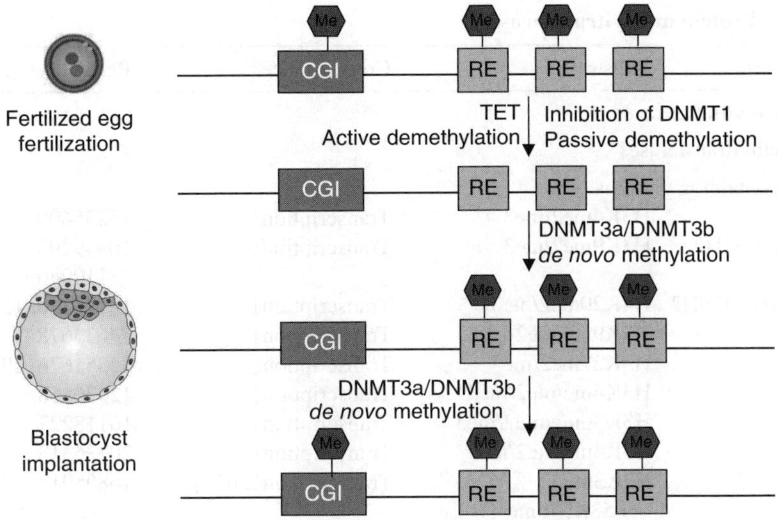

Figure 32.3 De novo methylation during embryonic development. Shortly after fertilization, methylation paternal genome and maternal genome are erased, via active demethylation by the 10–11 translocation (TET) family of enzymes or passive demethylation by inhibition of DNMT1. After implantation, the embryo undergoes de novo methylation by DNMT3a/DNMT3b that establishes a new methylation pattern in genome mostly in repetitive elements (RE). The CpG islands (CGIs) remain hypomethylated until later in development, when some of them become methylated.

germline specific [20], including pluripotency genes. This suggests that methylation is a crucial mechanism by which transcription of key genes are suppressed during differentiation.

PROTEIN METHYLATION

Protein methylation is conducted by protein methyltransferases which add a methyl group from the donor SAM to carboxyl groups of glutamate, leucine, and isoprenylated cysteine, or on the side-chain nitrogen atoms of lysine, arginine, and histidine residues [21]. Two families of enzymes have been identified that catalyze protein methylation: (i) histone methyltransferases and (ii) carboxyl methyltransferases. Their biochemical features and functions are briefly summarized in Table 32.1.

Histone Methylation

Histone methylation is the most studied mechanism in the field of protein methylation. It mostly occurs on arginine (R) or lysine (K) catalyzed by either lysine methyltransferases or arginine methyltransferases (RMT) (Table 32.1) [22]. Lysines can be monomethylated (me), dimethylated (me2), or trimethylated (me3), while arginines can be monomethylated, symmetrically dimethylated (me2s), or asymmetrically dimethylated (me2a). Furthermore, it should be noted that histidine methylation of histones can occur but it appears to be rare and has not been well characterized [23]. The most extensively studied sites of histone (H) lysine methylation lies in the N-termini of H3 and H4 histones. Methylation on H3K9, H3K27, H3K79, and H4K20 is usually within heterochromatin, which is tightly packed chromatin, genetically inactive, tightly coiled, and darkly stained. Methylation on H3K4 and H3K36 histones is usually associated with genes that are potentially activated in euchromatin, which is a lightly packed chromatin and genetically active, partially or fully uncoiled, and lightly stained [23]. Sites of arginine methylation include H3R2, H3R8, H3R17, H3R26, and H4R3. Originally, arginine methylation was

TABLE 32.1 Protein methyltransferases

Methyltransferase	Substrate	Consequence	References (PMID)
(1) Histone methyltransferases			
(A) Lysine methyltransferases			
SET domain-containing proteins			
SMYD3	H3K4me2/me3	Transcription↑	15235609
SUV39H1, SUV39H2	H3K9me2/me3	Transcription↓	10949293, 14690610, 11094092
SUV420H1, SUV420H2	H4K20me2/me3	Transcription↓	18296440, 15145825
PRDM2	H3K9me/me2/me3	Transcription↓	14633678
EZH1, EZH2	H3K27me2/me3	Transcription↓	12351676, 19026780
SETD1A	H3K4me/me2/me3	Transcription↑	12670868
SETD2	H3K36me/me2/me3	Transcription↑	16118227
MLL1,2,3,4,5	H3K4me/me2/me3	Transcription↑	20236313
SMYD2	H3K36me2, P53K370me	Transcription↓, p53↓	16805913, 17108971
SETD8	H4K20me, P53K382me	Transcription↑, p53↓	17562855, 17707234
SETD7/9	H3K4me, P53K372me	Transcription↑, P53↑	12086618, 15525938
DOT1-like proteins			
DOT1L	H3K79me/me2/me3	Transcription↓	12123582
(B) Protein arginine methyltransferases (PMRT)			
PRMT1	H4R3me/me2a	Transcription↑	11448779
CARM1(PRMT4)	H3R2me/me2a, H3R17me/me2a, H3R26me/me2a	Transcription↑	21074527
PRMT7	H3R2me/me2s, H4R3me/me2s	Transcription↓	15494416, 22231400
PRMT5	H3R8me/me2s, H4R3me/me2s	Transcription↓	15485929, 21074527
(2) Protein carboxyl methyltransferases			
ICMT	Small G-proteins	Trigger small G-proteins binding to cell membrane	19956441, 22189424
LCMT1	PP2A	PP2A holoenzyme assembly	21292165
PIMT	Proteins with abnormal aspartyl residues	Aged protein repairing	21204776

Symbol and abbreviations: ↑, activation; ↓, suppression; CARM1, coactivator-associated arginine methyltransferase-1; EZH2, enhancer of zeste homolog 2; ICMT, isoprenylcysteine carboxyl methyltransferase; LCMT1, leucine carboxyl methyltransferase-1; me, monomethylation; me2, dimethylation; me2a, asymmetric dimethylation; me2s, symmetric dimethylation; me3, trimethylation; MLL1, 2, 3, 4, 5, myeloid/lymphoid or mixed-lineage leukemia 1, 2, 3, 4, 5; PIMT, protein L-isoaspartyl (D-aspartyl) methyltransferase; PMRT, protein arginine methyltransferase; PP2A, protein phosphatase 2A; PRDM1, PR domain containing 1, with ZNF domain protein; SETD, SET domain-containing protein; SMYD, SET, and MYND domain-containing protein; SUV39H, suppressor of variegation 3–9 homolog; SUV420H, suppressor of variegation 4–20 homolog.

thought to be an activation mark because it was only found in chromatin where genes are actively transcribed [24]. Now, it is known that arginine methylation can either activate or repress gene transcription depending on the type of RMT involved [25]. As shown in Table 32.1, protein arginine *N*- methyltransferases (PRMT) 1 and 4 generate monomethyl-arginine and asymmetric dimethyl-arginine derivatives, which are involved in transcription activation. Meanwhile, PRMT5 and PRMT7 generate monomethyl-arginine and symmetric dimethyl-arginine derivatives, which are involved in repression [26].

Histone methylation on lysine or arginine sites can affect gene transcription via recruitment of proteins or domains to change chromatin structure. For example, H3K27me3 can be recognized by a transcriptionally repressive protein, *Drosophila* polycomb protein (PC) [27], whereas H3K4me2/me3 can recruit chromodomain helicase DNA-binding protein 1 (CHD1), which destabilizes nucleosomes and initiates transcription [28].

The SET domain-containing protein and DOT1-like proteins methylate lysines within the histone. Histone methyltransferases have been shown to methylate histones that are incorporated into chromatin and also free histones and nonhistone proteins [22]. Some members of the SET domain-containing protein family have been shown to regulate activity of p53, a transcription regulator that plays a central role in tumor suppression. Specifically, SMYD2-mediated p53K370me [29] and SETD8-mediated p53K382me [30] suppress p53. Oppositely, SETD7/9-mediated p53K372me activates p53 [31].

Protein Carboxyl Methylation

Unlike histone methylation, protein carboxyl methylation is a nonepigenetic posttranslational modification. In mammals, a few protein carboxyl methyltransferases have been found, such as isoprenylcysteine carboxyl methyltransferase (ICMT), protein L-isoaspartyl (D-aspartyl) methyltransferase (PIMT) [32], and leucine carboxyl methyltransferase-1 (LCMT1) [33]. ICMT-catalyzed methylation is the most intensively studied field for protein carboxyl methylation. ICMT is an endoplasmic reticulum (ER) resident enzyme which catalyzes the transfer of methyl groups from SAM to the α-carboxyl of isoprenylcysteine residues. The substrate of ICMT is a CAAX protein family, which contains a specific amino acid sequence at the C-terminal that directs posttranslational modification. The CAAX amino acid sequence is a cysteine residue followed by two aliphatic residues, and then any C-terminal amino acid residue. To date, more than 50 CAAX proteins have been identified. These proteins could be further classified into the small G-protein superfamily, including Ras, Rho, Rac, and CDC42; nuclear surface membrane proteins, such as lamin B1 and lamin B2; and intracellular membrane proteins, including HspxF, Cenp-F, and Cenp-E [34]. CAAX processing involves three steps: polyisoprenylation, proteolysis, and carboxyl methylation. At first, a protein farnesyltransferase (FTase) or geranylgeranyltransferase type I (GGTase I) attaches an isoprenoid lipid to the CAAX box. Next, Ras-converting enzyme 1 (RCE1) removes the AAX amino acids. Finally, ICMT catalyzes carboxyl methylation to significantly increase the hydrophobicity of the C-termini of CAAX-modified proteins to allow insertion into the cellular membrane, which is necessary for the biological activity of CAAX proteins [35].

ABERRANT METHYLATION AND DISEASES

Aberrant Methylation and Cancer

The cancer field has the most advanced DNA methylation research. Cancer cells are globally hypomethylated but are hypermethylated at site-specific CGI promoters [36]. This hypomethylation occurs at many noncoding regions, such as repetitive elements, retrotransposons, and introns [36]. In mouse models, hypomethylation has been shown to induce genomic instability and tumorigenesis [37, 38]. However, the causes of global hypomethylation in cancer cells remains poorly understood. It is suggested that upregulation of DNA demethylase may be involved in some cases [39]. The most recognized methylation mechanism in human tumors is the CGI promoter hypermethylation-associated silencing of tumor suppressor genes, such as cyclin-dependent kinase inhibitor 2A (CDKN2A), mutL homolog-1 (MLH1), breast cancer-associated-1 (BRCA1), and von Hippel–Lindau tumor suppressor (VHL) [40]. Cancer cells often show increased expression of DNMTs, especially DNMT1, which may cause the abnormal CGI hypermethylation. In fact, DNMT1 has been shown to be essential for the survival and proliferation of human cancer cells [41].

Because DNA methylation is reversible, many researchers focused on searching for antimethylation drugs, which may restore the normal methylation status in cancer cells by inhibiting DNA methyltransferase. In fact, the DNMT inhibitors, vidaza and decitabine (5- aza- and 5-aza-2′-deoxycytidine), have been approved by the FDA for patients with myelodysplastic syndrome who have few therapeutic options and develop acute leukemia. However, it should be noted that DNMT inhibitors may systemically suppress DNA methylation which would worsen the global hypomethylation status in cancer. Due to the lack of understanding of the mechanisms that lead to global hypomethylation, there is no drug to reverse genome-wide hypomethylation in cancer cells.

Aberrant Methylation Modifications in Atherosclerosis

In comparison to cancer, the atherosclerosis field is far less studied in terms of methylation. A search of the PubMed database using the terms "cancer and methylation" produced a total of 21,607 publications, whereas for "atherosclerosis and methylation" returned only a total of 180 publications as of May 2013. Two points could be highlighted from these publications. Firstly, genomic DNA isolated from human and mouse atherosclerotic lesions are globally hypomethylated. It is reported that DNA hypomethylation marks the area of vascular smooth muscle cell (VSMC) proliferation in advanced human atherosclerotic plaques where a 9% decrease in cytosine methylation has been detected [42]. Furthermore, similar hypomethylation is observed in the atheromas of $ApoE^{-/-}$ mice [43] and in the neointimal area in New Zealand White rabbits [42].

Secondly, aberrant methylation changes in DNA and histone have been found in the promoter regions of atherosclerosis-associated genes in EC, smooth muscle cells (SMC), and atherosclerotic plaques as described in Table 32.2.

Specific Methylation Changes in EC

EC proliferation inhibition and dysfunction lead to endothelium injury, which plays an important role in the atherosclerosis process. Hypomethylation has been found to contribute to the atherosclerotic phenotype of EC. CpG sites in the core promoter of endothelial nitric-oxide synthase (eNOS) were found to be hypomethylated in cultured human and murine ECs [45]. Meanwhile, human ECs are enriched in H3K4me2/3 at the eNOS core promoter when compared to other cell type [53]. Hypomethylation of CpG sites and enrichment of H3K4me2/3 at the eNOS core promoter both contribute to constitutive eNOS expression in EC to maintain endothelial function.

Risk factors of atherosclerosis inhibit EC growth and induce EC dysfunction via specific methylation changes in DNA. ECs are sensitive to hypomethylation regulation. At a clinically relevant concentration of Hcy, which is an independent risk factor of atherosclerosis, the intracellular levels of SAH, a general inhibitor for methyltransferase, are selectively increased in cultured EC, but not in VSMC [56]. Hcy induces hypomethylation in the promoter of cyclin A [13] at the CDE site which leads to the suppression of cyclin A expression, resulting in EC growth inhibition [13]. It is reported that Hcy causes hypomethylation of CpG sites 6 and 7 of P66shc promoter, which leads to increased P66shc expression, causing oxidative stress and endothelial dysfunction [44].

Specific Methylation Changes in VSMC

The proliferation and migration of VSMC into the intimal layer of the arterial conduit is a key step of the atherogenic process. Both

TABLE 32.2 (a) DNA methylation and (b) protein methylation related to atherosclerosis

Target	Inducer	Modification	Consequence	Year [Ref.]
DNA methylation				
Endothelial cell (EC)				
Cyclin A	Hcy	Promoter methylation↓	EC proliferation↓	2007 [13]
P66shc	Hcy	Promoter methylation↓	endothelium dysfunction	2011 [44]
eNOS	EC vs. non-EC	Promoter methylation↓	eNOS expression↑	2004 [45]
iNOS	EC vs. tumor cells	Promoter methylation↑	iNOS expression↓	2005 [46]
FGF2	Hcy	Promoter methylation↑	EC proliferation↓	2008 [47]
Vascular smooth muscle cell (VSMC)				
MMP2/9	OxLDL	Promoter methylation↓	VSMC migration↑	2011 [48]
ERα	Proliferating cells vs. differentiated cells	Promoter methylation↑	VSMC proliferation↑	2000 [49]
Atherosclerotic plaque (global DNA hypomethylation)				
ERβ	Atherosclerotic plaques	Promoter methylation↑	Proatherogenesis	2007 [50]
ERα	Atherosclerotic plaques	Promoter methylation↑	Proatherogenesis	1999 [51]
TFPI2	Atherosclerotic plaques	Promoter methylation↑	Lipids↓, Mϕ infiltration↓	2009 [52]
Histone methylation				
Endothelial cell (EC)				
eNOS	EC vs. non-EC	H3K4me2/3↑	eNOS expression↑	2005 [53]
iNOS	EC vs. tumor cells	H3K9me↑	iNOS induction↓	2005 [46]
ICAM1	Tumor EC	H3K4me↓	ICAM1↓	2006 [54]
Vascular smooth muscle cell (VSMC)				
miR29b	OxLDL	H3K4me↑, H3K9me↓	SMC migration↑	2012 [55]
Carboxyl methylation				
Endothelial cell				
$P21^{ras}$	Hcy	$P21^{ras}$ methylation↓	$P21^{ras}$ membrane association↓ EC proliferation↓	1997 [56]
Ras/RhoA	Hcy, ICMT inhibitor	Ras/Rho methylation↓	Ras/Rho activity↓ apoptosis↑	2003 [57], 2007 [58]
$BclX_L$	Overexpression of PIMT	Aged $BclX_L$ repairing	Apoptosis↓	2008 [59]

Reference numbers are consistent with the text and detailed in the Reference section. Abbreviation: $BclX_L$, B-cell lymphoma-extra large; EC, endothelial cells; eNOS, endothelial nitric-oxide synthase; ER, estrogen receptor; H3K4me, monomethylated H3 histone lysine 4; H3K4me2/3, di-/tri-methylated H3 histone lysine 4; H3K9me, monometylated H3 histone lysine 9; ICAM-1, intercellular adhesion molecule 1; ICMT, isoprenylcysteine-*O*-carboxyl methyltransferase; iNOS, inducible nitric-oxide synthase; FGF2, fibroblast growth factor 2; Hcy, homocysteine; Mϕ, macrophage; miR29b, microRNA miR29b; MMP2/9, matrix metalloproteinases 2/9; OxLDL, oxidized low density lipoprotein; PIMT, protein L-isoaspartyl (D-aspartyl) methyltransferase; Ras, Rat sarcoma; Rho, ras homolog; TFPI-2, tissue factor pathway inhibitor-2.

hypermethylation and hypomethylation changes are related with proliferation and migration in VSMC. For example, oxidized low-density lipoprotein was shown to increase cell migration via matrix metalloproteinases 2/9 (MMP2/9) induction which is associated with MMP2/9 promoter hypomethylation in cultured human aortic SMC [48]. It is suggested that oxLDL-induced MMP2/9 promoter hypomethylation is due to miR-29b-targeted DNMT3b suppression [48], which may be partially explained by reduced H3K9me and increased H3K4me in the promoter region of miR-29b [55].

In another example, estrogen has been shown to regulate cell proliferation both for cancer cells and for vascular cells. The promoter of estrogen receptor-α (ER-α) is hypermethylated in HASMC, which is linked to suppressed ER-α expression and increased SMC proliferation [49].

Aberrant Methylation Changes in Atherosclerotic Plaque

Specific methylation changes are also found in atherosclerotic plaques. Promoters of ER-α and ER-β are hypermethylated in human atherosclerotic plaques, which is associated with ER-α and ER-β reduction [50, 51].

Protein Carboxyl Methylation Associated with Atherosclerosis

Although most methylation changes associated with atherosclerosis are involved in epigenetic regulation, carboxyl methylation of proteins is found to play a role in the atherosclerosis process. Protein carboxyl methylation may affect EC proliferation and apoptosis. Oncoprotein $p21^{ras}$ is essential for cell growth. Its activity depends on its ability to associate with the cell membrane, which is mediated by a series of posttranslational modifications that include prenylation and carboxyl methylation of the $p21^{ras}$ C terminus [56]. We found that Hcy selectively decreased $p21^{ras}$ carboxyl methylation in EC and reduced membrane-associated $p21^{ras}$ protein, resulting in a suppressed $p21^{ras}$-mediated extracellular signal-regulated kinase (ERK) 1/2 signaling [56]. A following study confirmed this finding and reported that Hcy decreased $p21^{ras}$ carboxyl methylation and membrane association and promoted apoptosis in EC [57].

BIOCHEMICAL BASIS OF CELLULAR METHYLATION

Since all methyltransferases involved in DNA methylation, histone methylation, and protein carboxyl methylation require SAM as the methyl group donor, biochemical regulation of SAM plays a crucial role in cellular methylation and requires special attention.

SAM is generated from the methionine–homocysteine metabolic cycle (Figure 32.4) which can be described in five major steps: (1) SAM synthesis, (2) methyl transfer reactions, (3) SAH hydrolysis/synthesis, (4) Hcy remethylation/methionine synthesis, and (5) Hcy transsulfuration.

Briefly, SAM is derived from methionine and can be converted to S-adenosylhomocysteine (SAH) after donating its methyl group for cellular methylation. SAH is then hydrolyzed to adenosine and Hcy. Hcy can then be converted back to methionine via the remethylation pathway or then become cysteine via the transsulfuration pathway [60].

SAM Synthesis

Methionine adenosyltransferase (MAT) catalyzes the biosynthesis of SAM from methionine and ATP. Among the three forms of MAT identified in mammalian tissues, MAT I and MAT III are encoded by *MAT1A* gene and synthesized primarily in the liver. MAT II, on the other hand, is encoded by *MAT2A* and is found in the fetal liver as well as the kidney, brain, testis, and lymphocytes. The majority of the hypermethioninemia subjects with presumably MAT deficiency were clinically well, while only a few showed neurological deficiency with severe loss of MAT activity [61].

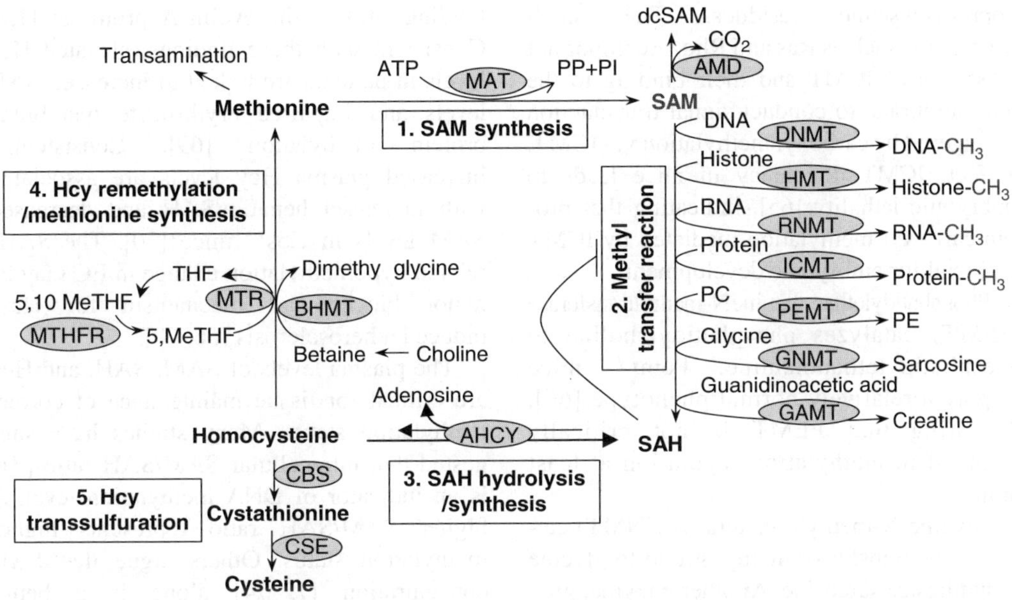

Figure 32.4 Methionine–homocysteine metabolic cycle and cellular methylation. SAM is the major methyl donor for cellular methylation and is an intermediate product in the methionine–homocysteine metabolism cycle, which has five key steps: (1) SAM synthesis; (2) methyl transfer reaction which transfers methyl group to DNA, histone, RNA, and protein as well as other molecules; (3) SAH hydrolysis/synthesis; SAH is a competitive inhibitor of SAM-dependent methyltransferases; SAH hydrolysis is a reversible reaction. Accumulation of Hcy leads to SAH production and inhibits cellular methylation, (4) Hcy remethylation to methionine, and (5) Hcy transsulfuration into cysteine. There are 15 enzymes involved in methionine–homocysteine metabolism. Abbreviations: 5,10MeTHF, 5-methylenetetrahydrofolate; 5MeTHF, 5-methyltetrahydrofolate; AHCY, S-adenosylhomocysteine hydrolase; AMD, S-adenosylmethionine decarboxylase; BHMT, betaine–homocysteine methyltransferase; CBS, cystathionine β-synthase; CSE, cystathionine γ-lyase; dcSAM, decarboxylated S-adenosylmethionine; DNMT, DNA methyltransferase; GAMT, guanidinoacetate *N*-methyltransferase; GNMT, glycine *N*-methyltransferase; Hcy, homocysteine; HMT, histone methyltransferase; ICMT, isoprenylcysteine carboxyl methyltransferases; MAT, methionine adenosyltransferase; Met, methionine; MTHFR, 5,10-methylenetetrahydrofolate reductase; MTR, 5-methyltetrahydrofolate–homocysteine methyltransferase; PC, phosphatidylcholine; PE, phosphatidylethanolamine; PEMT, phosphatidylethanolamine *N*-methyltransferase; RNMT, RNA methyltransferase; SAH, S-adenosylhomocysteine; SAM, S-adenosylmethionine; THF, tetrahydrofuran.

Methyl Transfer Reactions

SAM can be converted to SAH after donating its methyl group to DNA, RNA, protein, phospholipid, or other biochemical molecules through methyltransferases [62]. We profiled the tissue mRNA distribution of 12 enzymes involved in homocysteine metabolism and found that methyltransferases are differentially expressed in human and mouse tissues [63].

DNMT1 is important for maintenance of DNA methylation while DNMT3a and DNMT3b are essential for de novo methylation during development. HMTs mediate the histone methylation. Because DNMT and HMT deficiencies usually lead to embryonic lethal phenotype [3, 64] and severe development hindrance [1] in mice, it can be concluded that their function is more important in development than in metabolism and methylation regulation.

ICMT catalyzes the transfer of methyl groups from SAM to the α-carboxyl of isoprenylcysteine residues. The small G-proteins, such as Ras and Rho, are important substrates of ICMT and their binding to the cell membrane to conduct signal transduction is triggered by carboxyl methylation via ICMT. In fact, ICMT deficiency in mice leads to embryonic lethality [65], indicating that protein carboxyl methylation mediated by ICMT is crucial for embryonic development.

Phosphatidylethanolamine N-methyltransferase (PEMT) catalyzes phosphatidylcholine to phosphatidylethanolamine. Pemt$^{-/-}$ mice display a relatively normal phenotype [66], indicating that PEMT is not critically involved in methylation regulation at least in mice.

Glycine N-methyltransferase (GNMT) catalyzes the transfer of methyl group to glycine to synthesize sarcosine. Another enzyme, guanidinoacetate methyltransferase, catalyzes the transfer of methyl group to guanidinoacetate to form creatine. Among all methylation reactions, synthesis of creatine/creatinine normally consumes 75–80% of the total SAM generated in young adult [67].

SAH Hydrolysis/Synthesis

SAH is hydrolyzed to adenosine and Hcy by SAH hydrolase [68]. SAH hydrolysis serves not only to sustain the flux of methionine sulfur toward cysteine, but also plays a critical role in the regulation of methylation. This reaction is reversible but favors the SAH formation. Under physiologic conditions, adenosine and Hcy are rapidly removed, driving the reaction toward SAH hydrolysis. However, in pathological conditions such as during high levels of Hcy, this reaction is reversed, leading to SAH accumulation. Importantly, this reaction is regulated in a cell type specific manner. When EC and VSMC are both treated with pathologically relevant concentration of Hcy (50 μM), intracellular SAH levels are significantly elevated only in EC but not in VSMC [56]. This induction is associated with EC-specific growth inhibition via DNA hypomethylation on CDE (suppressor binding site) of the cyclin A promoter [13]. Consistent with these findings, elevated Hcy levels in patients are linked to increased SAH levels and impaired erythrocyte membrane protein methylation [69]. Consistently, increased plasma Hcy levels are associated with increased hepatic SAH and decreased SAM levels in Cbs$^{+/-}$ mice [70]. The SAH-related hypomethylation change in EC may be major biochemical mechanism for Hcy-induced atherosclerosis [56, 60].

The plasma levels of SAM, SAH, and Hcy are critical for tissue maintenance of correct methylation status. Many studies have suggested that intracellular SAM/SAH ratio [71] is an indicator of DNA methylation level. A higher SAM/SAH ratio represents higher methylation status. Others argue that SAH concentration [72–74] alone is a better indicator of DNA methylation based on the correlation between DNA methylation level and SAH concentration. In light of this, it is recommended to use both SAH concentration and SAH/SAM ratio to determine cellular methylation status.

Hcy Remethylation/Methionine Synthesis

Hcy remethylation can be catalyzed by 5-methyltetrahydrofolate–homocysteine S-methyltransferase (MTR), also called methionine synthase, which transfers a methyl group from 5-methyltetrahydrofolate to Hcy, generating tetrahydrofolate (THF) and methionine. MTR is extremely important in development as indicated by *Mtr* homozygous knockout embryos which die soon after implantation [75]. This lethality is probably due to its role in supplementing methionine synthesis and thus regulating the subsequent methyl transfer reaction. The 5,10-methylenetetrahydrofolate reductase (MTHFR) catalyzes the conversion of 5,10-methylenetetrahydrofolate (5,10MeTHF) to 5-methyltetrahydrofolate (5MeTHF), a cosubstrate for Hcy remethylation to methionine. Hcy remethylation is an essential biochemical process and highly regulated at the tissue level, with both MTR and

MTHFR being broadly expressed in most of human and mouse tissues [63]. In fact, $Mthfr^{-/-}$ mice have increased plasma Hcy and SAH levels, as well as DNA hypomethylation in the liver and brain [76]. It is strongly suggested that deficiency or decreased activity of MTHFR reduces the methyl transfer to methionine and contributes to global DNA hypomethylation.

In addition, betaine–homocysteine methyltransferase (BHMT) converts betaine to dimethylglycine while remethylating Hcy to methionine. This reaction mostly takes place in the liver, where BHMT is primarily expressed in human and mice [63]. In accordance, $Bhmt^{-/-}$ mice had increased hepatic Hcy (six to eightfold) and SAH (threefold) concentrations and decreased SAM (43%) [77].

Hcy Transsulfuration

The Hcy transsulfuration pathway involves two catabolic steps: (1) cystathionine β-synthase (CBS) converts Hcy to cystathionine, and (2) cystathionine γ-lyase (CSE) converts cystathionine to cysteine. The *CBS* gene is expressed in almost all human tissues [63]; therefore, humans possess a higher level of tissue-specific regulation of the CBS reaction. In mice however, CBS is predominantly expressed in the liver and therefore is rather heavily dependent on liver metabolism. In the clinic, deficiency of CBS leads to hyperhomocysteinemia (HHcy) ranging from 200 to 300 μM compared to approximately 10 μM in healthy individuals [78]. Patients with CBS homozygous mutation develop severe HHcy, CVD, thromboembolic diseases, lens dislocations, and osteoporosis and usually die before 30 years of age due to thromboembolism complications if not treated [79]. Similarly, the $Cbs^{-/-}$ mice have severe HHcy with concentrations of over 200 μmol/l and die within 5 weeks postnatal [80], suggesting an important role of CBS during the postnatal development.

CSE is expressed predominantly in the liver in both human and mice. In fact, the CSE reaction is likely liver dependent and not critical for development nor growth-related metabolism. $Cse^{-/-}$ mice do not exhibit any growth problem [81].

HHCY LINKS METHYLATION TO ATHEROSCLEROSIS

Cellular methylation is largely controlled by the methionine–homocysteine metabolic cycle, which produces the methyl donor via SAM and modulates methyltransferase activity by SAH. One of the most important intermediate products in this metabolic cycle is Hcy. Hcy has become one of the research focuses in the cardiovascular field since it was initially identified as an independent risk factor for cardiovascular and thrombotic diseases in the general population [82]. Several in vivo studies suggest that Hcy levels may induce global DNA hypomethylation. In healthy young women, increased levels of plasma Hcy were found to be associated with both increased SAH concentrations and DNA hypomethylation in lymphocytes [74]. Other studies also confirm this inverse relationship between Hcy plasma concentrations and DNA methylation patterns [83, 84].

We have performed a series of studies and demonstrated that Hcy selectively induces SAH accumulation and subsequent cellular hypomethylation in EC, leading to EC-specific growth inhibition [56]. Hcy also causes an EC-specific suppression of Ras protein carboxyl methylation and cyclin A promoter methylation [13, 56, 85]. Studies from other labs also indicate that Hcy inhibits EC growth by downregulating FGF2 via hypermethylation of the FGF2 promoter [47] and that Hcy promotes EC dysfunction by upregulating p66shc via hypomethylation of the p66shc promoter [44]. We propose that hypomethylation is a key biochemical mechanism [60] by which Hcy causes vascular diseases [13, 56, 86]. This hypothesis is supported by several studies [87]. In one such example, increased Hcy levels along with a reduced global lymphocyte DNA methylation pattern were reported in patients with renal functional impairment, a clinical condition that augments the risk for vascular diseases [88].

It is clear that HHcy represents the biochemical fundamentals of methylation deregulation and links methylation to atherosclerosis. Specific therapeutic approaches targeting the aberrant methylation patterns in vascular cells which are induced by risk factors, such as Hcy, may provide new possibilities for the more effective treatments of atherosclerosis.

REFERENCES

1. Okano M, Bell DW, Haber DA, Li E. (1999) DNA methyltransferases DNMT3a and DNMT3b are essential for de novo methylation and mammalian development. *Cell*. 99, 247–257.
2. Leonhardt H, Page AW, Weier HU, Bestor TH. (1992) A targeting sequence directs DNA methyltransferase to sites of DNA replication in mammalian nuclei. *Cell*. 71, 865–873.
3. Li E, Bestor TH, Jaenisch R. (1992) Targeted mutation of the DNA methyltransferase gene results in embryonic lethality. *Cell*. 69, 915–926.
4. Mayer W, Niveleau A, Walter J, Fundele R, Haaf T. (2000) Demethylation of the zygotic paternal genome. *Nature*. 403, 501–502.
5. Rougier N, Bourc'his D, Gomes DM, et al. (1998) Chromosome methylation patterns during mammalian preimplantation development. *Genes Dev*. 12, 2108–2113.
6. Bhutani N, Burns DM, Blau HM. (2011) DNA demethylation dynamics. *Cell*. 146, 866–872.
7. Wang Y, Leung FC. (2004) An evaluation of new criteria for CpG islands in the human genome as gene markers. *Bioinformatics*. 20, 1170–1177.
8. Hellman A, Chess A. (2007) Gene body-specific methylation on the active X chromosome. *Science*. 315, 1141–1143.
9. Zilberman D, Gehring M, Tran RK, Ballinger T, Henikoff S. (2007) Genome-wide analysis of arabidopsis thaliana DNA methylation uncovers an interdependence between methylation and transcription. *Nat Genet*. 39, 61–69.
10. Moarefi AH, Chedin F. (2011) Icf syndrome mutations cause a broad spectrum of biochemical defects in DNMT3b-mediated de novo DNA methylation. *J Mol Biol*. 409, 758–772.
11. Laurent L, Wong E, Li G, et al. (2010) Dynamic changes in the human methylome during differentiation. *Genome Res*. 20, 320–331.
12. Thai P, Loukoianov A, Wachi S, Wu R. (2008) Regulation of airway mucin gene expression. *Annu Rev Physiol*. 70, 405–429.
13. Jamaluddin MD, Chen I, Yang F, et al. (2007) Homocysteine inhibits endothelial cell growth via DNA hypomethylation of the cyclin a gene. *Blood*. 110, 3648–3655.
14. Jones PL, Veenstra GJ, Wade PA, et al. (1998) Methylated DNA and MeCP2 recruit histone deacetylase to repress transcription. *Nat Genet*. 19, 187–191.
15. Loebel DA, Tam PP. (2004) Genomic imprinting: Mice without a father. *Nature*. 428, 809–811.
16. Dean W, Santos F, Reik W. (2003) Epigenetic reprogramming in early mammalian development and following somatic nuclear transfer. *Semin Cell Dev Biol*. 14, 93–100.
17. Illingworth RS, Bird AP. (2009) CpG islands—a rough guide. *FEBS Lett*. 583, 1713–1720.
18. Farthing CR, Ficz G, Ng RK, et al. (2008) Global mapping of DNA methylation in mouse promoters reveals epigenetic reprogramming of pluripotency genes. *PLoS Genet*. 4, e1000116.
19. Challen GA, Sun D, Jeong M, et al. (2012) DNMT3a is essential for hematopoietic stem cell differentiation. *Nat Genet*. 44, 23–31.
20. Weber M, Hellmann I, Stadler MB, et al. (2007) Distribution, silencing potential and evolutionary impact of promoter DNA methylation in the human genome. *Nat Genet*. 39, 457–466.
21. Clarke S. (1993) Protein methylation. *Curr Opin Cell Biol*. 5, 977–983.
22. Greer EL, Shi Y. (2012) Histone methylation: A dynamic mark in health, disease and inheritance. *Nat Rev Genet*. 13, 343–357.
23. Bannister AJ, Kouzarides T. (2005) Reversing histone methylation. *Nature*. 436, 1103–1106.
24. Lee DY, Teyssier C, Strahl BD, Stallcup MR. (2005) Role of protein methylation in regulation of transcription. *Endocr Rev*. 26, 147–170.
25. Wysocka J, Allis CD, Coonrod S. (2006) Histone arginine methylation and its dynamic regulation. *Front Biosci*. 11, 344–355.
26. Di Lorenzo A, Bedford MT. (2011) Histone arginine methylation. *FEBS Lett*. 585, 2024–2031.

27. Margueron R, Trojer P, Reinberg D. (2005) The key to development: Interpreting the histone code? *Curr Opin Genet Dev*. 15, 163–176.
28. Berger SL. (2007) The complex language of chromatin regulation during transcription. *Nature*. 447, 407–412.
29. Huang J, Perez-Burgos L, Placek BJ, et al. (2006) Repression of p53 activity by Smyd2-mediated methylation. *Nature*. 444, 629–632.
30. Shi X, Kachirskaia I, Yamaguchi H. (2007) Modulation of p53 function by set8-mediated methylation at lysine 382. *Mol Cell*. 27, 636–646.
31. Chuikov S, Kurash JK, Wilson JR, et al. (2004) Regulation of p53 activity through lysine methylation. *Nature*. 432, 353–360.
32. Desrosiers RR, Fanelus I. (2011) Damaged proteins bearing l-isoaspartyl residues and aging: A dynamic equilibrium between generation of isomerized forms and repair by PIMT. *Curr Aging Sci*. 4, 8–18.
33. Stanevich V, Jiang L, Satyshur KA, et al. (2011) The structural basis for tight control of pp2a methylation and function by LCMT-1. *Mol Cell*. 41, 331–342.
34. Gao J, Liao J, Yang GY. (2009) CAAX-box protein, prenylation process and carcinogenesis. *Am J Transl Res*. 1, 312–325.
35. Ahearn IM, Haigis K, Bar-Sagi D, Philips MR. (2012) Regulating the regulator: Post-translational modification of RAS. *Nat Rev Mol Cell Biol*. 13, 39–51.
36. Esteller M. (2008) Epigenetics in cancer. *N Engl J Med*. 358, 1148–1159.
37. Gaudet F, Hodgson JG, Eden A, et al. (2003) Induction of tumors in mice by genomic hypomethylation. *Science*. 300, 489–492.
38. Eden A, Gaudet F, Waghmare A, Jaenisch R. (2003) Chromosomal instability and tumors promoted by DNA hypomethylation. *Science*. 300, 455.
39. Rai K, Sarkar S, Broadbent TJ, et al. (2010) DNA demethylase activity maintains intestinal cells in an undifferentiated state following loss of APC. *Cell*. 142, 930–942.
40. Rodriguez-Paredes M, Esteller M. (2011) Cancer epigenetics reaches mainstream oncology. *Nat Med*. 17, 330–339.
41. Chen T, Hevi S, Gay F, et al. (2007) Complete inactivation of DNMT1 leads to mitotic catastrophe in human cancer cells. *Nat Genet*. 39, 391–396.
42. Hiltunen MO, Turunen MP, Hakkinen TP, et al. (2002) DNA hypomethylation and methyltransferase expression in atherosclerotic lesions. *Vasc Med*. 7, 5–11.
43. Lund G, Andersson L, Lauria M, et al. (2004) DNA methylation polymorphisms precede any histological sign of atherosclerosis in mice lacking apolipoprotein e. *J Biol Chem*. 279, 29147–29154.
44. Kim CS, Kim YR, Naqvi A, et al. (2011) Homocysteine promotes human endothelial cell dysfunction via site-specific epigenetic regulation of p66shc. *Cardiovasc Res*. 92, 466–475.
45. Chan Y, Fish JE, D'Abreo C, et al. (2004) The cell-specific expression of endothelial nitric-oxide synthase: A role for DNA methylation. *J Biol Chem*. 279, 35087–35100.
46. Chan GC, Fish JE, Mawji IA, Leung DD, Rachlis AC, Marsden PA. (2005) Epigenetic basis for the transcriptional hyporesponsiveness of the human inducible nitric oxide synthase gene in vascular endothelial cells. *J Immunol*. 175, 3846–3861.
47. Chang PY, Lu SC, Lee CM, et al. (2008) Homocysteine inhibits arterial endothelial cell growth through transcriptional downregulation of fibroblast growth factor-2 involving g protein and DNA methylation. *Circ Res*. 102, 933–941.
48. Chen KC, Wang YS, Hu CY, et al. (2011) Oxldl up-regulates microrna-29b, leading to epigenetic modifications of MMP-2/MMP-9 genes: A novel mechanism for cardiovascular diseases. *FASEB J*. 25, 1718–1728.
49. Ying AK, Hassanain HH, Roos CM, et al. (2000) Methylation of the estrogen receptor-alpha gene promoter is selectively increased in proliferating human aortic smooth muscle cells. *Cardiovasc Res*. 46, 172–179.
50. Kim J, Kim JY, Song KS, et al. (2007) Epigenetic changes in estrogen receptor beta gene in atherosclerotic cardiovascular tissues and in-vitro vascular senescence. *Biochim Biophys Acta*. 1772, 72–80.
51. Post WS, Goldschmidt-Clermont PJ, Wilhide CC, et al. (1999) Methylation of the estrogen receptor gene is associated with aging and atherosclerosis in the cardiovascular system. *Cardiovasc Res*. 43, 985–991.
52. Zawadzki C, Chatelain N, Delestre M, et al. (2009) Tissue factor pathway inhibitor-2 gene methylation is associated with low expression in

carotid atherosclerotic plaques. *Atherosclerosis*. 204, e4–e14.

53. Fish JE, Matouk CC, Rachlis A, et al. (2005) The expression of endothelial nitric-oxide synthase is controlled by a cell-specific histone code. *J Biol Chem*. 280, 24824–24838.
54. Hellebrekers DM, Castermans K, Vire E, et al. (2006) Epigenetic regulation of tumor endothelial cell anergy: Silencing of intercellular adhesion molecule-1 by histone modifications. *Cancer Res*. 66, 10770–10777.
55. Chen KC, Liao YC, Hsieh IC, Wang YS, Hu CY, Juo SH. (2012) Oxldl causes both epigenetic modification and signaling regulation on the microrna-29b gene: Novel mechanisms for cardiovascular diseases. *J Mol Cell Cardiol*. 52, 587–595.
56. Wang H, Yoshizumi M, Lai K, et al. (1997) Inhibition of growth and p21ras methylation in vascular endothelial cells by homocysteine but not cysteine. *J Biol Chem*. 272, 25380–25385.
57. Kramer K, Harrington EO, Lu Q, et al. (2003) Isoprenylcysteine carboxyl methyltransferase activity modulates endothelial cell apoptosis. *Mol Biol Cell*. 14, 848–857.
58. Lu Q, Harrington EO, Newton J, Jankowich M, Rounds S. (2007) Inhibition of ICMT induces endothelial cell apoptosis through grp94. *Am J Respir Cell Mol Biol*. 37, 20–30.
59. Cimmino A, Capasso R, Muller F, et al. (2008) Protein isoaspartate methyltransferase prevents apoptosis induced by oxidative stress in endothelial cells: Role of bcl-xl deamidation and methylation. *PLoS ONE*. 3, e3258.
60. Lee ME, Wang H. (1999) Homocysteine and hypomethylation. A novel link to vascular disease. *Trends Cardiovasc Med*. 9, 49–54.
61. Mudd SH, Levy HL, Tangerman A, et al. (1995) Isolated persistent hypermethioninemia. *Am J Hum Genet*. 57, 882–892.
62. Niculescu MD, Zeisel SH. (2002) Diet, methyl donors and DNA methylation: Interactions between dietary folate, methionine and choline. *J Nutr*. 132, 2333S–2335S.
63. Chen NC, Yang F, Capecci LM, et al. (2010) Regulation of homocysteine metabolism and methylation in human and mouse tissues. *FASEB J*. 24, 2804–2817.
64. Yu BD, Hanson RD, Hess JL, Horning SE, Korsmeyer SJ. (1998) Mll, a mammalian trithorax-group gene, functions as a transcriptional maintenance factor in morphogenesis. *Proc Natl Acad Sci U S A*. 95, 10632–10636.
65. Bergo MO, Leung GK, Ambroziak P, et al. (2001) Isoprenylcysteine carboxyl methyltransferase deficiency in mice. *J Biol Chem*. 276, 5841–5845.
66. Zhu X, Song J, Mar MH, Edwards LJ, Zeisel SH. (2003) Phosphatidylethanolamine n-methyltransferase (PEMT) knockout mice have hepatic steatosis and abnormal hepatic choline metabolite concentrations despite ingesting a recommended dietary intake of choline. *Biochem J*. 370, 987–993.
67. Mudd SH, Poole JR. (1975) Labile methyl balances for normal humans on various dietary regimens. *Metabolism*. 24, 721–735.
68. Baric I, Fumic K, Glenn B, et al. (2004) S-adenosylhomocysteine hydrolase deficiency in a human: A genetic disorder of methionine metabolism. *Proc Natl Acad Sci U S A*. 101, 4234–4239.
69. Perna AF, Ingrosso D, Zappia V, Galletti P, Capasso G, De Santo NG. (1993) Enzymatic methyl esterification of erythrocyte membrane proteins is impaired in chronic renal failure. Evidence for high levels of the natural inhibitor s-adenosylhomocysteine. *J Clin Invest*. 91, 2497–2503.
70. Chen Z, Karaplis AC, Ackerman SL, et al. (2001) Mice deficient in methylenetetrahydrofolate reductase exhibit hyperhomocysteinemia and decreased methylation capacity, with neuropathology and aortic lipid deposition. *Hum Mol Genet*. 10, 433–443.
71. Kim JM, Hong K, Lee JH, Lee S, Chang N. (2009) Effect of folate deficiency on placental DNA methylation in hyperhomocysteinemic rats. *J Nutr Biochem*. 20, 172–176.
72. Caudill MA, Wang JC, Melnyk S, et al. (2001) Intracellular s-adenosylhomocysteine concentrations predict global DNA hypomethylation in tissues of methyl-deficient cystathionine beta-synthase heterozygous mice. *J Nutr*. 131, 2811–2818.
73. Ulrey CL, Liu L, Andrews LG, Tollefsbol TO. (2005) The impact of metabolism on DNA methylation. *Hum Mol Genet*. 14 Spec No 1, R139–147.
74. Yi P, Melnyk S, Pogribna M, Pogribny IP, Hine RJ, James SJ. (2000) Increase in plasma homocysteine associated with parallel increases in plasma s-adenosylhomocysteine and lymphocyte

DNA hypomethylation. *J Biol Chem*. 275, 29318–29323.

75. Swanson DA, Liu ML, Baker PJ, et al. (2001) Targeted disruption of the methionine synthase gene in mice. *Mol Cell Biol*. 21, 1058–1065.
76. Dayal S, Devlin AM, McCaw RB, et al. (2005) Cerebral vascular dysfunction in methionine synthase-deficient mice. *Circulation*. 112, 737–744.
77. Teng YW, Mehedint MG, Garrow TA, Zeisel SH. (2011) Deletion of betaine-homocysteine s-methyltransferase in mice perturbs choline and 1-carbon metabolism, resulting in fatty liver and hepatocellular carcinomas. *J Biol Chem*. 286, 36258–36267.
78. Wang L, Chen X, Tang B, Hua X, Klein-Szanto A, Kruger WD. (2005) Expression of mutant human cystathionine beta-synthase rescues neonatal lethality but not homocystinuria in a mouse model. *Hum Mol Genet*. 14, 2201–2208.
79. Kraus JP, Janosik M, Kozich V, et al. (1999) Cystathionine beta-synthase mutations in homocystinuria. *Hum Mutat*. 13, 362–375.
80. Watanabe M, Osada J, Aratani Y. (1995) Mice deficient in cystathionine beta-synthase: Animal models for mild and severe homocyst(e) inemia. *Proc Natl Acad Sci U S A*. 92, 1585–1589.
81. Yang G, Wu L, Jiang B, et al. (2008) H2S as a physiologic vasorelaxant: Hypertension in mice with deletion of cystathionine gamma-lyase. *Science*. 322, 587–590.
82. Stampfer MJ, Malinow MR, Willett WC, et al. (1992) A prospective study of plasma homocyst(e)ine and risk of myocardial infarction in us physicians. *JAMA*. 268, 877–881.
83. Friso S, Choi SW, Girelli D, et al. (2002) A common mutation in the 5,10-methylenetetrahydrofolate reductase gene affects genomic DNA methylation through an interaction with folate status. *Proc Natl Acad Sci U S A*. 99, 5606–5611.
84. Castro R, Rivera I, Blom HJ, Jakobs C, Tavares de Almeida I. (2006) Homocysteine metabolism, hyperhomocysteinaemia and vascular disease: An overview. *J Inherit Metab Dis*. 29, 3–20.
85. Wang H, Jiang X, Yang F, et al. (2002) Cyclin a transcriptional suppression is the major mechanism mediating homocysteine-induced endothelial cell growth inhibition. *Blood*. 99, 939–945.
86. Tan H, Jiang X, Yang F, et al. (2006) Hyperhomocysteinemia inhibits post-injury reendothelialization in mice. *Cardiovasc Res*. 69, 253–262.
87. Pogribny IP, Beland FA. (2009) DNA hypomethylation in the origin and pathogenesis of human diseases. *Cell Mol Life Sci*. 66, 2249–2261.
88. Ingrosso D, Cimmino A, Perna AF, et al. (2003) Folate treatment and unbalanced methylation and changes of allelic expression induced by hyperhomocysteinaemia in patients with uraemia. *Lancet*. 361, 1693–1699.

33

MICROMANAGING ATHEROGENESIS: ROLE OF MICRORNAS IN ATHEROSCLEROSIS

Shusheng Wang[1,2] and Chastain Anderson[1]

[1]Department of Cell and Molecular Biology, Tulane University, New Orleans, LA, USA

[2]Department of Ophthalmology, Tulane University, New Orleans, LA, USA

OVERVIEW OF ATHEROSCLEROSIS DEVELOPMENT

Atherosclerosis is a chronic inflammatory disease that leads to atherosclerotic lesion/plaque development and narrowing of the arterial lumen. Plaque rupture and thrombosis cause myocardial infarction (MI) and stroke, the leading causes of death and morbidity throughout the developed world [1]. Currently, no cure is available, and the cholesterol-lowering drugs "statins" are not sufficient to fully prevent atherosclerosis progression.

Atherosclerotic lesion development begins with endothelial dysfunction in response to disturbed laminar flow or injury at predilection sites of the arterial wall, which permits subendothelial accumulation of low-density lipoprotein (LDL) and infiltration of immune cells. Early "fatty streak" lesions are composed of cholesterol-laden monocytes and macrophages, as well as T cells. Lipids and LDLs in the intima are susceptible to oxidative modification. The oxidized LDL (oxLDL), no longer recognizable by normal LDL receptors, is taken up by macrophages in the intima through their scavenger receptors. When oxLDL levels reach the point that macrophages cannot properly dispose of the cholesterol, it accumulates within the cytoplasm and renders the macrophage a "foam cell."

The development from fatty streak to advanced lesion is characterized by a transition in vascular smooth muscle cell (VSMC) involvement and a strong inflammatory response. This immune response has both pro- and anti-atherosclerotic effects, but the net outcome is a worsening of the disease phenotype. VSMCs immigrate from the media to the intima where they proliferate, uptake modified LDLs, contribute to foam cell formation, and synthesize extracellular matrix proteins. This leads to the development of a 'fibrotic cap' that pushes out into the lumen of the arteries.

At this point, the plaque will remain clinically silent. Successive accumulation of apoptotic

Atherosclerosis: Risks, Mechanisms, and Therapies, First Edition. Edited by Hong Wang and Cam Patterson.

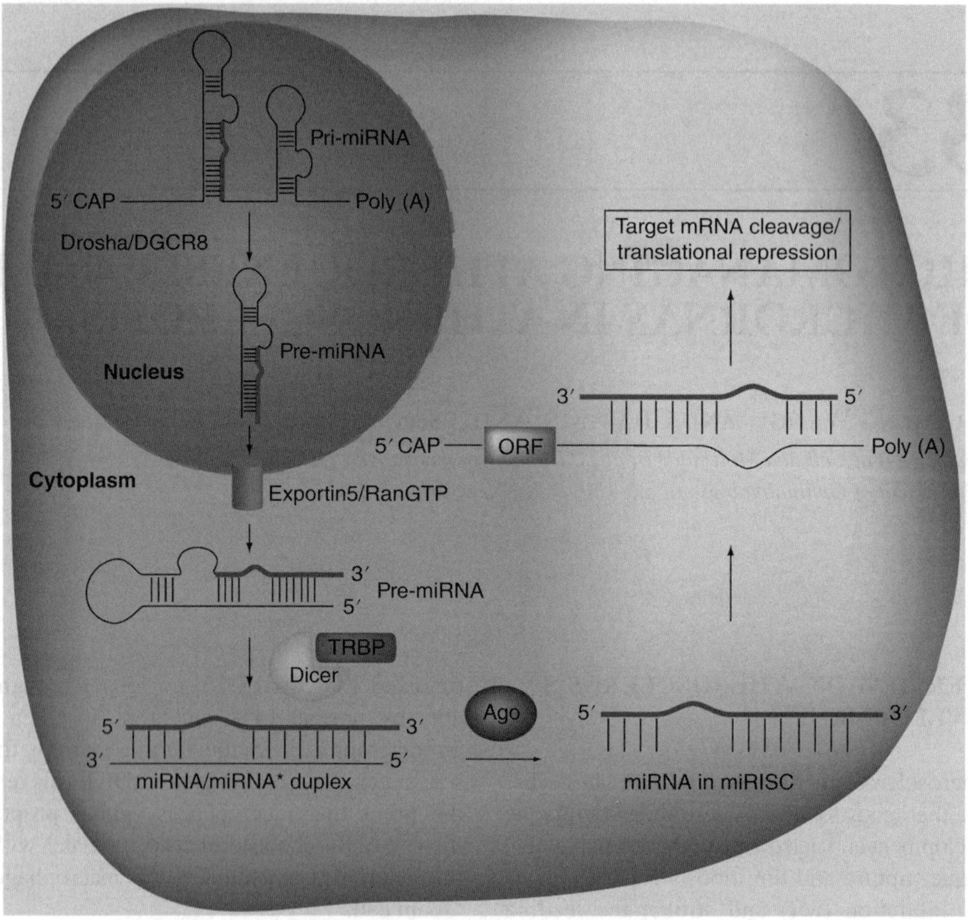

Figure 33.1 Canonical microRNA biogenesis pathway. Primary (pri)-miRNAs are transcribed by RNA Polymerase II or III, and are processed by Drosha/DGCR8 into precursor (pre)-miRNAs. The pre-miRNA is transported from the nucleus to the cytoplasm by Exportin-5 and Ran-GTP, where it is cut by a Dicer complex (Dicer/TRBP) into a miRNA/miRNA* duplex. One strand of the duplex is preferably incorporated into miRISC containing Ago proteins. The miRISC mediates sequence-specific gene repression by binding to the 3′-UTR of the target mRNAs and repressing translational or inducing mRNA degradation Ago, argonaute; DGCR8, DiGeorge syndrome critical region 8; ORF, open reading frame; TRBP, TAR RNA-binding protein. (Reprinted with permission from Wang et al. [4]. © 2012, Future Science Ltd).

cells, cholesterol crystals, and mixture of leukocytes forms a necrotic core under the fibrous cap. Continuous growth of the plaque occludes the artery and causes ischemia, rupture, and thrombosis, leading to MI and stroke. Although the determinants involved in plaque rupture have not been fully characterized, apoptosis, neovascularization, and subsequent hemorrhage within this necrotic core are major risk factors for lesion instability and rupture.

MICRORNA BIOGENESIS AND FUNCTIONAL MECHANISM

MicroRNAs (miRNAs or miRs) are small RNAs that function posttranscriptionally as negative regulators of gene expression [2]. Lin-4 was the first miRNA discovered by Victor Ambros' group in 1993. Lin-4 controls the timing of larva development in *Caenorhabditis elegans* through producing a pair of small

RNAs with complementarities to targeting sites in the 3′-untranslated region (3′-UTR) of the *lin-14* gene [3]. To date, more than 1000 miRNAs have been identified in humans and are predicted to directly regulate more than one third of the protein-coding genes. In the genome, miRNAs can be localized in the intron of protein-coding genes, or extragenically where they are subject to independent regulation. miRNAs are transcribed by RNA polymerase II or III into a primary transcript (pri-miRNA). In the canonical pathway (Figure 33.1), Drosha, an RNase III-type endoribonuclease localized in the nucleus, processes the pri-miRNA into the pre-miRNA (about 72 nucleotides in length). The pre-miRNA is then exported to the cytoplasm, where Dicer processes it into a miRNA/miRNA* duplex (sized about 22 nucleotides in each strand). Finally, one strand of the duplex is preferably incorporated into the miRNA-induced silencing complex (miRISC), which guides sequence-specific posttranscriptional gene repression. Typically, a short sequence of approximately six to seven nucleotides near the 5′-end of the miRNA, called the "seed region," binds to sequence(s) on the target mRNAs, usually in the 3′-UTR, and represses translation or induces target mRNA degradation. Because of the incomplete complementarity between miRNAs and their target sequences, one miRNA can target multiple mRNAs, and one mRNA can be targeted by multiple miRNAs, adding yet another layer of posttranscriptional gene regulation to the genome.

Recent studies have indicated that miRNA regulatory mechanisms play critical roles in numerous developmental and disease processes. Specifically, miRNAs have been shown to regulate many pathological or physiological processes, including lipid metabolism, immune regulation, inflammation, VSMC phenotypic switching, and angiogenesis, which are critically involved in atherogenesis. More excitingly, several miRNAs have shown powerful effects in several disease models, including atherosclerosis [5, 6]. These discoveries have resulted in the rapid advancement of miRNA inhibitors as a new class of therapeutics. In this chapter, we summarize the recent progress in the field of miRNA regulation and function in processes involved in atherosclerosis. We surmise that miRNAs may be viable therapeutic targets for atherosclerosis.

REGULATION OF MIRNAS IN ATHEROSCLEROSIS

The whole picture of miRNA regulation in atherosclerosis is still unclear. A recent study aimed to examine the expression level of about 900 miRNAs in human atherosclerotic plaque compared to that of the nonatherosclerotic thoracic artery using a miRNA microarray approach [7]. Of the ten miRNAs regulated in the array, miR-21, miR-210, miR-34a, miR-146a, and miR-146b-5p were confirmed to be highly upregulated in atherosclerotic plaques. Another study investigated a preselected set of miRNAs in the intima of arteriosclerosis obliterans (ASO), a form of atherosclerotic disease with narrowing or occlusion in peripheral arteries [8]. miR-21, miR-130a, miR-27b, let-7f, and miR-210 were found to be significantly upregulated, while miR-221 and miR-222 are significantly downregulated in the sclerotic samples compared to the normal samples. Consistently, miR-222 shows progressively decreased expression from early to advanced atherosclerotic lesion [9]. In an independent study, 20 miRNAs were found highly expressed, and three miRNAs are downregulated in plaque tissues compared to the healthy arteries [10]. Surprisingly, there is nearly no overlap in the miRNAs dysregulated in this study compared to the previous studies. These results suggest an involvement of miRNA in atherosclerosis, but highlight the complexity and variability of the components in atherosclerotic plaques. Increasing sample size, carefully characterizing the stage of plaques, and choosing appropriate controls are needed to complete a picture of miRNA regulation during atherogenesis. Of note, there were also scattered studies of individual miRNA regulation in atherosclerosis, which will be described during the course of the chapter.

MIRNAS IN LIPID METABOLISM AND ATHEROGENESIS

Although atherosclerosis is a multifactorial disease, high serum cholesterol level is unique among atherosclerotic risk factors in being sufficient to drive lesion development in the absence of other known factors [1]. This makes cholesterol regulation a prime target for therapeutic intervention. Modified LDLs promote atherogenesis at almost all stages, from recruitment of the inflammatory cells to the lesion to macrophage and VSMC foam cell formation. Conversely, high-density lipoproteins (HDL) transfer cholesterol from peripheral tissue to the liver for excretion and exert anti-inflammatory functions. Increasing plasma HDL has been shown to hinder atherosclerotic plaque progression and promote regression. Recent studies have discovered key roles of miRNAs in regulating lipid metabolism in atherosclerosis. miR-33a and miR-33b (simplified as miR-33) are intronic miRNAs cotranscribed with their host genes, sterol-response-element-binding protein genes *SREBF2* and *SREBF1*, respectively, and are found to be critical regulators of cholesterol and fatty acid homeostasis [11–13] (Figure 33.2). *SREBF1* and *SREBF2* genes code for transcriptional factors SREBP1 and

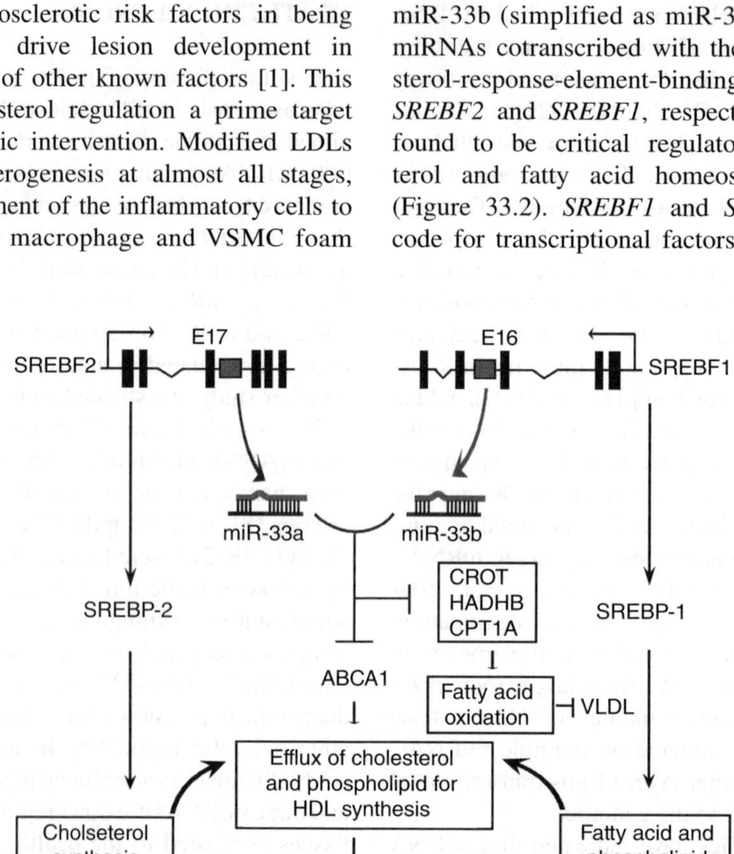

Figure 33.2 miR-33 regulates lipid metabolism in atherosclerosis. miR-33a and miR-33b exist as introns within the SREBF2 and SREBF1, respectively. The SREBFs encode proteins that activate cholesterogenic and lipogenic pathways. miR-33 represses cholesterol and phospholipid efflux for HDL synthesis through targeting ABCA1 protein and decreases fatty acid oxidation and therefore increases VLDL by targeting CROT, CPT1A, and HADHB. Inhibition of miR-33 protects from atherosclerosis by increasing HDL while decreasing VLDL levels. These exemplify the cooperation of miRNAs and their host genes to regulate cholesterol and fatty acid metabolism. Of note, miR-33a is present in all mammals, while miR-33b exists in mammals except rodents. ABCA1, ATP-binding cassette transporter; CPT1A, carnitine palmitoyltransferase 1A; CROT, carnitine octanoyltransferase; HADHB, hydroxyacyl-CoA dehydrogenase beta subunit; HDL, high-density lipoprotein; SREBF, sterol-response-element-binding protein gene; SREBP, sterol-response-element-binding protein; VLDL, very-low-density lipoprotein.

SREBP2, which have established roles in regulating cholesterol and fatty acid metabolism. miR-33 prevents cholesterol efflux by downregulating ATP-binding cassette transporter A1 (ABCA1) which is responsible for apolipoprotein A1-mediated transportation of cholesterol out of cells to generate nascent HDL particles (a process called reverse cholesterol transport or cholesterol efflux). miR-33 also decreases fatty acid oxidation and increases VLDL by targeting fatty acid oxidation genes *carnitine O-octanoyltransferase (CROT)*, *hydroxyacyl-CoA-dehydrogenase (HADHB)*, and *carnitine palmitoyltransferase 1A (CPT1A)*. Silencing of miR-33 using locked nucleic acid (LNA)-modified antisense inhibitors in both mice and nonhuman primates has been shown to increase hepatic ABCA1 and plasma HDL levels by as much as 40% [5, 11–13]. More excitingly, inhibiting miR-33 expression in mice has shown strong atheroprotective capability and has actually reversed the atherosclerotic phenotypes as indicated by reduction in plaque size and lipid content, increased markers of plaque stability, and decreased inflammatory gene expression [6]. These elegant discoveries strongly suggest miR-33 antagonists as potential therapeutics for atherosclerosis and related metabolic diseases.

Cholesterol metabolism is also broadly regulated by miR-122, the most abundant miRNA in the liver. Silencing of miR-122 in mice caused sustained reduction of total plasma cholesterol levels including both the LDL and HDL fractions [14, 15]. These results are consistent with its regulation of genes implicated in fatty acid synthesis and oxidation, including fatty acid synthase (FASN), acetyl-CoA carboxylase (ACC)-1 and -2. Further studies in African green monkeys showed that miR-122 silencing resulted in 20–30% reduction in total plasma cholesterol without apparent toxicity or histopathological changes in the liver [16]. Currently, anti-miR-122 agent miravirsen is on phase I clinical trial and shown to be well tolerated, offering promise for miRNA-based therapeutics for various diseases.

MIRNAS IN THE EARLY LESION DEVELOPMENT

The key events of early atherosclerotic lesion development include endothelial dysfunction, lipid accumulation in the arterial wall, and monocyte infiltration which leads to a profusion of cholesterol-laden macrophage foam cells. Atherosclerosis develops preferentially in arterial branches and curvatures where local flow is disturbed. Laminar shear flow tends to maintain endothelial cells (ECs) in a quiescent state, contributing to the atheroprotection in the straight portions of the arteries, while endothelial dysfunction induced by disturbed laminar flow or injury may lead to the upregulation of several cell adhesion molecules, including vascular cell adhesion molecule-1 (VCAM-1), E-selectin, and P-selectin, which are involved in leukocyte adhesion and infiltration into the arterial wall. miRNAs have been implicated in early atherosclerotic lesion development (Figure 33.3). A list of miRNAs has been shown to be significantly regulated in laminar flow-treated ECs [17, 18]. Among them, miR-19 and miR-23b were upregulated by laminar flow and contribute to shear stress-induced growth arrest [17, 18]. Downregulation of miR-10a in athero-prone endothelium in vivo could contribute to inflammation in ECs in the athero-prone arteries through upregulation of monocyte chemotactic protein-1 (MCP-1), interleukin-6 (IL-6), IL-8, VCAM-1, and E-selectin [19]. miR-92a is downregulated more in ECs exposed to atheroprotective pulsatile shear flow compared to athero-prone oscillatory shear flow, which in turn increases the expression of Kruppel-like factor 2 (KLF2), a critical mediator of shear stress and atheroprotective endothelial function [20]. miR-126, an EC-enriched miRNA, has been shown to be induced by KLF2 (at least in zebrafish) and repress leukocyte adhesion to ECs through targeting VCAM-1 [21, 22]. miR-21 is upregulated by both unilateral shear stress and athero-prone oscillatory shear stress [23, 24]. miR-21 upregulation by unilateral shear stress has been reported to protect EC apoptosis,

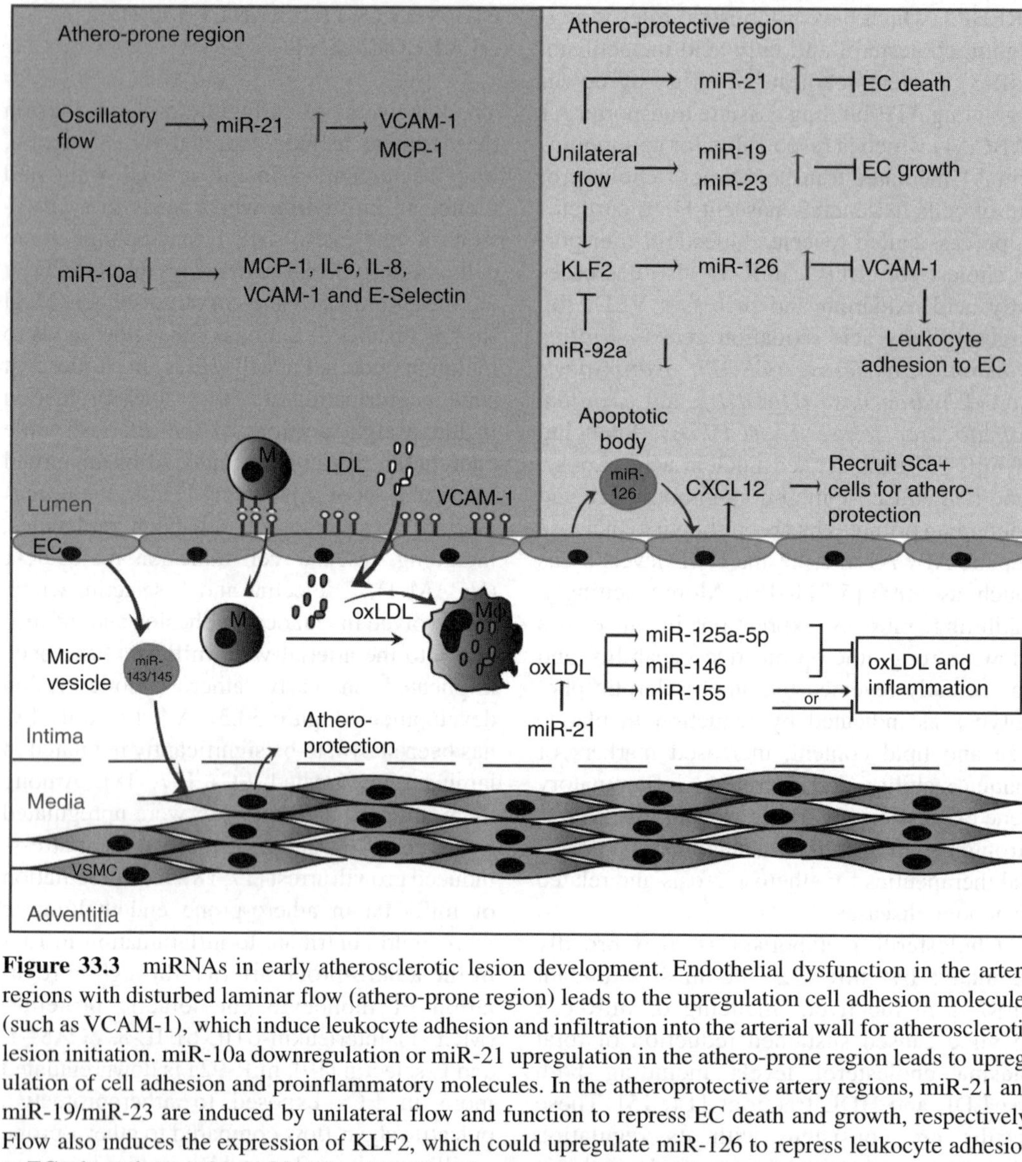

Figure 33.3 miRNAs in early atherosclerotic lesion development. Endothelial dysfunction in the artery regions with disturbed laminar flow (athero-prone region) leads to the upregulation cell adhesion molecules (such as VCAM-1), which induce leukocyte adhesion and infiltration into the arterial wall for atherosclerotic lesion initiation. miR-10a downregulation or miR-21 upregulation in the athero-prone region leads to upregulation of cell adhesion and proinflammatory molecules. In the atheroprotective artery regions, miR-21 and miR-19/miR-23 are induced by unilateral flow and function to repress EC death and growth, respectively. Flow also induces the expression of KLF2, which could upregulate miR-126 to repress leukocyte adhesion to ECs through targeting VCAM-1. miR-92a is downregulated in the atheroprotective region and represses KLF2 expression. miR-126 in apoptotic bodies secreted by ECs upregulates CXCL12 expression in the recipient cells, which functions to recruit Sca$^+$ progenitor cells for atheroprotection, while miR-143/miR-145 in microvesicles secreted by ECs can function on VSMCs to exert atheroprotection. oxLDL induces the expression of miR-125p, miR-146, and miR-155, which function as feedback mechanism to reduce oxLDL level and inflammation (miR-155 might be pro- or anti-inflammatory depending on cell type). Up and down arrows represent changes in miRNA expression level. EC, endothelial cell; LDL, low-density lipoprotein; M, monocytes; Mφ, macrophage; oxLDL, oxidized LDL; VCAM-1, vascular cell adhesion molecule-1; VSMC, vascular smooth muscle cell.

while miR-21 upregulation by athero-prone oscillatory shear stress increases adhesion of monocytes to ECs by enhancing the expression of VCAM-1 and MCP-1. Further, miR-21 has been implicated in the initial oxidation responsible for increasing oxLDL levels. Proteomic analysis suggests that elevated levels of miR-21 repress superoxide dismutase-2 (SOD2) and its target gene sprouty-2, thereby stimulating the formation of reactive oxygen species [25]. Tentatively, manipulating the level of these miRNAs, such as overexpression of miR-10 or miR-126, might be beneficial in protecting against atherogenesis at early stage.

Recent studies indicate that ECs can secrete membranous microvesicles, such as exosomes and apoptotic bodies, which confer atheroprotective function [26, 27]. miRNAs in the microvesicles can be transferred to the recipient ECs or VSMCs and regulate their behavior in atherosclerosis. miR-126 in the apoptotic bodies from ECs can be taken up by the adjacent ECs and upregulate the expression of chemokine CXCL12, which promotes the mobilization and incorporation of Sca-1$^+$ progenitor cells to the atherosclerotic plaque and exerts atheroprotective function [26]. miR-143/miR-145 from KLF2-expressing EC-derived extracellular vesicles can regulate smooth muscle gene expression and reduce atherosclerotic lesion formation in vivo [27]. These studies exemplify the importance of miRNAs in cell–cell communication in atherogenesis and suggest an atheroprotective function for miR-126 and miR-143/miR-145.

Once infiltrated into the arterial wall, monocytes can differentiate into macrophages, uptake oxLDL particles through scavenger receptors, and lead to the formation of cholesterol-laden macrophage foam cells. At this stage, the immune responses elicited by monocytes/macrophages and T cells establish a chronic inflammatory state in the lesion. miRNAs have been implicated in the regulation of immune cell function in atherosclerosis. miR-9, miR-125a-5p, miR-146b-5p, and miR-155 have been shown to be upregulated by oxLDL in human primary monocytes. Among them, miR-125a-5p, miR-146, and miR-155 seem to provide a negative feedback mechanism controlling excessive lipid uptake and inflammatory response [28–30]. However, miR-155 knockout mice were resistant to collagen-induced arthritis [31]. These results suggest a cell type-specific function and overall proinflammatory function of miR-155. miR-146a has been shown to inhibit oxLDL accumulation in THP-1 macrophage and negatively regulate IL-6 and IL-8 expression through targeting Toll-like receptor-4 (TLR4), TNF receptor-associated factor-6 (TRAF-6), and IL-1 receptor-associated kinase-1 (IRAK-1) [32, 33]. Therapeutic targeting of inflammation with candidate miRNAs, such as miR-146 mimics, should be an appealing approach for managing atherosclerosis.

MIRNAS IN THE ADVANCED ATHEROSCLEROTIC LESION/PLAQUE

The transition from the fatty streak to advanced lesion/plaque is characterized by the immigration of VSMCs from the media to the intima and the formation of a fibrous cap, followed by the formation of necrotic core through the accumulation of apoptotic and necrotic cholesterol-laden macrophages and VSMCs in the advanced lesion. VSMC phenotypic switching from a contractile state to a proliferative state is a significant contributor to neointima formation in atherosclerosis. Several specific miRNAs with a well-characterized role in VSMC phenotype alteration include miR-21, miR-26a, miR-133, miR-143/miR-145, and miR-221/miR-222 (Figure 33.4). miR-133 expression is inversely correlated with VSMC proliferation, and its overexpression reduces VSMC proliferation and migration in vitro and in vivo [34]. miR-26a was shown to be downregulated in two mouse models of abdominal aortic aneurysm (AAA) formation and functions to promote VSMC proliferation while inhibiting cellular differentiation and apoptosis [35]. Knockdown of miR-21 or miR-221/miR-222 and overexpression of miR-143/miR-145 each

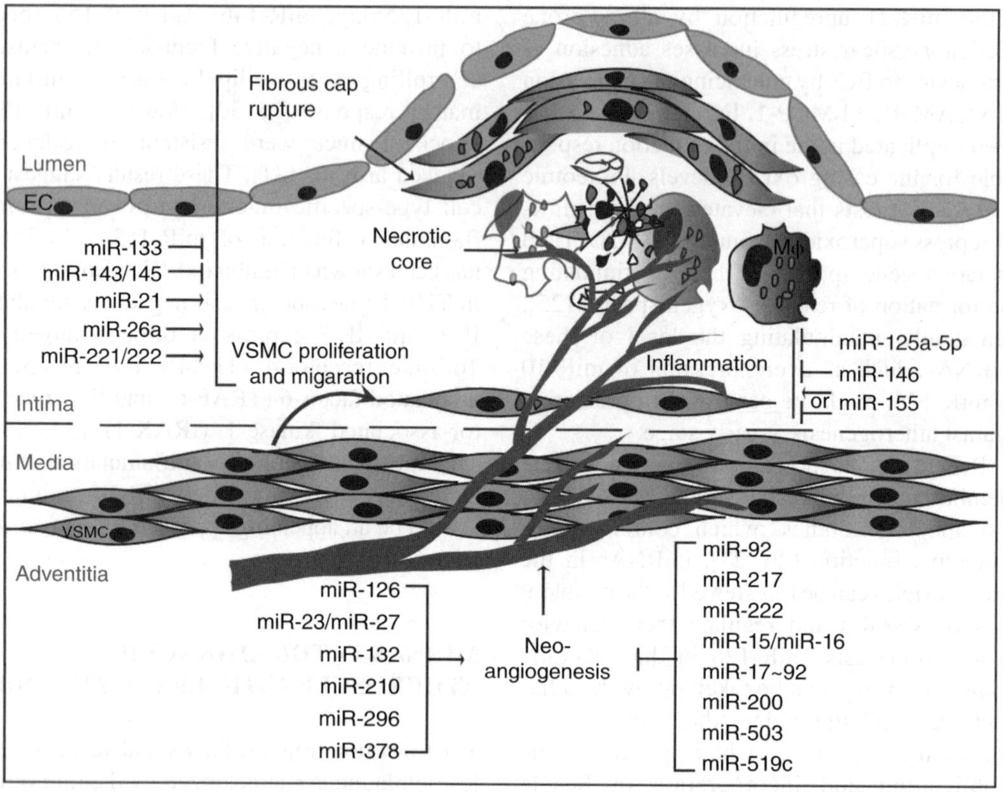

Figure 33.4 miRNAs in advanced lesion development and plaque stabilization/rupture. Excessive inflammation, VSMC phenotypic switching, and neoangiogenesis underlie advanced lesion development and plaque rupture in atherosclerosis. miR-125p, miR-146, and miR-155 could regulate inflammation involved in advanced lesion development and fibrous cap formation. miR-21, miR-26, and miR-221/miR-222 promote VSMC proliferation and migration, therefore contributing to advanced lesion formation and potentially fibrous cap rupture, while miR-133 and miR-143/miR-145 function oppositely to protect advanced lesion development and likely promote plaque stability. Numerous miRNAs enhance or repress angiogenesis and therefore could regulate advanced lesion development and plaque stabilization/rupture. EC, endothelial cell; Mϕ, macrophage; VSMC, vascular smooth muscle cell.

inhibited neointima formation in a balloon injury model [36–39]. *miR-143/miR-145* knockout mice developed neointima lesion in femoral arteries at 18 months of age but were surprisingly resistant to neointima formation in response to balloon injury. miR-143/miR-145 expression level was markedly decreased in the aorta of *ApoE*$^{-/-}$ atherosclerosis mouse model and was further reduced when the mice were fed on a high-fat diet. miR-21 promotes proliferation but represses apoptosis of VSMCs, an effect possibly mediated by its targets phosphatase and tensin homolog (PTEN) and B-cell lymphoma (BCL) 2 [36]. miR-26a promotes VSMC proliferation/migration while inhibiting differentiation by targeting SMAD-1 and SMAD-4 in the transforming growth factor (TGF)-β pathway [35]. miR-221 expression in VSMCs is induced by platelet-derived growth factor (PDGF), which is present in human atherosclerotic plaques and functions to recruit VSMCs to the intima [40]. miR-221/miR-222 enhances VSMC proliferation via their target proteins p27(Kip1) and p57(Kip2) [39]. miR-143/miR-145 controls VSMC plasticity by promoting differentiation and repressing proliferation of VSMCs [37, 41, 42]. Multiple genes involved in the regulation of serum response factor (SRF) activity and actin dynamics, including KLF4 and KLF5, are validated miR-143/miR-145 targets for regulating VSMC plasticity. The direct evidence for miRNA involvement in VSMC phenotypic switching in advanced atherosclerotic lesion formation is still lacking. Nevertheless, the identification of miRNAs like miR-143/miR-145 as major regulators of VSMC phenotypic switching may open a new avenue for attenuating arteriosclerotic pathogenesis.

MIRNAS IN PLAQUE STABILITY AND RUPTURE

Continuous development of the advanced atherosclerotic lesion leads to the narrowing of the vessel lumens and therefore ischemic symptoms, while plaque rupture and thrombosis result in MI and stroke. The parameters that determine the risk of plaque rupture include the thickness of the fibrous cap, the size of the necrotic core, and the presence of neovascularization and intraplaque hemorrhage. A stable, clinically silent plaque has a thick fibrous cap, smaller necrotic core, and little angiogenesis; an unstable plaque has a large necrotic core and a small fibrous cap in addition to (and partially because of) neovascularization. Because of the infancy of miRNA research, direct evidence for miRNAs involvement in plaque stabilization/rupture is still lacking. However, many miRNAs have been shown to regulate cell death, inflammation, VSMC phenotypic switching, and angiogenesis and are therefore potentially implicated in plaque stability/rupture (Figure 33.4). We have discussed several miRNAs that are involved in VSMC phenotypic switching. These miRNAs may also be modulated to keep VSMCs in a quiet state in the fibrous cap, therefore maintaining its stability. Matrix metalloproteinases (MMPs) are strongly implicated in atherosclerotic lesion development and plaque rupture. miR-29b has been shown to be upregulated by oxLDL in VSMCs, which leads to increased MMP-2 and MMP-9 expression through targeting DNA methyltransferase 3b (DNMT3b) [43]. This miRNA might have implications in plaque rupture. Neoangiogenesis in closely linked to plaque growth, destabilization, and rupture. An anti-angiogenic approach has been proposed as a potential strategy to normalize immature intraplaque vessels and stabilize rupture-prone plaques [44]. An increasing number of miRNAs have been shown to regulate angiogenesis via modulation of multiple angiogenic pathways [45]. Among them, miR-92 was shown to be upregulated by ischemic injury and functions to repress angiogenesis [46]. miR-126 is a well-characterized proangiogenic miRNA which has been shown to be required for maintaining vascular integrity and angiogenesis through targeting negative regulators of the pathways, including the Sprouty-related protein-1 and phosphoinositol-3 kinase regulatory subunit 2 (PIK3R2/p85-β) [47, 48]. miR-100 is downregulated after femoral artery occlusion in mice and functions as an endogenous repressor of the serine/threonine protein kinase mammalian target of rapamycin (mTOR) in ECs and VSMCs [49]. miR-217 is expressed and negatively correlated with silent information regulator 1 (SirT1) expression level in human atherosclerotic lesions in vivo and functions to repress angiogenesis but increase senescence by targeting SirT1 [50]. miR-222 expression is negatively correlated with lesion progression, while the expression of its target protein signal transducer and activator of transcription (STAT) 5A is positively correlated with lesion progression [9]. Overexpression of miR-222 blocked EC proliferation and migration in response to IL-3 and basic fibroblast growth factor (BFGF) via targeting STAT5A. These studies suggest a role for miR-222 in vascular remodeling and lesion progression in atherosclerosis. Although much work is needed, these aforementioned miRNAs merit further investigation regarding their function in atherosclerotic lesion progression and plaque stabilization. Other miRNAs reported to be involved in angiogenesis include but are not limited to the following: miR-15/miR-16, miR-17~92, miR-21, miR-23/miR-27, miR-132, miR-200, miR-210, miR-296, miR-378, miR-503, and miR-519c. Angiogenesis seems to be a double-edged sword in the progression of atherosclerosis. Although angiogenesis is a major risk factor for plaque rupture, it is a vital factor in survival following the infarction and ischemia that can result from the ruptured plaque. Therefore, anti-angiogenic approach must be carefully devised based on the stage of atherosclerosis. As an example, blocking miR-126 may improve plaque stabilization by repressing angiogenesis but might interfere the recovery after MI, since miR-126 knockout mice show severe compromised survival rate after MI compared to wild-type mice [47]. Conversely, miR-126 mimics may decrease leukocyte adhesion to ECs, promote Sca^+ progenitor cell-mediated vascular repair, and increase survival after MI, but it may increase angiogenesis in the atherosclerotic plaque.

MIRNA THERAPEUTICS IN ATHEROSCLEROSIS

The nascent field of miRNA therapeutics relies on modulating gene expression through miRNA mimics or miRNA antagonists. A miRNA mimic is a molecule that mimics the function of a particular miRNA thereby downregulating the expression of its target genes; conversely, a miRNA antagonist disrupts the function of a specific miRNA causing an upregulation in the expression of its target genes. Since an individual miRNA can regulate multiple genes in a pathway or several pathways simultaneously, although the effect of a single miRNA on a single gene is mild, simultaneous regulation of multiple targets especially those in the related pathways may have far-reaching biological effects. This "one drug/multiple target" model of miRNA therapeutics would provide a natural means of normalizing the expression of disease genes and achieve outstanding potency while potentially avoiding the toxicity, drug resistance, or activation of alternative pathways caused by switching a single target on or off. miRNA antagonists are typically chemically modified single stranded oligonucleotides with sequences complementary to the miRNAs of interest, which competitively bind to the endogenous miRNAs. The commonly used chemical modifications include cholesterol-conjugated 2′-*O*-methyl modification (so-called antagomiRs) and LNA modification (so-called LNA-antimiRs) [16, 51]. miRNA "sponges" contain complementary binding sites to miRNAs of interest and expressed in an expression vector. miRNA mimics can be chemically modified double-stranded miRNAs or miRNAs expressed by plasmid or viral vectors.

miRNA therapeutics for atherosclerosis are still in their infancy. However, as discussed in this chapter, a list of miRNAs has demonstrated promise as future therapeutic solutions for atherosclerosis. Inhibition of miR-33 in both mice and nonhuman primate was shown to increase plasma HDL levels by as much as 40%. Moreover, atheroprotection and even reversion of the atherosclerotic phenotype was observed by miR-33 silencing in mice. miR-122 silencing in monkeys resulted in 20–30% reduction in total plasma cholesterol without apparent toxicity. Currently, miR-122 antagonists are underway in clinical trials and are shown to be well tolerated in humans, offering promise for miRNA-based therapeutics for diseases including atherosclerosis (NCT00688012, NCT01200420, and NCT00979927). Besides these, miR-126, miR-143/miR-145, and miR-146 may protect atherogenesis through multiple mechanisms. These miRNAs should be prioritized as candidates for therapeutics in atherosclerosis.

Though promising, the field of miRNA therapeutics is not without challenges. The very thing that makes a miRNA regulator such an appealing pharmaceutical candidate, namely its ability to affect multiple genes, provides the greatest source of challenge: specificity. Each individual miRNA can have many gene targets. This is not a problem if the major targets of a miRNA are related and have a concerted effect, but becomes an issue of concern when a miRNA has both beneficial and adverse effects depending on cell types or biological processes (e.g., miR-126 and miR-21). Future work should be carefully designed to further define the role of the candidate miRNAs in different stages and processes of atherogenesis. Efficacy and safety profile of the miRNAs should be rigorously tested in atherosclerosis animal models before moving to human clinical trials.

CONCLUSION

The field of miRNA research in atherosclerosis has emerged in the last few years. miRNAs have been implicated in almost all processes involved in atherogenesis, from atherosclerotic lesion formation to plaque stabilization and rupture. These processes include lipid metabolism, endothelial dysfunction, immune response and inflammation, VSMC phenotypic switching, as well as neoangiogenesis. It is safe to say that a miRNA connection to atherosclerosis has been firmly established. Specifically, several miRNAs, including miR-33, miR-122, miR-126, miR-143/miR-145, and miR-146, promise to be potential therapeutic targets for atherosclerosis. We are cautiously optimistic about treating the multifactorial atherosclerosis disease by taking advantage of the multitargets and multimechanisms of miRNAs in the near future. Future work should focus on systematically identifying miRNAs involved in different aspects of atherosclerosis in animal models and human patients, with an emphasis on dissecting their functions. Since atherogenesis is an extreme long process in humans, establishment of a robust, specific and safe delivery system for miRNA agents is critical for atherosclerosis therapeutics in the future.

ACKNOWLEDGMENTS

SW was supported by a startup fund from Tulane University, President's Research Council, New Investigator Award from UT Southwestern Medical Center, NIH Grant EY021862, a career development award from the Research to Prevent Blindness foundation, and a Bright Focus Foundation Award in Age-related Macular Degeneration.

REFERENCES

1. Glass CK, Witztum JL. (2001) Atherosclerosis. The road ahead. *Cell* 104(4), 503–16.
2. Bartel DP. (2004) MicroRNAs: genomics, biogenesis, mechanism, and function. *Cell* 116(2), 281–97.
3. Lee RC, Feinbaum RL, Ambros V. (1993) The *C. elegans* heterochronic gene lin-4 encodes small RNAs with antisense complementarity to lin-14. *Cell* 75(5), 843–54.
4. Wang S, Koster K, He Y, Zhou Q. (2012) miRNAs as potential therapeutic targets for age-related macular degeneration. *Future Med Chem* 4(3) 277–87.
5. Rayner KJ, Esau CC, Hussain FN, et al. (2011) Inhibition of miR-33a/b in non-human primates raises plasma HDL and lowers VLDL triglycerides. *Nature* 478(7369), 404–7.
6. Rayner KJ, Sheedy FJ, Esau CC, et al. (2011) Antagonism of miR-33 in mice promotes reverse cholesterol transport and regression of atherosclerosis. *J Clin Invest* 121(7), 2921–31.
7. Raitoharju E, Lyytikainen LP, Levula M, et al. (2011) miR-21, miR-210, miR-34a, and miR-146a/b are up-regulated in human atherosclerotic plaques in the Tampere Vascular Study. *Atherosclerosis* 219(1), 211–7.
8. Li T, Cao H, Zhuang J, et al. (2011) Identification of miR-130a, miR-27b, and miR-210 as serum biomarkers for atherosclerosis obliterans. *Clin Chim Acta* 412(1–2), 66–70.
9. Dentelli P, Rosso A, Orso F, Olgasi C, Taverna D, Brizzi MF. (2010) MicroRNA-222 controls neovascularization by regulating signal transducer and activator of transcription 5A expression. *Arterioscler Thromb Vasc Biol* 30(8), 1562–8.
10. Bidzhekov K, Gan L, Denecke B, et al. (2012) microRNA expression signatures and parallels between monocyte subsets and atherosclerotic plaque in humans. *Thromb Haemost* 107(4), 619–25.
11. Marquart TJ, Allen RM, Ory DS, Baldan A. (2010) miR-33 links SREBP-2 induction to

repression of sterol transporters. *Proc Natl Acad Sci U S A* 107(27), 12228–32.

12. Najafi-Shoushtari SH, Kristo F, Li Y, et al. (2010) microRNA-33 and the SREBP host genes cooperate to control cholesterol homeostasis. *Science* 328(5985), 1566–9.
13. Rayner KJ, Suarez Y, Davalos A, et al. (2010) miR-33 contributes to the regulation of cholesterol homeostasis. *Science* 328(5985), 1570–3.
14. Elmen J, Lindow M, Silahtaroglu A, et al. (2008) Antagonism of microRNA-122 in mice by systemically administered LNA-antimiR leads to up-regulation of a large set of predicted target mRNAs in the liver. *Nucleic Acids Res* 36(4), 1153–62.
15. Esau C, Davis S, Murray SF, et al. (2006) miR-122 regulation of lipid metabolism revealed by *in vivo* antisense targeting. *Cell Metab* 3(2), 87–98.
16. Elmen J, Lindow M, Schutz S, et al. (2008) LNA-mediated microRNA silencing in non-human primates. *Nature* 452(7189), 896–9.
17. Qin X, Wang X, Wang Y, et al. (2010) microRNA-19a mediates the suppressive effect of laminar flow on cyclin D1 expression in human umbilical vein endothelial cells. *Proc Natl Acad Sci U S A* 107(7), 3240–4.
18. Wang KC, Garmire LX, Young A, et al. (2010) Role of microRNA-23b in flow-regulation of Rb phosphorylation and endothelial cell growth. *Proc Natl Acad Sci U S A* 107(7), 3234–9.
19. Fang Y, Shi C, Manduchi E, Civelek M, Davies PF. (2010) MicroRNA-10a regulation of proinflammatory phenotype in athero-susceptible endothelium *in vivo* and *in vitro*. *Proc Natl Acad Sci U S A* 107(30), 13450–5.
20. Wu W, Xiao H, Laguna-Fernandez A, et al. (2011) Flow-dependent regulation of Kruppel-like factor 2 is mediated by microRNA-92a. *Circulation* 124(5), 633–41.
21. Nicoli S, Standley C, Walker P, Hurlstone A, Fogarty KE, Lawson ND. (2010) MicroRNA-mediated integration of haemodynamics and Vegf signalling during angiogenesis. *Nature* 464(7292), 1196–200.
22. Harris TA, Yamakuchi M, Ferlito M, Mendell JT, Lowenstein CJ. (2008) MicroRNA-126 regulates endothelial expression of vascular cell adhesion molecule 1. *Proc Natl Acad Sci U S A* 105(5), 1516–21.
23. Zhou J, Wang KC, Wu W, et al. (2011) Micro RNA-21 targets peroxisome proliferators-activated receptor-alpha in an autoregulatory loop to modulate flow-induced endothelial inflammation. *Proc Natl Acad Sci U S A* 108(25), 10355–60.
24. Weber M, Baker MB, Moore JP, Searles CD. (2010) miR-21 is induced in endothelial cells by shear stress and modulates apoptosis and eNOS activity. *Biochem Biophys Res Commun* 393(4), 643–8.
25. Fleissner F, Jazbutyte V, Fiedler J, et al. (2010) Short communication: asymmetric dimethylarginine impairs angiogenic progenitor cell function in patients with coronary artery disease through a microRNA-21-dependent mechanism. *Circ Res* 107(1), 138–43.
26. Zernecke A, Bidzhekov K, Noels H, et al. (2009) Delivery of microRNA-126 by apoptotic bodies induces CXCL12-dependent vascular protection. *Sci Signal* 2(100), ra81.
27. Hergenreider E, Heydt S, Treguer K, et al. (2012) Atheroprotective communication between endothelial cells and smooth muscle cells through miRNAs. *Nat Cell Biol* 14(3), 249–56.
28. Chen T, Huang Z, Wang L, et al. (2009) microRNA-125a-5p partly regulates the inflammatory response, lipid uptake, and ORP9 expression in oxLDL-stimulated monocyte/macrophages. *Cardiovasc Res* 83(1), 131–9.
29. Huang RS, Hu GQ, Lin B, Lin ZY, Sun CC. (2010) microRNA-155 silencing enhances inflammatory response and lipid uptake in oxidized low-density lipoprotein-stimulated human THP-1 macrophages. *J Investig Med* 58(8), 961–7.
30. Yang K, He YS, Wang XQ, et al. (2011) MiR-146a inhibits oxidized low-density lipoprotein-induced lipid accumulation and inflammatory response via targeting toll-like receptor 4. *FEBS Lett* 585(6), 854–60.
31. Kurowska-Stolarska M, Alivernini S, Ballantine LE, et al. (2011) microRNA-155 as a proinflammatory regulator in clinical and experimental arthritis. *Proc Natl Acad Sci U S A* 108(27), 11193–8.
32. Taganov KD, Boldin MP, Chang KJ, Baltimore D. (2006) NF-kappaB-dependent induction of microRNA miR-146, an inhibitor targeted to signaling proteins of innate immune responses. *Proc Natl Acad Sci U S A* 103(33), 12481–6.
33. Bhaumik D, Scott GK, Schokrpur S, et al. (2009) microRNAs miR-146a/b negatively modulate the senescence-associated inflammatory mediators IL-6 and IL-8. *Aging (Albany NY)* 1(4), 402–11.

34. Torella D, Iaconetti C, Catalucci D, et al. (2011) microRNA-133 controls vascular smooth muscle cell phenotypic switch *in vitro* and vascular remodeling *in vivo*. *Circ Res* 109(8), 880–93.
35. Leeper NJ, Raiesdana A, Kojima Y, et al. (2011) microRNA-26a is a novel regulator of vascular smooth muscle cell function. *J Cell Physiol* 226(4), 1035–43.
36. Ji R, Cheng Y, Yue J, et al. (2007) MicroRNA expression signature and antisense-mediated depletion reveal an essential role of microRNA in vascular neointimal lesion formation. *Circ Res* 100(11), 1579–88.
37. Elia L, Quintavalle M, Zhang J, et al. (2009) The knockout of miR-143 and -145 alters smooth muscle cell maintenance and vascular homeostasis in mice: correlates with human disease. *Cell Death Differ* 16(12), 1590–8.
38. Cheng Y, Liu X, Yang J, et al. (2009) microRNA-145, a novel smooth muscle cell phenotypic marker and modulator, controls vascular neointimal lesion formation. *Circ Res* 105(2), 158–66.
39. Liu X, Cheng Y, Zhang S, Lin Y, Yang J, Zhang C. (2009) A necessary role of miR-221 and miR-222 in vascular smooth muscle cell proliferation and neointimal hyperplasia. *Circ Res* 104(4), 476–87.
40. Davis BN, Hilyard AC, Nguyen PH, Lagna G, Hata A. (2009) Induction of microRNA-221 by platelet-derived growth factor signaling is critical for modulation of vascular smooth muscle phenotype. *J Biol Chem* 284(6), 3728–38.
41. Boettger T, Beetz N, Kostin S, et al. (2009) Acquisition of the contractile phenotype by murine arterial smooth muscle cells depends on the Mir143/145 gene cluster. *J Clin Invest* 119(9), 2634–47.
42. Xin M, Small EM, Sutherland LB, et al. (2009) microRNAs miR-143 and miR-145 modulate cytoskeletal dynamics and responsiveness of smooth muscle cells to injury. *Genes Dev* 23(18), 2166–78.
43. Chen KC, Wang YS, Hu CY, et al. (2011) OxLDL up-regulates microRNA-29b, leading to epigenetic modifications of MMP-2/MMP-9 genes: a novel mechanism for cardiovascular diseases. *FASEB J* 25(5), 1718–28.
44. Jain RK, Finn AV, Kolodgie FD, Gold HK, Virmani R. (2007) Antiangiogenic therapy for normalization of atherosclerotic plaque vasculature: a potential strategy for plaque stabilization. *Nat Clin Pract Cardiovasc Med* 4(9), 491–502.
45. Wang S, Olson EN. (2009) AngiomiRs—key regulators of angiogenesis. *Curr Opin Genet Dev* 19(3), 205–11.
46. Bonauer A, Carmona G, Iwasaki M, et al. (2009) microRNA-92a controls angiogenesis and functional recovery of ischemic tissues in mice. *Science* 324(5935), 1710–3.
47. Wang S, Aurora AB, Johnson BA, et al. (2008) The endothelial-specific microRNA miR-126 governs vascular integrity and angiogenesis. *Dev Cell* 15(2), 261–71.
48. Fish JE, Santoro MM, Morton SU, et al. (2008) miR-126 regulates angiogenic signaling and vascular integrity. *Dev Cell* 15(2), 272–84.
49. Grundmann S, Hans FP, Kinniry S, et al. (2011) microRNA-100 regulates neovascularization by suppression of mammalian target of rapamycin in endothelial and vascular smooth muscle cells. *Circulation* 123(9), 999–1009.
50. Menghini R, Casagrande V, Cardellini M, et al. (2009) microRNA 217 modulates endothelial cell senescence via silent information regulator 1. *Circulation* 120(15), 1524–32.
51. Krutzfeldt J, Rajewsky N, Braich R, et al. (2005) Silencing of microRNAs *in vivo* with 'antagomirs'. *Nature* 438(7068), 685–9.

34

POSTTRANSLATIONAL MODIFICATIONS: ROLE IN ATHEROSCLEROSIS

Justyna Fert-Bober[1,2], Ronald J. Holewinski[1,2], Joel A. Kooren[1,2], and Jennifer E. Van Eyk[1,2,3]

[1] *Advanced Clinical Biosystems Research Institute, Heart Institute and Department of Medicine, Cedars-Sinai Medical Center, Los Angeles, CA, USA*

[2] *Department of Medicine, Johns Hopkins University, Baltimore, MD, USA*

[3] *Departments of Biomedical Engineering and Biological Chemistry, JHU-Bayview Proteomics Center, Johns Hopkins University, Baltimore, MD, USA*

INTRODUCTION, BACKGROUND, AND SIGNIFICANCE

The prevalence and mortality associated with atherosclerosis have motivated a great deal of research [1, 2]. Despite the advances in understanding of the pathophysiology of atherosclerosis, it remains as a leading cause of heart disease. Currently, proteomic investigations aim to maximize the number of biological agents of molecular mechanisms that underlie cardiac physiology and pathology. This requires not only analyzing the proteomes of cardiovascular diseases but the posttranslational modifications (PTMs) associated with these proteins as well. Knowledge of protein PTM sites is essential to understand phenotype as they can alter protein activity, cellular localization, turnover, and interaction with other macromolecules, which provides insights into the possible regulation of cellular physiology. However, the identification and quantification of specific PTMs is challenging, especially from crowded cellular environments. It has become clear that a protein can have a variety of PTMs at multiple amino acid residues. Furthermore, different PTMs may compete with each other for a particular residue [3]. Recent developments of PTM-specific enrichment techniques combined with advanced mass spectrometry (MS) methods have provided great insight into the global identification of multiple PTMs. Today, more than 400 different PTMs are recognized and described, although far fewer have been documented in higher organisms such as mammals [4]; and atherosclerosis is a good example of multifactorial disease that involves many PTMs of proteins, that is, oxidation, glycosylation, and phosphorylation [5–7]. The aim of this review is to give an overview of protein phosphorylation,

Justyna Fert-Bober and Ronald J. Holewinski are coauthors.

Atherosclerosis: Risks, Mechanisms, and Therapies, First Edition. Edited by Hong Wang and Cam Patterson.

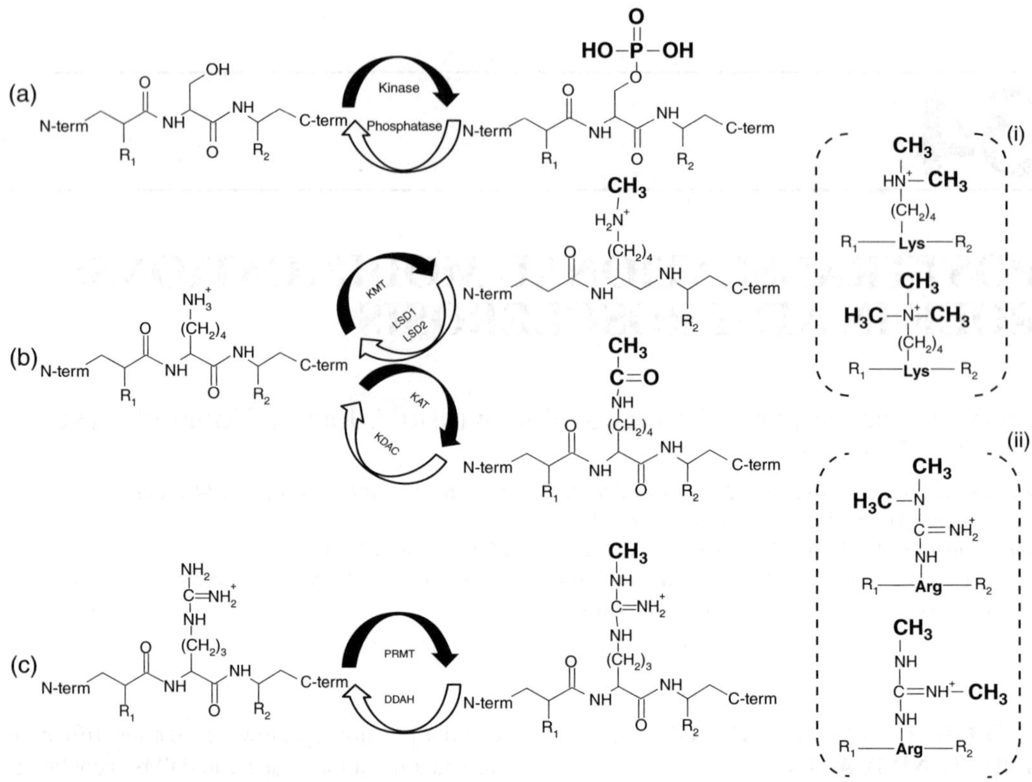

Figure 34.1 Schematic representations of enzyme-catalyzed posttranslational modifications (PTMs) of (a) Ser phosphorylation, (b) Lys methylation (top) and acetylation (bottom), and (c) Arg methylation. The enzymes with filled arrows represent the enzymes responsible for the addition of the modification and the ones with open arrows represent the enzymes responsible for the removal of the modification. (i) Dimethyl (top) and trimethyl (bottom) lysine and (ii) asymmetrical (top) and symmetrical (bottom) dimethyl arginine.

methylation, and acetylation and their role in atherosclerosis. We will take inventory of the methods available for the measurement of PTM and assess the potential contribution of new technologies for their clinical use.

Serine, Threonine, and Tyrosine Phosphorylation

The most studied modification is protein phosphorylation which is known to be involved in the regulation of many biological processes including cell growth, cell communication, proliferation, differentiation, transcriptional and translational regulation, homeostasis, and cell survival [8]. Abnormal alterations in protein phosphorylation status within a cell can lead to many adverse effects and has been shown to play a role in many disease settings such as cancer [9], neurodegenerative disease [10], cardiovascular disease [11], and atherosclerosis [6].

Protein phosphorylation is regulated by protein kinases, which transfer a phosphate group from a high-energy donor molecule, such as ATP, to the protein substrate, and protein phosphatases, which dephosphorylate the amino acid residue (Figure 34.1). Kinases are often grouped into three classes according to the amino acids they phosphorylate: serine (Ser) protein kinases, threonine (Thr) protein kinases, and tyrosine (Tyr) protein kinases; these are further divided into transmembrane receptor-linked kinases and cytoplasmic kinases and dual specificity protein kinases, which phosphorylate Ser, Thr, and Tyr (Table 34.1). Protein

TABLE 34.1 Examples of enzymes responsible for the addition and removal of protein phosphorylation, acetylation, and methylation

		Enzyme	Substrate	Example
Addition of modification	Phosphorylation	Serine (Ser)/threonine (Thr) kinases	Ser, Thr	Protein kinases A, C, G (PKA, PKC, PKG) Calcium/calmodulin-dependent kinases (CamKs) Mitogen-activated protein kinases (MAPKs)
		Tyrosine (Tyr) kinases	Tyr	Janus kinases (JAK) Epidermal growth factor receptor (EGFR) Platelet-derived growth factor receptor (PDGFR)
		Dual specificity kinases	Ser, Thr, Tyr	MAPK kinases (MEKs) Extracellular signal-regulating kinase (ERKs) Dual specificity protein kinase (TTK)
	Methylation and acetylation	Lysine (Lys) methyltransferases	Lys	N-Lys methyltransferase (SMYD2)
		Arginine (Arg) methyltransferases	Arg	Protein arginine methyltransferases (PRMTs) PRMT1,3,4,6,8—asymmetrical dimethyl addition PRMT5,7,9—symmetrical dimethyl addition
		Lys acetyltransferases	Lys	Histone acetyltransferase (KAT2A) Nuclear receptor coactivator 1 (NCOA1) General transcription factor 3C polypeptide 4 (GTF3C4)
Removal of modification	Phosphorylation	Ser/Thr phosphatases	Ser, Thr	Ser/Thr protein phosphatases (PP1, PP2A, PP2B, PP2C)
		Tyr phosphatases	Tyr	Tyr protein phosphatase nonreceptor type Receptor-type Tyr protein phosphatases
		Dual specificity phosphatases	Ser, Thr, Tyr	M-phase inducer phosphatases (CDC25A,B,C) Dual specificity phosphatases 1-28
	Methylation and acetylation	Lys demethylases	Lys	Lys-specific histone demethylases JmjC domain-containing proteins
		Arg demethylases	Arg	Protein Arg demethylase(s) unknown; dimethylarginine dimethylaminohydrolase (DDAH) has activity on free methyl Arg amino acids → citrulline
		Lys deacetylases	Lys	Class I—HDACs 1,2,3,8 Class II—HDACs 4,5,6,7,9,10 Class III—sirtuin protein family (NAD^+ dependent)

phosphatases are a group of enzymes that cleave the phosphoester bond to form free phosphate and a free alcohol on the side chain of the amino acid modified (Figure 34.1). Just as with protein kinases, protein phosphatases can be grouped based on their substrate specificity, with phosphatases having specificity for Ser/Thr, Tyr, or Ser/Thr/Tyr (dual specificity phosphatases). However, the number of phosphatases within the genome is much lower than kinases [12]. There is also considerable structural diversity between the phosphatase catalytic domains of each group, in addition to a variety of regulatory or targeting subunits and/or domains that determine the subcellular localization and substrates available to the catalytic domain.

Arginine Methylation and Lysine Methylation/Acetylation

Arginine (Arg) and lysine (Lys) modifications are emerging from the shadow of protein phosphorylation, largely due to the identification of the family of enzymes with methylation/acetylation activity and the development of new tools that allows for their detection in complex cellular protein mixtures. Arg and Lys are positively charged amino acid residues known to mediate hydrogen bonding and amino–aromatic interactions. The nitrogen of those amino acids within polypeptides can be posttranslationally modified, receiving a methyl groups on Arg/Lys or an acetyl group on Lys residues (Figure 34.1). The phosphorylation, methylation, and acetylation of proteins play an important role in a wide range of cardiovascular diseases [13–15].

Methylation usually occurs at Lys and Arg but has also been found on histidine, glutamic acid, and the carboxy terminus of proteins [16]. Arg can be methylated once (monomethylated Arg) or twice, with either both methyl groups on one terminal nitrogen (asymmetric dimethylated Arg) or one or both nitrogens (symmetric dimethylated Arg), by protein arginine methyltransferases (PRMTs). Currently, 11 mammalian PRMTs have been identified (Table 34.1). All members are able to monomethylate Arg, but they are grouped into two classes according to their type of dimethylation: type I (PRMT1, 3, 4, 6, and 8) asymmetrically dimethylate Arg and type II (PRMT5, 7, and 9) symmetrically dimethylate Arg [17]. To date, one enzyme, bifunctional Arg demethylase/lysyl-hydroxylase (JMJD6), was identified as a possible Arg demethylase [18], but further research has shown that it lacks such activity [19, 20] (Table 34.1). It was initially thought that peptidylarginine deiminase (PAD) enzymes could convert methylated arinines to citrulline [21]. However, it now appears that PAD enzymes lack demethylase activity [22]. However, dimethylarginine dimethylaminohydrolase (DDAH) catalyzes the conversion of N^G-monomethyl-L-arginine (L-NMMA) and asymmetrical N^G,N^G-dimethyl-L-arginine (ADMA) but not symmetrical N^G,N^G-dimethyl-L-arginine (SDMA) to citrulline and either methyl- or dimethylamine [23, 24]. DDAH converts free methylated Arg that arise from proteolytic digestion rather than methylated Arg residues in proteins.

Methylation of Lys involves the addition of one to three methyl groups on the amino acid's ε-amine group to form mono-, di- or trimethyllysine (Figure 34.1). This process is catalyzed by a group of Lys methyltransferases (KMT). These enzymes have been subgrouped into seven main families and named according to their founding member: SUV39, SET1, SET2, EZ, RIZ, SMYD, and SUV4-20 [25]. Although Lys methylation has long been considered to be an irreversible modification, the recent discovery of two groups of Lys demethylases has changed that perspective [26]. The first group of amine oxidase domain-containing enzymes is represented by LSD1 (also known as AOF2) and LSD2 (also known as AOF1). The second group of KMTs is represented by the Jumonji C (jmjC) domain-containing proteins; the members of this group are Fe(II)- and 2-oxoglutarate (2OG)-dependent oxygenases [21, 27, 28].

Acetylation refers to the addition of an acetyl group (CH_3CO) from acetyl-CoA to organic compounds. Lys acetylation is a reversible PTM that is tightly regulated by Lys acetyltransferases

(KATs) [29] and Lys deacetylases (KDACs) [30] in both prokaryotic and eukaryotic cells. Currently, there are three classes of human deacetylases: class I histone deacetylases (HDACs) are homologous to the yeast enzyme Rpd3 and include HDAC1, 2, 3, and 8; class II HDACs are homologous to yeast HDAC1 and are divided into class IIa (HDAC4, 5, 7, and 9) and class IIb (HDAC6 and 10) based on structure. The human class III HDACs includes the sirtuin (SIRT) family of NAD^+-dependent protein deacetylases. Hence, there is considerable diversity in the enzymes that catalyze these Arg/Lys modifications and these modifications may play a broad role in diverse cellular processes.

ROLE OF PHOSPHORYLATION, METHYLATION, AND ACETYLATION IN ATHEROSCLEROSIS AND VASCULAR DISEASE

Atherosclerosis is the result of the interaction of multiple diverse factors such as lipid metabolism, blood coagulation, hemodynamic stress, as well as PTMs, which alter the structural and biological properties of proteins. This terminology refers to a wide range of chemical, enzymatic, and nonenzymatic reactions that occur after the enzymatic processes of protein maturation and leads to progressive alteration of physiological functions at various levels [6, 9, 11, 13–15]. Here, we will examine some examples of phosphorylation, methylation, and acetylation modification in the context of atherosclerosis and vascular disease.

Protein Phosphorylation Regulation in Atherosclerosis

One of the key changes that occur in the cell after stress is ATP depletion and a coincident rise in the ratio of AMP/ATP. The enzyme AMP-activated protein kinase (AMPK), a highly conserved, multiple substrate Ser/Thr kinase involved in the regulation of cellular and whole organ metabolism, is able to make changes in energy demand and through signaling cascades can mediate the production of ATP. When ATP generation is impaired or cellular demand increased, AMPK is activated by phosphorylation at Thr172 by upstream AMPK kinases (LBK1, CAMKKβ, and TAK1) [31], leading to the activation of ATP-generating catabolic pathways and downregulation of ATP-consuming anabolic pathways. AMPK has been widely studied in the context of atherosclerosis and vascular disease and has been implicated primarily to have vasoprotective and antiatherosclerotic effects. Recent studies have reported that AMPK-activating agents, such as AICAR, metformin, and adiponectin, induce endothelium-dependent relaxation [6]. Studies have also shown that in addition to inducing vasorelaxation, AMPK activation can be a potent inhibitor of smooth muscle cell proliferation and can reduce inflammatory cell adhesion and migration, an essential step in the progression of atherosclerotic lesions [6]. The effect of AMPK activation on cell proliferation has also been shown to be mediated through the inhibition of the mTORC1 pathway, through the phosphorylation of tuberin (TSC2) by AMPK, resulting in reduction of protein synthesis and cell cycle arrest [31]. The beneficial effects that are observed through AMPK activation, either through vasorelaxation or inhibition of vascular smooth muscle cell (VSMC) proliferation, make this enzyme a potential therapeutic target for atherosclerosis and other vascular diseases. For example, a recent study by Han and coworkers showed that berberine, a drug that prevents and suppresses proinflammatory cytokines, causes a dose- and time-dependent activation of AMPK in cultured endothelial cells [32], likely via AMPK-dependent phosphorylation of endothelial nitric oxide synthase (eNOS) and modulation of nitric oxide (NO) production [32, 33].

Interestingly, AMPK activation has also been linked to alterations in gene expression through the phosphorylation and modulation of histone acetylases, which also play a role in regulating the endothelial proinflammatory response (discussed in following section). A specific example of this is the acetylation of

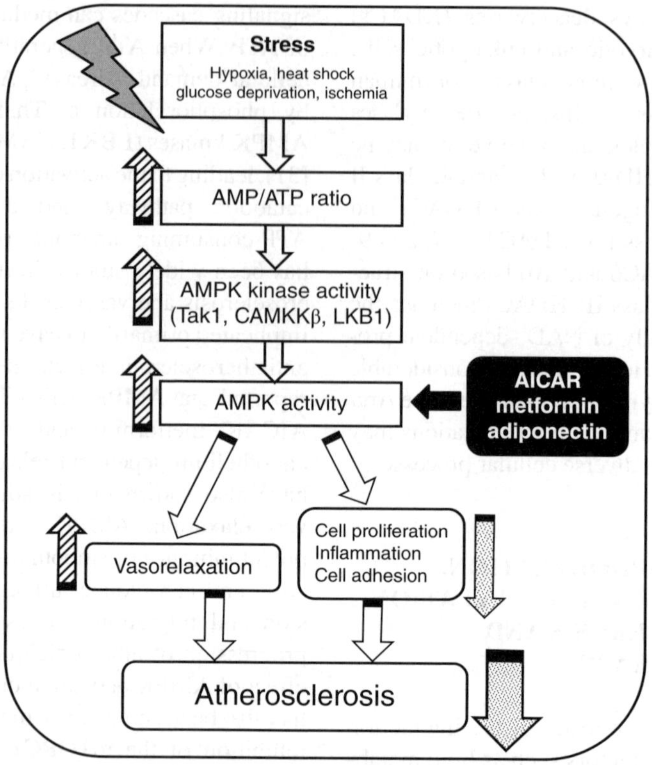

Figure 34.2 General schematic for mechanism of AMPK vasoprotection. AMPK exerts its vasoprotective effects through multiple mechanisms and pathways, which were excluded from the figure for simplicity. Details of these mechanisms can be found in the references cited in the text.

NF-κB, a key transcription factor involved in regulating the expression of proinflammatory mediators, such as VCAM-1, through TNF-α signaling, thereby playing a critical role in mediating inflammatory responses. Zhang and coworkers showed that acetylation of NF-κB p65 by the HAT p300 constitutes a regulatory mechanism for TNF-α-dependent VCAM-1 transcription and monocyte adhesion to endothelial cells (EC) [34]. However, pharmacological or genetic activation of AMPK was shown to inhibit the endothelial proinflammatory response observed in this setting. They concluded that the anti-inflammatory effects imposed by AMPK activation were the result of AMPK phosphorylation of p300 at Ser79, which resulted in p300 inactivation and subsequently decreased acetylation of NF-κB and inhibition of TNF-α inflammatory signaling pathways [34]. The AMPK/p300/NF-κB cascade is just one example of the cross talk that takes place between different PTMs within the cell and highlights how aberrations in one PTM can have downstream effects on other PTMs in biological pathways (Figure 34.2 outlines a general scheme for how AMPK exerts some of its protective effects).

Another example of how phosphorylation facilitates a protective role came from Fuster and coworkers, who investigated the phosphorylation of growth suppressor $p27^{Kip1}$. $p27^{Kip1}$ is a protein that binds to and inhibits cyclin-dependent kinases, preventing cell proliferation, and $p27^{Kip1}$ was previously reported to have an atheroprotective function in diet-induced atherosclerotic mice [35]. They also showed that deficient phosphorylation of the growth suppressor $p27^{Kip1}$ at Ser10 is a hallmark of human and mouse atherosclerosis and promotes disease progression in mice, presumably by limiting a proliferation-independent atheroprotective function of $p27^{Kip1}$ in macrophages [35].

Mitogen-activated protein kinase (MAPK) signaling cascades have recently been reported to have a role in atherosclerotic lesion formations, primarily involved in foam cell formation. Treatment of cultured macrophages with oxidized low-density lipoprotein (oxLDL) results in rapid activation of the MAPKs c-Jun N-terminal kinase (JNK), p38α MAPK, and extracellular signal-regulating kinase (ERK). However, it appears that foam cell formation is dependent upon activation of JNK2 and perhaps p38α MAPK, as treatment with ERK inhibitors does not block foam cell formation, whereas treatment with JNK or p38α MAPK inhibitors does block foam cell formation [36]. The exact mechanism by which JNK and p38α MAPK facilitate foam cell formation remains unclear, but it is possible that JNK or p38α MAPK activity regulates the expression or activity of proteins required for uptake of oxLDL prior to oxLDL binding to cell-surface receptors [36].

Wong et al. also showed an involvement of MAPKs in vascular inflammation, through the activation of angiotensin II (AngII)-induced cyclooxygenase-2 (COX2) expression, which in turn is related to generation of proinflammatory cytokines [37]. The AngII-induced COX2 expression appears to be dependent upon p38α MAPK and ERK activity, as inhibitors of these kinases independently reduced AngII-induced COX2 expression and when administered simultaneously produced an additive effect, suggesting a parallel involvement of p38α MAPK and ERK in mediating AngII-induced COX2 expression. They also pinpointed a novel role of protein kinase C delta (PKC_δ) in the AngII-induced COX2 expression, as PKC_δ inhibitors prevented activation of ERK, suggesting PKC_δ activation is upstream of ERK in COX2 induction. PKC_δ inhibitors did not have an effect on p38α MAPK activity, and other isoforms of PKC showed a little involvement in the AngII-induced COX2 expression [37].

Furthermore, protein phosphorylation can be used as a potential biomarker. An interesting set of studies by Martin-Ventura on heat shock protein 27 (Hsp27) revealed two Hsp27 isoforms corresponding to the nonphosphorylated (isoform 2) and the monophosphorylated state of Hsp27 at Ser82 (isoform 1) that have been shown to be a potential marker of atherosclerosis [38]. Despite these few examples, there are many more that implicate protein phosphorylation as an essential component of vascular disease pathology. However, to date, no large-scale phosphorylation proteomic screening has been carried out in the area of vascular smooth muscle or atherosclerosis.

Protein Methylation and Acetylation Regulation in Atherosclerosis

The methylation of proteins has been established to influence the activity of nuclear and nucleic acid binding proteins. However, less is known about the importance of protein methylation in the regulation of cytosolic proteins biological function and their role in cardiovascular disease. Indeed, protein methylation has been associated with increased levels of reactive oxygen species mainly through the ability of ADMA to inhibit nitric oxide synthase (NOS), resulting in decreased NO production [15] (Figure 34.3). Specifically, eNOS is important to help maintain the vasculature in the relaxed state, inhibit the adhesion of platelets, and suppress smooth muscle cell proliferation [39]. Studies have shown that decreased NO production arising from ADMA inhibition of NOS leads to hypertension and cardiac dysfunction [40]. In addition, it has been suggested that ADMA can be used as a risk marker as it has been shown that plasma ADMA levels are elevated in patients with systemic atherosclerosis, hypertension, renal failure, as well as other diseases [41]. Moreover, other clinical studies have suggested that ADMA may also play a role as a novel cardiovascular risk factor. Zoccali demonstrated that in patients undergoing hemodialysis, elevated ADMA is associated with increased risk of

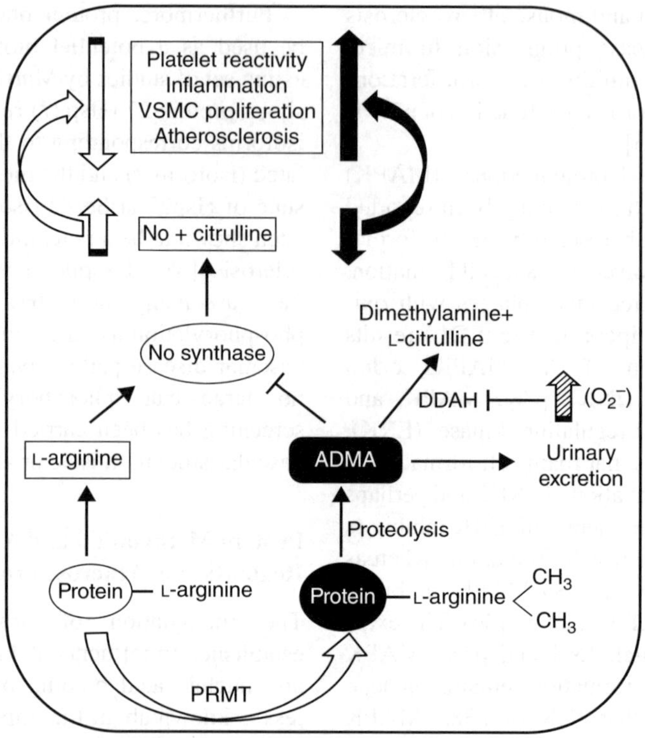

Figure 34.3 Schematic for the metabolism of ADMA and L-arginine resulting from proteolytic digestion of a protein with and without ADMA modification, respectively. ADMA is cleared through urinary secretion or enzymatic conversion of ADMA by DDAH to dimethylamine and citrulline. Inhibition of DDAH by superoxide anion, resulting from oxidative stress, leads to ADMA accumulation which inhibits NO synthase (by competing with L-arginine for active site) and decreased NO production.

future cardiovascular events and mortality [42]. Valkonen's group showed that elevated ADMA was associated with increased risk for acute coronary events in clinically healthy men [43]. By contrast, DDAH overexpression can significantly reduce ADMA concentration with effect on endothelial function [44]. Furthermore, several examples of Lys methylation of histones and nonhistone proteins have been discovered. In the case of the Lys methyltransferases, such as Set7/9, direct methylation of the farnesoid X receptor regulates the expression of its target genes, including p53, TAF10, p300/CBP-associated factor, estrogen receptor, androgen receptor, and the RelA subunit of NF-κB that are associated with atherosclerosis or can determine atherosclerosis risk [45].

Lys acetylation also plays a role in the pathogenesis of vascular disease, primarily through the acetylation and deacetylation of nuclear proteins, in particular histones, which have an effect on transcription and translation. Specifically, Kawahara et al. observed hypernuclear acetylation (HNA) in nuclei of VSMCs from atherosclerotic coronary lesions but not in normal coronary arteries, suggesting a role for HNA in the pathogenesis of this disease [14]. HNA in atherosclerosis is based on thrombin-activated transcription factors. Thrombin activates signal-dependent transcription factors, such as Ets-1, NF-κB, and AP-1, which interact with CREB- binding protein, which is a transcriptional coactivator that has intrinsic histone acetyltransferase activity. Furthermore, Findeisen et al. reported that pharmacological inhibition of HDACs using Scriptaid prevents VSMC proliferation and reduces neointima formation following vascular injury [13]. Interestingly, acetylation and deacetylation also target a variety of substrates outside of the nucleus (e.g., mitochondria) which have been strongly linked with aging and age-related diseases. Indeed, SIRTs have been shown to regulate the acetylation of mitochondrial proteins [46] in addition to the regulation of nuclear proteins such as the nuclear receptor LXT and transcriptional factor FOXO1 [47]. Most Lys-acetylated proteins identified in mitochondrial fractions were metabolic enzymes, for example, the antioxidant enzyme superoxide dismutase (SOD2) [48]. SIRT1 has also been shown to be an antiatherosclerosis factor by preventing the oxLDL-induced apoptosis, upregulating eNOS expression, and improving endothelium relaxation function [49]. Furthermore, hypoxia-induced redox stress stimulates SIRT1 activity, which in turn deacetylates and activates hypoxia-inducible factors, like vascular endothelial growth factor (VEGF), which promotes atherosclerosis [50]. These data suggest that SIRT may be a potent target for atherosclerosis prevention and treatment. As with phosphorylation, no large-scale proteomic study on these specific PTMs has been done in area of vascular smooth muscle or atherosclerosis.

PROTEOMIC METHODS: SEEING MOLECULAR DETAILS

Given the prevalence and diversity of protein PTMs, robust technical methods are needed for the analysis/study of modified proteins. Proteomics is capable of PTM identification and enabling site mapping and quantitation; however, while useful, these methods suffer from various shortcomings. First, highly sensitive methods are required for detection due to a low stoichiometry. Furthermore, since the covalent bond between the PTM and amino acid side chain is typically labile, it is difficult to maintain the peptide in its modified state during sample preparation. Also, PTMs are frequently transient in dynamic homeostasis. In this section, we review the utility and limitations of commonly used detection methods for the analysis of protein phosphorylation, methylation, and acetylation.

Detection of PTMs by Non-MS-Based Methods

In order to separate the various PTM forms of the same protein, two-dimensional electrophoresis (2DE) has been widely used. This method allows separation of different modified forms of the same protein into distinct spots on the gel, based on their charge and molecular weight [4]. For example, phosphorylation of a protein gives a series of spots with the same molecular weight but different pI values. Additionally, 2D fluorescence difference gel electrophoresis or separation of the proteins under nondenaturing conditions provides additional information regarding protein–protein interactions and the composition of protein complexes [51, 52]. Other classical methods of directly measuring protein PTMs involve specific antibody development [53], radioactive isotope-labeled substrates [54], Western blot analysis, and more recently diverse chemical tags peptide, where the detection tag can be either a fluorescent dye for visualization or an affinity tag for protein enrichment and MS identification [55–57]. Another valid approach is based on the specificity of PTM-specific enzymes, where by using peptide or protein arrays it's possible to identify sequence motifs for different enzymes and therefore identified new targets [58]. Nevertheless, PTM substrate candidates identified by any of these approaches require further validation by MS analysis of the purified endogenous proteins. In summary, despite technical advances in the past few decades, more efficient and sensitive bioanalytical technologies are needed to address key bottlenecks in the identification of PTM substrate proteins, in mapping PTM sites, and to investigate *in vivo* PTM dynamics.

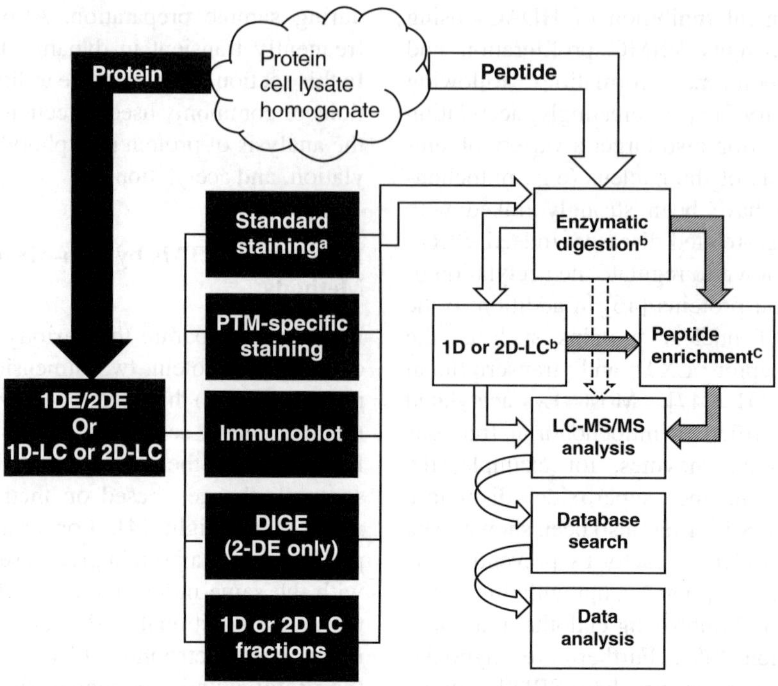

Figure 34.4 General schematic for proteomics workflow and analysis by gel-based methods (black boxes) or MS-based methods (white boxes). Proteins can be excised out of gels or membranes and enzymatically digested for MS analysis. [a]Examples of standard protein stains are Coomassie blue and silver stain. [b]Enzymatic digestion can be carried out in solution (directly from the protein, lysate, or homogenate) or after gel analysis or LC fractionation. Peptides can then be analyzed directly by LC–MS/MS or prefractionated using 1D or 2D LC before analysis. [c]For MS analysis of PTMs, peptide enrichment is routinely carried after digestion or peptide fractionation by 1D or 2D LC (gray arrows).

Direct Detection of PTMs by MS-Based Methods

MS has become the main analytical tool used for measuring the molecular mass of a sample and provides a powerful means of protein/peptide identification. MS can be used to characterize intact, nondigested proteins, with or without prior separation, and offers tools to interrogate proteins/peptides for PTMs. Most MS proteomics experiments employ a "bottom-up" approach, in which proteins are first digested with a protease and the corresponding peptides are analyzed by MS, with or without prior fractionation or PTM enrichment (Figure 34.4). Peptide sequences are determined by the parent ion (intact peptide) and fragment ion (fragments of the peptide) masses. The proteins are then inferred from the peptides that are detected. PTM-specific increase in the mass of a peptide and amino acid residue in mass spectra reveal the presence of PTMs. In some cases, the characteristic fragmentation properties of peptides containing PTMs, such as neutral losses or specific reporter masses, can be exploited (Table 34.2). As mentioned earlier, peptide mixtures derived from a complex protein sample are challenging to comprehensively analyze especially if PTMs are the focus. A major issue for the analysis of PTM proteins is the often substoichiometric nature of modified peptides compared to the unmodified form, which requires a selective enrichment step after digestion. Currently, three enrichment methods are commonly adopted for enhancing the identification of PTM peptides: (i) antibody-based affinity,

TABLE 34.2 Mass changes due to phosphorylation, methylation, and acetylation peptide modifications

Modification	Mass increase per instance*	Neutral loss mass*	Signature reporter ions*
Phosphorylation (Ser, Thr, Tyr)	80 Da	Phosphoric acid (H_3PO_4) 98 Da (Ser, Thr)† Metaphosphoric acid (HPO_3) 80 Da (Tyr)†	Immonium ion *m/z* 216 (Tyr)
Methylation (Lys)	14 Da (monomethyl) 28 Da (dimethyl) 42 Da (trimethyl)‡	Trimethylamine (NC_3H_9) 59 Da (trimethyl)	Immonium ion *m/z* 98 (monomethyl)
Methylation (Arg)	14 Da (monomethyl) 28 Da (dimethyl)	Monoethylamine (NH_2CH_3) 31 Da (MMA, SDMA) Dimethylamine (NHC_2H_6) 45 Da (ADMA) Dimethylcarbodiimidium 70 Da (ADMA, SDMA)	Immonium ions *m/z* 70, 112, 115 (MMA, DMA) Dimethylammonia *m/z* 46 Da (ADMA) Dimethylcarbodiimidium *m/z* 71 (ADMA, SDMA) Dimethylated guanidinium *m/z* 88 (ADMA, SDMA)
Acetylation (Lys)	42 Da‡	N/A	Immonium ion *m/z* 126§

*All masses reported in table are rounded to the nearest integer; exact monoisotopic masses (up to 3–4 decimal places) are used experimentally. Additionally, the masses indicated in the first 2 columns correspond to the total mass of chemical group, whereas the mass to charge ratio (*m/z*) is given for the signature ions.
†The loss of 98 Da from phosphopeptides upon fragmentation is due to the loss of phosphoric acid (H_3PO_4), most likely proceeding through a beta elimination of water and the phosphate group. However, a partial neutral loss of 80 Da (HPO_3) is more commonly observed with Tyr residues, whose phosphate group is much more stable than that of Ser and Thr.
‡Trimethyl Lys has an exact mass increase of 42.047 Da and Lys acetylation has an exact mass increase of 42.011 Da, which can be distinguished with high accuracy MS instruments.
§The *m/z* 126 immonium ion is characteristic of Lys acetylation and is another way to distinguish between trimethyl Lys and acetylated Lys, as trimethyl Lys does not produce the *m/z* 126 immonium ion.

(ii) ionic interaction-based, and (iii) chemical derivatization; however, the last two methods are used primarily for phosphopeptide enrichment. Isolation of PTM peptides from tryptic digests by immunoaffinity purification using a PTM-specific antibody is well recognized and has been successfully used for global analysis of protein Lys acetylation [46], Lys/Arg methylation [59], and Tyr phosphorylation [60]. In the second method, the unique ionic characteristic of the PTM groups are recognized. For example, phosphopeptides can be captured selectively through their negatively charged phospho group using immobilized metal affinity chromatography (IMAC) and titanium dioxide (TiO_2) chromatography [61]. Interestingly, the combined use of IMAC and TiO_2 (called SIMAC) seems able to separate monophosphorylated peptides from multiphosphorylated counterparts, thereby reducing sample complexity and leading to higher sensitivity [61]. In the third method, phosphate groups are chemically replaced by biotinylated moieties that can be used as affinity handles and enriched on avidin or streptavidin beads [62]. Regardless of the enrichment method utilized, liquid chromatography–mass spectrometry (LC–MS) or MS analysis is carried out to identify the site of modification.

Quantitation of these modifications can be done with or without adding a mass tag to the peptide (tag vs. label-free methods, respectively). Quantification of label-free methods is based on the signal intensity originated in the

MS (peak area) or on the number of times a peptide is observed [63]. Tag methods require stable isotope labeling by amino acids (SILAC) [64] or chemical derivatization (e.g., isobaric tag for relative and absolute quantification [iTRAQ]) [65]. Many other chemical labeling techniques have also been described, for example, labeling by heavy or light dimethyl groups is very economical and therefore also allows chemical labeling in cases where large amounts of starting material are required [66]. A newly adapted MS technology is multiple reaction monitoring (MRM) and its closely related method single/selective reactive monitoring (SRM). The biggest advantages with MRM are the unbiased quantification, the ability to distinguish between PTMs and other protein forms, and the ability to quantify site-specific PTMs within one or more proteins [67].

FUTURE PROSPECTIVE AND CONCLUDING REMARKS

Currently, atherosclerosis and vascular disease are serious health problems that are not fully understood. It is therefore important to search for factors that cause and contribute to the development of the processes involved in these diseases. PTMs greatly amplifying the diversity of proteins and have become an interest to atherosclerosis proteomes. Understanding protein phosphorylation, methylation, acetylation and their relationship with atherosclerosis models will allow a more in-depth understanding of pathogenic situations. Although the current procedures for PTMs analysis are greatly improved, there are still concerns regarding sample loss and inconsistencies in sample preparation, enrichment, and instrumentation. Consequently, new techniques for the selective enrichment, identification, and quantification of each modified amino acid residue are required. It is necessary to develop and use techniques that can assess multiple PTMs as well as identify and quantify each modified residue. The pathology of atherosclerosis is complex; therefore, characterization of the protein networks and their PTMs will help to reveal the current state and potential future directions of cardiovascular proteomics investigation.

ACKNOWLEDGMENT

This work was supported by 5RC1HL100021-02 (JVE) and NHLBI-HV-10-05 (2) (JVE). The authors thank Dr. Chris Murray for editing this manuscript.

REFERENCES

1. Lusis, A.J. (2000) Atherosclerosis. *Nature* 407(6801), 233–241.
2. Woolard, K.J. (2013) Immunological aspects of atherosclerosis. *Clin Sci* 125, 221–235.
3. Hu, P., Shimoji, S., Hart, G.W. (2010) Site-specific interplay between O-GlcNAcylation and phosphorylation in cellular regulation. *FEBS Lett* 584, 2526–2538.
4. Agnetti, G., Husberg, C., Van Eyk, J.E. (2011) Divide and conquer: The application of organelle proteomics to heart failure. *Circ Res* 108, 512–526.
5. Basta, G. (2008) Receptor for advanced glycation endproducts and atherosclerosis: From basic mechanisms to clinical implications. *Atherosclerosis* 196, 9–21.
6. Ewart, M.A., Kennedy, S. (2011) AMPK and vasculoprotection. *Pharmacol Ther* 131, 242–253.
7. Liu, S.X., Hou, F.F., Guo, Z.J., et al. (2006) Advanced oxidation protein products accelerate atherosclerosis through promoting oxidative stress and inflammation. *Arterioscler Thromb Vasc Biol* 26, 1156–1162.
8. Hunter, T. (2000) Signaling—2000 and beyond. *Cell* 100, 113–127.
9. Hoekstra, E., Peppelenbosch, M.P., Fuhler, G.M. (2012) The role of protein tyrosine phosphatases in colorectal cancer. *Biochim Biophys Acta* 1826, 179–188.
10. Salminen, A., Kaarniranta, K., Haapasalo, A., Soininen, H., Hiltunen, M. (2011) AMP-activated protein kinase: A potential player in Alzheimer's disease. *J Neurochem* 118, 460–474.

11. Kushnir, A., Shan, J., Betzenhauser, M.J., Reiken, S., Marks, A.R. (2010) Role of CaMKIIdelta phosphorylation of the cardiac ryanodine receptor in the force frequency relationship and heart failure. *Proc Natl Acad Sci U S A* 107, 10274–10279.
12. Depaoli-Roach, A.A., Park, I.K., Cerovsky, V., et al. (1994) Serine/threonine protein phosphatases in the control of cell function. *Adv Enzyme Regul* 34, 199–224.
13. Findeisen, H.M., Gizard, F., Zhao, Y., et al. (2011) Epigenetic regulation of vascular smooth muscle cell proliferation and neointima formation by histone deacetylase inhibition. *Arterioscler Thromb Vasc Biol* 31, 851–860.
14. Kawahara, K., Watanabe, S., Ohshima, T., et al. (1999) Hypernuclear acetylation in atherosclerotic lesions and activated vascular smooth muscle cells. *Biochem Biophys Res Commun* 266, 417–424.
15. Vallance, P., Leiper, J. (2004) Cardiovascular biology of the asymmetric dimethylarginine: Dimethylarginine dimethylaminohydrolase pathway. *Arterioscler Thromb Vasc Biol* 24, 1023–1030.
16. Grillo, M.A., Colombatto, S. (2005) S-adenosylmethionine and protein methylation. *Amino Acids* 28, 357–362.
17. Litt, M., Qiu, Y., Huang, S. (2009) Histone arginine methylations: Their roles in chromatin dynamics and transcriptional regulation. *Biosci Rep* 29, 131–141.
18. Chang, B., Chen, Y., Zhao, Y., Bruick, R.K. (2007) JMJD6 is a histone arginine demethylase. *Science* 318, 444–447.
19. Webby, C.J., Wolf, A., Gromak, N., et al. (2009) JmJd6 catalyses lysyl-hydroxylation of U2AF65, a protein associated with RNA splicing. *Science* 325, 90–93.
20. Mantri, M., Krojer, T., Bagg, E.A., et al. (2010) Crystal structure of the 2-oxoglutarate- and FE(II)-dependent lysyl hydroxylase JMJD6. *J Mol Biol* 401(2), 211–222.
21. Wang, Y., Wysocka, J., Sayegh, J., et al. (2004) Human PAD4 regulates histone arginine methylation levels via demethylimination. *Science* 306, 279–283.
22. Raijmakers, R., Zendman, A.J.W., Egberts, W.V., et al. (2007) Methylation of arginine residues interferes with citrullination by peptidylarginine deiminases *in vitro*. *J Mol Biol* 367, 1118–1129.
23. Leiper, J.M., Santa Maria, J., Chubb, A., et al. (1999) Identification of two human dimethylarginine dimethylaminohydrolases with distinct tissue distributions and homology with microbial arginine deiminases. *Biochem J* 343(Pt 1), 209–214.
24. Ogawa, T., Kimoto, M., Sasoka, K. (1989) Purification and properties of a new enzyme, N^G,N^G-dimethylarginine dimethylaminohydrolase, from rat kidney. *J Biol Chem* 264 10205–10209.
25. Dillon, S.C., Zhang, X., Trievel, R.C., Cheng, X. (2005) The SET-domain protein superfamily: Protein lysine methyltransferases. *Genome Biol* 6, 227.
26. Shi, Y., Lan, F., Matson, C., et al. (2004) Histone demethylation mediated by the nuclear amine oxidase homolog LSD1. *Cell* 119, 941–953.
27. Mosammaparast, N., Shi, Y. (2010) Reversal of histone methylation: Biochemical and molecular mechanisms of histone demethylases. *Annu Rev Biochem* 79, 155–179.
28. Cloos, P., Christiensen, J., Agger, K., Helin, K. (2008) Erasing the methyl mark: Histone demethylases at the center of cellular differentiation and disease. *Genes Dev* 22(9), 1115–1140.
29. Roth, S.Y., Denu, J.M., Allis, C.D. (2001) Histone acetyltransferases. *Annu Rev Biochem* 70, 81–120.
30. Bolden, J.E., Peart, M.J., Johnstone, R.W. (2006) Anticancer activities of histone deacetylase inhibitors. *Nat Rev Drug Discov* 5, 769–784.
31. Ferri, N. (2012) AMP-activated protein kinase and the control of smooth muscle cell hyperproliferation in vascular disease. *Vascul Pharmacol* 56, 9–13.
32. Han, Y., Wang, Q., Song, P., Zhu, Y., Zou, M.H. (2010) Redox regulation of the AMP-activated protein kinase. *PLoS ONE* 5, e15420.
33. Cheng, Z., Pang, T., Gu, M., et al. (2006) Berberine-stimulated glucose uptake in L6 myotubes involves both AMPK and p38 MAPK. *Biochim Biophys Acta* 1760, 1682–1689.
34. Zhang, Y., Qiu, J., Wang, X., Xia, M. (2011) AMP-activated protein kinase suppresses endothelial cell inflammation through phosphorylation of transcriptional coactivator p300. *Arterioscler Thromb Vasc Biol* 31, 2897–2908.
35. Fuster, J.J., Gonzalez-Navarro, H., Vinue, A., et al. (2011) Deficient p27 phosphorylation at

serine 10 increases macrophage foam cell formation and aggravates atherosclerosis through a proliferation-independent mechanism. *Arterioscler Thromb Vasc Biol* 31, 2455–2463.

36. Muslin, A.J. (2008) MAPK signalling in cardiovascular health and disease: Molecular mechanisms and therapeutic targets. *Clin Sci (Lond)* 115, 203–218.
37. Wong, S.L., Lau, C.W., Wong, W.T., et al. (2011) Pivotal role of protein kinase Cdelta in angiotensin II-induced endothelial cyclooxygenase-2 expression: A link to vascular inflammation. *Arterioscler Thromb Vasc Biol* 31, 1169–1176.
38. Martin-Ventura, J.L., Duran, M.C., Blanco-Colio, L.M., et al. (2004) Identification by a differential proteomic approach of heat shock protein 27 as a potential marker of atherosclerosis. *Circulation* 110, 2216–2219.
39. Napoli, C., de Nigris, F., Williams-Ignarro, S., Pignalosa, O., Sica, V., Ignarro, L.J. (2006) Nitric oxide and atherosclerosis: An update. *Nitric Oxide* 15, 265–279.
40. Achan, V., Broadhead, M., Malaki, M., et al. (2003) Asymmetric dimethylarginine causes hypertension and cardiac dysfunction in humans and is actively metabolized by dimethylarginine dimethylaminohydrolase. *Arterioscler Thromb Vasc Biol* 23, 1455–1459.
41. Boger, R.H. (2006) Asymmetric dimethylarginine (ADMA): A novel risk marker in cardiovascular medicine and beyond. *Ann Med* 38, 126–136.
42. Zoccali, C. (2006) Asymmetric dimethylarginine (ADMA): A cardiovascular and renal risk factor on the move. *J Hypertens* 24, 611–619.
43. Valkonen, V.P., Paiva, H., Salonen, J.T., et al. (2001) Risk of acute coronary events and serum concentration of asymmetrical dimethylarginine. *Lancet* 358, 2127–2128.
44. Dayoub, H., Rodionov, R.N., Lynch, C., et al. (2008) Overexpression of dimethylarginine dimethylaminohydrolase inhibits asymmetric dimethylarginine-induced endothelial dysfunction in the cerebral circulation. *Stroke* 39, 180–184.
45. Balasubramaniyan, N., Ananthanarayanan, M., Suchy, F.J. (2012) Direct methylation of FXR by Set7/9, a lysine methyltransferase, regulates the expression of FXR target genes. *Am J Physiol Gastrointest Liver Physiol* 302, G937–G947.
46. Kim, S.C., Sprung, R., Chen, Y., et al. (2006) Substrate and functional diversity of lysine acetylation revealed by a proteomics survey. *Mol Cell* 23, 607–618.
47. Potente, M., Ghaeni, L., Baldessari, D., et al. (2007) SIRT1 controls endothelial angiogenic functions during vascular growth. *Genes Dev* 21, 2644–2658.
48. Westphal, C.H., Dipp, M.A., Guarente, L. (2007) A therapeutic role for sirtuins in diseases of aging? *Trends Biochem Sci* 32, 555–560.
49. Yu, W., Fu, Y.C., Chen, C.J., Wang, X., Wang, W. (2009) SIRT1: A novel target to prevent atherosclerosis. *J Cell Biochem* 108, 10–13.
50. Dioum, E.M., Chen, R., Alexander, M.S., et al. (2009) Regulation of hypoxia-inducible factor 2alpha signaling by the stress-responsive deacetylase sirtuin 1. *Science* 324, 1289–1293.
51. Burre, J., Wittig, I., Schagger, H. (2009) Non-classical 2-D electrophoresis. *Methods Mol Biol* 564, 33–57.
52. Husberg, C., Agnetti, G., Holewinski, R.J., Christensen, G., Van Eyk, J.E. (2012) Dephosphorylation of cardiac proteins *in vitro*—A matter of phosphatase specificity. *Proteomics* 12, 973–978.
53. Bannister, A.J., Schneider, R., Kouzarides, T. (2002) Histone methylation: Dynamic or static? *Cell* 109, 801–806.
54. Lin, Y.Y., Lu, J.Y., Zhang, J., et al. (2009) Protein acetylation microarray reveals that NuA4 controls key metabolic target regulating gluconeogenesis. *Cell* 136, 1073–1084.
55. Kinoshita, E., Kinoshita-Kikuta, E., Koike, T. (2012) Phos-tag SDS-PAGE systems for phosphorylation profiling of proteins with a wide range of molecular masses under neutral pH conditions. *Proteomics* 12, 192–202.
56. Su, H.C., Hutchison, C.A., 3rd, Giddings, M.C. (2007) Mapping phosphoproteins in *Mycoplasma genitalium* and *Mycoplasma pneumoniae*. *BMC Microbiol* 7, 63.
57. Yang, Y.Y., Ascano, J.M., Hang, H.C. (2010) Bioorthogonal chemical reporters for monitoring protein acetylation. *J Am Chem Soc* 132, 3640–3641.
58. Rathert, P., Dhayalan, A., Murakami, M., et al. (2008) Protein lysine methyltransferase G9a acts on non-histone targets. *Nat Chem Biol* 4, 344–346.
59. Iwabata, H., Yoshida, M., Komatsu, Y. (2005) Proteomic analysis of organ-specific

post-translational lysine-acetylation and -methylation in mice by use of anti-acetyllysine and -methyllysine mouse monoclonal antibodies. *Proteomics* 5, 4653–4664.

60. Rikova, K., Guo, A., Zeng, Q., et al. (2007) Global survey of phosphotyrosine signaling identifies oncogenic kinases in lung cancer. *Cell* 131, 1190–1203.
61. Thingholm, T.E., Jensen, O.N., Larsen, M.R. (2009) Analytical strategies for phosphoproteomics. *Proteomics* 9, 1451–1468.
62. Oda, Y., Nagasu, T., Chait, B.T. (2001) Enrichment analysis of phosphorylated proteins as a tool for probing the phosphoproteome. *Nat Biotechnol* 19, 379–382.
63. Wang, M., You, J., Bemis, K.G., Tegeler, T.J., Brown, D.P. (2008) Label-free mass spectrometry-based protein quantification technologies in proteomic analysis. *Brief Funct Genomic Proteomic* 7, 329–339.
64. Ong, S.E., Blagoev, B., Kratchmarova, I., et al. (2002) Stable isotope labeling by amino acids in cell culture, SILAC, as a simple and accurate approach to expression proteomics. *Mol Cell Proteomics* 1, 376–386.
65. Wiese, S., Reidegeld, K.A., Meyer, H.E., Warscheid, B. (2007) Protein labeling by iTRAQ: A new tool for quantitative mass spectrometry in proteome research. *Proteomics* 7, 340–350.
66. Hsu, J.L., Huang, S.Y., Chow, N.H., Chen, S.H. (2003) Stable-isotope dimethyl labeling for quantitative proteomics. *Anal Chem* 75, 6843–6852.
67. Cox, D.M., Zhong, F., Du, M., Duchoslav, E., Sakuma, T., McDermott, J.C. (2005) Multiple reaction monitoring as a method for identifying protein post-translational modifications. *J Biomol Tech* 16, 83–90.

35

PHOSPHORYLATION REGULATION BY KINASES AND PHOSPHATASES IN ATHEROSCLEROSIS

Katherine J. Elliott and Satoru Eguchi

Department of Physiology, Cardiovascular Research Center, Temple University School of Medicine, Philadelphia, PA, USA

INTRODUCTION

Phosphorylation of proteins is a critical posttranslational modification involved in regulating signal transduction cascades, which results in altered nuclear transcription factor activity eventually leading to changes in cellular functions such as cell proliferation, differentiation, and apoptosis. Specific phosphorylation events are tightly regulated by kinases and phosphatases, and an understanding of these events has contributed to uncovering the basis of many disease processes. Although misregulation of phosphorylation by mutant, oncogenic, or viral kinases is known to lead to neoplasms in humans, less is known regarding protein phosphorylation events involved in cardiovascular diseases. However, recent data suggest a critical involvement of protein phosphorylation leading to or enhancing atherosclerosis. This chapter deals with the current understanding of protein phosphorylation events regulated by protein kinases and phosphatases and the mechanisms potentially relevant for atherosclerosis.

SERINE/THREONINE KINASES

Mitogen-Activated Protein Kinase Cascades and the Substrates

The mitogen-activated protein kinase (MAPK) signaling cascades are a universal signaling system used by eukaryotic cells. The MAPK cascade consists of a set of kinases that are activated in series to transduce an external signal into the nucleus (Figure 35.1). Initially an MAPK kinase kinase (MAPKKK) is activated by phosphorylation after an external stimulus such as growth factors or cytokines binds to cell surface receptors. Next, an MAPK kinase (MAPKK) is phosphorylated, and finally, a terminal MAPK is activated by phosphorylation, prompting a biologic response by the cell. Additionally, cross talk can occur

Atherosclerosis: Risks, Mechanisms, and Therapies, First Edition. Edited by Hong Wang and Cam Patterson.

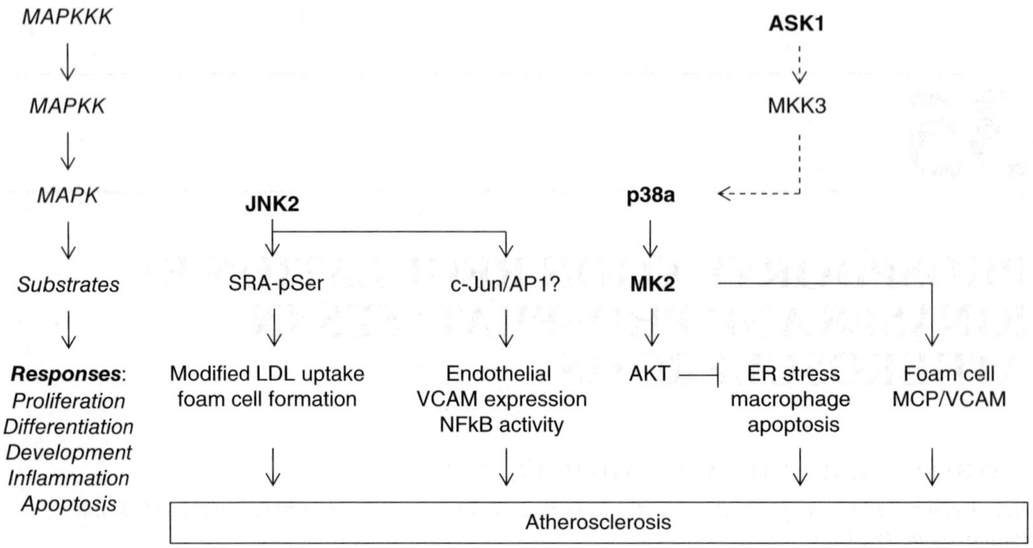

Figure 35.1 MAPK cascades involved in atherosclerosis. The column to the left indicates a generalized pathway. On the right are members of the MAPK pathway thought to be involved in atherosclerosis. Defined connections are indicated by solids lines, less well-established connections are indicated by dashed lines. Proteins discussed in the text are in bold.

between the MAPK pathways within a cell and is used as a way for the cell to coordinate multiple signal inputs. How these pathways can maintain specificity while containing significant overlap has been a source of confusion in the field. Currently, data suggest that scaffolding proteins are used by the cell to localize specific components of the cascade in order to facilitate select signal transduction [1]. Regarding atherosclerosis, MAPK pathways have been implicated in foam cell formation, endothelial cell activation, and vascular smooth muscle migration and proliferation. As such, phosphorylation substrates of the MAPK pathways are of importance to vascular biology as potential specific and selective therapeutic targets [2]. By performing knockout experiments in mice lacking low-density lipoprotein receptor (LDLr) or apolipoprotein E (ApoE), researchers have shown that several components and targets of the MAPK signaling cascade are critical for atherogenesis. Table 35.1 contains a summary of the literature reporting critical kinases and phosphatases and their substrates, subsequent functions, and cell-type specificity involved in regulation of atherogenesis.

Jun N-Terminal Kinases

JNK2 is an isoform of c-Jun N-terminal kinases (JNKs) and a member of the MAPK family. JNK2 is widely expressed but not required for normal mouse development as evidenced by the ability of JNK knockout mice to develop normally. (These mice are immunocompromised however due to defects in T cell function.) To establish the importance of JNK in atherosclerotic plaque formation, double knockout ApoE$^{-/-}$ JNK1$^{-/-}$ and ApoE$^{-/-}$ JNK2$^{-/-}$ mice were fed a high-fat diet. ApoE$^{-/-}$ JNK2$^{-/-}$ mice developed considerably smaller plaques than ApoE$^{-/-}$ control mice, whereas loss of JNK1 had no effect. The decrease in plaque size appears to be due to decreased foam cell formation rather than effects on endothelial or T cells in this model. Scavenger receptor-A (SR-A) is involved in receptor-mediated uptake of modified LDL and is necessary for foam cell formation. The loss of

TABLE 35.1 Critical kinases and phosphatases and their substrates and cell specificity in atherosclerosis

Cell type	Kinase	Direct/*downstream* substrates	Athero-promoting (+)/athero-suppressing (–) key functions	Reference
EC	Akt	eNOS–Ser1177	(–) Prevent apoptosis	[3]
	IKK	IkB	(+) Cytokine expression	[4]
	PKCβ	p66Shc, *JNK*	(+) Egr1/MMP2 induction	[5]
	IR	IRS1/2, *Akt*, *eNOS*	(–) NO production	[6]
	JAK	STAT3	(+) IL-8 induction	[7]
Macrophage	JNK2	SR-A, c-Jun	(+) Foam cell formation	[8, 9]
	p38a	MK2, *Akt*	(–) Macrophage apoptosis	[10]
	ASK1	JNK, p38	(–) Macrophage apoptosis	[11]
	IRAK4	IRAK1	(+) NFκB activation	[12]
	Rip2	*ERK*	(–) Macrophage lipid uptake	[13]
	ROCK1	ERM	(+) Chemotaxis	[14]
	Ddr1	Src, *ERK1/2*	(+) Macrophage adhesion	[15]
	IR	IRS1/2	(+) Helps cytokine induction	[16]
VSMC	Ddr1	*ERK1/2*	(+) Plaque stability	[17]
Cell type	**Phosphatase**	**Direct/downstream substrate**	**Effect**	**Reference**
Macrophage	MKP-1	ERK, JNK, p38	(+) Macrophage spreading/ migration	[18]
EC	MKP-1	p38, JNK	(–) Suppress inflammation	[19]

EC, endothelial cell; VSMC, vascular smooth muscle cell.

JNK2 leads to a decrease in SR-A cytoplasmic tail serine phosphorylation and consequently a decrease in internalization of modified LDL and finally fewer foam cells, suggesting that SR-A is a substrate of JNK2 [8].

Further evidence that JNK is involved in atherosclerosis can be found in studies conducted with a pharmacologic inhibitor of JNK. ApoE$^{-/-}$ mice were fed a high-fat diet, and blood flow was disturbed by a cast placed around the carotid artery in order to mimic low/disturbed flow-induced plaque formation. These studies specifically looked at the involvement of JNK in endothelial cells by examining the effects of shear stress on plaque formation. The addition of SP600125, a pharmacologic JNK inhibitor, significantly reduced the incidence of plaque formation in regions of low/disturbed flow. The reduced plaque formation caused by JNK inhibition was correlated with decreased vascular cell adhesion molecule 1 (VCAM1) expression and attenuation of nuclear factor kappa-light-chain-enhancer of activated B cells (NFκB) activity [20].

p38 MAPKs

In contrast to JNK2, p38a, a member of the terminal MAPKs may play a protective role in atherosclerosis. In macrophages, p38a appears to be important for suppressing apoptosis. ApoE$^{-/-}$ mice deficient in macrophage p38a were fed a high-fat diet, and the extent of plaque formation was assessed. Loss of p38a led to increased macrophage apoptosis in advanced atherosclerotic plaques, suggesting an atheroprotective role of p38a [10]. The data indicate that p38a suppresses endoplasmic reticulum (ER) stress-induced macrophage apoptosis via Akt activation (note that p38 forms a complex with MAPK-activated protein kinase-2 (MK2) to activate and phosphorylate Akt at Ser473). However, a similar study by a different group found that loss of p38a in either macrophages or endothelial cells in ApoE$^{-/-}$ mice fed a high-fat diet did not affect plaque formation or composition. This aptly illustrates the complexity of studying the signaling systems involved in atherosclerosis [21].

Apoptosis Signal-Regulating Kinase 1

Apoptosis signal-regulating kinase 1 (ASK1) is an MAPKKK, which is activated by various cytotoxic stressors and activates JNK and p38. ASK1 is known to regulate cardiomyocyte and macrophage apoptosis. To study the importance of macrophage apoptosis in atherosclerosis double knockout mice, ApoE$^{-/-}$ ASK1$^{-/-}$ mice were fed a high-fat diet, and the level of atherosclerosis was assessed. ASK1$^{-/-}$ ApoE$^{-/-}$ mice had accelerated hyperlipidemia-induced atherosclerosis, which was characterized by less apoptosis of macrophages and fewer necrotic areas. The decrease in macrophage apoptosis correlated with a decrease in caspase-3 and an increase in bcl-2 expression. Although likely, whether this atheroprotective role of ASK is mediated through p38 remains undetermined [11].

MK2

Another aspect of p38 MAPK signaling cascade to explore is at the substrate level of terminal MAPKs. MK2 is a direct substrate of p38a and is an important regulator of proinflammatory signals. To establish the importance of MK2 in atherosclerosis, LDLr$^{-/-}$ mice were bred with MK2$^{-/-}$ mice to produce a double knockout mouse. A significant decrease in atherosclerosis was observed in the MK2$^{-/-}$ mice fed a high-fat diet compared with control mice. MK2 deficiency specifically resulted in reduced foam cell formation (which likely involves reduced SR-A expression) and decreased endothelial monocyte chemotactic protein-1 (MCP-1) and VCAM1 expression. This is because MK2 increases the transcript stability of inflammatory mediators via targeting AU-rich elements located in the 3′-untranslated region by phosphorylating a zinc finger protein tristetraprolin which mediates degradation of the transcript [22]. Taken together, the atheroprotective roles of p38a may not be explained by its ability to activate MK2, since MK2 appears to promote atherosclerosis.

MAPK Phosphatase

Significant time and resources have been devoted to study various protein kinases involved in MAPK signaling cascades and cardiovascular diseases [2]. Much less is known regarding the involvement of protein phosphatases. One phosphatase studied is MAPK phosphatase-1 (MKP-1), which is known to act as a dual specificity phosphatase to remove phosphates from extracellular signal-regulated kinase (ERK), JNK, and p38 at tyrosine and serine/threonine (S/T) residues, resulting in their inactivation. Deletion of MKP-1 in ApoE$^{-/-}$ mice leads to a reduction in atherosclerosis, which is associated with decreased levels of interleukin-1α and tumor necrosis factor-α (TNFα) [18]. The reduction in atherosclerotic lesions may be due to macrophage defects in spreading/migration and unexpected suppression of ERK signaling or activation of p38a. Reduced MCP-1 production from macrophages is also associated with these mice [23].

In contrast, MKP-1 expression in endothelial cells has been associated with an atheroprotective phenotype. MKP-1 is preferentially expressed in endothelial cells located in shear stress-protected regions of the aorta and was shown to be a negative regulator of p38 and JNK [19]. These conflicting data indicate that MKP-1 may have cell-specific effects.

Activator Protein-1

Although much of the research into MAPK pathway involvement in atherosclerosis has been performed in the knockout mouse models, it is important to confirm the significance of these players in human disease scenarios as well. To that end, correlative studies looking at expression of MAPK components at various stages of atherosclerotic disease progression in human samples are one way to complement the

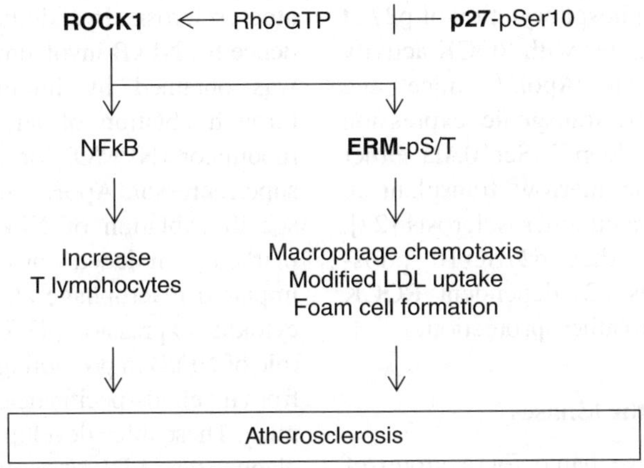

Figure 35.2 Rho kinase/ROCK pathway involvement in atherosclerosis. Proteins discussed in the text are in bold.

animal studies. Activator protein-1 (AP-1) is a common inflammatory transcription factor which functions as a heterodimer of c-Fos/c-Jun. c-Jun is phosphorylated and activated by JNK and is considered as an important terminal phosphorylation target of the MAPK pathway. Activated c-Jun was found to be increased in human aortic wall samples of atherosclerotic disease compared with normal samples, indicating that the JNK MAPK pathways are potentially important in atherosclerosis in humans [9].

OTHER RELEVANT S/T KINASES AND THEIR SUBSTRATES

Rho-kinase/ROCK and p27

Rho-kinase/ROCK, a downstream effector of Rho small G protein, is an important mediator of pathophysiological cardiovascular states, including atherosclerosis, through its effects on various cell types [24] (Figure 35.2). The first indication of ROCK's involvement in early atherogenesis *in vivo* came from studies of LDLr$^{-/-}$ mice fed a high-fat diet and treated with a ROCK inhibitor, Y-27632. Mice treated with the inhibitor showed significant decrease in lesion size compared to control mice, which was associated with reduced T lymphocyte accumulation and NFκB cascade activation. A major limitation of inhibitor studies is the inability to determine in which cell type the loss of function is most important as well as the potential for inhibiting other S/T kinases. Bone marrow derived from ROCK1$^{-/-}$ mice was transplanted into irradiated LDLr$^{-/-}$ mice to determine the effect of specific loss of ROCK1 on atherosclerosis. The data indicate that ROCK1 is critically important for macrophage chemotaxis, uptake of modified LDL, and foam cell formation [14]. Therefore, loss of ROCK1 in bone marrow protects these mice from atherosclerosis. The role of ROCK2 in atherosclerosis as well as the ROCK1 target substrate in macrophages remains unknown; however Y-27632 was shown to inhibit ezrin–radixin–moesin (ERM) protein phosphorylation (Thr567 of ezrin, Thr564 of radixin, and Thr558 of moesin) in the atherosclerotic plaques [25], suggesting that the ERM phosphorylation may mediate macrophage activity in the plaques. Interestingly, patients with coronary artery disease treated with fasudil, a ROCK inhibitor, show significant improvement in endothelium-dependent vasodilation compared with healthy individuals [26]. Thus, endothelial ROCKs may also promote vascular diseases by reducing endothelial nitric oxide synthase (eNOS) activity.

A cyclin-dependent kinase inhibitor, p27, is known as an endogenous negative regulator of ROCK by binding to RhoA. Interestingly, it has been shown that phosphorylation of p27 at Ser10 inversely correlates with ROCK activity and atherosclerosis in ApoE$^{-/-}$ mice and humans. Significantly, transgenic expression of nonphosphorylatable p27 Ser10Ala either globally or via bone marrow transplant in ApoE$^{-/-}$ mice accelerated atherosclerosis [27], providing evidence that p27Ser10 phosphorylation mediates p27-dependent ROCK suppression leading to atheroprotection.

NFκB cascade and the kinases

NFκB is the collective name for a group of ubiquitous transcription factors that regulate cellular responses to multiple stimuli, primarily by initiating transcription of proinflammatory genes. NFκB is sequestered in an inactive state in the cytosol via binding to an inhibitory partner protein, IκBalpha. Upon cell stimulation, IκB is phosphorylated by IκB kinase complex (IKK) leading to degradation of the inhibitor, leaving NFκB free to translocate into the nucleus and induce transcription (Figure 35.3). Incorrect regulation of NFκB is involved in inflammatory diseases, including

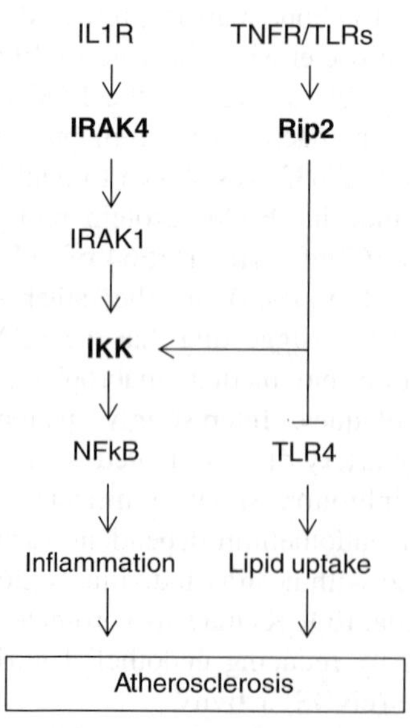

Figure 35.3 NFkB pathway involvement in atherosclerosis. Proteins discussed in the text are in bold.

atherosclerosis. In endothelial cells, *in vivo* evidence for NFκB involvement in atherosclerosis was obtained by inhibiting NFκB activity through ablation of an IKK subunit, NFκB modulator (NEMO), or expression of an IκB superrepressor. ApoE$^{-/-}$ mice with endothelial specific ablation of NFκB activity displayed markedly reduced plaque formation with impaired macrophage attachment and reduced cytokine expression [4]. Further supporting the role of NFκB in promoting atherogenesis is data from myeloid-specific deletion of IκB in LDLr$^{-/-}$ mice. These mice develop larger atherosclerotic plaques than LDLr$^{-/-}$ control mice [28].

While IκB and NEMO are directly upstream of NFκB activation, kinases further upstream in the NFκB pathway may also be involved in atherosclerosis. IL-1 receptor-associated kinase 4 (IRAK4) is an upstream kinase of the NFκB cascade, which appears to be involved in modified LDL-induced NFκB activation. IRAK4 is known to phosphorylate IRAK1, leading to the activation of NFκB and a member of the MAPK cascade, JNK. Macrophages deficient in IRAK4 kinase activity have reduced IκBα phosphorylation and therefore decreased NFκB activity. ApoE$^{-/-}$ mice with a functional deficiency in IRAK4 kinase activity show decreased atherosclerotic lesion formation, reduced proinflammatory mRNA expression, but no change in LDL uptake or foam cell formation [12].

Receptor-interacting protein 2 (Rip2) is an S/T kinase associated with membrane-bound receptors such as TNF receptor 1 (TNFR1) and toll-like receptor (TLR). Rip2 is also associated with multiple NFκB activation pathways. Surprisingly, myeloid-specific deletion of Rip2 in LDLr$^{-/-}$ mice resulted in increased atherosclerosis even while exhibiting impaired inflammation associated with decreased NFκB activation. Increased lipid accumulation and uptake by the Rip2$^{-/-}$ macrophages indicate an uncoupling of inflammation and lipid accumulation during atherogenesis. The increased lipid uptake by Rip2-deficient macrophages was associated with increased TLR4 expression, suggesting that Rip2 may normally function as a negative regulator or TLR4 [13].

Akt

Akt1 is known to be involved in endothelial cell proliferation and migration, vascular permeability, apoptosis, and angiogenesis. Akt1 is an S/T kinase activated in endothelial cells by shear stress and Vascular Endothelial Growth Factor (VEGF). Upon activation, Akt1 can phosphorylate Ser1177 of eNOS for maintenance of vascular tone (see also insulin receptor section). Knockout of Akt1 in ApoE$^{-/-}$ mice leads to increased atherosclerosis. As expected, eNOS phosphorylation was reduced and endothelial cells from these mice showed greater levels of apoptosis [3]. The forkhead box transcription factor O (FoxO) is phosphorylated and negatively regulated by Akt1, leading to enhanced cell survival. Loss of Akt1 could promote cell death in endothelial cells due to the loss of negative regulation of FoxO.

Loss of Akt1 also results in greater levels of vascular smooth muscle cell (VSMC) apoptosis as well as reduced proliferation and migration. In ApoE Akt1 double knockout mice, these changes in VSMC physiology are correlated with increased plaque vulnerability characterized by plaques with relatively larger necrotic core area, smaller cap area, and reduced collagen composition [29].

Protein kinase C

Protein kinase C (PKC) refers to a family of S/T kinases that phosphorylate a wide array of substrates and is involved in regulating cell growth and transcription factor activation, among other activities. PKC has been shown to be a regulator of early growth response protein (Egr-1) in response to vascular stress. Mice lacking PKCβ and ApoE show decreased atherosclerosis, as well as decreased levels of Egr-1 and matrix metalloproteinase 2 (MMP2) mRNAs. PKCβ inhibitor also prevented atherosclerosis. These effects appear to be mediated through the PKCβ-JNK pathway in endothelial cells [5]. While the downstream substrate of PKCβ linking it to JNK activation needs to be elucidated, p66 Shc may be the substrate downstream of JNK [30].

AMP-activated protein kinase

AMP-activated protein kinase (AMPK) functions as a regulator of cellular energy homeostasis at both the cellular and whole-body level. AMPK is activated in response to cell stresses that deplete cellular ATP levels such as low glucose and hypoxia. Activated AMPK can phosphorylate a number of target proteins, which inhibit anabolic pathways to conserve ATP or activate catabolic pathways to generate more ATP. In vascular cells, AMPK has been shown to have protective effects by inhibiting inflammation, oxidant production, and VSMC proliferation. One mechanism for these actions may be through an ability to inhibit errant ER stress response. Specifically, double knockout LDLr$^{-/-}$ AMPKα2$^{-/-}$ mice have increased levels of ER stress and accelerated atherosclerosis [31]. In human cultured endothelial cells, AMPK mediates the phosphorylation of p300 Hat via atypical PKC activation [32]. Another AMPK substrate is sterol regulatory element-binding protein (SREBP-1c). In hepatic cells, AMPK phosphorylates SREBP-1c at Ser372, resulting in inactivation and decreased lipogenic transcription [33].

MLCK

Phosphorylation of myosin light chain (MLC) by non-muscle MLC kinase (nmMLCK) in endothelial cells is implicated in endothelial dysfunction. nmMLCK knockout in ApoE$^{-/-}$ mice resulted in decreased atherosclerotic lesion formation due in part to reduced monocyte migration and endothelial hyperpermeability. These effects may be mediated through Src interaction with nmMLCK [34].

RECEPTOR TYROSINE KINASES

Receptor tyrosine kinases (RTKs) are one of the major ways that environmental information is transmitted into a cell. These receptors are critical for normal growth and development; consequently, disruption or enhancement of their signaling can lead to pathophysiological states. Several RTKs, in a variety of cells, have been shown to be important for the development of atherosclerosis.

DDR1

The RTK discoidin domain receptor 1 (Ddr1) is activated by binding to triple helical collagens and controls cell proliferation, differentiation, and migration. Ddr1 clearly plays a proatherogenic role in disease progression. The $LDLr^{-/-}$ mouse model of atherosclerosis has shown that loss of Ddr1 resulted in decreased plaque development and changes in plaque composition; specifically, increased extracellular matrix content and decreased macrophage accumulation were observed [35]. In order to analyze the role of Ddr1 in macrophages during plaque development, bone marrow transplants were performed with $Ddr^{-/-}$ or $Ddr^{+/+}$ mice into $LDLr^{-/-}$ mice, and the mice were fed a high-fat diet. Loss of Ddr1 in bone marrow led to decreased atherosclerosis with a reduction in the number of donor-derived macrophages in plaques. *In vitro* studies on the Ddr-deficient macrophages indicate that Ddr1 is important for macrophage adhesion and chemotactic invasion through type IV collagen [15].

In addition to being expressed in macrophages, Ddr1 is expressed in VSMC in the fibrous cap of advanced atherosclerotic lesions, indicating that Ddr1 is important in SMC and may have a potential role in influencing plaque stability [17]. In support of the importance of Ddr1 in SMC, the kinase domain of Ddr1 was shown to be critical for attachment and migration of SMC *in vitro* [36]. The downstream signaling pathways from Ddr1 are beginning to be elucidated. VSMC from $Ddr1^{-/-}$ mice overexpressing human Ddr1 were cultured *in vitro* and activated with type I collagen. Upon activation, Ddr1 can be coimmunoprecipitated with Src and phosphorylation of ERK1/2 occurs. Inhibition of Src by pharmacologic treatments disrupts Ddr1 signaling and SMC migration [37], suggesting that Src may be one of the responsible substrates of Ddr1 to promote atherosclerosis.

Tie1

Tyrosine kinase with immunoglobulin-like and EGF-like domains 1 (Tie1) is one of several RTKs whose expression appears to be necessary for atherosclerotic plaque formation. Tie1 is expressed primarily on endothelial cells and acts as an effector of shear stress. It is necessary for embryonic vessel development as Tie1 knockout mice are not viable. $Tie1^{+/-}$ mice were bred with $ApoE^{-/-}$ mice in order to assess the role of Tie1 in atherosclerosis. Tie1 attenuation reduces murine atherosclerosis in a dose-dependent and shear stress-specific manner [38]. However, the downstream substrates phosphorylated by Tie1 to promote atherosclerosis remain to be identified.

Fibroblast Growth Factor Receptor and Other RTKs

Activated fibroblast growth factor receptor (FGFR) signals through the MAPK, PKC, and Akt pathways and is involved in promoting cell proliferation and differentiation. Regarding atherosclerosis, activated FGFR has proatherogenic properties. The effects of FGFR tyrosine kinase inhibitor SU5402 on atherosclerosis in $ApoE^{-/-}$ mice fed a high-fat diet were examined. Inhibition of FGFR phosphorylation resulted in decreased lesion size with fewer macrophages and VSMC [39]. $ApoE^{-/-}$ mice expressing a constitutively active FGFR2 conditionally targeted to endothelial cells exhibited greater plaque formation when fed a high-fat diet than control mice. Expression of constitutively active FGFR2 in endothelial cells led to increased synthesis of p21/cip1, a cell cycle-dependent kinase inhibitor. Increases in p21 and platelet-derived growth factor (PDGF) are hypothesized to mediate the proatherogenic outcomes of FGFR2 [40]. In addition to the aforementioned RTKs, both PDGF receptor and epidermal growth factor (EGF) receptor have been implicated in atherosclerosis [41, 42]. Among the PDGF receptor substrates, LDL receptor-related protein 1 was found to be atheroprotective by negatively regulating the expression and induction of PDGF receptor in VSMC [41].

Insulin Receptor (IR) and the Substrates

In contrast to proatherogenic properties of the aforementioned RTKs, activation of IR may be antiatherogenic. Although IR signaling and its relationship to atherosclerosis are complicated due to the IR being ubiquitously expressed and having effects both systemically and locally, data indicate that IR signaling in endothelium is atheroprotective (Figure 35.4). ApoE$^{-/-}$ mice with a conditional knockout of IR in endothelial cells were used to assess the importance of endothelial IR for plaque formation. Mice lacking endothelial IR were more prone to disease initiation and progression and had more complex advanced lesions than control mice. Insulin-stimulated phosphorylations of Akt at Ser473 and eNOS Ser1177 were lost in IR-negative aortic endothelial cells [6]. In addition, LDLr$^{-/-}$ mice with IR deficient bone marrow recipient developed larger and more complex atherosclerotic lesions [43]. However, this understanding of IR signaling as antiatherogenic is hampered by the data that indicate IR signaling in macrophages is rather proatherogenic. Macrophage-specific IR deletion in an ApoE$^{-/-}$ mouse fed a high-fat diet produces significantly smaller

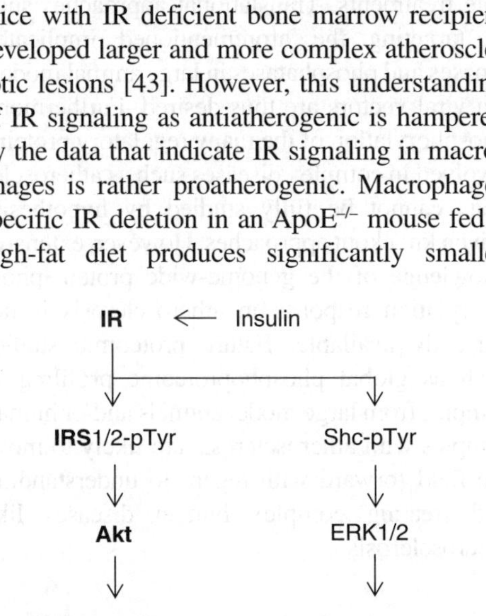

Figure 35.4 Insulin receptor signal transduction involvement in atherosclerosis. Proteins discussed in the text are in bold.

plaque size compared to control mice. This effect seems to be mediated by the IR substrate, IRS2 in macrophages, while systemic IR activation and subsequent IRS2 phosphorylation appear to be atheroprotective [16].

IRS1, IRS2, and Shc are major substrates for IR. Studies looking at decreased systemic expression (IRS$^{+/-}$ mice) of either IRS1 or IRS2 show increased levels of atherosclerosis in ApoE$^{-/-}$ mice, indicating an atheroprotective role for IR. In mice with decreased levels of IRS2, the levels of Akt were also reduced [44]. In mice with decreased levels of IRS1, vascular eNOS phosphorylation was reduced, whereas vascular ERK phosphorylation was enhanced. Thus, the vascular IR substrates seem to shift in favor of Shc under an insulin-resistant state [45].

MER Receptor Tyrosine Kinase Receptor

MER receptor tyrosine kinase (Mertk) is another antiatherogenic RTK. The function of Mertk receptor is to facilitate efferocytosis by binding to phosphatidylserine binding protein growth arrest-specific 6, which acts as a bridge between apoptotic cells and macrophages. To determine the effects of Mertk receptor on advanced atherosclerotic lesions, kinase-defective Mertk receptor mice were bred with ApoE$^{-/-}$ mice fed a high-fat diet, and the number of apoptotic cells was analyzed. The level of apoptosis was significantly increased in Mertk-defective plaques when compared with control plaques, indicating a decrease in efferocytosis in the mutant cells [46]. The level of apoptotic cells is relevant because it is a marker for the stability of the plaque; however, responsible downstream signaling events remain less clear.

NONRECEPTOR KINASES

Janus Kinase and the Substrate Signal Transducers and Activators of Transcription

The Janus kinase/signal transducers and activators of transcription (JAK/STAT) pathway is a ubiquitous and critical pathway for transduction of cytokine and proinflammatory signals. Upon cytokine binding to cell surface receptor, JAK tyrosine kinase activity is induced resulting in tyrosine phosphorylation of the receptor to create SH2 binding sites. The SH2 binding sites recruit STAT proteins to the complex, resulting in STAT phosphorylation, dimerization, and nuclear translocation, culminating in activation of STAT responsive genes. Conditional knockout of STAT3 in endothelial cells of mice results in smaller atherosclerotic lesion size, indicating the importance of the JAK/STAT pathway in atherosclerosis. In addition, samples of atherosclerotic lesions of human carotid artery show STAT3 is activated in the endothelium and monocytes [7].

Proline-Rich Tyrosine Kinase 2

Proline-rich tyrosine kinase 2 (PYK2) has been shown to be activated by many proatherogenic signals. Indeed, PYK2$^{-/-}$ mice show attenuated levels of atherosclerosis. PYK2 is activated by a high-cholesterol diet resulting in oxidative stress and TNFα induction, leading to gene expression of proatherogenic proteins such as MCP-1 [47]. Src has been shown to be a key downstream substrate and an effector of PYK2.

PROTEIN TYROSINE PHOSPHATASES

Although significantly less research, especially mechanistic, has been performed to understand the role of phosphatases in atherosclerosis, a few human studies have suggested a link between protein tyrosine phosphatase nonreceptor type 22 (PTPN22) and autoimmune diseases, including atherosclerosis. PTPN22 encodes lymphoid-specific tyrosine phosphatase and regulates T cell receptor signaling via dephosphorylation of lymphocyte-specific protein tyrosine kinase. The single nucleotide polymorphism at PTPN22+1858 is correlated with an increased risk for autoimmune disease and has also been associated with an increased risk for atherosclerosis [48]. This mutation is likely to be a gain-of-function mutation with increased phosphatase activity [49] and is hypothesized to increase disease risk due to a failure to delete autoreactive T cells [50].

PERSPECTIVES AND FUTURE DIRECTIONS

This chapter has provided much information about knockout mouse experiments involving protein kinases implicated in pathophysiological signal transduction and has demonstrated critical roles for these kinases and the downstream protein phosphorylation events which are likely to be involved in atherosclerosis in humans. However, clear limitations exist with these types of experiments. Specifically, animal models of disease do not always correlate with human disease and thus may not lead to clinically workable treatments. Translational approaches such as targeting the aforementioned implicated kinases and phosphatases in large animal models via viral vectors are thus desired. Furthermore, phosphorylation of the many regulatory proteins involved in complex diseases such as atherosclerosis cannot be fully studied by hypothesis-driven knockout approaches. However, extensive knowledge of the genome-wide protein phosphorylation response in atherosclerosis is not currently available. Future proteomic studies such as global phosphoproteome profiling in samples from large model animals and/or human samples with atherosclerosis are likely to move the field forward with regard to understanding and treating complex human diseases like atherosclerosis.

REFERENCES

1. Cargnello M, Roux PP. Activation and function of the MAPKs and their substrates, the MAPK-activated protein kinases. *Microbiol Mol Biol Rev* 2011; **75**(1): 50–83.
2. Muslin AJ. MAPK signalling in cardiovascular health and disease: Molecular mechanisms and

therapeutic targets. *Clin Sci (Lond)* 2008; **115**(7): 203–218.

3. Fernandez-Hernando C, Ackah E, Yu J, et al. Loss of Akt1 leads to severe atherosclerosis and occlusive coronary artery disease. *Cell Metab* 2007; **6**(6): 446–457.
4. Gareus R, Kotsaki E, Xanthoulea S, et al. Endothelial cell-specific NF-kappaB inhibition protects mice from atherosclerosis. *Cell Metab* 2008; **8**(5): 372–383.
5. Harja E, Chang JS, Lu Y, et al. Mice deficient in PKCbeta and apolipoprotein E display decreased atherosclerosis. *FASEB J* 2009; **23**(4): 1081–1091.
6. Rask-Madsen C, Li Q, Freund B, et al. Loss of insulin signaling in vascular endothelial cells accelerates atherosclerosis in apolipoprotein E null mice. *Cell Metab* 2010; **11**(5): 379–389.
7. Gharavi NM, Alva JA, Mouillesseaux KP, et al. Role of the Jak/STAT pathway in the regulation of interleukin-8 transcription by oxidized phospholipids in vitro and in atherosclerosis in vivo. *J Biol Chem* 2007; **282**(43): 31460–31468.
8. Ricci R, Sumara G, Sumara I, et al. Requirement of JNK2 for scavenger receptor A-mediated foam cell formation in atherogenesis. *Science* 2004; **306**(5701): 1558–1561.
9. Meijer CA, Le Haen PA, van Dijk RA, et al. Activator protein-1 (AP-1) signalling in human atherosclerosis: Results of a systematic evaluation and intervention study. *Clin Sci (Lond)* 2012; **122**(9): 421–428.
10. Seimon TA, Wang Y, Han S, et al. Macrophage deficiency of p38alpha MAPK promotes apoptosis and plaque necrosis in advanced atherosclerotic lesions in mice. *J Clin Invest* 2009; **119**(4): 886–898.
11. Yamada S, Ding Y, Tanimoto A, et al. Apoptosis signal-regulating kinase 1 deficiency accelerates hyperlipidemia-induced atheromatous plaques via suppression of macrophage apoptosis. *Arterioscler Thromb Vasc Biol* 2011; **31**(7): 1555–1564.
12. Kim TW, Febbraio M, Robinet P, et al. The critical role of IL-1 receptor-associated kinase 4-mediated NF-kappaB activation in modified low-density lipoprotein-induced inflammatory gene expression and atherosclerosis. *J Immunol* 2011; **186**(5): 2871–2880.
13. Levin MC, Jirholt P, Wramstedt A, et al. Rip2 deficiency leads to increased atherosclerosis despite decreased inflammation. *Circ Res* 2011; **109**(11): 1210–1218.
14. Wang HW, Liu PY, Oyama N, et al. Deficiency of ROCK1 in bone marrow-derived cells protects against atherosclerosis in LDLR$^{-/-}$ mice. *FASEB J* 2008; **22**(10): 3561–3570.
15. Franco C, Britto K, Wong E, et al. Discoidin domain receptor 1 on bone marrow-derived cells promotes macrophage accumulation during atherogenesis. *Circ Res* 2009; **105**(11): 1141–1148.
16. Baumgartl J, Baudler S, Scherner M, et al. Myeloid lineage cell-restricted insulin resistance protects apolipoprotein E-deficient mice against atherosclerosis. *Cell Metab* 2006; **3**(4): 247–256.
17. Ferri N, Carragher NO, Raines EW. Role of discoidin domain receptors 1 and 2 in human smooth muscle cell-mediated collagen remodeling: Potential implications in atherosclerosis and lymphangioleiomyomatosis. *Am J Pathol* 2004; **164**(5): 1575–1585.
18. Shen J, Chandrasekharan UM, Ashraf MZ, et al. Lack of mitogen-activated protein kinase phosphatase-1 protects ApoE-null mice against atherosclerosis. *Circ Res* 2010; **106**(5): 902–910.
19. Zakkar M, Chaudhury H, Sandvik G, et al. Increased endothelial mitogen-activated protein kinase phosphatase-1 expression suppresses proinflammatory activation at sites that are resistant to atherosclerosis. *Circ Res* 2008; **103**(7): 726–732.
20. Wang J, An FS, Zhang W, et al. Inhibition of c-Jun N-terminal kinase attenuates low shear stress-induced atherogenesis in apolipoprotein E-deficient mice. *Mol Med* 2011; **17**(9–10): 990–999.
21. Kardakaris R, Gareus R, Xanthoulea S, Pasparakis M. Endothelial and macrophage-specific deficiency of P38alpha MAPK does not affect the pathogenesis of atherosclerosis in ApoE$^{-/-}$ mice. *PLoS ONE* 2011; **6**(6): e21055.
22. Jagavelu K, Tietge UJ, Gaestel M, Drexler H, Schieffer B, Bavendiek U. Systemic deficiency of the MAP kinase-activated protein kinase 2 reduces atherosclerosis in hypercholesterolemic mice. *Circ Res* 2007; **101**(11): 1104–1112.
23. Imaizumi S, Grijalva V, Priceman S, et al. Mitogen-activated protein kinase phosphatase-1 deficiency decreases atherosclerosis in apolipoprotein E null mice by reducing monocyte chemoattractant protein-1 levels. *Mol Genet Metab* 2010; **101**(1): 66–75.

24. Zhou Q, Gensch C, Liao JK. Rho-associated coiled-coil-forming kinases (ROCKs): Potential targets for the treatment of atherosclerosis and vascular disease. *Trends Pharmacol Sci* 2011; **32**(3): 167–173.
25. Rekhter M, Chandrasekhar K, Gifford-Moore D, et al. Immunohistochemical analysis of target proteins of Rho-kinase in a mouse model of accelerated atherosclerosis. *Exp Clin Cardiol* 2007; **12**(4): 169–174.
26. Nohria A, Grunert ME, Rikitake Y, et al. Rho kinase inhibition improves endothelial function in human subjects with coronary artery disease. *Circ Res* 2006; **99**(12): 1426–1432.
27. Fuster JJ, Gonzalez-Navarro H, Vinue A, et al. Deficient p27 phosphorylation at serine 10 increases macrophage foam cell formation and aggravates atherosclerosis through a proliferation-independent mechanism. *Arterioscler Thromb Vasc Biol* 2011; **31**(11): 2455–2463.
28. Goossens P, Vergouwe MN, Gijbels MJ, et al. Myeloid IkappaBalpha deficiency promotes atherogenesis by enhancing leukocyte recruitment to the plaques. *PLoS ONE* 2011; **6**(7): e22327.
29. Fernandez-Hernando C, Jozsef L, Jenkins D, Di Lorenzo A, Sessa WC. Absence of Akt1 reduces vascular smooth muscle cell migration and survival and induces features of plaque vulnerability and cardiac dysfunction during atherosclerosis. *Arterioscler Thromb Vasc Biol* 2009; **29**(12): 2033–2040.
30. Shi Y, Cosentino F, Camici GG, et al. Oxidized low-density lipoprotein activates p66Shc via lectin-like oxidized low-density lipoprotein receptor-1, protein kinase C-beta, and c-Jun N-terminal kinase kinase in human endothelial cells. *Arterioscler Thromb Vasc Biol* 2011; **31**(9): 2090–2097.
31. Dong Y, Zhang M, Liang B, et al. Reduction of AMP-activated protein kinase alpha2 increases endoplasmic reticulum stress and atherosclerosis in vivo. *Circulation* 2010; **121**(6): 792–803.
32. Zhang Y, Qiu J, Wang X, Xia M. AMP-activated protein kinase suppresses endothelial cell inflammation through phosphorylation of transcriptional coactivator p300. *Arterioscler Thromb Vasc Biol* **31**(12): 2897–2908.
33. Li Y, Xu S, Mihaylova MM, et al. AMPK phosphorylates and inhibits SREBP activity to attenuate hepatic steatosis and atherosclerosis in diet-induced insulin-resistant mice. *Cell Metab* **13**(4): 376–388.
34. Sun C, Wu MH, Yuan SY. Non-muscle myosin light-chain kinase deficiency attenuates atherosclerosis in apolipoprotein E-deficient mice via reduced endothelial barrier dysfunction and monocyte migration. *Circulation* 2011; **124**(1): 48–57.
35. Franco C, Hou G, Ahmad PJ, et al. Discoidin domain receptor 1 (ddr1) deletion decreases atherosclerosis by accelerating matrix accumulation and reducing inflammation in low-density lipoprotein receptor-deficient mice. *Circ Res* 2008; **102**(10): 1202–1211.
36. Hou G, Vogel WF, Bendeck MP. Tyrosine kinase activity of discoidin domain receptor 1 is necessary for smooth muscle cell migration and matrix metalloproteinase expression. *Circ Res* 2002; **90**(11): 1147–1149.
37. Lu KK, Trcka D, Bendeck MP. Collagen stimulates discoidin domain receptor 1-mediated migration of smooth muscle cells through Src. *Cardiovasc Pathol* 2011; **20**(2): 71–76.
38. Woo KV, Qu X, Babaev VR, et al. Tie1 attenuation reduces murine atherosclerosis in a dose-dependent and shear stress-specific manner. *J Clin Invest* 2011; **121**(4): 1624–1635.
39. Raj T, Kanellakis P, Pomilio G, Jennings G, Bobik A, Agrotis A. Inhibition of fibroblast growth factor receptor signaling attenuates atherosclerosis in apolipoprotein E-deficient mice. *Arterioscler Thromb Vasc Biol* 2006; **26**(8): 1845–1851.
40. Che J, Okigaki M, Takahashi T, et al. Endothelial FGF receptor signaling accelerates atherosclerosis. *Am J Physiol Heart Circ Physiol* 2011; **300**(1): H154–H161.
41. Andrae J, Gallini R, Betsholtz C. Role of platelet-derived growth factors in physiology and medicine. *Genes Dev* 2008; **22**(10): 1276–1312.
42. Shafi S, Lamb D, Modjtahedi H, Ferns G. Periadventitial delivery of anti-EGF receptor antibody inhibits neointimal macrophage accumulation after angioplasty in a hypercholesterolaemic rabbit. *Int J Exp Pathol* 2010; **91**(3): 224–234.
43. Han S, Liang CP, DeVries-Seimon T, et al. Macrophage insulin receptor deficiency increases ER stress-induced apoptosis and necrotic core formation in advanced atherosclerotic lesions. *Cell Metab* 2006; **3**(4): 257–266.

44. Gonzalez-Navarro H, Vinue A, Vila-Caballer M, et al. Molecular mechanisms of atherosclerosis in metabolic syndrome: Role of reduced IRS2-dependent signaling. *Arterioscler Thromb Vasc Biol* 2008; **28**(12)**:** 2187–2194.
45. Galkina EV, Butcher M, Keller SR, et al. Accelerated atherosclerosis in Apoe$^{-/-}$ mice heterozygous for the insulin receptor and the insulin receptor substrate-1. *Arterioscler Thromb Vasc Biol* 2012; **32**(2)**:** 247–256.
46. Thorp E, Cui D, Schrijvers DM, Kuriakose G, Tabas I. Mertk receptor mutation reduces efferocytosis efficiency and promotes apoptotic cell accumulation and plaque necrosis in atherosclerotic lesions of apoe$^{-/-}$ mice. *Arterioscler Thromb Vasc Biol* 2008; **28**(8)**:** 1421–1428.
47. Katsume A, Okigaki M, Matsui A, et al. Early inflammatory reactions in atherosclerosis are induced by proline-rich tyrosine kinase/reactive oxygen species-mediated release of tumor necrosis factor-alpha and subsequent activation of the p21Cip1/Ets-1/p300 system. *Arterioscler Thromb Vasc Biol* 2011; **31**(5)**:** 1084–1092.
48. Pertovaara M, Raitala A, Juonala M, et al. Autoimmunity and atherosclerosis: Functional polymorphism of PTPN22 is associated with phenotypes related to the risk of atherosclerosis. The cardiovascular risk in young Finns study. *Clin Exp Immunol* 2007; **147**(2)**:** 265–269.
49. Vang T, Congia M, Macis MD, et al. Autoimmune-associated lymphoid tyrosine phosphatase is a gain-of-function variant. *Nat Genet* 2005; **37**(12)**:** 1317–1319.
50. Saccucci P, Banci M, Cozzoli E, et al. Atherosclerosis and PTPN22: A study in coronary artery disease. *Cardiology* 2011; **119**(1)**:** 54–56.

36

EFFECT OF MMPs ON CARDIOVASCULATURE AND BLOOD FLOW: ROLE OF ROS, MITOPHAGY, AND ANTIOXIDANTS

Thomas P. Vacek, Diana Neamtu, and Suresh C. Tyagi
Department of Physiology and Biophysics, School of Medicine, University of Louisville, Louisville, KY, USA

INTRODUCTION

Homocysteine (Hcy) is a sulfur-containing, nonproteinogenic amino acid formed during the metabolism of methionine [1]. Elevated levels of total plasma Hcy are known as hyperhomocysteinemia (HHcy). Hcy is a well-known molecule that has been thoroughly researched but continues to leave us with further questions. It has been well documented to be an independent risk factor for heart disease [2, 3]. Several dismiss this data and say that Hcy must just be an innocent bystander. However, when feeding mice or rats Hcy or in mice/rats lacking cystathionine-β-synthase (CBS) gene, these animals do proceed with developing multisystem organ failure such as chronic kidney disease as well as heart failure [3, 4]. This tends to occur in the higher ranges of blood Hcy levels, however.

A mechanistic role for this has also been discovered using cells, and Hcy is known to act as a ligand on specific receptors such as NMDA-R1 receptor that increases calcium influx upstream and promotes an oxidizing environment within the cell [5]. Hcy is also known to bind to PPARγ receptor and downregulate this beneficial receptor, thereby creating an oxidizing environment in the cell with its downregulation [6]. Activation of proteins such as calcium-dependent calpain-1 and increase of the mitochondrial permeability transition pore (MPTP) within mitochondria have been documented in destabilizing this organelle and promoting an increase of radicals within the cell [7]. These radicals can proceed to cleave matrix metalloproteinase (MMP) enzyme to activate this protein. Overactivation of this protein has been shown in various disease profiles such as atherosclerosis,

A part of this study was supported by NIH grants: HL-71010; HL-74185; HL-88012 and NS-51568.

Atherosclerosis: Risks, Mechanisms, and Therapies, First Edition. Edited by Hong Wang and Cam Patterson.

osteoporosis, kidney disease, neurodegenerative disease, and so on [8–11].

Moreover, there is a link of Hcy to the autophagic process by which mitochondria is degraded [12, 13]. The damage to mitochondria, incurred by decoupling mitochondria to electron transport chain via disruption of the intermitochondrial membrane (IMM), encourages the autophagic/mitophagic process of destroying this organelle [14]. There is still some debate of whether autophagy or mitophagy is protective or detrimental [15]. The role of Hcy is obvious when considering how it increases ROS within cells [16]. Other more subtle roles involve Hcy acting to modulate proteins involved in the multimetric machinery that is used in the actual autophagic process. For instance, Hcy has been shown to decrease expression levels of Bcl-2 protein that is involved in the balance of autophagy versus retention preservation of organelles [17].

MMP expression is known to occur in various pathological profiles such as atherosclerosis and osteoporosis. Hence, the increased expression of MMPs under HHcy is of special concern since ameliorating this process may lead to resolution of the disease state within patients. Some molecules that have been proposed to do this are antioxidants such as STS, an H_2S donor, as well as folate [5, 18]. Further studies with other antioxidant molecules as well as vitamins like B12 and folate are warranted to discover possible routes by which Hcy levels or the oxidation state within cells may be lowered or neutralized.

MMPs IN ATHEROSCLEROSIS AND CORONARY VASCULAR DISEASE

MMPs have a very interesting role in atherosclerosis with different MMPs known to have different roles in promoting vessel remodeling and plaque evolution/stability. One study indicated that the drug atorvastatin stabilized plaques in acute coronary syndrome and that this was correlated with a decrease in MMP-9 and hsCRP [19]. Another study concluded that higher MMP-9 levels were observed in ventricle of patients with previous myocardial infarction. Another study has confirmed that the increase of MMP-9 and MMP-2 activity are correlated with plaque instability [20]. The role of MMP-2 and MMP-9 in plaque evolution has shown that they cleave fibrillar collagen that is intact as well as nonfibrillar and fragmented interstitial collagen [21, 22]. Another study has shown that MMP-3 is involved in neointima formation and migration of smooth muscle cells (SMC). In fact, neointima was reduced by 75% in mice with a knockout for this MMP [23]. Moreover, MMP-3 and MMP-9 appeared to act in synergy to promote this neointima formation within atherosclerotic vessels. It also appeared that MMP-12 played a redundant role in doing this [23]. Another study has shown that increase in MMP-9 was associated with pulmonary vascular resistance, lending further evidence to its destructive role in atherosclerosis when overactivated [24]. It is very interesting to note the different roles that each MMP has in the evolution of an atherosclerotic plaque.

It is easy to believe that all MMPs are villains when it comes to atherosclerosis after being presented with the results of these studies. However, some MMPs have been shown to be beneficial in cases of carotid atherosclerosis. Patients were more prone to plaque occurrence when carrying polymorphisms for the genes associated with producing MMP-8. MMP-8 is involved in cleaving collagen type 1. Hence, the absence of this MMP-8 was shown to be detrimental and made patients more prone to having plaques [25]. Another study has shown that the inhibition of MMP-13 results in collagen accumulation in plaques [26]. This feature in humans has been associated with a resistance to rupture. Hence, this MMP-13 decrease in activity by inhibitor was shown to be beneficial by allowing more collagen to accumulate in the plaque, thereby stabilizing the plaque [26]. The stabilization of the plaque prevents plaque rupture.

The clinical application of this is endless when it comes to discerning prognosis of patients who already are experiencing atherosclerotic changes. Moreover, it is also very

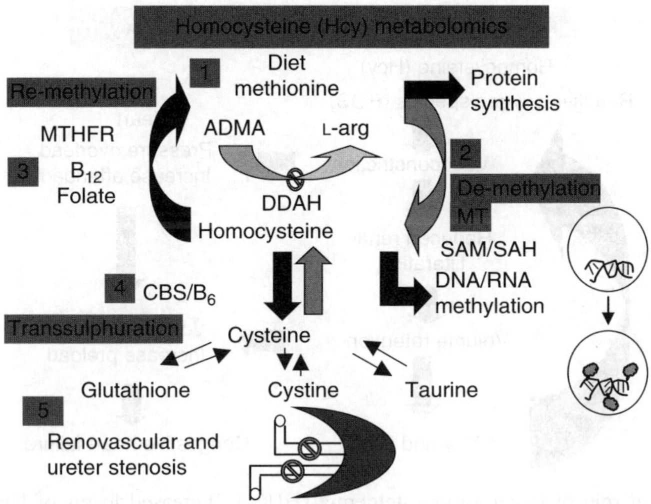

Figure 36.1 Homocysteine (Hcy) metabolomics. There are five pathways involved in Hcy metabolism: increased methionine in diet, demethylation of methionine via SAM/SAH enzyme, conversion of Hcy to methionine via MTHFR enzyme B12 and folate, conversion of Hcy to cysteine via CBS enzyme and vitamin B6, as well as renovascular reabsorption of the molecule.

useful in determining future risk of patients. When presented with these individual unique risks, perhaps the excuse that "John Smith lived to be 90 years old and smoked two packs a day with no problems" will no longer suffice. It's clear that John Smith most likely had a robust genetic profile for handling these sort of free radical insults on his vessels. Most others are not so lucky.

HCY METABOLOMICS

Hcy is an amino acid that consists of cysteine with an additional methylene group. High levels of Hcy can occur through various metabolic blockades of Hcy metabolomics. The first way is via an increase in unutilized protein in the diet that increases methionine levels. Hence, it is proposed that the excess protein in the diet may potentially be detrimental to the body and may promote increased levels of Hcy. There is a demethylation of methionine that is catalyzed by enzymes SAM/SAH, thereby producing Hcy. Hcy can also be converted back to methionine via the enzyme MTHFR using the cofactors B12 and folate. Furthermore, Hcy can be converted to cysteine via CBS enzyme with its vitamin B6 cofactor. If the CBS enzyme is damaged, such as in the heterozygote mouse, $CBS^{+/-}$, then this can produce a state of HHcy. Moreover, cysteine can have subtle molecular modifications that can change this molecule to glutathionine, cystine, and taurine. This is illustrated in Figure 36.1.

INTEGRAL ROLE OF HHCY

There is an integral role of HHcy in causing congestive heart failure. Hcy has been shown to incur an increase in reactive oxygen species (ROS) that leads to vasoconstriction [27]. An increase in vasoconstriction decreases the amount of blood that is delivered to the kidney via the renal arteries. A reduced renal blood flow also reduces the renal filtration rate or glomerular filtration rate (GFR). The GFR is used as a general marker of kidney function. This results in volume retention that produces a state of volume overload whereby there is too much fluid within the cardiovascular system. More blood also reaches the heart resulting in an increase of preload, or increase of volume

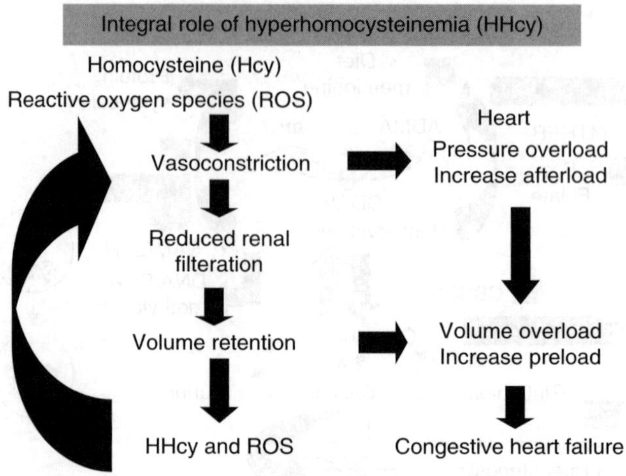

Figure 36.2 Integral role of hyperhomocysteinemia (HHcy). Increased levels of Hcy lead to increased ROS that leads to vasoconstriction, reduced renal filteration, and volume retention, leading to further increase in Hcy and ROS and further vasoconstriction. Vasoconstriction leads to pressure overload on the heart with increased afterload. Volume retention leads to volume overload and increased preload. This leads to congestive heart failure.

in the left ventricle before contraction. An increase in preload of blood can lead to congestive heart failure as blood backs up into the left atrium and lungs. This back pressure leads to fluid seeping through lung alveoli and congestive heart failure. An increase in ROS from Hcy also damages the vasculature, thereby promoting atherosclerosis. Atherosclerosis is a narrowing of the arteries that functions physiologically in the same way as a constriction of the arteries since the passage by which blood flows becomes less and flow can become less. This produces a state of increased afterload or increased pressure back on the heart that it must overcome in order to pump blood systemically. This can lead to hypertrophic cardiomyopathy and heart failure. This process is illustrated in Figure 36.2.

EFFECT OF HCY AND VOLUMETRIC LOAD ON BODY MMPs

It is known that the increased presence of ROS is influenced by NADPH oxidase, superoxide dismutase (SOD), peroxiredoxin (Prx), and catalase. Reactive nitrogen species (RNS) are modulated by inducible nitrous oxide enzyme (iNOS), mitochondrial nitrous oxide (mtNOS), endothelial nitrous oxide (eNOS), and neuronal nitric oxide synthase (nNOS). Reactive thiosulfate (RTS) is modulated by thioredoxin (Trx) and GSSG/GSH oxidase. Moreover, ROS, RNS, and RTS are all increased under conditions of increased volumetric load within the cardiovasculature, diabetes, Hcy, and renin–angiotensin–aldosterone system. This increases MMP 1/13, MMP 2/7/9/12 as well as membrane type matrix metalloproteinase (MT-MMP), metalloproteinase (ADAM), and meprin (MMP-1/MMP-13). This influences angiostatin (plasminogen), endostatin (collagen-18), and parstatin (PAR-1). This influence results in antiangiogenesis and chronic heart failure (CHF) from increased afterload and increased blood backing up through the heart to the lungs. MMP-2, MMP-7, MMP-9, and MMP-12 influence TGF-β and VEGF that results in angiogenesis. Tissue inhibitors of MMP-1 inhibit MMP-1 and MMP-13 as well as MMP-2, MMP-7, MMP-9, and MMP-12. Tissue inhibitor of metalloproteinase (TIMP-4) inhibits MMP-2, MMP-7, MMP-9, and MMP-12, while that of TIMP-3

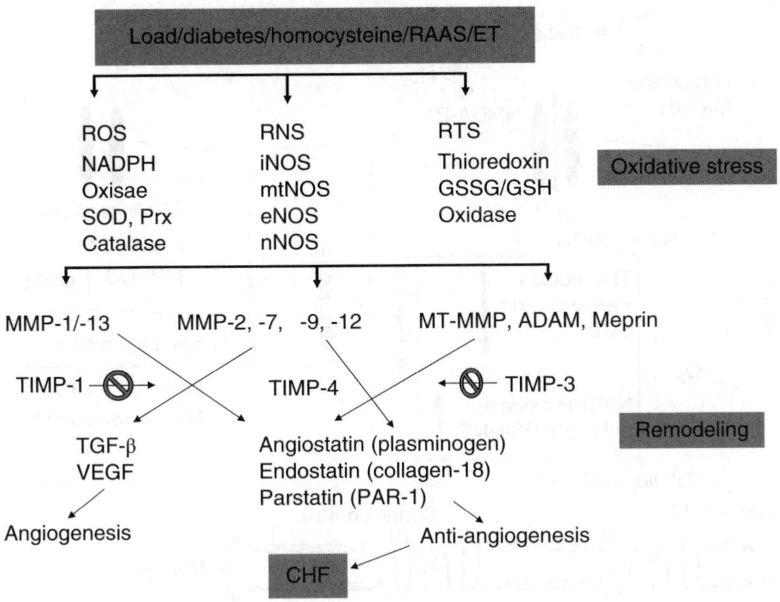

Figure 36.3 Effects of load/diabetes/Hcy/RAAS/ET. Increased load, diabetes, Hcy, RAAS dysregulation, and ET increase ROS, RNS, and RTS. ROS are influenced by NADPH oxidase, SOD, Prx, and catalase. RNS is influenced by iNOS, mtNOS, eNOS, and nNOS. RTS is influenced by Trx, GSSG/GSH, and oxidase. This leads to increase MMP-1/-MMP13, increase MMP-2/MMP-7/MMP-9/MMP-12, and increase MT-MMP, ADAM, and meprin. TIMP-1, TIMP-4, and TIMP-3 are involved in inhibiting these MMPs. The MMPs in combination with TGF-β and VEGF influence angiogenesis. MT-MMP, ADAM, and meprin can act together with angiostatin, endostatin, and parstatin to influence antiangiogenesis and CHF.

inhibits MT-MMP, ADAM, and meprin. This is illustrated in Figure 36.3.

The activation of MMPs via RTS, RNS, and ROS occurs by cleaving a particular molecular group on the latent MMP enzyme combined with NO molecule and TIMP enzyme that inhibits a particular MMP, resulting in the production of NO_2 and tyrosine or NO_2 and arginine. Hence, the MMP/NO/TIMP combination is converted to an oxidized TIMP+MMP active whereby the TIMP and NO are no longer attached.

The balance of MMP-9 and TIMP-4 is disrupted by Hcy by binding to such receptors as PAR-1, leading to an increase of MMP-9 and a decrease of TIMP-4. This disrupted balance leads to contractile dysfunction. Moreover, one mechanism by which HHcy can cause arrhythmia is via Hcy binding to NMDA-R1 receptor, resulting in an influx of calcium. Dizocilpine MK-801 is an NMDA-R1 blocker that can inhibit the effects of Hcy agonist on NMDA-R1 receptor. Using this receptor-specific blocker, we were able to determine that Hcy acts as an agonist to NMDA-R1 receptor. The increase in calcium can also lead to the increase of iNOS and production of NO. Within mitochondria, there is a decrease of Prx, Trx, and SOD and an increase in NADPH oxidase, mtNOS, mitochondrial reactive oxygen species (mtROS), and mitochondrial reactive nitrogen species (mtRNS). This leads to the production of O^{2-} molecules. Together with NO, this produces peroxynitrite ($ONOO^-$) that leads to activation of MMP and disruption of molecules called connexins. Connexins are channels between myocytes that allow myocytes to communicate electrically and to transmit a concordant electrical signal without being uncoupled. Hence, the destruction of these connexins (e.g., connexin-43) leads to a proarrhythmic state. This is illustrated in Figure 36.4.

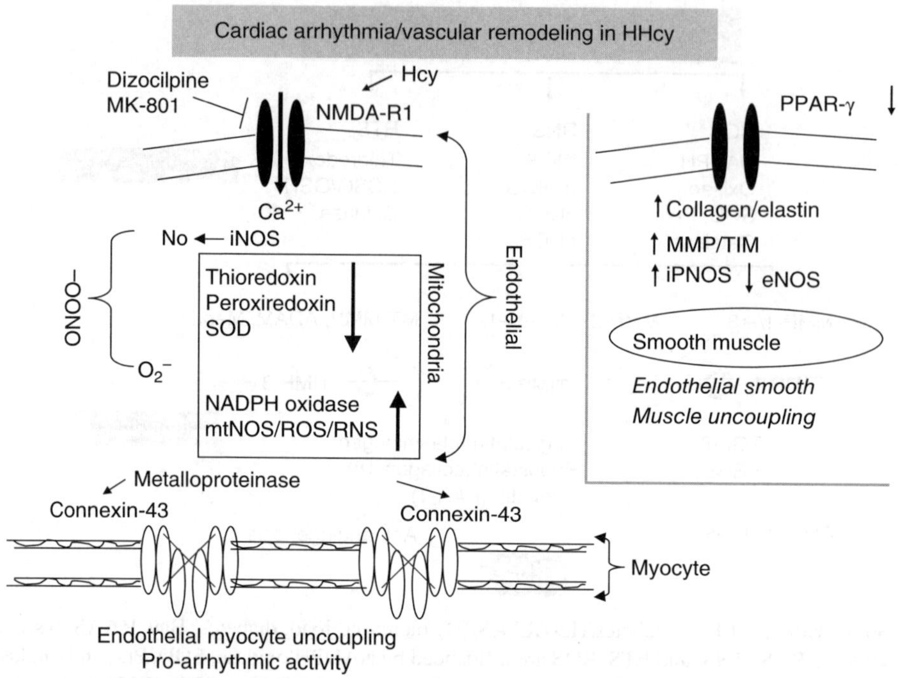

Figure 36.4 Mechanism of cardiac arrhythmia in HHcy. Hcy acts as agonist to NMDA-R1, and dizocilpine MK-801 acts as a blocker of NMDA-R1 receptor. Activation of receptor increases calcium transients. Increase iNOS leads to NO and superoxide activation of MMPs. Within mitochondria, there is a decrease in Trx, Prx, and SOD. There is an increase in NADPH oxidase, mtNOS, ROS, and RNS. MMPs degrade connexin-43 within the myocyte. These lead to endothelial–myocyte uncoupling and a proarrhythmic state. Hcy decrease expression of PPARγ receptor. This leads to a decrease in Trx and Prx, increase in NADH oxidase, and increase in ROS/RNS/mtNOS. From this cellular state, iNOS increases, eNOS decreases, MMP and TIMP increase, and collagen and elastin increase. There then becomes an endothelial–smooth muscle uncoupling.

OTHER MECHANISMS OF HCY ROS PRODUCTION

Hcy has been known to act as a ligand on PPARγ that is actually known to produce a reducing environment [28]. However, the continual agonizing of this receptor actually will result in a downregulation of this reducing environment, leading to an oxidizing environment within the cell. In fact, the expression of PPARγ is decreased with increasing exposure to Hcy. An activated PPARγ increases the expression of SOD and catalase by decreasing NAD/NADPH oxidase [29]. Agonists of PPAR decrease oxidative stress and MMP activity within the cell. This results in an intramitochondrial oxidative stress [29].

An increase in plasma levels or tissue can lead to a decrease in PPARγ receptor that is known to promote a reducing environment. Hence, this leads to a decrease in Trx and Prx. There is also an increase in NADH oxidase as well as ROS, RNS, and mtNOS within mitochondria. This results in an increase in iNOS, decrease in eNOS, increase in MMP/TIMP, and increase in collagen/elastin of smooth muscle. This leads to an endothelial–smooth muscle uncoupling. When there is an uncoupling, there can be an invasion of the vessel by molecules like LDL to promote the inflammatory process of atherosclerosis. This process is illustrated in Figure 36.4.

HCY AND BONE BLOOD FLOW

A proper blood flow is essential to so many systemic processes that any kind of alteration of blood flow also alters the ability of organ systems to function properly. Several studies have shown that HHcy can alter the functionality of vasculature to dilate since it decreases NO bioavailability [30]. This increases vascular resistance, thereby allowing for a decrease in blood flow and a decrease of nutrients to the organs supplied [30].

Our lab has shown that there is a decrease in bone blood flow of tibia in rats under HHcy conditions with rats having been fed significant concentrations of Hcy in drinking water before developing HHcy [27, 31]. Kumar et al. has also shown a decrease in carotid artery blood flow in CBS heterozygote mice with elevated plasma levels of Hcy in those mice having caused an increase in arterial remodeling due to an increase in collagen/elastin ratio [32]. The increase in collagen/elastin ratio led to an increase in resistance and a measurable decrease in carotid artery blood flow [32]. Moreover, it was shown in other studies that Hcy decreased the ability of the body to create parallel circuits of arteries to reduce vascular resistance. Hence, this increase in resistance occurs as a consequence of decrease in angiogenesis [33].

In a follow-up study on bone blood flow, our lab has discovered that there was an increased iNOS level as well as a decrease eNOS level [34]. Blood flow was examined with three different agonists in CBS heterozygotes versus control mice: acetylcholine (an endothelial-dependent vasodilator), nitroprusside (an endothelial-independent vasodilator), and phenylephrine (an alpha-adrenergic vasoconstrictor) [34]. There was a significant decrease in tibial artery flux with CBS heterozygotes compared to CT mice [34]. This provided another experimental model under which a decrease in bone blood flow was noted. The mechanism of this is proposed to be via a decrease in NO bioavailability and the ability of the vessel to vasodilate. Hence, a decrease in bone blood flow resulted from these conditions.

FAILING HEART AND NO AND AUTOPHAGY

Several studies have suggested that mitochondria play a role in inducing the autophagic process by which a cell will consume itself. It has been suggested that the placement of the MPTP that permanently depolarizes the mitochondria and doesn't allow it to act as a capacitor may be an upstream event inducing autophagy of the cardiomyocyte in ischemia/reperfusion [13]. Aged cardiomyocytes have shown an accumulation of supersized mitochondria [13]. An aging mitochondria has been known to exhibit such qualities as swollen appearance, disrupted cristae, destruction of intramitochondrial enzymes, as well as mutations in the mitochondrial DNA. This has also been known to disrupt the energy-making process; moreover, the disruption of the electron transport chain results in a decrease in ATP production—the currency that the body uses for energy [13].

A failing heart has been shown to produce large amounts of iNOS, which decreases heart performance and contractility. In fact, our lab has shown that Hcy acts as an upstream event to increase ROS expression, increase iNOS, and decrease eNOS. This results in a decrease of the NO bioavailability to the cell. NO was also shown to play an important role in protection during ischemia/reperfusion. During ischemia, there is activation of nitric oxide synthase logically since the body would prefer the vessel to be free flowing and NO helps to facilitate this by dilating the vessel. Hence, there is an activation of guanylyl cyclase, production of cGMP, and activation of protein kinase G [35]. It remains unclear whether NO plays some direct role in initiating the autophagic process. The use of an NO donor, SIN-1, has been used in studies to determine this. Administration of this molecule increased several proapoptotic caspases with no change in genes associated with autophagy: *beclin-1*, *Atg51*, and *Atg* [35]. It was concluded from this study that NO was not a trigger for autophagy and does not play a role in autophagy-influenced heart failure.

Autophagy and the formation of the preautophagosome are dependent on several proteins in the Atg class labeled by different numbers for slightly different permutations. These proteins form a multimetric protein that allows autophagy to take place. Some major players include the following proteins: Atg9, Atg17, Atg8, Atg12, and Atg16 [35]. Some studies have found that mice that are deficient in several of these proteins actually develop hypertrophic cardiomyopathy as senescent cells begin to accumulate large mitochondria and other intracellular remnants [35].

Moreover, Hcy is thought to play a role in the autophagic process of heart via decrease of Bcl-2, a protein that serves in the balance between apoptosis of cell and autophagy [36]. It has been shown that administration of Hcy actually decreased the expression of Bcl-2, thereby altering this balance of apoptosis versus autophagy [36]. A decrease of Bcl-2 might allow autophagy to proceed by allowing another protein involved in autophagy, beclin-1, to continue in its actions. It is unclear whether this is beneficial or maladaptive. A study has shown that disrupting the gene encoding beclin-1 served to decrease autophagy in myocytes and also even diminished pressure-induced pathological remodeling [35]. Autophagy that is pressure load induced has been shown to be detrimental. Moreover, autophagy has been suggested to be protective during ischemia but destructive during the reperfusion stage [35]. So there is still some debate on whether it is beneficial or destructive. However, it does appear that Hcy has some role in beginning this process with the documented effect in expression of the autophagy protein Bcl-2.

There is a clear link of an increase in serum Hcy also increasing ROS production via binding of receptors that facilitate intracellular production of ROS. ROS includes the free radicals themselves, like superoxide or other molecules that are capable of creating an oxidized state. Hcy, for example, is a known inducer of iNOS that generates radicals [37]. There appears also to be a role of the MAPK cascade that Hcy is known to affect. In fact, inhibitors of MAPK cascade have been shown to decrease ROS and decrease the incidence of autophagy [35]. Autophagy, as it appears in some circumstances, acts to decrease the presence of ROS. Mitochondria are known to produce ROS, and the autophagic process of consuming mitochondria may actually be compensatory and beneficial to the cell as a means of neutralizing radicals or ridding the cells of damaged protein products.

Hcy has been known to elevate calcium levels by acting as a ligand to NMDA-R1 receptor, which induces Ca^{2+} and K^{+} currents [38]. This allows for an influx of calcium intracellularly. An increase in cytosolic calcium is proapoptotic—inducing autophagy through activation of calmodulin-dependent kinase [39]. An increase in intracellular calcium and intramitochondrial calcium also results in an increase of ROS[40]. An increase in ROS is a signal for autophagy to occur as intracellular damage accumulates. The rate at which mitochondria accumulate radicals is affected by the coupling state of mitochondria, of which calcium has a great influence. With increased calcium, there is an increased production of superoxide that leads to mitochondria damage and promotes the autophagic process [40].

MMPs REMODELING AND MITOCHONDRIA

The cardiovasculature is a very dynamic system that is able to adapt to new stresses. Not only can the vasculature dilate or constrict in response to chemical agonists, but it is also capable of a physical change in size or composition. For instance, when constructing a venous graft, the venous graft will actually remodel itself with the newly imposed pressures placed on the graft. The graft will then begin to appear more like an artery with a greater proportion of smooth muscle layer [41]. Unfortunately, some remodeling can be pathological with excessive stresses placed on the system leading to such inflammatory processes as atherosclerosis that narrows arteries and decreases blood flow. This involves the migration of SMCs from the media to intima and endothelial smooth muscle [42].

MMPs play a major role in this pathological remodeling process. In fact, it was found that inhibitors of MMPs actually inhibited the SMC and intimal thickening [43]. Since activation of MMPs is known to have this pathological role, one could also say that Hcy is also detrimental by this measure since Hcy is known to activate MMPs. MMPs are found everywhere within and around the cell: intracellularly, extracellularly, intranuclearly, and intramitochondrialy. The complex pathways by which MMPs are activated and translocated from one location to another are beginning to be elucidated. For instance, our lab has discovered that Hcy binding to NMDA-R1 receptor causes an influx of calcium ions [44]. This influx of calcium has been shown to activate a calcium-dependent enzyme calpain-1 that proceeds with disrupting IMM [45]. In fact, using a calpain inhibitor actually reduced the expression of MMP-2 and MMP-9, lending more evidence to its role in this process [45]. This disruption has been shown to increase ROS within mitochondria in association with activation of MMPs and extrusion of MMP intracellularly. Activation of ERK 1/2 has also been shown to be involved in this process [45]. This is not a mutually exclusive pathway, however, since its activation also involves calcium influx into the cell. The calcium influx will then allow for phosphorylation and activation of ERK 1/2 as shown by our lab. Of course, the initial influx of calcium via activation of NMDA-R1 is also crucial for beginning this cascade with a final result of activating MMPs. To determine this, the NMDA-R1 blocker, MK-801, was used [45]. Upon blocking NMDA-R1, the MPTP within mitochondria was not put into place.

HYDROGEN SULFIDE PROTECTS AGAINST VASCULAR REMODELING

The vasculature must overcome some very great pressures and shearing forces on a daily basis to function properly. H_2S is a molecule that can act as an antioxidant as well as to dilate the vessels and decrease the stresses that are imposed on them. The protein SOD is involved in neutralizing free radicals within the cell. Hence, an increased expression of SOD is viewed as a beneficial thing for the cell so that the cell may not be exposed to harmful molecules. TIMPs directly bind to MMP molecule and act to inhibit the MMP molecule until the complex is cleaved and the MMP will be allowed to degrade matrix. A balance exists between these two molecules such that if active MMPs are higher, TIMPs are found in lower concentration within the cell. This makes sense since for the MMP molecule to be active, the TIMP molecule must no longer be bound to the MMP molecule.

Our lab has demonstrated that with AVF heart failure, there is an increased expression of the MMP-2 and MMP-9. Moreover, this expression of MMP-2 and MMP-9 was reduced to control levels with administration of H_2S donor STS [46]. We hypothesized in our lab that physical damage of the vessel via plastic catheter, or even damage by volume and pressure overload caused by HHcy, would result in a pathological condition within the cell. This condition includes increased ROS and increased expression of MMP-2 as well as MMP-9. Moreover, we proposed that the antioxidant and vasodilator H_2S would restore normal remodeling [47]. These results were confirmed by our lab. This provides some promising data for certain compounds to administer to patients under pathological conditions or interventional stress caused by intra-arterial catheters that are so often used with patients.

Dosage of H_2S is a controversial matter, however, since low concentrations of H_2S have been shown to cause vasoconstriction due to the blocked vasoprotective effect of NO and nitrosothiol [48]. Moreover, some studies have suggested that H_2S can be damaging in certain concentrations, or oxidizing, while acting as an antioxidant at other concentrations. H_2S may prove to be a practical molecule to be used under certain pathological conditions such as pressure or volume overload, as well as under interventional stress. However, concentration must be closely monitored if it is to have a beneficial role in ameliorating damage or preventing future damage to occur via inflammatory process of atherosclerosis and SMC migration. The experiments conducted by our lab still suggest a role for certain antioxidant agents in ameliorating damage from interventional stress or Hcy-induced pathological changes within the cardiovasculature.

CONCLUSION

Hcy continues to be a fascinating mystery in its role of being present under disease states as well as proving to be a disease-causing factor itself. The role of MMPs in modifying cardiovasculature and causing system-wide pathologies is of special concern to us. The mechanisms by which Hcy incurs disease are still being discovered as well as the routes to try to alleviate these disease states via antioxidants including H_2S, folate, and B12. The role of Hcy and autophagy/mitophagy is yet another realm to delve into. It is still unclear whether this process is beneficial or detrimental to the cell, but there appears to be a few starting points, including protein Bcl-2 of which Hcy has been shown to decrease the expression.

REFERENCES

1. Perry TL, Hansen S, Bar HP, Macdougall L. Homocystinuria: Excretion of a new sulfur-containing amino acid in urine. *Science*. 1966; 152:776–778.
2. Vizzardi E, Bonadei I, Zanini G, et al. Homocysteine and heart failure: An overview. *Recent Patents on Cardiovascular Drug Discovery*. 2009;4:15–21.
3. Sen U, Munjal C, Qipshidze N, Abe O, Gargoum R, Tyagi SC. Hydrogen sulfide regulates homocysteine-mediated glomerulosclerosis. *American Journal of Nephrology*. 2010;31:442–455.
4. Walker E, Black J, Parris C, et al. Effect of experimental hyperhomocysteinemia on cardiac structure and function in the rat. *Annals of Clinical and Laboratory Science*. 2004;34: 175–180.
5. Tyagi N, Mishra PK, Tyagi SC. Homocysteine, hydrogen sulfide (H_2S) and NMDA-receptor in heart failure. *Indian Journal of Biochemistry & Biophysics*. 2009;46:441–446.
6. Gillespie W, Tyagi N, Tyagi SC. Role of PPAR gamma, a nuclear hormone receptor in neuroprotection. *Indian Journal of Biochemistry & Biophysics*. 2011;48:73–81.
7. Moshal KS, Kumar M, Tyagi N, et al. Restoration of contractility in hyperhomocysteinemia by cardiac-specific deletion of NMDA-R1. *American Journal of Physiology. Heart and Circulatory Physiology*. 2009;296: H887–H892.
8. Moshal KS, Metreveli N, Frank I, Tyagi SC. Mitochondrial MMP activation, dysfunction and arrhythmogenesis in hyperhomocysteinemia. *Current Vascular Pharmacology*. 2008;6: 84–92.
9. Guo LJ, Luo XH, Wu XP, et al. Serum concentrations of MMP-1. MMP-2, and TIMP-1 in Chinese women: Age-related changes, and the relationships with bone biochemical markers, bone mineral density. *Clinica Chimica Acta*. 2006;371:137–142.
10. Sen U, Basu P, Abe OA, et al. Hydrogen sulfide ameliorates hyperhomocysteinemia-associated chronic renal failure. *American Journal of Physiology. Renal Physiology*. 2009;297: F410–F419.
11. Miller A, Mujumdar V, Shek E, et al. Hyperhomocyst(e)inemia induces multiorgan damage. *Heart and Vessels*. 2000;15:135–143.
12. Tyagi N, Vacek JC, Givvimani S, Sen U, Tyagi SC. Cardiac specific deletion of n-methyl-D-aspartate receptor 1 ameliorates mtMMP-9 mediated autophagy/mitophagy in hyperhomocysteinemia. *Journal of Receptor and Signal Transduction Research*. 2010;30:78–87.
13. Vacek TP, Vacek JC, Tyagi SC. Mitochondrial mitophagic mechanisms of myocardial matrix metabolism and remodelling. *Archives of Physiology and Biochemistry*. 2012;118:31–42.
14. Kurihara Y, Kanki T, Aoki Y, et al. Mitophagy plays an essential role in reducing mitochondrial production of reactive oxygen species and mutation of mitochondrial DNA by maintaining mitochondrial quantity and quality in yeast. *The Journal of Biological Chemistry*. 2012;287: 3265–3272.

15. De Meyer GR, De Keulenaer GW, Martinet W. Role of autophagy in heart failure associated with aging. *Heart Failure Reviews*. 2010;15: 423–430.
16. Tyagi N, Qipshidze N, Sen U, Rodriguez W, Ovechkin A, Tyagi SC. Cystathionine beta synthase gene dose dependent vascular remodeling in murine model of hyperhomocysteinemia. *International Journal of Physiology, Pathophysiology, and Pharmacology*. 2011;3: 210–222.
17. Dong F, Zhang X, Wold LE, Ren Q, Zhang Z, Ren J. Endothelin-1 enhances oxidative stress, cell proliferation and reduces apoptosis in human umbilical vein endothelial cells: Role of ETB receptor, NADPH oxidase and caveolin-1. *British Journal of Pharmacology*. 2005;145:323–333.
18. Tyagi N, Moshal KS, Sen U, et al. H_2S protects against methionine-induced oxidative stress in brain endothelial cells. *Antioxidants & Redox Signaling*. 2009;11:25–33.
19. Yu DQ, Lin SG, Chen JY, et al. Effect of atorvastatin therapy on borderline vulnerable lesions in patients with acute coronary syndrome. *Archives of Medical Science*. 2011;7:433–439.
20. Heo SH, Cho CH, Kim HO, et al. Plaque rupture is a determinant of vascular events in carotid artery atherosclerotic disease: Involvement of matrix metalloproteinases 2 and 9. *Journal of Clinical Neurology*. 2011;7:69–76.
21. Aimes RT, Quigley JP. Matrix metalloproteinase-2 is an interstitial collagenase—inhibitor-free enzyme catalyzes the cleavage of collagen fibrils and soluble native type-I collagen generating the specific 3/4-length and 1/4-length fragments. *The Journal of Biological Chemistry*. 1995;270:5872–5876.
22. Shah PK, Galis ZS. Matrix metalloproteinase hypothesis of plaque rupture—players keep piling up but questions remain. *Circulation*. 2001;104:1878–1880.
23. Johnson JL, Dwivedi A, Somerville M, George SJ, Newby AC. Matrix metalloproteinase (MMP)-3 activates MMP-9 mediated vascular smooth muscle cell migration and neointima formation in mice. *Arteriosclerosis, Thrombosis, and Vascular Biology*. 2011;31:E35–U23.
24. Kawut SM, Barr RG, Johnson WC, et al. Matrix metalloproteinase-9 and plasminogen activator inhibitor-1 are associated with right ventricular structure and function: The MESA-RV study. *Biomarkers*. 2010;15:731–738.
25. Djuric T, Stankovic A, Koncar I, et al. Association of MMP-8 promoter gene polymorphisms with carotid atherosclerosis: Preliminary study. *Atherosclerosis*. 2011;219:673–678.
26. Quillard T, Tesmenitsky Y, Croce K, et al. Selective inhibition of matrix metalloproteinase-13 increases collagen content of established mouse atherosclerosis. *Arteriosclerosis, Thrombosis, and Vascular Biology*. 2011;31: 2464–U2276.
27. Tyagi N, Kandel M, Munjal C, et al. Homocysteine mediated decrease in bone blood flow and remodeling: Role of folic acid. *Journal of Orthopaedic Research*. 2011;29: 1511–1516.
28. Mishra PK, Tyagi N, Sen U, Joshua IG, Tyagi SC. Synergism in hyperhomocysteinemia and diabetes: Role of PPAR gamma and tempol. *Cardiovascular Diabetology*. 2010;9:49.
29. Tyagi N, Moshal KS, Sen U, Lominadze D, Ovechkin AV, Tyagi SC. Ciglitazone ameliorates homocysteine-mediated mitochondrial translocation and matrix metalloproteinase-9 activation in endothelial cells by inducing peroxisome proliferator activated receptor-gamma activity. *Cell Mol Biol (Noisy-le-grand)*. 2006;52:21–27.
30. Banfi G, Iorio EL, Corsi MM. Oxidative stress, free radicals and bone remodeling. *Clinical Chemistry and Laboratory Medicine*. 2008;46: 1550–1555.
31. Tyagi N, Vacek TP, Fleming JT, Vacek JC, Tyagi SC. Hyperhomocysteinemia decreases bone blood flow. *Vascular Health and Risk Management*. 2011;7:31–35.
32. Kumar M, Tyagi N, Moshal KS, et al. Homocysteine decreases blood flow to the brain due to vascular resistance in carotid artery. *Neurochemistry International*. 2008;53: 214–219.
33. Zhu JH, Wang XX, Fu GS, Shang YP, Zhang FR, Chen JZ. Reduced number and activity of circulating endothelial progenitor cells in patients with idiopathic pulmonary arterial hypertension. *Respiratory Medicine*. 2008;102: 1073–1079.
34. Tyagi N, Kandel M, Munjal C, et al. Homocysteine mediated decrease in bone blood flow and remodeling: Role of folic acid. *Journal of Orthopaedic Research*. 2011;29:1511–1516.
35. Vacek TP, Vacek JC, Tyagi N, Tyagi SC. Autophagy and heart failure: A possible role for homocysteine. *Cell Biochemistry and Biophysics*. 2012;62:1–11.

36. Dong F, Zhang XC, Wold LE, Ren Q, Zhang ZJ, Ren J. Endothelin-1 enhances oxidative stress, cell proliferation, and reduces apoptosis in human umbilical vein endothelial cells: Role of ETB receptor, NADPH oxidase, and caveolin-1. *British Journal of Pharmacology*. 2005;145: 323–333.
37. Steed MM, Tyagi N, Sen U, Schuschke DA, Joshua IG, Tyagi SC. Functional consequences of the collagen/elastin switch in vascular remodeling in hyperhomocysteinemic wild-type, eNOS(–/–), and iNOS(–/–) mice. *American Journal of Physiology. Lung Cellular and Molecular Physiology*. 2010;299:L301–L311.
38. Moshal KS, Tipparaju SM, Vacek TP, et al. Mitochondrial matrix metalloproteinase activation decreases myocyte contractility in hyperhomocysteinemia. *American Journal of Physiology. Heart and Circulatory Physiology*. 2008;295: H890–H897.
39. Hoyer-Hansen M, Bastholm L, Szyniarowski P, et al. Control of macroautophagy by calcium, calmodulin-dependent kinase kinase-beta, and Bcl-2. *Molecular Cell*. 2007;25:193–205.
40. Dalton TD, Shertzer HG, Puga A. Regulation of gene expression by reactive oxygen. *Annual Review of Pharmacology and Toxicology*. 1999; 39:67–101.
41. Shormanov SV, Pshenisnov KP. Changes in the walls of autologous venous grafts after transplantation into arteries. *Arkhiv Anatomii, Gistologii I Embriologii*. 1987;92:19–25.
42. Geer JC, Haust MD. Smooth muscle cells in atherosclerosis. *Monographs on Atherosclerosis*. 1972;2:1–140.
43. Bendeck MP, Irvin C, Reidy MA. Inhibition of matrix metalloproteinase activity inhibits smooth muscle cell migration but not neointimal thickening after arterial injury. *Circulation Research*. 1996;78:38–43.
44. Moshal KS, Kumar M, Tyagi N, et al. Restoration of contractility in hyperhomocysteinemia by cardiac-specific deletion of NMDA-R1. *American Journal of Physiology. Heart and Circulatory Physiology*. 2009;296:H887–H892.
45. Moshal KS, Singh M, Sen U, et al. Homocysteine-mediated activation and mitochondrial translocation of calpain regulates MMP-9 in MVEC. *American Journal of Physiology. Heart and Circulatory Physiology*. 2006;291:H2825–H2835.
46. Sen U, Vacek TP, Hughes WM, et al. Cardioprotective role of sodium thiosulfate on chronic heart failure by modulating endogenous H_2S generation. *Pharmacology*. 2008;82: 201–213.
47. Vacek TP, Gillespie W, Tyagi N, Vacek JC, Tyagi SC. Hydrogen sulfide protects against vascular remodeling from endothelial damage. *Amino Acids*. 2010;39:1161–1169.
48. Ali MY, Ping CY, Mok YY, et al. Regulation of vascular nitric oxide in vitro and in vivo: A new role for endogenous hydrogen sulphide? *British Journal of Pharmacology*. 2006;149:625–634.

SECTION V

CLINICAL MANAGEMENT

37

IMPACT OF PREVENTIVE THERAPIES ON CLINICAL MANAGEMENT AND OUTCOMES

DANNY EAPEN[1], AHSAN ACHTCHI[2], REYNARIA NIEVA[3], KIRAN VALIANI[2], FARESA ZARREEN[2], AALOK PATEL[4], ALLEN DOLLAR[1], IJEOMA ISIADINSO[2], SUSMITA PARASHAR[2], JEFFERSON BAER[1], KRETON MAVROMATIS[2], AND LAURENCE SPERLING[1]

[1]*Department of Medicine, The Center for Heart Disease Prevention, Emory University School of Medicine, Atlanta, GA, USA*
[2]*Emory University School of Medicine, Atlanta, GA, USA*
[3]*University of Colorado, Denver, CO, USA*
[4]*Mayo Clinic, Rochester, MN, USA*

INTRODUCTION

An ounce of prevention is worth a pound of cure —Benjamin Franklin

Cardiovascular disease (CVD) death rates decreased by 30% from 1998 to 2008, in part because of the significant impact of preventive therapies [1]. Despite this, CVD remains the leading cause of death for both men and women in the United States and worldwide. Many of the risk factors for CVD, such as diabetes, hypertension, physical inactivity, and smoking, are targets for effective prevention efforts. Extensive research has shown that among patients at risk for or with coronary heart disease (CHD), risk factor modification can improve outcomes. It has also been well documented that the use of specific therapies after myocardial infarction (MI), such as aspirin, beta-adrenergic antagonists, angiotensin-converting enzyme (ACE) inhibitors, and statins, results in a decrease in cardiovascular (CV) mortality.

In 2011, the American Heart Association (AHA) and the American College of Cardiology (ACC) published an update of the guidelines on secondary prevention for coronary and atherosclerotic vascular disease. These recommendations outline target goals for management of CVD risk factors, as well as evidence-based medical therapies for patients with CHD [2]. While secondary preventive efforts are readily accepted by most clinicians, the most effective and cost-efficient strategy in the long term is primary prevention.

This chapter will focus on the impact that medical therapy, diet, and psychosocial factors have on CVD. Current clinical data related to the prevention and medical management of CVD will be reviewed. Emerging therapies

Atherosclerosis: Risks, Mechanisms, and Therapies, First Edition. Edited by Hong Wang and Cam Patterson.

will be highlighted as well. Increased acceptance and knowledge of a comprehensive approach to CV prevention by health professionals can translate into further decreases in CV morbidity and mortality.

EPIDEMIOLOGIC AND ECONOMIC IMPACT OF CVD PREVENTIVE MEASURES

According to recent estimates by the AHA, the annual combined cost of CVD and stroke in the United States is approximately $297.7 billion [1]. CVD tops the list of diagnoses responsible for direct healthcare spending in the United States, accounting for 16% of total health-related expenditures. It is expected that the aging of the U.S. population will cause an increase in the prevalence of CVD, and in turn, an increase in future CVD-related expenditures. By 2030, 40.5% of the US population is projected to have some form of CVD [1]. In view of such predictions there has been increasing interest in the primary and secondary prevention of atherosclerosis to reduce the societal and economic burden of CVD.

The AHA and ACC joint guidelines recommend a series of secondary prevention measures to combat the progression of disease in patients with known atherosclerosis. These guidelines encourage physical activity and tobacco cessation, and set optimal goals for blood pressure, lipids, and hemoglobin A1C [2]. Interventions with potentially high value include the use of statins in hyperlipidemia, diuretics and ACE inhibitors for hypertension, ACE inhibitors and beta-blockers for post-MI treatment, implantable cardiac defibrillators for patients at high risk for sudden death, antiplatelet agents for patients with known CVD, and warfarin to prevent future stroke in patients with nonvalvular atrial fibrillation. Regular exercise, cardiac rehabilitation, and smoking cessation programs are also cost-effective [3, 4].

Primary prevention, which focuses on the modification of risk factors before the development of CVD, is also a growing area of interest. Risk factors that account for the largest share of the disease burden include tobacco consumption, high blood pressure, obesity, and physical inactivity [5]. Few validated trials exist on primary prevention in part because, in contrast to curative care, prevention has no identifiable beneficiaries and is commonly characterized by immediate costs and delayed benefits. Assessing the value of prevention is also difficult because the time span before an apparently healthy individual manifests signs of disease may be many decades, which makes follow-up difficult.

Primary prevention should not be minimized, however, as it has the potential to greatly reduce the economic burden of CVD by both preventing early death and limiting morbidity. A recent report by Trust for America's Health showed that an investment of $10 per person per year in proven community-based prevention programs could save the U.S. more than $16 billion annually within 5 years based on medical cost savings alone. Additional unmeasured savings could also be assumed from increased worker productivity or quality-of-life measures [5].

Efforts must also be directed toward reducing health disparities, as lower socioeconomic and educational status are established risk factors for CVD. Ethnicity also plays a role, as risk factors for CVD are disproportionally prevalent in certain populations, including non-Hispanic blacks, American Indians, Hispanics/Latinos, and Pacific Islanders compared to non-Hispanic whites [5].

ANTIHYPERTENSIVE THERAPY

Hypertension remains the most important modifiable risk factor for the prevention of atherosclerosis in both men and women. A meta-analysis of a million adults in 61 prospective studies and decades of epidemiologic evidence demonstrates that CV risk increases consistently with increased blood pressure, beginning as low as 115/75 mmHg and increasing risk nearly twofold per 20 mmHg increase in systolic blood pressure (SBP) in patients 40–69 years of age [6]. Treatment of hypertension with primary mitigation of CV risk has proven to be substantially more cost-effective than expensive secondary interventions after the occurrence of CV events, including thrombolysis and coronary artery bypass [7].

Beta-adrenergic Antagonists

Beta-adrenergic antagonists (beta-blockers) are effective in reducing blood pressure through their negative inotropic effects and their inhibition of renin–angiotensin–aldosterone system (RAAS) as well as beta-1 blockade with agents such as carvedilol. The Seventh Report of the Joint National Committee on Prevention, Detection, Evaluation, and Treatment of High Blood Pressure (JNC VII) [8], however, advises the use of beta-blockers as second-line agents for the treatment of uncomplicated hypertension, as other antihypertensive agents offer significantly better protection from adverse CV events in the setting of primary prevention. Beta-blockers are first-line agents, however, in patients with concomitant ischemic heart disease or congestive heart failure [8]. Beta-blockers may slow the progression of atherosclerosis and help in the stabilization of potentially vulnerable coronary artery plaques, thereby reducing the risk of initial CV events and recurrent events in patients with known atherosclerosis [9]. In addition, they improve exercise tolerance [10], reduce anginal symptoms [11], prevent ventricular arrhythmias [2] and sudden cardiac death [12], and reduce CV and all-cause mortality in individuals with prior ST elevation myocardial infarction (STEMI) [13]. The ACC/AHA guidelines [6] recommend long-term use of a beta-adrenergic antagonist in all patients without contraindications who are not considered to be low risk post-STEMI (class I/level of evidence A) [14].

ACE Inhibitors and Angiotensin Receptor Blockers

The JNC VII guidelines recommend ACE inhibitors and angiotensin receptor blockers (ARBs) as second-line therapy in uncomplicated hypertension after thiazide diuretics or as first-line therapy in combination with thiazide diuretics in patients requiring more than one agent. ACE inhibitors are indicated as first-line therapy in the setting of many significant comorbid conditions, including congestive heart failure, post-MI, chronic renal disease, diabetes mellitus, and stroke. A meta-analysis of 29,805 subjects at risk for or with established CHD indicated that ACE inhibitors significantly reduced all-cause and CV mortality, nonfatal MI, all stroke, heart failure, and need for coronary artery bypass surgery [15].

In contrast to ACE inhibitors, ARBs appear to have less impact on prevention of recurrent MI or primary events in high-risk patients. The rates of these outcomes are no higher than placebo, however, and the antihypertensive effects are equivalent to ACE inhibitors, which are modest [16]. In patients with symptomatic heart failure, left ventricular ejection fraction less than 40%, comorbid diabetes mellitus or kidney disease, and an inability to tolerate ACE inhibitors, ARBs are a reasonable alternative for RAAS blockade, alone or in combination with aldosterone receptor antagonists [8].

Calcium Antagonists

Calcium antagonists are second-line therapy in uncomplicated hypertension after thiazide diuretics, or as first-line therapy in combination with thiazide diuretics in patients requiring more than one agent. When used in the primary prevention of atherosclerosis via treatment of hypertension, they demonstrate benefit in patients at high risk for future events [17].

LIPID THERAPIES

Lipid targets for the primary and secondary prevention of CV events have dramatically evolved over the last four decades. Initial epidemiologic evidence showed a positive correlation between total cholesterol and coronary events. Subsequent randomized trials have demonstrated a clear benefit in lowering low-density lipoprotein cholesterol (LDL-C) in reducing CV events and mortality. Although elevated triglycerides are associated with CVD, a causal relationship has not been demonstrated when controlling for other lipid risk factors, such as LDL-C and HDL-C. In contrast, high-density lipoprotein cholesterol (HDL-C) has a strong inverse relationship with CV outcomes. However, clinical trials with HDL-C-raising therapies have shown mixed results, and currently, this approach to CVD prevention is unclear.

HMG CoA Reductase Inhibitors (Statins)

Statins competitively inhibit HMG CoA reductase, the rate-limiting enzyme in the hepatic cholesterol synthesis pathway. Potency in reducing total cholesterol and LDL-C varies among different statins. Total cholesterol may be reduced anywhere between 14–40% and LDL-C by 17–63%. Statin therapy has potential pleiotropic activities, such as decreased inflammation, oxidation, and endothelial injury, all crucial effects for combating atherosclerotic heart disease.

Large randomized trials have demonstrated significant CV risk reduction in both primary and secondary prevention. In secondary prevention trials, risk is decreased by 23–44% in patients with both high and relatively low LDL-C [18]. Similar CV risk reduction has been shown in primary prevention studies, such as the WOSCOPS, AFCAPS/TexCAPS, and JUPITER trials. Statin use in patients with established vascular disease, in addition to individuals with increased vascular risk, is effective and is first-line lipid therapy.

Niacin

Niacin, or nicotinic acid, reduces total cholesterol, LDL-C, and triglycerides, and increases HDL-C through several mechanisms. Niacin is the most potent HDL-C-raising drug currently available, and acts to inhibit HDL-C hepatic uptake, thereby decreasing HDL-C catabolism and increasing HDL-C serum concentration. *In vitro* studies have shown that niacin inhibits vascular inflammation, LDL oxidation, and vascular adhesion of macrophages and monocytes [19].

As a monotherapy and in combination with statins, niacin has been shown to decrease atherosclerosis and CV events [20]. Despite these and other studies showing clinical benefit of niacin, the AIM-HIGH trial, a recent randomized study, did not show clinical benefit to patients with stable CHD who were already medically optimized. The Drug Safety and Monitoring Board stopped AIM-HIGH prematurely because of lack of efficacy and futility; in addition, there were several related methodological problems, including a concern that it was underpowered. Results of the HPS2-THRIVE study may better define the role of niacin as a potential addition to statin therapy in patients with stable CHD.

Fibrates

Hypertriglyceridemia has been associated with CVD, but this correlation is attenuated when controlling for LDL-C and HDL-C [21]. Though fibrates such as gemfibrozil and fenofibrate typically have only mild LDL-C-lowering effects, they can lower triglycerides by 35–50% and raise HDL-C by 20%.

Clinical studies related to fibrates have shown mixed results. In the VA-HIT study, gemfibrozil treatment of patients with a history of CVD and low HDL-C and low LDL-C levels led to a significant reduction in CV events. Findings in subsequent trials, such as BIP, ACCORD, and FIELD, have largely failed to show significant reduction in primary end points of MI, stroke, or death from a CV cause. In most of these trials, the average triglyceride level was well below 200 mg/dl. Post hoc analysis of BIP and ACCORD studies did show a significant reduction in primary end points for patients with triglycerides ≥200 mg/dl. Thus, combined treatment with statin and fibrate may play a more significant role in a select group of patients with higher baseline triglyceride levels.

Cholesterol Absorption Inhibitors

Cholesterol absorption inhibitors such as ezetimibe limit the intestinal absorption of exogenous cholesterol and thereby lower LDL-C. In the ENHANCE trial, treatment with ezetimibe combined with simvastatin failed to show a beneficial effect on carotid artery intima–media thickness when compared to treatment with simvastatin alone. The IMPROVE-IT trial may better define the effect of ezetimibe in combination with a statin on CV outcomes.

Bile Acid Sequestrants

Bile acid sequestrants (BAS) prevent bile acid reabsorption in the small intestines, lowering LDL-C by up to 10–24% [22]. The Lipid Research Clinics Coronary Primary Prevention Trial (LRL-CPPT) and the National Heart, Lung, and Blood Institute (NHLBI) Type II Coronary Intervention study conducted in the 1970s were the earliest landmark studies showing a significant reduction in LDL-C and CV risk in patients treated with cholestyramine compared to placebo [22]. The most common side effects of BAS are gastrointestinal such as constipation, dyspepsia, and nausea. Colesevelam was developed in the 1990s and has better tolerability than cholestyramine and colestipol. Used as an adjunct to statin therapy, BAS may help further decrease LDL-C in patients who are not yet at goal.

LDL Apheresis

LDL apheresis is most commonly used in the treatment of familial hypercholesterolemia (FH). It also has a role in patients suffering from severe dyslipidemias (LDL-C >300 or >200 with CHD) who are either treatment resistant or intolerant to maximal antilipid therapy [23].

Postapheresis reduction in LDL-C can be as significant as 80%. The frequency at which patients must undergo apheresis depends on the degree of hyperlipidemia. Usually, patients with homozygous FH must undergo apheresis every 7–10 days, while those with heterozygous FH must undergo FH every 14 days. The average cost of an apheresis session ranges from $2500 to $4500, making it a very expensive treatment modality. LDL-C reduction and improvement in CV mortality have been largely studied in patients with FM in uncontrolled settings. Further investigation is ongoing.

EMERGING LIPID THERAPIES

Introduction

Investigation of emerging lipid therapies is an active area of development. Several of these therapies are outlined in this section.

Cholesteryl Ester Transfer Protein Inhibitors

Cholesteryl ester transfer protein (CETP) plays a major role in HDL-C metabolism and mediates the transfer of cholesteryl esters from HDL-C- to triglyceride-rich lipoproteins such as LDL-C and Very Low Density Lipoprotein Cholesterol (VLDL). Exchange of cholesteryl ester for triglyceride yields lower HDL-C levels and increased plasma concentrations of LDL-C and VLDL. Patients with CETP gene mutations resulting in low levels of serum CETP were shown to have a lower CV risk than age-match controls [24]. Although the first CETP inhibitor, torcetrapib, was effective in increasing HDL-C by 61%, it was associated with increased mortality compared to controls, likely due to off-target effects such as hypertension and metabolic alkalosis [25]. Second-generation CETP inhibitors like dalcetrapib and anacetrapib are also potent HDL-C-raising agents that thus far seem effective and safe. Large clinical trials are underway to evaluate clinical and CV outcomes for dalcetrapib and anacetrapib.

Mipomersen

Mipomersen, a stabilized antisense oligonucleotide, is administered as a weekly injection to target the messenger RNA for apolipoprotein B, thus inhibiting the translation of the molecule that is essential for LDL-C production. The drug is currently under phase 3 investigation. Phase 2 data has demonstrated an additional 21–52% reduction in LDL-C in patients with hypercholesterolemia already on stable statin therapy treated with mipomersen [26]. More recent phase 3 trials demonstrate that mipomersen offers promise in patients with statin intolerance and patients with FM, as this may be a more cost-effective alternative to LDL apheresis. Furthermore, it has been shown to reduce lipoprotein(a) [27]. Further investigation on the long-term safety profile of mipomersen is still ongoing.

Proprotein Convertase Subtilisin/Kexin Type 9 Inhibitors

Proprotein convertase subtilisin/kexin type 9 (PCSK9) is a circulating zymogen protein that is responsible for the degradation of LDL receptors, thereby leading to a decreased clearance and increased serum levels of LDL-C. Most data regarding the effects of PCSK9 in humans come from observational studies in individuals with loss-of-function mutations in PCSK9 that lead to lower levels of LDL-C with subsequent reduction in CV mortality without observable deleterious effects. Conversely, families with gain-of-function mutations and associated higher levels of LDL-C are subject to increased levels of CV morbidity and mortality [28].

PCSK9 inhibition represents a novel strategy for lowering LDL-C levels. Two major categories of PCKS9 inhibitors currently under investigation include agents that either (i) suppress PCSK9 formation via induction of mRNA degradation or (ii) interfere with the interaction of PCSK9 and LDL receptors via neutralizing antibodies against PCSK9 [28].

ANTIPLATELET THERAPY

Introduction

Acute atherothrombosis occurs rapidly and, in most cases, begins with the process of plaque rupture or erosion. This results in platelet activation (via exposure to vascular wall components such as tissue factor and collagen) and aggregation (via glycoprotein IIb–IIIa complex and fibrinogen binding). These platelets then release the cytokines thromboxane A2 and adenosine diphosphate (ADP), causing more platelet activation/aggregation, leading to thrombus formation. Thus, inhibiting platelet activation is a mainstay of the prevention of atherothrombotic events.

Aspirin

Primary prevention of atherothrombotic events with aspirin in individuals at high risk for atherosclerosis, but without an established diagnosis, is also beneficial. Aspirin therapy may be used in patients at moderate to high risk for a nonfatal MI. Individual cardiac risk assessment may be performed with tools such as the Framingham Risk Calculator. Generally, patients with a 10-year risk greater than 10% benefit from aspirin therapy.

Use of aspirin for secondary prevention in patients with known atherosclerosis has been well studied. Meta-analyses have shown that aspirin reduces the risk of CV events by 18% ($P<0.0001$) [29]. Similar risk reductions are seen with doses ranging from 75 to 600 mg/day [30]. However, such studies also demonstrate that the risk of gastrointestinal bleeding increases incrementally with aspirin dose. Hence, in patients with a history of CVD, ongoing low-dose (75–100 mg/day) aspirin therapy is preferentially recommended [31].

ADP Receptor (P2Y12) Inhibitors

This class of antiplatelet agents inhibits ADP-mediated platelet activation by binding to the ADP P2Y12 receptor. The thienopyridines—ticlopidine, clopidogrel, and prasugrel—are prodrugs that require hepatic activation via cytochrome P450-dependent processes and then bind irreversibly to the P2Y12 receptor [32]. Ticagrelor is a nonthienopyridine P2Y12 inhibitor that does not require hepatic activation and reversibly binds P2Y12.

Clopidogrel has been compared to aspirin in patients with known atherosclerotic disease and was found to yield a small relative risk reduction (8.7%) in CV events ($P=0.043$). Therefore, clopidogrel is recommended for secondary prevention of CV events in patients with aspirin intolerance [33].

Compared to aspirin alone, dual antiplatelet therapy with clopidogrel plus aspirin reduces the risk of recurrent CV events in patients with acute coronary syndromes [34]. Both prasugrel and ticagrelor have been studied in combination with aspirin in such patients, and found to offer benefit in reducing recurrent clinical CV events compared with clopidogrel plus aspirin [35, 36]. They have not, however, been well studied as stand-alone agents or in combination with aspirin in stable atherosclerotic disease.

VITAMINS AND HERBAL SUPPLEMENTS

Introduction

In the 2002 National Health Interview Survey, 36% of 10,572 respondents with CVD used complementary and alternative medicine in the past 12 months. The most common therapy used was herbal supplementation, with 10% of respondents using it specifically for CV conditions.

Omega-3 Polyunsaturated Fatty Acid

Omega-3 polyunsaturated fatty acid (PUFA) supplements have demonstrated pleiotropic effects on the CV system, from decreasing inflammation and oxidative stress to improving lipid profiles. In particular, omega-3 PUFA significantly decrease serum triglycerides at doses of at least 3–4 g/day [37]. In some studies, supplementation with lower doses of omega-3 PUFA has been shown to decrease the incidence of sudden cardiac death, arrhythmia, hypertension, and heart failure in patients with CVD. Some randomized studies have shown benefit with consumption of omega-3 PUFA, while others have not shown any significant benefit in CV outcomes and mortality (Table 37.1). Despite the mixed evidence, the AHA has recommended 1 g of omega-3 PUFA in the form of DHA and EPA for secondary prevention [38].

Multivitamins, Antioxidants, and B Vitamins

Multivitamins are the most common supplements consumed in the United States. Some observational studies have demonstrated a reduction in CV risk with multivitamin supplementation. Other studies have shown no clinical benefit or suggested a potentially harmful effect. The same holds true for vitamins with antioxidant properties (including vitamins E and C and beta-carotene). While hyperhomocysteinemia has been independently associated with increased CV risk and with B vitamins playing an integral role in homocysteine metabolism, B-vitamin supplementation has not demonstrated a reduction in CV risk. Currently, there is insufficient evidence to support the use of multivitamins, antioxidants, or B vitamins in either the primary or secondary prevention of CVD [39].

Vitamin D

Vitamin D is a fat-soluble vitamin, and its intracellular receptors are present in many tissues, including cells in the immune system, cardiomyocytes, vascular smooth muscles, and endothelium. Reduced levels have been associated with hypertension, atherosclerosis, and CVD. In the Framingham Offspring Study, subjects with low vitamin D levels had an associated hazard ratio of 1.62 ($P = 0.01$) for development of a CV event [40].

A meta-analysis, however, did not demonstrate benefit with vitamin D supplementation [41]; keeping the Framingham results in mind, this finding suggests that vitamin D deficiency might serve as a marker for other lifestyle habits such as decreased exercise and poor nutrition. Recommendations may be influenced by the ongoing VITAL trial.

DIETARY APPROACHES TO CVD

Current AHA recommendations are to balance caloric intake with physical activity in order to maintain body weight so that body mass index falls in the range of 18.5–24.9 kg/m^2. These recommendations (Table 37.2) are especially stressed when eating outside the home, where large portion sizes and high caloric density are common features [42].

TABLE 37.1 Omega-3 PUFA trials

Title	Formulation and dose	Number of subjects	Subject characteristics	Follow-up	Findings
Positive trials					
COMBOS trial	Lovaza 4 g/day	254	Patients on stable statin therapy Mean fasting TG > 200 Mean LDL-C < 10% of NCEP ATP III goal	8 weeks	Reduction of TG Increase in HDL-C No difference in adverse reactions
Diet and Reinfarction Trial (DART)	Dietary advice	2,033	Men with recent MI	2 years	29% reduction in all-cause mortality
GISSI Prevenzione Trial	EPA and DHA 850 mg	11,324	Patients with recent < 3 months of MI	3.5 years	10% relative risk decrease in CV events
Japan EPA Lipid Intervention Study (JELIS)	EPA 1800 mg	18,645	Patients with total cholesterol 6.5 mmol/l or greater	5 years	19% relative risk reduction of coronary events ($P = 0.011$) In patients with no history of CHD, nonsignificant difference in CV event
Negative trials					
Lack of benefit of dietary advice to men with angina	Dietary advice, two portions of fatty fish a week, or 3 g of fish oil supplement	3,114	Men with history of angina	3–9 years	No difference in all-cause mortality Subgroup taking fish oil capsules had increased risk of SCD that was attributed to compensational behaviors
Effects of B vitamins and omega-3 fatty acids on cardiovascular diseases	EPA and DHA 600 mg	2,501	Patients with history of MI, CVA or unstable angina	4.7 years	No significant difference in CV events (nonfatal MI, CVA, or CV death)
Alpha Omega Trial	EPA and DHA 400 mg	4,873	Patients with history of MI	40 months	No difference in primary end point of CV event
Omega Trial	EPA and DHA 1 gm	3,851	Patients admitted with STEMI or NSTEMI	1 year	No difference in all end points of CV events, death from CV cause, all-cause mortality, and revascularization

TABLE 37.2 American Heart Association recommendations for a healthy diet

Dietary modification	AHA recommendation	Comments
Fruits and vegetables (rich in nutrients, low in calories, and high in fiber)	Choose fruits and vegetables that are deeply colored (e.g., spinach, carrots, peaches, and berries) Prepare fruits and vegetables in a way to preserve the natural nutrient content without adding unnecessary calories, saturated/trans fat, sugar, or salt	Shown to decrease BP and improve CVD risk factors
Whole-grain and high-fiber foods	At least half of grain intake should come from whole grains	Modest reduction in LDL-C and CVD risk and slows progression of CVD in high-risk individuals. Soluble fiber increases short-chain fatty acid synthesis, thereby reducing endogenous cholesterol production
Fish (long-chain omega-3 PUFAs: eicosapentaenoic acid (EPA) and docosahexaenoic acid (DHA))	Consume fish, especially oily fish, at least twice a week	2 servings (8 oz) per week associated with decreased risk of sudden death and death from CHD Note: children and pregnant women should avoid fish with high mercury content
Saturated and trans fatty acids	Limit intake of saturated fat to < 7% of daily energy, trans fat to < 1% of daily energy, and cholesterol to < 300 mg/day by choosing lean meats and vegetable alternatives (beans and fish) and fat-free or low-fat dairy products. Also minimize intake of partially hydrogenated fats	This decreases CVD largely through effects on LDL-C cholesterol
"Added sugar" (e.g., sucrose, corn syrup, and high-fructose corn syrup)	Minimize beverages and foods with added sugar	Important to decrease total caloric intake and weight gain
Salt	Up to 2.3 g/day	Salt sensitivity is more important in blacks, middle aged, and elderly and those with HTN, DM, or CKD. Decrease in salt intake leads to an age-adjusted decline in blood pressures; this effect is found to be dose responsive (lower is better)
Alcohol	≤2 drinks a day for men and ≤1 drink per day for women, ideally with meals. Drink equivalent = 12 oz bottle of beer, 4 oz glass of wine, 1.5 oz shot of 80-proof spirits all contain the same amount of alcohol (one half ounce)	Note that alcohol can increase caloric intake and has the potential for addiction

Source: Adapted from Lichtenstein et al. [42]. © Lippincott Williams & Wilkins.

Whereas there is consensus that a healthy diet to reduce CVD risk should be low in saturated fat, there is debate about which macronutrient should be the predominant source of calories: carbohydrate, protein, or unsaturated fat. The Optimal Macronutrient Intake Trial to Prevent Heart Disease (OmniHeart) demonstrated that three separate diets—a carbohydrate-rich diet similar to the Dietary Approaches to Stop Hypertension (DASH) diet, a diet rich in protein (almost half from plant sources), and a diet rich in unsaturated, mostly monounsaturated fat—all improved blood pressure and lipid profiles [43].

Effect of Diet on Hypertension

Current dietary recommendations for the patient with prehypertension or frank hypertension focus on fresh fruits, vegetables, and low-fat dairy, as outlined in the DASH diet. The DASH diet recommends 4–5 servings of fruits, 4–5 servings of vegetables per day, 2–3 servings of low-fat dairy per day, and <25% fat. This diet can rapidly (within 2 weeks) and modestly lower blood pressure [44].

Sodium restriction is another important dietary modification for the control of blood pressure. Clinical trials have documented that a low sodium diet can prevent hypertension, with a relative risk reduction of 20% with or without concomitant weight loss [45]. Of note, African Americans and the elderly are specifically sensitive to sodium (as well as potassium) controlled diets. Other dietary measures that modulate hypertension include potassium supplementation, moderate alcohol intake, and vegetarian diet [42].

Effect of Diet on Dyslipidemia

Mediterranean diet In a meta-analysis of 50 clinical trials that collectively represents nearly 535,000 participants, the Mediterranean diet demonstrated a protective role by ameliorating risk-associated components of metabolic syndrome, including waist circumference, low HDL-C, high triglyceride levels, high blood pressure, and elevated blood glucose [46]. Greater improvements in lipid and inflammatory markers were also seen in the Mediterranean diet when compared to a low-fat diet (per AHA recommendations) [47].

Popular/fad diets Popular fad diets often promote low-carbohydrate and/or high-protein meal plans for weight loss, weight maintenance, and heart disease risk reduction. However, sustained individual compliance with these diets remains a problem. In a randomized controlled trial to evaluate effectiveness and adherence with four different dietary regimens (Weight Watchers diet, Atkins diet, Zone diet, and Ornish diet), 160 patients were randomized into four dietary patterns and followed for 12 months. Each diet showed modest declines in body weight, waist circumference, blood pressure, lipid panel, and blood sugar. Interestingly, there were no significant differences among the four diets. Overall, adherence to each diet was low, with the highest discontinuation rates in the Ornish and Atkins diet. The most common reasons cited for discontinuation were difficulty of maintaining the regimen and insufficient weight loss [48].

PSYCHOSOCIAL FACTORS IN CVD: STRESS AND DEPRESSION

Psychological stress and depression are associated with increased risk of MI [49]. Depression is a common condition in CVD patients, with rates as high as 25% in patients after MI or with congestive heart failure. Furthermore, data suggests that young women may be at particularly high risk for depression after acute MI [50]. Depression is a strong and independent risk factor for mortality, recurrent cardiac events, and lower functional status in several CHD patient groups, including patients with MI, unstable angina, and CHF. The importance of routine screening and treatment of depression in CVD patients is reflected in the 2011 AHA/ACC prevention guidelines [2].

CONCLUSION

The significant reduction in mortality from CVD over the past decade has in large part been related to the impact of preventive therapies on clinical management and outcomes. Competing with the gains largely achieved by more aggressive and widespread use of medical treatments, however, is the public and vascular health challenge related to the global obesity epidemic. Comprehensive efforts to address lifestyle and behavioral factors affecting CV risk are greatly needed. Importantly, several novel therapies currently in development may provide strategies for further risk reduction.

REFERENCES

1. Roger VL, Go AS, Lloyd-Jones DM, et al. Heart disease and stroke statistics–2012 update: A report from the American Heart Association. *Circulation*. 2012;125(1):e2–e220.
2. Smith SC, Jr, Benjamin EJ, Bonow RO, et al. AHA/ACCF secondary prevention and risk reduction therapy for patients with coronary and other atherosclerotic vascular disease: 2011 update. A guideline from the American Heart Association and American College of Cardiology Foundation. *Circulation*. 2011;124(22):2458–73.
3. Jeffrey LP. How cost-effective are new preventive strategies for cardiovascular disease? *Am J Cardiol*. 2003;91(10 Suppl. 1):22–7.
4. Braithwaite RS, Mentor SM. Identifying favorable-value cardiovascular health services. *Am J Managed Care*. 2011;17(6):431–8.
5. Weintraub WS, Daniels SR, Burke LE, et al. Value of primordial and primary prevention for cardiovascular disease. *Circulation*. 2011; 124(8):967–90.
6. Lewington S, Clarke R, Qizilbash N, Peto R, Collins R. Age-specific relevance of usual blood pressure to vascular mortality: A meta-analysis of individual data for one million adults in 61 prospective studies. *Lancet*. 2002;360(9349): 1903–13.
7. Gaziano TA. Cardiovascular disease in the developing world and its cost-effective management. *Circulation*. 2005;112(23):3547–53.
8. Chobanian AV, Bakris GL, Black HR, et al. Seventh report of the Joint National Committee on prevention, detection, evaluation, and treatment of high blood pressure. *Hypertension*. 2003;42(6):1206–52.
9. Sipahi I, Tuzcu EM, Wolski KE, et al. Beta-blockers and progression of coronary atherosclerosis: Pooled analysis of 4 intravascular ultrasonography trials. *Ann Intern Med*. 2007; 147(1):10–8.
10. Harris FJ, Low RI, Paumer L, Amsterdam EA, Mason DT. Antianginal efficacy and improved exercise performance with timolol. Twice-daily beta-blockade in ischemic heart disease. *Am J Cardiol*. 1983;51(1):13–8.
11. Aronow WS, Turbow M, Van Camp S, Lurie M, Whittaker K. The effect of timolol vs placebo on angina pectoris. *Circulation*. 1980;61(1): 66–9.
12. Anderson JL, Rodier HE, Green LS. Comparative effects of beta-adrenergic blocking drugs on experimental ventricular fibrillation threshold. *Am J Cardiol*. 1983;51(7): 1196–202.
13. Solomon SD, Zelenkofske S, McMurray JJ, et al. Sudden death in patients with myocardial infarction and left ventricular dysfunction, heart failure, or both. *N Engl J Med*. 2005; 352(25):2581–8.
14. Antman EM, Hand M, Armstrong PW, et al. 2007 focused update of the ACC/AHA 2004 guidelines for the management of patients with ST-elevation myocardial infarction: Report of the American College of Cardiology/American Heart Association Task Force on practice guidelines. *J Am Coll Cardiol*. 2008;51(2):210–47.
15. Dagenais GR, Pogue J, Fox K, Simoons ML, Yusuf S. Angiotensin-converting-enzyme inhibitors in stable vascular disease without left ventricular systolic dysfunction or heart failure: A combined analysis of three trials. *Lancet*. 2006;368(9535):581–8.
16. Bangalore S, Kumar S, Wetterslev J, Messerli FH. Angiotensin receptor blockers and risk of myocardial infarction: Meta-analyses and trial sequential analyses of 147020 patients from randomised trials. *BMJ*. 2011;342:d2234.
17. Waal-Manning HJ, Paulin JM, Simpson FO. MRC trial of treatment of mild hypertension. *Brit Med J (Clin Res Ed)*. 1985;291(6500):972.
18. Randomised trial of cholesterol lowering in 4444 patients with coronary heart disease: The Scandinavian Simvastatin Survival Study (4S). *Lancet*. 1994;344(8934):1383–9.

19. Kamanna VS, Kashyap ML. Mechanism of action of niacin. *Am J Cardiol.* 2008;101(8A): 20B–6B.
20. Brown BG, Zhao XQ, Chait A, et al. Simvastatin and niacin, antioxidant vitamins, or the combination for the prevention of coronary disease. *N Engl J Med.* 2001;345(22):1583–92.
21. Austin MA, Hokanson JE, Edwards KL. Hypertriglyceridemia as a cardiovascular risk factor. *Am J Cardiol.* 1998;81(4A):7B–12B.
22. Yamaoka-Tojo M, Tojo T, Izumi T. Beyond cholesterol lowering: Pleiotropic effects of bile acid binding resins against cardiovascular disease risk factors in patients with metabolic syndrome. *Curr Vasc Pharmacol.* 2008;6(4): 271–81.
23. Thompson GR. LDL apheresis. *Atherosclerosis.* 2003;167(1):1–13.
24. Thompson A, Di Angelantonio E, Sarwar N, et al. Association of cholesteryl ester transfer protein genotypes with CETP mass and activity, lipid levels, and coronary risk. *JAMA.* 2008; 299(23):2777–88.
25. Nissen SE, Tardif JC, Nicholls SJ, et al. Effect of torcetrapib on the progression of coronary atherosclerosis. *N Engl J Med.* 2007;356(13): 1304–16.
26. Akdim F, Stroes ES, Sijbrands EJ, et al. Efficacy and safety of mipomersen, an antisense inhibitor of apolipoprotein B, in hypercholesterolemic subjects receiving stable statin therapy. *J Am Coll Cardiol.* 2010;55(15):1611–8.
27. Ricotta DN, Frishman W. Mipomersen: A safe and effective antisense therapy adjunct to statins in patients with hypercholesterolemia. *Cardiol Rev.* 2012;20(2):90–5.
28. Cariou B, Le May C, Costet P. Clinical aspects of PCSK9. *Atherosclerosis.* 2011;216(2): 258–65.
29. Baigent C, Blackwell L, Collins R, et al. Aspirin in the primary and secondary prevention of vascular disease: Collaborative meta-analysis of individual participant data from randomised trials. *Lancet.* 2009;373(9678):1849–60.
30. Antiplatelet Trialists' Collaboration. Collaborative overview of randomised trials of antiplatelet therapy-I: Prevention of death, myocardial infarction, and stroke by prolonged antiplatelet therapy in various categories of patients. *BMJ.* 1994;308(6921):81–106.
31. Mehta SR, Bassand JP, Chrolavicius S, et al. Dose comparisons of clopidogrel and aspirin in acute coronary syndromes. *N Engl J Med.* 2010;363(10):930–42.
32. Mega JL, Close SL, Wiviott SD, et al. Cytochrome p-450 polymorphisms and response to clopidogrel. *N Engl J Med.* 2009;360(4): 354–62.
33. Gibbons RJ, Chatterjee K, Daley J, et al. ACC/AHA/ACP-ASIM guidelines for the management of patients with chronic stable angina: Executive summary and recommendations. A Report of the American College of Cardiology/American Heart Association Task Force on Practice Guidelines (Committee on Management of Patients with Chronic Stable Angina). *Circulation.* 1999;99(21):2829–48.
34. Yusuf S, Zhao F, Mehta SR, Chrolavicius S, Tognoni G, Fox KK. Effects of clopidogrel in addition to aspirin in patients with acute coronary syndromes without ST-segment elevation. *N Engl J Med.* 2001;345(7):494–502.
35. Montalescot G, Sideris G, Cohen R, et al. Prasugrel compared with high-dose clopidogrel in acute coronary syndrome. The randomised, double-blind ACAPULCO study. *Thromb Haemostasis.* 2010;103(1):213–23.
36. Wallentin L, Becker RC, Budaj A, et al. Ticagrelor versus clopidogrel in patients with acute coronary syndromes. *N Engl J Med.* 2009; 361(11):1045–57.
37. Davidson MH, Stein EA, Bays HE, et al. Efficacy and tolerability of adding prescription omega-3 fatty acids 4 g/d to simvastatin 40 mg/d in hypertriglyceridemic patients: An 8 week, randomized, double-blind, placebo-controlled study. *Clin Therapeutics.* 2007; 29(7): 1354–67.
38. Wenger NK. 2011 ACCF/AHA focused update of the guidelines for the management of patients with unstable angina/non-ST-elevation myocardial infarction (updating the 2007 guideline): Highlights for the clinician. *Clin Cardiol.* 2012;35(1):3–8.
39. Honarbakhsh S, Schachter M. Vitamins and cardiovascular disease. *Brit J Nutrit.* 2009; 101(8):1113–31.
40. Wang TJ, Pencina MJ, Booth SL, et al. Vitamin D deficiency and risk of cardiovascular disease. *Circulation.* 2008;117(4):503–11.
41. Autier P, Gandini S. Vitamin D supplementation and total mortality: A meta-analysis of randomized controlled trials. *Arch Intern Med.* 2007;167(16):1730–7.

42. Lichtenstein AH, Appel LJ, Brands M, et al. Summary of American Heart Association Diet and Lifestyle Recommendations revision 2006. *Arterioscler Thromb Vasc Biol.* 2006;26(10):2186–91.
43. Appel LJ, Sacks FM, Carey VJ, et al. Effects of protein, monounsaturated fat, and carbohydrate intake on blood pressure and serum lipids: Results of the OmniHeart randomized trial. *JAMA*. 2005;294(19):2455–64.
44. Sacks FM, Svetkey LP, Vollmer WM, et al. Effects on blood pressure of reduced dietary sodium and the Dietary Approaches to Stop Hypertension (DASH) diet. DASH-Sodium Collaborative Research Group. *N Engl J Med.* 2001;344(1):3–10.
45. Appel LJ, Brands MW, Daniels SR, Karanja N, Elmer PJ, Sacks FM. Dietary approaches to prevent and treat hypertension: A scientific statement from the American Heart Association. *Hypertension*. 2006;47(2):296–308.
46. Kastorini CM, Milionis HJ, Esposito K, Giugliano D, Goudevenos JA, Panagiotakos DB. The effect of Mediterranean diet on metabolic syndrome and its components: A meta-analysis of 50 studies and 534,906 individuals. *J Am Coll Cardiol*. 2011;57(11):1299–313.
47. Estruch R, Martinez-Gonzalez MA, Corella D, et al. Effects of a Mediterranean-style diet on cardiovascular risk factors: A randomized trial. *Ann Intern Med*. 2006;145(1):1–11.
48. Dansinger ML, Gleason JA, Griffith JL, Selker HP, Schaefer EJ. Comparison of the Atkins, Ornish, Weight Watchers, and Zone diets for weight loss and heart disease risk reduction: A randomized trial. *JAMA*. 2005; 293(1):43–53.
49. Yusuf S, Hawken S, Ôunpuu S, et al. Effect of potentially modifiable risk factors associated with myocardial infarction in 52 countries (the INTERHEART study): Case-control study. *Lancet*. 2004;364(9438):937–52.
50. Mallik S, Spertus JA, Reid KJ, et al. Depressive symptoms after acute myocardial infarction: Evidence for highest rates in younger women. *Arch Intern Med*. 2006;166(8):876–83.

38

RISK STRATIFICATION IN CLINICAL PRACTICE

PHIL MENDYS[1], GOLSA JOODI[2], AND ROSS J. SIMPSON, JR.[3]

[1] *Cardiology Division, University of North Carolina, Chapel Hill, NC, USA*
[2] *Lipid and Prevention Clinic, University of North Carolina at Chapel Hill, Chapel Hill, NC, USA*
[3] *Cardiology Section, Lipid and Prevention Clinic, University of North Carolina at Chapel Hill, Chapel Hill, NC, USA*

INTRODUCTION

As practitioners in a cardiology-based lipid and prevention service, we aggressively treat lipids and traditional risk factors in patients with evident vascular disease. Our emphasis is based on the assertion that low-density lipoprotein (LDL) cholesterol is a necessary and sufficient cause for atherosclerotic heart disease [1]. We also treat patients who have traditional risk factors, such as a positive family history, or the so-called emerging risk factors, such as an elevated high-sensitivity C-reactive protein (hs-CRP), who may not have been identified with a diagnosis of coronary disease. In these latter patients, we may obtain a more comprehensive lipid analysis, search for additional nontraditional risk factors such as Lp(a), or perform cardiac or vascular imaging or stress testing to better assess their risk of cardiovascular disease (CVD). We incorporate this information for each of our patients to assess the individual's risk for future cardiovascular events and to guide us in formulating the goals of patient's lipid management. In this chapter, we will review the basic model we follow in estimating a patient's risk for cardiovascular events and describe how we use this risk estimate to guide therapy [2].

We will review a composite of all risk factors, both traditional and emerging, and outline how we use this information to choose appropriate targets for lipid management. We also consider two examples of potentially high-risk, commonly encountered conditions—male erectile dysfunction (ED) and rheumatoid arthritis—that seem to increase the risk of cardiovascular events, and whose associated risk may not be fully appreciated in routine clinical practice. Last, we summarize strategies for patient care based on the existing clinical evidence.

Disclosures: Dr Simpson is a consultant for Merck, Pfizer, and LipoScience. Dr Mendys is an employee of Pfizer.

Atherosclerosis: Risks, Mechanisms, and Therapies, First Edition. Edited by Hong Wang and Cam Patterson.

CLINICAL RISK FACTORS, RISK MARKERS, AND THE CAUSE OF ATHEROSCLEROSIS

It is important to conceptualize the proximate cause of atherosclerosis as abnormal levels of blood LDL cholesterol. The evidence to support this view is based on the observational and experimental evidence developed over decades—from animal, pathological, genetic, and population studies and from multiple randomized clinical trials. Moreover, recent genetic and epidemiological studies indicate that moderate lifelong reduction in the plasma level of LDL cholesterol is associated with a substantial reduction in the incidence of coronary events, even in populations with a high prevalence of non-lipid-related cardiovascular risk factors [1].

Based on these data, on the one hand, we use the term *risk factor* to refer to the small number of patient characteristics that seem causal for atherosclerosis and account for the vast majority of incident myocardial infarctions. These include traditional factors such as the patient's age, gender, total cholesterol, blood pressure, and smoking status. Over time, this list of evidence-based risk factors has expanded to include diabetes as a described CVD risk "equivalent," such that those patients who present with a diagnosis of diabetes should be managed as aggressively as those who have already had a cardiovascular event [3].

On the other hand, we use the term *risk marker* to refer to characteristics that have been associated with atherosclerosis in some studies but not in others, or for which a benefit based on their control has not yet been successfully demonstrated in randomized trials. These factors may also be referred to as nontraditional or "emerging risk factors," as they are not fully endorsed by evidence-based treatment guidelines [4]. Risk markers encompass a long list of individual patient characteristics that includes inflammatory markers (e.g., hs-CRP), newer lipid assays (including LDL cholesterol particle number), environmental characteristics (e.g., air quality), behavioral habits (e.g., heavy exercise), and social considerations (economic well-being). Moreover, new imaging modalities that show components of atherosclerosis in arteries, including coronary artery calcium (CAC) scores and cardiac stress tests that characterize cardiac ischemia, can add to estimates of the probability of future heart attacks or strokes, and may be incorporated in our estimate of risk.

The traditional risk factors are often treated as independent "causes" of atherosclerosis. Treatment of some of these, particularly smoking and hypertension, can dramatically lower cardiovascular risk, and we independently target these factors for intervention. Risk "markers," on the contrary, may not cause atherosclerosis directly, and it is currently uncertain that specific treatment to modify these markers will help prevent the late complications of atherosclerosis. For example, at this point in time, due to the absence of clinical trials showing that treatments to increase low high-density lipoprotein (HDL) cholesterol improve risk, we regard low HDL cholesterol as a risk marker—and our response to its identification in a given patient is to intensify efforts to decrease LDL cholesterol, in keeping with current guidelines. Similarly, risk markers, such as lipoprotein(a) (Lp(a)), hs-CRP, and microalbuminuria, among others, are only used to better define a patient's risk, and are not generally targeted independently for control [5].

Not all of the risk markers are candidates for inclusion in risk prediction. Of the emerging risk factors, hs-CRP may be the best candidate for formal inclusion in risk assessment. However, even here the benefits of its inclusion are still controversial. Lp(a) and homocysteine levels are less consistent or validated [4, 5]. For example, a recent comprehensive analysis of Lp(a) identified a continuous, independent, and specific association of Lp(a) concentration with vascular outcomes, consistent with the existence of a causal relationship. Pending additional research, it is not clear if Lp(a) should be considered a potential therapeutic target, or just included as a risk modifier. At present, as is the case with hs-CRP, treatment directed at Lp(a) is uncertain, and we recommend that Lp(a) be measured to provide additional data related to potential risk. Elevated Lp(a) levels may potentially be used to recalibrate the intensity of LDL cholesterol management, but they are not targeted independently for lowering [6].

NONINVASIVE IMAGING AND RISK EVALUATION

As stated earlier, traditional risk factors are good predictors of future events for patients. However, for patients at intermediate or undetermined risk for CVD, noninvasive markers of subclinical atherosclerosis may improve risk classification and thus clinical decision-making. The 2010 ACCF/AHA guidelines for assessment of cardiovascular risk in asymptomatic adults recommend the use of carotid intima-media thickness (CIMT) and CAC scores, as class IIa indications, for risk classification of asymptomatic patients with an intermediate risk of events [7]. CIMT and CAC score improve conventional risk prediction, especially in patients in the intermediate risk group [8–10]. The extent of value added by CIMT varied among different studies, whereas CAC measurement showed more consistency in risk assessment [9]. Many studies are heterogeneous based on study population, duration of follow-up, event rates, categories of risk, CAC thresholds, end points, and risk factors selected. Therefore, meta-analyses encompassing large populations, economic evaluation of lifetime costs and benefits, and cost-effectiveness studies are needed before widespread introduction of CAC in clinical practice is justified [8]. In addition, careful operator training and experience in performance in assessing CIMT is particularly necessary [7].

RHEUMATOID ARTHRITIS

Patients with rheumatoid arthritis suffer a high rate of cardiovascular events, and coronary artery disease (CAD) is a common cause of death and disability in these patients. Traditional CVD risk factors account partially for this risk, and it is believed that the chronic inflammation that accompanies rheumatoid arthritis accelerates endothelial dysfunction and atherosclerosis in these patients [11]. Accordingly, control of the underlying inflammation by immunosuppressive or biologic therapies may improve CVD risk, and is advocated [12].

In our experience, the traditional risk factors for CVD in patients with rheumatoid arthritis are often underappreciated and poorly controlled. This may be not only due to focus on the chronic underlying disease out of proportion to CVD risk factors but also due to concern about potential drug interactions [13].

Based on the best available evidence, the European League against Rheumatism (EULAR) recommends that rheumatoid arthritis patients be considered at high risk for CVD, and it also emphasizes that the existing risk score models may underestimate the CVD risk in these patients [14, 15]. EULAR recommendations include controlling the primary rheumatologic disease to lower this risk, as well as following national cardiovascular guidelines for risk reduction.

ERECTILE DYSFUNCTION

ED is a common clinical syndrome that affects approximately 18 million men in the United States. Like cardiovascular risk, ED increases with age and shares many of the same risk factors, such as hypertension, high cholesterol, diabetes, and smoking [16, 17].

ED is a powerful marker of cardiovascular risk—it may confer risk similar to that seen in patients with established coronary disease, and its presence must alert the physician about the potential for a high rate of future cardiovascular events. A diagnosis of ED is particularly useful in predicting CVD events in men aged below 60 years. If we consider the composite effect that the presence of ED typically precedes the onset of angina by 2–3 years and cardiovascular events by 3–5 years, and that men with ED have increased mortality-driven primarily by increased cardiovascular death, it is reasonable to consider ED as a CVD risk equivalent [18].

Regardless of the precise designation, men who present with ED should have a comprehensive medical assessment, including testosterone, fasting lipids, fasting glucose, and blood pressure measurement. Following this baseline assessment, select patients should be stratified according to the risk of future cardiovascular events, and those at high risk of CVD should be evaluated by stress testing with the use of computed tomography or coronary angiography. Prior to initiating therapy directed at the ED, men should undergo exercise tolerance testing, and the stability of any cardiovascular conditions should be ensured [19]. Aggressive management of traditional risk factors, including blood pressure reduction, LDL cholesterol management, diabetes control, and smoking cessation should be employed, bearing in mind that therapies should not interfere with management of ED. Lifestyle modifications that target CVD risk factors also improve ED, and should be likewise encouraged. Ultimately, the important messages for both clinicians and patients are that (i) the presence of ED represents increased risk for cardiovascular events and (ii) the documented interval between the onset of ED and symptomatic CAD allows for timely risk factor management.

DIABETES MELLITUS

Current treatment guidelines support the position that diabetes confers a risk status to a patient equivalent to the presence of known coronary disease [3]. While there is continued debate about the wisdom of this recommendation and whether it applies to all patients with diabetes, large-scale epidemiological analysis confirms that a fasting blood sugar greater than 126 mg/dl confers a twofold independent risk. However, the relationship between impaired fasting glucose appears to be nonlinear, and the level of blood sugar is only modestly predictive of vascular disease. For example, in an analysis of 400 older men with diabetes, a 10-year duration of diabetes identified those patients with a clear coronary heart disease risk equivalent [20]. Diabetes presents challenges for the clinician and the patient because of the multiple layers of interventions—dietary counseling and exercise recommendations, as well as medical management for the lipids, blood pressure, and blood sugar. The recent ACCORD (Action to Control Cardiovascular Risk in Diabetes) study provides helpful guidance to help reduce this complexity by showing that overly intensive glycemic and blood pressure control may not be helpful and that there may be limited value in adding a fibrate to statin-based therapy [21]. By contrast, benefit of statins in patients with diabetes is clearly established [22].

Summary recommendations for glycemic, blood pressure, and lipid control in adults with diabetes

Treatment approach	Goal	Comments
Hemoglobin A_{1c}	<7.0%	More or less stringent glycemic goals may be appropriate for individual patients. Goals should be individualized based on duration of diabetes, age/life expectancy, comorbid conditions, known CVD or advanced microvascular complications, hypoglycemia unawareness, individual, and patient considerations
Blood pressure	<130/80 mmHg	Based on patient characteristics and response to therapy, higher or lower systolic blood pressure targets may be appropriate
Lipids (LDL cholesterol)	<100 mg/dl	In individuals with overt CVD, a lower LDL cholesterol goal of 70 mg/dl (1.8 mmol/l), using a high dose of a statin, is an option

Source: American Diabetes Association [23].

LIPID MEASUREMENTS

Since our focus of therapy is LDL cholesterol, the standard, fasting lipid profile that we measure in all patients includes total cholesterol, HDL cholesterol, and triglycerides; we also utilize an estimated LDL cholesterol value, given that LDL cholesterol accounts for almost 70% of the total serum cholesterol burden in patients with normal triglycerides [3].

When fasting triglycerides exceed 200 mg/dl, we consider that additional atherogenic lipoproteins are contained in very-low-density lipoprotein (VLDL) cholesterol, and estimate their contribution to risk by subtracting HDL cholesterol from total cholesterol to calculate the non-HDL cholesterol level. This value seems a reasonable estimate of the apolipoprotein B (ApoB)-containing lipoproteins. A typical goal for non-HDL cholesterol is a level that is 30 mg/dl higher than the LDL cholesterol goal. Recent analyses have commented on the additional value of non-HDL cholesterol measurement among statin-treated patients, in which the risk of future cardiovascular events had a more significant strength of association for non-HDL cholesterol [24]. Both elevated triglycerides and low HDL cholesterol are common in patients who are overweight, physically inactive, smoke, and consume very high levels of carbohydrates (>60% of total energy intake). We do not routinely measure direct LDL cholesterol, as we feel that this value may often underestimate the other atherogenic ApoB-containing lipoproteins [25]. We do not routinely measure ApoB or LDL particle number, but may do so once the LDL cholesterol approaches what seems an appropriate goal from our basic risk predictions.

High triglycerides in excess of 400 mg/dl need to be treated to prevent pancreatitis or eruptive xanthomas. We have had good success in treating patients with elevated triglycerides by treating underlying diabetes or hypothyroidism aggressively, eliminating certain medications, changing to a low simple carbohydrate/low-fat diet, and instituting a regular exercise program. Addition of high doses of fish oil or the occasional need for fibrates or niacin may also be considered, although the evidence that such combination therapy reduces cardiovascular risk is lacking. We also find that if the underlying causes are not addressed and patients do not maintain a low-carbohydrate diet and exercise program, fibrates and niacin are not very effective. We limit the use of omega-3 fatty acids to avoid potential drug interactions, as many of these patients also need statins to treat associated elevated LDL cholesterol.

RISK ASSESSMENT AND RISK MODELS

The ability to assess a patient's risk is a critical skill for all clinicians. Detecting the presence of risk factors and markers in patients requires accurate assessment of historical and laboratory data, and the clinician must separate risk factors that are well supported by clinical evidence from emerging or biologically plausible but unsupported risk markers. In assessing which risk factors are important, we consider the guidelines from the Adult Treatment Panel (ATP) III as a framework of care, and supplement those with what has been learned from recent, well-conducted clinical trials and epidemiologic and physiologic studies. We follow the ATP guidelines. These guidelines, first published in 1988, were updated in the ATP III in 2003 and have been prescient in predicting the results of clinical trials and epidemiologic research [2, 3]. Other, newer risk equations and classification systems are available—these seem as valid, and may be as useful for other practices and in other settings [26]. The foundation of any practical guideline is that risk assessment is the means to guide therapy, and that the focus of therapy to prevent and modify the course of atherosclerosis is measurement and treatment of LDL cholesterol.

Whichever guideline is followed, the model is similar. Traditional risk factors and risk markers or the presence of established atherosclerotic disease are used to establish a general risk model to guide the management goals and targets for LDL cholesterol.

In our practice, we calculate a modified Framingham risk estimate for patients following the ATP III guidelines. This equation enters the patient's age, gender, total cholesterol, HDL cholesterol, smoking status, and blood pressure to estimate the 10-year probability for fatal and nonfatal myocardial infarction. Patients are then grouped by their risk: those at less than 10% risk over 10 years are generally considered as low risk, those between 10 and 20% as intermediate, and those with risk of over 20% as high risk. Those patients at the highest risk, that is, those patients with known atherosclerosis and at least one other major risk factor, are classified separately for the most aggressive therapy for LDL cholesterol reduction. Patients with diabetes are treated as if they have established atherosclerosis [3]. Some factors such as obesity, physical inactivity, and an atherogenic diet contribute to risk, but do NOT impact the risk equations, as each of these factors have multiples effects that are already included in the risk models. A thrombogenic state is also recognized as a nonlipid risk factor based on the observation that antiplatelet and anticoagulant therapies have been shown to reduce CVD, but do not weigh into the consideration for risk. Age and gender are nonmodifiable risk characteristics and are included in the initial overall patient assessment. Other characteristics, such as a strong family history of premature coronary disease, may not be directly reflected in the assessment, but may be considered as "treatment modifiers" to help set our LDL cholesterol targets.

The ATP III risk equations are based on the 10-year probability of a patient developing a cardiovascular event. The limitation to this approach is that, as mentioned, it does not include many of the risk markers we have described above. Some of these may be important to assess the risk in individual patients, and the model may underestimate or overestimate risk in some groups of patients. Of more concern is that the model has a short, 10-year projection of risk in a disease that lasts a lifetime and does not encompass all of the complications or harm from atherosclerosis. Estimating risk over a longer time frame, for example, 20 years or a lifetime, would seem to reflect better our current knowledge of the benefits of lifestyle interventions and drug safety [27]. Extending risk predictions to 20 years or an age predicted lifetime risk would seem to offer the potential to treat atherosclerosis in younger patients, before the disease enters its more advanced stages. However, at the present time, we follow a shorter time window to balance the benefits and harm of aggressive therapy to lower risk. As data emerge, we expect to modify our approach to reflect risk predictions that encompass longer time frames.

SUMMARY RECOMMENDATIONS FOR RISK MANAGEMENT

The therapy to lower LDL cholesterol should be directed to achieve a specific goal LDL cholesterol according to guidelines, as informed by evidence from well-controlled randomized trials. For high- and very high-risk patients, for example, patients with CAD or diabetes and one additional risk factor, we strive to achieve LDL cholesterol levels of 70 mg/dl. For less high-risk patients, we may aim for LDL cholesterol levels of 100 mg/dl or less; while for still lower risk patients, the LDL cholesterol target is 130 mg/dl. Each of these approaches is consistent with the current treatment guidelines.

The most critical principle is that LDL cholesterol can be impacted by therapeutic changes in lifestyle, diet, and exercise, as well as medications. In the clinical setting, therapeutic lifestyle changes are not greatly effective in lowering LDL cholesterol, yet dietary modification and exercise are vital to our success at lipid management. Limiting consumption of simple carbohydrates, such as sugary drinks and alcohol; reducing dietary saturated and trans fats; and increasing activities of daily living, especially walking seem to improve patients' sense of well-being and aid in the effectiveness of drug therapy.

Statins are central to evidence-based drug therapy of hypercholesterolemia. If we are not successful in reaching LDL cholesterol goals on statins alone, we may add ezetimibe to the therapeutic regimen; occasionally, we use such drugs alone in patients who cannot tolerate statins. Adherence to cholesterol therapy is always a concern. Although poor patient compliance with statins is a common and complex problem, the benefits of medication adherence from a multidisciplinary team approach are clear.

Patient concerns about drug safety, drug expense, limited understanding of the benefits and safety of statins, and complexity of follow-up programs can be major barriers to good adherence. However, we find that these barriers can be surmounted by a positive attitude, shared by physicians and the health care team, about the importance of statin-based drug therapy.

REFERENCES

1. Cohen JC, Boerwinkle E, Mosley TH, Jr., Hobbs HH. (2006) Sequence variations in PCSK9, low LDL, and protection against coronary heart disease. *N Engl J Med.* 354(12):1264–1272.
2. Grundy SM, Cleeman JI, Merz CN, et al. (2004) Implications of recent clinical trials for the National Cholesterol Education Program Adult Treatment Panel III guidelines. *Circulation.* 110(2):227–239.
3. Expert Panel on Detection, Evaluation, and Treatment of High Blood Cholesterol in Adults. (2001) Executive summary of the third report of the National Cholesterol Education Program (NCEP) Expert Panel on Detection, Evaluation, and Treatment of High Blood Cholesterol in Adults (Adult Treatment Panel III). *JAMA.* 285(19):2486–2497.
4. Hlatky MA, Greenland P, Arnett DK, et al. (2009) Criteria for evaluation of novel markers of cardiovascular risk: a scientific statement from the American Heart Association. *Circulation.* 119(17):2408–2416.
5. Helfand M, Buckley DI, Freeman M, et al. (2009) Emerging risk factors for coronary heart disease: a summary of systematic reviews conducted for the U.S. Preventive Services Task Force. *Ann Intern Med.* 151(7):496–507.
6. Myers GL, Christenson RH, Cushman M, et al. (2009) National Academy of Clinical Biochemistry Laboratory Medicine Practice guidelines: emerging biomarkers for primary prevention of cardiovascular disease. *Clin Chem.* 55(2):378–384.
7. Greenland P, Alpert JS, Beller GA, et al. (2010) 2010 ACCF/AHA guideline for assessment of cardiovascular risk in asymptomatic adults: executive summary: a report of the American College of Cardiology Foundation/American Heart Association Task Force on Practice Guidelines. *Circulation.* 122(25): 2748–2764.
8. Peters SA, Bakker M, den Ruijter HM, Bots ML. (2011) Added value of CAC in risk stratification for cardiovascular events: a systematic review. *Eur J Clin Invest.* 42(1):110–116.
9. Peters SA, den Ruijter HM, Bots ML, Moons KG. (2012) Improvements in risk stratification for the occurrence of cardiovascular disease by imaging subclinical atherosclerosis: a systematic review. *Heart.* 98(3):177–184.
10. Rana JS, Gransar H, Wong ND, et al. (2012) Comparative value of coronary artery calcium and multiple blood biomarkers for prognostication of cardiovascular events. *Am J Cardiol.* 109(10):1449–1453.
11. Turiel M, Sitia S, Atzeni F, et al. (2010) The heart in rheumatoid arthritis. *Autoimmun Rev.* 9(6):414–418.
12. Murdaca G, Colombo BM, Cagnati P, Gulli R, Spano F, Puppo F. (2012) Endothelial dysfunction in rheumatic autoimmune diseases. *Atherosclerosis.* 224(2):309–317.
13. Chung CP, Giles JT, Petri M, et al. (2012) Prevalence of traditional modifiable cardiovascular risk factors in patients with rheumatoid arthritis: comparison with control subjects from the multi-ethnic study of atherosclerosis. *Semin Arthritis Rheum.* 41(4):535–544.
14. Crowson CS, Matteson EL, Roger VL, Therneau TM, Gabriel SE. (2012) Usefulness of risk scores to estimate the risk of cardiovascular disease in patients with rheumatoid arthritis. *Am J Cardiol.* 110(3):420–424.
15. Peters MJ, Symmons DP, McCarey D, et al. (2010) EULAR evidence-based recommendations for cardiovascular risk management in patients with rheumatoid arthritis and other

forms of inflammatory arthritis. *Ann Rheum Dis.* 69(2):325–331.
16. Schwartz BG, Kloner RA. (2011) Clinical cardiology: physician update: erectile dysfunction and cardiovascular disease. *Circulation.* 123(1): 98–101.
17. Dong JY, Zhang YH, Qin LQ. (2011) Erectile dysfunction and risk of cardiovascular disease: meta-analysis of prospective cohort studies. *J Am Coll Cardiol.* 58(13):1378–1385.
18. Jackson G, Boon N, Eardley I, et al. (2010) Erectile dysfunction and coronary artery disease prediction: evidence-based guidance and consensus. *Int J Clin Pract.* 64(7):848–857.
19. Montorsi P, Ravagnani PM, Galli S, et al. (2006) Association between erectile dysfunction and coronary artery disease: matching the right target with the right test in the right patient. *Eur Urol.* 50(4):721–731.
20. Sarwar N, Gao P, Seshasai SR, et al. (2010) Diabetes mellitus, fasting blood glucose concentration, and risk of vascular disease: a collaborative meta-analysis of 102 prospective studies. *Lancet.* 375(9733):2215–2222.
21. Gerstein HC, Miller ME, Genuth S, et al. (2011) Long-term effects of intensive glucose lowering on cardiovascular outcomes. *N Engl J Med.* 364(9):818–828.
22. Ginsberg HN, Elam MB, Lovato LC, et al. (2010) Effects of combination lipid therapy in type 2 diabetes mellitus. *N Engl J Med.* 362(17):1563–1574.
23. American Diabetes Association. (2012) Standards of medical care in diabetes—2012. *Diabetes Care.* 35(Suppl 1):S11–S63.
24. Ramjee V, Sperling LS, Jacobson TA. (2011) Non-high-density lipoprotein cholesterol versus apolipoprotein B in cardiovascular risk stratification: do the math. *J Am Coll Cardiol.* 58(5): 457–463.
25. Baruch L, Agarwal S, Gupta B, et al. (2010) Is directly measured low-density lipoprotein clinically equivalent to calculated low-density lipoprotein? *J Clin Lipidol.* 4(4):259–264.
26. Cook NR, Paynter NP, Eaton CB, et al. (2012) Comparison of the Framingham and Reynolds Risk scores for global cardiovascular risk prediction in the multiethnic Women's Health Initiative. *Circulation.* 125(14):1748–1756, S1–S11.
27. Elward KS, Simpson RJ, Jr., Mendys P. (2012) Improving cardiovascular risk reduction for primary prevention—utility of lifetime risk assessment. *Postgrad Med.* 122(4): 192–199.

39

MOLECULAR IMAGING

ARUN K. THUKKANI[1] AND FAROUC A. JAFFER[2,3]

[1] *Cardiovascular Division, Brigham and Women's Hospital, Harvard Medical School, Boston, MA, USA*

[2] *Cardiovascular Research Center, Cardiology Division, Department of Medicine, Massachusetts General Hospital, Harvard Medical School, Boston, MA, USA*

[3] *Department of Radiology, Massachusetts General Hospital, Boston, MA, USA*

INTRODUCTION

As a major cause of morbidity and mortality, atherosclerosis is a public health concern no longer limited to Western nations. Although pharmacotherapy and revascularization for atherosclerosis have produced tangible mortality benefits [1], healthcare expenditures related to atherosclerosis continue to rise.

While traditional methods of assessing and treating atherosclerosis rely on measures of luminal stenosis defined by angiography, several lines of evidence have together suggested that this approach may not be ideal. Early on, as atherosclerotic lesions develop, lesions can undergo positive remodeling—a process by which luminal area is preserved by counterbalancing outward expansion [2]. Consequently, these relatively advanced lesions with high plaque burden often escape detection by lumen-focused angiography. As autopsy series have demonstrated that the majority of culprit lesions responsible for fatal myocardial infarction are of only moderate angiographic severity [3], it is evident that angiography is inadequate in detecting vulnerable plaques.

Coupled with clinical studies demonstrating no mortality benefit offered by intervening upon angiographically significant lesions [4], it is clear that new approaches to identify vulnerable plaques are needed. To address this, a number of structurally based approaches engineered to detect vulnerable plaque have been developed including intravascular ultrasound (IVUS)-virtual histology (VH), optical coherence tomography, and near-infrared spectroscopy. Recently, a large clinical trial based on IVUS-VH demonstrated limited predictive power to identify vulnerable plaques, suggesting that a new diagnostic paradigm is necessary [5].

Spurred by advances in vascular molecular biology, material science, and diagnostic imaging, the field of cardiovascular molecular imaging has emerged to potentially address these myriad challenges. By utilizing sensitive

Atherosclerosis: Risks, Mechanisms, and Therapies, First Edition. Edited by Hong Wang and Cam Patterson.

molecular and cellular imaging agents detected by various imaging platforms, biological features of vulnerable plaques can be assessed *in vivo*. Vulnerable plaque lesion progression and rupture have been assessed in preclinical animal studies and early human trials. This chapter will initially review the concepts fundamental to atherosclerotic molecular imaging, with an emphasis on the integration of atherosclerosis biology and molecular imaging targets. Next, clinical inflammation molecular imaging approaches, such as nanoparticle-enhanced magnetic resonance imaging (MRI) and ^{18}F-fluorodeoxyglucose positron emission tomography (^{18}F-FDG PET), will be discussed. Last, novel molecular imaging platforms ideally suited to interrogate coronary atherosclerosis will be highlighted.

MOLECULAR IMAGING OF ATHEROSCLEROSIS: CONCEPT AND TARGETS

Biological Framework

Atherosclerosis is a chronic inflammatory process initiated and driven by lipid deposition. Mature atherosclerotic lesions comprise (i) cholesterol and cholesterol ester deposits; (ii) fibrocellular elements, including extracellular matrix, fibrin, and calcium; and (iii) inflammatory cells, such as macrophages, as well as platelets and smooth muscle cells [6]. In response to subendothelial low-density lipoprotein (LDL) deposition, lipid oxidation and endothelial dysfunction ensue; as lipoproteins continue to accumulate, this response-to-retention reaction promotes the expression of inflammatory chemokines and cytokines [7]. Activated endothelia, expressing cell adhesion molecules including vascular cell adhesion molecule-1, intercellular adhesion molecule-1, and P-selectin, next recruit circulating monocytes that, localizing to these areas, transform into activated plaque macrophages and further amplify the inflammatory cascade by secreting cytokines and matrix metalloproteinases [8]. Fatty streak lesions develop as macrophages ingest oxidized LDL and evolve into foam cells. Foam cells may undergo necrosis and apoptosis, releasing intracellular lipids, cytokines, and proteases, thus promoting expansion of the necrotic core. Enhanced proteolytic activity, particularly in the shoulder regions of the fibrous cap, greatly increases the risk for rupture and thrombotic occlusion. In response to hypoxia and inflammatory stimuli, neovessels develop from the vasa vasorum. This immature neovascular network not only augments macrophage influx into the core, but is itself prone to rupture, leading to intraplaque hemorrhage and plaque expansion. Together, these structural and biological hallmarks characterize a range of vulnerable atherosclerotic lesions that collectively typify the thin-capped fibroatheroma (TCFA). Believed to immediately precede overt plaque rupture, the unstable, vulnerable TCFA is generally believed to be responsible for clinically significant thrombotic complications including myocardial infarction and stroke.

Molecular Imaging

By coupling clinical imaging modalities including MRI, PET, single-photon emission computed tomography (SPECT), and fluorescence imaging with specialized molecular imaging agents, *in vivo* readouts of plaque inflammation, angiogenesis, tissue proteolysis, and apoptotic burden can be obtained. Potential advantages of atherosclerosis molecular imaging include (i) improving *in vivo* knowledge regarding atheroma biology, (ii) assessing the efficacy of biotherapeutics targeting specific antiatherosclerotic pathways, (iii) improving identification of plaques at high risk of rupture producing acute coronary syndromes and stroke, (iv) providing an *in vivo* molecular basis for selection of specific antiatherosclerosis therapies, and (v) developing platforms to enable integrated imaging and therapy (theranostic agents) [9].

The following discussion will review specific molecular imaging approaches currently used to detect atherosclerotic vascular inflammation, with emphasis on clinically available or clinically promising approaches.

USPIO-ENHANCED MOLECULAR MR IMAGING OF INFLAMMATORY PLAQUE MACROPHAGES

Inflammation plays a central role in the initiation, progression, and rupture of atherosclerotic plaques. As a source of detrimental reactive oxygen species and proteolytic enzymes, macrophages are the key cellular mediator of inflamed plaque and, therefore, represent an important target for imaging.

One of the first clinical imaging approaches to robustly detect plaque macrophages *in vivo* leveraged ultrasmall superparamagnetic iron oxide (USPIO) nanoparticles. USPIOs consist of monocrystalline magnetic cores with a dextran coating, typically resulting in a full particle size of less than 50 nm in diameter [10]. When internalized by macrophages, USPIOs generate T2/T2*-weighted signal loss on MR images (i.e., negative contrast). Landmark preclinical MRI studies of atheroma-bearing rabbits demonstrated that USPIOs internalized within plaque macrophages *in vivo* [11]. The recognition that USPIOs might target inflamed human atheroma *in vivo* was incidentally noted when signal loss was observed in the large arteries of patients undergoing MR imaging for oncologic staging [12, 13].

Clinical Studies with USPIO-Enhanced MRI

Landmark studies have expanded upon these initial observations and together support the utility of USPIOs to detect activated macrophages in inflamed, rupture-prone, symptomatic carotid plaque *in vivo*. In the first systematic study of symptomatic patients undergoing carotid endarterectomy, Kooi et al. [14] obtained serial T2*-weighted carotid plaque MR images before and after USPIO administration. The resulting images were then divided into quadrants that were analyzed for signal change. At 24 hours after the administration of USPIOs, 54% of all quadrants evaluated exhibited a 24% signal loss relative to baseline. *Ex vivo* analyses of the resected tissue demonstrated USPIOs not only within plaque macrophages in 10 of the 11 patients, but also in 75% of those lesions exhibiting overt rupture or structural alterations suggesting vulnerability. Only 7% of the stable lesions contained USPIOs. Findings were validated histologically by staining immunohistochemically for macrophages as well as for iron. Collectively, these results stress the potential of USPIOs to identify rupture-prone plaque.

In a second key clinical study, Tang et al. [15] evaluated 20 symptomatic patients with bilateral carotid atherosclerosis to determine whether subclinical plaques are also inflamed. Angiography demonstrated an average stenosis on the symptomatic and contralateral, asymptomatic sides of 77 and 46%, respectively. MR imaging was conducted before and after USPIO administration. USPIO-enhanced MRI suggested that symptomatic carotid lesions were associated with systemic inflammation, analogous to acute coronary syndromes [16]. Though all symptomatic plaques demonstrated signal loss, perhaps unexpectedly, 95% of contralateral asymptomatic plaques also exhibited signal loss indicating active plaque inflammation. Moreover, conventional MR imaging revealed that the asymptomatic plaques largely lacked the structural alterations typifying TCFAs. Taken together, it is then plausible that clinically silent but inflamed lesions may in fact be at increased risk for rupture and cerebrovascular events. Further prospective clinical studies are required to better delineate the prognostic implications associated with USPIO-defined inflammation in asymptomatic plaques.

In addition, the prospect of extending this technique to evaluate other arterial territories is likely. Inflammation associated with aortic aneurysms was readily detected by USPIOs [17]. When similarly sized aneurysms were analyzed, those demonstrating USPIO accumulation were associated with accelerated growth rates [18]. These results suggest that MR imaging with USPIOs may better articulate the future risk of aneurysmal expansion and rupture when compared with standard techniques.

Pharmaceutical Evaluation with USPIO-MRI

A recent study exemplified the use of USPIOs to evaluate the efficacy of anti-inflammatory pharmaceuticals designed to treat atherosclerosis. In the Atorvastatin Therapy: Effects on Reduction of Macrophage Activity (ATHEROMA) study, the effects of low- and high-dose atorvastatin on carotid plaque inflammation as assessed by USPIO-enhanced MRI were investigated (Figure 39.1) [19].

Patients with carotid stenosis who demonstrated signal dropout (inflammation) on baseline USPIO-enhanced MR imaging were randomized to either 10 or 80 mg of atorvastatin for 12 weeks. Serial MRI then assessed the reduction in USPIO-defined inflammation at both 6 and 12 weeks. The patients on high-dose statin therapy demonstrated significant reductions in carotid inflammation when compared with baseline at both time points; this trend correlated with reductions in LDL cholesterol levels and cerebral microemboli burden. At 12 weeks, a statistically significant greater reduction in plaque inflammation was detected in the high-dose statin versus the low-dose statin group. Thus, USPIO-enhanced MRI may represent a means to evaluate the anti-inflammatory potency of investigational therapeutics. Molecular imaging of inflamed lesions could therefore be used to gauge the potential of novel pharmaceuticals, and thus expedite clinical trials.

Limitations

Lacking ionizing radiation, high-resolution MRI with USPIOs is a promising molecular imaging approach. Though these reports are encouraging, prospective, longitudinal studies validating diagnostic accuracy and prognostic implications are required. However, several limitations must be overcome. Partial volume effects and other *in vivo* artifacts producing inherent T2* effects may limit the accuracy of USPIO-defined inflammation [20]. Serial imaging is technically difficult and depends on the reproducibility of coil positioning and image co-registration. Diverse methods of quantifying signal dropout have been described, and a consensus regarding a standard method has yet to be achieved. This may be aided by new gradient-echo acquisition protocols that generate positive contrast from USPIOs *in vivo*. Though these techniques have been utilized to detect inflammation in preclinical animal models [21], they have not yet been reported feasible clinically. Last, the long-term effects of USPIOs on atherosclerosis need further evaluation since chronic administration may augment plaque iron content, theoretically promoting oxidative stress and lesion progression [22].

NUCLEAR MOLECULAR IMAGING OF PLAQUE INFLAMMATION

^{18}F-FDG-enhanced PET is a powerful, rapidly disseminating approach to noninvasively characterize plaque inflammation. This system relies on the accumulation of ^{18}F-FDG, a glucose analogue, in metabolically active cells. ^{18}F-FDG competes with glucose for uptake into cells by glucose transporter proteins (GLUTs). While glucose and ^{18}F-FDG are both phosphorylated by hexokinase to glucose-6-phosphate and ^{18}F-FDG-6-phosphate, respectively, the latter cannot be metabolized further and accumulates within cells (metabolic trapping). Only negligible ^{18}F-FDG-6-phosphate undergoes dephosphorylation due to low intracellular levels of phosphatases [23]. *In vitro* studies have shown that GLUT-1 and GLUT-3 expression, glycolytic flux, and hexokinase activity all increase in activated macrophages [24]. Compared with other modalities, ^{18}F-FDG PET imaging is exquisitely sensitive, capable of detecting

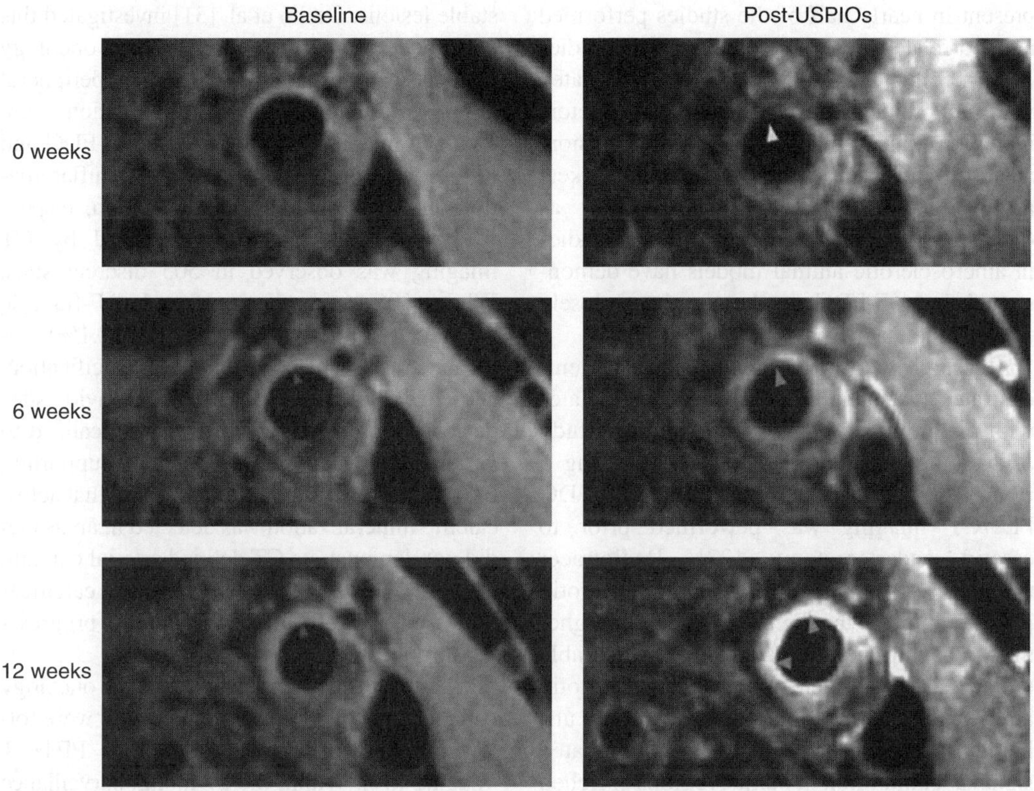

Figure 39.1 USPIO-enhanced MR images of the left common carotid artery in a patient treated with high-dose atorvastatin from the ATHEROMA study. Serial T2*-weighted MR images before (left panels) and after (right panels) USPIO administration at 0, 6, and 12 weeks, respectively, are depicted. Signal void is appreciated in the carotid plaque after USPIO infusion at 0 week (arrowhead at 12 o'clock, upper right panel). At 6 weeks, imaging demonstrates no retention of USPIOs at baseline (arrowhead at 12 o'clock). Following USPIO administration, plaque enhancement (arrowhead at 12 o'clock) rather than a signal void indicates the absence of USPIO uptake and plaque macrophages. Similar results are apparent when imaging was later performed at 12 weeks. USPIO, ultrasmall superparamagnetic iron oxide (Modified from Tang et al. [19]. With permission. © Elsevier).

picomolar levels of radiotracer *in vivo*, and uniquely quantifiable through measuring standard uptake values (SUVs). However, the spatial resolution of PET is relatively poor (3–5 mm) compared with other modalities, and thus provides only limited anatomical information. Thus, hybrid PET/CT imaging systems have been developed integrating PET with high-resolution CT imaging in order to precisely localize radiotracer uptake. Hybrid PET/MR imaging systems are also actively being developed.

Clinical Vascular ^{18}F-FDG PET Imaging

^{18}F-FDG PET/CT imaging is primarily used in oncology to detect metastases. However, the detection of ^{18}F-FDG aortic uptake in patients with Takayasu's arthritis suggested that PET-based vascular imaging may be feasible [25]. Subsequent reports suggesting that plaque inflammation could be detected as ^{18}F-FDG accumulation were noted in the large arteries of patients undergoing ^{18}F-FDG PET/CT oncologic imaging [26]. Retrospective analyses of these patients demonstrated that this uptake was present in nearly half of the studies performed and correlated with age. Subsequent studies confirmed that ^{18}F-FDG uptake also correlated with traditional cardiovascular risk factors including male gender, diabetes, hypertension, and hypercholesterolemia, as well as biomarkers associated with atherosclerosis such as C-reactive protein [27]. Last, preclinical studies in atherosclerotic animal models have demonstrated that ^{18}F-FDG uptake correlates closely with plaque macrophage density [28].

Research within the past decade has demonstrated that ^{18}F-FDG PET signal associates with plaque macrophages. In a seminal study of patients with carotid stenosis suffering a recent transient ischemic attack, ^{18}F-FDG PET/CT imaging was performed prior to carotid endarterectomy [27]. Radiotracer uptake was seen in all symptomatic carotid plaques at levels that were significantly higher than in asymptomatic plaques. No measurable radioactivity was detected in normal carotid arteries. *Ex vivo* autoradiography studies utilizing tritiated deoxyglucose demonstrated signal accumulation in plaque regions enriched with macrophages. Similar studies not only confirmed these findings but also reported a correlation between quantified PET signal and plaque macrophage content [29].

More recently, the ability of ^{18}F-FDG PET/CT to predict future cardiovascular risk has been studied. Paulmier et al. [30] retrospectively compared two groups of patients, matched for cardiovascular risk factors, with either high or low arterial ^{18}F-FDG uptake in larger peripheral arteries. Cardiovascular events were more frequent in the high uptake group. While this group was associated with more vascular calcification, ^{18}F-FDG activity did not localize to calcified lesions. Analyses demonstrated that high ^{18}F-FDG uptake was associated with cardiovascular events within the prior 6 months, while calcification was related to older events. Together, this would imply that ^{18}F-FDG accumulation and vascular calcification reflect different stages of atherosclerosis with the former associated with inflamed plaques at greater risk for rupture and the latter with older, stable lesions. Derlin et al. [31] investigated this concept further. In a novel study of oncology patients, CT scans of the larger peripheral arteries to detect vascular calcification were obtained in conjunction with both ^{18}F-FDG and ^{18}F-fluoride PET to detect vascular inflammation and active calcific mineralization, respectively. Vascular calcification derived by CT imaging was observed in 505 discrete sites. While 105 areas demonstrated ^{18}F-fluoride accumulation, only 81 of these (77.1%) co-localized to CT-identified vascular calcification. Moreover, of the 124 ^{18}F-FDG avid sites detected, only 18 areas (14.5%) co-localized to vascular calcium. While generally supporting the findings of Paulmier et al., the fact that active calcific mineralization was detected in areas that did not demonstrate CT-defined arterial calcium suggests that the process of vascular calcification itself is likely a dynamic process progressing through distinct phases as well.

To study this concept further, 932 oncology patients without cardiovascular disease were followed after undergoing ^{18}F-FDG PET/CT imaging [32]. Within the 29-month surveillance period, 15 subjects suffered a cardiovascular event. Compared with controls, this cohort demonstrated significantly higher levels of baseline arterial ^{18}F-FDG uptake. These findings have been corroborated by others [33]. In contradistinction to the results of Paulmier et al., a more accurate assessment of clinical risk was realized when the extent of vascular calcification on baseline imaging was integrated with the PET data. The BioImage study, a prospective multimodal imaging study of high-risk patients using MRI to identify plaques, CT to assess for arterial calcification, and ^{18}F-FDG PET/CT to measure vascular inflammation will hopefully better characterize the interaction between arterial inflammation and calcification, as well as their relative influences on future cardiovascular risk [34].

Coronary ^{18}F-FDG PET Imaging

Though challenging due to myocardial ^{18}F-FDG uptake and cardiac motion, coronary ^{18}F-FDG PET/CT imaging has been performed. However, due to high background myocardial radiotracer uptake, evaluation of many coronary segments, including the left anterior descending coronary artery, is inadequate [35]. Currently, accurate quantitative coronary ^{18}F-FDG PET imaging appears to be limited to assessing the larger, relatively less mobile left main coronary artery. Dunphy et al. [36] retrospectively analyzed ^{18}F-FDG PET/CT images of patients undergoing oncologic staging. Coronary ^{18}F-FDG uptake, when detected, most frequently localized to the left main coronary artery. Using high-fat, low-carbohydrate loading to limit myocardial glucose uptake, others have corroborated these findings [37]. Further, in a study of patients following coronary stenting, high levels of ^{18}F-FDG accumulation were detected in the left main coronary arteries of patients presenting with unstable angina (Figure 39.2) [38].

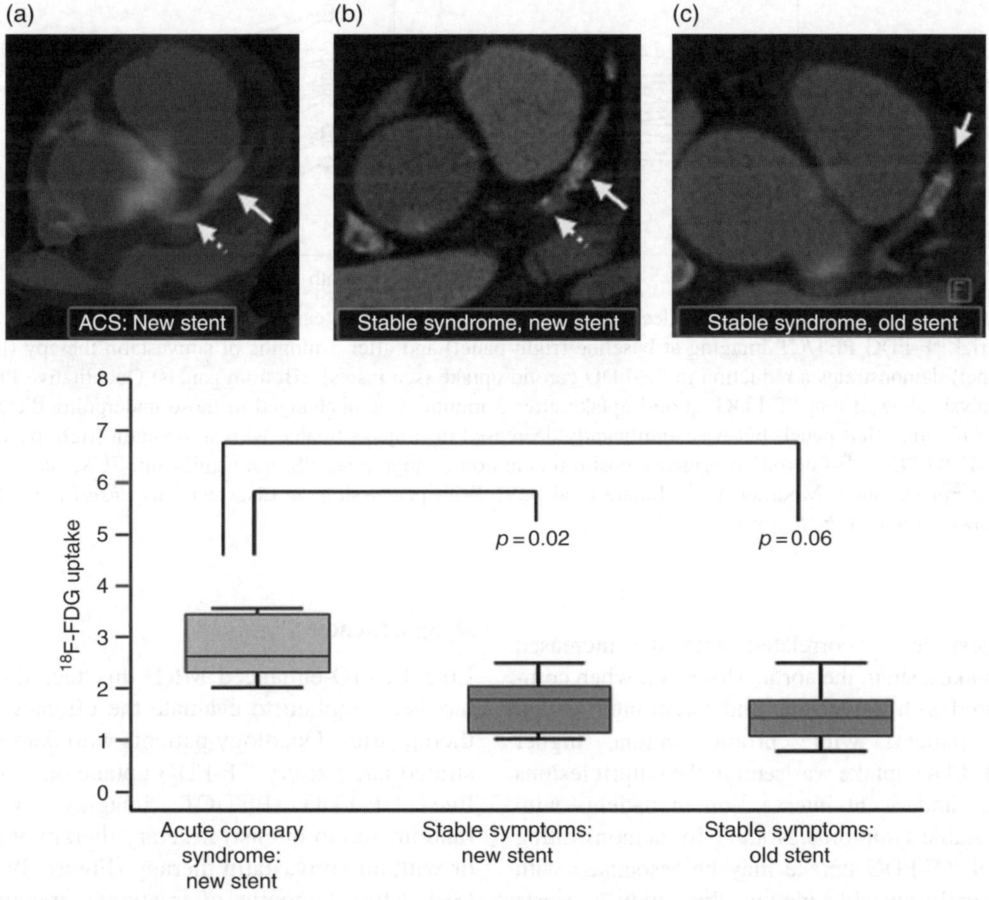

Figure 39.2 ^{18}F-FDG PET/CT coronary imaging (top panels) and FDG target-to-background ratios (bottom graph) in patients with acute coronary syndrome (panel a) and stable angina (panels b and c), respectively, that underwent percutaneous intervention. (a) ^{18}F-FDG PET/CT imaging demonstrated ^{18}F-FDG uptake in the left main (hatched arrows) and recently stented left anterior descending artery culprit lesion (solid arrows). (b and c) In patients with stable chest pain syndromes, significantly lower levels of ^{18}F-FDG uptake were associated with culprit lesions recently or previously stented, respectively. ^{18}F-FDG PET, ^{18}F-fluorodeoxyglucose positron emission tomography (Modified from Rogers et al. [38]. With permission. © Elsevier). (*See insert for color representation of the figure.*)

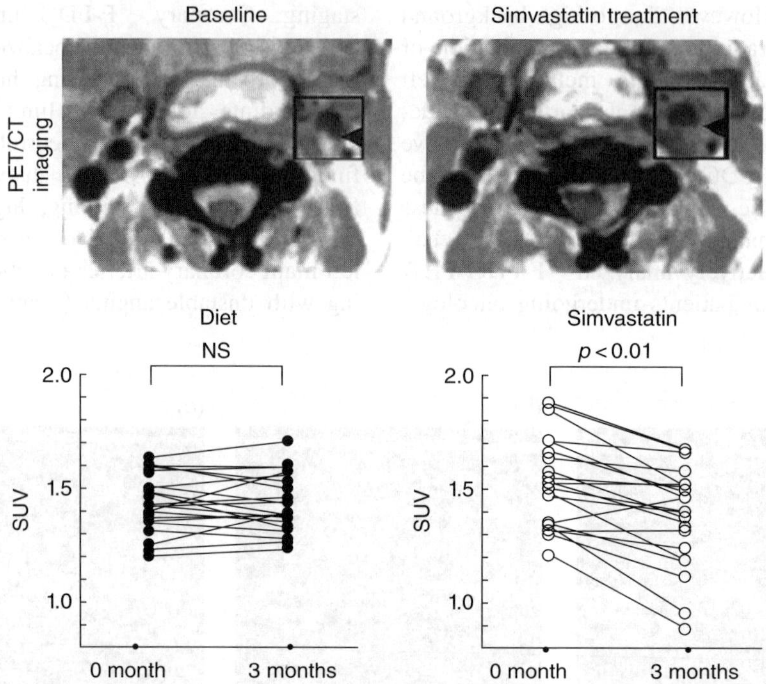

Figure 39.3 Simvastatin therapy decreases ^{18}F-FDG accumulation in carotid atherosclerosis. (Top panels) Serial ^{18}F-FDG PET/CT imaging at baseline (right panel) and after 3 months of simvastatin therapy (left panel) demonstrates a reduction in ^{18}F-FDG carotid uptake (see insets). (Bottom panels) Quantitative PET analysis showed that ^{18}F-FDG carotid uptake after 3 months was unchanged in those undergoing dietary modification (left panel) but was significantly decreased in subjects treated with simvastatin (right panel). ^{18}F-FDG PET, ^{18}F-fluorodeoxyglucose positron emission tomography; NS, not significant; SUV, standardized uptake value (Modified from Tahara et al. [39]. With permission. © Elsevier). (*See insert for color representation of the figure.*)

These levels correlated with the increased uptake seen in the aorta. Moreover, when compared with lesions that underwent intervention in patients with chronic angina, higher ^{18}F-FDG uptake was seen in the culprit lesions that underwent intervention in patients with unstable symptoms. Notable for demonstrating that ^{18}F-FDG uptake may be associated with actively unstable plaque, this study suggests that aortic and coronary uptake may correlate, thus offering an alternative imaging strategy if reproducible ^{18}F-FDG PET/CT coronary imaging is not feasible. Further investigation corroborating these results and optimizing the adjunctive techniques necessary for robust PET-based coronary imaging is still required.

Drug Efficacy

Like USPIO-enhanced MRI, this technique has been applied to evaluate the efficacy of therapeutics. Oncology patients who demonstrated large artery ^{18}F-FDG uptake on baseline ^{18}F-FDG PET/CT imaging were randomized to intensive dietary therapy with or without simvastatin therapy (Figure 39.3) [39]. After 3 months of treatment, imaging demonstrated that only in those patients treated with simvastatin, plaque radiotracer uptake had diminished significantly. This finding correlated with increased serum high-density lipoprotein (HDL) cholesterol levels but intriguingly not LDL levels. In another study of 60 healthy patients, baseline

^{18}F-FDG PET/CT imaging showed 352 ^{18}F-FDG positive lesions in large arteries [40]. After 17 months of intensive lifestyle modification, only 124 ^{18}F-FDG-avid lesions were later detected, and this reduction correlated with an increase in serum HDL cholesterol as well. The dal-PLAQUE study, a recent multicenter clinical trial examining the safety and efficacy of dalcetrapib, a novel cholesteryl ester transfer protein inhibitor, perhaps best exemplifies how this molecular imaging technology can greatly benefit pharmaceutical development [41]. A total of 130 patients with either a history of cardiovascular disease or at high cardiovascular risk were randomized to either dalcetrapib or placebo treatment for 24 months. Notable for the use of imaging indices as primary endpoints, serial MRI and ^{18}F-FDG PET/CT imaging studies were conducted to assess changes in plaque structure and plaque inflammation, respectively. When compared with the placebo group, dalcetrapib therapy for 6 months resulted in decreased carotid artery ^{18}F-FDG uptake. At 24 months, MRI showed that dalcetrapib was also associated with less increase in total vessel area, as well as a decrease in total wall area compared with placebo. These effects mediated by dalcetrapib correlated with increases in serum HDL cholesterol. The findings underscore the link between plaque inflammation and lesion progression, while suggesting that inflammation-driven vascular remodeling may be curtailed by pharmacologically enhancing HDL cholesterol levels. With confirmatory trials studying clinical endpoints ongoing [42, 43], this landmark imaging trial highlights the utility of multimodal imaging in accelerating drug development: by first assessing the direct *in vivo* structural and functional effects in plaques initially; the efficacy of experimental therapies can be validated expeditiously in smaller but still adequately powered Phase II trials, limiting study costs and fast-tracking those initially promising therapeutics for Phase III clinical outcomes evaluation.

Limitations

Though these studies together demonstrate the feasibility of ^{18}F-FDG PET/CT imaging to detect inflamed plaque, several issues must be addressed. Serial imaging will be limited given the cumulative radiation exposure. Further, studies evaluating the kinetics of ^{18}F-FDG plaque accumulation are required to determine the optimal imaging window following radiotracer administration, since prior studies have used divergent protocols. Reproducibly measuring SUVs in diabetic patients—a key demographic at high cardiovascular risk—is likely to be difficult as high serum glucose levels are associated with widely variable ^{18}F-FDG arterial uptake [29]. ^{18}F-FDG PET/CT coronary imaging is associated with additional challenges. Relatively lower radiotracer uptake in small coronary plaques in conjunction with partial volume effects will compromise signal-to-background ratios. High background myocardial ^{18}F-FDG uptake is an additional difficulty. While dietary manipulation has been described, no standard protocol has been universally adopted. Cardiac motion is also an obstacle that could be addressed with cardiac gating but would further diminish sensitivity. Despite these challenges, ^{18}F-FDG PET/CT remains a major area of investigation in the field of atherosclerotic molecular imaging. Moreover, SPECT molecular imaging agents targeting other biopathologic processes of inflamed plaque are being developed. Studies have described detecting apoptosis in subjects with symptomatic carotid atherosclerosis using ^{99}technetium-labeled Annexin V [44]. However, the application of ^{18}F-FDG PET/CT will also likely expand to other cardiovascular pathologies, including detecting inflammation associated with aortic stenosis and aneurysmal disease [45, 46].

EMERGING MOLECULAR IMAGING PLATFORMS

Recently, advancements in molecular biology, material science, chemistry, and radiochemistry have together led to the development of novel molecular imaging platforms. Including various nanoparticle and lipoprotein-based agents among others, these imaging agents have not only been functionalized with multimodal reporter molecules but have also been engineered to convey therapeutic payloads (theranostic agents). A complete discussion is beyond the scope of this chapter, but extensive reviews are available [47]. A particularly notable example of an innovative molecular imaging platform to detect inflammation in atherosclerosis was recently reported by Jaffer et al. [48] (Figure 39.4).

In conjunction with a near-infrared fluorescence (NIRF) imaging agent previously shown to detect cysteine protease activity, a rotational

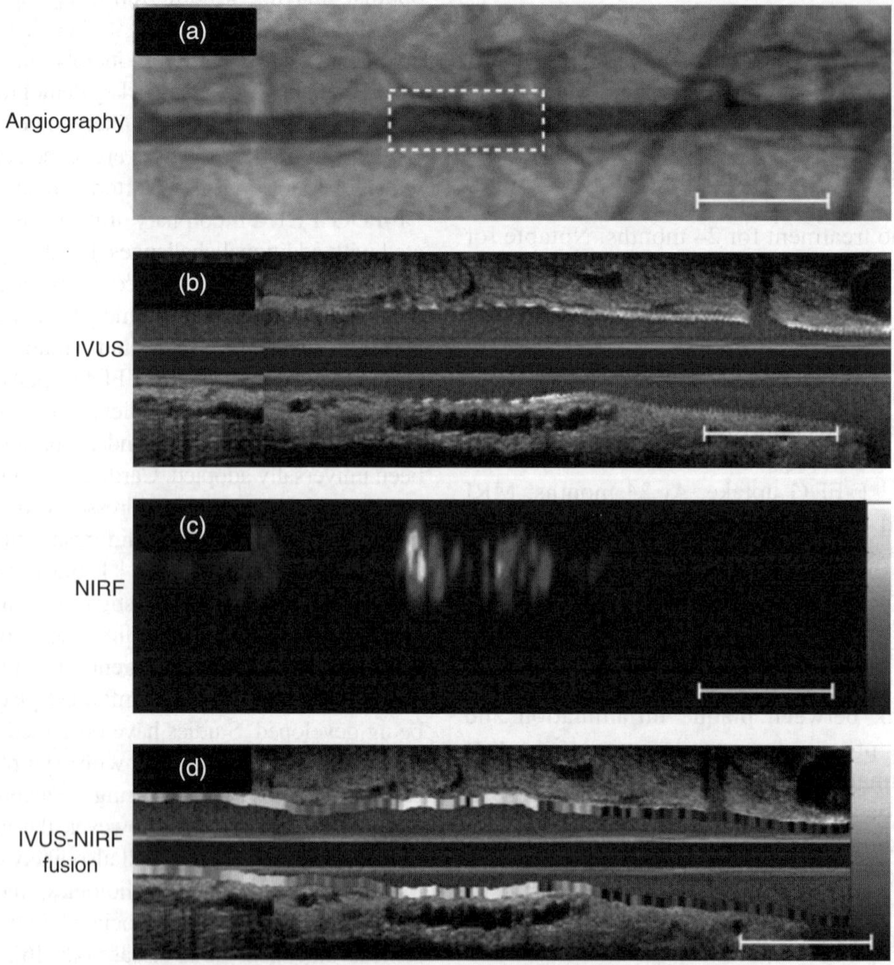

Figure 39.4 Intravascular two-dimensional NIRF imaging of plaque inflammation in coronary-sized arteries through blood. The NIRF molecular imaging catheter detects vascular inflammation (cathepsin protease activity) 7 days following stent implantation in rabbits. (a) The dashed box denotes the position of a stent deployed in the abdominal aorta. (b and c) Longitudinal and rotational IVUS (panel b) and NIRF (panel c) interrogation facilitated by automated pullback demonstrates increased proteolytic activity localized to the stented region. (d) Longitudinal NIRF–IVUS fusion image (yellow/white color denotes strongest intensity NIRF signal; red/black color denotes lowest intensity NIRF signal). IVUS, intravascular ultrasound; NIRF, near-infrared fluorescence (Modified from Jaffer et al. [48]. With permission. © Elsevier). (*See insert for color representation of the figure.*)

two-dimensional imaging catheter was used to evaluate coronary-sized plaques and stent-induced arterial injury in the aortas of hyperlipidemic rabbits. This intravascular imaging strategy not only detected arterial inflammation with high signal-to-background ratios, but also delineated the protease activity associated with stent implantation. The detection of atherosclerotic inflammation in rabbit aortas that approximate the size of human coronary arteries suggests that catheter-based approaches may be a viable alternative strategy, particularly if noninvasive modalities are not feasible for robust coronary molecular imaging. Moreover, the use of invasive molecular imaging platforms may represent an adjunctive tool complementing conventional angiography. A second-generation NIRF imaging catheter engineered with optical frequency domain imaging capability to assess vessel microstructure has also very recently been described and shows promise for quantitative high-resolution intravascular NIRF imaging [49].

CONCLUSION

By detecting the cardinal biopathologic processes of actively inflamed atherosclerotic plaque, the field of cardiovascular molecular imaging seeks to better articulate future cardiovascular risk beyond traditional methods. The use of specific molecular imaging approaches such as USPIO-enhanced MRI and ^{18}F-FDG PET/CT imaging to evaluate large artery atherosclerosis has demonstrated the feasibility of imaging plaque inflammation in clinical subjects. While prospective clinical trials are still required, these modalities may possibly enhance currently utilized clinical risk paradigms. Moreover, by more accurately identifying patients harboring asymptomatic yet vulnerable plaques, individualized pharmacotherapy in conjunction with more judicious implementation of percutaneous intervention may not only bolster clinical outcomes but also help contain costs. By serially detecting alterations in both plaque composition and physiology, these technologies will also accelerate the development of novel pharmaceuticals designed to treat atherosclerosis. Specifically, by directly evaluating its mechanism of action *in vivo*, initially promising candidate pharmaceuticals can be readily identified for Phase III clinical outcomes studies. In addition, site-directed drug delivery systems may be developed as targeted molecular imaging agents are engineered to also convey high-capacity pharmaceutical payloads. These therapeutic-diagnostic agents (theranostic agents) can potentially be more efficacious than traditional pharmaceuticals by delivering higher effective quantities of active drug directly to *in vivo* cellular and molecular targets while simultaneously minimizing off-target toxicities. Clearly, the development, preclinical evaluation, translation, and clinical application of molecular imaging strategies based on USPIO-enhanced MRI, ^{18}F-FDG PET/CT, next-generation nanoparticle systems, as well as innovative catheter-based imaging platforms, will continue to be a major focus of atherosclerotic molecular imaging investigation in the near future.

REFERENCES

1. Krumholz HM, Wang Y, Chen J, et al. Reduction in acute myocardial infarction mortality in the United States: Risk-standardized mortality rates from 1995–2006. *JAMA*. 2009;302: 767–773.
2. Glagov S, Weisenberg E, Zarins CK, Stankunavicius R, Kolettis GJ. Compensatory enlargement of human atherosclerotic coronary arteries. *N Engl J Med*. 1987;316:1371–1375.
3. Virmani R, Burke AP, Farb A, Kolodgie FD. Pathology of the vulnerable plaque. *J Am Coll Cardiol*. 2006;47:C13–C18.
4. Adamu U, Knollmann D, Alrawashdeh W, et al. Results of interventional treatment of stress positive coronary artery disease. *Am J Cardiol*. 2010;105:1535–1539.
5. Stone GW, Maehara A, Lansky AJ, et al. A prospective natural-history study of coronary atherosclerosis. *N Engl J Med*. 2011;364: 226–235.

6. Fuster V, Fayad ZA, Moreno PR, Poon M, Corti R, Badimon JJ. Atherothrombosis and high-risk plaque: Part II: Approaches by noninvasive computed tomographic/magnetic resonance imaging. *J Am Coll Cardiol*. 2005;46: 1209–1218.
7. Libby P, Ridker PM, Hansson GK. Progress and challenges in translating the biology of atherosclerosis. *Nature*. 2011;473(7347): 317–325.
8. Virmani R, Burke AP, Farb A, Kolodgie FD. Pathology of the unstable plaque. *Prog Cardiovasc Dis*. 2002;44:349–356.
9. Jaffer FA, Libby P, Weissleder R. Molecular imaging of cardiovascular disease. *Circulation*. 2007;116:1052–1061.
10. Trivedi RA, U-King-Im JM, Graves MJ, Kirkpatrick PJ, Gillard JH. Noninvasive imaging of carotid plaque inflammation. *Neurology*. 2004;63:187–188.
11. Ruehm SG, Corot C, Vogt P, Kolb S, Debatin JF. Magnetic resonance imaging of atherosclerotic plaque with ultrasmall superparamagnetic particles of iron oxide in hyperlipidemic rabbits. *Circulation*. 2001;103:415–422.
12. Schmitz SA, Taupitz M, Wagner S, Wolf KJ, Beyersdorff D, Hamm B. Magnetic resonance imaging of atherosclerotic plaques using superparamagnetic iron oxide particles. *J Magn Reson Imaging*. 2001;14:355–361.
13. Schmitz SA, Taupitz M, Wagner S, et al. Iron-oxide-enhanced magnetic resonance imaging of atherosclerotic plaques: Postmortem analysis of accuracy, inter-observer agreement, and pitfalls. *Invest Radiol*. 2002;37:405–411.
14. Kooi ME, Cappendijk VC, Cleutjens KB, et al. Accumulation of ultrasmall superparamagnetic particles of iron oxide in human atherosclerotic plaques can be detected by in vivo magnetic resonance imaging. *Circulation*. 2003;107:2453–2458.
15. Tang T, Howarth SP, Miller SR, et al. Assessment of inflammatory burden contralateral to the symptomatic carotid stenosis using high-resolution ultrasmall, superparamagnetic iron oxide-enhanced MRI. *Stroke*. 2006;37:2266–2270.
16. Buffon A, Biasucci LM, Liuzzo G, D'Onofrio G, Crea F, Maseri A. Widespread coronary inflammation in unstable angina. *N Engl J Med*. 2002;347:5–12.
17. Sadat U, Taviani V, Patterson AJ, et al. Ultrasmall superparamagnetic iron oxide-enhanced magnetic resonance imaging of abdominal aortic aneurysms—A feasibility study. *Eur J Vasc Endovasc Surg*. 2011;41:167–174.
18. Richards JM, Semple SI, MacGillivray TJ, et al. Abdominal aortic aneurysm growth predicted by uptake of ultrasmall superparamagnetic particles of iron oxide: A pilot study. *Circ Cardiovasc Imaging*. 2011;4:274–281.
19. Tang TY, Howarth SP, Miller SR, et al. The atheroma (atorvastatin therapy: Effects on reduction of macrophage activity) study. Evaluation using ultrasmall superparamagnetic iron oxide-enhanced magnetic resonance imaging in carotid disease. *J Am Coll Cardiol*. 2009;53:2039–2050.
20. Hyafil F, Laissy JP, Mazighi M, et al. Ferumoxtran-10-enhanced MRI of the hypercholesterolemic rabbit aorta: Relationship between signal loss and macrophage infiltration. *Arterioscler Thromb Vasc Biol*. 2006;26:176–181.
21. Mani V, Briley-Saebo KC, Itskovich VV, Samber DD, Fayad ZA. Gradient echo acquisition for superparamagnetic particles with positive contrast (grasp): Sequence characterization in membrane and glass superparamagnetic iron oxide phantoms at 1.5T and 3T. *Magn Reson Med*. 2006;55:126–135.
22. Stadler N, Lindner RA, Davies MJ. Direct detection and quantification of transition metal ions in human atherosclerotic plaques: Evidence for the presence of elevated levels of iron and copper. *Arterioscler Thromb Vasc Biol*. 2004; 24:949–954.
23. Kubota R, Yamada S, Kubota K, Ishiwata K, Tamahashi N, Ido T. Intratumoral distribution of fluorine-18-fluorodeoxyglucose in vivo: High accumulation in macrophages and granulation tissues studied by microautoradiography. *J Nucl Med*. 1992;33:1972–1980.
24. Taneja N, Coy PE, Lee I, Bryson JM, Robey RB. Proinflammatory interleukin-1 cytokines increase mesangial cell hexokinase activity and hexokinase II isoform abundance. *Am J Physiol Cell Physiol*. 2004;287: C548–C557.
25. Theron J, Tyler JL. Takayasu's arteritis of the aortic arch: Endovascular treatment and correlation with positron emission tomography. *Am J Neuroradiol*. 1987;8:621–626.
26. Yun M, Yeh D, Araujo LI, Jang S, Newberg A, Alavi A. F-18 FDG uptake in the large arteries:

A new observation. *Clin Nucl Med*. 2001; 26:314–319.

27. Rudd JH, Warburton EA, Fryer TD, et al. Imaging atherosclerotic plaque inflammation with [18F]-fluorodeoxyglucose positron emission tomography. *Circulation*. 2002;105: 2708–2711.
28. Vallabhajosula S, Fuster V. Atherosclerosis: Imaging techniques and the evolving role of nuclear medicine. *J Nucl Med*. 1997;38:1788–1796.
29. Tawakol A, Migrino RQ, Hoffmann U, et al. Noninvasive in vivo measurement of vascular inflammation with F-18 fluorodeoxyglucose positron emission tomography. *J Nucl Cardiol*. 2005;12:294–301.
30. Paulmier B, Duet M, Khayat R, et al. Arterial wall uptake of fluorodeoxyglucose on PET imaging in stable cancer disease patients indicates higher risk for cardiovascular events. *J Nucl Cardiol*. 2008;15:209–217.
31. Derlin T, Toth Z, Papp L, et al. Correlation of inflammation assessed by ^{18}F-FDG PET, active mineral deposition assessed by ^{18}F-fluoride PET, and vascular calcification in atherosclerotic plaque: A dual-tracer PET/CT study. *J Nucl Med*. 2011;52:1020–1027.
32. Rominger A, Saam T, Wolpers S, et al. ^{18}F-FDG PET/CT identifies patients at risk for future vascular events in an otherwise asymptomatic cohort with neoplastic disease. *J Nucl Med*. 2009;50:1611–1620.
33. Arauz A, Hoyos L, Zenteno M, Mendoza R, Alexanderson E. Carotid plaque inflammation detected by ^{18}F-fluorodeoxyglucose-positron emission tomography. Pilot study. *Clin Neurol Neurosurg*. 2007;109:409–412.
34. Muntendam P, McCall C, Sanz J, Falk E, Fuster V. The BioImage Study: Novel approaches to risk assessment in the primary prevention of atherosclerotic cardiovascular disease—study design and objectives. *Am Heart J*. 2010;160:49–57.e1.
35. Saam T, Rominger A, Wolpers S, et al. Association of inflammation of the left anterior descending coronary artery with cardiovascular risk factors, plaque burden and pericardial fat volume: A PET/CT study. *Eur J Nucl Med Mol Imaging*. 2010;37:1203–1212.
36. Dunphy MP, Freiman A, Larson SM, Strauss HW. Association of vascular ^{18}F-FDG uptake with vascular calcification. *J Nucl Med*. 2005;46:1278–1284.
37. Wykrzykowska J, Lehman S, Williams G, et al. Imaging of inflamed and vulnerable plaque in coronary arteries with ^{18}F-FDG PET/CT in patients with suppression of myocardial uptake using a low-carbohydrate, high-fat preparation. *J Nucl Med*. 2009;50:563–568.
38. Rogers IS, Nasir K, Figueroa AL, et al. Feasibility of FDG imaging of the coronary arteries: Comparison between acute coronary syndrome and stable angina. *JACC Cardiovasc Imaging*. 2010;3:388–397.
39. Tahara N, Kai H, Ishibashi M, et al. Simvastatin attenuates plaque inflammation: Evaluation by fluorodeoxyglucose positron emission tomography. *J Am Coll Cardiol*. 2006;48: 1825–1831.
40. Lee SJ, On YK, Lee EJ, Choi JY, Kim BT, Lee KH. Reversal of vascular ^{18}F-FDG uptake with plasma high-density lipoprotein elevation by atherogenic risk reduction. *J Nucl Med*. 2008;49:1277–1282.
41. Fayad ZA, Mani V, Woodward M, et al. Safety and efficacy of dalcetrapib on atherosclerotic disease using novel non-invasive multimodality imaging (dal-PLAQUE): A randomised clinical trial. *Lancet*. 2011;378:1547–1559.
42. Tardif J-C, Lesage F, Harel F, Romeo P, Pressacco J. Imaging biomarkers in atherosclerosis trials. *Circ Cardiovasc Imaging*. 2011;4:319–333.
43. Schwartz GG, Olsson AG, Ballantyne CM, et al. Rationale and design of the dal-OUTCOMES trial: Efficacy and safety of dalcetrapib in patients with recent acute coronary syndrome. *Am Heart J*. 2009;158:896–901.e3.
44. Kietselaer BL, Reutelingsperger CP, Heidendal GA, et al. Noninvasive detection of plaque instability with use of radiolabeled annexin A5 in patients with carotid-artery atherosclerosis. *N Engl J Med*. 2004;350:1472–1473.
45. Marincheva-Savcheva G, Subramanian S, Qadir S, et al. Imaging of the aortic valve using fluorodeoxyglucose positron emission tomography increased valvular fluorodeoxyglucose uptake in aortic stenosis. *J Am Coll Cardiol*. 2011;57:2507–2515.
46. Reeps C, Essler M, Pelisek J, Seidl S, Eckstein HH, Krause BJ. Increased ^{18}F-fluorodeoxyglucose uptake in abdominal aortic aneurysms in positron emission/computed tomography is associated with inflammation, aortic wall instability, and acute

symptoms. *J Vasc Surg*. 2008;48:417–423; discussion 424.
47. Osborn EA, Jaffer FA. The year in molecular imaging. *JACC Cardiovasc Imaging*. 2012;5:317–328.
48. Jaffer FA, Calfon MA, Rosenthal A, et al. Two-dimensional intravascular near-infrared fluorescence molecular imaging of inflammation in atherosclerosis and stent-induced vascular injury. *J Am Coll Cardiol*. 2011;57:2516–2526.
49. Yoo H, Kim JW, Shishkov M, et al. Intra-arterial catheter for simultaneous microstructural and molecular imaging in vivo. *Nat Med*. 2011;17:1680–1684.

40

PHARMACOLOGICAL THERAPY IN THE MANAGEMENT OF ACUTE CORONARY SYNDROMES

Andrew J. Sampson, Timir Paul, and George A. Stouffer
Division of Cardiology, University of North Carolina, Chapel Hill, NC, USA

INTRODUCTION

Atherosclerotic plaque rupture and erosion are the primary initiating events in most cases of acute coronary syndromes (ACS). Once a plaque ruptures, intravascular hemorrhage and thrombus formation occur. The thrombus can extend into the lumen and be occlusive or nonocclusive depending on thrombogenic and hemodynamic factors. Treatment with platelet inhibitors and anticoagulants form the cornerstone of medical therapy for ACS, and the use of these agents is a Class 1 recommendation in the American College of Cardiology/American Heart Association (ACC/AHA) and the European Society of Cardiology (ESC) guidelines for unstable angina (UA), non-ST segment elevation myocardial infarction (NSTEMI) and ST-elevation MI (STEMI) [1–4].

Aspirin inhibits platelet aggregation via irreversible inhibition of cyclo-oxygenase, the enzyme that converts arachidonic acid to thromboxane A2. Aspirin reduces mortality in patients with UA or acute MI [5–7]. The Second International Study of Infarct Survival (ISIS-2) was a pivotal trial that demonstrated the benefit of aspirin in 17,187 patients with suspected MI who were randomized to treatment with aspirin, intravenous streptokinase, combination, or placebo [5]. Aspirin therapy reduced 5-week vascular mortality by 23% compared to placebo. The combination of aspirin and streptokinase resulted in a 42% reduction in mortality compared to the use of neither agent (8% vs. 13.2%, $P<0.001$). Further, aspirin therapy was associated with significant reduction in the incidence of nonfatal reinfarction (1.0% vs. 2.0%) without any increase risk of major bleeding or hemorrhagic stroke. Aspirin is also effective in reducing the transition from stable to unstable vascular disease: a meta-analysis of 287 studies involving 135,000 patients found that antiplatelet therapy reduced the occurrence of vascular events by 25–33% in patients with history of MI, transient ischemic attack, stable angina, or peripheral vascular disease [6, 8].

Atherosclerosis: Risks, Mechanisms, and Therapies, First Edition. Edited by Hong Wang and Cam Patterson.

Full-dose aspirin (e.g., 300–325 mg) was used in most studies, but lower doses may be equally effective in treatment of ACS. The Clopidogrel and Aspirin Optimal Dose Usage to Reduce Recurrent Events—Seventh Organization to Assess Strategies in Ischemic Syndromes (CURRENT-OASIS 7) trial found no significant difference in the combined outcome of cardiovascular death, MI, or stroke at 30 days between higher (300–325 mg) versus lower doses (75–100 mg) of aspirin in 25,086 patients with ACS (Table 40.1). Although there was no significant difference in the rate of major bleeding, the rates of minor bleeding and gastrointestinal bleeding were higher in patients who received the higher dose of aspirin [9].

Thrombin is generated at sites of atherosclerotic plaque rupture, and numerous studies have shown that anticoagulant therapy, when used in addition to aspirin, reduces the incidence of recurrent MI and death in patients with ACS. The four agents recommended for use in ACS (Figure 40.1) are unfractionated heparin, enoxaparin (a low-molecular-weight heparin), fondaparinux (an inhibitor of factor Xa), and bivalirudin (a direct thrombin inhibitor). Adenosine diphosphate (ADP) release during thrombosis amplifies the platelet response and stimulates platelet aggregation. ADP acts on platelets via two G-protein-coupled receptors: $P2Y_1$ and $P2Y_{12}$. $P2Y_{12}$ is the predominant receptor involved in ADP-stimulated platelet activation of the glycoprotein (GP) IIb/IIIa receptor, and inhibitors of $P2Y_{12}$ have protective effects in patients with ACS. The most widely used ADP receptor antagonists are thienopyridines (e.g., ticlopidine, clopidogrel, and prasugrel). These prodrugs are metabolized to active agents that covalently bind the $P2Y_{12}$ receptor, resulting in irreversible platelet inhibition. Although all of the thienopyridines require CYP450 metabolism for the generation of active metabolites, the pathways differ between the drugs. High-affinity allosteric inhibitors of the $P2Y_{12}$ receptor (e.g., cangrelor and ticagrelor) that do not require metabolism and result in reversible, concentration-dependent inhibition of the receptor have more recently entered clinical use [10].

Clopidogrel is the most extensively studied $P2Y_{12}$ antagonist. When used in conjunction with aspirin, clopidogrel reduces major adverse cardiovascular events (MACEs) in patients with ACS, acute MI treated with thrombolytic agents, stable vascular disease, and in patients who receive an intracoronary stent. Clopidogrel use increases the risk of bleeding. For example, in the Clopidogrel in Unstable Angina to Prevent Recurrent Events (CURE) trial, clopidogrel plus aspirin was associated with a significant increase in major (3.7% vs. 2.7%), minor (5.1% vs. 2.4%), and gastrointestinal (1.3% vs. 0.7%) bleeding, but not life-threatening bleeding compared to aspirin alone (Table 40.2) [11]. Discontinuation of clopidogrel or ticagrelor for approximately 5 days and prasugrel for 7 days is recommended in most patients undergoing coronary artery bypass graft surgery to minimize the risk of bleeding.

Since it is difficult, if not impossible, to effectively block all agonists that cause platelet activation, therapeutic agents were developed that inhibit platelet aggregation by binding to GPIIb/IIIa, the receptor in the final common pathway of platelet aggregation [12]. This transmembrane protein is a member of the integrin family that undergoes a conformational change that enables it to bind fibrinogen (i.e., platelet activation). Pharmacological agents that inhibit fibrinogen binding to GPIIb/IIIa have the theoretical potential to completely abolish platelet aggregation, and are more effective at inhibiting *ex vivo* platelet aggregation than aspirin or ADP receptor antagonists. Four intravenous (IV) GPIIb/IIIa inhibitors—abciximab, eptifibatide, tirofiban, and lamifiban—have been shown to be effective in ACS and are currently in clinical use (lamifiban is not approved for use in the United States). All of these agents are delivered as a bolus followed by continuous infusion, and result in rapid platelet inhibition and prolongation of bleeding times when administered at therapeutic doses. The major pharmacological

TABLE 40.1 Details of trials that are discussed in the text

Trial	*N*	Inclusion criteria	Therapy	Primary endpoints	Secondary endpoints
ISIS-2 (1988)	17,187	Suspected acute MI	Aspirin (ASA), streptokinase (SK), ASA + SK or placebo	Vascular death at 5 months SK alone and ASA alone each produced a highly significant reduction in 5-week vascular mortality The combination of SK + ASA was significantly better than either agent alone.	Aspirin significantly reduced nonfatal reinfarction (1.0% vs. 2.0%) and nonfatal stroke (0.3% vs. 0.6%), and was not associated with any significant increase in cerebral hemorrhage or in bleeds requiring transfusion.
CURRENT-OASIS 7 (2010)	25,086	ACS who were referred for an invasive strategy	2-by-2 factorial design with either double-dose or standard-dose clopidogrel and either higher-dose or lower-dose aspirin	CV death, MI, or stroke at 30 days 4.2% of patients assigned to double dose as compared with 4.4% assigned to standard-dose clopidogrel ($P = 0.30$) 4.2% of patients assigned to higher dose as compared with 4.4% assigned to lower dose aspirin ($P = 0.61$). ($P = 0.001$)	Major bleeding at 30 days 2.5% in the double-dose clopidogrel group and 2.0% in the standard-dose group ($P = 0.01$) No difference in aspirin groups (2.3% vs. 2.3%; $P = 0.90$)
COMMIT-CCS 2 (2005)	45,825	Hospitalized with ACS <24 hours from symptom onset	ASA + clopidogrel vs. ASA + placebo	Death, MI, or stroke prior to discharge ASA + clopidogrel = 9.2% ASA + placebo = 10.1% ($P = 0.002$)	Major bleeding Clopidogrel 0.58% Placebo 0.55% ($P = 0.59$)
			IV the oral metoprolol vs. placebo	Death, MI, or cardiac arrest prior to discharge Metoprolol = 9.4% Placebo = 9.9% ($P = 0.1$)	Death Metoprolol = 7.7% Placebo = 7.8% ($P = 0.7$) Metoprolol was associated with less reinfarction and ventricular fibrillation but more cardiogenic shock
PCI-CLARITY (2005)	1,863	STEMI treated with ASA + fibrinolytic and then PCI	Clopidogrel vs. placebo	CV death, MI, or stroke after PCI Clopidogrel = 3.6% Placebo = 6.2% ($P = 0.008$)	CV death, MI, or stroke at 30 days Clopidogrel = 7.5% Placebo = 12% ($P = 0.001$)

(Continued)

TABLE 40.1 (Continued)

Trial	N	Inclusion criteria	Therapy	Primary endpoints	Secondary endpoints
ON-TIME 2	4,162	Hospitalized for ACS in past 10 days with STEMI, NSTEMI, and UA	Atorvastatin 80 mg vs. pravastatin 40 mg	Death, MI, UA, revascularization, or CVA at 2 years	Death from CHD or MI
				Atorvastatin group = 22.4%	Atorvastatin = 8.3%
				Pravastatin group = 26.3% ($P = 0.005$)	Pravastatin = 10% ($P = 0.06$)
BRAVE 3	19,134	Admitted <24 hours of acute MI (STEMI/NSTEMI)	Lisinopril, nitroglycerin, lisinopril + NTG, placebo	All-cause mortality at 6 weeks	Death, symptomatic HF, EF < 35%, or dyskenesis
				Controls = 7.2% Lisinopril = 6.3% Nitrates = 7.0% Lisinopril + NTG = 6.0% P value not stated	Controls = 17.0% Lisinopril = 16.4% Nitrates = 17.0% Lisinopril + NTG = 14.8%
HORIZONS-AMI	16,027	Patients within 12 hours of acute MI	Atenolol vs. placebo	Vascular mortality at 7 days Atenolol = 3.8% Placebo = 4.6% ($P < 0.04$)	Vascular mortality at 1 year Atenolol = 10.7% Placebo = 12.8% ($P < 0.01$)
OASIS-5 (2006)	20,078	ACS	Fondaparinux or enoxaparin for a mean of 6 days	Death, myocardial infarction, or refractory ischemia at 9 days	Major bleeding at 9 days was lower with fondaparinux (2.2% vs. 4.1%; $P < 0.001$)
				Fondaparinux = 5.8%	Mortality was lower with fondaparinux at 30 days ($P = 0.02$) and at 180 days ($P = 0.05$)
				Enoxaparin = 5.7% (P = NS)	
OASIS-6 (2006)	12,092	STEMI of which ~75% received reperfusion	Fondaparinux vs. heparin/control	Death or reinfarction at 30 days 11.2% in the control group and 9.7% in the fondaparinux group ($P = 0.008$)	Fondaparinux was beneficial in patients receiving fibrinolytic therapy but not in those treated with primary PCI
FINESSE (2008)	2,452	STEMI within 6 hours of symptom onset who were undergoing PCI	Half-dose reteplase + abciximab prior to PCI, abciximab prior to PCI or abciximab at the time of PCI	Composite of death from all causes, ventricular fibrillation occurring more than 48 hours after randomization, cardiogenic shock, and congestive heart failure at 90 days	Mortality at 90 days 5.2% in reteplase + abciximab prior to PCI 5.5% in abciximab prior to PCI 4.5% in abciximab at time of PCI ($P = 0.49$)
				9.8% in reteplase + abciximab prior to PCI	Early ST segment resolution 43.9% in reteplase + abciximab prior to PCI
				10.5% in abciximab prior to PCI 10.7% in abciximab at time of PCI ($P = 0.55$)	33.1% in abciximab prior to PCI 31.0% in abciximab at time of PCI ($P = 0.01$ and $P = 0.003$, respectively)

SYNERGY (2004)	10,027	NSTEMI	Unfractionated heparin vs. enoxaparin	Death or MI at 30 days	No differences in ischemic events during PCI between enoxaparin and unfractionated heparin.
				The primary end point occurred in 14.0% of patients assigned to enoxaparin and 14.5% of patients assigned to unfractionated heparin (P=NS)	Increased TIMI major bleeding with enoxaparin (9.1% vs. 7.6%, P=0.008)
ACUITY (2006)	13,819	"Moderate or high-risk ACS" undergoing invasive evaluation	Unfractionated heparin or enoxaparin + GP IIb/IIIa inhibitor, bivalirudin + GP IIb/IIIa inhibitor, or bivalirudin alone	Ischemia endpoint = death, MI, or unplanned revascularization at 30 days Net clinical endpoint = ischemia endpoints + major bleeding at 30 days Bivalirudin + GP IIb/IIIa inhibitor vs. heparin + GP IIb/IIIa inhibitor—noninferior rates of composite ischemia end point, major bleeding, and net clinical outcome end point Bivalirudin alone vs. heparin + GP IIb/IIIa inhibitor—noninferior rate of the composite ischemia end point and significantly reduced rates of major bleeding and the net clinical outcome end point	
SAVE (1992)	2,331	Age 21–80, EF < 40%, survive first 3 days after MI	Captopril vs. placebo	All-cause mortality at 42 months Captopril group = 20.4% Placebo group = 24.6% (P=0.019)	CV mortality Captopril = 16.9% Placebo = 21.0% (P=0.014)

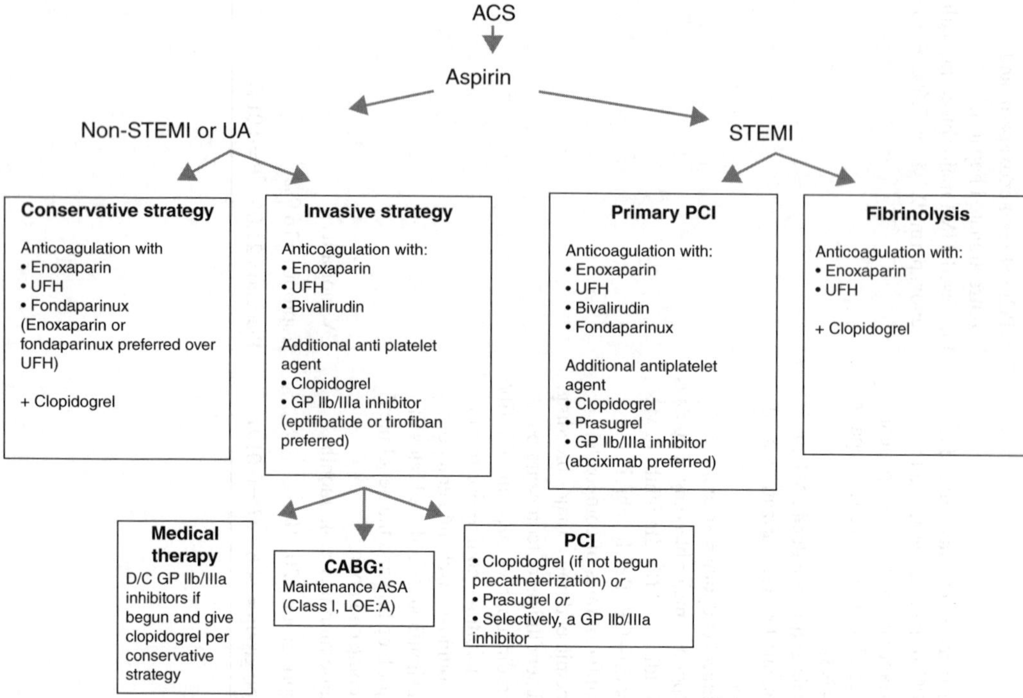

Figure 40.1 Brief overview of antiplatelet and anticoagulation therapy recommended for ACS.

differences between these agents are the derivation of the binding site, specificity for GPIIb/IIIa, and biological half-life.

STEMI

Introduction

STEMI accounts for approximately 20–30% of ACS. Coronary artery reperfusion with either primary percutaneous coronary intervention (PCI) or fibrinolysis improves clinical outcomes in patients with STEMI (for the purpose of this discussion, STEMI also includes an MI with a new left bundle branch block). While initially controversial, it is now well established that primary PCI is the reperfusion strategy of choice in hospitals with availability of 24-hour interventional cardiology facilities. A large meta-analysis demonstrated a significant reduction in short-term (≤6 weeks) mortality with primary PCI compared to fibrinolysis in randomized controlled trials (34% reduction) and in observational studies (23% reduction) [13]. Pharmacotherapy with antiplatelet agents plays a key role in improving clinical outcomes in patients with STEMI receiving treatment with fibrinolytic agents or with primary PCI. Aspirin should be started immediately upon presentation with STEMI and continued indefinitely.

Fibrinolytic Therapy

There are many hospitals that do not have on-site cardiac catheterization facilities, and fibrinolysis is an effective treatment in this setting. The benefit of fibrinolysis increases the earlier it is given after the onset of symptoms [14]. TIMI 3 flow (i.e., complete perfusion) is achieved in 50–60% of patients who received fibrinolytic therapy versus 93–96% of patients with primary PCI. Potential complications of fibrinolytic therapy include early recurrence of ischemia in 20–30% of patients, major hemorrhagic complications in 2–3% of patients, and intracranial hemorrhage in 1% of patients. Absolute contraindications include active

TABLE 40.2 Trials evaluating P2Y$_{12}$ receptor antagonists in ACS

Trials	*N*	Adjunctive therapy	Inclusion criteria	Primary endpoints	Results
CURE (2001)	12,562	Aspirin	Hospitalized with NSTEMI ≤24 hours	CV death, MI, or stroke	Clopidogrel vs. placebo (9.3% vs. 11.4%; $P<0.001$)
PCI-CURE (2001)	2,658	Aspirin	NSTEMI, PCI with BMS	CV death, MI, or urgent TVR at 30 days	Clopidogrel vs. placebo (4.5% vs. 6.4%; $P=0.03$)
CREDO* (2002)	2,116	Aspirin; GPIIb/IIIa inhibitors (40%)	Elective PCI	Incidence of death, MI, or stroke at 1 year	Clopidogrel vs. placebo (8.5% vs. 11.5%; $P=0.02$)
CURRENT-OASIS 7 (2010)	25,086	Aspirin	ACS (STEMI 29.2%; UA/NSTEMI 70.8%)	CV death, MI, or stroke at 30 days	Clopidogrel double vs. regular dose (4.2% vs. 4.4 %, $P=0.3$) High dose aspirin vs. low dose aspirin (4.2% vs. 4.4 %, $P=0.61$)
TRITON-TIMI 38 (2009)	13,608	Aspirin	ACS (STEMI 26%; UA/NSTEMI 74%)	CV death, MI, or stroke at 15 months	Prasugrel vs. clopidogrel (9.7% vs. 11.9%; $P=0.0001$)
PLATO (2007)	18,624	Aspirin	ACS (STEMI 48.8%; UA/NSTEMI 51.2%)	CV death, MI, or stroke	Ticagrelor vs. clopidogrel (9.0% vs. 10.7%; $P=0.0025$)
CLARITY-TIMI 28 (2005)	3,491	Aspirin; heparin if received fibrin-specific lytic	STEMI with chest pain ≤12 hours, age <75 years	Occluded infarct-related artery, death, or recurrent MI at time of angiography	Clopidogrel vs. placebo (15.0% vs. 21.7%; $P=0.001$)
COMMIT/CCS-2 (2005)	45,582	Aspirin; fibrinolytics (54%); anticoagulant (75%)	Acute MI (STEMI/new LBBB)	All-cause mortality	Clopidogrel vs. placebo (7.5% vs. 8.1%; $P=0.001$)
				Death, reinfarction, or stroke	Clopidogrel vs. placebo (9.3% vs. 10.1%; $P=0.0002$)

*CREDO, Clopidogrel for the Reduction of Events during Observation. Refer to the text for an explanation of other abbreviations.

bleeding, recent stroke, history of intracranial bleeding, closed head or facial trauma within 3 months, suspected aortic dissection, uncontrolled high blood pressure, and/or structural cerebral vascular lesions. Efficacy of fibrinolytic therapy has not been demonstrated in patients with cardiogenic shock [15].

The use of aspirin as an adjunct to fibrinolytic therapy in patients with acute MI reduced the combined endpoint of subsequent nonfatal MI, nonfatal stroke, or vascular death by 30% in studies involving more than 19,000 patients [6]. Heparin should be used in all patients treated with a fibrin-specific fibrinolytic agent such as alteplase, tenecteplase, or reteplase. The choice of unfractionated heparin versus enoxaparin in patients with STEMI treated with fibrin-specific fibrinolytic agents was examined in a meta-analysis of over 27,000 patients [16]. This study found that the primary endpoint of death, MI, or major bleed at 30 days was significantly reduced (11.1% vs. 12.9%), but that major bleeding was significantly higher in patients receiving enoxaparin. There is no benefit in the use of heparin in conjunction with streptokinase [16].

Early administration of clopidogrel to STEMI patients treated with fibrinolytic therapy plus heparin plus aspirin improves outcomes [17, 18]. The Clopidogrel and Metoprolol in Myocardial Infarction Trial (COMMIT/CCS-2) trial randomized 45,582 patients with a suspected acute MI to clopidogrel (75 mg/day without a loading dose) or placebo [18]. Of these patients, 93% had evidence of ST elevation or left bundle branch block, and 54% received fibrinolytic therapy, 75% received an anticoagulant, and all were treated with aspirin. The primary endpoints of all-cause mortality and the composite of death, reinfarction, or stroke were significantly lower in the patients treated with clopidogrel, without any increase in major bleeding.

In summary, all patients with STEMI treated with fibrinolytic therapy should receive aspirin and clopidogrel, and those treated with a fibrin-specific fibrinolytic should also receive heparin [14]. At present, there is no evidence that support the use of GPIIb/IIIa inhibitors in the setting of fibrinolytic therapy. Prasugrel and ticagrelor have not been evaluated in STEMI patients treated with fibrinolysis.

Adjunctive Therapy for Primary PCI

Aspirin and $P2Y_{12}$ inhibitors are standard treatment with PCI, and thus no randomized trials have been conducted to evaluate the efficacy of aspirin or aspirin plus $P2Y_{12}$ antagonists in STEMI patients treated with primary PCI. A subgroup analysis of patients who underwent PCI in the Clopidogrel as Adjunctive Reperfusion Therapy—TIMI 28 trial (PCI-CLARITY), a study of 3491 patients with STEMI and age ≤75 years who received aspirin plus fibrinolytic therapy plus unfractionated heparin or low-molecular-weight heparin and were randomized to either clopidogrel or placebo, indicated that clopidogrel was associated with a significant reduction in the primary endpoint of cardiovascular death, MI, or stroke (3.6% vs. 6.2%) at 30 days after PCI [19].

Studies of GPIIb/IIIa inhibitors in STEMI treated with primary PCI have produced differing results [12]. The Ongoing Tirofiban in Myocardial Infarction Evaluation (On-TIME) two trial compared the prehospital use of tirofiban to placebo in nearly 1400 STEMI patients undergoing primary PCI treated with aspirin, heparin, and clopidogrel. There was a significant reduction in the rate of major adverse cardiovascular events at 30 days with tirofiban (5.8% vs. 8.6%), without a significant increase in the rate of bleeding [20]. The Bavarian Reperfusion Alternatives Evaluation-3 (BRAVE-3) trial compared abciximab with placebo, administered in the intensive care unit, in 800 patients with STEMI treated with 600 mg of clopidogrel [21]. There was no significant difference in death, recurrent MI, stroke, or urgent revascularization of the infarct-related artery at 30 days.

Optimal anticoagulation during primary PCI was studied in the Harmonizing Outcomes with Revascularization and Stents in Acute Myocardial Infarction (HORIZONS AMI) trial, which compared bivalirudin to heparin plus GPIIb/IIIa inhibitor during primary PCI in 3006 patients with STEMI [22]. The composite endpoint of death, reinfarction, target vessel revascularization, or stroke was similar at 30 days in the two treatment groups, while major bleeding was significantly lower in patients treated with bivalirudin. At 3 years, bivalirudin was associated with lower rates of all-cause mortality, cardiac mortality, reinfarction, and major bleeding not related to bypass graft surgery, with no significant differences in ischemia-driven target vessel revascularization or stent thrombosis [23].

The use of fondaparinux in STEMI was evaluated in OASIS-6, a randomized double-blind study of 12,092 patients from 447 hospitals in 41 countries. Patients in the control arm were treated with heparin. Death or reinfarction at 30 days was significantly reduced in patients treated with fondaparinux, but the benefit was limited to those treated with fibrinolytic therapy, and there was no benefit in patients treated with primary PCI [24]. There was no difference in bleeding or stroke incidence. Fondaparinux, while effective in STEMI, has been associated with a risk of catheter-related thrombosis, and thus supplemental anticoagulation is needed during PCI [25].

Facilitated PCI

Facilitated primary PCI for STEMI is defined as the use of pharmacological agents, in particular half-dose thrombolytic therapy and/or GPIIb/IIIa inhibitors, before PCI. The Facilitated Intervention with Enhanced Reperfusion Speed to Stop Events (FINESSE) trial of 2452 patients found no significant difference between abciximab plus reduced dose reteplase prior to primary PCI, early administration of abciximab alone, or abciximab given at the time of primary PCI in the composite endpoint of death from all causes, ventricular fibrillation occurring more than 48 hours after randomization, cardiogenic shock, and heart failure at 90 days. There was a significant increase in TIMI major and minor bleeding in the "facilitated" (reduced dose reteplase, early abciximab) groups compared to the group receiving abciximab at the time of PCI [26].

A meta-analysis of 17 trials including 4504 patients found that the facilitated approach resulted in a greater number of patients with TIMI 3 flow on the initial angiogram (37% vs. 15%), but not on the final angiogram (89% vs. 88%). Patients assigned to the facilitated approach were more likely to die, have a nonfatal reinfarction, need urgent target vessel revascularization, and/or have major bleeding. The increase in adverse events occurred primarily with thrombolytic-therapy-based regimens [27].

UNSTABLE ANGINA AND NSTEMI

Antiplatelet Therapy

Antiplatelet agents have been shown to improve clinical outcomes in patients with UA and NSTEMI. A meta-analysis involving 12 trials of over 5000 patients with NSTEMI demonstrated that treatment with aspirin resulted in a significant reduction in the combined endpoint of subsequent nonfatal MI, nonfatal stroke, or vascular death (8.0% vs. 13.3%) [6]. Evidence suggests that there is a beneficial effect of $P2Y_{12}$ antagonists that is additive to aspirin. In the CURE trial, combination therapy with aspirin and clopidogrel was associated with a reduction in the primary endpoint (9.3% vs. 11.4%), mostly due to a decrease in MI incidence [11]. Whether patients were treated medically or invasively, clopidogrel showed similar relative risk reduction in low-, intermediate-, and high-risk patients as defined by the TIMI risk score [28]. The benefit began to emerge within 24 hours and gradually increased in magnitude during the first 30 days. The use of GPIIb/IIIa inhibitors in patients with NSTEMI has decreased, but there is some evidence that these agents can be beneficial during PCI, especially in patients not receiving a loading dose of ADP receptor antagonist [12].

Anticoagulants

Anticoagulant therapy should be started as soon as possible in addition to antiplatelet therapy in patients with UA/NSTEMI. Anticoagulants along with antiplatelet agents reduce the incidence of MI, death, and need for revascularization in patients with UA/NSTEMI. Both unfractionated heparin and enoxaparin are effective. The largest direct comparison was the Superior Yield of the New Strategy of Enoxaparin, Revascularization and Glycoprotein IIb/IIIa Inhibitors (SYNERGY) trial, which enrolled 10,027 high-risk patients with NSTEMI [29]. There was no difference in efficacy (a composite end point of all-cause death or nonfatal MI during the first 30 days), but enoxaparin was associated with increased major bleeding. A meta-analysis of 21,946 patients found a very modest reduction in the incidence of death or nonfatal MI in patients treated with enoxaparin without any increase in bleeding [30].

Different anticoagulant strategies were evaluated in the Acute Catheterization and Urgent Intervention Triage Strategy (ACUITY) trial of almost 14,000 moderate-to-high-risk ACS patients undergoing PCI [31]. Bivalirudin alone was noninferior to unfractionated heparin or enoxaparin plus a GPIIb/IIIa inhibitor for the rate of ischemic complications at 30 days; however, bivalirudin-treated group had a significantly reduced rate of major bleeding. Fondaparinux, in comparison with enoxaparin in NSTEMI in the OASIS-5 trial, was associated with less bleeding but a small but important increase in catheter-related thrombosis [25, 32].

The ACC/AHA 2007 guidelines and 2011 focused update on UA/NSTEMI recommended the following in regard to anticoagulation in patients with non-STEMI (Figure 40.1) [1]. Among patients in whom a conservative strategy is selected, either enoxaparin or unfractionated heparin can be used. Bivalirudin alone is as efficacious as unfractionated heparin or enoxaparin plus GPIIb/IIIa inhibitor in moderate-to-high-risk ACS patients undergoing PCI, and is associated with a reduced rate of bleeding. When bivalirudin is selected as the anticoagulant, it is reasonable to omit a GPIIb/IIIa inhibitor before diagnostic angiography and PCI, as long as clopidogrel (at least 300 mg) was given at least 6 hours earlier.

NEWER ANTIPLATELET AGENTS

The Trial to Assess Improvement in Therapeutic Outcomes by Optimizing Platelet Inhibition with Prasugrel (TRITON)-TIMI 38 trial evaluated the efficacy of prasugrel compared to clopidogrel in 13,608 moderate-to-high-risk ACS patients undergoing PCI, including 3534 patients with STEMI [33]. Prasugrel was associated with fewer MACE (death from cardiovascular causes, nonfatal MI, or nonfatal stroke; 9.9% vs. 12.1%), but with an increase in major bleeding and fatal bleeding. Patients with STEMI treated with prasugrel had significantly lower rates of cardiovascular death, nonfatal MI, or nonfatal stroke at 15 months follow-up [34].

The Platelet Inhibition and Patient Outcomes (PLATO) trial enrolled 18,624 ACS patients (38% with STEMI) and compared the efficacy of ticagrelor with clopidogrel. The primary end point (MI, stroke, or cardiovascular death) at 12 months was significantly lower with ticagrelor compared to clopidogrel (9.8% vs. 11.7%). Ticagrelor was associated with a lower rate of death from any cause but a higher rate of major bleeding not related to coronary-artery bypass grafting (4.5% vs. 3.8%, $P = 0.03$) [35]. Unexpectedly, clopidogrel was associated with a nonsignificant trend toward better outcomes in North America, whereas ticagrelor was associated with better outcomes in the rest of the world. The investigators felt that the regional interaction could arise from chance, or from an interaction with aspirin maintenance dose. When using ticagrelor, the lowest risk of cardiovascular death, MI, or stroke was associated with low-maintenance doses of aspirin [36].

ADJUNCTIVE THERAPY

Reperfusion strategies, antiplatelet drugs, and antithrombotic agents remain the cornerstone of treatment of ACS. Beyond these interventions, current ACC/AHA guidelines focus on adjuvant pharmacologic treatments of ACS directed at reducing myocardial oxygen demand and malignant arrhythmias (beta-blockers), minimizing ischemia (supplemental O_2 and nitrates), plaque stabilization (statins), and regulation of remodeling (angiotensin-converting enzyme inhibitors (ACEIs)).

Nitrates, which may be administered via IV, sublingual, aerosol, or transdermal delivery, are routinely used in ACS, and have a Class I, level of evidence (LOE) C indication. Nitrates induce smooth muscle relaxation via a cyclic GMP-mediated intracellular signaling cascade that decreases intracellular calcium and reduces contraction. Smooth muscle relaxation and dilation of coronary arteries and arterioles improves perfusion, and has been shown to decrease infarct size in animal models. Vasodilation leads to increased capacitance in the venous system, effectively lowering preload and reducing filling pressures, which is especially advantageous in the setting of pulmonary edema. Despite these hemodynamically desirable effects, there is no conclusive evidence to support a mortality benefit with nitrates in ACS [37]. In addition, the decision to use nitrates needs to be carefully weighed against their possible deleterious effects. Nitrates are contraindicated in patients with systemic hypotension (SBP < 90), severe bradycardia (HR < 50), or tachycardia (HR < 100) in the absence of heart failure, or in those in whom right ventricular infarction is suspected. Hemodynamic collapse may be precipitated in these individuals. Further, nitrates should not be administered in the setting of low blood pressure where it would prevent the use of other agents with well-established mortality benefit, such as ACEI [37]. Nitrates should not be administered to individuals who have recently taken a phosphodiesterase inhibitor.

The use of supplemental oxygen remains the standard of care for ACS patients with shortness of breath, heart failure, and/or oxyhemoglobin concentrations <90% [37]. These recommendations were driven largely by studies performed in the 1970s. More recently, however, the routine use of supplemental O_2 in patients without these clinical indications has been called into question, and a 2010 Cochrane review examining randomized control trials in ACS found a trend toward harm with routine use of supplemental oxygen, although the number of patients was small [38]. Several hypotheses exist for this detrimental effect, including hyperoxia-induced coronary vasoconstriction and free radical production. In a study of the effects of hyperoxia on the coronary circulation using intracoronary Doppler flow wire and coronary angiography, hyperoxia decreased coronary blood flow by 29% and raised coronary vascular resistance by 41% [39]. The most recent ACC/AHA guidelines conclude that there is insufficient data to support the routine use of supplemental oxygen in uncomplicated ACS in the absence of hypoxemia or heart failure [1, 2].

The potential beneficial effects of beta-adrenergic blockade include decreased oxygen demand, decreased malignant arrhythmias, improved diastolic coronary perfusion time, reduction in cardiac remodeling, and improved left ventricular function. Early studies of IV beta-blockers demonstrated reduced mortality and infarct size when administered early in the course of ACS [40]. The results from the COMMIT/CCS2 trial, however, called for a more nuanced approach to early IV beta-blocker therapy [41]. Patients with hemodynamic instability treated with IV beta-blockers in this study incurred increased mortality, while hemodynamically stable patients showed a trend toward benefit. As a result, current ACC/AHA guidelines recommend initiation of oral beta-blocker therapy in those patients without signs of hemodynamic compromise or significant heart block within 24 hours of medical contact (Class I, LOE A) [1, 2]. In hemodynamically stable ACS patient with significant hypertension and tachycardia unrelated to cardiogenic shock, IV beta-blockade remains a reasonable therapeutic choice and is a class IIa recommendation. Though much has been made about the risk of beta-blockers in patients with chronic obstructive pulmonary disease (COPD) or reactive airway disease given the theoretical possibility of precipitating bronchospasm, a number of studies have consistently shown a decrease in mortality in COPD patients treated with beta-blockers.

Under normal physiological circumstances, the healthy myocardium utilizes fatty acids as the primary metabolic substrate for energy production. In the absence of adequate perfusion, the myocardium relies on anaerobic metabolism, and fatty acids become toxic to the myocardial cells [42]. The concept of metabolic modulation with glucose–insulin–potassium (GIK) was therefore predicated on the following ideas: (i) insulin would shift the primary metabolic substrate of the ischemic myocardium from free fatty acids to glucose; (ii) glucose would provide metabolic substrate for anaerobic glycolysis; and (iii) potassium would prevent insulin-mediated hypokalemia while optimizing intramyocellular potassium levels. Initial results in small, randomized clinical trials were promising, but subsequent larger randomized clinical trials failed to show a significant benefit [43]. As a result, GIK therapy remains a rarely used modality and Class IIB, LOE C recommendation.

ACEIs reduce mortality rates in patients with acute or recent MI [44]. This benefit appears greatest in those patients with left ventricular dysfunction or heart failure, with benefit also seen in diabetics and in patients with high-risk coronary disease. ACEIs positively influence left ventricular remodeling, and trials of ACEI initiated shortly after MI (24 hours to 16 days) have demonstrated an improvement in left ventricular ejection fraction at 1 year. The benefits of ACEI appear to be durable over time, with preservation of mortality benefit up to 12 years in the Trandopril Cardiac Evaluation (TRACE) trial [45]. In light of this data, current ACC/AHA guidelines recommend that ACEI be given and continued indefinitely for patients recovering from ACS with heart failure, left ventricular dysfunction (LVEF$<$0.40), hypertension, or diabetes mellitus, unless contraindicated. Angiotensin receptor blockade (ARB) with candesartan proved to be as effective as ACEI in patients whose post-MI course was complicated by heart failure, left ventricular dysfunction, or both [44], and thus for those unable to tolerate ACEI, an ARB should be initiated (Class I, LOE A).

Statins reduce cardiovascular morbidity and mortality in secondary prevention of coronary heart disease [46]. Accordingly, current AHA/ACC guidelines recommend that patients with ACS should receive statin therapy prior to discharge, regardless of baseline LDL level, and treatment should be continued indefinitely in the absence of contraindications (Class I, LOE A) [1, 2]. Given the available data, it is unclear whether the benefit seen with statin therapy is a class effect, or if treatment with more potent statins confers a greater benefit. Results of the Pravastatin or Atorvastatin Evaluation and Infection Therapy–Thrombolysis in Myocardial Infarction 22 (PROVE IT-TIMI 22) trial would suggest that atorvastatin may be superior to simvastatin, and possibly to other lower potency statins [47].

PROSPECTS FOR THE FUTURE

There are several areas under active study that may change the medical management of ACS in the future. Various strategies evaluating different antiplatelet medications at different doses and utilizing different measurements of platelet function are being evaluated to determine the best approach to ensuring adequate platelet inhibition in every patient. Blockade of the protease-activated receptor 1 on platelets represents a promising target to inhibit platelet function. Factor IXa represents a new and interesting target for anticoagulation in patients with ACS. Various algorithms are being developed to help predict bleeding complications, thus enabling tailored anticoagulant therapy in patients with ACS. Among different approaches to modify lipids in patients with ACS, medications designed to increase HDL levels are subjects of extensive study.

REFERENCES

1. Wright RS, Anderson JL, Adams CD, et al. 2011 ACCF/AHA focused update incorporated into the ACC/AHA 2007 guidelines for the management of patients with unstable angina/non-ST-elevation myocardial infarction: a report of the American College of Cardiology Foundation/American Heart Association Task Force on Practice Guidelines developed in collaboration with the American Academy of Family Physicians, Society for Cardiovascular Angiography and Interventions, and the Society of Thoracic Surgeons. *J Am Coll Cardiol.* 2011;57:e215–e367.

2. Kushner FG, Hand M, Smith SC, Jr., et al. 2009 focused updates: ACC/AHA guidelines for the management of patients with ST-elevation myocardial infarction (updating the 2004 guideline and 2007 focused update) and ACC/AHA/SCAI guidelines on percutaneous coronary intervention (updating the 2005 guideline and 2007 focused update) a report of the American College of Cardiology Foundation/American Heart Association Task Force on Practice Guidelines. *J Am Coll Cardiol*. 2009;54:2205–2241.
3. Hamm CW, Bassand JP, Agewall S, et al. ESC guidelines for the management of acute coronary syndromes in patients presenting without persistent ST-segment elevation: the Task Force for the management of acute coronary syndromes (ACS) in patients presenting without persistent ST-segment elevation of the European Society of Cardiology (ESC). *Eur Heart J*. 2011;32:2999–3054.
4. Van De WF, Bax J, Betriu A, et al. Management of acute myocardial infarction in patients presenting with persistent ST-segment elevation: the Task Force on the management of ST-segment elevation acute myocardial infarction of the European Society of Cardiology. *Eur Heart J*. 2008;29:2909–2945.
5. ISIS Collaborative Group. Randomized trial of intravenous streptokinase, oral aspirin, both or neither among 17187 cases of suspected acute myocardial infarction: ISIS-2. *Lancet*. 1988;2:349–360.
6. Antithrombotic Trialists' Collaboration. Collaborative meta-analysis of randomised trials of antiplatelet therapy for prevention of death, myocardial infarction, and stroke in high risk patients. *BMJ*. 2002;324:71–86.
7. Baigent C, Blackwell L, Collins R, et al. Aspirin in the primary and secondary prevention of vascular disease: collaborative meta-analysis of individual participant data from randomised trials. *Lancet*. 2009;373:1849–1860.
8. Antiplatelet Trialists' Collaboration. Collaborative overview of randomised trials of antiplatelet therapy—I: prevention of death, myocardial infarction, and stroke by prolonged antiplatelet therapy in various categories of patients [published erratum appears in *BMJ*. 1994;308(6943):1540]. *BMJ*. 1994;308:81–106.
9. Mehta SR, Bassand JP, Chrolavicius S, et al. Dose comparisons of clopidogrel and aspirin in acute coronary syndromes. *N Engl J Med*. 2010;363:930–942.
10. Wallentin L. P2Y(12) inhibitors: differences in properties and mechanisms of action and potential consequences for clinical use. *Eur Heart J*. 2009;30:1964–1977.
11. Yusuf S, Zhao F, Mehta SR, Chrolavicius S, Tognoni G, Fox KK. Effects of clopidogrel in addition to aspirin in patients with acute coronary syndromes without ST-segment elevation. *N Engl J Med*. 2001;345:494–502.
12. Starnes HB, Patel AA, Stouffer GA. Optimal use of platelet glycoprotein IIb/IIIa receptor antagonists in patients undergoing percutaneous coronary interventions. *Drugs*. 2011;71: 2009–2030.
13. Keeley EC, Boura JA, Grines CL. Primary angioplasty versus intravenous thrombolytic therapy for acute myocardial infarction: a quantitative review of 23 randomised trials. *Lancet*. 2003;361:13–20.
14. Morse MA, Todd JW, Stouffer GA. Optimizing the use of thrombolytics in ST-segment elevation myocardial infarction. *Drugs*. 2009;69: 1945–1966.
15. Kiernan TJ, Gersh BJ. Thrombolysis in acute myocardial infarction: current status. *Med Clin North Am*. 2007;91:617–637.
16. Eikelboom JW, Quinlan DJ, Mehta SR, Turpie AG, Menown IB, Yusuf S. Unfractionated and low-molecular-weight heparin as adjuncts to thrombolysis in aspirin-treated patients with ST-elevation acute myocardial infarction: a meta-analysis of the randomized trials. *Circulation*. 2005;112:3855–3867.
17. Sabatine MS, Cannon CP, Gibson CM, et al. Addition of clopidogrel to aspirin and fibrinolytic therapy for myocardial infarction with ST-segment elevation. *N Engl J Med*. 2005;352: 1179–1189.
18. Chen ZM, Jiang LX, Chen YP, et al. Addition of clopidogrel to aspirin in 45,852 patients with acute myocardial infarction: randomised placebo-controlled trial. *Lancet*. 2005;366: 1607–1621.
19. Sabatine MS, Cannon CP, Gibson CM, et al. Effect of clopidogrel pretreatment before percutaneous coronary intervention in patients with ST-elevation myocardial infarction treated with fibrinolytics: the PCI-CLARITY study. *JAMA*. 2005;294:1224–1232.
20. Ten Berg JM, van 't Hof AW, Dill T, et al.Effect of early, pre-hospital initiation of high bolus dose tirofiban in patients with ST-segment

elevation myocardial infarction on short- and long-term clinical outcome. *J Am Coll Cardiol.* 2010;55:2446–2455.
21. Mehilli J, Kastrati A, Schulz S, et al. Abciximab in patients with acute ST-segment-elevation myocardial infarction undergoing primary percutaneous coronary intervention after clopidogrel loading: a randomized double-blind trial. *Circulation.* 2009;119:1933–1940.
22. Stone GW, Witzenbichler B, Guagliumi G, et al. Bivalirudin during primary PCI in acute myocardial infarction. *N Engl J Med.* 2008;358: 2218–2230.
23. Stone GW, Witzenbichler B, Guagliumi G, et al. Heparin plus a glycoprotein IIb/IIIa inhibitor versus bivalirudin monotherapy and paclitaxel-eluting stents versus bare-metal stents in acute myocardial infarction (HORIZONS-AMI): final 3-year results from a multicentre, randomised controlled trial. *Lancet.* 2011;377: 2193–2204.
24. Yusuf S, Mehta SR, Chrolavicius S, et al. Effects of fondaparinux on mortality and reinfarction in patients with acute ST-segment elevation myocardial infarction: the OASIS-6 randomized trial. *JAMA.* 2006;295:1519–1530.
25. Mehta SR, Boden WE, Eikelboom JW, et al. Antithrombotic therapy with fondaparinux in relation to interventional management strategy in patients with ST- and non-ST-segment elevation acute coronary syndromes: an individual patient-level combined analysis of the Fifth and Sixth Organization to Assess Strategies in Ischemic Syndromes (OASIS 5 and 6) randomized trials. *Circulation.* 2008;118:2038–2046.
26. Ellis SG, Tendera M, de Belder MA, et al. Facilitated PCI in patients with ST-elevation myocardial infarction. *N Engl J Med.* 2008;358: 2205–2217.
27. Keeley EC, Boura JA, Grines CL. Comparison of primary and facilitated percutaneous coronary interventions for ST-elevation myocardial infarction: quantitative review of randomised trials. *Lancet.* 2006;367:579–588.
28. Budaj A, Yusuf S, Mehta SR, et al. Benefit of clopidogrel in patients with acute coronary syndromes without ST-segment elevation in various risk groups. *Circulation.* 2002;106: 1622–1626.
29. Ferguson JJ, Califf RM, Antman EM, et al. Enoxaparin vs. unfractionated heparin in high-risk patients with non-ST-segment elevation acute coronary syndromes managed with an intended early invasive strategy: primary results of the SYNERGY randomized trial. *JAMA.* 2004;292:45–54.
30. Petersen JL, Mahaffey KW, Hasselblad V, et al. Efficacy and bleeding complications among patients randomized to enoxaparin or unfractionated heparin for antithrombin therapy in non-ST-segment elevation acute coronary syndromes: a systematic overview. *JAMA.* 2004;292:89–96.
31. Stone GW, McLaurin BT, Cox DA, et al. Bivalirudin for patients with acute coronary syndromes. *N Engl J Med.* 2006;355:2203–2216.
32. Jolly SS, Faxon DP, Fox KA, et al. Efficacy and safety of fondaparinux versus enoxaparin in patients with acute coronary syndromes treated with glycoprotein IIb/IIIa inhibitors or thienopyridines: results from the OASIS 5 (Fifth Organization to Assess Strategies in Ischemic Syndromes) trial. *J Am Coll Cardiol.* 2009;54: 468–476.
33. Wiviott SD, Braunwald E, McCabe CH, et al. Prasugrel versus clopidogrel in patients with acute coronary syndromes. *N Engl J Med.* 2007;357:2001–2015.
34. Montalescot G, Wiviott SD, Braunwald E, et al. Prasugrel compared with clopidogrel in patients undergoing percutaneous coronary intervention for ST-elevation myocardial infarction (TRITON-TIMI 38): double-blind, randomised controlled trial. *Lancet.* 2009;373:723–731.
35. Wallentin L, Becker RC, Budaj A, et al. Ticagrelor versus clopidogrel in patients with acute coronary syndromes. *N Engl J Med.* 2009;361:1045–1057.
36. Mahaffey KW, Wojdyla DM, Carroll K, et al. Ticagrelor compared with clopidogrel by geographic region in the Platelet Inhibition and Patient Outcomes (PLATO) trial. *Circulation.* 2011;124:544–554.
37. O'Connor RE, Brady W, Brooks SC, et al. Part 10: acute coronary syndromes: 2010 American Heart Association guidelines for cardiopulmonary resuscitation and emergency cardiovascular care. *Circulation.* 2010;122:S787–S817.
38. Cabello JB, Burls A, Emparanza JI, Bayliss S, Quinn T. Oxygen therapy for acute myocardial infarction. *Cochrane Database Syst Rev.* 2010; (6):CD007160.
39. McNulty PH, King N, Scott S, et al. Effects of supplemental oxygen administration on coronary blood flow in patients undergoing cardiac

catheterization. *Am J Physiol Heart Circ Physiol*. 2005;288:H1057–H1062.

40. First International Study of Infarct Survival Collaborative Group. Randomised trial of intravenous atenolol among 16 027 cases of suspected acute myocardial infarction: ISIS-1. *Lancet*. 1986;2:57–66.
41. Chen ZM, Pan HC, Chen YP, et al. Early intravenous then oral metoprolol in 45,852 patients with acute myocardial infarction: randomised placebo-controlled trial. *Lancet*. 2005;366: 1622–1632.
42. Kloner RA, Nesto RW. Glucose-insulin-potassium for acute myocardial infarction: continuing controversy over cardioprotection. *Circulation*. 2008;117:2523–2533.
43. Zhao YT, Weng CL, Chen ML, et al. Comparison of glucose-insulin-potassium and insulin-glucose as adjunctive therapy in acute myocardial infarction: a contemporary meta-analysis of randomised controlled trials. *Heart*. 2010;96: 1622–1626.
44. Flather MD, Yusuf S, Kober L, et al. Long-term ACE-inhibitor therapy in patients with heart failure or left-ventricular dysfunction: a systematic overview of data from individual patients. ACE-Inhibitor Myocardial Infarction Collaborative Group. *Lancet*. 2000;355: 1575–1581.
45. Buch P, Rasmussen S, Abildstrom SZ, Kober L, Carlsen J, Torp-Pedersen C. The long-term impact of the angiotensin-converting enzyme inhibitor trandolapril on mortality and hospital admissions in patients with left ventricular dysfunction after a myocardial infarction: follow-up to 12 years. *Eur Heart J*. 2005;26: 145–152.
46. Baigent C, Keech A, Kearney PM, et al. Efficacy and safety of cholesterol-lowering treatment: prospective meta-analysis of data from 90,056 participants in 14 randomised trials of statins. *Lancet*. 2005;366: 1267–1278.
47. Ray KK, Cannon CP, McCabe CH, et al. Early and late benefits of high-dose atorvastatin in patients with acute coronary syndromes: results from the PROVE IT-TIMI 22 trial. *J Am Coll Cardiol*. 2005;46:1405–1410.

41

TARGETED MOLECULAR THERAPEUTICS FOR ATHEROSCLEROSIS

LANTZ C. MACKEY AND JONATHON W. HOMEISTER

Department of Pathology and Laboratory Medicine, McAllister Heart Institute, University of North Carolina Medical School, Chapel Hill, NC, USA

INTRODUCTION

Traditional risk factors for atherosclerosis (see Section I) include lifestyle features such as diet, exercise, and smoking, as well as certain medical conditions including hypertension and diabetes. Many of the preventive therapies for atherosclerosis discussed in Chapter 38 are aimed at modifying these risk factors. Unfortunately, these therapies generally produce modest, at best, results in preventing the long-term sequelae of the disease. Some of the earlier discussed, existing preventive medical therapies for atherosclerosis, such as the use of statins, nicotinic acid, and antiplatelet agents, can be considered targeted molecular therapeutics because we understand at least, in part, the molecular and cellular biology processes that they modulate. However, even with these medical treatments, the morbidity and mortality associated with atherosclerosis remain high. Fortunately, recent advances in cellular and molecular biology and genomics are leading to a better understanding of the pathogenesis of atherosclerosis. As these advancements help us unravel the molecular and cellular mechanisms of the disease, we identify new molecular therapeutic targets that can potentially be manipulated to help prevent or treat it. This chapter includes a brief introduction to genetic and molecular techniques to identify and manipulate new molecular targets to treat atherosclerosis. It examines select molecular targets with therapeutic potential that have been uncovered by cellular, molecular, and genomic studies in animals and humans, with emphasis on targets identified by human genome-wide association studies (GWAS). These targets can be grouped broadly into those that modulate lipid metabolism, endothelial function, cell growth and migration, extracellular matrix turnover, or innate and adaptive immunity.

Atherosclerosis: Risks, Mechanisms, and Therapies, First Edition. Edited by Hong Wang and Cam Patterson.

GENETIC AND MOLECULAR THERAPEUTIC STRATEGIES

Recent advancements in genomic tools and techniques, coupled with advanced proteomic and metabolomics studies, are leading to the identification of unrecognized molecular pathways critical to the pathogenesis of atherosclerosis. Together, these studies will lead to the discovery and validation of new therapeutic targets for the disease.

A detailed discussion of the genetic and genomic techniques useful to identify genes involved in complex diseases such as atherosclerosis is beyond the scope of this chapter, but they have been reviewed recently [1]. These techniques include gene expression studies, genetic linkage analysis, and genetic association (either candidate gene or whole genome type) studies. Each of these study types has inherent advantages and disadvantages, but GWAS are especially suited for the discovery of genes and pathways previously not known to be involved in a particular disease process [1]. A number of recent GWAS have identified 34 genetic loci significantly associated with myocardial infarction (MI) and coronary artery disease (CAD) [2]. Some of the genes identified at these loci are discussed later in this chapter.

Once a gene at a particular locus is confirmed to participate in the pathogenesis of atherosclerosis, it becomes a valid target for therapeutic manipulation by traditional pharmacologic, biologic, or molecular techniques. Enhanced or inhibited expression of the gene may be desirable depending on its function and role in atherogenesis. Molecular methods for enhanced expression of a gene (a "gain-of-function" strategy) require somatic gene therapy, which involves overexpression of a DNA sequence encoding the gene product that is lacking or deficient. A number of different types of vectors have been developed for delivery of the genetic material, and multiple preclinical therapeutic studies have been performed with genes whose functions are relevant to vascular and myocardial ischemic disease [3]. A few early phase clinical trials of gene therapy have been directed at angiogenic therapy for ischemic disease and cytostatic therapy for neointima formation; but in general, the widespread use of gene therapy awaits the development of safer and more efficacious methods of gene delivery [3].

Several molecular techniques are available to inhibit gene expression (a "loss-of-function" strategy), including the use of antisense oligonucleotides, ribozymes, or RNA interference. These techniques are described in detail elsewhere [4, 5], but all result in regulation of gene expression by causing degradation, or blocking the translation, of messenger RNA. Antisense oligonucleotides and ribozymes have been used in preclinical studies to decrease gene expression. A few examples have shown successful use of these technologies in clinical trials for treatment of cancer, infectious diseases, and lipid metabolism. However, the Food and Drug Administration has yet to approve any of these therapies. RNA interference can be achieved with short-interfering RNA, short-hairpin RNA, or micro-RNA (miRNA). MiRNAs are small endogenously produced RNAs that bind to messenger RNAs (mRNAs) and facilitate their degradation. Recent findings demonstrate that miRNAs regulate the expression of genes involved in atherosclerotic processes such as monocyte maturation, cellular cholesterol transport, angiogenesis, leukocyte adhesion, and smooth muscle cell proliferation, differentiation, and apoptosis [6]. In addition, circulating miRNAs may be useful biomarkers of coronary artery disease [7]. These findings show that miRNAs are key regulators of processes involved in atherogenesis, and suggest that miRNAs hold the potential to be developed as new molecular therapies for atherosclerosis [6, 8].

MODULATION OF LIPID METABOLISM

Lipids including cholesterol and fatty acids are transported in the bloodstream as components of lipoproteins, and it is clear that lipoprotein metabolism is intimately related to the development of atherosclerosis. High plasma concentrations of low-density lipoprotein

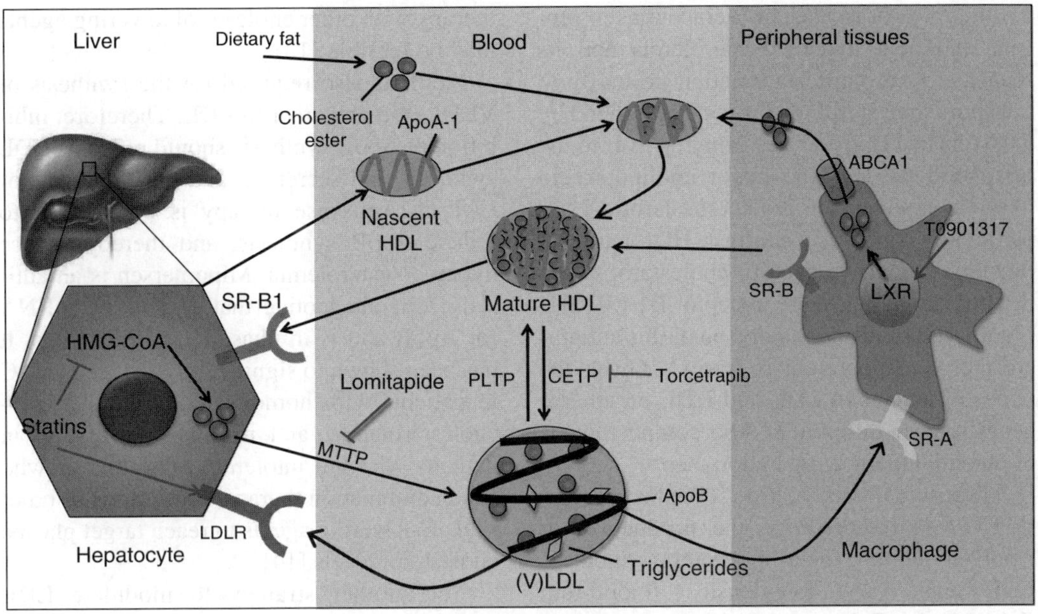

Figure 41.1 Targeted therapies for lipid metabolism. Decreased plasma HDL-cholesterol or increased plasma (V)LDL cholesterol are known risk factors for atherosclerosis. Some traditional pharmacologic and newer molecular therapies targeting specific aspects of lipoprotein metabolism, with the goal of reducing (V)LDL- and increasing HDL-cholesterol, are depicted. Newer strategies include blockade of MTTP and CETP function with lomitapide and torcetrapib, respectively, and enhancement of LXR activity with T0901317. See text for detailed explanations (Adapted by permission from Rader and Daugherty [9]. © 2008, Nature).

(LDL) cholesterol promote atherosclerosis, whereas high concentrations of high-density lipoprotein (HDL) cholesterol are protective against atherosclerosis. The general relationships between the lipoproteins and the tissues involved in their metabolism are shown in Figure 41.1, along with therapeutic targets discussed in this chapter. It is important to note the two primary sources of plasma cholesterol as (1) dietary, via gut uptake, and (2) hepatic, via de novo synthesis by the hepatocyte.

Cholesterol Metabolism

The recognition of plasma cholesterol, and in particular LDL cholesterol (LDL-C), as a risk factor for atherosclerosis triggered intense study of cholesterol and lipoprotein function and metabolism. The enzyme 3-hydroxy-3-methylglutaryl coenzyme A (HMG-CoA) reductase was identified as the catalyst for the rate-limiting step in the hepatic biosynthesis of cholesterol, and subsequently HMG-CoA reductase inhibitors (statins; Figure 41.1) were developed. The statins include atorvastatin (Lipitor), simvastatin (Zocor), lovastatin (Mevacor), and others. They inhibit hepatic cholesterol biosynthesis, leading to increased LDL receptor expression and thus increased clearance of plasma LDL [10]. Statins are widely used to reduce plasma cholesterol, particularly in the LDL fraction, with concomitant protection against the sequelae of atherosclerosis. They also have atheroprotective actions apart from their ability to modulate cholesterol synthesis and LDL metabolism; these include raising HDL concentrations and inhibiting inflammation [11].

Apolipoprotein B Function and Low-Density Lipoprotein Metabolism

The overwhelming clinical success of the statins underscores the importance of knowing the molecular mechanisms of lipoprotein assembly, function, and metabolism for the development of molecular therapeutics to prevent and treat hypercholesterolemia and its sequelae. LDL particles function primarily to transport cholesterol (dietary or synthesized by the liver) and triglycerides from the liver to the peripheral tissues. The major apolipoprotein (Apo) component of proatherosclerotic LDL particles is Apo B. By contrast, HDL particles function primarily to return cholesterol to the hepatocyte via scavenger receptor B1 (SR-B1; Figure 41.1), and the major Apo of this antiatherosclerotic lipoprotein is ApoA1. Given the opposing effects of LDL and HDL on atherogenesis, their differential Apo composition is of potential therapeutic importance.

Microsomal triglyceride transfer protein (MTTP) is required for the production of ApoB-containing beta-lipoproteins such as LDL and very low-density lipoprotein (VLDL). ApoB incorporated into these lipoproteins serves as the ligand for cellular LDL receptors (LDLRs). Binding to LDLRs facilitates (V)LDL internalization and degradation in the lysosomal compartment. Mutations of the *MTTP* gene cause abetalipoproteinemia, and mutations of the *APOB* gene cause hypobetalipoproteinemia. The former manifests as severe disease caused by reduced absorption of dietary fat, fat-soluble vitamins, and cholesterol, and reduced synthesis of beta-lipoproteins. Both of these genetic diseases are characterized by low plasma concentrations of LDL, making these molecules attractive potential therapeutic targets to modulate plasma cholesterol. An orally active, small-molecule MTTP inhibitor (Lomitapide; Figure 41.1) is being developed to reduce high LDL-C in patients with genetic deficiency of LDLR (homozygous familial hypercholesterolemia), and to treat high plasma triglycerides in patients with familial lipoprotein lipase deficiency [12]. Analogous to genetic MTTP deficiency in abetalipoproteinemia, MTTP inhibition causes reduced gastrointestinal absorption of fat. This side effect may limit widespread lomitapide monotherapy for the treatment of elevated LDL-C in the general population; however low-dose combination therapy with other cholesterol lowering agents may be feasible [13].

ApoB is also required for the synthesis of VLDL, the precursor to LDL. Therefore, inhibition of ApoB synthesis should reduce VLDL synthesis and secretion, and plasma levels of LDL-C. Antisense therapy is being used to reduce ApoB synthesis, and thereby reduce hypercholesterolemia. Mipomersen is an antisense oligonucleotide that targets the mRNA for ApoB and is in phase III clinical trials. It has been shown to significantly reduce LDL-C in patients with homozygous familial hypercholesterolemia, and may also be useful in patients who are intolerant of statins or who need combination therapy with statins or other LDL-C-lowering agents to reach target plasma cholesterol levels [14].

Yet another strategy to modulate LDL metabolism is to target the LDLR. GWAS confirm earlier genetic studies in mice and humans that associate the protease proprotein convertase subtilisin/kexin type 9 (PCSK9) with coronary heart disease [2]. PCSK9 is a protease that binds to the LDLR causing its degradation, thereby reducing removal of plasma LDL by the liver. Thus, PCSK9 is an attractive stand-alone therapeutic target whose inhibition should enhance cell surface LDLR expression and removal of plasma LDL. In addition, PCSK9 expression is sterol-responsive and is increased by statin treatment. Low-dose statins markedly lower LDL-C, while higher doses further drop plasma LDL-C only modestly. The sterol responsiveness may explain the modest increase in effect associated with higher statin doses, and suggests that PCSK9 inhibition may be a useful adjunct treatment to statin therapy. PCSK9 inhibition has been achieved by both small interfering RNA [5] and antibody targeting strategies that reduce LDL-C in preclinical models. A phase II clinical trial using an anti-PCSK9 antibody achieved exciting, expected, and good results with few adverse side effects [15]. Additional trials will determine the therapeutic potential for this and similar agents.

Apolipoprotein A1 and HDL Function

GWAS have confirmed that a locus containing genes for several Apos including ApoA1 is highly associated with risk for CAD [2]. ApoA1 (see Figure 41.1) is the major Apo component of HDL particles, where it facilitates the uptake of cholesterol into the lipoprotein, and promotes the efflux of cholesterol from nonhepatic tissue such as atherosclerotic plaque and redirects it to the liver for excretion. These ApoA1-dependent HDL-mediated atheroprotective actions of HDL are well known and are termed "reverse cholesterol transport" (RCT).

Targeted therapeutics that enhance RCT and other HDL functions are an active research focus. HDL-mediated RCT can be enhanced using inhibitors of cholesterol ester transfer protein (CETP) to block the transfer of cholesterol from HDL to (V)LDL. A recent clinical trial of an oral CETP inhibitor, torcetrapib (see Figure 41.1), revealed off-target side effects that will prevent its use, but development of new small-molecule CETP inhibitors or inhibition of CETP by molecular methods such as antisense therapy remain promising therapeutic strategies [16].

ApoA1-dependent cholesterol uptake will also enhance HDL-mediated RCT, and HDL ApoA1 content and function are therefore targets for therapeutic manipulation. A small-molecule enhancer of ApoA1 mRNA production and protein synthesis, RVX-208, has been shown in a phase II clinical trial to modestly increase plasma ApoA1 and HDL content. However, longer-term studies are needed for a better assessment of its therapeutic potential [17].

Therapeutic augmentation of HDL function using ApoA1-mimetics is also being explored [18]. These molecules are short polypeptides without sequence homology to ApoA1, but with a similar secondary structure that mediates lipid binding and other ApoA1 properties in a manner similar to endogenous ApoA1. Studies show that no single mimetic completely recapitulates all functions of ApoA1. Instead, different mimetics possess varying ApoA1 activities (i.e., cholesterol uptake, anti-inflammatory effects, and improved insulin resistance) that may be superior to endogenous ApoA1 activity. Therefore, combination mimetic therapy may achieve desirable clinical results. Preclinical trials of a number of ApoA1 mimetics have been performed, and early phase clinical trials with two different mimetics have promising results [18].

Treatment with recombinant ApoA1 (rApoA1) is also being investigated as a means of augmenting HDL function [19]. This approach utilizes treatment with rApoA1 that is either wild-type, or possesses the Milano mutation (ApoA1$_M$). Patients with ApoA1$_M$ are protected from atherosclerosis, despite having low plasma concentrations of HDL and elevated plasma triglyceride. Various formulations of rApoA1, some that include phospholipids (termed rHDL), have beneficial effects on plasma lipid profiles, and on plaque histology in preclinical models. Several phase I and II clinical trials have shown that various formulations of rHDL mediate beneficial changes in plasma lipids, RCT, glucose handling, endothelial function, platelet function, and atheroma characteristics [19].

Lipoprotein Lipase

Lipoprotein lipase (LPL) is widely expressed on endothelial cells, adipocytes, and heart and skeletal muscle where it hydrolyzes triglycerides (TGs) found in TG-rich VLDL, LDL, and chylomicrons, leading to decreased plasma TG and increased HDL. Previous studies have demonstrated that LPL has an atheroprotective role in animals and humans, and an association between LPL variants and heart disease has been confirmed in a recent GWAS [20]. Lipoprotein lipase deficiency (LPLD) is a rare but debilitating autosomal-recessive disorder characterized by a lack of functional LPL causing hyperchylomicronemia, and leading to eruptive xanthomas, lipemia retinalis, hepatosplenomegaly, pancreatitis, and premature atherosclerosis. *LPL* gene replacement therapy is being developed as a treatment for LPLD but is conceivably also useful as an adjunct to treatment to reduce hypertriglyceridemia and the concomitant risk of atherosclerosis. *LPL* gene therapy uses adeno-associated virus-1 (AAV-1)-mediated delivery of a *LPL* "gain-of-function" gene variant, LPL^{S447X} ($AAV1\text{-}LPL^{S447X}$), via intramuscular injection. Proof of concept has been demonstrated in preclinical animal models, and several phase I and II clinical trials in patients with LPLD have shown short- and long-term beneficial alterations in TG metabolism with an excellent safety profile and few side effects [21].

Scavenger Receptors

Scavenger receptors (SR) including SR-A, SR-B, CD36, LOX-1, and others, are a group of membrane bound proteins that are expressed on macrophages, platelets, endothelial cells, epithelial cells, and adipocytes where they bind modified LDL particles and polyanionic ligands. Ligand binding results in receptor internalization and intracellular signaling that lead to apoptosis, endothelial dysfunction, and foam cell formation. Several scavenger receptors modulate the pathogenesis of atherosclerosis. Scavenger receptor A (SR-A) and CD36 both contribute to foam cell formation and atherosclerosis, but SR-B may reduce atherosclerosis by promoting reverse cholesterol transport from macrophages. LOX-1 leads to endothelial cell dysfunction. A number of preclinical gene therapy studies using forms of these SRs (summarized by Stephen et al. [22]) demonstrated the therapeutic potential of this technique to modulate SR-dependent modulation of atherogenesis. However, they suffered from a lack of stable long-term gene expression, and similar clinical gene therapy studies await improved gene delivery strategies [22].

Liver X Receptors

The liver X receptors (LXR; see Figure 41.1) α and β are transcription factors that belong to the nuclear hormone receptor superfamily that are expressed in the liver, intestine, kidney, and macrophages. Activation of LXRs by their oxysterol ligands induces heterodimerization of LXRs with 9-*cis*-retinoid X receptors (RXR) and facilitates DNA binding to LXR-responsive elements within the promoter sequences of target genes [23]. In macrophages, LXRs control many elements of RCT by increasing the expression of ATP-binding cassette (ABC) transporters A1, ABCG1, and ABCG4, which facilitate the transport of intracellular cholesterol to HDL particles for clearance by the liver. Additionally, LXRs induce the expression of LPL and phospholipase transfer protein (PLTP), which increases HDL production by transporting phospholipids from LDL particles to developing HDLs. In addition to regulating cholesterol handling, activation of LXR reduces inflammation by suppressing nuclear factor-κB (NF-κB) signaling. Small-molecule agonists of LXRs (i.e., T0901317 and GW3965; Figure 41.1) have been developed to modulate multiple aspects of cholesterol metabolism, and to dampen the inflammatory response. These agonists have been shown to reduce atherosclerosis in preclinical studies [23, 24]. There is a single report of a phase I trial that showed favorable effects on ABCG1 and ABCA1 expression in peripheral blood cells, but also significant neurological side effects [25].

Other Potential Molecular Targets for Lipid Metabolism

Recent GWAS have identified 34 loci that are strongly associated with MI and CAD [2]. A subset of these loci contains genes whose products modulate lipid metabolic pathways that are known to affect atherogenesis; and in the earlier sections, we discussed molecular therapies relevant to these genes and their products. However, other loci contain lipid metabolism genes (summarized in Table 41.1) not previously associated with atherosclerosis. Studies to validate the contribution of these genes to the pathogenesis of atherosclerosis are in progress, and validated genes are likely future targets for molecular therapies to prevent or treat atherosclerosis.

TABLE 41.1 Lipid metabolism genes not previously associated with atherosclerosis, but at loci now associated with myocardial infarction and coronary artery disease by recent genome-wide association studies

Gene symbol	Gene name	Gene information*
PPAP2B	Phosphatidic acid phosphatase type 2B	Plasma membrane enzyme that converts phosphatidic acid to diacylglycerol
SORT1	Sortilin 1	A multifunctional trans-Golgi network protein that modulates hepatic VLDL secretion
ABCG5/8	ATP-binding cassette G5 and G8	Adjacent genes for ATP-binding cassette transporters involved in sterol absorption and secretion
LIPA	Lipase A	A lysosomal lipase that hydrolyzes cholesterol esters and triglycerides from endocytosed lipoproteins
TRIB1	Tribbles homolog 1	A modulator of MAPK activity. Polymorphisms modulate plasma LDL-C, HDL, and TG

*Summarized from information at GeneCards®: www.genecards.org.

TREATMENT FOR ENDOTHELIAL DYSFUNCTION

Endothelial dysfunction is a critical event in the pathogenesis of atherosclerosis. It represents a transformation of the endothelial cell phenotype from anti-inflammatory and antithrombotic, to proinflammatory and prothrombotic [26]. Dysfunction is broadly characterized by decreased endothelial-dependent vasorelaxation, and increased expression of proinflammatory molecules including leukocyte adhesion molecules, cytokines, chemokines, immunomodulatory molecules, and growth factors. Thus, the state of dysfunction represents altered expression of many gene products, each of which represents a potential therapeutic target. However, because many genes are involved, modulation of any particular single gene or gene product using molecular therapy may yield a small effect, if any, on the general state of dysfunction.

There is evidence that oxidative stress is a significant contributor to the pathogenesis of dysfunction. Unfortunately, there are many triggers and sources of oxygen species that alter the redox state of the cell, making targeted therapy difficult. Treatments with antioxidant vitamins to nonspecifically reduce oxidative stress have generally not been beneficial, further diminishing the drive to develop targeted therapies to reduce oxidative stress [27].

NF-κB is a transcription factor that is another major contributor to inflammation and endothelial dysfunction [28, 29]. Translocation of NF-κB into the nucleus is a common molecular event triggered by a number of factors associated with atherosclerosis, and results in gene expression of numerous proinflammatory molecules including leukocyte adhesion molecules, cytokines, growth factors, apoptosis-related molecules, and matrix metalloproteinases. Its central role in integrating and mediating proinflammatory signals makes it an attractive therapeutic target. A number of compounds of varied type, including the statins, proteasome inhibitors, steroids, and immunosuppressant agents, have been demonstrated *in vitro* and *in vivo* to nonspecifically inhibit NF-κB activity. The potential clinical utility of these agents as NF-κB inhibitors, specifically, has not been evaluated to date. The usefulness of specific NF-κB inhibitors may be limited by the central role of NF-κB in maintaining homeostasis, and may require development of molecular therapies that allow for temporally and spatially controlled modulation of its activity [28, 29].

TARGETED REGULATION OF VASCULAR CELL GROWTH

Atherosclerotic lesions are characterized in part by abnormal intimal accumulations of several cell types. Early lesions contain abundant macrophage-derived foam cells, and later stage lesions contain various additional inflammatory cell types, fibroblasts, and smooth muscle cells. Numerous studies have investigated the role of growth factors in orchestrating the proliferation and migration of fibroblasts, smooth muscle cells, and endothelial cells during atherogenesis [30]. Growth factors that modulate the pathogenesis of atherosclerosis include vascular endothelial growth factor (VEGF), platelet-derived growth factor (PDGF), epidermal growth factor (EGF), transforming growth factor-beta (TGF-β), fibroblast growth factor (FGF), and others. Factors such as VEGF and PDGF are known to contribute to plaque formation, while others including TGF-β appear to have both protective and detrimental effects on atherogenesis. Targeted therapy to modulate expression of growth factors or their receptors has potential to treat atherosclerosis and arteriosclerosis. Interestingly, a locus containing the gene for platelet-derived growth factor D has been associated with CAD in multiple recent GWAS [2]. However, given the number of involved growth factors and the complexity of the disease processes, therapy targeted at a single factor may be ineffective at controlling the proliferation of multiple cell types that contribute to the progression of a chronic process such as atherosclerosis [31]. Instead, recent clinical trials based on growth factors have generally focused on growth factor (primarily VEGF and FGF) gene delivery and expression for proangiogenic therapy to treat the consequences of coronary artery and peripheral artery disease [32, 33].

The atherosclerotic proliferative response involves a number of individual growth factors and cytokines that may trigger common or separate signaling pathways, which may have similar or opposing effects on target cells. Therefore, in lieu of targeting the growth factors, cytokines, and their receptors, a more direct and potentially feasible approach to control the proliferative response is to target the downstream molecular cell cycle machinery that orchestrates cell proliferation, such as cyclin-dependent kinases. These kinases govern the entry and progression of the cell through phases of the cell cycle. The success of drug-eluting stents that release inhibitors of cell proliferation, such as paclitaxel and sirolimus, to prevent postangioplasty neointima formation demonstrate the validity and efficacy of this approach. The likely relevance of this approach to atherosclerosis is highlighted by findings in recent GWAS. The 9p21 locus was the first to be associated with CAD by GWAS, and this association has been reproduced in multiple studies [2]. It is not yet entirely clear which gene(s) at the locus underlie the association, but cyclin-dependent kinase inhibitors 2A (*CDKN2A*) and the homologue *CDKN2B* reside at the locus. Evidence suggests that mutations in the region affect neighboring gene expression and vascular cell proliferation in a manner involving *CDKN2A* and *CDKN2B* [34]. Additional work is needed to understand fully how genes at this important locus impact atherogenesis, but this understanding will likely identify new targets for molecular therapies that have great potential to prevent and treat atherosclerotic disease.

EXTRACELLULAR MATRIX TURNOVER AND PLAQUE RUPTURE

Extracellular matrix (EM) degradation and remodeling is a prominent feature in the development of atherosclerotic plaque, and also plays a role in plaque destabilization and rupture [35]. The EM comprises many types of proteins, and a diverse array of proteinases is involved in matrix remodeling. Most of these proteases are either of the serine, cysteine, or metalloproteinase subtypes. The matrix metalloproteinases (MMPs) are particularly relevant to atherosclerotic disease because of their presence in the diseased arterial wall, and their ability to degrade all extracellular matrix components. Cysteine cathepsins also contribute to extracellular matrix degradation under pathophysiologic conditions.

Studies of a number of individual MMPs in animal models have demonstrated that MMP-2, -8, -9, and -12 play a major role in plaque development, and that MMP-1, -2, -8, -12, and -14 may contribute to vulnerable plaque formation. However, there are also studies that implicate some of these MMPs in plaque stabilization, and show that MMP-3 and -11 may limit plaque size [36]. Inhibition of MMP activity can be achieved through nonspecific small-molecule inhibitors, or by overexpression of endogenous tissue inhibitors of MMPs (TIMPs). Neither of these approaches has been pursued clinically in the setting of atherosclerosis due to the complexity of the disease process, limited understanding of the role of each of the MMPs individually and in concert, lack of specific small-molecule inhibitors, and unavailability of robust gene therapy techniques. Recently, genetic studies have demonstrated associations between risk factors for MI and CAD and single nucleotide polymorphisms (SNPs) in MMP-1, MMP-2, and MMP-9. However, another SNP in MMP-1 and an SNP in MMP-8 are associated with reduced risk for CAD [36]. Additional human genetic studies will help us to identify potential therapeutic targets relevant to EM turnover in atherosclerosis, and subsequent preclinical research will be necessary to verify these targets and develop appropriate molecular therapies for these targets.

A number of cathepsins, including subtypes F, K, L, and S, are expressed in atherosclerotic plaques, primarily by macrophages, smooth muscle cells, and endothelial cells. Studies in mice show that loss of cathepsin K or S results in decreased atherosclerotic lesion size and progression. By contrast, deficiency of the endogenous inhibitor of cathepsins, cystatin C, increases plaque size, collagen content, and elastin breaks. The cathepsins also degrade LDL and reduce cholesterol efflux from macrophages. Because of their actions in atherosclerosis and other diseases, cathepsins K and S are valuable therapeutic targets. Small molecule inhibitors of cathepsins K and S have been developed and are being evaluated in clinical trials for the treatment of osteoporosis, psoriasis, and osteoarthritis. Clinical trials of pharmacologic inhibitors, and molecular approaches to modify activity of the cathepsins or cystatin C, have not been pursued for reasons similar to those discussed for the MMPs. Expression patterns of MMPs and cathepsins are being evaluated and show promise as markers for cardiovascular disease [37, 38].

A locus at 15q25.1 containing the *ADAMTS7* gene is strongly associated with CAD [2]. A disintegrin and metalloproteinase with thrombospondin motifs 7 (ADAMTS7) degrades cartilage oligomeric matrix protein (COMP), a noncollagenous EM protein that interacts with other EM proteins. ADAMTS7 activity is implicated in smooth muscle cell migration and neointima formation, and therefore atherosclerotic plaque development [39]. Additional studies are needed to validate the role for ADAMTS7 in atherogenesis and elucidate the mechanisms whereby it contributes to its pathogenesis. These studies may verify ADAMTS7 as a target for future molecular therapies to treat the disease.

MODULATION OF INNATE AND ADAPTIVE IMMUNITY IN ATHEROSCLEROSIS

Our current understanding of atherosclerosis as an inflammatory disease is based on numerous studies that show both the innate and adaptive immune responses contribute greatly to the pathogenesis of the disease [40]. A wide variety of molecules that orchestrate inflammatory and immune responses have been implicated in the disease process. These molecules can be broadly grouped into (i) cytokines and their receptors, (ii) chemokines and their receptors, (iii) leukocyte adhesion molecules and their receptors, and (iv) biologically active lipids and their receptors. The contributions of some of these molecules to the pathogenesis of atherosclerosis are discussed elsewhere in this book. Discussion of the contributions of the large number of remaining molecules is beyond the scope of this chapter. However, the macrophage is generally viewed as the inflammatory cell most central to the pathogenesis of atherosclerosis. Recent publications review the preclinical and early clinical (if available) studies that address the value of therapeutic agents targeting the cytokines, chemokines, adhesion molecules, and biologically active lipids that induce, or result from, macrophage activation during atherogenesis [41–44]. Some of the individual molecules within these broad classes have been targeted in preclinical trials. Few have been targeted in early phase clinical studies, either pharmacologically or with molecular techniques, most likely because of the complex nature of the inflammatory response and the lack of good techniques to quantify plaque inflammation, size, and regression *in vivo* [43].

GWAS have identified two separate loci strongly associated with CAD that contain the *IL5* (interleukin 5) and *CXCL12* (chemokine ligand 12) genes, respectively [2]. IL-5 stimulates B-cell growth, immunoglobulin secretion, and eosinophil activation. CXCL12 is a chemotactic factor for lymphocytes and monocytes, and is angiogenic. Neither molecule were known to participate in atherogenesis prior to the association studies, but are now being investigated as possible future targets for molecular therapies.

CONCLUSION

The complex pathogenesis of atherosclerosis involves many cell types, including leukocytes, endothelial cells, smooth muscle cells, fibroblasts, and hepatocytes. It also involves many biologic processes including lipid metabolism, inflammation, and cell proliferation, differentiation, and apoptosis. Our understanding of the molecular components of these processes and their role in the pathogenesis of the disease is incomplete. However, research has identified a number of therapeutic targets that modulate these processes. This is particularly evident with respect to lipid metabolism, but molecular targets that modulate the other processes have also been identified.

The recent application of genomic studies, particularly GWAS, to atherosclerosis has identified a number of genes and biologic pathways previously unknown to be associated with the disease. These studies have provided new avenues for the discovery of additional therapeutic targets that may be useful to prevent and treat atherosclerosis.

The currently approved therapeutic agents that modulate these targets are traditional pharmacologic agents. However, newer molecular techniques to modulate the expression of genes involved in the pathogenesis of atherosclerosis are being investigated as possible treatments for the disease. These techniques include various forms of gene therapy to increase gene expression, and the use of antisense oligonucleotides, ribozymes, and RNA inhibition to decrease gene expression. With rare exception, the use of these molecular techniques for the treatment of atherosclerosis has been limited to preclinical and early phase clinical trials. However, more efficient, effective, and safer methods for the delivery of molecular therapeutics are being developed, and will allow for the widespread clinical use of targeted molecular therapy for traditional targets as well as new targets identified by recent genomic studies.

REFERENCES

1. Miller DT, Ridker PM, Libby P, Kwiatkowski DJ. Atherosclerosis: The path from genomics to therapeutics. *J Am Coll Cardiol*. 2007;49: 1589–1599.
2. Stylianou IM, Bauer RC, Reilly MP, Rader DJ. Genetic basis of atherosclerosis: Insights from mice and humans. *Circ Res*. 2012;110: 337–355.
3. Melo LG, Pachori AS, Gnecchi M, Dzau VJ. Genetic therapies for cardiovascular diseases. *Trends Mol Med*. 2005;11:240–250.

4. Rayburn ER, Zhang R. Antisense, RNAi, and gene silencing strategies for therapy: Mission possible or impossible? *Drug Discov Today*. 2008;13:513–521.
5. Watts JK, Corey DR. Silencing disease genes in the laboratory and the clinic. *J Pathol*. 2012;226:365–379.
6. Martin K, O'Sullivan JF, Caplice NM. New therapeutic potential of microRNA treatment to target vulnerable atherosclerotic lesions and plaque rupture. *Curr Opin Cardiol*. 2011;26:569–575.
7. Fichtlscherer S, De Rosa S, Fox H, et al. Circulating microRNAs in patients with coronary artery disease. *Circ Res*. 2010;107: 677–684.
8. Weber C, Schober A, Zernecke A. MicroRNAs in arterial remodelling, inflammation and atherosclerosis. *Curr Drug Targets*. 2010;11: 950–956.
9. Rader DJ, Daugherty A. Translating molecular discoveries into new therapies for atherosclerosis. *Nature*. 2008;451:904–913.
10. Maron DJ, Fazio S, Linton MF. Current perspectives on statins. *Circulation*. 2000;101: 207–213.
11. Lardizabal JA, Deedwania P. Lipid-lowering therapy with statins for the primary and secondary prevention of cardiovascular disease. *Cardiol Clin*. 2011;29:87–103.
12. Lomitapide. *Am J Cardiovasc Drugs*. 2011;11:347–352.
13. Samaha FF, McKenney J, Bloedon LT, Sasiela WJ, Rader DJ. Inhibition of microsomal triglyceride transfer protein alone or with ezetimibe in patients with moderate hypercholesterolemia. *Nat Clin Pract Cardiovasc Med*. 2008;5: 497–505.
14. Ricotta DN, Frishman W. Mipomersen: A safe and effective antisense therapy adjunct to statins in patients with hypercholesterolemia. *Cardiol Rev*. 2012;20:90–95.
15. McKenney JM, Koren MJ, Kereiakes DJ, Hanotin C, Ferrand AC, Stein EA. Safety and efficacy of a monoclonal antibody to proprotein convertase subtilisin/kexin type 9 serine protease, SAR236553/REGN727, in patients with primary hypercholesterolemia receiving ongoing stable atorvastatin therapy. *J Am Coll Cardiol*. 2012;59(25):2344–2353.
16. Duffy D, Rader DJ. Update on strategies to increase HDL quantity and function. *Nat Rev Cardiol*. 2009;6:455–463.
17. Nicholls SJ, Gordon A, Johansson J, et al. Efficacy and safety of a novel oral inducer of apolipoprotein A-I synthesis in statin-treated patients with stable coronary artery disease a randomized controlled trial. *J Am Coll Cardiol*. 2011;57:1111–1119.
18. Getz GS, Reardon CA. Apolipoprotein A-I and A-I mimetic peptides: A role in atherosclerosis. *J Inflamm Res*. 2011;4:83–92.
19. Smith JD. Apolipoprotein A-I and its mimetics for the treatment of atherosclerosis. *Curr Opin Investig Drugs*. 2010;11:989–996.
20. Samani NJ, Erdmann J, Hall AS, et al. Genomewide association analysis of coronary artery disease. *N Engl J Med*. 2007;357: 443–453.
21. Gaudet D, de Wal J, Tremblay K, et al. Review of the clinical development of alipogene tiparvovec gene therapy for lipoprotein lipase deficiency. *Atheroscler Suppl*. 2010;11:55–60.
22. Stephen SL, Freestone K, Dunn S, et al. Scavenger receptors and their potential as therapeutic targets in the treatment of cardiovascular disease. *Int J Hypertens*. 2010;2010: 646929.
23. Zhu Y, Li Y. Liver X receptors as potential therapeutic targets in atherosclerosis. *Clin Invest Med*. 2009;32:E383–E394.
24. Im SS, Osborne TF. Liver X receptors in atherosclerosis and inflammation. *Circ Res*. 2011;108:996–1001.
25. Katz A, Udata C, Ott E, et al. Safety, pharmacokinetics, and pharmacodynamics of single doses of LXR-623, a novel liver X-receptor agonist, in healthy participants. *J Clin Pharmacol*. 2009;49:643–649.
26. Endemann DH, Schiffrin EL. Endothelial dysfunction. *J Am Soc Nephrol*. 2004;15: 1983–1992.
27. Sugamura K, Keaney JF, Jr. Reactive oxygen species in cardiovascular disease. *Free Radic Biol Med*. 2011;51:978–992.
28. Pamukcu B, Lip GY, Shantsila E. The nuclear factor-kappa B pathway in atherosclerosis: A potential therapeutic target for atherothrombotic vascular disease. *Thromb Res*. 2011; 128:117–123.
29. Dabek J, Kulach A, Gasior Z. Nuclear factor kappa-light-chain-enhancer of activated B cells (NF-kappaB): A new potential therapeutic target in atherosclerosis? *Pharmacol Rep*. 2010;62:778–783.

30. Kapoor M, Nomiyama T, Bruemmer D, Kojima F, Crofford LJ. Growth factors and cytokines: Emphasis on their role in wound healing and atherosclerosis. *Curr Anaesth Crit Care*. 2006;17:13–20.
31. Dzau VJ, Gibbons GH, Mann M, Braun-Dullaeus R. Future horizons in cardiovascular molecular therapeutics. *Am J Cardiol*. 1997; 80:33I–39I.
32. Williams PD, Kingston PA. Plasmid-mediated gene therapy for cardiovascular disease. *Cardiovasc Res*. 2011;91:565–576.
33. Williams PD, Ranjzad P, Kakar SJ, Kingston PA. Development of viral vectors for use in cardiovascular gene therapy. *Viruses*. 2010;2: 334–371.
34. Visel A, Zhu Y, May D, et al. Targeted deletion of the 9p21 non-coding coronary artery disease risk interval in mice. *Nature*. 2010; 464:409–412.
35. Newby AC. Dual role of matrix metalloproteinases (matrixins) in intimal thickening and atherosclerotic plaque rupture. *Physiol Rev*. 2005; 85:1–31.
36. Ketelhuth DF, Back M. The role of matrix metalloproteinases in atherothrombosis. *Curr Atheroscler Rep*. 2011;13:162–169.
37. Liu J, Sukhova GK, Sun JS, Xu WH, Libby P, Shi GP. Lysosomal cysteine proteases in atherosclerosis. *Arterioscler Thromb Vasc Biol*. 2004;24:1359–1366.
38. Lutgens SP, Cleutjens KB, Daemen MJ, Heeneman S. Cathepsin cysteine proteases in cardiovascular disease. *FASEB J*. 2007;21: 3029–3041.
39. Reilly MP, Li M, He J, et al. Identification of ADAMTS7 as a novel locus for coronary atherosclerosis and association of ABO with myocardial infarction in the presence of coronary atherosclerosis: Two genome-wide association studies. *Lancet*. 2011;377:383–392.
40. Hansson GK, Hermansson A. The immune system in atherosclerosis. *Nat Immunol*. 2011; 12:204–212.
41. Saha P, Modarai B, Humphries J, et al. The monocyte/macrophage as a therapeutic target in atherosclerosis. *Curr Opin Pharmacol*. 2009;9:109–118.
42. Tabas I. Consequences and therapeutic implications of macrophage apoptosis in atherosclerosis: The importance of lesion stage and phagocytic efficiency. *Arterioscler Thromb Vasc Biol*. 2005;25:2255–2264.
43. Choudhury RP, Lee JM, Greaves DR. Mechanisms of disease: Macrophage-derived foam cells emerging as therapeutic targets in atherosclerosis. *Nat Clin Pract Cardiovasc Med*. 2005;2:309–315.

42

HIGH-DENSITY LIPOPROTEIN THERAPIES—THEN AND NOW

Henry J. Pownall, Corina Rosales, Baiba K. Gillard, and Antonio M. Gotto, Jr.

Atherosclerosis & Lipoprotein Research Laboratory, Cardiology Department, Houston Methodist Research Institute, Houston, TX, USA

INTRODUCTION

Annually in the United States, there are about 500,000 heart attack deaths and over 700,000 stroke incidents that exact great personal pain from patients and their families with an attendant economic burden on families, health care organizations, and governments. The underlying cause of most heart attacks and strokes is atherosclerosis, the accumulation of cholesterol and other compounds in arteries in a way that blocks circulation directly or through the formation of vulnerable sites that attract clot formation. Statins reduce plasma low-density lipoprotein-cholesterol (LDL-C) levels, a risk factor for atherosclerosis. Yet, there is an unfulfilled need for more potent agents to complement statins.

REVERSE CHOLESTEROL TRANSPORT

High-density lipoproteins (HDLs) are thought to be antiatherogenic through their role in RCT, that is, the transfer of cholesterol from macrophages in the subendothelial space of the arterial wall to HDL carriers to the liver for disposal (Figure 42.1). RCT is initiated by macrophage cholesterol efflux via the ATP-binding cassette transporters ABCA1 and ABCG1 and by spontaneous desorption, all of which yield nascent forms of HDL that are modified by a second step, esterification of their free cholesterol by lecithin: cholesterol acyltransferase (LCAT). The third and final step is the selective removal of cholesteryl esters (CE) by the hepatic scavenger receptor class B, Type I

Atherosclerosis: Risks, Mechanisms, and Therapies, First Edition. Edited by Hong Wang and Cam Patterson.

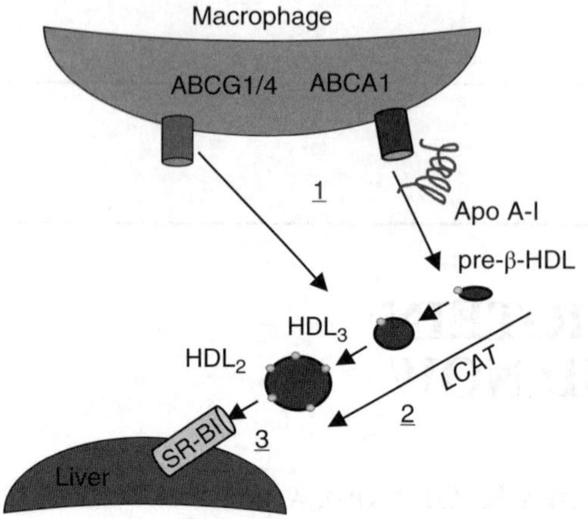

Figure 42.1 Schematic representation of RCT comprising (1) cholesterol efflux from macrophages, (2) its esterification by LCAT, and (3) uptake of HDL-cholesteryl esters by hepatic SR-BI.

(SR-BI). It is widely thought that these three steps, when efficient, are atheroprotective.

PLASMA HDL-C—HIGHER IS BETTER

Numerous observational and prospective studies support the hypothesis that raising HDL-C is a rationale therapeutic goal. On the basis of a 10-year prospective study of male Livermore Radiation Laboratory employees, Gofman et al. first reported [1] that men who developed ischemic heart disease had lower HDL2 and HDL3, and higher LDL, IDL, and small very-low-density lipoprotein (VLDL). A recent 29-year follow-up to the Livermore study [1] showed that total incident coronary heart disease (CHD) was inversely related to HDL2- and HDL3-mass, and concordantly related to LDL-mass, IDL-mass, and small and large VLDL-mass concentrations. The hazard reduction per mg/dL was greater for HDL2 than HDL3. Risk for premature CHD was greater in men within the lowest HDL2 and HDL3 quartiles. Other observational and interventional trials also support the hypothesis that raising HDL-C would have a cardioprotective effect. The oft-cited Framingham Heart Study begun in 1948 identified many CHD risk factors. Among them, high HDL-C levels were associated with a lower death rate [2]. In some studies, interventions that increased HDL-C reduced CHD events. In both the Helsinki Heart Study [3] and the Veterans Administration HDL Intervention Trial [4], increased HDL was associated with reduced CHD events. Finally, regular, modest alcohol consumption is associated with reduced CHD death [5], an effect that is mediated mainly by the attendant increases in HDL-C [6], although this has never been tested in a prospective placebo-controlled study and likely will not be.

THE CASE AGAINST HDL-RAISING THERAPIES

High plasma levels of HDL-C and low rates of CHD events are not axiomatically linked; many patients with low HDL-C do not present with CHD, and vice versa. In mouse models, enhanced (vs. reduced) hepatic HDL-C uptake and attendant lower (higher) plasma HDL-C leads to reduced (increased) atherosclerosis [7, 8]. Cholesteryl ester transfer protein (CETP) is a plasma protein that exchanges the CE of HDL for the triglycerides of VLDL and chylomicrons. This effect is dose-dependent with respect to plasma VLDL concentrations [9], and as a consequence the HDL CE content is reduced, thereby lowering HDL-C without changing HDL particle number. Patients with CETP deficiency have high plasma HDL-C levels but have less cardioprotection than CETP-competent patients with the same HDL-C levels [10, 11]. The plasma of patients with similar HDL-C levels, whether high or low, can have very different capacities for cellular cholesterol efflux [12], the initial step in reverse cholesterol transport. Patients in the Copenhagen City Heart Study with genetically elevated levels of apo A-I and HDL-C did not have a reduced risk for ischemic heart disease or myocardial infarction [13]. Low plasma HDL-C levels due to heterozygous loss-of-function *ABCA1* mutations were not associated with an increased risk of CHD [13]. According to a Mendelian randomization study, an HDL-raising gene (LIPG Asn396Ser) is not associated with reduced myocardial infarction [14]. In the Copenhagen City Heart Study and the Copenhagen General Population Study, a common LCAT variant identified in individuals with the lowest 2% HDL-C was associated with a 13% decrease in HDL cholesterol ($n=85$), but not with increased risk of myocardial infarction or other ischemic end points [15]. Comparison of LCAT activity in cardiovascular disease (CVD) cases and controls revealed that high plasma LCAT activity did not predict reduced CVD risk [16]. In the EPIC-Norfolk and the IMPROVE [17] studies, plasma LCAT levels showed weak or no association with future cardiovascular events. These findings in CVD patients do not support high LCAT activity as being antiatherogenic. Finally, a 2011 study of the Cohort of Norway Population showed that controlling for HDL-C does not affect the magnitude of the relationship between alcohol intake and CVD death [18]. In summary, many mechanisms that increase HDL are not atheroprotective; enhancing RCT via improved cholesterol efflux or esterification is not clearly atheroprotective. Only the final RCT step, cholesterol disposal to liver via SR-BI, seems to atheroprotect, and it does so by reducing plasma HDL-C concentrations.

IMPROVING HDL FUNCTIONALITY VIA ENHANCED RCT

The RCT pathway (Figure 42.1) provides three steps that have been targets of improvement. Studies with cultured macrophages have shown that cholesterol efflux is mediated by ABCA1 and ABCG1. In the former case, lipid-free apolipoproteins (apos) microsolubilize plasma membrane-free cholesterol and phospholipids to form nascent HDL that, being rich in free cholesterol, provides a target for esterification by LCAT. The efflux step can be mediated by apos A-I, C, E, and several of their peptides, and reaches maximum efflux rate at approximately 50 μg/ml [19]. Cholesterol efflux can also occur by spontaneous desorption or via ABCG1 to lipidated apos such as reconstituted (r) HDL. There are several barriers to enhancement of macrophage ABCA1 efflux *in vivo*. First, Tangier patients, who are ABCA1-deficient, do not have a profoundly atherogenic profile. Second, plasma levels of lipid-free apo A-I, the major ABCA1 ligand, are immeasurably low. Other studies of infused cholesterol acceptors have given mixed results. Despite early promising studies [20], subsequent studies showed that $rHDL_{Milano}$ infusion remodels the arterial external elastic membrane with no change in the lumen dimensions [21]. Infusions of delipidated HDL plasma increased preβ-like HDL and reduced α-HDL, a cardioprotective profile. However, intravascular ultrasound showed only a trend in atheroregression [22]. Synthetic peptide analogs of apos also mediate cholesterol efflux [23]. A problem with any infused agent that removes cholesterol from cells, be they peptides, rHDL, delipidated HDL, or lipid vesicles, is that they lack directionality and would likely remove cholesterol from all tissue sites, including the liver, the target of therapeutic disposal.

There have been various attempts to potentiate the second RCT step, cholesterol esterification via LCAT. This seems to be a sound approach, because unlike free cholesterol, its ester cannot transfer back to macrophages by a spontaneous mechanism. The ester is the preferred form of cholesterol storage, because it is less cytotoxic and proinflammatory than free cholesterol. Moreover, in humans, CETP transfers cholesterol from HDL to the much more rapidly metabolized LDL, which should improve RCT. Mice over expressing LCAT have elevated plasma HDL-C, but more atherosclerosis [24]. However, recombinant LCAT normalizes the plasma lipoprotein profiles of LCAT-deficient mice, and the combination of rHDL and LCAT infusions has some early promise [25].

Enhancement of the final RCT step, hepatic cholesterol disposal via SR-BI, would seem to be a promising target. This step provides directionality to RCT, and in mouse models of enhanced (vs. reduced) hepatic HDL-C uptake and attendant lower plasma HDL-C, atherosclerosis is reduced (vs. increased) [7, 8, 26, 27]. There are several options. One could develop agents that alter the interaction of HDL with hepatic receptors; one could identify components of diet that alter HDL composition and its hepatic uptake; last, one could identify mechanisms to alter HDL structure in a way that enhances cholesterol hepatic disposal. Of these, the last has seen some success *in vitro* and in mouse models of lipoprotein metabolism. Courtney et al. [28] discovered that a streptococcal virulence factor, serum opacity factor (SOF), disrupts the structure of HDL. Subsequently, Gillard et al. [29] showed that a recombinant (r) SOF catalyzes the partial disproportionation of HDL into a cholesteryl ester-rich microemulsion (CERM) and a new HDL-like particle, neo HDL, with the concomitant release of lipid-free (LF)-apo A-I. Opacification is unique; rSOF transfers apo E and nearly all neutral lipids of greater than 100,000 HDL particles into a collection of CERM with sizes that increase with HDL-CE content (r~100–250 nm), leaving a neo-HDL that is phospholipid- and protein-rich, especially apo A-II rich, but cholesterol-poor. rSOF is potent; within 30 minutes at 37 °C, 10 nM rSOF opacifies 4 μM HDL. CERM formation and apo A-I release have similar kinetics, which suggests parallel or rapid sequential steps. According to reaction products and chemical kinetics, rSOF is a heterodivalent fusogenic protein that uses a docking site to displace apo A-I and bind to exposed CE surfaces on HDL; the resulting rSOF-HDL complex recruits additional HDL with its binding-delipidation site and through multiple fusion steps forms a CERM. Gillard et al. speculated that rSOF may be a clinically useful and novel modality for improving reverse cholesterol transport. With apo E and a high CE content, CERM could transfer large amounts of cholesterol to the liver for disposal via the LDL receptor; neo-HDL is likely a better acceptor of cellular cholesterol than HDL; LF-apo A-I could enhance efflux via the ATP-binding cassette transporter ABCA1. These speculations have been largely supported by subsequent data. rSOF treatment of whole plasma made it a better acceptor of cholesterol efflux from THP-1 macrophages and a better LCAT substrate [30]. The underlying cause of the improvements was due to the rSOF-mediated conversion of HDL to neo-HDL, which is cholesterol-poor and phospholipid-rich. As expected with improved cholesterol efflux, reduction in the secretion of the inflammatory markers TNF-α and IL-6 by THP-1 macrophages mediated by neo-HDL was twice that of HDL. rSOF also improved the terminal RCT step, hepatic CE uptake [31]. [^{3}H] CE uptake by Huh7 cells was 4.4 times faster from rSOF-treated HDL than from control HDL. Hepatic [^{3}H]CE uptake was inhibited by both unlabeled LDL and HDL, thereby implicating both the LDL-receptor and SR-BI in CE uptake. Complementary studies in CHO cells expressing different levels of LDLR and SR-BI showed that the rate of HDL-[^{3}H]CE uptake was increased by preincubation with rSOF, and that both receptors mediate the uptake of CERM-[^{3}H]CE. Additional tests with inhibitors also implicated the LDL-receptor-related protein-1 and syndecan-1 in the CERM-[^{3}H]CE uptake. Finally, Rosales et al. [32] tested the effects of low-dose (4 μg) injection of rSOF into wild-type (WT) mice and showed that rSOF activity *in vivo* formed the same products—neo-HDL, CERM, and lipid-free apo A-I—that were formed *in vitro*. At the same time, rSOF reduced plasma total cholesterol approximately 40% in 3 hours due to preferential hepatic uptake of CERM. Comparison of WT plasma cholesterol kinetics with those in apo E-deficient or LDLR-deficient mice revealed that CERM clearance requires apo E and that most of the hepatic clearance of CERM occurs via the LDLR. These studies set the stage for future experiments testing whether rSOF enhances RCT and atheroregression via improved cholesterol efflux, cholesterol esterification, and reduction of plasma total cholesterol. Thus, rSOF catalyzes a novel reaction that might be developed as a new therapy that prevents or reverses atherosclerosis via improved reverse cholesterol transport.

LOOKING BEYOND RCT

In addition to lipoprotein and lipid risk factors, for example, high LDL-C and low HDL-C, the Framingham study also associated obesity and a sedentary life style with increased CHD risk. Although the Helsinki Heart Study revealed that reduced CVD correlated better with increased HDL-C than with reduced TG, it is important to recall that gemfibrozil is a potent TG-lowering agent. As suggested by Patsch et al., "HDL-C is not an independent parameter of lipid transport, but is strongly affected by triglyceride metabolism through lipolytic enzymes, as exemplified by postprandial lipemia, that affect both composition and plasma levels of HDL2" [33]. Further to this point, intense exercise (marathon training) increases HDL-C and reduces fasting and postprandial lipemia, effects mediated by enhanced lipoprotein lipase (LPL) activity [34]. These reports illustrate the importance of efficient energy metabolism in atheroprotection.

Current National Cholesterol Education Program (NCEP) guidelines recognize the presence of metabolic syndrome (MetS) if three of five of the following are present: hypertension, low HDL-C, hypertriglyceridemia, hyperglycemia, and hyperinsulinemia. Notably, the latter three of these are determinants of insulin resistance and/or energy metabolism and its regulation, and low HDL-C is a function hypertriglyceridemia. On the basis of literature reports, this list could be expanded as shown in Table 42.1.

TABLE 42.1 Expanded definition of MetS

Hypertriglyceridemia*	Elevated CETP activity
High plasma NEFA	Low lipoprotein lipase
Intense, persistent postprandial lipemia	High hepatic lipase
Alcoholic lipemia	Hyperinsulinemia*
Low HDL-C*	Hyperglycemia*
No HDL2	Hypertension*
Small, dense LDL	Abdominal obesity*

*NCEP guideline definition for MetS.

These analytes can be rationally incorporated into a global model of lipid metabolism by identifying underlying factor(s). Coexistence of hyperglycemia and hyperinsulinemia is reflective of insulin resistance, one of the hallmarks of which is impaired insulin-mediated uptake of nonesterified fatty acids (NEFAs), which leads to increased hepatic extraction of NEFA that are converted to glycerolipids, mostly triglycerides, which are in turn secreted as VLDL, producing a hypertriglyceridemic state. In this state, the transfer of TG from VLDL to HDL and LDL and CE from HDL to VLDL via CETP lowers the HDL- and LDL-CE content and correspondingly raises their triglyceride content [9]. Importantly, CETP activity is increased by elevated plasma NEFA levels. The increased triglyceride content of HDL and LDL makes them substrates for hepatic lipase, which hydrolyzes the triglycerides, thereby reducing both HDL and LDL to their smaller denser counterparts; in the case of HDL, this is a triglyceride-rich HDL3. Notably, CETP activity in type 2 diabetic subjects is profoundly higher than that of controls [34a]. Lipoprotein lipase and adipose tissue lipolysis are reciprocally regulated so that during postprandial periods of high lipoprotein lipase activity, adipose tissue lipolysis is down regulated so that adipocytes can take up and esterify the NEFA released by lipoprotein lipase. By the same token, during periods of impaired insulin-mediated NEFA uptake, lipoprotein lipase activity is downregulated, further exacerbating hypertriglyceridemia. In parallel, downregulation of lipoprotein lipase is associated with the upregulation of hepatic lipase [33]. This study suggests that HDL-cholesterol is not an independent parameter of lipid transport, but is strongly affected by triglyceride metabolism through lipolytic enzymes, as exemplified by postprandial lipemia that affects both composition and plasma levels of HDL2 [33, 34]. Elevated plasma NEFA also accumulate in skeletal muscle, where they induce a lipotoxic state, the consequences of which are impaired glucose disposal. Comparison of intramyocellular lipid (IMCL) content, as assessed by proton nuclear magnetic resonance, and glucose disposal measured by a hyperinsulinemic euglycemic clamp, shows that glucose disposal declines as IMCL increases [35].

Although the underlying connection between MetS and hypertension has not been clearly established, there is a growing consensus about the role of central obesity in MetS. The importance of fat distribution in obesity has been well recognized through the studies of Vague [36] and Kissebah and Peiris [37], whose seminal studies established a correlation between central fat accumulation and type 2 diabetes. Later studies began to look at the association of both central and noncentral fat depots with markers for CVD risk and insulin resistance. According to the Quebec Family Study, a narrow waist and large hips may both protect against CVD [38]. A large waist circumference (adjusted for age, BMI, and hip circumference) was associated with low HDL-C concentrations and high fasting triglyceride, insulin, and glucose concentrations. A narrow hip circumference (adjusted for age, BMI, and waist circumference) was associated with low HDL-C and high glucose concentrations in men and high triglyceride and insulin concentrations in men and women. The Hoorn Study confirmed this; and even if trunk fat is taken into account, accumulation of fat in the legs seems to protect against a disturbed glucose metabolism, particularly in women [39]. Other more recent studies support the concept that defective fat storage and/or lean body mass in femoral-gluteal depots are a risk factor for MetS, obesity-linked diabetes, and CVD [40]. Finally, data from the LookAhead Study showed that obese patients with type 2 diabetes not only have more visceral fat and less subcutaneous adipose tissue than healthy control subjects but also more intermuscular adipose tissue. Differences in body fat distribution are also associated with differences in lipid and lipoprotein risk factors for cardiovascular disease. According to one study, waist-to-hip (WHR) ratio correlates with atherogenic plasma lipid and lipoprotein profiles [41]. WHR positively correlated with plasma concentrations of triglycerides, total cholesterol, and LDL-C and VLDL-C, small LDL, intermediate density lipoprotein (IDL), and VLDL, but it negatively correlated with HDL-C and HDL2. However, another study [42] using dual-energy x-ray absorptiometry and abdominal computed tomography revealed a favorable effect of leg fat on fasting insulin, insulin sensitivity, triglycerides, and HDL-C after controlling for trunk fat. After controlling for either visceral or subcutaneous abdominal adiposity, however, triglyceride was the only risk factor with a favorable effect of leg fat.

Some investigators have assigned dysregulation of energy metabolism to that component of waist circumference that includes intra-abdominal adiposity because of portal drainage of the intraperitoneal depot, whose high lipolytic activity and unique venous drainage may underlie dysregulated lipid metabolism and attendant cardiovascular disease [42a]. In this model, hypersecretion of metabolites from intraperitoneal fat, particularly NEFA, reduces hepatic insulin sensitivity while increasing hepatic glucose output and triglyceride synthesis. Moreover, according to Randle et al. [43], systemic increases in NEFA concentrations induce peripheral insulin resistance by inhibiting skeletal muscle glucose uptake. However, kinetic studies of regional NEFA flux [44] do not support this hypothesis. It is difficult to reconcile the idea of hypersecretion of NEFA by intraperitoneal adipose tissue when this is the site that has an increased accumulation of fat in patients with central obesity.

The issue was clarified by a study of nondiabetic, middle-aged men with varying degrees of obesity, which showed that variation in subcutaneous abdominal fat mass was a better predictor of insulin sensitivity (Rd value) than that of intraperitoneal fat mass, and that the sum of truncal skin fold thickness also better predicted insulin resistance than intraperitoneal, retroperitoneal, or peripheral subcutaneous fat [45]. Further, posterior subcutaneous abdominal fat mass better predicted insulin sensitivity than anterior subcutaneous abdominal fat mass [46]. Studies in men with type 2 diabetes mellitus also showed a strong relationship between subcutaneous truncal fat and glucose disposal [47].

Despite the inclusion of central obesity as a component of MetS, the data clearly support the concept that there are profound differences in fat metabolism and its regulation in adipose tissue that are a function of the anatomical site. Dysregulated energy metabolism and attendant lipoatrophy in noncentral fat depots, particularly the femoral-gluteal depots, is linked mechanistically to the cluster of risk factors that define metabolic syndrome including central obesity [48, 49]. Although these differences have not been adequately explored on a depot-by-depot basis, they do provide the basis for understanding the mechanistic link between body fat distribution and the other biomarkers of deranged lipid and lipoprotein metabolism that are associated with obesity-linked type 2 diabetes.

Although most studies using state-of-the-art technologies show that central obesity and not lower body obesity is more strongly correlated with risk factors for cardiovascular disease, the underlying pathways that are the mechanistic links likely originate from dysregulated energy metabolism in noncentral fat depots, particularly femoral-gluteal depots. According to one model, MetS is a manifestation of systemic steatosis stemming from excess lipolysis via adipose triglyceride lipase (ATGL) and hormone-sensitive lipase (HSL) and failure of femoral-gluteal fat depots to sequester, esterify, and store NEFA (Figure 42.2). As a consequence, the excess NEFA circulate to other sites—to the liver, where the NEFA are

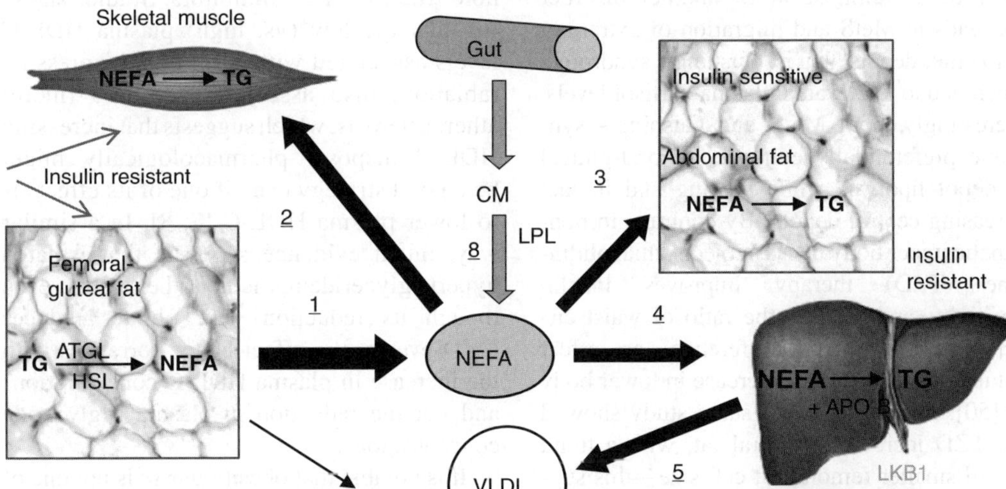

Figure 42.2 A mechanistic model for MetS. (1) Insulin resistance in femoral-gluteal fat depots increases lipolysis that raises the plasma NEFA concentration. The plasma NEFA are transported to other sites—(2) skeletal muscle, (3) abdominal fat depots, and (4) liver, which (5) hypersecretes NEFA-derived VLDL triglyceride. The CETP-mediated exchange of VLDL-triglyceride for HDL and LDL CE (6) makes them TG-rich substrates for hepatic lipase (7), which converts them to their smaller, denser analogs. Hydrolysis of chylomicron-triglyceride (8) would raise the plasma NEFA concentration still more thereby exacerbating the hyper-NEFAnemic state.

ultimately incorporated into VLDL-triglycerides that when secreted produce a hypertriglyceridemic state, to skeletal muscle, where they induce an insulin-resistant state that leads to hyperglycemia, and to central fat depots producing the MetS waist line. Ultimately, CETP exchanges the triglycerides of the expanded pool of VLDL to HDL and LDL in exchange for their cholesteryl esters. The now triglyceride-rich HDL and LDL are substrates for hepatic lipase, which reduces the triglyceride content and thereby produces the smaller, denser counterparts, HDL3 and small, dense LDL. However, the effects are far more profound than the failure to store NEFA, because the femoral-gluteal depots diminish in size, indicating increased lipolysis. This condition is reminiscent of another disorder that leads to MetS and migration of extremity fat to other depots, that is, Cushing's syndrome, which is due to elevated plasma cortisol levels. Interestingly, both MetS and Cushing's syndrome preferentially enhance femoral-gluteal fat depot lipolysis while sparing and in fact increasing central depots. By contrast, in non-diabetic upper body obese subjects, thiazolidindone (TZD) therapy improves insulin sensitivity, and reduces the ratio of waist circumference-to-hip circumference—an effect that is due to a selective increase in lower body fat [50]. Interestingly, the same study showed that TZD increased femoral fat, with a trend toward smaller femoral fat cell size—this suggests that adipocyte proliferation rather than hypertrophy underlies the leg fat gain, a finding that is consistent with the effects of TZDs on preadipocytes. The effects of cortisol, which enhances femoral-gluteal lipolysis, and of TZDs, which increase femoral-gluteal adipogenesis, support the concept that femoral-gluteal depots are more sensitive than other depots to pharmacological perturbations. The underlying mechanisms and structure responsible for the depot-specific effects are not known, but their identification would certainly shed light on possible therapeutic targets.

What are the atherogenic components of MetS? Although all the factors in Table 42.1 have been associated with CVD risk, mechanistic connections to disease have been more difficult to establish than those related to each of the elements of Table 42.1. HDL in and of itself is likely not a direct antiatherogenic; and on the basis of the above discussion, it might be safer to say, "A high plasma HDL-C concentration is not as important as how it got to that level; that is, mechanism is important." Raising plasma HDL-C by exercise, alcohol consumption, and fibrate and niacin therapy are cardioprotective, even though the mechanisms have not been clearly established. CETP deficiency and endothelial lipase gain of function do not produce the levels of cardioprotection commensurate with the attendant increases in plasma HDL-C; this may also hold true for CETP inhibitors. Studies showing that the low (vs. high) plasma HDL-C levels associated with SR-BI overexpression (ablation) also associate with less (more) atherosclerosis, which suggests that increasing HDL-C disposal pharmacologically might be a good strategy even if one of its effects is to lower plasma HDL-C [7, 8]. In a similar way, much evidence suggests that isolated hypertriglyceridemia is not atherogenic; even though its reduction with fibrates reduces CVD events, the effect better correlates with the increase in plasma HDL-C concentrations and not the reduction in plasma triglyceride concentrations.

It is notable that obesity *per se* is not one of the criteria of the NCEP definition of MetS, an omission that we have retained in our expanded criteria (Table 42.1). However, the inclusion of central obesity in NCEP guidelines goes closer to the underlying cause of MetS, but does not complete the model, which requires the simultaneously decreased fat storage from other sites, especially the prominent femoral-gluteal sites. It is possible that none of the analytes of Table 42.1 have anything to do with atherosclerosis, and that they are simply biomarkers for underlying derangements in energy metabolism, which are derived from insulin resistance in femoral-gluteal fat depots. The usually strong correlation of high HDL-C concentrations with atheroprotection may merely indicate that this analyte is a more robust marker of insulin sensitivity than other analytes.

SUMMARY

This discussion brings us to two major points. First, a low plasma HDL-C concentration is associated with increased cardiovascular disease. However, elevation of plasma HDL-C concentrations is not axiomatically linked to reduced cardiovascular events, but rather depends on the specific mechanism(s) involved, some being cardioprotective while others are not. Given the atheroprotective effects of SR-BI over expression in a mouse model, future efforts might best focus on increasing HDL functionality, particularly hepatic HDL-C disposal. Second, central fat is strongly associated with increased cardiovascular disease. However, the underlying cause is more likely associated with dysregulated energy metabolism in noncentral depots, especially femoral-gluteal depots. The future here lies in the discovery of compounds superior to PPARγ agonists that promote fat storage in noncentral fat depots without promoting weight gain, and the identification of the underlying cause of age-dependent insulin resistance in noncentral fat depots.

ACKNOWLEDGMENT

This work was supported in part by a National Institutes of Health grant to HJP (HL056865); CR was supported by a National Institutes of Health Training Grant (HL007812).

REFERENCES

1. Williams PT, Feldman DE. Prospective study of coronary heart disease vs. HDL2, HDL3, and other lipoproteins in Gofman's Livermore cohort. *Atherosclerosis*. 2011;214:196–202.
2. Wilson PW, Abbott RD, Castelli WP. High density lipoprotein cholesterol and mortality. The Framingham Heart study. *Arteriosclerosis*. 1988;8:737–741.
3. Frick MH, Elo O, Haapa K, et al. Helsinki Heart study: Primary-prevention trial with gemfibrozil in middle-aged men with dyslipidemia. Safety of treatment, changes in risk factors, and incidence of coronary heart disease. *N Engl J Med*. 1987;317:1237–1245.
4. Robins SJ, Collins D, Wittes JT, et al. Relation of gemfibrozil treatment and lipid levels with major coronary events: Va-hit: A randomized controlled trial. *JAMA*. 2001;285:1585–1591.
5. Thun MJ, Peto R, Lopez AD, et al. Alcohol consumption and mortality among middle-aged and elderly U.S. adults. *N Engl J Med*. 1997; 337:1705–1714.
6. Langer RD, Criqui MH, Reed DM. Lipoproteins and blood pressure as biological pathways for effect of moderate alcohol consumption on coronary heart disease. *Circulation*. 1992;85:910–915.
7. Braun A, Trigatti BL, Post MJ, et al. Loss of SR-BI expression leads to the early onset of occlusive atherosclerotic coronary artery disease, spontaneous myocardial infarctions, severe cardiac dysfunction, and premature death in apolipoprotein e-deficient mice. *Circ Res*. 2002;90:270–276.
8. Arai T, Wang N, Bezouevski M, Welch C, Tall AR. Decreased atherosclerosis in heterozygous low density lipoprotein receptor-deficient mice expressing the scavenger receptor B-I transgene. *J Biol Chem*. 1999;274:2366–2371.
9. Pownall HJ, Brauchi D, Kilinc C, et al. Correlation of serum triglyceride and its reduction by omega-3 fatty acids with lipid transfer activity and the neutral lipid compositions of high-density and low-density lipoproteins. *Atherosclerosis*. 1999;143:285–297.
10. Bruce C, Sharp DS, Tall AR. Relationship of HDL and coronary heart disease to a common amino acid polymorphism in the cholesteryl ester transfer protein in men with and without hypertriglyceridemia. *J Lipid Res*. 1998;39: 1071–1078.
11. Zhong S, Sharp DS, Grove JS, et al. Increased coronary heart disease in Japanese-American men with mutation in the cholesteryl ester transfer protein gene despite increased HDL levels. *J Clin Invest*. 1996;97:2917–2923.

12. de la Llera-Moya M, Drazul-Schrader D, Asztalos BF, Cuchel M, Rader DJ, Rothblat GH. The ability to promote efflux via ABCA1 determines the capacity of serum specimens with similar high-density lipoprotein cholesterol to remove cholesterol from macrophages. *Arterioscler Thromb Vasc Biol*. 2010;30:796–801.
13. Haase CL, Tybjaerg-Hansen A, Grande P, Frikke-Schmidt R. Genetically elevated apolipoprotein A-I, high-density lipoprotein cholesterol levels, and risk of ischemic heart disease. *J Clin Endocrinol Metab*. 2010;95:E500–E510.
14. Voight BF, Peloso GM, Orho-Melander M, et al. Plasma HDL cholesterol and risk of myocardial infarction: A Mendelian randomisation study. *Lancet*. 2012;380:572–580.
15. Haase CL, Tybjaerg-Hansen A, Qayyum AA, Schou J, Nordestgaard BG, Frikke-Schmidt R. Lcat, HDL cholesterol and ischemic cardiovascular disease: A Mendelian randomization study of HDL cholesterol in 54,500 individuals. *J Clin Endocrinol Metab*. 2012;97:E248–E256.
16. Dullaart RP, Perton F, van der Klauw MM, Hillege HL, Sluiter WJ. High plasma lecithin: Cholesterol acyltransferase activity does not predict low incidence of cardiovascular events: Possible attenuation of cardioprotection associated with high HDL cholesterol. *Atherosclerosis*. 2010;208:537–542.
17. Calabresi L, Baldassarre D, Simonelli S, et al. Plasma lecithin: Cholesterol acyltransferase and carotid intima-media thickness in European individuals at high cardiovascular risk. *J Lipid Res*. 2011;52:1569–1574.
18. Magnus P, Bakke E, Hoff DA, et al. Controlling for high-density lipoprotein cholesterol does not affect the magnitude of the relationship between alcohol and coronary heart disease. *Circulation*. 2011;124:2296–2302.
19. Gillotte KL, Zaiou M, Lund-Katz S, et al. Apolipoprotein-mediated plasma membrane microsolubilization. Role of lipid affinity and membrane penetration in the efflux of cellular cholesterol and phospholipid. *J Biol Chem*. 1999;274:2021–2028.
20. Nissen SE, Tsunoda T, Tuzcu EM, et al. Effect of recombinant ApoA-I Milano on coronary atherosclerosis in patients with acute coronary syndromes: A randomized controlled trial. *JAMA*. 2003;290:2292–2300.
21. Nicholls SJ, Tuzcu EM, Sipahi I, et al. Relationship between atheroma regression and change in lumen size after infusion of apolipoprotein A-I Milano. *J Am Coll Cardiol*. 2006;47:992–997.
22. Waksman R, Torguson R, Kent KM, et al. A first-in-man, randomized, placebo-controlled study to evaluate the safety and feasibility of autologous delipidated high-density lipoprotein plasma infusions in patients with acute coronary syndrome. *J Am Coll Cardiol*. 2010;55:2727–2735.
23. Mendez AJ, Anantharamaiah GM, Segrest JP, Oram JF. Synthetic amphipathic helical peptides that mimic apolipoprotein A-I in clearing cellular cholesterol. *J Clin Invest*. 1994;94:1698–1705.
24. Berard AM, Foger B, Remaley A, et al. High plasma HDL concentrations associated with enhanced atherosclerosis in transgenic mice overexpressing lecithin-cholesteryl acyltransferase. *Nat Med*. 1997;3:744–749.
25. Rousset X, Vaisman B, Auerbach B, et al. Effect of recombinant human lecithin cholesterol acyltransferase infusion on lipoprotein metabolism in mice. *J Pharmacol Exp Ther*. 2010;335:140–148.
26. Kozarsky KF, Donahee MH, Glick JM, Krieger M, Rader DJ. Gene transfer and hepatic overexpression of the HDL receptor SR-BI reduces atherosclerosis in the cholesterol-fed LDL receptor-deficient mouse. *Arterioscler Thromb Vasc Biol*. 2000;20:721–727.
27. Mardones P, Quinones V, Amigo L, et al. Hepatic cholesterol and bile acid metabolism and intestinal cholesterol absorption in scavenger receptor class B type I-deficient mice. *J Lipid Res*. 2001;42:170–180.
28. Courtney HS, Zhang YM, Frank MW, Rock CO. Serum opacity factor, a streptococcal virulence factor that binds to apolipoproteins A-I and A-II and disrupts high density lipoprotein structure. *J Biol Chem*. 2006;281:5515–5521.
29. Gillard BK, Courtney HS, Massey JB, Pownall HJ. Serum opacity factor unmasks human plasma high-density lipoprotein instability via selective delipidation and apolipoprotein A-I desorption. *Biochemistry*. 2007;46:12968–12978.
30. Tchoua U, Rosales C, Tang D, et al. Serum opacity factor enhances HDL-mediated cholesterol efflux, esterification and anti-inflammatory effects. *Lipids*. 2010;45:1117–1126.
31. Gillard BK, Rosales C, Pillai BK, Lin HY, Courtney HS, Pownall HJ. Streptococcal serum opacity factor increases the rate of hepatocyte

uptake of human plasma high-density lipoprotein cholesterol. *Biochemistry*. 2010;49: 9866–9873.

32. Rosales C, Tang D, Gillard BK, Courtney HS, Pownall HJ. Apolipoprotein E mediates enhanced plasma high-density lipoprotein cholesterol clearance by low-dose streptococcal serum opacity factor via hepatic low-density lipoprotein receptors in vivo. *Arterioscler Thromb Vasc Biol*. 2011;31:1834–1841.
33. Patsch JR, Prasad S, Gotto AM, Jr., Patsch W. High density lipoprotein2. Relationship of the plasma levels of this lipoprotein species to its composition, to the magnitude of postprandial lipemia, and to the activities of lipoprotein lipase and hepatic lipase. *J Clin Invest*. 1987;80:341–347.
34. Patsch JR, Karlin JB, Scott LW, Smith LC, Gotto AM, Jr. Inverse relationship between blood levels of high density lipoprotein subfraction 2 and magnitude of postprandial lipemia. *Proc Natl Acad Sci U S A*. 1983;80:1449–1453.

34a. Bagdade JD, Ritter MC, Subbaiah PV. Accelerated cholesteryl ester transfer in patients with insulin-dependent diabetes mellitus. *Eur J Clin Invet*. 1991;21:161–167.

35. McGarry JD. Banting lecture 2001: Dysregulation of fatty acid metabolism in the etiology of type 2 diabetes. *Diabetes*. 2002;51:7–18.
36. Vague J. Importance of the measurement of fat distribution in pathology [in French]. *Bull Mem Soc Med Hop Paris*. 1950;66:1572–1574.
37. Kissebah AH, Peiris AN. Biology of regional body fat distribution: Relationship to non-insulin-dependent diabetes mellitus. *Diabetes Metab Rev*. 1989;5:83–109.
38. Seidell JC, Perusse L, Despres JP, Bouchard C. Waist and hip circumferences have independent and opposite effects on cardiovascular disease risk factors: The Quebec Family study. *Am J Clin Nutr*. 2001;74:315–321.
39. Snijder MB, Dekker JM, Visser M, et al. Larger thigh and hip circumferences are associated with better glucose tolerance: The Hoorn study. *Obes Res*. 2003;11:104–111.
40. Yim JE, Heshka S, Albu JB, Heymsfield S, Gallagher D. Femoral-gluteal subcutaneous and intermuscular adipose tissues have independent and opposing relationships with CVD risk. *J Appl Physiol*. 2008;104:700–707.
41. Terry RB, Stefanick ML, Haskell WL, Wood PD. Contributions of regional adipose tissue depots to plasma lipoprotein concentrations in overweight men and women: Possible protective effects of thigh fat. *Metabolism*. 1991;40:733–740.
42. Van Pelt RE, Jankowski CM, Gozansky WS, Schwartz RS, Kohrt WM. Lower-body adiposity and metabolic protection in postmenopausal women. *J Clin Endocrinol Metab*. 2005;90:4573–4578.

42a. Bjorntorp P. "Portal" adipose tissue as a generator of risk factors for cardiovascular disease and diabetes. *Arteriosclerosis*. 1990;10: 493–496.

43. Randle PJ, Garland PB, Hales CN, Newsholme EA. The glucose fatty-acid cycle. Its role in insulin sensitivity and the metabolic disturbances of diabetes mellitus. *Lancet*. 1963;1: 785–789.
44. Basu A, Basu R, Shah P, Vella A, Rizza RA, Jensen MD. Systemic and regional free fatty acid metabolism in type 2 diabetes. *Am J Physiol Endocrinol Metab*. 2001;280: E1000–E1006.
45. Abate N, Garg A, Peshock RM, Stray-Gundersen J, Grundy SM. Relationships of generalized and regional adiposity to insulin sensitivity in men. *J Clin Invest*. 1995;96: 88–98.
46. Misra A, Garg A, Abate N, Peshock RM, Stray-Gundersen J, Grundy SM. Relationship of anterior and posterior subcutaneous abdominal fat to insulin sensitivity in nondiabetic men. *Obes Res*. 1997;5:93–99.
47. Abate N, Garg A, Peshock RM, Stray-Gundersen J, Adams-Huet B, Grundy SM. Relationship of generalized and regional adiposity to insulin sensitivity in men with NIDDM. *Diabetes*. 1996;45:1684–1693.
48. Okura Y, Takao M, Zhang B, Nakashima Y, Saku K. Cardiovascular risk factor profiles and endothelial function in coronary artery disease patients treated with statins. *Hypertens Res*. 2004;27:723–729.
49. Balasubramanyam A, Sekhar RV, Jahoor F, Jones PH, Pownall HJ. Pathophysiology of dyslipidemia and increased cardiovascular risk in HIV lipodystrophy: A model of "systemic steatosis". *Curr Opin Lipidol*. 2004;15:59–67.
50. Shadid S, Jensen MD. Effects of pioglitazone versus diet and exercise on metabolic health and fat distribution in upper body obesity. *Diabetes Care*. 2003;26:3148–3152.

43

STEM CELL THERAPEUTICS

Mei Mei Wong and Qingbo Xu
Cardiovascular Division, King's College London, London, UK

INTRODUCTION

Vascular cell damage or injury is a key event in the initiation of vascular diseases, for example, native atherosclerosis, angioplasty-induced restenosis, transplant arteriosclerosis, vein bypass graft atherosclerosis, and aneurysm formation [1]. Dysfunction of the endothelium is thought to be one of the earliest abnormalities that leads to atherosclerosis and is intimately related to modifiable risk factors such as high cholesterol levels, high blood pressure, smoking, and diabetes [1]. The past 30 years have witnessed several major therapeutic achievements in prevention and treatment of atherosclerosis. One of these is the discovery of the statin class of drugs that effectively lower cholesterol, a finding based on the understanding of mechanisms of lipid metabolism [2]. Other significant advances include the development of antithrombosis and antiplatelet agents, which are now widely used and important therapies for vascular diseases and arteriosclerosis [3]. In terms of prevention, smoking cessation and adoption of a healthy lifestyle that includes exercise and low-salt and low-energy intake can both reduce the morbidity and mortality of vascular disease. Despite these advances, cardiovascular disease remains a major cause of disability and death worldwide. While revascularization therapies such as thrombolysis, primary angioplasty, and coronary artery bypass grafting can save lives in the setting of acute myocardial infarction, postinfarction survivors are susceptible to vascular complications due to restenosis, progression of atherosclerotic disease at other sites, and inadequate revascularization due to unfavorable lesion characteristics or lack of vascular graft material. These factors can all contribute to the development of congestive heart failure due to impaired systolic function and to the susceptibility to life-threatening cardiac arrhythmias. Consequently, the search continues to identify new therapies for vascular disease that can overcome current problems

Atherosclerosis: Risks, Mechanisms, and Therapies, First Edition. Edited by Hong Wang and Cam Patterson.

such as atherosclerosis, in-stent restenosis, and vein graft arteriosclerosis.

Recent studies provide evidence that stem cell therapy has significant promise for treatment of vascular diseases. The identification of stem/progenitor cells in the vessel wall indicates that endogenous (adult) stem cells have the ability to repair the endothelium, thereby restoring the integrity of the vessel and importantly improving critical endothelial functions [4]. Secondly, embryonic stem cells (ESCs), a population of undifferentiated cells that can develop into any type of cells in the body including endothelial cells and smooth muscle cells (SMCs), occur as a second source for stem cell therapy [5]. Finally, a recent progress in somatic cell reprogramming, that is, induced pluripotent stem cells (iPSCs), suggests even greater potential for the field of vascular stem cell therapeutics than could be envisioned just a few years ago [6]. The advent of iPSCs derived from specific individuals promises to circumvent some of the important limitations of ESCs and adult stem cells for regenerative medicine, including ethical issues regarding the source of ESCs, scalable production of desired cell lineages (for adult progenitors), and immunogenicity and allograft rejection (for both). Such concepts have generated high-profile publicity for the use of stem cells as tools for therapeutic intervention in vascular disease. However, potential limitations of iPSCs persist, including the use of genetic manipulation for efficient reprogramming of cells via ectopic expression of a select group of transcription factors (combinations of Sox2, LIN28, c-Myc, Oct3/4, KLF4, and Nanog), which raises concern for potential oncogenicity in stem cell recipients. In the present review, we will focus on the application of stem cells to treat vascular disease, discuss the mechanisms of stem cell differentiation, and consider limitations of stem cell therapy in these applications.

ATHEROSCLEROSIS AND ENDOTHELIAL REPAIR

Atherosclerosis is an inflammatory condition of large- and medium-sized muscular arteries and can be characterized by endothelial dysfunction that results in delayed or impaired endothelialization. Once the integrity of the endothelium is interrupted, lipid penetration and mononuclear cell adhesion processes are initiated. Evidence shows that the overall rates of cell turnover in the artery are very low [7] and clusters of increased cell replication in some areas correlate with increased permeability to plasma proteins [8, 9]. In animal models, endothelial cells in atherosclerosis-resistant areas of the artery have a lifespan of about 12 months, whereas cells at lesion-prone sites live for weeks or sometimes even less [10]. In a recent study of endothelial cell turnover in the whole aortas, our group demonstrated that the numbers of dead and proliferating cells in aortas of *Apoe*$^{-/-}$ mice differ significantly from those of wild-type animals [11]. Importantly, lesion-prone areas display a higher rate of endothelial cell turnover, as indicated by BrdU incorporation assays. Furthermore, our data strongly suggest that endothelial progenitor cells are involved in cellular repair and turnover that follow endothelial death in the specific regions.

Several groups have reported that circulating progenitor cells contribute to endothelial repair and suggested further that impairment of such cells promotes disease progression. For instance, Hill et al. [12] described a reduction in endothelial progenitor colony-forming units in healthy individuals with risk factors for cardiovascular disease. Vasa et al. [13] documented a similar inverse correlation, suggesting that decreases in endothelial progenitor numbers and function might serve as novel biomarkers of cardiovascular disease. Data from chimeric animal models of vascular grafting showed that the regenerated endothelial cells of vein grafts originated, at least in part, from the recipients' circulating blood, while the remaining endothelial cells came from the donor vessel [4]. One of the most important observations in this study was the complete loss of endothelium in vein grafts within 3 days of surgery, which indicates that certain surface areas of the vein graft lack a monolayer of endothelial cells for an initial period of time. Such exposure of blood to the subendothelial matrix promotes thrombus formation. Various strategies to restore the integrity of the endothelium using endothelial progenitors have therefore been investigated, either by increasing progenitor cell number in vivo using cytokines or by increasing progenitor cell number ex vivo prior to reinfusion to repair the damaged endothelium (Figure 43.1). However, recent studies indicate some concerns on the nature of endothelial progenitor cells in the blood (see following text).

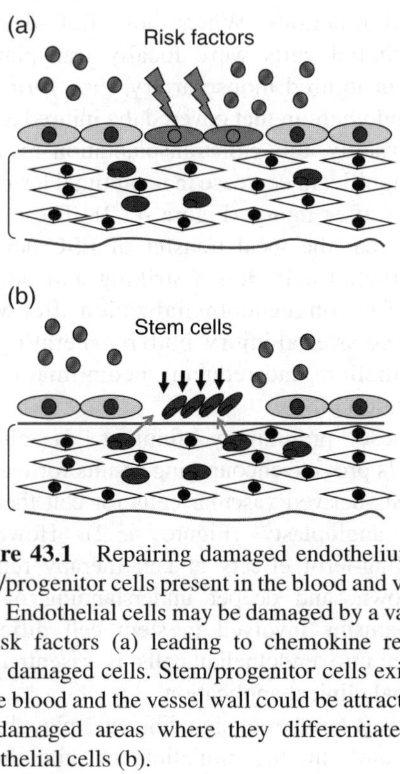

Figure 43.1 Repairing damaged endothelium by stem/progenitor cells present in the blood and vessel wall. Endothelial cells may be damaged by a variety of risk factors (a) leading to chemokine release from damaged cells. Stem/progenitor cells existing in the blood and the vessel wall could be attracted to the damaged areas where they differentiate into endothelial cells (b).

The strategy using endothelial progenitors to repair the endothelium and restore its function is to increase the number of progenitors by transfusing preprepared cells derived from the blood or bone marrow. A global improvement in endothelial function was seen with a systemic transfusion of vascular progenitor cells in the study by Wassmann et al. [14] Intravenous transfusion of spleen-derived mononuclear cells isolated from wild-type mice on 3 consecutive days significantly improved endothelium-dependent vasodilation in *Apoe*$^{-/-}$ mice maintained on a high-cholesterol diet. Endothelial progenitors cultured from splenic mononuclear cells and transfused intravenously were found to home to the site of injury in a mouse model, thereby causing enhanced reendothelialization and associated reduction of neointima formation [15]. Furthermore, this study showed that cells administered intravenously localized exclusively to the injury site when homing to the host organ was prevented by splenectomy. If preferential homing to the spleen could be "switched off" in some way in humans, this might enhance homing of these cells to the site of injury in clinical practice.

However, recent data demonstrate that "endothelial progenitor cells" described by many papers are in fact monocytes/macrophages. Findings from different groups have shown that classic "endothelial progenitor cells" fail to incorporate into the vasculature and do not differentiate into endothelial cells [16]. New terms that reflect these distinctions, such as circulating angiogenic cells, early angiogenic cells, early outgrowth cells, and others, are currently being introduced. Consequently, it is now widely accepted that more stringent criteria are needed for endothelial progenitor studies, as outlined by Hirschi [17]. Thus, endothelial progenitor cells in circulation could exist in a small proportion, but one has to be cautious in the interpretation of the papers earlier published in the literature.

Recent evidence shows that vascular progenitor cells are present in the vessel wall [18]. Furthermore, human embryonic aorta studies suggest the presence of progenitor cells with wide lineage potential [19, 20]. Studies of adult human veins and arteries show that the vessel wall contains progenitor cells with clonogenic potential like that of circulating endothelial progenitor cells [21]. Progenitor cells with endothelial potential have also been found in the media–adventitia border of vessels from a range of organs [22]. Endothelial cells of neovessels in cardiac transplant arteriopathy were shown to have a myeloid/progenitor cell lineage, similar to the pattern of lineage markers expressed in endothelial outgrowth cells derived from healthy volunteers [23]. Taken together, these data suggest that from development through adulthood, endothelial progenitor cells are present in the vessel walls of both normal and diseased tissues (Figure 43.1). However, the extent to which resident vascular progenitor cells are responsible for endothelial repair and angiogenesis after vessel injury is not clear. Thus, further investigation on the physiological functions of local stem and progenitor cells in the vessel wall will be needed in the near future.

ENDOTHELIAL REPAIR AFTER ANGIOPLASTY

Angioplasty is a technique that involves the mechanical expansion of a blood vessel stenosis or obstruction, which is typically the result of atherosclerosis [24, 25]. In this process, a catheter is introduced percutaneously into the vasculature and advanced to the site of vessel obstruction. At this location, a small balloon inflated within the vessel lumen crushes fatty deposits and transiently stretches the vascular wall, thereby opening up the blood vessel lumen to improve flow. A stent may or may not be inserted at the time of balloon inflation to ensure that the vessel remains open. After angioplasty, although relief of atherosclerotic symptoms in patients can occurs rapidly, injury to the inner surface of the vessel wall may cause thrombus formation. Damage to the endothelium is unavoidable, and despite systemic and/or local administration of anticoagulant and antiproliferative drugs, neointimal lesions develop in the treated vessels [24, 25]. If lost endothelium could be efficiently replaced by exogenous stem cells, reduced thrombosis and neointimal formation might be achieved.

In preclinical studies along these lines, our previous data showed that stem cell antigen-1 (Sca-1) can be used as a sorting marker of vascular progenitors to isolate these cells from adult adventitial tissues [26]. After intensive searching, we found that Sca-1+ progenitors isolated from ESC cultures can serve as a common vascular progenitor for both endothelial cells and SMCs [27, 28]. ESC-derived endothelial cells can form capillary-like networks on Matrigel™ in vitro and vascular-like structures within Matrigel™ plugs in vivo, which indicates that these cells have endothelial functions. When these ESC-derived endothelial cells were locally transplanted into an injured mouse artery, they formed a neoendothelium that covered the injured areas. Importantly, this cell transplantation resulted in a significant decrease in neointimal lesions 2 weeks after injury (Figure 43.2). These data show that the local transfer of ESC-derived endothelial cells has a striking and beneficial effect on reendothelialization after wire-induced arterial injury, both regenerating the endothelium and reducing neointimal lesion formation [28].

These preliminary findings in animal models provide encouraging results for the use of ESC-derived vascular cells for cell therapy after angioplasty (Figure 43.2). However, the long-term effects of cell therapy remain unknown, and deeper understanding of the mechanisms involved in stem cell differentiation into endothelial cells is essential for optimal clinical application.

A vast number of signaling cues play different roles in the initiation of endothelial differentiation. The release of signaling protein molecules such as vascular endothelial growth

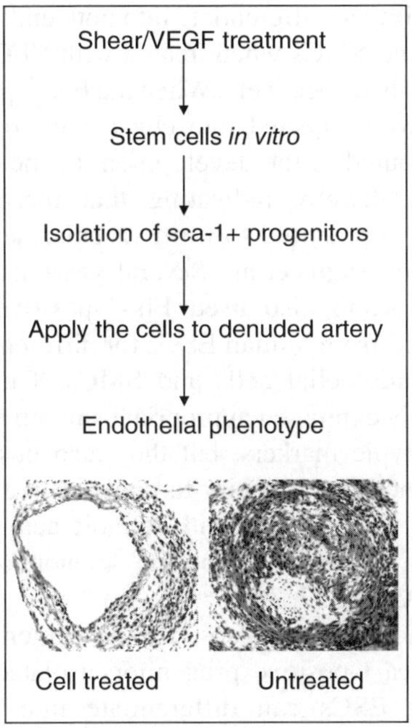

Figure 43.2 Stem cell-based therapy for injured vessels. Shear stress or VEGF treatment results in ESC differentiation into Sca1+ progenitors, which were isolated with microbeads. These cells were applied to injured arteries locally resulting in reduced lesion formation. The lumen size of injured artery with Sca1+ cell treatment is significantly larger than that without cell treatment.

factor (VEGF), Wnts, or bone morphogenetic proteins (BMPs) from distinct cell populations triggers multiple intracellular signaling cascades in neighboring cells. Thorough understanding of the mechanisms that control signaling events pertaining to endothelial cell development in differentiating ESC populations can also offer insights into how one might manipulate these signals for therapeutic advantages.

So far, two factors that have been found to initiate endothelial differentiation are VEGFs and mechanical force generated by the intrinsic flow of the blood, that is, laminar shear stress [29, 30]. VEGF and/or shear stress significantly induces the expression of endothelial cell-specific markers such as CD31, von Willebrand factor (vWF), and VE-cadherin while changing the cells from a fibroblast-like appearance to a more cobblestone shape at the same time [29].

Interestingly, laminar flow also appears to activate molecular pathways that lead to histone modifications, including the stabilization and activation of histone deacetylases (HDACs). ESCs undergo complex gene-specific and functionally important remodeling of chromatin structure that is necessary for differentiation into endothelial cells. Overexpression of HDAC3 has been shown to increase protein levels of endothelial cell lineage-specific markers in VEGF-induced endothelial differentiation. Furthermore, Zeng et al. [30] also provided evidence that shear or VEGF stimulation activates HDAC3 by posttranslational stabilization through fetal liver kinase-1–phosphatidylinositol 3-kinase–Akt (Flk-1–PI3K–Akt) signaling pathways, thus providing mechanistic insight into shear-induced stem cell differentiation toward endothelial cells. Taken together, it is conceivable that several signaling pathways coordinately regulate stem cell differentiation toward an endothelial phenotype.

TISSUE-ENGINEERED VESSELS FROM STEM CELLS

Vascular bypass grafting is commonly performed to supplant the function of severely diseased arteries and restore perfusion of ischemic tissues and organs (Figure 43.3). However, a majority of patients do not have healthy vessels that are suitable for use in such replacement procedures [31]. Additionally, the use of small-caliber artificial vascular prostheses as coronary artery bypass grafts or in arteriosclerosis obliterans below the knee leads to extremely high failure rates that are generally attributed to thrombus formation and occlusion.

Tissue engineering, a relatively new discipline, offers the possibility of creating replacement structures with stem cells derived from autologous or nonautologous sources [32]. Early studies described the production of

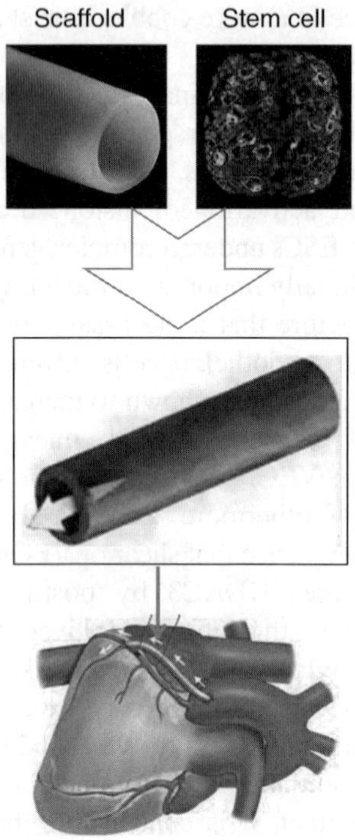

Figure 43.3 Schematic representation for vascular tissue engineering. Scaffold and stem cells could interact in a bioreactor where they form a blood vessel, which can be grafted to coronary artery to replace severely diseased vessel. (*See insert for color representation of the figure.*)

completely biological tissues and organs with surprisingly high mechanical strength and without the need for any exogenous plastic scaffold [33]. Such techniques can produce completely biological vessels that fulfill fundamental requirements for grafting, such as high resistance to bursting (burst strength), ease of surgical manipulation, and appropriate endothelial function [33]. Such vessels have already been used successfully as arteriovenous fistulas for hemodialysis access [34].

Recently, stem cells have been widely used to produce vascular cells for vascular tissue engineering [35]. Such approaches require the production of large numbers of both endothelial cells and vascular SMCs, and several methods have been established. Yamashita and colleagues [36] demonstrated that Flk-1-positive cells derived from mouse ESCs had the ability to differentiate into both endothelial cells and SMCs when treated with VEGF and PDGF-β, respectively. When the Flk-1-positive cells were injected into chick embryos, they contributed to the development of the vasculature, thereby indicating that these cells have the potential to form vascular networks in tissue engineering. Several years later, the same group also used Flk-1-positive cells isolated from human ESCs for differentiation into endothelial cells and SMCs. The latter not only expressed appropriate smooth muscle phenotypic markers, but they also possessed the functional capacity to contract like native SMCs when treated with retinoic acid and/or dibutyryl cyclic adenosine monophosphate (db-cAMP) [37, 38].

Xiao and colleagues [27, 28] demonstrated that Sca-1-positive progenitors isolated from mouse ESCs can differentiate into highly purified cultures of endothelial cells or SMCs in the presence of VEGF or PDGF-β, respectively. When Sca-1-positive progenitors were cultured for several passages in the presence of VEGF, the cells displayed cobblestone morphologies, and most expressed high levels of endothelial markers (CD31, CD106, CD144, Flk-1, Flt-1, and vWF). Similarly, they also showed that Sca-1-positive progenitors isolated from mouse ESCs, when cultured on type IV collagen in the presence of PDGF, differentiated into highly pure (>95%) populations of SMCs that expressed smooth muscle markers such as smooth muscle α-actin (α- SMA), SM-MHC, and calponin. Their studies also showed that the integrin (α1, β1, α5)-FAK/paxillin-PI3K-MEK-ERK/JNK pathway is associated with the type IV collagen and PDGF-β signaling-mediated SMC differentiation [27]. Therefore, in light of the need for high-purity, ESC-derived vascular cells, this study provides a valuable differentiation technique within the field of vascular engineering.

Current reports have also demonstrated the involvement of chromatin-related proteins in the signaling cascades that promote SMC differentiation. Such studies are ongoing in many laboratories around the world, but here we describe two examples of recent advances in understanding of chromatin-based molecular mechanisms that contribute to the control of SMC differentiation.

Heterogeneous nuclear ribonucleoproteins (hnRNPs) are a set of nuclear proteins associated with heterogeneous nuclear RNA (hnRNA) during the posttranscriptional process, up to the point when mature mRNA is transported into the cytoplasm to participate in protein translation [39]. hnRNPs can range in size from 30 to 120 kDa in humans and are composed of six major nuclear core proteins known as A1, A2, B1, B2, C1, and C2, as well as another 16 proteins, or groups of proteins, designated as D1, D2, I, and U [40]. To date, although not all the hnRNP functions and mechanisms are known, it could be postulated that hnRNPs can affect the structure of hnRNAs. Therefore, they may act importantly to facilitate or hinder the interaction of hnRNA sequences with other nuclear factors in processing of pre-mRNAs and nucleocytoplasmic transport of mRNAs. In addition, they may function in other cellular processes—for example, these hnRNPs have also been shown to inhibit serine-/arginine-rich proteins binding to intronic or exonic enhancers to attenuate their splicing [41]. Taken together, the interactions between different hnRNPs within the hnRNA–hnRNP complex and with other proteins will promote or inhibit the splicing between specific pairs of splice sites, thus contributing to alternative splicing [42]. In a study of the potential importance of hnRNP A2/B1 in ESC–SMC differentiation, the expression of hnRNP A2/B1 was found to be upregulated during ESC differentiation to SMC [43]. Additionally, this study found that hnRNP A2/B1 can activate SMC differentiation by direct binding and activation of the promoters of the SMC differentiation marker genes encoding α-SMA and SM22α via regulation of chromobox protein homolog genes (Cbx).

Cbxs are highly conserved molecules that are involved in retinoic acid-induced primitive endoderm differentiation [44, 45]. Recently, our group found that Cbx3 is crucial for SMC differentiation from stem cells via a mechanism that requires the Cbx3 chromoshadow domain [46]. This study showed that four amino acids (165–168) within the chromoshadow domain of Cbx3 were key elements for Cbx3 interaction with diaphanous homolog 1 (mDia1) and for Cbx3-induced SMC differentiation. Interestingly, it was recently reported that mDia1 is highly expressed in a number of SMC-containing organs, including the aorta [47], and accumulating evidence suggests that mDia1 can act as an effector of the small GTPase RhoA, which is known to be involved in actin polymerization and SMC differentiation [47]. Interestingly, the C-terminal protein sequence of mDia1 bears a conserved pentapeptide PxVxL motif (pevql) that may mediate interaction with the Cbx3 chromoshadow domain, making it highly likely that mDia1 is a partner protein for Cbx3 in the regulation of stem cell differentiation into SMCs. Taken together, these results confirm the crucial role of Cbx3 in SMC differentiation.

To elucidate the mechanisms involved in vascular lineage differentiation, a bioreactor could be designed to optimize the conditions required to create a tissue-engineered vessel (Figure 43.3). A bioreactor refers to a system that can simulate a physiological environment to allow or induce certain behavior in cells, tissues, or organs. Ideally, the bioreactor should be capable of supplying oxygen and nutrition and removing accumulated products of cellular metabolism while providing the appropriate physical stimulation to cells. In vascular tissue engineering, the grafts should encounter hemodynamic conditioning within an artificial circulation system that provides biochemical and physiological stimulation to the tissue-engineered vessels before implantation in vivo (Figure 43.3). To date, bioreactors have been shown to facilitate cell seeding, differentiation, maturation, adhesion, and protein secretion [48, 49] and to enhance the mechanical properties of tissue-engineered vessels [50–52]. Overall, the application of stem cell-derived vascular cells in a bioreactor for vascular tissue engineering could lead to a graft in which the resultant morphological and mechanical properties closely resemble those of a native vessel.

SUMMARY AND PERSPECTIVE

As described earlier, the therapeutic potential of stem and progenitor cells in vascular disease is an exciting and important area of cardiovascular research. In adults, there are stem/progenitor cells in the circulating blood and within the vessel walls that are able to differentiate into endothelial cells and SMCs. The therapeutic strategy envisioned for atherosclerosis-related disease would ideally involve enhanced stem cell homing to areas of damage, where these therapeutic cells can repair the vessel and retard further lesion progression. For the problems associated with percutaneous vascular interventions, the major goal for prevention of restenosis would be the stimulation of arterial reendothelialization. Stem cells have the potential to meet the demands of repair and regeneration in diseased vessels by replacing cells that have been lost or damaged following catheter-based interventions, thereby restoring vessel integrity and function. In combination with other technologies such as tissue engineering, it is plausible that these cells can be directed to grow ex vivo into highly organized structures, closely resembling blood vessels, that can be implanted into patients. For vascular regeneration and repair, as for many other biomedical fields, the recent advances in iPSC technology offer potential solutions to major problems. iPSC techniques, through which terminally differentiated somatic cells can be reprogrammed to a pluripotent state, may provide unlimited sources of pluripotent stem cells that can be used in cell replacement therapy without the ethical concerns associated with ESCs. Recent research highlights include the direct differentiation of mature cells that bypass the pluripotent state during somatic cell reprogramming, thus accelerating the production of a desired cell type in a direct and more efficient manner [53]. Further embellishments include the use of therapeutic iPSCs or somatic cell reprogramming, which may 1 day make it possible to generate autologous cells that are genetically compatible with patient tissue. These advancements can obviate concerns of immune system rejection of the stem cell transplant. If successful, stem cell therapy via therapeutic cloning or other techniques would contribute greatly to personalized medicine.

ACKNOWLEDGMENTS

The authors would like to acknowledge all the collaborators who contributed to the work summarized in the review. This work was supported by grants from the British Heart Foundation and Oak Foundation.

REFERENCES

1. Ross R. Atherosclerosis—an inflammatory disease. *N Engl J Med*. 1999;340:115–126.
2. Goldstein JL, Kita T, Brown MS. Defective lipoprotein receptors and atherosclerosis. Lessons from an animal counterpart of familial hypercholesterolemia. *N Engl J Med*. 1983;309: 288–296.
3. Katz JN, Turer AT, Becker RC. Cardiology and the critical care crisis: A perspective. *J Am Coll Cardiol*. 2007;49:1279–1282.
4. Xu Q. The impact of progenitor cells in atherosclerosis. *Nat Clin Pract Cardiovasc Med*. 2006;3: 94–101.
5. Margariti A, Zeng L, Xu Q. Stem cells, vascular smooth muscle cells and atherosclerosis. *Histol Histopathol*. 2006;21:979–985.
6. Takahashi K, Yamanaka S. Induction of pluripotent stem cells from mouse embryonic and adult fibroblast cultures by defined factors. *Cell*. 2006; 126:663–676.
7. Schwartz SM, Benditt EP. Cell replication in the aortic endothelium: A new method for study of the problem. *Lab Invest*. 1973;28:699–707.
8. Schwartz SM, Benditt EP. Clustering of replicating cells in aortic endothelium. *Proc Nat Acad Sci U S A*. 1976;73:651–653.
9. Caplan BA, Schwartz CJ. Increased endothelial cell turnover in areas of in vivo Evans blue

uptake in the pig aorta. *Atherosclerosis*. 1973; 17:401–417.

10. Schwartz SM, Benditt EP. Aortic endothelial cell replication. I. Effects of age and hypertension in the rat. *Circ Res*. 1977;41:248–255.
11. Foteinos G, Hu Y, Xiao Q, Metzler B, Xu Q. Rapid endothelial turnover in atherosclerosis-prone areas coincides with stem cell repair in apolipoprotein e-deficient mice. *Circulation*. 2008;117:1856–1863.
12. Hill JM, Zalos G, Halcox JP, et al. Circulating endothelial progenitor cells, vascular function, and cardiovascular risk. *N Engl J Med*. 2003; 348:593–600.
13. Vasa M, Fichtlscherer S, Aicher A, et al. Number and migratory activity of circulating endothelial progenitor cells inversely correlate with risk factors for coronary artery disease. *Circ Res*. 2001;89:E1–E7.
14. Wassmann S, Werner N, Czech T, Nickenig G. Improvement of endothelial function by systemic transfusion of vascular progenitor cells. *Circ Res*. 2006;99:e74–e83.
15. Werner N, Junk S, Laufs U, et al. Intravenous transfusion of endothelial progenitor cells reduces neointima formation after vascular injury. *Circ Res*. 2003;93:e17–e24.
16. Zampetaki A, Kirton JP, Xu Q. Vascular repair by endothelial progenitor cells. *Cardiovasc Res*. 2008;78:413–421.
17. Hirschi KK. Vascular precursors: Origin, regulation and function. *Arterioscler Thromb Vasc Biol*. 2010;30:1078–1079.
18. Torsney E, Xu Q. Resident vascular progenitor cells. *J Mol Cell Cardiol*. 2011;50:304–311.
19. Alessandri G, Girelli M, Taccagni G, et al. Human vasculogenesis ex vivo: Embryonal aorta as a tool for isolation of endothelial cell progenitors. *Lab Invest*. 2001;81:875–885.
20. Fang B, Li Y, Song Y, Li N. Isolation and characterization of multipotent progenitor cells from the human fetal aorta wall. *Exp Biol Med (Maywood)*. 2010;235:130–138.
21. Ingram DA, Mead LE, Moore DB, Woodard W, Fenoglio A, Yoder MC. Vessel wall-derived endothelial cells rapidly proliferate because they contain a complete hierarchy of endothelial progenitor cells. *Blood*. 2005;105:2783–2786.
22. Zengin E, Chalajour F, Gehling UM, et al. Vascular wall resident progenitor cells: A source for postnatal vasculogenesis. *Development*. 2006;133:1543–1551.
23. Liu C, Wang S, Metharom P, Caplice NM. Myeloid lineage of human endothelial outgrowth cells circulating in blood and vasculogenic endothelial-like cells in the diseased vessel wall. *J Vasc Res*. 2009;46:581–591.
24. Gray BH, Sullivan T. The treatment of iliac artery atherosclerosis with angioplasty and intravascular stents. *Vasc Med*. 1996;1:287–291.
25. Kuntz RE. Importance of considering atherosclerosis progression when choosing a coronary revascularization strategy: The diabetes-percutaneous transluminal coronary angioplasty dilemma. *Circulation*. 1999;99:847–851.
26. Hu Y, Zhang Z, Torsney E, et al. Abundant progenitor cells in the adventitia contribute to atherosclerosis of vein grafts in Apoe-deficient mice. *J Clin Invest*. 2004;113:1258–1265.
27. Xiao Q, Zeng L, Zhang Z, Hu Y, Xu Q. Stem cell-derived sca-1+ progenitors differentiate into smooth muscle cells, which is mediated by collagen IV-integrin alpha1/beta1/alphav and PDGF receptor pathways. *Am J Physiol Cell Physiol*. 2007;292:C342–C352.
28. Xiao Q, Zeng L, Zhang Z, et al. Sca-1+ progenitors derived from embryonic stem cells differentiate into endothelial cells capable of vascular repair after arterial injury. *Arterioscler Thromb Vasc Biol*. 2006;26:2244–2251.
29. Wang H, Riha GM, Yan S, et al. Shear stress induces endothelial differentiation from a murine embryonic mesenchymal progenitor cell line. *Arterioscler Thromb Vasc Biol*. 2005;25: 1817–1823.
30. Zeng L, Xiao Q, Margariti A, et al. Hdac3 is crucial in shear- and VEGF-induced stem cell differentiation toward endothelial cells. *J Cell Biol*. 2006;174:1059–1069.
31. Hoenig MR, Campbell GR, Rolfe BE, Campbell JH. Tissue-engineered blood vessels: Alternative to autologous grafts? *Arterioscler Thromb Vasc Biol*. 2005;25:1128–1134.
32. Campbell GR, Campbell JH. Development of tissue engineered vascular grafts. *Curr Pharm Biotechnol*. 2007;8:43–50.
33. L'Heureux N, Paquet S, Labbe R, Germain L, Auger FA. A completely biological tissue-engineered human blood vessel. *FASEB J*. 1998;12: 47–56.
34. L'Heureux N, Dusserre N, Marini A, Garrido S, de la Fuente L, McAllister T. Technology insight: The evolution of tissue-engineered vascular grafts—from research to clinical

practice. *Nat Clin Pract Cardiovasc Med.* 2007;4:389–395.

35. Adams B, Xiao Q, Xu Q. Stem cell therapy for vascular disease. *Trends Cardiovasc Med.* 2007;17:246–251.
36. Yamashita J, Itoh H, Hirashima M, et al. Flk1-positive cells derived from embryonic stem cells serve as vascular progenitors. *Nature*. 2000; 408:92–96.
37. Drab M, Haller H, Bychkov R, et al. From totipotent embryonic stem cells to spontaneously contracting smooth muscle cells: A retinoic acid and db-cAMP in vitro differentiation model. *FASEB J*. 1997;11:905–915.
38. Huang H, Zhao X, Chen L, et al. Differentiation of human embryonic stem cells into smooth muscle cells in adherent monolayer culture. *Biochem Biophys Res Comm*. 2006;351: 321–327.
39. Dreyfuss G, Matunis MJ, Pinol-Roma S, Burd CG. HnRNP proteins and the biogenesis of mRNA. *Annu Rev Biochem*. 1993;62: 289–321.
40. Ball EE, Rehm EJ, Goodman CS. Cloning of a grasshopper cDNA coding for a protein homologous to the a1, a2/b1 proteins of mammalian hnRNP. *Nucleic Acids Res*. 1991;19:397.
41. He Y, Smith R. Nuclear functions of heterogeneous nuclear ribonucleoproteins A/B. *Cell Mol Life Sci*. 2009;66:1239–1256.
42. Martinez-Contreras R, Cloutier P, Shkreta L, Fisette JF, Revil T, Chabot B. HnRNP proteins and splicing control. *Adv Exp Med Biol*. 2007; 623:123–147.
43. Wang G, Xiao Q, Luo Z, Ye S, Xu Q. Functional impact of heterogeneous nuclear ribonucleoprotein a2/b1 in smooth muscle differentiation from stem cells and embryonic arteriogenesis. *J Biol Chem*. 2012;287:2896–2906.
44. Cammas F, Oulad-Abdelghani M, Vonesch JL, Huss-Garcia Y, Chambon P, Losson R. Cell differentiation induces Tif1beta association with centromeric heterochromatin via an HP1 interaction. *J Cell Sci*. 2002;115:3439–3448.
45. Cammas F, Herzog M, Lerouge T, Chambon P, Losson R. Association of the transcriptional corepressor TIF1beta with heterochromatin protein 1 (HP1): An essential role for progression through differentiation. *Genes Dev*. 2004;18:2147–2160.
46. Xiao Q, Wang G, Yin X, et al. Chromobox protein homolog 3 is essential for stem cell differentiation to smooth muscles in vitro and in embryonic arteriogenesis. *Arterioscler Thromb Vasc Biol*. 2011;31:1842–1852.
47. Staus DP, Blaker AL, Taylor JM, Mack CP. Diaphanous 1 and 2 regulate smooth muscle cell differentiation by activating the myocardin-related transcription factors. *Arterioscler Thromb Vasc Biol*. 2007;27:478–486.
48. McFetridge PS, Abe K, Horrocks M, Chaudhuri JB. Vascular tissue engineering: Bioreactor design considerations for extended culture of primary human vascular smooth muscle cells. *ASAIO J*. 2007;53:623–630.
49. Williams C, Wick TM. Perfusion bioreactor for small diameter tissue-engineered arteries. *Tissue Eng*. 2004;10:930–941.
50. Jeong SI, Kim SY, Cho SK, et al. Tissue-engineered vascular grafts composed of marine collagen and PLGA fibers using pulsatile perfusion bioreactors. *Biomaterials*. 2007;28: 1115–1122.
51. Barron V, Lyons E, Stenson-Cox C, McHugh PE, Pandit A. Bioreactors for cardiovascular cell and tissue growth: A review. *Ann Biomed Eng*. 2003;31:1017–1030.
52. Isenberg BC, Williams C, Tranquillo RT. Small-diameter artificial arteries engineered in vitro. *Circ Res*. 2006;98:25–35.
53. Margariti A, Winkler B, Karamariti E, et al. Direct reprogramming of fibroblasts into endothelial cells capable of angiogenesis and reendothelisation in tissue-engineered vessels. *Proc Natl Acad Sci U S A*. 2012;109:13793–13798.

44

ADVANCES IN SURGICAL REVASCULARIZATION

Stephen J. Forest, Ahmed Khan, and Joseph J. DeRose, Jr.

Department of Cardiovascular and Thoracic Surgery, Montefiore-Einstein Heart Center, Bronx, NY, USA

INTRODUCTION

The first successful coronary artery bypass graft (CABG) was performed in 1960 by Dr. Robert Goetz, when he placed a pedicled right internal thoracic artery (RITA) to the right coronary artery using a tantalum stent on the beating heart [1]. Dr. Vasilii I. Kolesov described the first use of the left internal thoracic artery (LITA) to the left anterior descending (LAD) artery on the beating heart using a suture technique through a left thoracotomy 4 years later in six patients [2]. However, it was not until 8 years later, when saphenous vein grafts (SVGs) [3] and bilateral internal thoracic arteries (BITAs) [4] were used to bypass obstructed coronary arteries on the arrested heart using cardiopulmonary bypass (CPB), that the modern-day CABG operation was born. For over four decades, CABG has been one of the most commonly performed major operations in the world, with nearly 450,000 cases performed in the United States yearly.

The longevity of CABG and its use as a durable and reliable intervention for the prognostic and symptomatic treatment of coronary artery disease (CAD) have been validated by over 30 years of prospective randomized study. The wealth of prospective randomized research directed at CABG as it compares to both medical therapy and the myriad of interventional therapies that have been employed over this time period gives us a solid guidepost for recommending this therapy in selected patients.

TECHNIQUES OF CABG

The underlying premise of CABG is to reroute the flow of blood in a coronary artery around a proximal obstruction in order to provide maximum oxygen delivery to the myocardial territory in question. The three basic requirements for CABG are an inflow source, a conduit, and an outflow bed. The inflow typically comes from the left subclavian artery for the

Atherosclerosis: Risks, Mechanisms, and Therapies, First Edition. Edited by Hong Wang and Cam Patterson.

LITA and the aorta for free grafts, which include reverse SVGs, radial artery grafts, and the RITA. The outflow vascular bed is critically important for maintaining the patency of the constructed bypass graft. Diffuse distal disease, small vessel caliber, and distal obstructions are all unfavorable characteristics for maintaining graft patency.

ARTERIAL AND VENOUS CONDUITS

The LITA

The most important development in CABG surgery has been the use of the LITA to bypass the LAD [5]. In early studies comparing conventional CABG to medical therapy [6–9], as well as later studies comparing conventional CABG to percutaneous coronary intervention [10–12], the most powerful predictor conferring a survival advantage to CABG has been the presence of a patent LITA–LAD bypass graft. The patency of this bypass graft has been time tested, and it is the gold standard against which all other revascularization strategies of the anterior wall have been measured. Studies in which routine 6-month and 1-year angiographies have been performed following LITA–LAD bypass grafting during conventional CABG reveal a short-term patency rate of 96–98% [13–15]. Longer-term analysis of LITA–LAD patency out to 10 and 20 years reveals that the durability of this graft is excellent, with an overall patency rate of 90–95% [16].

The LITA rarely has native atherosclerosis upon harvest and almost never develops atherosclerotic changes when placed into the coronary circulation. The reproducible and durable patency of the LITA–LAD graft likely owe both to this resistance of the LITA to atherosclerosis, the excellent outflow of the LAD, and the technical ease of grafting this anterior coronary artery.

The Reversed Saphenous Veins

The patency rate of the second most commonly used bypass graft, the reversed saphenous vein, is much lower than that of the LITA. The predisposition of the SVG to develop ongoing atherosclerosis when placed in the arterial circulation contributes to its failure in both early and late time frames after surgery. The patency of SVGs to all coronary targets is 75–85% at 1 year [13, 15, 17, 18] and approximately 50% at 10 years [14, 16].

Classically, the saphenous vein conduit has been harvested from the leg with an open or open bridging incision technique. The major complications associated with such a harvest technique have been wound infection and leg edema, especially in diabetic patients. The advent of endoscopic vein harvesting techniques in 1996 delivered a new paradigm for avoidance of the morbid wound complications associated with open vein harvesting.

Endoscopic vein harvesting is performed through a small incision near the knee. An endoscope and endoscopic cautery are used to remove the vein through a small tunnel created along both the proximal and distal course of the saphenous vein. This has resulted in a markedly decreased incidence of wound infection, postoperative pain, and lower extremity complications when compared to open harvesting [19]. According to the Society of Thoracic Surgeons' National Database, endoscopic vein harvesting was used in 70% of CABG procedures performed in the United States in 2008.

However, some studies have suggested that the patency of endoscopically harvested vein grafts may be lower than those harvested in a conventional open manner [18, 20]. In a substudy of the PREVENT IV study which evaluated the effect of ex vivo vein graft engineering via transfection, the short- and long-term patencies of saphenous veins harvested via conventional open techniques and endoscopic techniques were compared [18]. Routine 1-year angiography revealed a vein graft failure rate of 46.7% in patients undergoing endoscopic vein harvesting, versus a 38% vein graft failure rate in patients undergoing conventional open vein harvesting techniques. More concerning was the fact that endoscopic vein harvesting was associated with a significantly higher combined rate of death, myocardial infarction, and repeat revascularization at 3 years of follow-up (20.2% vs. 17.4%). A similar subanalysis of the Randomized On/Off Bypass (ROOBY) trial evaluated 894 of the 1471 patients who underwent routine 1-year angiography [20]. Again, a statistically significant decrease in patency was observed in the endoscopic vein harvest group compared to the open vein harvesting group (74.5% vs. 85.2%). Although the 1-year composite outcome of death, myocardial infarction, and revascularization was no different between endoscopic and open techniques (8.2% vs. 4.9%, $P = 0.061$), there was a significant difference in the need for repeat revascularization (6.7% vs. 3.4%, $P = 0.033$).

Excessive traction causing intimal injury and cautery-mediated adventitial trauma have been invoked as causative factors in possible damage incurred during endoscopic vein harvesting [21]. Technical aspects of endoscopic vein harvesting should not be underestimated, and the degree of injury to the saphenous vein may be related to operator experience [22]. Nonetheless, it is evident that careful saphenous vein harvesting with minimal traction, distention, and cautery damage can positively affect the long-term results of this conduit.

The BITAs

The superior patency rates of the LITA compared to SVGs have led many investigators to conclude that the use of the RITA to non-LAD targets might yield similar advantageous results. Multiple observational studies and several meta-analyses have shown the RITA patency to be better than SVGs to left-sided coronary targets, but not as good as the LITA–LAD patency [16, 23–25].

The RITA can be used as an arterial conduit in three ways: (1) as a pedicled graft, with its proximal connection to the right subclavian artery intact; (2) as a free graft sewn to the pedicled LITA in an end-to-side fashion to construct a "y" or "t" graft; and (3) as a free graft sewn directly to the aorta or a vein hood for its proximal connection. All three arrangements have their advantages and disadvantages. The pedicled RITA has the advantage of keeping its proximal connection undisturbed but suffers from a lack of length to reach all left-sided targets comfortably. The free RITA can be used with adequate length to most circumflex targets, but the proximal aortic RITA connection does offer some technical challenges that can affect long-term patency. The "t" or "y" connection allows for the most length and complete left-sided revascularization, but some surgeons are reticent to base the entire revascularization strategy on inflow to the LITA.

As early as 1999, surgeons at the Cleveland Clinic demonstrated improved survival, decreased incidence of reintervention, and a decreased need for reoperation over a 10-year follow-up period in patients undergoing BITA grafting [23]. In a propensity-matched study, these same investigators showed that BITA grafting produced improved survival compared with single internal thoracic artery (SITA) grafting during the second postoperative decade, and the magnitude of that benefit increased through 20 postoperative years [16, 24] (Figure 44.1).

Despite the improved long-term survival and patency demonstrated with BITA grafting when compared to SITA grafting, its adoption by most surgeons has been slow, with only 5% of CABG cases performed in the United States conducted with BITAs. BITA grafting is time-consuming and increases the technical difficulty of the operation, and in some patient subsets, it has been associated with an increased risk of serious sternal wound complications. The increased risk of wound infection and mediastinitis can be overcome with a number of different surgical strategies, including the use of skeletonized ITA grafts and the harvest of a small piece of RITA to be used as a "t" graft, instead of harvesting the entire vessel.

The Radial Artery

In order to avoid the technical challenges of BITA grafting, many surgeons have used the radial artery as a second arterial conduit. The

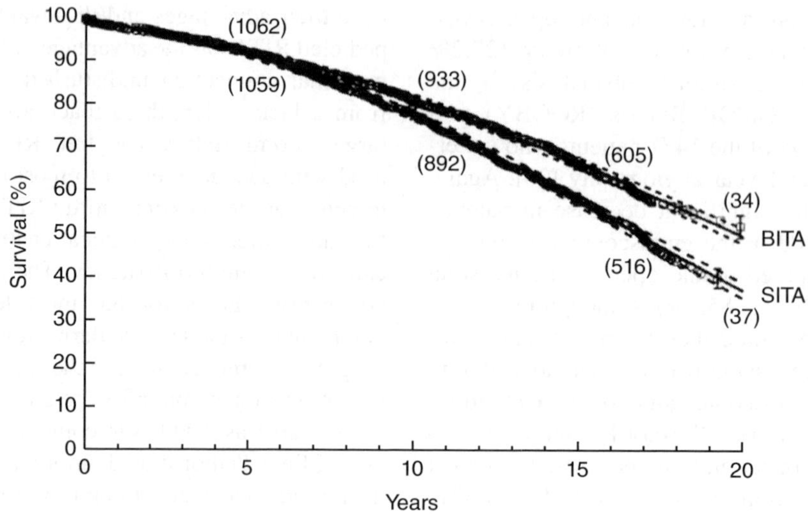

Figure 44.1 Survival of propensity-matched BITA versus SITA patients with one standard deviation confidence intervals over 20 years. Number of patients alive is noted in the parentheses. BITA, bilateral internal thoracic artery; SITA, single internal thoracic artery (Reprinted with permission from Elsevier, from Lytle et al. [16]. © Elsevier).

radial artery is not as resistant to atherosclerosis as the RITA, and it is a thick-walled vessel that is prone to spasm. Prior to harvest, an evaluation of the palmar arch is critical in order to ensure adequate collateral flow to the hand. A duplex Doppler study of the radial artery with an Allen test is the most reliable evaluation of the safety of radial artery harvest. It can also identify areas of calcification and stenosis of the radial artery, which can be especially important in diabetic patients and patients with renal failure.

Radial artery patency is very dependent upon the degree of stenosis present in the bypassed coronary artery. Noncritical stenoses (<75%) can cause competitive flow between the native coronary and the vasospastic radial artery, resulting in a radial artery "string sign" and early graft failure [26].

Nonetheless, when evaluated and used appropriately, radial artery patency can rival the patency of RITA grafts to non-LAD, left-sided coronary targets. In two recent randomized prospective studies in which routine angiography was performed at 1 year, the patency of radial artery grafts was 89–92% [13, 17]. A recent propensity-matched retrospective study of 14-year follow-up has shown an improved survival with radial artery grafting compared to SITA grafting [27].

ON-PUMP VERSUS OFF-PUMP CABG

Conventional CABG is performed through a sternotomy incision with the heart arrested and the patient supported on the CPB machine. This allows for delicate coronary anastomoses to be performed on a dry and motionless surgical field. However, during CPB, blood is exposed to the foreign surfaces of the bypass machine tubing and the oxygenator, resulting in a systemic inflammatory response. In selected patients, this inflammatory response can result in organ dysfunction [28]. Furthermore, the aortic manipulation which occurs during cannulation, bypass, and cross-clamping can increase a patient's risk of stroke [29]. In an attempt to reduce the potential for stroke and end-organ damage associated with on-pump CABG, off-pump CABG (OpCAB) was performed through a sternotomy on the beating heart.

During OpCAB, special stabilizers are used to elevate and position the heart, as well as to limit the motion of the coronary target during anastomosis. With the use of "t" grafting and special devices for anastomoses to the aorta, a "no touch" technique can be used on the aorta to minimize stroke risk. The technique is technically more demanding than on-pump CABG, and its utilization has reached a plateau in which only 22% of all CABGs performed in the United States are performed off pump.

Over the past decade, randomized trials have shown OpCAB to be well tolerated and effective when performed by operators with significant OpCAB experience. However, in the largest randomized studies of mostly low-risk patients, no difference in 30-day mortality or early morbidity has been demonstrated with OpCAB [15, 30–33]. Nonetheless, in the highest-risk patients, OpCAB may have a benefit as compared to conventional CABG [34]. In a large observational analysis of 14,766 patients, Puskas and colleagues [34] demonstrated no significant risk reduction for OpCAB in the lower three risk quartiles but a significant reduction for OpCAB patients in the highest risk quartile (Figure 44.2).

Although low-risk patients have not had a risk reduction in stroke with OpCAB, there does appear to be growing evidence that OpCAB can reduce stroke in high-risk patients with severe aortic atheroma [35].

The technical demands of OpCAB have brought into question the quality and patency of bypass grafts constructed off pump. Completeness of revascularization and patency of bypass grafts to lateral wall targets may be compromised with OpCAB [15]. The patency of LITA–LAD grafts appears to be equally excellent in all prospective randomized studies comparing OpCAB and conventional CABG [15, 31–36].

Most surgeons agree that OpCAB has a role in high-risk patients with multiple comorbidities, particularly when these include an extensive aortic atheromatous burden. However, a significant experience with OpCAB appears to be critical to obtaining revascularization results that are equivalent to the time-tested procedure of conventional on-pump CABG [37].

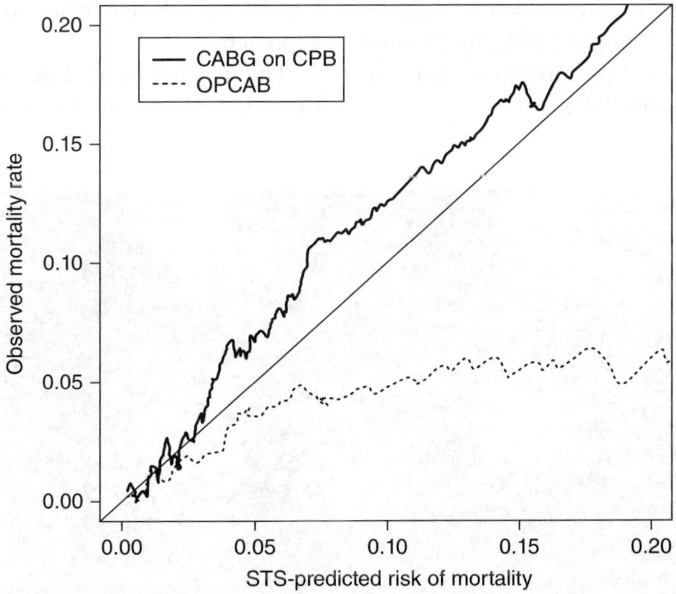

Figure 44.2 A comparison of coronary artery bypass grafting with or without cardiopulmonary bypass by observed mortality rate versus predicted risk of mortality. (Reprinted with permission from Elsevier, from Puskas et al. [34]).

MINIMALLY INVASIVE CABG

In most surgical specialties, minimally invasive surgery refers to the access the operator takes to the surgical field. Minimally invasive cardiac surgery, however, aims to ameliorate two potentially invasive surgical components: the CPB machine and the sternotomy incision.

Most minimally invasive CABG procedures focus on LITA–LAD revascularization through a sternum-sparing incision with the operation performed off pump. A minimally invasive direct coronary artery bypass (mid-CAB) is one such operation performed on the beating heart through a small left-sided thoracotomy in the fourth or fifth intercostal space [38]. Costal cartilage removal or disarticulation is necessary, and a special chest wall retractor is used to allow for open LITA takedown. Cardiac stabilization can be accomplished with a stabilizer, which is delivered directly through the wound. Mid-CAB has the advantage of not requiring any special endoscopic or robotic skills to master the LITA takedown. Although single-lung ventilation improves exposure, chest cavity insufflation is not necessary. However, the degree of chest wall retraction necessary to allow for open LITA mobilization is quite extensive, and postoperative pain control can be a challenge [39].

Endoscopic atraumatic coronary artery bypass (endo-ACAB) refers to the robotic identification of the LAD and complete mobilization of the LITA without compromising the integrity of the chest wall [40, 41]. A directed, nonrib-spreading, or limited rib-spreading thoracotomy then allows for a hand-sewn LITA–LAD anastomosis on the beating heart (Figure 44.3). This approach takes advantage of facilitating robotic technology to allow for limited chest wall retraction and improved visualization and accuracy during LITA takedown.

Despite the excellent results with robotic endo-ACAB and mid-CAB, some authors have pursued the "holy grail" of totally endoscopic coronary revascularization. The first TECAB surgeries were performed robotically on the arrested heart during CPB [42]. Peripheral CPB and an intra-aortic balloon occluder for cardioplegic arrest are critical components to the arrested heart TECAB. The complications associated with intra-aortic balloon occlusion and the inflammatory response of CPB led most minimally invasive surgeons to opt for beating-heart off-pump revascularization in lieu of an arrested heart totally endoscopic approach.

The beating-heart TECAB was the next extension of the totally endoscopic arrested

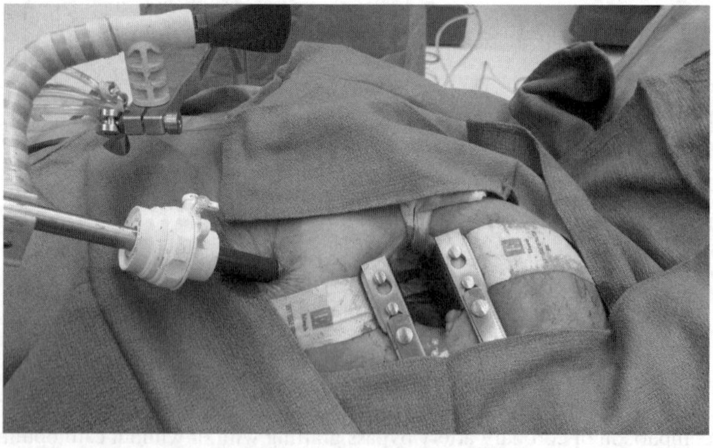

Figure 44.3 Intraoperative view of a nonrib-spreading or limited rib-spreading thoracotomy in the left fourth intercostal space during endo-ACAB procedure with endoscopic stabilizer in the left arm port.

heart CABG. The beating-heart TECAB is performed in an identical way as the robotic endo-ACAB, with the exception of the 4 cm thoracotomy for open, hand-sewn anastomosis. Instead, the anastomosis is performed intracorporeally with the robot. The TECAB has proved to be an incredibly challenging operation that only a few have mastered. Early results are encouraging from a small number of skilled operators, but widespread adoption of this operation had not occurred [43].

A learning curve exists for all of these minimally invasive approaches, and specialized training and experience are necessary to master them. Nonetheless, in high-volume centers with highly skilled operators, the patency results of these procedures have been shown to be equal to conventional CABG performed on bypass. The decrease in morbidity and rapid recovery associated with minimally invasive CABG promise to retain the durability of the LITA–LAD anastomosis while simultaneously decreasing the trauma of surgery.

HYBRID CORONARY REVASCULARIZATION

Hybrid coronary revascularization (HCR) utilizes a combination of minimally invasive CABG and percutaneous coronary intervention (PCI) techniques for revascularization in the setting of multivessel CAD [44]. While CABG confers a survival advantage by a patent LITA graft to the LAD artery, PCI to additional non-LAD targets offers a lower level of invasiveness, quicker recovery, and, in certain patients, fewer complications. Furthermore, minimally invasive surgical techniques for isolated LITA–LAD revascularization have served to reduce the incidence of adverse outcomes when compared to conventional CABG by sparing a sternal incision, avoiding aortic manipulation, or eliminating CPB. The rationale of HCR is to minimize the morbidity without compromising vessel patency and patient survival in this combined approach.

There are three different approaches for HCR: PCI before CABG, PCI after CABG, or as a simultaneous procedure with CABG in a hybrid operative suite. Each approach offers unique advantages. When PCI is performed first, the revascularization of the non-LAD targets minimizes the risk of ischemia during minimally invasive CABG. This approach is especially useful when the patient presents with an acute coronary syndrome in a non-LAD distribution. The drawbacks of a PCI-first strategy center mostly on the need for surgery sometime shortly after fresh stent placement, which may increase the risk of stent thrombosis even in the face of uninterrupted dual antiplatelet therapy. Proceeding with CABG first permits continuous antiplatelet therapy after PCI, routine angiography of the LITA–LAD anastomosis, and PCI of high-risk, non-LAD lesions after LAD revascularization. Finally, a simultaneous procedure allows for complete revascularization and angiography of the LITA–LAD anastomosis in a single step. This approach entails some disadvantages, which include hemorrhage with continuous antiplatelet therapy, an increased risk of stent thrombosis secondary to the inflammatory response of surgery, and potential negative economic consequences for the hospital when two costly procedures are compensated as one.

Controversy remains for the optimal approach for HCR. At this point in time, each patient requires an individualized evaluation that weighs the potential advantages and disadvantages of each procedure. Angiographic evidence of LITA–LAD patency rates in minimally invasive CABG compares favorably to conventional CABG. The reintervention rate of LITA–LAD varies from 1 to 3% in HCR studies. While short- and midterm results appear promising, empiric recommendations are limited by the variability of HCR technique, the timing of the combined procedures, the percentage of drug-eluting stents (DES) versus bare-metal stents used, and the lack of a well-controlled, randomized study in the literature.

LONG-TERM RESULTS OF CABG

Over 30 years of prospective randomized studies have provided evidenced-based indications for the appropriate use of CABG. The utility of CABG for coronary revascularization must be weighed against the alternatives of optimal medical therapy (OMT) and PCI. When choosing a therapeutic option, the data of prospective randomized studies must also be evaluated in light of the medical comorbidities of each patient in order to identify an appropriate individualized treatment plan.

The earliest randomized studies evaluating CABG were conducted in the early 1980s and compared OMT and conventional CABG. These earliest studies included the Coronary Artery Surgery Study (CASS), the Veterans Administration Coronary Artery Bypass Surgery Cooperative Study Group, and the European Coronary Surgery Study [6–9]. All three of these prospective randomized studies demonstrated improved 5- and 10-year survival in patients undergoing CABG versus OMT alone in the setting of (i) left main disease and (ii) severe triple-vessel disease with a decreased left ventricular ejection fraction. Patients who had preserved myocardial function, regardless of their coronary disease, showed an equivalent 10-year survival with OMT and CABG. However, in these intention-to-treat studies, nearly 40% of patients treated with medical therapy alone eventually had CABG. It also has to be recognized that OMT in the 1980s did not encompass some important therapies that have now become standard, including glycoprotein IIb/IIIa platelet inhibitors, targeted anticholesterol therapy, and routine beta-blockade.

The first studies comparing PCI to CABG evaluated percutaneous transluminal coronary angioplasty (PTCA) (without stenting) to conventional CABG. The Bypass Angioplasty Revascularization Investigation (BARI) compared 1829 randomized patients to either CABG or PTCA as initial therapy for their CAD requiring revascularization [10]. These investigators found an 89.3% 5-year survival rate in the CABG group versus 86.3% in the PTCA group. These subjects also had similar rates of 5-year freedom from myocardial infarction at 80.4 and 78.6%, respectively. There was, however, a significant difference in the repeat intervention rate between the two groups. BARI showed a 54% repeat revascularization rate in the patients who underwent PTCA as initial modality for their revascularization, compared to 8% in the CABG arm. Thirty-one percent of the PTCA group who required further intervention underwent a CABG within a 5-year follow-up period. In addition to the studied endpoints of survival and mortality, BARI found that diabetes mellitus was a significant predictor of improved survival with CABG versus PTCA. Among patients with controlled diabetes mellitus (19% of the overall cohort), 5-year survival with PTCA was 65.5%, versus 80.6% with CABG.

Studies evaluating CABG with bare-metal stenting have similarly shown a survival advantage in patients undergoing CABG with triple-vessel disease and a decreased left ventricular ejection fraction and in patients with diabetes mellitus and triple-vessel disease. In these studies, restenosis rates with bare-metal stents (~30%) were improved over those with PTCA (~50%), but still were not good enough to provide survival advantages over CABG.

Restenosis rates with PCI have steadily improved, and the advent of DES spawned the SYNTAX trial, which compared modern-day CABG with DES in a prospective randomized trial of 1800 "real-world" patients with triple-vessel and/or left main disease [11, 12]. The primary endpoint was major adverse cardiac and cerebral events (MACCE), including all-cause death and the incidence of repeat revascularization, at 1 year following initial intervention. SYNTAX was designed as a noninferiority trial, with a difference of 5.9% between the two groups designated as positive. This noninferiority was not met, as MACCE rates were significantly higher in the PCI group compared to the CABG group (17.8% vs. 12.4% for CABG, $P=0.002$), in large part

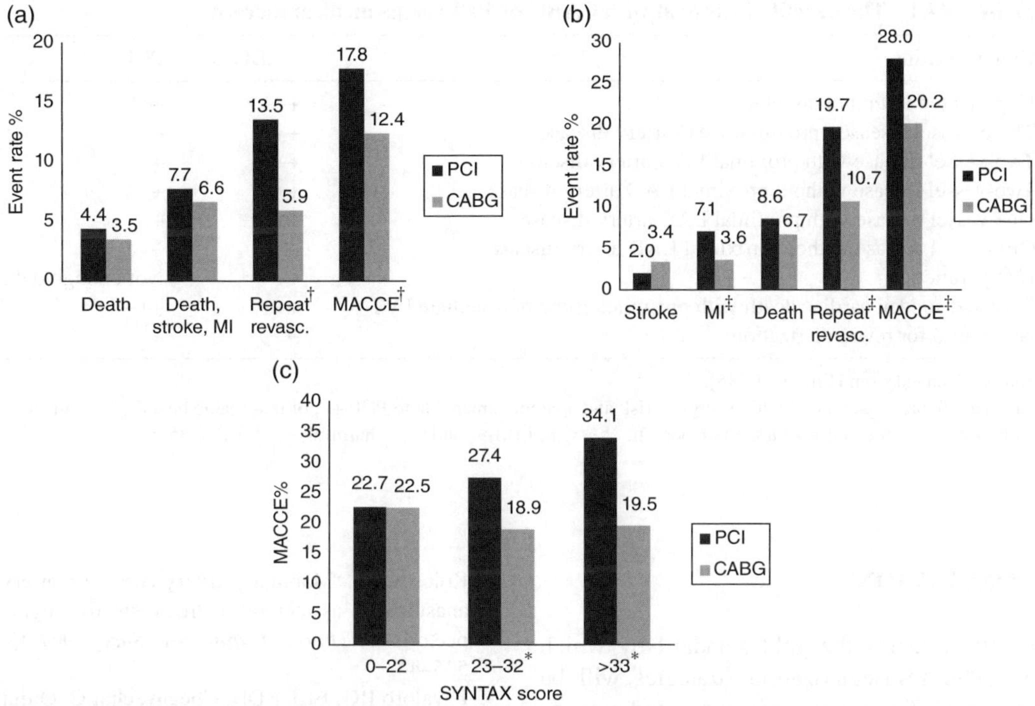

Figure 44.4 Results of the SYNTAX trial and 3-year follow-up. (a) The cumulative rates for events in the SYNTAX trial at 12 months—death from any cause, death from any cause plus stroke and MI, repeat revascularization, and MACCE. †There was a statistically significant increased rate of repeat revascularization ($P < 0.001$) and MACCE ($P = 0.002$) in the PCI group. (b) The cumulative rates for events in the SYNTAX trial at 3-year follow-up—stroke, MI, death from any cause, repeat revascularization, and MACCE. ‡There was a statistically significant increased rate of MI ($P = 0.007$), repeat revascularization ($P < 0.001$), and MACCE ($P < 0.001$) in the PCI group. (c) The MACCE at 3-year follow-up of the SYNTAX trial stratified by patient SYNTAX score category. *There was a statistically significant higher rate of MACCE in patients with more severe coronary artery disease who received PCI when SYNTAX score was 23–32 ($P = 0.02$) and >33 ($P < 0.001$) (Adapted from Serruys et al. [11] and Kappetein et al. [12]).

due to the significant difference in repeat revascularization (5.9% vs. 13.5%, $P<0.0001$) (Figure 44.4a). At 1 year of follow-up, all-cause death rates were similar (3.5%, CABG and 4.3%, PCI), but the incidence of stroke was slightly higher in the CABG group (2.2% CABG and 0.6% PCI, $P=0.003$).

In a 3-year follow-up of the SYNTAX trial [11], MACCE (CABG 20.2% vs. PCI 28.0%, $P=0.001$), repeat revascularization (CABG 10.7% vs. PCI 19.7%, $P=0.001$), and myocardial infarction (CABG 3.6% vs. PCI 7.1%, $P=0.002$) were all elevated in the PCI group (Figure 44.4b). Interestingly, the 3-year rate of stroke was equal between CABG and PCI (3.4 vs. 2.0%, $P=0.07$). Based on these studies, a SYNTAX scoring system of CAD severity was developed in order to guide future therapeutic recommendations. In the 3-year follow-up, patients with the highest two tertiles of coronary disease based on SYNTAX score had the greatest improvement in MACCE, whereas in the lowest tertile SYNTAX scores, the results were equivalent (Figure 44.4c).

Based on the data of these randomized controlled trials, the American College of Cardiology and the American Heart Association have made consensus recommendations about when CABG and PCI are indicated in patients with CAD (Table 44.1).

TABLE 44.1 The benefit of survival with CABG or PCI versus medical therapy

Clinical setting	CABG	PCI
Unprotected left main disease	+++	++*
Three-vessel disease ± proximal LAD artery disease	+++	+
Two-vessel disease with proximal LAD artery disease	+++	+
Two-vessel disease without proximal LAD artery disease	++**	+
One-vessel disease with proximal LAD artery disease	++***	+
One-vessel disease without proximal LAD artery disease	–	–
LV dysfunction	++****	Not enough data
Survived sudden cardiac death with presumed ischemia-mediated VT	+++	+++
No criteria for revascularization	–	–

Source: Adapted from Hillis et al. [45].
+++, beneficial; *, patient with high surgical risk and anatomy amenable to PCI; ++, of reasonable benefit; **, patient with extensive ischemia; +, of uncertain benefit; ***, with LIMA conduit; –, harmful; ****, EF of 35–50%.

CONCLUSIONS

CABG remains the gold standard by which all other revascularization strategies will be measured. The survival advantage and patency conferred by LIMA–LAD revascularization have yet to be replicated. However, the advent of minimally invasive and robotic approaches for surgical revascularization provides opportunity for less morbid operations with comparable outcomes. Furthermore, patients with multivessel disease can now undergo a combination approach to coronary revascularization using percutaneous intervention and robotic endoscopic atraumatic coronary artery bypass termed HCR. In the near future, we can look forward to the long-term results of combined revascularization techniques. Nonetheless, the optimal care of the patient with critical CAD relies on a multidisciplinary team in which the best treatment options are uniquely designed for each particular patient.

REFERENCES

1. Goetz RH, Rohman M, Haller JD, Dee R, Rosenak SS. Internal mammary-coronary artery anastomosis. A nonsuture method employing tantalum ring. *J Thorac Cardiovasc Surg* 1961; 41:378–86.
2. Kolesov VI. Mammary artery-coronary artery anastomosis as method of treatment for angina pectoris. *J Thorac Cardiovasc Surg* 1967;45: 535–44.
3. Favaloro RG, Effler DB, Cheanvechai C, Quint RA, Sones FM Jr. Acute coronary insufficiency (impending myocardial infarction and myocardial infarction): surgical treatment by the saphenous vein graft technique. *Am J Cardiol* 1971;55:598–607.
4. Green GE, Stertzer SH, Reppert EH. Coronary arterial bypass grafts. *Ann Thorac Surg* 1968; 5:443–50.
5. Loop FD, Lytle BW, Cosgrove DM, et al. Influence of the internal-mammary-artery graft on 10-year survival and other cardiac events. *N Engl J Med* 1986;314:1–6.
6. Passamani E, Davis KB, Gillespie MJ, Killip T, and the CASS Principal Investigators and Their Associates. A randomized trial of coronary artery bypass surgery: survival of patients with a low ejection fraction. *N Engl J Med* 1985;312:1665–71.
7. The Veterans Administration Coronary Artery Bypass Surgery Cooperative Study Group. Eleven-year survival in the Veterans Administration randomized trial of coronary bypass surgery for stable angina. *N Engl J Med* 1984;311:1333–9.
8. Varnauskas E, and the European Coronary Surgery Study Group. Twelve-year follow-up of survival in the randomized European coronary surgery study. *N Engl J Med* 1988;319: 332–7.

9. Alderman EL, Bourassa MG, Cohen LS. Ten-year follow-up of survival and myocardial infarction in the randomized coronary artery surgery study. *Circulation* 1990;82:1629–46.
10. The Bypass Angioplasty Revascularization Investigation (BARI) Investigators. Comparison of coronary bypass surgery with angioplasty in patients with multivessel disease. *N Engl J Med* 1996;335:217–25.
11. Serruys PW, Morice MC, Kappetein P, et al. Percutaneous coronary intervention versus coronary artery bypass grafting for severe coronary artery disease. *N Engl J Med* 2009; 360:961–72.
12. Kappetein AP, Feldman TO, Mack MJ, et al. Comparison of coronary bypass surgery with drug-eluting stenting for the treatment of left main and/or three-vessel disease: 3 year follow-up of the SYNTAX trial. *Eur Heart J* 2011; 32:2125–34.
13. Goldman S, Sethy GK, Holman W, et al. Radial artery grafts vs. saphenous vein grafts in coronary artery bypass surgery. A randomized trial. *JAMA* 2011;305:167–74.
14. Sabik JF, Lytle BW, Blackstone EH, Houghtaling PL, Cosgrove DM. Comparison of saphenous vein and internal thoracic artery graft patency by coronary system. *Ann Thorac Surg* 2005;79:544–51.
15. Shroyer AL, Grover FL, Hattler B, et al. On-pump versus off-pump coronary artery bypass surgery. *N Engl J Med* 2009;361: 1827–37.
16. Lytle BW, Blackstone EH, Sabik JF, Houghtaling P, Lop FD, Cosgrove DM. The effect of bilateral internal thoracic artery grafting on survival during 20 post-operative years. *Ann Thorac Surg* 2004;78:2005–14.
17. Desai ND, Cohen EA, Naylor CD, Fremes SE. Radial Artery Patency Study Investigators. A randomized comparison of radial-artery and saphenous-vein coronary bypass grafts. *N Engl J Med* 2004;351:2302–9.
18. Lopes RD, Hafley GE, Allen KB, et al. Endoscopic versus open vein-graft harvesting in coronary-artery bypass surgery. *N Engl J Med* 2009;361(3):235–44.
19. Markar SR, Kutty R, Edmonds L, Sadat U, Nair S. A meta-analysis of minimally invasive versus traditional open vein harvest for coronary artery bypass graft surgery. *Interact Cardiovasc Thorac Surg* 2010;10:266–70.
20. Zenati MA, Shrover AL, Collins JF, et al. Impact of endoscopic versus open saphenous vein harvest technique on late coronary artery bypass grafting patient outcomes in the ROOBY (Randomized On/Off Bypass) Trial. *J Thorac Cardiovasc Surg* 2011;141:338–44.
21. Rousou LJ, Taylor KB, Lu XG, et al. Saphenous vein conduits harvested by endoscopic technique exhibit structural and functional damage. *Ann Thorac Surg* 2009;87:62–70.
22. Desai P, Kiani S, Thiruvanthan N, et al. Impact of the learning curve for endoscopic vein harvest on conduit quality and early graft patency. *Ann Thorac Surg* 2011;91:1385–92.
23. Lytle BW, Blackstone EH, Loop FD, et al. Two internal thoracic artery grafts are better than one. *J Thorac Cardiovasc Surg* 1999;117: 855–72.
24. Sabik JF, Blackstone EH, Gillinov AM, Smedira NG, Lytle BW. Occurrence and risk factors for reintervention after coronary artery bypass grafting. *Circulation* 2006;114(Suppl I):I-454–60.
25. Hwang HY, Kim JS, Cho KR, Kim KB. Five year angiographic patency and long-term clinical outcomes of bilateral internal thoracic artery in situ versus y-composite grafting. *Ann Thorac Surg* 2011;92:579–86.
26. Desai ND, Naylor CD, Kiss A, et al. Radial Artery Patency Study Investigators. Impact of patient and target-vessel characteristics on arterial and venous bypass graft patency. *Circulation* 2007;115(6):684–91.
27. Tranbaugh RF, Dimitrova KR, Friedmann P, et al. Radial artery conduits improve long term survival after coronary artery bypass grafting. *Ann Thorac Surg* 2010;90:1165–72.
28. Edmunds LH Jr. Inflammatory response to cardiopulmonary bypass. *Ann Thorac Surg* 1998;66(Suppl):S12–16.
29. Taylor KM. Brain damage during cardiopulmonary bypass. *Ann Thorac Surg* 1998; 65(Suppl):S20–6.
30. Halkos ME, Puskas JD. Off-pump coronary surgery: where do we stand in 2010? *Curr Opin Cardiol* 2010;25:583–8.
31. Khan NE, De Souza A, Mister R, et al. A randomized comparison of off-pump and on-pump multivessel coronary-artery bypass surgery. *N Engl J Med* 2004;350:21–8.
32. Nathoe HM, van Dijk D, Jansen EW, et al. A comparison of on-pump and off pump coronary

bypass surgery in low-risk patients. *N Engl J Med* 2003;348:394–402.

33. Moller CH, Perko MJ, Lund JT, et al. No major differences in 30-day outcomes in high-risk patients randomized to off-pump versus on-pump coronary bypass surgery. The Best Bypass Surgery Trial. Circulation 2010;121:498–504.
34. Puskas JD, Thourani VH, Kilgo P, et al. Off-pump coronary artery bypass disproportionately benefits high-risk patients. *Ann Thorac Surg* 2009;88:1142–7.
35. Brizzio ME, Zapolanski A, Shaw RE, et al. Stroke-related mortality in coronary surgery is reduced by the off-pump approach. *Ann Thorac Surg* 2010;89:19–23.
36. Puskas J, Cheng D, Knight J, et al. Off-pump versus conventional coronary bypass grafting: a meta-analysis and consensus statement from the 2004 ISMICS Consensus Conference. *Innovations* 2005;1:3–27.
37. Konety SH, Rosenthal GE, Vaughan-Sarrazin MS. Surgical volume and outcomes of off-pump coronary artery bypass graft surgery: does it matter? *J Thorac Cardiovasc Surg* 2009;137: 1116–23.e1.
38. Holzhey DM, Jacobs S, Mochalski M, et al. Seven-year follow-up after minimally invasive direct coronary artery bypass: experience with more than 1300 patients. *Ann Thorac Surg* 2007;83:108–14.
39. Bucerius J, Metz S, Walther T. Endoscopic internal thoracic artery dissection leads to significant reduction of pain after minimally invasive direct coronary artery bypass graft surgery. *Ann Thorac Surg* 2002;73:1180–4.
40. Vassiliades TA Jr., Reddy VS, Puskas JD, Guyton RA. Long-term results of the endoscopic atraumatic coronary artery bypass. *Ann Thorac Surg* 2007;83:979–84.
41. Halkos ME, Vassiliades TA, Douglas JS, et al. Hybrid coronary revascularization versus off-pump coronary artery bypass grafting for the treatment of multivessel coronary artery disease. *Ann Thorac Surg* 2011;92: 1695–701.
42. de Canniere D, Wimmer-Greinecker G, Cichon R, et al. Feasibility, safety, and efficacy of totally endoscopic coronary artery bypass grafting: multicenter European experience. *J Thorac Cardiovasc Surg* 2007;134:710–6.
43. Srivastava S, Gadasalli S, Agusala M, et al. Robotically assisted beating heart totally endoscopic coronary artery bypass (TECAB). Is there a future? *Innovations* 2008;3:52–5.
44. DeRose JJ Jr. Current state of integrated hybrid revascularization. *Semin Cardiovasc Thorac Surg* 2009;21:229–36.
45. Hillis LD, Smith PK, Anderson JL, et al. 2011 ACCF/AHA guideline for coronary artery bypass graft surgery: executive summary: a report of the American College of Cardiology Foundation/American Heart Association Task Force on Practice Guidelines. *J Thorac Cardiovasc Surg* 2012;143:4–34.

INDEX

Note: Page numbers in *italic* refers figure pages and page numbers in **bold** refers to table pages.

Atherosclerosis: Risks, Mechanisms and Therapies, First Edition. Edited by Hong Wang and Cam Patterson.